Advances in Mathematical Fluid Mechanics

Series editors

Giovanni P. Galdi, Pittsburgh, USA
John G. Heywood, Vancouver, Canada
Rolf Rannacher, Heidelberg, Germany

Dieter Bothe • Arnold Reusken

Editors

Transport Processes at Fluidic Interfaces

Editors
Dieter Bothe
Fachbereich Mathematik
Technische Universität Darmstadt
Darmstadt, Germany

Arnold Reusken
Institut für Geometrie und Praktische
  Mathematik
RWTH Aachen University
Aachen, Germany

ISSN 2297-0320 ISSN 2297-0339 (electronic)
Advances in Mathematical Fluid Mechanics
ISBN 978-3-319-56601-6 ISBN 978-3-319-56602-3 (eBook)
DOI 10.1007/978-3-319-56602-3

Library of Congress Control Number: 2017942773

Mathematics Subject Classification (2010): 35JXX, 76D05, 76T99, 65M08, 65M60, 65N08, 65N30

This book is published under the trade name Birkhäuser, www.birkhauser-science.com
The registered company is Springer International Publishing AG
The registered company address is: Gewerbestrasse 11, 6330 Cham, Switzerland

Preface

In multiphase processes, interfaces play a prominent role as the surface at which the bulk phases are in contact. If the bulk phases are fluids, i.e., gas or liquid, these interfaces are often free to move and deform; hence geometric properties such as the shape and curvature, but also physical quantities such as surface tension, are of fundamental importance. In many cases, such fluid interfaces exhibit further physico chemical properties stemming from partial mass being accumulated at the interface, like inherent surface viscosities. Typically, such adsorbed species are surface-active agents, so-called surfactants, and the interplay between sorption and transport processes then leads to inhomogeneous surface tension, causing surface stresses. These few facts already show that the interface between fluids often comes as a phase in itself, and, for this reason, we call this a "fluidic interface" rather than a fluid interface. The behavior of a multiphase flow system is determined by the complex interplay between the bulk phases and the interface, mediated by the transport of mass, momentum and energy at the interface as well as the transfer of these quantities across it. This includes the dynamics of the interfaces themselves, heat and mass transport between the fluids, adsorption effects at the interface and transport of species on the interface, variable interface properties, and phase change. In general, these processes are strongly coupled and properties of the interface play a dominant role. Therefore, a rigorous understanding of the behavior of such complex flow problems must be based on physically sound mathematical models accounting especially for local processes at the interface. The realization of this requires interdisciplinary research with expertise from Applied Analysis, Numerical Mathematics, Interface Physics, and Chemistry as well as relevant research areas in the Engineering Sciences.

Against this background, the German Research Foundation (Deutsche Forschungsgemeinschaft, DFG) has funded a targeted research program in the form of a national priority program (SPP) on "Transport Processes at Fluidic Interfaces," running from 2010 to 2016, to further develop and expand such models, to analyze their mathematical properties, and to develop and advance numerical methods for the rigorous simulation of these mathematical models. Important goals of this SPP 1506 have been: (1) to derive and expand mathematical models that describe relevant physico-chemical interface phenomena, (2) to improve and deepen the understanding of mechanisms and phenomena occurring at fluidic interfaces by means of rigorous mathematical analysis of the underlying systems of partial differential equations, (3) to develop and analyze numerical methods for the simulation of multiphase flow problems which resolve the local processes at the interface, and (4) to validate these models and the numerical simulation methods by means of specifically designed experiments. This book is the closing report of this priority program SPP 1506 "Transport Processes at Fluidic Interfaces." It summarizes the major findings obtained in 22 research projects with 32 principal investigators from 20 German universities and research centers, covering a wide range of research topics from mathematical analysis of relevant classes of partial differential equations, to development, analysis, and application of tailor-made numerical methods, to specifically designed high-resolution measurement techniques, to detailed studies of particular interfacial processes. These topics and the researchers that participated in the priority program are from several different disciplines, fostering the strongly interdisciplinary nature of the work done in the SPP 1506.

The authors and the editors of this book would like to thank the Deutsche Forschungsgemeinschaft for funding the SPP 1506 "Transport Processes at Fluidic Interfaces." Our special thanks go to Dr. Frank Kiefer (DFG) for his continued support. We are also very grateful to the international referees, who evaluated new and renewal proposals during the duration of the priority program. We thank Prof. Akio Tomiyama (Kobe University) for acting as a research fellow within this program, by supporting the guiding measure on Taylor bubbles and Taylor flow within numerous discussions and by providing valuable data. Finally, we thank the editors of this book series, especially Paolo Galdi (University of Pittsburgh) for the possibility to include this closing report in the series on "Advances in Mathematical Fluid Mechanics," and Springer Verlag, Heidelberg, for the constructive and pleasant collaboration in preparing this volume.

Darmstadt, Germany   Dieter Bothe
Aachen, Germany   Arnold Reusken
February 2017

Contents

Part II   Analysis and Simulation of Diffusive Interface Models

**Part III   Experimental and Numerical Investigation of Interfacial
            Phenomena**

List of Contributors

Helmut Abels Fakultät für Mathematik, Universität Regensburg, Regensburg, Germany

Rufat Abiev Department of Optimization of Chemical and Biotechnological Equipment, St. Petersburg State Institute of Technology, St. Petersburg, Russia

Sebastian Aland Institute of Scientific Computing, TU Dresden, Dresden, Germany

Faculty of Informatics/Mathematics, HTW Dresden, Dresden, Germany

Eberhard Bänsch Department Mathematics, Friedrich-Alexander Universität Erlangen-Nürnberg, Erlangen, Germany

Stephan Boden Institute of Fluid Dynamics, Experimental Fluid Dynamics Division, Helmholtz-Zentrum Dresden-Rossendorf, Dresden, Germany

Institute of Power Engineering, AREVA Endowed Chair of Imaging Techniques in Energy and Process Engineering, Technische Universität Dresden, Dresden, Germany

Thomas Boeck Institute of Thermodynamics and Fluid Mechanics, TU Ilmenau, Ilmenau, Germany

Stefan Bommer Experimental Physics, Saarland University, Saarbrücken, Germany

Dieter Bothe Mathematical Modeling and Analysis, Fachbereich Mathematik, Technische Universität Darmstadt, Darmstadt, Germany

Oleg Boyarkin Institute of Mathematics, University of Augsburg, Augsburg, Germany

Stefan Burger Institute of Physics, University of Augsburg, Augsburg, Germany

Kerstin Eckert Institute of Process Engineering, TU Dresden, Dresden, Germany

Institute of Fluid Dynamics, Helmholtz-Zentrum Dresden-Rossendorf, Dresden, Germany

Carlos Falconi Institute for Catalysis Research and Technology (IKFT), Karlsruhe Institute of Technology (KIT), Karlsruhe, Germany

Thomas Franke Institute of Physics, University of Augsburg, Augsburg, Germany

Nanosystems Initiative Munich, München, Germany

Chair of Biomedical Engineering, School of Engineering, College of Science & Engineering, University of Glasgow, Glasgow, UK

Thomas Fraunholz Institute of Mathematics, University of Augsburg, Augsburg, Germany

Heinrich Freistühler Department of Mathematics and Statistics, University of Konstanz, Konstanz, Germany

Sashikumaar Ganesan Department of Computational and Data Sciences, Indian Institute of Science, Bangalore, India

Harald Garcke Fakultät für Mathematik, Universität Regensburg, Regensburg, Germany

Stephan Gerber Department of Mathematics and Computer Science, Mathematics, Freie Universität Berlin, Berlin, Germany

Michael Griebel Institute for Numerical Simulation, University of Bonn, Bonn, Germany

Fraunhofer SCAI, Schloss Birlinghoven, Augustin, Germany

Günther Grün Department Mathematik, Friedrich-Alexander Universität Erlangen-Nürnberg, Erlangen, Germany

Mohammadreza Haghnegahdar Institute of Fluid Dynamics, Experimental Fluid Dynamics Division, Helmholtz-Zentrum Dresden-Rossendorf, Dresden, Germany

Andreas Hahn Institut für Analysis und Numerik, Otto-von-Guericke Universität, Magdeburg, Germany

Uwe Hampel Institute of Fluid Dynamics, Experimental Fluid Dynamics Division, Helmholtz-Zentrum Dresden-Rossendorf, Dresden, Germany

Institute of Power Engineering, AREVA Endowed Chair of Imaging Techniques in Energy and Process Engineering, Technische Universität Dresden, Dresden, Germany

Michael Hintermüller Weierstraß-Institut, Berlin, Germany

Institut für Mathematik, Humboldt-Universität zu Berlin, Berlin, Germany

Michael Hinze Fachbereich Mathematik, Universität Hamburg, Hamburg, Germany

Marko Hoffmann Institute of Multiphase Flows, Hamburg University of Technology, Hamburg, Germany

Ronald H.W. Hoppe Institute of Mathematics, University of Augsburg, Augsburg, Germany

Department of Mathematics, University of Houston, Houston, TX, USA

Sebastian Jachalski Weierstraß Institute, Berlin, Germany

Christian Kahle Chair of Optimal Control, Center for Mathematical Sciences, Garching, Germany

Talmira Kairaliyeva Max Planck Institute of Colloids and Interfaces, Potsdam, Germany

Christina Kallendorf Institute of Fluid Dynamics, TU Darmstadt, Darmstadt, Germany

Sven Kastens Institute of Multiphase Flows, Hamburg University of Technology, Hamburg, Germany

Tobias Keil Weierstraß-Institut, Berlin, Germany

Institut für Mathematik, Humboldt-Universität zu Berlin, Berlin, Germany

Simon Kirschler Institute of Physics, University of Augsburg, Augsburg, Germany

Rupert Klein Department of Mathematics and Computer Science, Mathematics, Freie Universität Berlin, Berlin, Germany

Thomas Köllner Institute of Fluid Dynamics, Helmholtz-Zentrum Dresden-Rossendorf, Dresden, Germany

Matthias Kotschote Department of Mathematics and Statistics, University of Konstanz, Konstanz, Germany

Florian Kummer Institute of Fluid Dynamics, TU Darmstadt, Darmstadt, Germany

Kei Fong Lam Fakultät für Mathematik, Universität Regensburg, Regensburg, Germany

Christoph Lehrenfeld Institut für Numerische und Angewandte Mathematik, University of Göttingen, Göttingen, Germany

Kevin Lindner Institute of Physics, University of Augsburg, Augsburg, Germany

YuNing Liu NYU-ECNU Institute of Mathematical Sciences at NYU Shanghai, Shanghai, China

Holger Marschall Mathematical Modeling and Analysis, Fachbereich Mathematik, Technische Universität Darmstadt, Darmstadt, Germany

Stefan Metzger Department Mathematik, Friedrich-Alexander Universität Erlangen-Nürnberg, Erlangen, Germany

Christoph Meyer Institute of Multiphase Flows, Hamburg University of Technology, Hamburg, Germany

Reinhard Miller Max Planck Institute of Colloids and Interfaces, Potsdam, Germany

Björn Müller Institute of Fluid Dynamics, TU Darmstadt, Darmstadt, Germany

Ingo Nitschke Institute of Scientific Computing, TU Dresden, Dresden, Germany

Robert Nürnberg Department of Mathematics, Imperial College London, London, UK

Martin Oberlack Institute of Fluid Dynamics, TU Darmstadt, Darmstadt, Germany

Michael Oevermann Department for Applied Mechanics, Combustion, Chalmers University of Technology, Gothenburg, Sweden

Dirk Peschka Weierstraß Institute, Berlin, Germany

Chiara Pesci Mathematical Modeling and Analysis, Fachbereich Mathematik, Technische Universität Darmstadt, Darmstadt, Germany

Malte A. Peter Institute of Mathematics, University of Augsburg, Augsburg, Germany

Augsburg Centre for Innovative Technologies, University of Augsburg, Augsburg, Germany

Arnold Reusken Institut für Geometrie und Praktische Mathematik, RWTH Aachen University, Aachen, Germany

Sebastian Reuther Institute of Scientific Computing, TU Dresden, Dresden, Germany

Christian Rieger Institute for Numerical Simulation, University of Bonn, Bonn, Germany

Alexander Schier Institute for Numerical Simulation, University of Bonn, Bonn, Germany

Michael Schlüter Institute of Multiphase Flows, Hamburg University of Technology, Hamburg, Germany

Andreas Schöttl Fakultät für Mathematik, Universität Regensburg, Regensburg, Germany

Karin Schwarzenberger Institute of Process Engineering, TU Dresden, Dresden, Germany

Ralf Seemann Experimental Physics, Saarland University, Saarbrücken, Germany

Kristin Simon Institut für Analysis und Numerik, Otto-von-Guericke Universität, Magdeburg, Germany

Florian Strobl Institute of Physics, University of Augsburg, Augsburg, Germany

Nanosystems Initiative Munich, München, Germany

Lutz Tobiska Institut für Analysis und Numerik, Otto-von-Guericke Universität, Magdeburg, Germany

Vamseekrishna Ulaganathan Max Planck Institute of Colloids and Interfaces, Potsdam, Germany

Thomas Utz Institute of Fluid Dynamics, TU Darmstadt, Darmstadt, Germany

Axel Voigt Institute of Scientific Computing, TU Dresden, Dresden, Germany

Dresden Center for Computational Materials Science (DCMS), TU Dresden, Dresden, Germany

Center for Systems Biology Dresden (CSBD), Dresden, Germany

Barbara Wagner Weierstraß Institute, Berlin, Germany

Matthias Waidmann Department of Mathematics and Computer Science, Mathematics, Freie Universität Berlin, Berlin, Germany

Josef Weber Fakultät für Mathematik, Universität Regensburg, Regensburg, Germany

Stephan Weller CD-adapco, Nürnberg Office, Nürnberg , Germany

Achim Wixforth Institute of Physics, University of Augsburg, Augsburg, Germany

Augsburg Centre for Innovative Technologies, University of Augsburg, Augsburg, Germany

Nanosystems Initiative Munich, München, Germany

Martin Wörner Institute for Catalysis Research and Technology (IKFT), Karlsruhe Institute of Technology (KIT), Karlsruhe, Germany

Part I
Numerical Methods for Sharp Interface Models

The first part reports on developments of numerical methods which are specifically designed for the efficient and accurate simulation of transport processes at fluid interfaces based on sharp interface continuum models. This includes contributions concerning the numerical treatment of two-phase hydrodynamics, where capillarity, i.e. surface tension, is an important issue. Additional transport processes involve the transfer of mass across the interface, leading to evolving discontinuous, as well as the transfer of mass from the bulk to the interface and the transport of adsorbed mass (surfactants) on the moving and deforming fluid interface. Both interface tracking and interface capturing techniques for the numerical representation of the interface are considered. Finite Element (conforming and nonconforming DG) methods and Finite Volume discretization methods are studied in the various contributions, and spatial as well as time discretization methods are considered. There are two contributions on the discrete exterior calculus (DEC) technique for the spatial discretization of surface partial differential equations.

In this part on numerical methods for sharp interface models there are the following contributions:

Chapter 1. S. Ganesan, A. Hahn, K. Simon and L. Tobiska, *ALE-FEM for Two-Phase and Free Surface Flows with Surfactants*.

Chapter 2. C. Lehrenfeld and A. Reusken, *High Order Unfitted Finite Element Methods for Interface Problems and PDEs on Surfaces*.

Chapter 3. T. Utz, C. Kallendorf, F. Kummer, B. Müller and M. Oberlack, *An Extended Discontinuous Galerkin Framework for Multiphase Flows*.

Chapter 4. M. Waidmann, S. Gerber, M. Oevermann and R. Klein, *Building Blocks for a Strictly Conservative Generalized Finite Volume Projection Method for Zero Mach Number Two-Phase Flows*.

Chapter 5. S. Weller and E. Bänsch, *Time Discretization for Capillary Flow: Beyond Backward Euler*.

Chapter 6. M. Griebel, C. Rieger and A. Schier, *Upwind Schemes for Scalar Advection-Dominated Problems in the Discrete Exterior Calculus*.

Chapter 7.   I. Nitschke, S. Reuther and A. Voigt, *Discrete Exterior Calculus (DEC) for the Surface Navier-Stokes Equation.*

We outline the main topics of these contributions. In the first three chapters *finite element* discretization methods for two-phase flow problems are treated. A standard sharp interface two-phase flow model is used.

In Chap. 1 a *fitted* finite element technique is used, based on the arbitrary Lagrangian-Eulerian (ALE) approach, which results in triangulations that are aligned to the interface. Hence, this method can be classified as an interface tracking method. Besides the bulk flow problem also the transport of soluble surfactants is considered. The latter is modeled by coupled convection-diffusion equations in the bulk phases and on the interface. Special features of the method presented in Chap. 1 are an isoparametric high order discretization of the surface PDE and a local projection stabilization in the surface finite element method that results in a stable discretization also for convection dominated surfactant transport on the interface. Different options for constructing the ALE mapping are studied. Results of numerical experiments with these finite element methods applied to three-dimensional axisymmetric free surface problems are presented. Simulation results of coupled fluid dynamics and surfactant transport for a pending droplet problem and a freely oscillating droplet are presented and certain physical phenomena caused by surfactant transport are studied.

In Chap. 2 *unfitted* finite element methods for the discretization of mass and surfactant transport equations in two-phase flows are treated. Due to the fact that a level set (LS) method is used for interface capturing, the computational grids are *not* aligned to the interface. So-called unfitted finite element methods (also called CutFEM in the literature) are studied, both for the discretization of a convection-diffusion equation which models mass transport between the bulk phases and for the discretization of the surfactant transport equation on the interface. The Henry condition that occurs in the former is handled by the Nitsche method. Special features of the finite element methods presented are an isoparametric unfitted FEM, which results in higher order discretization accuracy, and space-time variants of the unfitted finite element and Nitsche methods. Optimal order discretization error bounds, resulting from an error analysis, are included. Results of numerical experiments in which these methods are applied to three dimensional two-phase flow problems with mass transport and insoluable surfactants are presented.

In Chap. 3 several aspects related to *discontinuous* Galerkin (DG) finite element methods applied to two-phase flow problems are treated. As in Chap. 2, a level set method is used for interface capturing and thus the triangulations are not aligned to the interface. This automatically leads to so-called cut cells, which result as subdomains in finite elements that are intersected by the interface. An accurate quadrature on such cut cells, which can have a very complex geometry, is a key component in an accurate (high order) two-phase flow solver. In Chap. 3 such an accurate quadrature method for cut cells, based on the hierarchical moment fitting method, is treated. The DG finite element spaces for two-phase flows, which have similarities with the unfitted FE spaces used in Chap. 2, are studied. Certain stabilization techniques, e.g. an agglomeration strategy for elimination of cut-cells that are very small and a filtering procedure for a stable curvature approximation, are introduced. Related to surfactant transport on the interface, an extension procedure is used, in which the surface PDE is extended to a narrow band and then discretized by DG based finite element method.

In Chap. 4 a *finite volume* method for the discretization of immiscible zero Mach number two-phase flows is presented. A sharp interface model in conservative integral form and with appropriate interface conditions is considered. As in Chaps. 2 and 3, a level set method is applied for interface capturing and thus the Cartesian grid, which is the basis for the finite volume discretization, is not aligned to the interface. A mass conserving variant of the level set method is used, which is based on a combination of the level set technique with a volume-of-fluid algorithm. This contribution focuses on the systematic design of a conservative sharp interface finite volume projection method on Cartesian grids with explicit representation of discontinuities and an asymptotics based efficient solution strategy for the variable coefficient Poisson projection equation for stationary interface problems with arbitrarily large ratio of the coefficients. A strategy for obtaining a well-balanced and conservative sharp interface finite volume discretization of surface tension forces is treated.

The contribution in Chap. 5 focuses on *time discretization* of capillary free surface flows. A standard sharp interface model for one- or two-phase capillary flows is considered. For space discretization an ALE method similar to the one treated in Chap. 1 is used. This results in an interface tracking method with moving meshes, which are aligned to the interface or free surface. The method of lines approach is applied and several time discretization methods are studied and compared, based on important criteria such as order of accuracy, stability properties, and amount of dissipativity. The class of time discretization methods considered comprises implicit Euler, BDF2, Rosenbrock methods (e.g., ROS3P, ROS34PW2) and Discontinuous Galerkin methods (dG) in time. Results of numerical experiments for an oscillating droplet (i.e., one phase free surface flow), rising droplet and a Taylor bubble flow problem are presented. A quantitative comparison, involving experimental order of convergence and numerical dissipativity, between the different classes of time discretization methods is made.

In Chaps. 6 and 7 the *discrete exterior calculus* (DEC) technique is studied for the spatial discretization of surface partial differential equations. In Chap. 6 a DEC method for the discretization of a convection-diffusion equation, e.g., the surfactant equation in a two-phase flow, is developed. The conventional DEC approach suffers from lack of stability if it is applied to convection-diffusion problems with strong convection. In this contribution a modification of the DEC is developed which leads to a stable discretization, also in cases with strong convection. This modification is based on an equivalence of the discrete exterior calculus discretization with certain standard finite volume or finite difference schemes. The modified DEC scheme with upwinding is applied to several test problems such as pure advection, advection of Zalesak's disc and an advection-diffusion problem on a flat regular mesh. Also an experiment with an advection-diffusion equation on a curved mesh is considered.

The results show the stabilization effect and the numerical convergence properties of the modified discrete exterior calculus method.

In Chap. 7 a DEC method is applied for the spatial discretization of a *surface (time-dependent) incompressible Navier-Stokes* equation. A surface variant of the Navier-Stokes equations is considered, in which the covariant derivative, the surface divergence and the surface Laplace-DeRham differential operator occur. The unknowns are a surface pressure and a tangential velocity field. The main topic of this contribution is the development of a DEC based approach for the spatial discretization of this flow problem on a stationary surface. It is shown how the DEC framework can be used to realize the concept of discrete parallel transport. The discretization is based on the covariant form of the Navier-Stokes equations and uses a discrete version of the Hodge $*$ and the Stokes theorem for the exterior derivative. Results of numerical experiments are presented in which the DEC discretization is compared with a vorticity-stream function approach. The examples demonstrate the interplay between topology, geometry and flow properties.

Chapter 1
ALE-FEM for Two-Phase and Free Surface Flows with Surfactants

Sashikumaar Ganesan, Andreas Hahn, Kristin Simon, and Lutz Tobiska

Abstract We study two-phase and free surface flows with soluble and insoluble surfactants. A numerical analysis of the contained convection-diffusion equations is carried out. The surface equation is stabilized by Local Projection Stabilization. The benefit of Local Projection Stabilization on surfaces is shown by a numerical example. An advanced finite element method that allow for a robust and accurate numerical simulation is presented. The arbitrary Langrangian-Eulerian framework is utilized to capture the moving surface. This allows the usage of a fitted finite element mesh. A decoupling strategy is used to divide the origin problem into subproblems easier to solve. Different time discretizations are considered and the problem of spurious velocities for the spatial discretization is discussed. Numerical examples in 2d and 3d illustrate the potential of the proposed algorithm. The comparison to mathematical predicted values validates the obtained results.

1.1 Introduction

The influence of surface active agents (surfactants) on the deformation of droplets and on the dynamics of the surrounding flow field is an active research area with numerous applications [12, 43]. In weak flows a nearly uniform concentration of the surfactants on the surface takes place and the behaviour of the flow is (after a suitable scaling) similar to flows without surfactants, but with reduced surface tension. Suppose that the flow field becomes stronger, thinning effects appear due to the stretching of the surface. This changes the surface tension locally. The convective transport induced by the flow field generates a local accumulation of surfactants and the resulting Marangoni forces may lead to a destabilization of the interface with essential consequences on the flow dynamics. This is a complex process, whose

S. Ganesan
Indian Institute of Science, Bangalore 560012, India
e-mail: sashi@cds.iisc.ac.in

A. Hahn • K. Simon • L. Tobiska (✉)
Otto von Guericke Universität Magdeburg, Fakultät für Mathematik, Universitätsplatz 2, 39016 Magdeburg, Germany
e-mail: andreas.hahn@ovgu.de; kristin.simon@ovgu.de; lutz.tobiska@ovgu.de

© Springer International Publishing AG 2017

D. Bothe, A. Reusken (eds.), *Transport Processes at Fluidic Interfaces*,
Advances in Mathematical Fluid Mechanics, DOI 10.1007/978-3-319-56602-3_1

tailored use in applications requires a fundamental understanding of the mutual interplay.

We present advanced finite element methods that allow for a robust and accurate numerical solution of the underlying system of partial differential equations. A complete model of two-phase and free surface flows with surfactants is stated in Sect. 1.2. We consider adsorption and desorption of surfactants at the sharp interface between two fluids and their transport and diffusion in the fluid phases and along the interface. One way to solve two-phase problems is to use a diffusive interface and consider the limit when the thickness of the interface goes to zero. This type of approach will not be considered here. For a detailed discussion we refer to [1, 16, 17]. Alternatively, sharp interface models can be discretized by fitted or unfitted finite elements. An overview on the general framework of unfitted or so-called CutFEM has recently been given in [10]. Using the CutFEM approach is convenient since the same finite element spaces defined on a background mesh can be used for solving the partial differential equations in the bulk and on the surface. However, a drawback is that the finite element matrices become arbitrarily ill-conditioned depending on the position of the surface in the background mesh. Therefore, although the fact that moving meshes have to be handled, we prefer fitted finite element discretizations, in which the interface is aligned with the mesh [19, 21, 24, 25].

In Sect. 1.3 we summarize the principles of discretizing partial differential equations on surfaces. The numerical analysis for the solution of a scalar convection-diffusion equation on a surface is discussed. We also consider the Local Projection Stabilization (LPS) to suppress spurious oscillations in convection dominated cases. Finally, we summarize our studies on a system of convection-diffusion equations in bulk domains coupled with a convection-diffusion equation on an embedded fixed surface as a model for the surfactant transport.

In Sect. 1.4 we compare interface capturing and interface tracking methods that handle partial differential equations on moving domains. The Arbitrary Langrangian–Eulerian (ALE) method is described and some techniques to retain the mesh quality without remeshing are discussed. We use discretizations of second order in space and time and propose a semi-implicit splitting of the two-phase flow problem with surfactants into smaller problems, a Navier–Stokes type problem and a problem for the transport of surfactants.

Selected test examples demonstrate in Sect. 1.5 the challenges and the potential of numerical simulations for a better understanding of the mutual interplay of different phenomena at fluidic interfaces.

1.2   Two-Phase and Free Surface Flows

We state a system of partial differential equations modeling two-phase and free surface flows. The weak formulation of the problem is given for a fixed domain.

1.2.1   Mathematical Model

We consider an incompressible two-phase flow with surfactants in a fixed bounded domain $\Omega \subset \mathbb{R}^d$, $d = 2, 3$. Let the liquid filling the domain $\Omega_1(t)$ at time $t \in [0, T]$ be completely surrounded by another liquid filling $\Omega_2(t) = \Omega \setminus \Omega_1(t)$. We assume that the two liquids are immiscible and separated by the sharp interface $\Gamma(t) = \partial \Omega_1(t)$. The model consists of the time-dependent incompressible Navier-Stokes equation in each phase

$$\rho_i \left(\partial_t \mathbf{u} + \mathbf{u} \cdot \nabla \mathbf{u} \right) - \nabla \cdot \mathbb{S}(\mathbf{u}, p) = \rho_i g \mathbf{e}, \qquad \nabla \cdot \mathbf{u} = 0 \quad \text{in } \Omega_i(t) \times (0, T], \quad (1.1)$$

for $i = 1, 2$, the initial conditions

$$\Omega_i(0) = \Omega_{i,0}, \qquad \mathbf{u}(\cdot, 0) = \mathbf{u}_0, \qquad\qquad (1.2)$$

the kinematic and force balance conditions

$$\mathbf{w} \cdot \mathbf{n} = \mathbf{u} \cdot \mathbf{n}, \quad [\mathbf{u}] = \mathbf{0}, \quad [\mathbb{S}(\mathbf{u}, p)] \cdot \mathbf{n} = \sigma(c_\Gamma) \mathscr{K} \mathbf{n} + \nabla_\Gamma \sigma(c_\Gamma) \quad \text{on } \Gamma(t), \quad (1.3)$$

and homogeneous Dirichlet-type boundary conditions on the fixed (in time) boundary $\partial \Omega$. Here, $\mathbf{u} = (u_1, \ldots, u_d)$ denotes the fluid velocity, p is the pressure, ρ_i, $i = 1, 2$, are the densities of the corresponding fluid phases, g is the gravitational constant, and $\mathbf{e}$ is a unit vector in the direction of the gravitational force. Further, $\mathbf{w}$ on $\Gamma(t)$ denotes the velocity of the interface, $\mathbf{n}$ is the outer unit normal vector (on $\Gamma(t)$ directed outward of $\Omega_1(t)$), $\mathscr{K}$ is the sum of principal curvatures, $[\cdot]$ denotes the jump across the interface $\Gamma(t)$, c_Γ is the surfactant concentration on the interface, $\sigma(c_\Gamma)$ is the surface tension coefficient depending on c_Γ, and ∇_Γ is the surface gradient.

The stress tensor $\mathbb{S}(\mathbf{u}, p)$ for a Newtonian incompressible fluid and the velocity deformation tensor $\mathbb{D}$ are given by

$$\mathbb{S}_i(\mathbf{u}, p) = 2\mu_i \mathbb{D}(\mathbf{u}) - p\mathbb{I}, \qquad \mathbb{D}(\mathbf{u}) = \frac{1}{2} \left(\nabla \mathbf{u} + (\nabla \mathbf{u})^T \right), \quad i = 1, 2,$$

where μ_i denotes the dynamic viscosity of the corresponding fluid phase, and $\mathbb{I}$ is the identity tensor.

In case of a free surface flow, we assume that the effect of the flow field in the surrounding phase is negligible and thus, $\Omega(t) := \Omega_1(t)$. The kinematic force balance condition (1.3) becomes

$$\mathbf{w} \cdot \mathbf{n} = \mathbf{u} \cdot \mathbf{n}, \quad \mathbb{S}(\mathbf{u}, p) \cdot \mathbf{n} = \sigma(c_\Gamma) \mathscr{K} \mathbf{n} + \nabla_\Gamma \sigma(c_\Gamma) \quad \text{on } \Gamma(t). \qquad (1.4)$$

Note that in case of a two-phase flow (in contrast to a free surface flow) the pressure is determined only up to an additive constant. For a detailed discussion on the uniqueness of the pressure, we refer to [22].

The model has to be completed by a set of equations that describe the surfactant transport in the bulk phases and on the interface. Let us assume that the surfactant is soluble in both phases. Then, for the surfactant concentration c_i, $i = 1, 2$, in the bulk phases we have

$$\partial_t c_i + \mathbf{u} \cdot \nabla c_i = \nabla \cdot (D_i \nabla c_i) \quad \text{in } \Omega_i(t) \times (0, T], \tag{1.5}$$

completed by the initial conditions

$$c_i \big|_{t=0} = c_{i,0} \qquad \text{in } \Omega_i(0), \tag{1.6}$$

and the boundary conditions

$$(-1)^i \mathbf{n} \cdot (D_i \nabla c_i) = S_i(c_i, c_\Gamma) \quad \text{on } \Gamma(t), \quad \mathbf{n} \cdot (D_2 \nabla c_2) = 0 \quad \text{on } \partial\Omega. \tag{1.7}$$

Here, D_i is the diffusion coefficient in phase i and S_i is the source term. The surfactant concentration c_Γ on the evolving surface satisfies the initial condition $c_\Gamma \big|_{t=0} = c_{\Gamma,0}$ and the partial differential equation

$$\dot{c}_\Gamma + (\nabla_\Gamma \cdot \mathbf{w}) c_\Gamma + \nabla_\Gamma \cdot [c_\Gamma (\mathbf{u} - \mathbf{w})] = \nabla_\Gamma \cdot (D_\Gamma \nabla_\Gamma c_\Gamma) + \sum_{i=1}^{2} S_i(c_i, c_\Gamma) \tag{1.8}$$

on $\Gamma(t) \times (0, T]$, where $\dot{c}_\Gamma$ denotes the material derivative with respect to $\mathbf{w}$ and D_Γ is the surface diffusion coefficient. It is assumed that the interface Γ is a closed surface, therefore no boundary condition is needed for (1.8).

Suppose the surfactant is soluble only in one phase or is even insoluble. Then, the corresponding equations in the other phase or in both phases are removed from the model (1.5)–(1.8).

Finally, the effect of the surfactant on the surface tension and the transport of surfactant between the interface and the bulk phase is modelled [15]. We consider the linear Henry equation of state

$$S_i(c_i, c_\Gamma) = k_{a,i} C_\Gamma^\infty c_i - k_{d,i} c_\Gamma \quad \text{and} \quad \sigma(c_\Gamma) = \sigma_0 \left(1 + E \left(1 - \frac{c_\Gamma}{C_\Gamma^\infty} \right) \right), \tag{1.9}$$

or the nonlinear Langmuir equation of state

$$S_i(c_i, c_\Gamma) = k_{a,i} (C_\Gamma^\infty - c_\Gamma) c_i - k_{d,i} c_\Gamma \quad \text{and} \quad \sigma(c_\Gamma) = \sigma_0 \left(1 + E \ln \left(1 - \frac{c_\Gamma}{C_\Gamma^\infty} \right) \right). \tag{1.10}$$

Here, $k_{a,i}$ and $k_{d,i}$ are the adsorption and desorption constants, respectively, σ_0 a reference surface tension, E the surface elasticity constant, and C_Γ^∞ a reference surfactant concentration (linear case) or the maximum surface surfactant packing concentration (nonlinear case).

1.2.2   Dimensionless Weak Formulation in Fixed Domains

We introduce dimensionless variables by setting

$$\tilde{\mathbf{x}} = \frac{\mathbf{x}}{L}, \ \tilde{\mathbf{u}} = \frac{\mathbf{u}}{U_\infty}, \ \tilde{\mathbf{w}} = \frac{\mathbf{w}}{U_\infty}, \ \tilde{t} = \frac{U_\infty t}{L}, \ \tilde{p} = \frac{p}{\rho_1 U_\infty^2}, \ \tilde{c}_\Gamma = \frac{c_\Gamma}{C_\Gamma^\infty}, \ \tilde{c}_i = \frac{c_i}{C_\infty},$$

where L is a characteristic length, U_∞ is a characteristic velocity, and C_∞ is a characteristic surfactant concentration in the bulk regions. In order to simplify the notations we drop the tilde afterwards. Then, the model equations are parameterised by the following dimensionless numbers: the Reynolds number Re, the Weber number We, the Froude number Fr, the Peclet numbers for the bulk surfactant transport Pe_i, the Peclet number for the surface surfactant transport Pe_Γ, the Biot numbers Bi_i, the Damköhler numbers Da_i, the surface elasticity E, the surfactant scaling β, and the dimensionless scaling factors ρ_2/ρ_1 and μ_2/μ_1. In more detail, the dimensionless numbers are given by

$$\mathrm{Re} = \begin{cases} Re_1 & \mathbf{x} \in \Omega_1 \\ Re_1\mu_1/\mu_2 & \mathbf{x} \in \Omega_2 \end{cases}, \quad \text{where } Re_1 = \frac{\rho_1 U_\infty L}{\mu_1},$$

$$\mathrm{We} = \frac{\rho_1 U_\infty^2 L}{\sigma_0}, \quad \mathrm{Fr} = \frac{U_\infty^2}{Lg}, \quad \mathrm{Pe}_i = \frac{U_\infty L}{D_i}, \quad \mathrm{Pe}_\Gamma = \frac{U_\infty L}{D_\Gamma},$$

$$\mathrm{Bi}_i = \frac{k_{d,i} L}{U_\infty}, \quad \mathrm{Da}_i = \frac{k_{a,i} C_\Gamma^\infty}{U_\infty}, \quad \beta = \frac{C_\Gamma^\infty}{L C_\infty},$$

$$\text{with } i = 1, 2, \text{ and } \rho = \begin{cases} 1 & \mathbf{x} \in \Omega_1 \\ \rho_2/\rho_1 & \mathbf{x} \in \Omega_2 \end{cases}.$$

The weak formulation of the problem is derived as usual. Multiplying with test functions $\mathbf{v} \in V(\Omega) := H_0^1(\Omega)^d$, $q \in Q(\Omega) := L_0^2(\Omega)$, $\varphi \in G(\Omega) := H^1(\Omega_1) \times H^1(\Omega_2)$, $\psi \in M(\Gamma) := H^1(\Gamma)$, applying integration by parts to remove the highest order of differentiation, and incorporating the boundary conditions, we obtain the weak formulation of the two-phase problem:

Problem 1 Find $(\mathbf{u}, p, c, c_\Gamma) \in V \times Q \times G \times M$ such that for all $(\mathbf{v}, q, \varphi, \psi) \in V \times Q \times G \times M$

$$(\rho \partial_t \mathbf{u}, \mathbf{v}) + a(\mathbf{u}, \mathbf{u}, \mathbf{v}) - b(p, \mathbf{v}) + b(q, \mathbf{u}) = f(c_\Gamma, \mathbf{v}),$$

$$(\partial_t c, \varphi) + a_c(\mathbf{u}, c, \varphi) = f_c(c, c_\Gamma, \varphi),$$

$$\frac{d}{dt} \int_\Gamma c_\Gamma \psi \, d\gamma + a_\Gamma(\mathbf{u} - \mathbf{w}, c_\Gamma, \psi) = f_\Gamma(c, c_\Gamma, \psi).$$

For the case of a free surface flow we set $\Omega(t) = \Omega_1(t)$, $Q(\Omega) := L^2(\Omega)$, and the equations on $\Omega_2(t)$ are omitted. The above forms are given by

$$a(\mathbf{z}, \mathbf{u}, \mathbf{v}) = \left(\frac{2}{\text{Re}} \mathbb{D}(\mathbf{u}), \mathbb{D}(\mathbf{v}) \right) + (\rho(\mathbf{z} \cdot \nabla)\mathbf{u}, \mathbf{v})$$

$$b(q, \mathbf{v}) = (q, \nabla \cdot \mathbf{v})$$

$$f(c_\Gamma, \mathbf{v}) = \frac{1}{\text{Fr}} (\rho\mathbf{e}, \mathbf{v}) - \frac{1}{\text{We}} \langle \sigma(c_\Gamma)(\mathbb{I} - \mathbf{n} \otimes \mathbf{n}), \nabla_\Gamma \mathbf{v} \rangle_\Gamma$$

$$a_c(\mathbf{z}, c, \varphi) = \sum_{i=1}^{2} \left[\frac{1}{\text{Pe}_i} (\nabla c_i, \nabla \varphi_i)_{\Omega_i} + (\mathbf{z} \cdot \nabla c_i, \varphi_i)_{\Omega_i} \right]$$

$$f_c(c, c_\Gamma, \varphi) = -\beta \sum_{i=1}^{2} \langle S_i(c_i, c_\Gamma), \varphi_i \rangle_\Gamma$$

$$a_\Gamma(\mathbf{z}, c_\Gamma, \psi) = \frac{1}{\text{Pe}_\Gamma} \langle \nabla_\Gamma c_\Gamma, \nabla_\Gamma \psi \rangle_\Gamma - \langle c_\Gamma \mathbf{z}, \nabla_\Gamma \psi \rangle_\Gamma$$

$$f_\Gamma(c, c_\Gamma, \psi) = \sum_{i=1}^{2} \langle S_i(c_i, c_\Gamma), \psi \rangle_\Gamma .$$

Further, $(\cdot, \cdot)$, $(\cdot, \cdot)_{\Omega_i}$, and $\langle \cdot, \cdot \rangle_\Gamma$ denote the inner product in $L^2(\Omega)$, $L^2(\Omega_i)$, and $L^2(\Gamma)$, respectively, as well as its vector- and tensor-valued versions. The dimensionless surface tension law and the dimensionless source terms are given by

$$\sigma(c_\Gamma) = 1 + E(1 - c_\Gamma), \qquad S(c_i, c_\Gamma) = \frac{\text{Da}_i}{\beta} c_i - \text{Bi}_i c_\Gamma$$

in case of the Henry sorption isotherm and by

$$\sigma(c_\Gamma) = 1 + E \ln(1 - c_\Gamma), \qquad S(c_i, c_\Gamma) = \frac{\text{Da}_i}{\beta} c_i(1 - c_\Gamma) - \text{Bi}_i c_\Gamma$$

for the Langmuir sorption isotherm.

1.3 Finite Element Methods on Fixed Surfaces

The weak formulation of the two-phase or free surface flow contains an convection-diffusion type surface equation. Whereas finite element methods for bulk equations are well studied, the extension of these techniques to curved surfaces is an area of current research. The development of finite elements on surfaces started with the study of the Laplace-Beltrami equation on a fixed surface in [13]. An overview of the recent state of the art including different approaches and moving surfaces is given in [14].

In this section we present a finite element method for surface equations of convection-diffusion type. Additionally, we introduce a LPS technique for surface equations to reduce oscillations occurring at interior layers. The last part of this section handles a coupled convection-diffusion problem in the bulk phases and on the surface. Discretization techniques, solvability and error estimates are given.

1.3.1 Advection-Diffusion Problem

We study a steady state convection-diffusion equation on a given closed surface Γ instead of the surface equation (1.8) of the origin problem. By partial integration of the diffusion term we get the following problem

Problem 2 Find $u \in H^1(\Gamma)$ such that for all $v \in H^1(\Gamma)$

$$\varepsilon \langle \nabla_\Gamma u, \nabla_\Gamma v \rangle_\Gamma + \langle \nabla_\Gamma u, \mathbf{w} v \rangle_\Gamma + \langle \sigma u, v \rangle_\Gamma = \langle f, v \rangle_\Gamma .$$

Here, $\mathbf{w}$ and f are the given velocity field and right hand side, and ε the diffusion coefficient. We set $\sigma = \nabla_\Gamma \cdot \mathbf{w} + c$ with c being the reaction coefficient. Since Γ is not moving in time, we assume $\mathbf{w} \cdot \mathbf{n} = 0$ on Γ. Further, we suppose $0 < \sigma_0 \leq \sigma - \frac{1}{2}\nabla_\Gamma \cdot \mathbf{w} \leq \sigma_\infty$ to get unique solvability of the problem.

1.3.2 Surface Approximation and Discrete Problem

In the discretization of the surface equation in Problem 2 using finite elements several additional particularities have to be taken into account.

The given surface Γ is discretized by Γ_h, such that all nodes of Γ_h are on Γ. The geometric error introduced by the different integration domains and surface operators is estimated in [14]. We consider isoparametric surface approximation of order $k \geq 1$. In the simplest case, $k = 1$, the isoparametric surface Γ_h is a linear interpolation of Γ by flat simplices. In the general case, Γ_h is the union of curved simplices K, which are given as the image of a bijective mapping $F_K \in \mathbb{P}_k(\widehat{K})^d$ of the reference simplex $\widehat{K} \subset \mathbb{R}^{d-1}$. Thereby, $\mathbb{P}_k(\widehat{K})$ is the space of polynomials of degree less than or equal to k. Hence, F_K is a parametrisation of the element K over the reference triangle $\widehat{K}$, which has to be taken into account. Additionally, we need to extend the quantities given on Γ to a neighbourhood U of Γ including Γ_h. The extension operator $.^e$ is given as the constant extension along the surface normal.

Introducing a mapped Lagrangian surface finite element space of order k via

$$V_h = \{v_h \in H^1\left(\Gamma_h\right) \mid v_h \circ F_K \in \mathbb{P}_k(\widehat{K}) \text{ for all elements } K \text{ in } \Gamma_h\} \qquad (1.11)$$

the discretized problem reads:

Problem 3 Find $u_h \in V_h$ such that for all $v_h \in V_h$

$$a_h\left(u_h, v_h\right) = f_h\left(v_h\right)$$

with

$$a_h\left(u_h, v_h\right) = \varepsilon \left\langle \nabla_{\Gamma_h} u_h, \nabla_{\Gamma_h} v_h \right\rangle_{\Gamma_h} + \left\langle \nabla_{\Gamma_h} u_h, \mathbf{w}^e v_h \right\rangle_{\Gamma_h} + \left\langle \sigma^e u_h, v_h \right\rangle_{\Gamma_h}$$

$$f_h\left(v_h\right) = \left\langle f^e, v_h \right\rangle_{\Gamma_h}.$$

We want to point out that σ is discretized by $\sigma^e = \left(\nabla_\Gamma \cdot \mathbf{w}\right)^e + c^e$ instead of $\nabla_{\Gamma_h} \cdot \mathbf{w}^e + c^e$. This transfers the assumptions from σ on Γ to σ^e on Γ_h. However, an estimate of the difference between the discrete divergence $\nabla_{\Gamma_h} \cdot \mathbf{w}^e$ and the discretized divergence $\left(\nabla_\Gamma \cdot \mathbf{w}\right)^e$ can be shown

$$\left\| \nabla_{\Gamma_h} \cdot \mathbf{w}^e - \left(\nabla_\Gamma \cdot \mathbf{w}\right)^e \right\|_{\infty, \Gamma_h} \leq C h^k \left\| \nabla_\Gamma \mathbf{w} \right\|_{\infty, \Gamma}. \qquad (1.12)$$

1.3.3 Standard Galerkin FE Method

Using (1.12) the coercivity of the bilinear form a_h in V_h equipped with the standard H^1-norm is proven for h being small enough uniformly in ε and unique solvability follows. Considering convection dominated problems one is interested in error estimates uniformly in ε. We get coercivity of a_h in V_h equipped with the norm

$$\|v\| := \left(\varepsilon \|\nabla_{\Gamma_h} v\|_{0,\Gamma_h}^2 + \sigma_0 \|v\|_{0,\Gamma_h}^2\right)^{1/2}.$$

with a constant independent of ε.

The solution $u : \Gamma \to \mathbb{R}$ of the continuous problem and the solution $u_h : \Gamma_h \to \mathbb{R}$ are defined on different domains, thus cannot be compared directly. Therefore, the error between u_h and the extension of u onto Γ_h is evaluated. Using an isoparametric Galerkin approach of order k we obtain

$$\|u^e - u_h\| \leq C \left(h^{k+1}\left(\|u\|_{k,\Gamma} + \|f\|_{0,\Gamma}\right) + h^k \|u\|_{k+1,\Gamma}\right)$$

with a geometric error of order $k + 1$ and a finite element error of order k.

1.3.4 Local Projection Stabilization

For convection-diffusion equations in the bulk, it is well known that standard Galerkin finite element methods can lead to unphysical oscillations, if convection dominates diffusion and layers are unresolved by the mesh. Several stabilization techniques as Streamline Upwind Petrov Galerkin [29, 35], Continuous Interior Penalty [8, 9], and LPS [5, 33] have been developed for bulk equations. However, for surface equations only the Streamline Upwind Petrov Galerkin method on unfitted finite elements has been studied so far [38]. We present a LPS technique for fitted finite elements on surfaces.

LPS is based on an additional control over the gradient of the solution by adding a stabilization term to the discrete bilinear form a_h. To define the stabilization term a discontinuous projection space D_h is needed. The restriction of D_h to one element K is denoted by D_K. We introduce an elementwise L^2 projection $\pi_K : L^2(K) \to D_K$ into the projection space. We define the fluctuation operator $\kappa_K : L^2(K) \to L^2(K)$ as $\kappa_K = id - \pi_K$ and set the stabilization term to

$$LPS(u_h, v_h) = \sum_{K \in \Gamma_h} \gamma_K \langle \kappa_K (\nabla_{\Gamma_h} u_h), \kappa_K (\nabla_{\Gamma_h} v_h) \rangle_K.$$

Here γ_K are the stabilization parameters. The error analysis of the method provides that γ_K has to be chosen as $\gamma_K = \gamma^2 h_K$ with a fixed γ to get the optimal convergence order. The best value of γ is problem dependent and has to be found empirically. The stabilized problem reads:

Problem 4 Find $u_h \in V_h$ such that for all $v_h \in V_h$

$$a_s(u_h, v_h) := a_h(u_h, v_h) + LPS(u_h, v_h) = f_h(v_h).$$

According to the bilinear form a_s we introduce the s-triple norm

$$\|\|v\|\|_s = \left(\varepsilon \|\nabla_{\Gamma_h} v\|_{0,\Gamma_h}^2 + \sigma_0 \|v\|_{0,\Gamma_h}^2 + \sum_K \gamma_K \|\kappa_K (\nabla_{\Gamma_h} v)\|_{0,K}^2 \right)^{1/2}.$$

By construction we get the coercivity of a_s in V_h equipped with the s-triple norm under the same restriction for h as in the standard Galerkin case. Using the Lax-Milgram lemma the unique solvability of Problem 4 is shown.

We can improve the error estimate from Sect. 1.3.3 utilize the stronger s-triple norm following the idea from [33]. To this end, we assume the existence of a D_h-orthogonal interpolator $j_h : H^2(\Gamma_h) \to V_h$ with interpolation order k. The existence of such an interpolator depends on the choice of V_h and D_h and is related to a local inf-sup condition. Common pairs on triangles are $V_h = \mathbb{P}_k^+$, e.g. $\mathbb{P}_k$ enriched by $b \cdot \mathbb{P}_{k-1}$ where $b|_K \in H_0^1(K)$, $\forall K \in \Gamma_h$, is a piecewise cubic function, and $D_h = \mathbb{P}_{k-1}^{\text{disc}}$ [33].

For an isoparametric approach, e.g. $V_h = \mathbb{P}_k^+$, $D_h = \mathbb{P}_{k-1}^{\mathrm{disc}}$ and surface approximation with polynomials of degree k, we get the following convergence estimate

$$\||u^e - u_h\||_s \leq Ch^{k+1}\left(\|u\|_{r,\Gamma} + \|f\|_{0,\Gamma}\right) + Ch^k\left(\varepsilon^{1/2} + \gamma h^{1/2}\right)\|u\|_{k+1,\Gamma}.$$

Thus, for LPS we have stability in the stronger s-triple norm and an improved $k+1/2$ order of convergence in the convection dominated case $\varepsilon < h$.

1.3.5 Coupled Bulk-Surface Transport Problem

The next step is to couple the investigated surface equation on Γ to steady-state convection-diffusion equations in the bulk domains Ω_i. We study the problem

$$- \varepsilon_i \Delta u_i + \mathbf{w} \cdot \nabla u_i = f_i \qquad \text{in } \Omega_i, \ i = 1, 2, \quad (1.13)$$

$$-\varepsilon_\Gamma \Delta_\Gamma u_\Gamma + \mathbf{w} \cdot \nabla_\Gamma u_\Gamma = L\sum_{i=1}^2 S_i\left(u_i, u_\Gamma\right) + f_\Gamma \quad \text{on } \Gamma, \qquad (1.14)$$

$$(-1)^i \varepsilon_i \frac{\partial u_i}{\partial \mathbf{n}} = S_i\left(u_i, u_\Gamma\right) \qquad \text{on } \Gamma, \ i = 1, 2, \quad (1.15)$$

$$\varepsilon_2 \frac{\partial u_2}{\partial \mathbf{n}} = 0 \qquad \text{on } \partial\Omega \qquad (1.16)$$

using the linear Henry source term, compare Sect. 1.2.1. We assume that the velocity field $\mathbf{w}$ satisfies $\nabla \cdot \mathbf{w} = 0$ in the bulk phases Ω_i, $i = 1, 2$, $\nabla_\Gamma \cdot \mathbf{w} = 0$ on the interface Γ, $\mathbf{w} \cdot \mathbf{n} = 0$ on $\Gamma \cup \partial\Omega$ and $\mathbf{f} = (f_1, f_2, f_\Gamma)$ fulfilling the solvability condition

$$\int_{\Omega_1} f_1 \, dx + \int_{\Omega_2} f_2 \, dx + \frac{1}{L}\int_\Gamma f_\Gamma \, d\gamma = 0.$$

The solution $\mathbf{u} = (u_1, u_2, u_\Gamma)$ of this problem is only fixed up to an additive constant $\lambda \mathbf{k}$, with $\mathbf{k} = (k_1, k_2, k_\Gamma)$ and $S_i\left(k_i, k_\Gamma\right) = 0$, $i = 1, 2$. To get uniqueness the condition of a given mass M is added.

We build the weak formulation summing up scaled weak formulations of the single equations such that the terms coming from the source term become symmetric in u and v. We end up with the bilinear form

$$a(\mathbf{u}, \mathbf{v}) := \sum_{i=1}^2 \alpha_i \left[\varepsilon_i \left(\nabla u_i, \nabla v_i\right)_{\Omega_i} + \left(\mathbf{w} \cdot \nabla u_i, v_i\right)_{\Omega_i}\right]$$

$$+\alpha_\Gamma \left[\varepsilon_\Gamma \langle\nabla_\Gamma u_\Gamma, \nabla_\Gamma v_\Gamma\rangle_\Gamma + \langle\mathbf{w} \cdot \nabla_\Gamma u_\Gamma, v_\Gamma\rangle_\Gamma\right]$$

$$+k_{d,2} \langle S_1\left(u_1, u_\Gamma\right), S_1\left(v_1, v_\Gamma\right)\rangle_\Gamma + k_{d,1} \langle S_2\left(u_2, u_\Gamma\right), S_2\left(v_2, v_\Gamma\right)\rangle_\Gamma,$$

where we set $\alpha_1 := k_{a,1}k_{d,2}$, $\alpha_2 := k_{a,2}k_{d,1}$ and $\alpha_\Gamma := k_{d,1}k_{d,2}/L$. One can easily check that under the additional mass condition the weak formulation is uniquely solvable.

For discretization we use an interface fitted triangulation of Ω. Let Γ_h be the piecewise linear approximation of Γ as described in Sect. 1.3.2. The polygonal approximations $\Omega_{1,h}$ and $\Omega_{2,h}$ of Ω_1 and Ω_2 are aligned to Γ_h, e.g. $\overline{\Omega}_{1,h} \cap \overline{\Omega}_{2,h} = \Gamma_h$. The linear surface approximation leads to a geometric error of order two for the surface terms but for bulk terms we get only an order of $3/2$.

The approximation space V_h is build by linear continuous finite elements in the bulk and on the surface. Then, introducing a discretized mass condition the discretized problem reads:

Problem 5 Find $\mathbf{u}_h \in V_h$ fulfilling the discretized mass condition such that

$$a_h\left(\mathbf{u}_h, \mathbf{v}_h\right) = f\left(\mathbf{v}_h\right) := \sum_{i=1}^{2} \alpha_i \left(f_i, v_{h,i}\right)_{\Omega_i} + \alpha_\Gamma \langle f_\Gamma, v_{h,\Gamma} \rangle_\Gamma \qquad \text{for all } \mathbf{v}_h \in V_h$$

with

$$a_h\left(\mathbf{u}_h, \mathbf{v}_h\right) := \sum_{i=1}^{2} \alpha_i \varepsilon_i \left(\nabla u_{h,i}, \nabla v_{h,i}\right)_{\Omega_{h,i}} + \alpha_\Gamma \varepsilon_\Gamma \langle \nabla_{\Gamma_h} u_{h,\Gamma}, \nabla_{\Gamma_h} v_{h,\Gamma} \rangle_{\Gamma_h}$$

$$+ \sum_{i=1}^{2} \frac{\alpha_i}{2} \left[\left(\mathbf{w}^e \cdot \nabla u_{h,i}, v_{h,i}\right)_{\Omega_{h,i}} - \left(\mathbf{w}^e \cdot \nabla v_{h,i}, u_{h,i}\right)_{\Omega_{h,i}} \right]$$

$$+ \frac{\alpha_\Gamma}{2} \left[\langle \mathbf{w}^e \cdot \nabla_{\Gamma_h} u_{h,\Gamma}, v_{h,\Gamma} \rangle_{\Gamma_h} - \langle \mathbf{w}^e \cdot \nabla_{\Gamma_h} v_{h,\Gamma}, u_{h,\Gamma} \rangle_{\Gamma_h} \right]$$

$$+ k_{2,d} \langle S_1\left(u_{h,1}, u_{h,\Gamma}\right), S_1\left(v_{h,1}, v_{h,\Gamma}\right) \rangle_{\Gamma_h}$$

$$+ k_{1,d} \langle S_2\left(u_{h,2}, u_{h,\Gamma}\right), S_2\left(v_{h,2}, v_{h,\Gamma}\right) \rangle_{\Gamma_h}.$$

In the space V_h restricted to the functions fulfilling the discretized mass condition and equipped with the corresponding energy norm coercivity of a_h is shown. Unique solvability follows directly.

Unfortunately, for this coupled bulk surface transport problem we do not have L^2-control uniformly in the diffusion parameters ε_1, ε_2, ε_Γ as for the Problem 2. Nevertheless, numerical tests show that the LPS improves stability and for fixed ε first order convergence can be established.

In [28] a slightly different weak formulation of (1.13)–(1.16), leading to a non-coercive bilinear form, has been studied for an unfitted finite element method without stabilization. In addition to energy-type estimates an optimal L^2-estimate has been established.

1.4   Finite Element Methods in Moving Domains

We describe the numerical scheme developed for the two-phase model (1.1)–(1.10). One of the main characteristic is the application of the ALE method to handle unknown moving interfaces.

1.4.1   Interface Capturing and Tracking Methods

The interface position has to be determined by the solution of the model equations. Two main classes of methods have been developed in the past, the interface capturing methods and the interface tracking methods.

Both methods have their advantages and disadvantages. The interface capturing methods can handle topological changes, like break-up and coalescence of phases, well and automatically. On the other hand, the handling of the marker function and the reconstruction of the interface induces several difficulties, like the accurate incorporation of surface tension forces, due to the implicit nature of the method. On the contrary, handling topological changes is very difficult with the tracking methods, but incorporation of surface properties, like surface tension, is quite easy.

In the class of tracking methods, the distinction in unfitted and fitted schemes gives further choice with correlated advantages and disadvantages. The unfitted methods are easy to implement, since the marker grid is independent from the bulk grid, hence existing code can be extended easily. But, since unfitted methods cut cells, they introduce discontinuities in the interior of a mesh cell. Special methods, like XFEM, had to be developed to handle those discontinuities. The fitted schemes do not cut cells, hence discontinuities appear only across cell boundaries and are easy to handle. But, special treatment is need for the bulk mesh, otherwise the alignment is lost. The bulk grid can no longer be a fixed grid, therefore special methods, like the ALE method, had to be developed.

1.4.2   Arbitrary Lagrangian-Eulerian Method

The ALE method is an interface tracking method [36]. The tracking grid, i.e. the discrete interface, is part of the bulk grid itself, consisting of faces of it.

At each instance of time t, the evolving domain is described by a mapping $\mathscr{A}_t : \hat{\Omega} \rightarrow \Omega(t)$ from a reference domain $\hat{\Omega}$ onto the domain $\Omega(t)$. A point $\mathbf{x}$ in the domain $\Omega(t)$ is given by $\mathbf{x} = \mathscr{A}_t(\hat{\mathbf{x}})$. Functions $\hat{v}(t, \hat{\mathbf{x}})$ on the reference domain and functions $v(t, \mathbf{x})$ on the evolving domain are related through $\hat{v}(t, \hat{\mathbf{x}}) = v(t, \mathscr{A}_t(\hat{\mathbf{x}}))$.

Time derivatives can be expressed in a Lagrangian frame, respective the domain velocity $\mathbf{w}$, such that a differential equation can be given regarding the evolving domain. In case of a convection-diffusion equation, the ALE form is given as

$$(\partial_t \hat{c}) \circ \mathscr{A}_t^{-1} + D\Delta c + (\mathbf{u} - \mathbf{w}) \cdot \nabla c = 0.$$

Similarly, the Navier-Stokes equations are transformed as

$$(\partial_t \hat{\mathbf{u}}) \circ \mathscr{A}_t^{-1} + (\mathbf{u} - \mathbf{w}) \cdot \nabla \mathbf{u} - \nabla \cdot \mathbb{S}(\mathbf{u}, p) = g\mathbf{e}$$

In both cases an additional convection term with the domain velocity $\mathbf{w}$ occurs. The time derivative is now a material derivative with respect to $\mathbf{w}$. Note that the equations are evaluated on the evolving domain, also called Eulerian frame. The ALE form of the partial differential equations becomes the starting point for our finite element discretization.

The domain velocity is, to a certain degree, arbitrary, since the model prescribes only the normal component on the interface. The choice for the extension of $\mathbf{w}$ into the bulk domain is of importance for the stability of the scheme. Different choices and their advantages and disadvantages are discussed in the next subsection.

1.4.3 Computation of the Domain Velocity

During the evolution of the domain the grid quality may undergo a strong deterioration. Since the grid quality is important for the accuracy of the finite element solution, this has to be avoided. If the grid quality becomes poor, a remeshing has to be done, which is essentially a restart of the simulation. A new grid with good quality is generated from the old grid, the current solution is transferred to the new grid, and the simulation continues. This step is very costly and introduces additional errors. Thus, it is best to have as few remeshing steps as possible.

The freedom in the domain velocity, introduced by the ALE method, is used to decrease the number of remeshing steps, or to avoid them at all. Two choices are important, how to advance of the interface and how to extend the interface velocity to the bulk domain velocity.

A simple method to extend the domain velocity is to do a harmonic extension of the interface velocity. The harmonic extension is very fast, since it has a special block structure, which allows to solve for each component of $\mathbf{w}$ with one smaller system. Further, it can be shown that the harmonic extension is the method which shows minimal distortion in a L^2-sense [37].

A second method, that was used, is the linear elastic extension, where the grid is treated as a linear elastic body. The solution of the resulting system is more expensive, due to the coupling of the velocity components. But, under certain circumstances, the method shows better results in regions of large deformations [32].

In both cases, the additional problem of the domain velocity extension is solved by a first order finite element method. Since the velocity extension step has to be done in each nonlinear iteration step of the Navier-Stokes solver, the performance of the velocity extension step is crucial. We prefer the harmonic extension method, due to its features: being fast and overall good.

More important for the grid quality, than the velocity extension into the bulk, is the choice of the interface velocity. Since the interface velocity is coupled to the fluid motion, most of the domain deformation happens there. However, the freedom of choice for the tangential surface velocity can be used to optimize the point distribution and grid cell shape on the interface.

An obvious method to advance the interface is by moving the grid nodes in a classic way, as a function of the fluid velocity

$$\dot{\mathbf{x}} = \mathbf{w}|_\Gamma = \mathbf{F}(\mathbf{u}) \quad \text{on } \Gamma. \tag{1.17}$$

$\mathbf{F}$ can be chosen to be, either the full fluid velocity $\mathbf{F}(\mathbf{u}) = \mathbf{u}|_\Gamma$, or the normal component of the fluid velocity $\mathbf{F}(\mathbf{u}) = (\mathbf{n} \cdot \mathbf{u}|_\Gamma)\mathbf{n}$. The classical discretization of this equation, e.g. with an Euler method, results in a stepping scheme for the n-th grid node $\mathbf{x}_n^i$ at time step i

$$\mathbf{x}_n^{i+1} = \mathbf{x}_n^i + \Delta t \mathbf{F}(\mathbf{u}_n^{i+1}).$$

In the continuous setting, both choices for $\mathbf{F}$ fulfill the kinematic interface condition of the model and result in identical interfaces, since they distinguish in the tangential component only. In the discrete setting, moving in normal direction only, can prevent local accumulation or coarsening of nodes. On the other hand, a local accumulation is often appreciated in order to resolve the local structure of the interface, hence full velocity has to be favored in those cases. The decision, which version is best, is not obvious and has to be taken from case to case. However, both versions suffer from the general problem that there is absolute no control of the node distribution and of the shape of surface cells.

In order to get more control over the node distribution, a new method has been developed by Barrett et al. [4]. A weak formulation of (1.17) with $\mathbf{F}(\mathbf{u}) = (\mathbf{n} \cdot \mathbf{u}|_\Gamma)\mathbf{n}$ was used and completed with an equation for the curvature using the weak form of Laplace-Beltrami identity $\Delta_\Gamma \mathrm{id} = \mathscr{K}\mathbf{n}$.

Problem 6 For given $\mathbf{x}^n \in \mathbb{P}_k(\Gamma_h)^d$ find $(\mathbf{x}^{n+1}, \mathscr{K}) \in \mathbb{P}_k(\Gamma_h)^d \times \mathbb{P}_k(\Gamma_h)$ such that for all $(\eta, \xi) \in \mathbb{P}_k(\Gamma_h)^d \times \mathbb{P}_k(\Gamma_h)$

$$\left\langle \mathbf{x}^{n+1} - \mathbf{x}^n, \xi\mathbf{n} \right\rangle_\Gamma^h - \Delta t_n \left\langle \mathbf{u}, \xi\mathbf{n} \right\rangle_\Gamma = 0,$$

$$\left\langle \mathscr{K}\mathbf{n}, \eta \right\rangle_\Gamma^h + \left\langle \nabla_\Gamma \mathbf{x}^{n+1}, \nabla_\Gamma \eta \right\rangle_\Gamma = 0.$$

By using piecewise linear finite elements and a lumped version $\langle \cdot, \cdot \rangle^h$ of the bilinear form of certain terms it can be shown that the semi-discrete, time continuous problem has the property of equi distributed nodes in the 2d case. The fully discrete scheme reaches equi distribution of nodes after a few steps [3].

This approach has been extended to piecewise quadratic isoparametric finite elements. Unfortunately, it turned out that the good properties are not transferring. The equi distribution of nodes is lost or takes an unrealistic amount of iteration steps and therefore cannot keep up with the fluid motion. A solution to keep the very good properties of the scheme for piecewise linear elements for isoparametric elements too, is to apply the scheme on a refined piecewise linear grid defined by all nodes of the isoparametric grid and piecewise linear elements. Further, using the full bilinear forms instead of the lumped ones, the nodes distribute according to the curvature of the interface. In regions of high curvature a higher number of nodes accumulate than in regions of low curvature, which is a beneficial property, since it uses the existing nodes efficiently for problems with varying curvature.

By using a linear combination of the lumped and non-lumped scheme, the user gets a parameter to control the node distribution from equi distributed to curvature dependent.

1.4.4 Discretization in Time and Space

We consider the time discretization now. Let $0 = t_0 < t_1 < \cdots < t_N = T$ be a decomposition of a time interval $[0, T]$ and $\Delta t_n = t_{n+1} - t_n$ be the time step size from time t_n to t_{n+1}. Discrete functions and domains at time t_n, $n = 0, \ldots, N$ get a superscript n, e.g. $\mathbf{u}^n$ and Ω^n. To emphasize the integration domain the forms get a superscript too. A generalized semi-discrete in time scheme of the two-phase flow problem with soluble surfactants reads:

Problem 7 For given Ω_k^n, $k = 1, 2$, $\mathbf{u}^n$, $\mathbf{w}^n$, p^n, c^n and c_Γ^n find $(\mathbf{u}^{n+1}, p^{n+1}, c^{n+1}, c_\Gamma^{n+1}) \in V(\Omega^{n+1}) \times Q(\Omega^{n+1}) \times G(\Omega^{n+1}) \times M(\Gamma^{n+1})$ such that for all $(\mathbf{v}, p, \varphi, \psi) \in V(\Omega^{n+1}) \times Q(\Omega^{n+1}) \times G(\Omega^{n+1}) \times M(\Gamma^{n+1})$

$$\left(\rho \mathbf{u}^{n+1}, \mathbf{v}\right)^{n+1} + \alpha \Delta t_n \left[a^{n+1}\left(\mathbf{u}^{n+1} - \mathbf{w}^{n+1}, \mathbf{u}^{n+1}, \mathbf{v}\right)_{\Omega^{n+1}} - f^{n+1}(c_\Gamma^{n+1}, \mathbf{v})\right]$$
$$+ \Delta t_n \left[b^{n+1}(q, \mathbf{u}^{n+1}) - b^{n+1}(p^{n+1}, \mathbf{v})\right]$$
$$= \left(\rho \mathbf{u}^n, \mathbf{v}\right)^n - \beta \Delta t_n \left[a^n\left(\mathbf{u}^n - \mathbf{w}^n, \mathbf{u}^n, \mathbf{v}\right) - f^n(c_\Gamma^n, \mathbf{v})\right] \qquad (1.18)$$

$$\left(c^{n+1}, \varphi\right)^{n+1} + \alpha \Delta t_n \left[a_c^{n+1}\left(\mathbf{u}^{n+1} - \mathbf{w}^{n+1}, c^{n+1}, \varphi\right) - f_c^{n+1}(c^{n+1}, c_\Gamma^{n+1}, \varphi)\right]$$
$$= \left(c^n, \varphi\right)^n - \beta \Delta t_n \left[a_c^n\left(\mathbf{u}^n - \mathbf{w}^n, c^n, \varphi\right) - f_c^n(c^n, c_\Gamma^n, \varphi)\right] \qquad (1.19)$$

$$\left(c_\Gamma^{n+1}, \psi\right)^{n+1} + \alpha \Delta t_n \left[a_\Gamma^{n+1}\left(\mathbf{u}^{n+1} - \mathbf{w}^{n+1}, c_\Gamma^{n+1}, \psi\right) - f_\Gamma^{n+1}(c^{n+1}, c_\Gamma^{n+1}, \psi)\right]$$
$$= \left(c_\Gamma^n, \psi\right)^n - \beta \Delta t_n \left[a_\Gamma^n\left(\mathbf{u}^n - \mathbf{w}^n, c_\Gamma^n, \psi\right) - f_\Gamma^n(c^n, c_\Gamma^n, \psi)\right] \qquad (1.20)$$

Table 1.1 Coefficients for time stepping schemes

	Forward Euler	Backward Euler	Crank-Nicolson	Fractional step-θ		
				Substep 1	Substep 2	Substep 3
α	0	1	0.5	0.585786	0.414214	0.585786
β	1	0	0.5	0.414214	0.585786	0.414214

For different coefficients α and β, different time stepping methods are obtained. In particular, the backward Euler scheme, the forward Euler scheme and the Crank-Nicolson scheme, see Table 1.1. By combining three time steps, of size θ, $1 - 2\theta$, and θ, to one step $(t_{3n}, t_{3(n+1)})$ and with $\theta = 1 - \sqrt{2}/2$ and choosing the coefficient according to Table 1.1 the fractional-step-θ scheme is obtained [42].

The equation system given in Problem 7 is highly nonlinear. Decoupling strategies are developed in order to split the problem into several smaller and simpler problems. Considering c_Γ explicit in the Navier-Stokes equations, they decouple from the surfactant transport equations. Further, the integration domains Ω and Γ are taken explicit in the Navier-Stokes step, but the domain velocity $\mathbf{w}$ is updated. The resulting problem is a standard Navier-Stokes problem with a modified convective velocity $\mathbf{u}^{n+1} - \mathbf{w}^{n+1}$.

Problem 8 For given Ω_k^n, $k = 1, 2$, $\mathbf{u}^n$, $\mathbf{w}^n$, p^n, and c_Γ^n find $(\mathbf{u}^{n+1}, p^{n+1}) \in V(\Omega^n) \times Q(\Omega^n)$ such that for all $(\mathbf{v}, p) \in V(\Omega^n) \times Q(\Omega^n)$

$$
\left(\rho\mathbf{u}^{n+1}, \mathbf{v}\right)^n + \alpha\Delta t_n \left[a^n \left(\mathbf{u}^{n+1} - \mathbf{w}^{n+1}, \mathbf{u}^{n+1}, \mathbf{v}\right) - f^n(c_\Gamma^n, \mathbf{v})\right]
$$
$$
+ \Delta t_n \left[b^n(q, \mathbf{u}^{n+1}) - b^n(p^{n+1}, \mathbf{v})\right]
$$
$$
= \quad (\rho\mathbf{u}^n, \mathbf{v})^n - \beta\Delta t_n \left[a^n \left(\mathbf{u}^n - \mathbf{w}^n, \mathbf{u}^n, \mathbf{v}\right) - f^n(c_\Gamma^n, \mathbf{v})\right]. \tag{1.21}
$$

Having the solution of the Navier-Stokes step given in Problem 8, the grid can be updated to Ω_i^{n+1}, $i = 1, 2$, and Γ^{n+1}. The functions $\mathbf{u}^{n+1}$, $\mathbf{w}^{n+1}$ and p^{n+1} are transferred implicit to the new grid.

After the Navier-Stokes step the surfactant transport is solved on the new grid Ω^{n+1} and Γ^{n+1}, using the updated velocity fields $\mathbf{u}^{n+1}$ and $\mathbf{w}^{n+1}$. The surfactant transport equations for the bulk and the surface are decoupled likewise. For each equation the remaining information is taken explicit. Although, this would not be necessary for the linear Henry sorption law, in the case of the non-linear Langmuir sorption law a nonlinear iteration is required anyway.

Problem 9 For given Ω^{n+1}, Γ^{n+1}, $\mathbf{u}^{n+1}$, $\mathbf{w}^{n+1}$, $c_0^{n+1} = c^n$, $c_{\Gamma,0}^{n+1} = c_\Gamma^n$ find $(c_j^{n+1}, c_{\Gamma j}^{n+1}) \in G(\Omega^{n+1}) \times M(\Gamma^{n+1})$, such that for all $(\varphi, \psi) \in G(\Omega^{n+1}) \times M(\Gamma^{n+1})$ and $i = 1, 2, \ldots, M$

$$
\left(c_j^{n+1}, \varphi\right)^{n+1} + \alpha\Delta t_n \left[a_c^{n+1} \left(\mathbf{u}^{n+1} - \mathbf{w}^{n+1}, c_j^{n+1}, \varphi\right) - f_c^{n+1}(c_j^{n+1}, c_{\Gamma,j-1}^{n+1}, \varphi)\right]
$$
$$
= \quad (c^n, \varphi)^n - \beta\Delta t_n \left[a_c^n \left(\mathbf{u}^n - \mathbf{w}^n, c^n, \varphi\right) - f_c^n(c^n, c_\Gamma^n, \varphi)\right] \tag{1.22}
$$

$$\left\langle c_{\Gamma j}^{n+1}, \psi \right\rangle^{n+1} + \alpha \Delta t_n \left[a_\Gamma^{n+1}(\mathbf{u}^{n+1} - \mathbf{w}^{n+1}, c_{\Gamma j}^{n+1}, \psi) - f_\Gamma^{n+1}(c_j^{n+1}, c_{\Gamma j}^{n+1}, \psi) \right]$$

$$= \left\langle c_\Gamma^n, \varphi \right\rangle^n - \beta \Delta t_n \left[a_\Gamma^n(\mathbf{u}^n - \mathbf{w}^n, c_\Gamma^n, \psi) - f_\Gamma^n(c^n, c_\Gamma^n, \psi) \right]. \qquad (1.23)$$

Here, j is the nonlinear iteration step and M the number of nonlinear iterations, resulting from the stopping criterion.

Finally, we take a closer look to the spatial discretization. Pressure discontinuities across the interface are an inherent property of two-phase flows with surface stresses. A finite element space that resolves this discontinuities is essential to get an acceptable approximation order and to suppress spurious velocities and oscillations at the interface [23].

In the unfitted approach the finite element space is enriched by the XFEM finite element method [27] in order to resolve discontinuities in pressure and kinks in the velocity fields. If the grid is fitted to the interface velocity kinks are resolved automatically and pressure jumps are resolved by having either a classical element-wise discontinuous pressure space or by taking a continuous pressure but allow jumps across the interface. By the so called node doubling, where the degrees of freedom are not identified at the interface, the pressure can have different values for each phase at the interface.

For a polyhedral domain it can be shown that node doubling the pressure space at the interface results in an inf-sup stable pair of finite element spaces, as long the underlying finite element pair is inf-sup stable in each phase. For the standard Taylor-Hood finite element family this is the case, under the usual assumptions to the subdomains and their triangulations. For a smooth interface a corresponding result is still left open.

Apart from spurious velocities and a poor approximation, not allowing a discontinuous pressure also results in a poor mass conservation. In fact, the mass conservation can only be guaranteed having a phase-wise discontinuous pressure. If this is not the case, it can happen that one phase vanishes.

We use the standard isoparametric Taylor-Hood element $\mathbb{P}_2/\mathbb{P}_1$ with node doubling in the pressure space, which can avoid the above-mentioned problems.

1.5 Numerical Results

The codes for numerical studies are all in-house developments. For the fully coupled time dependent Problem 1, a code based on MooNMD was developed. MooNMD is an in-house finite element code base written in C++. While the less demanding 2d and 3d axisymmetric computations are done with serial codes, the fully 3d computations are done with a parallelized version of the code base. The parallelization is done by domain decomposition and using the MPI standard. A numerical study of the fully coupled two-phase Problem 1 is given in [2] and will not repeated here. The fully coupled numerical studies presented in the following are free surface flows.

For the static shape computations in Sect. 1.5.2 a standalone code was developed in `Matlab`. This code is fully independent from the code developed for the fully coupled time dependent problems.

For the studies of surface equation in Sect. 1.5.1 a code in the language `Julia` [7], was developed. The code handles general differential equations on hypersurfaces in two or three dimensions. Stabilization by Local Projection is implemented.

1.5.1 Stabilized Advection-Diffusion Transport on a Surface

This example is inspired by Example 2 from [20], where a circular counter-clockwise flow in the domain $[0, 1]^2$ is simulated. The problem is transferred to a curved surface, and Problem 2 is solved.

Given a cylinder of radius 0.5 around the axis $x = 0.5$ parallel to the *y-axis* with height one. We set Γ to the half of the cylinder, where $z \geq 0$. The velocity field $(-y, x)$ in the plane case [20] is transferred to the curved surface resulting in

$$\mathbf{w} = \left(-\pi yz, \frac{1}{\pi} \arccos(1 - 2x), -\frac{\pi}{2}(1 - 2x) \right)^T .$$

The right hand side, diffusion and reaction coefficient are set to $f = 0$, $\varepsilon = 10^{-8}$ and $c = 0$, respectively. Along the outflow boundary, $\{(0, y, 0) \,|\, y \in (0, 1)\}$, we impose homogeneous Neumann conditions. On the remaining part of the boundary the following discontinuous Dirichlet-type boundary condition is prescribed

$$u_D(x, y, z) := \begin{cases} 1, & \text{if } y = 0 \text{ and } x \in \left(\frac{1}{3}, \frac{2}{3}\right) \\ 0, & \text{else} \end{cases} .$$

For the standard Galerkin approximation $\mathbb{P}_1$ surface finite element space is used. This is compared to the LPS stabilized formulation using $V_h = \mathbb{P}_1^+$ and $D_h = \mathbb{P}_0^{\mathrm{disc}}$ on the same mesh. The stabilization parameter is chosen as $\gamma^2 = 1.0$ (high), $\gamma^2 = 0.1$ (optimal), and $\gamma^2 = 0.01$ (low).

The discontinuous Dirichlet-type boundary conditions lead to interior layers of the solution. As observed for bulk equations numerical oscillations around the layers occur for the unstabilized Galerkin method, compare Fig. 1.1 (left). LPS can reduce and localize these oscillations, see Fig. 1.1 (right). The influence of the stabilization parameter is visualized in Fig. 1.2. For small ε we expect the outflow profile to approximate a step function. We observe a tendency of smearing out for the (too) low stabilization parameter and increasing oscillations for the (too) high stabilization parameter [41].

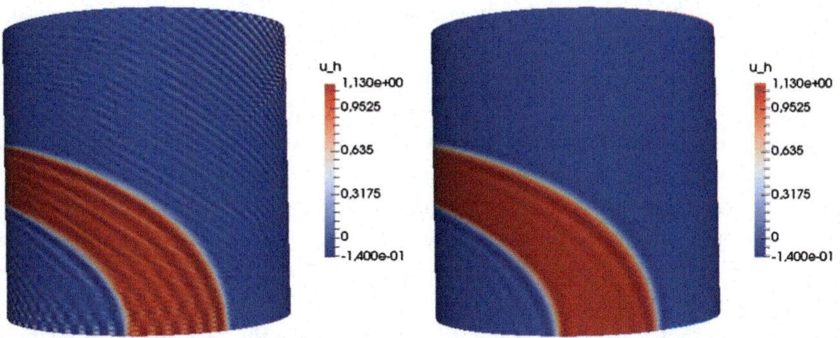

Fig. 1.1 $\mathbb{P}_1$ part of the solutions of the transport problem on a surface for the standard $\mathbb{P}_1$ (*left*) and the LPS-stabilized FEM with optimal stabilization parameter (*right*). View on x-y-plane

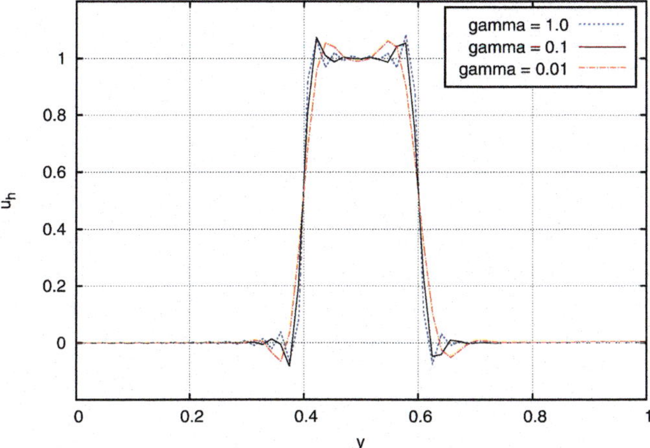

Fig. 1.2 $\mathbb{P}_1$ part of the solutions of the transport problem on a surface at the outflow boundary for the LPS-stabilized FEM with low, optimal and high stabilization parameter

1.5.2   Equilibrium Shape of a Pending Drop

In order to study the accuracy of the implemented algorithm we evaluate the shape of axisymmetric pendant droplets at a tip of a capillary. This is a typical situation in profile analysis tensiometry (PAT) for measuring the surface tension of liquids. In PAT a pendant drop is growing up to a certain volume at a capillary tip by a dosing system. The profile is then extracted from a picture of the drop by image processing and fitted to a solution of the Young-Laplace equation. It turns out that the shape of a dynamically generated drop differs considerably from the static shape, see e.g. [30]. We consider here the relaxation of clean and contaminated pendant drops at a tip of a capillary to the equilibrium state. For this two numerical schemes and are developed.

In [26] four different methods have been proposed to compute the static shape of axisymmetric drops at a tip of a capillary for a given drop volume and a given Bond number Bo = We/Fr. The method used in this section is based on an equivalent formulation of the problem as a constrained minimization problem: Find the surface of minimal energy which encloses a given fixed drop volume Vol. Applying the method of Lagrange multiplier and using a discretization by continuous, piecewise linear finite elements, we have to solve a nonlinear algebraic system of equations. As the initial guess for the Newton method we chose a piecewise linear approximation of a spherical shape enclosing the given volume. More details on this method and alternatives can be found in [26].

A 3d axisymmetric version of the 3d code based on the techniques described in Sects. 1.2–1.4 has been developed. Instead of starting with the strong axisymmetric form of the partial differential equation and deriving a weak formulation in suitable weighted Sobolev spaces we follow the approach in [18] where the 3d axisymmetric forms are developed directly from the 3d Cartesian forms. One advantage of this technique is that the boundary conditions at the artificial boundary along the symmetry axis appear in a natural way.

The geometric configuration of our test problem is as follows. The dimensionless radius of the capillary tip is equal to one its height equals also one. The axisymmetric pendant drop is assumed to have a pinned contact line (circle of radius one). The volume of the drop below the capillary is chosen to be Vol = 60. We start our test series with a surfactant free drop in equilibrium for Bo = 0.006708. As expected the drop stays in equilibrium and the bottom position remains unchanged, compare Table 1.2. Now we assume that the surface tension is suddenly reduced (uniformly over the surface Γ) leading to the Bond numbers Bo=0.013416 and Bo=0.03354, for two different values of surface tension. Since the drop is no longer in equilibrium, it starts to oscillate. In the dynamic computations with the 3d axisymmetric code, we chose Re = Fr = 1 such that We = Bo. We observe a perfect fit of the bottom position (Table 1.2 and Fig. 1.3) and the shape of the drop (Fig. 1.4 left) computed by the static code and the dynamic computations for $t \to \infty$. Note that larger amplitudes appear for the larger Bond number (larger reduction of surface tension). For the largest Bond number of Bo=0.067080 there exist a static equilibrium, however, due to the large amplitudes the dynamic computations predict the detachment of the drop. Our interpretation is that the static equilibrium is not stable with respect to large perturbations. We close this test series with a surfactant free droplet in equilibrium for Bo = 0.006708 and Vol = 60. Two uniform initial states for the bulk surfactant concentration are considered: $(c_0, c_{\Gamma,0}) = (1, 0)$ and $(c_0, c_{\Gamma,0}) = (3, 0)$. The dimensionless parameters are Re = Fr = 1, Pe = 1, Pe_Γ = 10, Bi = 0, Da = 10, E = 0.5, β = 100, and the Henry sorption law

Table 1.2 Bottom positions in equilibrium for different Bond numbers

Bo	0.006708	0.013416	0.033540	0.067080
$y_{\min}$	−4.7222	−4.7968	−5.0477	−5.6648
$y_{\min}^{dyn}$	−4.7222	−4.7969	−5.0477	–

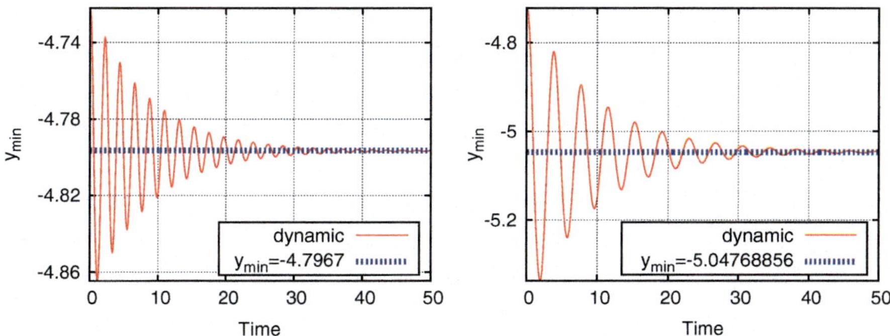

Fig. 1.3 Bottom position of a clean drop with suddenly reduced surface tension. Dynamic computations for Re = Fr = 1, We = Bo = 0.013416 (*left*) and We = Bo = 0.033540 (*right*)

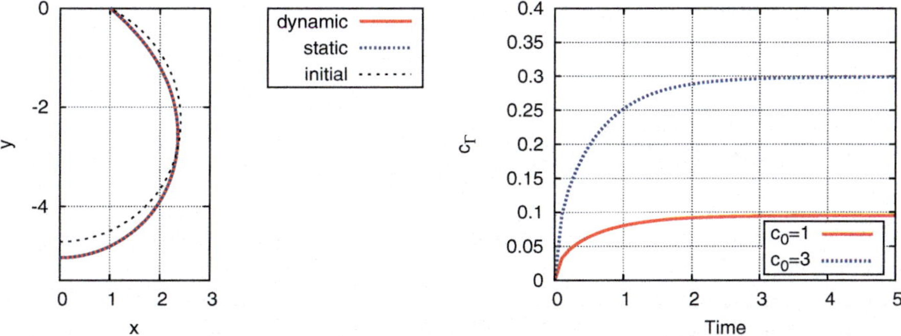

Fig. 1.4 *Left*: Clean drop with suddenly reduced surface tension. Initial, equilibrium, and final shape after relaxation for Re = Fr = 1, We = Bo = 0.033540. *Right*: Uniform surface surfactant concentration over time for a clean drop and two different initial bulk concentrations

is used. Since Bi = 0 we have no desorption and only adsorption takes place. The increase of the surface surfactant concentration c_Γ in the time interval $[0, 5]$ can be seen in Fig. 1.4 right.

The surfactant at the surface leads to a reduction in the surface tension and the drop is no longer in equilibrium. As it can be seen from Figs. 1.4 and 1.5 the time scale for the adsorption is much shorter than the time scale to attain the equilibrium. We see from Fig. 1.5 and Table 1.3 that even for a very small change in the surface tension, due to surfactant, our numerical scheme accurately captures the dynamics and the equilibrium bottom position. We do not compare the shapes of the relaxed drop with the shape of the equilibrium in an extra figure since they are practically identical.

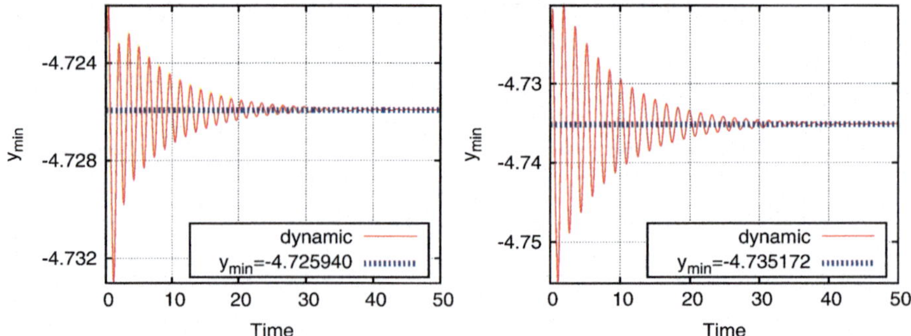

Fig. 1.5 Relaxation of a drop with soluble surfactant with different initial bulk concentrations. $c_0 = 1$ (*left*) and $c_0 = 3$ (*right*)

Table 1.3 Influence of adsorption of surfactants at the interface

	c_Γ	We	y_{min}	y_{min}^{dyn}
$c_0 = 1$	0.094106	0.007039	-4.7258	-4.7259
$c_0 = 3$	0.298672	0.007886	-4.7350	-4.7351

1.5.3 Freely Oscillating Droplet

A wide variety of analytical investigations of oscillating droplets make it an excellent choice for validation of numerical schemes and codes [31, 34, 40]. Here we will compare the theoretical models of oscillating droplets with soluble surfactant [40] and insoluble surfactant [34] to our fully three dimensional numerical scheme.

In the numerical computations gravity is neglected and the initial drop is in rest. The shape is given by a sphere of radius one perturbed with a spherical harmonic of second order and an amplitude of $a_2 = 0.1$. The drop is not in equilibrium and starts to oscillate. The Weber number is fixed to We $= 0.0081$ in all examples, which will induce roughly five oscillations in the time interval $[0, 1]$. The timestep size is $\Delta t = 10^{-4}$, i.e. 10,000 steps per computation.

We consider two examples, an insoluble case with a Reynolds number of Re $=$ 10.684 and a soluble case with Re $= 1.0684$. The Reynolds numbers are chosen such that the dimensionless viscosity $\sqrt{\text{We}}\text{Re}^{-1} < 0.1$, a bound given in [6, 39] where nonlinear viscous effects become negligible and the analytic approximations are valid. In both examples, the bulk and surface Peclet, the Damköhler and Biot numbers are Pe $= 1$, Pe$_\Gamma = 1$, Da $= 1$ and Bi $= 1$, respectively. The surface surfactant scaling factor β is set to one. The bulk surfactant concentration and the equilibrium surface surfactant concentration for the initial droplet is set to $c^{eq} = 0.1111$ and $c_\Gamma^{eq} = 0.1$, respectively.

From the linearized theory for small-amplitude oscillations of a drop, we can predict the angular frequency and the damping rate of the oscillation [6, 11, 31, 39]. The results are given in Table 1.4, where $\omega_{0,2}$ is the Lamb frequency, ω_2 the angular frequency, and δ_2 the damping rate of the second mode.

Table 1.4 Lamb frequency, angular frequency and damping rate for different Reynolds numbers

	Re	$\omega_{0,2}$	ω_2	δ_2
Soluble	1.0684	31.4270	31.0766	4.6799
Insoluble	10.684	31.4270	31.4235	0.4682

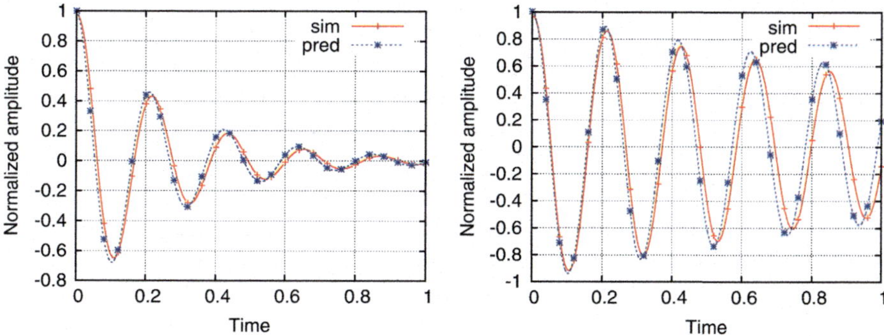

Fig. 1.6 Comparison of the course of the second mode of an freely oscillation drop with soluble surfactant (*left*) and insoluble surfactant (*right*), between the numerical simulation (sim) and the analytic prediction (pred), for $(\mathrm{Re}, E) = (1.0684, 1.0)$ and $(\mathrm{Re}, E) = (10.684, 1.0)$, respectively

Theoretical results for an oscillating droplet with insoluble surfactants are derived in [34]. A set of differential equations for the amplitude a_l of the l-th mode of shape oscillation and amplitude g_l of the l-th mode of the surface surfactant concentration are given. In order to compare the insoluble theory with the results here, the differential equations given in [34] are solved numerically with a Matlab code using the `ode45` routine.

Results for higher viscosities and a soluble surfactant case are given in [40]. For the dimensionless complex frequency α an first order approximation $\alpha = i(1 + \varepsilon + \mathcal{O}(\varepsilon^2))$, is given, where ε is specified by an explicit correlation from the material properties. The damping rate is given by the real part, and the frequency by the imaginary part of $\omega_{0,2}\alpha$

In Fig. 1.6 (left) a comparison of the normalized amplitudes of the shape oscillation in the soluble case is shown. Normalized means $a_2(t)$ is scaled with $a_2(0)^{-1}$ such that the graph starts at one. In the figure, the graph (sim) is the shape oscillation by our numerical computation and the graph (pred) is the shape oscillation obtain in [40]. We see a good agreement, although the prediction runs a little ahead.

In Fig. 1.6 (right) a comparison of the normalized amplitudes of the shape oscillation in the insoluble case is shown. The graph (pred) is the prediction of the shape oscillation obtained by solving the equations in [34]. We see a good agreement, the prediction runs ahead again and shows less damping.

In Fig. 1.7 (left) the damping rates versus different surface elasticities for the numerical simulation (sim) and the prediction (pred) after [40] is shown. In Fig. 1.7 (right) the same is shown for the frequencies. We see a quite good agreement in the frequencies over the considered range of surface elasticities. The agreement

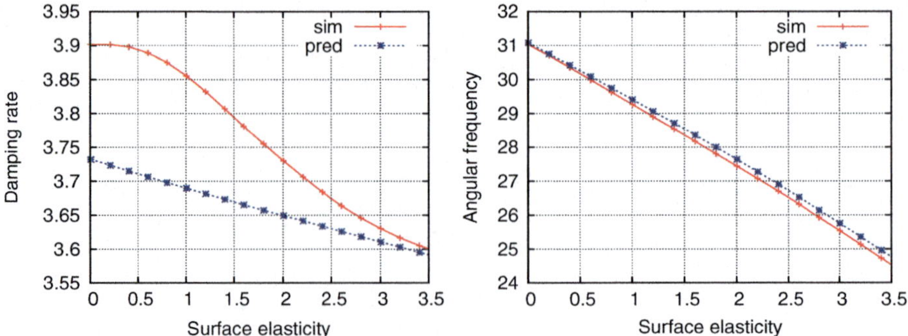

Fig. 1.7 Damping rate versus surface elasticity (*left*) and angular frequency versus surface elasticity (*right*) for Re $= 1.0684$ and soluble surfactant (*left*), for the numerical simulation (sim) and the analytic prediction (pred)

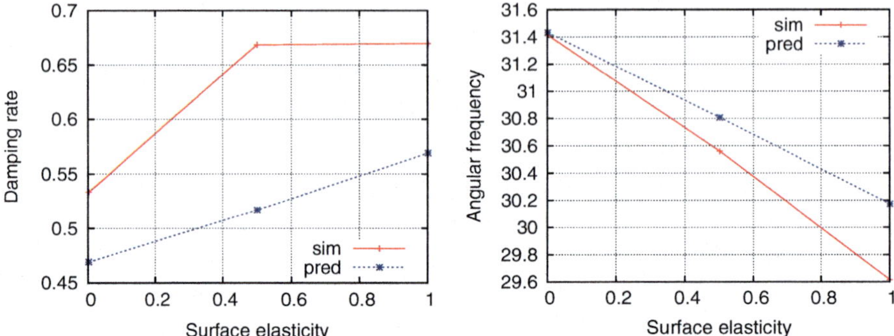

Fig. 1.8 Damping rate versus surface elasticity (*left*) and angular frequency versus surface elasticity (*right*) for Re $= 10.684$ and insoluble surfactant, for the numerical simulation (sim) and the analytic prediction (pred)

gets better for low surface elasticities. Contrary, we see an increasing disagreement in the damping rates for lower surface elasticities and a better agreement for higher surface elasticities.

In the case of insoluble surfactant, shown in Fig. 1.8, one gets a better agreement for the damping rates at lower surface elasticities. The angular frequencies are in good agreement for lower elasticities. Both, the mismatch in damping rate and the frequency increases with higher surface elasticities.

This numerical example confirms, as expected, that the linear theory by Lamb [31] fails to predict damping rates and frequencies for the case of viscous fluids and fluids with surfactants present.

In both cases the numerical simulation overestimates the damping. The disagreement increases with the surface elasticity in the insoluble case, what one might expect, since the theory chosen from [34] is for small surface elasticities. Also, the backward Euler scheme used in the numerical simulation introduces numerical damping. Contrary, in the soluble case, the disagreement increases with lower surface elasticities, thus we expect a problem with the numerical damping, and a lower time step size or a time discretization which introduce less numerical damping could be necessary.

1.6   Summary

We summarize our main contributions within the SPP 1506:

- Numerical analysis and implementation of a surface finite element method for convection-diffusion-reaction equations stabilized by local projection.
- Extension of the local projection stabilization to a coupled bulk-surface transport problem. Error analysis and implementation.
- Development and implementation of a higher order, fitted finite element method for two-phase and free surface flows with soluble and insoluble surfactants in 2d, 3d, and axisymmetric 3d cases.
- Application and validation of the developed algorithms in drop profile analysis tensiometry.
- Validation of different schemes within the SPP 1506 based on the Taylor flow problem. Comparison of numerical data and measurements of Taylor bubbles [2].

Acknowledgements   The authors wish to thank the Council of Scientific Research in India (CSIR) for financial support within the project 25(0228)/14/EMR-II and the German Research Foundation (DFG) for financial support within the Priority Programm SPP 1506 "Transport Processes at Fluidic Interfaces" with the project To143/11-2 and within the graduate program Micro-Macro-Interactions in Structured Media and Particle Systems (GK 1554).

References

1. Aland, S.: Modelling of two-phase flow with surface active particles. Ph.D. thesis, TU Dresden (2012)
2. Aland, S., Hahn, A., Kahle, C., Nürnberg, R.: Comparative simulations of Taylor-flow with surfactants based on sharp- and diffuse-interface methods. In: Bothe, D., Reusken, A. (eds.) Transport Processes at Fluidic Interfaces. Advances in Mathematical Fluid Mechanics. Springer (2017)
3. Barrett, J.W., Garcke, H., Nürnberg, R.: A parametric finite element method for fourth order geometric evolution equations. J. Comput. Phys. **222**(1), 441–467 (2007)
4. Barrett, J.W., Garcke, H., Nürnberg, R.: A stable parametric finite element discretization of two-phase Navier-Stokes flow. J. Sci. Comput. **63**(1), 78–117 (2015)
5. Becker, R., Braack, M.: A finite element pressure gradient stabilization for the Stokes equations based on local projections. Calcolo **38**(4), 173–199 (2001)
6. Becker, E., Hiller, W.J., Kowalewski, T.A.: Experimental and theoretical investigation of large-amplitude oscillations of liquid droplets. J. Fluid Mech. **231**, 189–210 (1991)

7. Bezanson, J., Edelman, A., Karpinski, S., Shah, V.B.: Julia: a fresh approach to numerical computing. CoRR **abs/1411.1607** (2014). http://arxiv.org/abs/1411.1607
8. Burman, E., Hansbo, P.: Edge stabilization for Galerkin approximations of convection-diffusion-reaction problems. Comput. Methods Appl. Mech. Eng. **193**(15–16), 1437–1453 (2004)
9. Burman, E., Fernández, M.A., Hansbo, P.: Continuous interior penalty finite element method for Oseen's equations. SIAM J. Numer. Anal. **44**(3), 1248–1274 (2006)
10. Burman, E., Claus, S., Hansbo, P., Larson, M.G., Massing, A.: CutFEM: discretizing geometry and partial differential equations. Int. J. Numer. Methods Eng. **104**(7), 472–501 (2015)
11. Chandrasekhar, S.: Hydrodynamic and Hydromagnetic Stability. The International Series of Monographs on Physics. Clarendon Press, Oxford (1961)
12. de Gennes, P.G., Brochard-Wyart, F., Quere, D.: Capillarity and Wetting Phenomena, Drops, Bubbles, Pearls, Waves. Springer, Berlin (2004)
13. Dziuk, G.: Finite elements for the Beltrami operator on arbitrary surfaces. In: Partial Differential Equations and Calculus of Variations. Lecture Notes in Mathematics, vol. 1357, pp. 142–155. Springer, Berlin (1988)
14. Dziuk, G., Elliott, C.M.: Finite element methods for surface PDEs. Acta Numer. **22**, 289–396 (2013)
15. Eastoe, J., Dalton, J.: Dynamic surface tension and adsorption mechanisms of surfactants at the air–water interface. Adv. Colloid Interf. Sci. **85**(2), 103–144 (2000)
16. Elliott, C.M., Stinner, B.: Analysis of a diffuse interface approach to an advection diffusion equation on a moving surface. Math. Models Methods Appl. Sci. **19**(5), 787–802 (2009)
17. Elliott, C., Stinner, B., Styles, V., Welford, R.: Numerical computation of advection and diffusion on evolving diffuse interfaces. IMA J. Numer. Anal. **31**, 786–812 (2011)
18. Ganesan, S., Tobiska, L.: An accurate finite element scheme with moving meshes for computing 3D-axisymmetric interface flows. Int. J. Numer. Methods Fluids **57**(2), 119–138 (2008)
19. Ganesan, S., Tobiska, L.: A coupled arbitrary Lagrangian-Eulerian and Lagrangian method for computation of free surface flows with insoluble surfactants. J. Comput. Phys. **228**(8), 2859–2873 (2009)
20. Ganesan, S., Tobiska, L.: Stabilization by local projection for convection-diffusion and incompressible flow problems. J. Sci. Comput. **43**(3), 326–342 (2010)
21. Ganesan, S., Tobiska, L.: Arbitrary Lagrangian-Eulerian finite-element method for computation of two-phase flows with soluble surfactants. J. Comput. Phys. **231**(9), 3685–3702 (2012)
22. Ganesan, S., Tobiska, L.: Finite Elements: Theory and Algorithms. Cambridge-IISc Series. Cambridge University Press, Cambridge (2017)
23. Ganesan, S., Matthies, G., Tobiska, L.: On spurious velocities in incompressible flow problems with interfaces. Comput. Methods Appl. Mech. Eng. **196**(7), 1193–1202 (2007)
24. Ganesan, S., Hahn, A., Held, K., Tobiska, L.: An accurate numerical method for computation of two-phase flows with surfactants. In: Eberhardsteiner, E., et al. (eds.) European Congress on Computational Methods in Applied Sciences and Engineering (ECCOMAS 2012) Vienna, Sept 10–14. CD-ROM. ISBN:978-3-9502481-9-7 (2012)
25. Ganesan, S., Hahn, A., Simon, K., Tobiska, L.: Finite element computations for dynamic liquid-fluid interfaces. In: Rahni, M., Karbaschi, M., Miller, R. (eds.) Computational Methods for Complex Liquid-Fluid Interfaces. Progress in Colloid and Interface Science, vol. 5, pp. 331–351. CRC Press Taylor & Francis Group, Boca Raton (2016)
26. Gille, M., Gorbacheva, Y., Hahn, A., Polevikov, V., Tobiska, L.: Simulation of a pending drop at a capillary tip. Commun. Nonlinear Sci. Numer. Simul. **26**, 137–151 (2015)
27. Groß, S., Reusken, A.: An extended pressure finite element space for two-phase incompressible flows with surface tension. J. Comput. Phys. **224**(1), 40–58 (2007)
28. Gross, S., Olshanskii, M.A., Reusken, A.: A trace finite element method for a class of coupled bulk-interface transport problems. ESAIM Math. Model. Numer. Anal. **49**(5), 1303–1330 (2015)

29. Hughes, T.J.R., Brooks, A.: A multidimensional upwind scheme with no crosswind diffusion. In: Finite Element Methods for Convection Dominated Flows (Papers, Winter Annual Meeting American Society of Mechanical Engineers, New York, 1979), AMD, vol. 34, pp. 19–35. American Society of Mechanical Engineers, New York (1979)
30. Karbaschi, M., Bastani, D., Javadi, A., Kovalchuk, V., Kovalchuk, N., Makievski, A., Bonaccurso, E., Miller, R.: Drop profile analysis tensiometry under highly dynamic conditions. Colloids Surf. A Physicochem. Eng. Asp. **413**, 292–297 (2012)
31. Lamb, H.: Hydrodynamics, 6th edn. Cambridge Mathematical Library. Cambridge University Press, Cambridge (1993). With a foreword by R. A. Caflisch [Russel E. Caflisch]
32. Matthies, G.: Finite element methods for free boundary value problems with capillary surfaces. Ph.D. thesis, Otto-von-Guericke-Universität, Fakultät für Mathematik, Magdeburg (2002)
33. Matthies, G., Skrzypacz, P., Tobiska, L.: A unified convergence analysis for local projection stabilisations applied to the Oseen problem. M2AN Math. Model. Numer. Anal. **41**(4), 713–742 (2007)
34. Nadim, A., Rush, B.: Determination of interfacial rheological properties through microgravity oscillations of bubbles and drops. Technical Report 20000120383, NASA (2000)
35. Nävert, U.: A finite element method for convection-diffusion problems. Chalmers Tekniska Högskola/Göteborgs Universitet. Department of Computer Science (1982)
36. Nobile, F.: Numerical approximation of fluid-structure interaction problems with application to haemodynamics. Ph.D. thesis, SB, Lausanne (2001). doi:10.5075/epfl-thesis-2458
37. Nochetto, R.H., Walker, S.W.: A hybrid variational front tracking-level set mesh generator for problems exhibiting large deformations and topological changes. J. Comput. Phys. **229**(18), 6243–6269 (2010)
38. Olshanskii, M.A., Reusken, A., Xu, X.: A stabilized finite element method for advection-diffusion equations on surfaces. IMA J. Numer. Anal. **34**(2), 732–758 (2014)
39. Prosperetti, A.: Free oscillations of drops and bubbles: the initial-value problem. J. Fluid Mech. **100**(2), 333–347 (1980)
40. Tian, Y., Holt, R.G., Apfel, R.E.: Investigations of liquid surface rheology of surfactant solutions by droplet shape oscillations: theory. Phys. Fluids **7**, 2938 (1995)
41. Tobiska, L.: On the relationship of local projection stabilization to other stabilized methods for one-dimensional advection-diffusion equations. Comput. Methods Appl. Mech. Eng. **198**(5–8), 831–837 (2009)
42. Turek, S.: Efficient Solvers for Incompressible Flow Problems. Lecture Notes in Computational Science and Engineering, vol. 6. Springer, Berlin (1999)
43. Wörner, M.: Numerical modeling of multiphase flows in microfluidics and micro process engineering: a review of methods and applications. Microfluid. Nanofluid. **12**(6), 841–886 (2012)

Chapter 2
High Order Unfitted Finite Element Methods for Interface Problems and PDEs on Surfaces

Christoph Lehrenfeld and Arnold Reusken

Abstract In this contribution we treat a special class of recently developed higher order unfitted finite element methods for the discretization of mass and surfactant transport equations. To achieve higher order accuracy for such PDEs on stationary geometries we combine standard techniques for numerical integration and a discretization with a special mesh transformation. This results in a new class of isoparametric unfitted finite element methods. For the treatment of such PDEs on evolving geometries we apply space-time variational formulations. These unfitted finite element techniques result in robust and accurate discretization methods for mass and surfactant transport problems in realistic two-phase flow simulations based on a sharp interface formulation. We present these finite element discretization methods, give theoretical error bounds for classes of model problems and present results of numerical simulations both for such model problems and for challenging two-phase flow applications.

2.1 Introduction

In this contribution we treat a special class of recently developed unfitted finite element methods for the discretization of mass and surfactant transport equations in incompressible two-phase flow problems. For the two-phase flow problem we restrict to a *sharp interface* model for the fluid dynamics, which consists of the Navier-Stokes equations for the bulk fluids with an interfacial surface tension force term in the momentum equation. In case of solute transport this Navier-Stokes equation is coupled with a convection-diffusion equation. If surfactants are present, a convection-diffusion equation on the (evolving) interface is used for modeling the

C. Lehrenfeld
Institut für Numerische und Angewandte Mathematik, University of Göttingen, 37083 Göttingen, Germany
e-mail: lehrenfeld@math.uni-goettingen.de

A. Reusken (✉)
Institut für Geometrie und Praktische Mathematik, RWTH Aachen University, Templergraben 55, 52056 Aachen, Germany
e-mail: reusken@igpm.rwth-aachen.de

© Springer International Publishing AG 2017

D. Bothe, A. Reusken (eds.), *Transport Processes at Fluidic Interfaces*, Advances in Mathematical Fluid Mechanics, DOI 10.1007/978-3-319-56602-3_2

surfactant transport. We refer to, e.g., [20] for a derivation and discussion of these models. In this contribution we do not consider the Navier-Stokes equations for the fluid dynamics and restrict to the transport equations for solute and surfactants which will be introduced below.

A key difficulty in the numerical simulation of two-phase flow problems is an accurate numerical approximation of the interface. For this different techniques have been developed in the literature, e.g., volume of fluid (VOF) and level set methods. In this contribution we restrict to the level set (LS) method. Furthermore, as discretization method for the mass and surfactant equations we restrict to finite element methods (FEM). In such a setting with a (standard) LS method for interface capturing, the underlying computational grids are typically *not* fitted to the (evolving) interface and thus one needs special finite element methods that can deal with such *un*fitted triangulations. Recently, significant progress has been made in the construction, analysis and application of so-called *unfitted FEM*, see for instance the papers [6, 7, 14, 18, 21, 40] and the references therein. In the literature different names for unfitted FEM for the discretization of the solute transport equation are also used, namely extended FEM (XFEM) and CutFEM. Unfitted FEM are also used for the discretization of PDEs on (evolving) manifolds, e.g., [39] and the references therein. While most of the work on unfitted discretizations has been on piecewise linear (unfitted) finite elements, many unfitted discretizations have a natural extension to higher order finite element spaces, see for instance [1, 24, 36, 44]. However, new techniques are required to obtain also higher order accuracy when errors due to the geometry approximation are considered.

In this contribution we present the main results on unfitted FEM obtained in the Priority Program 1506. More detailed treatments of these results are given in the papers [27–35, 39, 41] and the preprint [17]. We outline our main new results at the end of this chapter, cf. Sect. 2.5.

The structure of the paper is as follows. In Sect. 2.2 we treat higher order unfitted FEM for elliptic and parabolic interface problems. In Sect. 2.3 higher order unfitted FEM for elliptic and parabolic partial differential equations on (evolving) surfaces are discussed. These general techniques have rather straightforward applications to the mass and surfactant equations that occur in two-phase flows. This is addressed in Sect. 2.4. Finally, in Sect. 2.5 we outline our main new results and discuss some topics which we consider to be of interest for further study.

2.2 Unfitted FEM for Interface Problems

Below, in Sect. 2.2.1 we first restrict to a model interface problem with a *stationary* interface and then in Sect. 2.2.2 extend the approach to *evolving* interfaces. The discretizations of the interface problems are based on the unfitted Nitsche method from the seminal paper [21] and extended to higher order methods and evolving domains. The key ideas of the new finite element method are explained, optimal theoretical error bounds are discussed and results of a numerical example, which illustrates the behavior of the method, are included. The results in this section are based on [29, 30, 33–35].

2.2.1 A Model Elliptic Problem with a Stationary Interface

On a bounded connected polygonal domain $\Omega \subset \mathbb{R}^d$, $d = 2, 3$, we consider the model interface problem

$$-\text{div}(\alpha_i \nabla u) = f_i \quad \text{in } \Omega_i, \ i = 1, 2, \tag{2.1a}$$

$$[\![-\alpha \nabla u]\!]_\Gamma \cdot \mathbf{n}_\Gamma = 0, \quad [\![u]\!]_\Gamma = 0 \quad \text{on } \Gamma, \tag{2.1b}$$

$$u = 0 \quad \text{on } \partial\Omega. \tag{2.1c}$$

Here, $\Omega_1 \cup \Omega_2 = \Omega$ is a nonoverlapping partitioning of the domain, $\Gamma = \overline{\Omega}_1 \cap \overline{\Omega}_2$ is the interface, $[\![\cdot]\!]_\Gamma$ denotes the usual jump operator across Γ and $\mathbf{n}_\Gamma$ denotes the unit normal at Γ pointing from Ω_1 into Ω_2. f_i, $i = 1, 2$ are domain-wise described sources. In the remainder we will also use the source term f on Ω which we define as $f|_{\Omega_i} = f_i$, $i = 1, 2$. The diffusion coefficient α is assumed to be piecewise constant, i.e. it has a constant value $\alpha_i > 0$ on each sub-domain Ω_i. In the following we use the notation $u_i = u|_{\Omega_i}$, $i = 1, 2$.

The first interface condition in (2.1b) results from the conservation of mass principle while the second condition ensures continuity of the solution across the interface. Later, in Sect. 2.2.2 we replace the second condition $[\![u]\!] = 0$ with the more general—and in the context of two-phase flow applications more relevant—*Henry jump condition* $[\![\beta u]\!] = 0$ where β is strictly positive and piecewise constant. In contrast to the problem considered in Sect. 2.2.2, a problem as in (2.1) with the condition $[\![u]\!]_\Gamma = 0$ replaced by the Henry condition $[\![\beta u]\!] = 0$, $\beta_1 \neq \beta_2$ (which has a jump discontinuity in the solution u), can be transformed to a problem of the form of (2.1) using the variable $w = \beta u$. As the problem with the interface condition $[\![u]\!] = 0$ is a standard model problem in the literature we consider only this condition in the remainder of this section. We however note that the methods presented in the following can also be applied to the more general Henry condition, cf. [30, 48].

The weak formulation of the problem (2.1) is as follows: determine $u \in H_0^1(\Omega)$ such that

$$\int_\Omega \alpha \nabla u \cdot \nabla v \, dx = \int_\Omega f v \, dx \quad \text{for all } v \in H_0^1(\Omega). \tag{2.2}$$

We assume simplicial triangulations of Ω which are *not* fitted to Γ. Furthermore, the interface is characterized as the zero level of a given level set function ϕ. The numerically challenging aspect of the problem in (2.1) stems from the fact that the diffusion coefficient α can be discontinuous across the *unfitted* interface. Hence, the solution can have discontinuities in the gradient which are located inside individual elements. This lack of regularity of the solution introduces difficulties in the accurate approximation. Standard piecewise polynomial finite element spaces are no longer appropriate and lead to sub-optimal results, especially when aiming at higher order convergence. To overcome the approximation problem special finite element spaces are typically designed. Below, in Sect. 2.2.1.1 we consider an established choice, an unfitted finite element space as it is used in XFEM or CutFEM discretizations. One drawback of the application of this kind of finite element spaces is that none of the interface conditions in (2.1b) can easily be implemented in the finite element space as an essential condition. A suitable finite element method with these adapted finite element spaces needs a variational formulation which includes the interface conditions at least in a weak sense. Another important difficulty in the numerical treatment of (2.1) is in the accurate handling of geometries, which are only implicitly defined through level set functions. In particular the realization of integrals which have discontinuous integrands is difficult, for instance the weak Laplacian in (2.2). Especially for higher order methods, dealing with this difficulty is a challenging task. In the next section we introduce an unfitted finite element method which addresses all the aforementioned problems. Afterwards, in Sects. 2.2.1.2 and 2.2.1.3 we give a priori error bounds and numerical results. We restrict to the pure diffusion problem (2.1). In [30] a variant of the unfitted FEM, based on a streamline diffusion stabilization technique, is treated which is suitable for the discretization of elliptic convection dominated interface problems.

2.2.1.1　An Unfitted FEM

In [29] a new approach has been introduced to obtain higher order accurate approximations on domain and surface integrals of implicitly described geometries. The fundamental idea is the introduction of a parametric mapping Θ_h of the underlying mesh from a geometrical reference configuration to a final configuration.

Let $\mathscr{T}$ denote the simplicial triangulation of Ω and V_h^k denote the standard finite element space of continuous piecewise polynomials up to degree k. The nodal interpolation operator in V_h^k is denoted by I_k. It is assumed that a high order accurate finite element approximation $\phi_h = \phi_h^k$ of the level set function ϕ is known. Based on its piecewise linear interpolation $\hat{\phi}_h = I_1 \phi_h$ the reference configuration with the subdomains $\Omega_i^{\text{lin}} = \{\hat{\phi}_h \lessgtr 0\}$, $i = 1, 2$ and the interface $\Gamma^{\text{lin}} = \{\hat{\phi}_h = 0\}$ is defined. For this reference configuration a robust and accurate realization of numerical integration is fairly simple. Inside each element the interface and the subdomains are convex polytopes for which quadrature rules can easily be obtained, cf. (among others) [28, Chap. 4], [37], [38, Chap. 5]. This kind of strategy is used in many simulation codes that are based on unfitted finite elements, e.g. [7, 9, 12, 19, 45]. Note that this piecewise planar approximation of the geometry is only second order accurate and thus its application is limited to low order methods. However, with a suitable choice of the parametric mapping Θ_h this approximation can be improved to a higher order accurate one. In [29] we introduced such a mapping which is easy

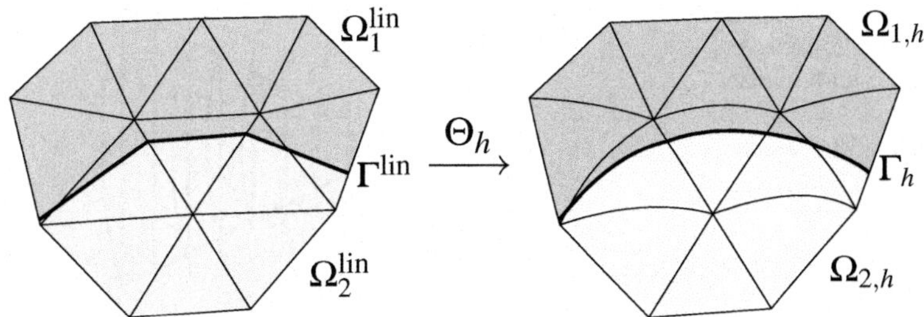

Fig. 2.1 Basic idea of the method in [29]: The zero level Γ^{lin} of the piecewise linear interpolation $\hat{\phi}_h$ is mapped approximately to the implicit interface $\{\phi_h = 0\}$ using a mesh transformation Θ_h

to construct and realizes the mapping from the reference geometries to a higher order accurate approximation. A sketch is given in Fig. 2.1.

The discretization approach consists of two steps. First, a (higher order) finite element discretization is constructed with respect to the reference configuration. Afterwards, applying the transformation Θ_h to this space and the geometries in the variational formulation results in a new unfitted finite element discretization with an accurate treatment of the geometries. The mapping renders the finite element spaces *isoparametric finite element spaces*. As usual in isoparametric finite element discretizations, volume and interface integrals that occur in the implementation of the method can be formulated in terms of integrals on the reference configuration, i.e. on convex polytopes. This allows to reuse the established strategy for numerical integration discussed before.

We outline the construction of Θ_h, cf. [29] for more details. For ease of presentation we assume quasi-uniformity of the mesh, s.t. h denotes a characteristic mesh size with $h \sim h_T := \mathrm{diam}(T),\ T \in \mathscr{T}$. All elements in the triangulation $\mathscr{T}$ which are cut by Γ^{lin} are collected in the set $\mathscr{T}^\Gamma := \{T \in \mathscr{T}, T \cap \Gamma^{\mathrm{lin}} \neq \emptyset\}$. The corresponding domain is $\Omega^\Gamma := \{x \in T, T \in \mathscr{T}^\Gamma\}$ and the restriction of the finite element space to Ω^Γ is $V_h^k(\Omega^\Gamma) = V_h^k|_{\Omega^\Gamma}$. The extended set which includes all direct neighbors to elements in $\mathscr{T}^\Gamma$ is $\mathscr{T}_+^\Gamma := \{T \in \mathscr{T}, \mathrm{meas}_{d-1}(T \cap \Omega^\Gamma) \neq 0\}$ with the corresponding domain $\Omega_+^\Gamma := \{x \in T, T \in \mathscr{T}_+^\Gamma\}$.

For the construction of the isoparametric mapping Θ_h we first introduce a mapping Ψ_h, which is then postprocessed to obtain the mapping Θ_h. The mapping Ψ_h is based on an approximate solution of the problem

$$\text{For every } x \in T \in \mathscr{T}^\Gamma \text{ find } y \in \Omega, \text{ such that } \hat{\phi}_h(x) = \phi(y). \tag{2.3a}$$

Such a mapping $x \to y$ maps each point on Γ^{lin} to a point on Γ. Clearly, such a y is not unique, since infinitely many points y exist with the level set value $\phi(y) = c = \hat{\phi}_h(x)$. Therefore, we restrict the "search direction" along which we seek for a point y. We take the normal direction $G(x) = \nabla\phi$ and obtain the new problem

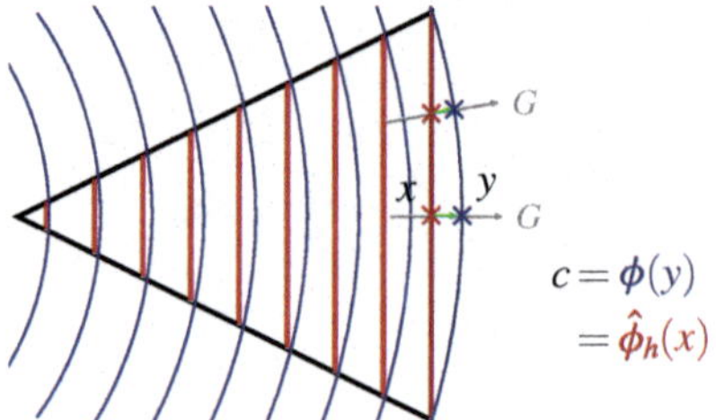

Fig. 2.2 Property of an ideal mapping as in (2.3b): Every point $x \in \Omega^{\Gamma}$ with (linear) level set value $c = \hat{\phi}_h(x)$ is mapped (along the search direction $G(x)$) onto the location $y \in \Omega$ with $\phi(y) = c$

For every $x \in T \in \mathcal{T}^{\Gamma}$ find the smallest (in absolute value) number $d \in \mathbb{R}$,

$$\text{such that } y = x + dG(x) \in \Omega \text{ solves } \hat{\phi}_h(x) = \phi(y). \tag{2.3b}$$

In Fig. 2.2 a sketch of this "ideal" mapping is given. Usually, the exact level set function ϕ is not known, and thus we construct Ψ_h based on approximations of ϕ and G. As an approximation of the search direction G we use $G_h = \nabla\phi_h$, but other options are also possible, see the discussion in [29]. For the approximation of ϕ in (2.3b) we do not directly take the finite element approximation ϕ_h, since this has the following two drawbacks. First, ϕ_h typically has weak discontinuities (kinks) across element interfaces which would lead to kinks in Ψ_h (located within elements). Second, for $x \in T$, the problem in (2.3b) with ϕ replaced by ϕ_h often requires evaluations of ϕ_h outside of T which can introduce significant computational overhead especially in parallel simulation environments. Instead we use $\mathcal{E}_T\phi_h$, the polynomial extension of $\phi_h|_T$. Thus we obtain the following definition of Ψ_h, which corresponds to an approximate solution of the problem in (2.3b),

For every $x \in T \in \mathcal{T}^{\Gamma}$ find the smallest (in absolute value) number $d_h \in \mathbb{R}$,

$$\text{such that } \Psi_h(x) := y = x + d_h G_h(x) \in \Omega \text{ solves } \hat{\phi}_h(x) = \mathcal{E}_T\phi_h(y). \tag{2.3c}$$

Note that Ψ_h has the property $\Psi_h(x_i) = x_i$ for all *vertices* x_i of $T \in \mathcal{T}^{\Gamma}$ as $\hat{\phi}_h(x_i) = \mathcal{E}_T\phi_h(x_i)$. This mapping is smooth only inside each element $T \in \mathcal{T}^{\Gamma}$, but can be discontinuous across element interfaces. We only have $\Psi_h \in C(\mathcal{T}^{\Gamma})^d$ with $C(\mathcal{T}^{\Gamma}) = \bigoplus_{T \in \mathcal{T}^{\Gamma}} C(T)$. Hence, we apply a postprocessing on Ψ_h to obtain a globally continuous mapping which is furthermore defined on whole Ω. To this end we make use of a projection $P_h : C(\mathcal{T}^{\Gamma})^d \to (V_h^k)^d$ and define the final mapping

$$\Theta_h := P_h\Psi_h = \text{id} + Q_h(d_h G_h). \tag{2.4}$$

Here $Q_h := Q_h^2 Q_h^1$ where Q_h^1 ensures continuity in Ω^{Γ}, $Q_h^1 : C(\mathcal{T}^{\Gamma})^d \to V_h^k(\Omega^{\Gamma})^d$ and $Q_h^2 : C(\Omega^{\Gamma})^d \to C(\Omega)^d$ realizes the continuous transition to zero in $\Omega_+^{\Gamma} \setminus \Omega^{\Gamma}$, cf. the sketch in Fig. 2.3. Note that Θ_h is a finite element (vector) function.

Note that since $\Psi_h(x_i) = \Theta_h(x_i) = x_i$ for all *vertices* x_i, we have $\Psi_h = \Theta_h = \text{id}$ for $k = 1$ in which case the mesh remains unchanged. For the projection operator

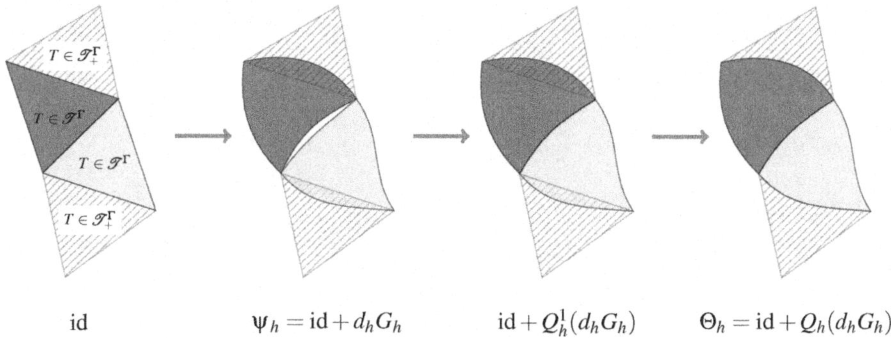

Fig. 2.3 Construction steps of the transformation Θ_h. In the first step, Ψ_h (only in Ω^Γ, pointwise, discontinuous across element interfaces) is constructed. In a second step, the discontinuities are removed through averaging (only in Ω^Γ). Finally, a continuous extension to the exterior is realized

Q_h^1 we consider a simple quasi-interpolation which averages function values across element interfaces, cf. also [43, Eqs. (25)–(26)] and [13]. For Q_h^2 we consider an operator which keeps the degrees of freedom in $\mathscr{T}^\Gamma$ unchanged and extends the discrete functions (element-wise) smoothly to zero on neighboring elements. For details on the projection P_h we refer to [29] and [33].

Remark 1 The mapping Θ_h should be a bijection on Ω and the transformed simplices $\Theta_h(T)$, $T \in \mathscr{T}$, should have some shape regularity property. One important result in [33] is that $\|D\Theta_h - I\|_{\infty,T} \lesssim h$, i.e. for sufficiently small mesh sizes the transformation becomes an arbitrary small perturbation of the identity and shape regularity of $\Theta_h(\mathscr{T})$ is inherited from the shape regularity of $\mathscr{T}$, cf. [33] for details. In cases where h is not sufficiently small, to guarantee shape regularity the transformation has to be adapted. We refer to [29] for a possible way to achieve this.

We now define the isoparametric Nitsche unfitted FEM as a transformed version of the original Nitsche unfitted FE discretization [21] with respect to the interface approximation $\Gamma_h = \Theta_h(\Gamma^{\text{lin}})$. We introduce some further notation. The standard unfitted space w.r.t. Γ^{lin} is denoted by

$$V_h^\Gamma := V_{h|\Omega_1^{\text{lin}}}^k \oplus V_{h|\Omega_2^{\text{lin}}}^k. \tag{2.5}$$

In the literature, a finite element method based on such a space is often called CutFEM, cf. [7] or extended FEM (XFEM), cf. [2, 14, 46]. To simplify the notation we do not explicitly express the polynomial degree k in V_h^Γ. The *isoparametric* unfitted FE space is defined as

$$V_{h,\Theta}^\Gamma := \{\, v_h \circ \Theta_h^{-1} \mid v_h \in V_h^\Gamma \,\} = \{\, \tilde{v}_h \mid \tilde{v}_h \circ \Theta_h \in V_h^\Gamma \,\}. \tag{2.6}$$

Based on this space we formulate a discretization of (2.1) using the unfitted Nitsche technique [21] with $\Gamma_h = \Theta_h(\Gamma^{\text{lin}})$ and $\Omega_{i,h} = \Theta_h(\Omega_i^{\text{lin}})$ as numerical approximation of the geometries: determine $u_h \in V_{h,\Theta}^{\Gamma}$ such that

$$A_h(u_h, v_h) := a_h(u_h, v_h) + N_h(u_h, v_h) = f_h(v_h) \quad \text{for all } v_h \in V_{h,\Theta}^{\Gamma} \tag{2.7}$$

with the bilinear forms

$$a_h(u, v) := \sum_{i=1}^{2} \alpha_i \int_{\Omega_{i,h}} \nabla u \cdot \nabla v \, dx, \tag{2.8a}$$

$$N_h(u, v) := N_h^c(u, v) + N_h^c(v, u) + N_h^s(u, v), \tag{2.8b}$$

$$N_h^c(u, v) := \int_{\Gamma_h} \{\!\!\{-\alpha \nabla v\}\!\!\} \cdot \mathbf{n}_{\Gamma_h} \, [\![u]\!] \, ds, \quad N_h^s(u, v) := \bar{\alpha} \frac{\lambda}{h} \int_{\Gamma_h} [\![u]\!][\![v]\!] \, ds, \tag{2.8c}$$

for $u, v \in V_{h,\Theta}^{\Gamma} + V_{\text{reg},h}$ with $V_{\text{reg},h} := H^1(\Omega) \cap H^2(\Omega_{1,h} \cup \Omega_{1,h})$.

Here, $\mathbf{n}_{\Gamma_h}$ denotes the outer normal of $\Omega_{1,h}$ and $\bar{\alpha} = \frac{1}{2}(\alpha_1 + \alpha_2)$ the mean diffusion coefficient. For the averaging operator $\{\!\!\{\cdot\}\!\!\}$ there are different possibilities. We use $\{\!\!\{w\}\!\!\} := \kappa_1 w_{|\Omega_{1,h}} + \kappa_2 w_{|\Omega_{2,h}}$ with a "Heaviside" choice where $\kappa_1 = 1$ if $|T_1| > \frac{1}{2}|T|$ and $\kappa_1 = 0$ if $|T_1| \leq \frac{1}{2}|T|$, $\kappa_2 = 1 - \kappa_1$. Here, $T_i = T \cap \Omega_i^{\text{lin}}$, i.e. the cut configuration on the undeformed mesh is used. This choice in the averaging renders the scheme in (2.7) stable (for sufficiently large λ) for arbitrary polynomial degrees k, independent of the cut position of Γ, cf. [33, Lemma 5.1]. A different choice for the averaging which also results in a stable scheme is $\kappa_i = |T_i|/|T|$.

In order to define the right-hand side functional f_h we first assume that the source term $f_i : \Omega_i \to \mathbb{R}$ in (2.1a) is (smoothly) extended to $\Omega_{i,h}$, such that $f_i = f_{i,h}$ on Ω_i holds. This extension is denoted by $f_{i,h}$. We define

$$f_h(v) := \sum_{i=1,2} \int_{\Omega_{i,h}} f_{i,h} v \, dx, \quad v \in V_{h,\Theta}^{\Gamma} + V_{\text{reg},h}. \tag{2.9}$$

We define f_h on Ω by $f_{h|\Omega_{i,h}} := f_{i,h}$, $i = 1, 2$.

For the implementation of this method, in the integrals we apply a transformation of variables $y := \Theta_h^{-1}(x)$. For example, the bilinear form $a_h(u, v)$ then results in

$$a_h(u, v) := \sum_{i=1,2} \alpha_i \int_{\Omega_i^{\text{lin}}} D\Theta_h^{-T} \nabla u \cdot D\Theta_h^{-T} \nabla v \, \det(D\Theta_h) \, dy. \tag{2.10}$$

Based on this transformation the implementation of integrals is carried out as for the case of the piecewise planar interface Γ^{lin}. The additional variable coefficients $D\Theta_h^{-T}$, $\det(D\Theta_h)$ are easily and efficiently computable using the property that Θ_h is a finite element (vector) function.

2.2.1.2   Optimal Order Error Bound

Optimal discretization error bounds, both in the H^1- and L^2-norm, for the isoparametric unfitted FEM presented above are derived in [33, 34]. We only present the H^1-norm error bound. We assume that the approximation $\phi_h \in V_h^k$ of the level set function ϕ satisfies the error estimate

$$\max_{T \in \mathscr{T}} |\phi_h - \phi|_{m,\infty,T \cap U} \lesssim h^{k+1-m}, \quad 0 \le m \le k+1. \tag{2.11}$$

Here $|\cdot|_{m,\infty,T \cap U}$ denotes the usual semi-norm on the Sobolev space $H^{m,\infty}(T \cap U)$. Here and in the remainder we use the notation $\lesssim$, which denotes an inequality with a constant that is independent of h and of how the interface Γ intersects the triangulation $\mathscr{T}$. This constant may depend on ϕ and on the diffusion coefficient α, cf. (2.1a). A main theorem proved in [33] is the following. In the theorem we use a certain piecewise smooth function u^e, which is smooth on the subdomain approximations $\Omega_{i,h}$, $i = 1, 2$, and defined by a smooth extension of $u_{|\Omega_i}$ (cf. [33] for details).

Theorem 1 *Let u be the solution of (2.1) and $u_h \in V_{h,\Theta}^{\Gamma}$ the solution of (2.7). We assume that $u \in H^{3,\infty}(\Omega_1 \cup \Omega_2)$ if $k = 2$, $u \in H^{k+1}(\Omega_1 \cup \Omega_2)$ if $k \ge 3$, and $f \in H^{1,\infty}(\Omega_1 \cup \Omega_2)$. Furthermore the data extension f_h satisfies the condition $\|f_h\|_{H^{1,\infty}(\Omega_{1,h} \cup \Omega_{2,h})} \lesssim \|f\|_{H^{1,\infty}(\Omega_1 \cup \Omega_2)}$. Then the following holds:*

$$|u^e - u_h|_{H^1(\Omega_{i,h} \cup \Omega_{i,h})} \lesssim h^k (\|u\|_{H^{k+1}(\Omega_1 \cup \Omega_2)} + \|f\|_{H^{1,\infty}(\Omega_1 \cup \Omega_2)})$$

Hence, the method has the optimal h^k error bound in the H^1-norm, under optimal smoothness assumptions on u.

2.2.1.3   Results of Numerical Experiments

We give a numerical example to illustrate the performance of the previously introduced method. On the domain $\Omega = [-1.5, 1.5]^2$ we prescribe the interface Γ by the level set function $\phi(x) = \|x\|_4 - 1$ where $\|x\|_4$ denotes the usual 4-norm on $\mathbb{R}^d$, $\Gamma = \{\phi(x) = 0\}$. The interface is a smoothed square, c.f. the sketch in Fig. 2.4, the level set function ϕ is equivalent to a signed distance function. The diffusion parameters are taken as $(\alpha_1, \alpha_2) = (1, 2)$, Dirichlet boundary data and right-hand side term f are chosen such that the solution is $u(x) = u_1(x) = 1 + \frac{\pi}{2} - \sqrt{2} \cdot \cos(\frac{\pi}{4} \|x\|_4^4)$ for $x \in \Omega_1$ and $u(x) = u_2(x) = \frac{\pi}{2} \|x\|_4$ for $x \in \Omega_2$.

The solution is continuous across the interface, but has a kink at the interface, $u_1|_\Gamma = u_2|_\Gamma$ and $\alpha_1 \nabla u_1 \cdot \mathbf{n}_\Gamma = \alpha_2 \nabla u_2 \cdot \mathbf{n}_\Gamma$ on Γ.

Based on a finite element approximation ϕ_h of ϕ which is obtained by higher order interpolation and the search direction $G_h = \nabla \phi_h$, we construct the mapping Θ_h as in (2.4). Starting from an initial simplicial mesh which resolves the interface sufficiently well such that we have shape regularity of the mesh after transformation with Θ_h, we repeatedly apply uniform refinements. The stabilization parameter in Nitsche's method is chosen as $\lambda = 20 \cdot k^2$. We note that in our experience the results of Nitsche's method depend only mildly on the choice of λ (as long as it is not chosen too small).

In Fig. 2.4 the convergence results of the error $u^e - u_h$ in the H^1 semi-norm for polynomial degrees $k = 1, 2, .., 5$ are shown. Here, u^e is the natural extension of the solution to the discrete domains $\Omega_{i,h}$. We observe optimal order of convergence. For details on this numerical example we refer to [33]. We note that optimal order, i.e. $\mathcal{O}(h^{k+1})$ convergence for the error is also observed in the L^2 norm. These results confirm the error analyses in [33] (H^1 norm) and [34] (L^2 norm).

Remark 2 (Dominating Convection) In many applications, especially in two-phase flows, problems of the form (2.1) with an additional dominating convection term $\mathbf{w} \cdot \nabla u$ are relevant where $\mathbf{w}$ is the convecting flow field. In this case additional stabilization becomes necessary for the unfitted discretization introduced before. In [30] we derived and analysed a discretization which combines Streamline Diffusion stabilization with an unfitted finite element formulation for piecewise linears for the case of an elliptic interface problem with dominating convection.

Remark 3 (Conditioning Stiffness Matrix) A disadvantage of the type of unfitted finite element method presented above is the fact that the stiffness matrix can be extremely ill-conditioned. In particular this condition number depends not only on the mesh size h, but also on how the interface intersects the triangulation $\mathscr{T}$. This topic is addressed in [35] for the case of *linear* finite elements. In that paper an additive subspace preconditioner is introduced which is optimal in the sense that the condition number of the preconditioned stiffness matrix is independent of *both*

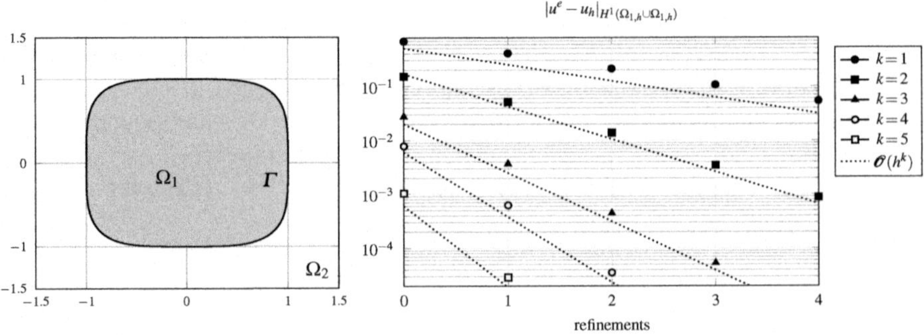

Fig. 2.4 Geometrical configuration and convergence results for the example in Sect. 2.2.1.3

the mesh size h *and* the interface position. Furthermore it is shown that already the simple diagonal scaling of the stiffness matrix results in a condition number that is bounded by ch^{-2}, with a constant c that does not depend on the location of the interface.

2.2.2 A Model Parabolic Problem with an Evolving Interface

We consider a mass transport problem with an *evolving* interface $\Gamma(t)$ which extends the previously considered model problem by adding convection and a temporal evolution of the concentration *and* the interface. We assume a given sufficiently smooth velocity field $\mathbf{w}$, with $\mathrm{div}\,\mathbf{w} = 0$, and assume that the transport of the interface is determined by this velocity field, in the sense that $V_\Gamma = \mathbf{w} \cdot \mathbf{n}_\Gamma$ holds. Here V_Γ is the normal velocity of the interface. We consider a standard model which describes the transport of a solute in a two-phase flow problem. In strong formulation this model is as follows:

$$\frac{\partial u}{\partial t} + \mathbf{w} \cdot \nabla u - \mathrm{div}(\alpha \nabla u) = f \quad \text{in } \Omega_i(t), \ i = 1, 2, \ t \in [0, T], \qquad (2.12a)$$

$$[\![-\alpha \nabla u]\!]_\Gamma \cdot \mathbf{n}_\Gamma = 0, \quad [\![\beta u]\!]_\Gamma = 0 \quad \text{on } \Gamma, \qquad (2.12b)$$

$$u(\cdot, 0) = u_0 \quad \text{in } \Omega_i(0), \ i = 1, 2, \qquad (2.12c)$$

$$u(\cdot, t) = 0 \quad \text{on } \partial\Omega, \ t \in [0, T]. \qquad (2.12d)$$

In (2.12a) we have standard parabolic convection-diffusion equations in the two subdomains Ω_1 and Ω_2. The coefficients α, β are assumed to be piecewise constant. We note that in contrast to the problem treated in Sect. 2.2.1.1 we now consider the general case $\alpha_1 \neq \alpha_2$ *and* $\beta_1 \neq \beta_2$. The second relation is the so-called *Henry condition*, cf. [3, 5, 22, 50, 51] and describes a jump discontinuity at the interface due to different solubilities within the respective fluid phases. Hence, the solution u is *discontinuous across the evolving interface*. In this section we treat an unfitted finite element method for this problem, introduced and analyzed in [31]. The method is based on a well-posed *space-time weak formulation* of (2.12) (which we do not present here). Compared to the unfitted finite element method presented in Sect. 2.2.1.1 there are two important differences. Firstly, we restrict to *linear* finite elements and a *linear* interface reconstruction, i.e., we do not use an isoparametric mapping (for getting a higher order approximation). Hence, the best we can obtain is a second order method. Secondly, we treat the evolution by using a *space-time* finite element approach. We will introduce an unfitted space-time FEM and combine this with a space-time Nitsche technique. The method and results that we present are from [31]. From the error analysis, cf. Sect. 2.2.2.2, it follows that the method is second order in space *and* time. We are not aware of any other Eulerian type discretization method for this class of parabolic interface problems which has a guaranteed (i.e., based on a-priori error bounds) second order convergence.

2.2.2.1 An Unfitted FEM

The space-time unfitted FEM that we introduce in this section has the form of a variational problem in a certain space-time finite element space. The same space is used for both trial and test functions. We introduce the method for the case of piecewise *bilinear* space-time functions (linear in space and linear in time). In Remark 5 we comment on generalizations. We introduce notation. The space-time domain is denoted by $Q = \Omega \times (0, T] \subset \mathbb{R}^{d+1}$. A partitioning of the time interval is given by $0 = t_0 < t_1 < \ldots < t_N = T$, with a uniform time step $\Delta t = T/N$. This assumption of a uniform time step is made to simplify the presentation, but is not essential for the method. Corresponding to each time interval $I_n := (t_{n-1}, t_n]$ we assume a given shape regular simplicial triangulation $\mathscr{T}_n$ of the spatial domain Ω. In general this triangulation is *not fitted* to the interface $\Gamma(t)$. In this section we assume that the implicit geometries can be handled without introducing additional errors, cf. Remark 4 below for the general case where geometries have to be approximated in order to obtain a realization of the discretization.

The triangulation $\mathscr{T}_n$ may vary with n. Let V_n be the finite element space of continuous piecewise linear functions on $\mathscr{T}_n$ with zero boundary values on $\partial\Omega$. The spatial mesh size parameter corresponding to V_n is denoted by h_n. Corresponding space-time finite element spaces on the time slab $Q^n := \Omega \times I_n$ are given by

$$W_n := \{ v : Q^n \to \mathbb{R} \mid v(x,t) = \phi_0(x) + t\phi_1(x), \quad \phi_0, \phi_1 \in V_n \}, \quad (2.13a)$$

$$W := \{ v : Q \to \mathbb{R} \mid v_{|Q^n} \in W_n \}. \quad (2.13b)$$

In the time slab Q^n we define the subdomains $Q_i^n := \{(x,t) \in Q^n, x \in \Omega_i(t)\}$, $i = 1, 2$, and also $Q_i := \{(x,t) \in Q, x \in \Omega_i(t)\}$, $i = 1, 2$. We will also use the notation $v_i := v|_{Q_i}$. The space-time *unfitted* FE spaces are given by

$$W_n^\Gamma := W_n|_{Q_1^n} \oplus W_n|_{Q_2^n}, \quad (2.14a)$$

$$W^{\Gamma*} := \{ v : Q \to \mathbb{R} \mid v_{|Q^n} \in W_n^\Gamma \} = W|_{Q_1} \oplus W|_{Q_2}. \quad (2.14b)$$

The symbol Γ_*^n denotes the space-time interface in Q^n, i.e., $\Gamma_*^n := \{(x,t) \in Q^n, x \in \Gamma(t)\}$, and $\Gamma_* := \{(x,t) \in Q, x \in \Gamma(t)\}$. The finite element spaces defined in (2.14) are natural space-time generalizations of the "cut" finite element space defined in (2.5). We treat the Henry condition $[\![\beta u]\!]_\Gamma = 0$ using a natural space-time generalization of the Nitsche technique used in Sect. 2.2.1.1. For this we need a suitable average across $\Gamma(t)$, denoted by $\{\!\!\{v\}\!\!\}_{\Gamma(t)} = \kappa_1(t)v|_{\Omega_1(t)} + \kappa_2(t)v|_{\Omega_2(t)}$. Take $t \in I_n$ and $T \in \mathscr{T}_n$ with $T_i := T \cap \Omega_i(t) \neq \emptyset$. We define the weights $\kappa_i(t) := |T_i|/|T|$. Note that those weights only depend on the *spatial* configuration at a given time t and there holds $\kappa_1(t) + \kappa_2(t) = 1$. Other choices for the averaging are possible, for instance the "Heaviside" choice that we also used in Sect. 2.2.1.1 for the stationary case.

In the discontinuous Galerkin method we need jump terms across the end points of the time intervals $I_n = (t_{n-1}, t_n]$. We define $u_+^{n-1}(\cdot) := \lim_{\varepsilon \downarrow 0} u(\cdot, t_{n-1} + \varepsilon)$ and introduce the notation

$$v^n(x) := v(x, t_n), \quad [v]^n(x) := v_+^n(x) - v^n(x), \quad 0 \leq n \leq N-1, \quad \text{with } v^0(x) := 0.$$

On the cross sections $\Omega \times \{t_n\}, 0 \leq n \leq N$, of Q we use a weighted L^2 scalar product

$$(u, v)_{0,t_n} := \int_\Omega \beta(\cdot, t_n) uv \, dx = \sum_{i=1}^{2} \beta_i \int_{\Omega_i(t_n)} uv \, dx.$$

This scalar product is uniformly (w.r.t. n and N) equivalent to the standard scalar product in $L^2(\Omega)$. Note that we use a weighting with β in this scalar product, which is not reflected in the notation.

The notation introduced above is used to define a bilinear form $B(\cdot, \cdot)$, which consists of three parts, namely a term $a(\cdot, \cdot)$ that directly corresponds to the partial differential equation, a term $d(\cdot, \cdot)$ which weakly enforces continuity with respect to t at the time interval end points t_k, and a term $N_{\Gamma_*}(\cdot, \cdot)$ which enforces in a weak sense the Henry condition $[\![\beta u]\!]_{\Gamma_*} = 0$. These terms are defined per time slab Q^n, i.e. $a(\cdot, \cdot)$ is of the form $a(u, v) = \sum_{n=1}^{N} a^n(u, v)$ and similarly for the other two terms.

We now define the bilinear forms corresponding to each time slab Q^n, see also Fig. 2.5 for a sketch of these contributions. For sufficiently smooth u, v and $1 \leq n \leq N$ we define

$$a^n(u, v) := \sum_{i=1}^{2} \int_{Q_i^n} \left(\frac{\partial u_i}{\partial t} + \mathbf{w} \cdot \nabla u_i\right)\beta_i v_i + \alpha_i \beta_i \nabla u_i \cdot \nabla v_i \, dx \, dt, \tag{2.15a}$$

$$d^n(u, v) := ([u]^{n-1}, v_+^{n-1})_{0,t_{n-1}}, \tag{2.15b}$$

$$N_{\Gamma_*}^n(u, v) := \int_{t_{n-1}}^{t_n} N_h(t; u, v) \, dt, \tag{2.15c}$$

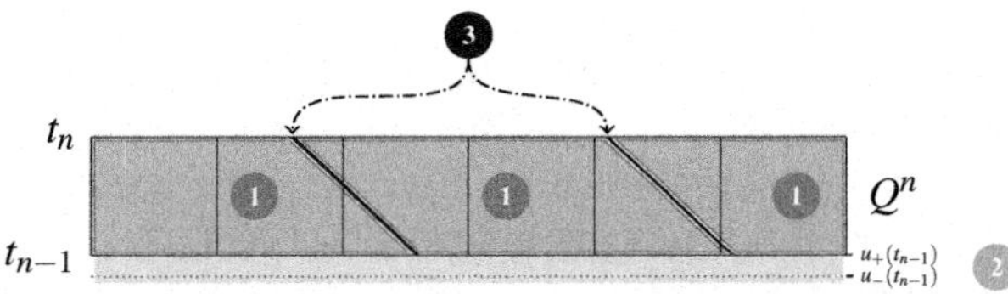

Fig. 2.5 Sketch of bilinear form contributions to space-time formulation for one time slab: (1) a^n: PDE (interior part), (2) d^n: temporal consistency, (3) $N_{\Gamma_*}^n$: interface conditions

with the *spatial* Nitsche bilinear form $N_h(t; u, v)$ for a fixed time t defined analogously to (2.8b) and (2.8c) as

$$N_h(t; u, v) := N_h^c(t; u, v) + N_h^c(t; v, u) + N_h^s(t; u, v), \tag{2.15d}$$

$$N_h^c(t; u, v) := \int_{\Gamma(t)} \{\!\!\{-\alpha \nabla u\}\!\!\}_{\Gamma(t)} \cdot \mathbf{n}_{\Gamma(t)} \, [\![\beta v]\!]_{\Gamma(t)} \, ds, \tag{2.15e}$$

$$N_h^s(t; u, v) := \bar{\alpha} \frac{\lambda}{h} \int_{\Gamma(t)} [\![\beta u]\!]_{\Gamma(t)} [\![\beta v]\!]_{\Gamma(t)} \, ds, \quad \lambda \geq 0. \tag{2.15f}$$

Finally, we introduce a right-hand side functional given by

$$f^1(v) = (u_0, v_+^0)_{0, t_0} + \int_{Q^1} f\beta \, v \, dx \, dt, \quad f^n(v) = \int_{Q^n} f\beta \, v \, dx \, dt, \quad 2 \leq n \leq N,$$

where u_0 is the initial condition from (2.12c) and f the source term in (2.12a). Corresponding global (bi)linear forms are obtained by summing over the time slabs:

$$q(u, v) = \sum_{n=1}^{N} q^n(u, v), \text{ for } q \in \{a, d, N_{\Gamma_*}\}, \quad f(v) = \sum_{n=1}^{N} f^n(v).$$

These bilinear forms and the functional f are well-defined on the space-time unfitted space W^{Γ_*}. The space-time unfitted FE discretization is defined as follows. Determine $u_h \in W^{\Gamma_*}$ such that

$$B(u_h, v_h) = f(v_h) \quad \text{for all } v_h \in W^{\Gamma_*},$$
$$B(u_h, v_h) := a(u_h, v_h) + d(u_h, v_h) + N_{\Gamma_*}(u_h, v_h). \tag{2.16}$$

Note that this formulation still allows to solve the space-time problem time slab by time slab.

Remark 4 (Quadrature in Space-Time) We assumed that no geometry errors are introduced in the discretization which implies the assumption that in each time slab integration on the space-time subdomains Q_i^n and on the space-time interface Γ_*^n can be done exactly. In practice this assumption is not realistic and an approximation of the geometries is necessary to construct quadrature rules. While piecewise linear interface approximations, which have second order accuracy, are standard in up to *three* dimensions, cf. (among others) [28, Chap. 4], [37], [38, Chap. 5], the numerical integration on geometries resulting from piecewise linear interface reconstructions in *four* dimensions (three space and one time dimension) required new strategies. In [27] we presented decomposition rules which allow to extend the previously known strategies from lower dimensions to four dimensional space-time geometries.

Remark 5 (Higher Order Methods)   We comment on a generalization to a higher order method. On Q, instead of the bilinear space W as in (2.13), a higher order space-time finite element space can be defined in an obvious manner, cf. [53]. A corresponding higher order unfitted finite element space is then defined as in (2.14) and the higher order discretization is obtained by the variational problem (2.16) with W^{Γ_*} replaced by this higher order unfitted finite element space. We conclude that the method (2.16) has a straightforward generalization to a higher order method. From an implementation point of view there is an important difference between such a higher order method and the bilinear method introduced above. In order to benefit from the higher order accuracy, one needs sufficiently accurate quadrature rules. For the case with an evolving interface such accurate approximations of the space-time integrals are difficult to realize. So far only second order quadrature rules have been developed for implicitly described geometries in space-time, cf. Remark 4. An extension of the higher order accurate geometry handling as explained in Sect. 2.2.1.1 to the space-time setting is a topic of current research.

2.2.2.2   Second Order Error Bounds

Below in Theorem 2 we present a main result of the a-priori error analysis of the unfitted space-time method. For this we first introduce some preliminaries. We need anisotropic Sobolev spaces. By $H^{k,l}(Q_1^n \cup Q_2^n)$ we denote the Sobolov space of functions on the domain $Q_1^n \cup Q_2^n$ with spatial partial derivatives up to degree k and temporal derivatives up to degree l in $L^2(Q_1^n \cup Q_2^n)$. The subscript 0 is used for the subspace of functions in $H^{k,l}(Q_1^n \cup Q_2^n)$ with zero value on $\partial\Omega \times (0, T)$ (in the trace sense), cf. [31] for details.

In order to derive second order bounds in the $L^2(\Omega(T))$ norm, we make use of duality arguments. This requires regularity assumptions for the following homogeneous backward problem (with data $\hat{v}_T$):

$$-\frac{\partial \hat{v}}{\partial t} - \mathbf{w} \cdot \nabla \hat{v} - \operatorname{div}(\alpha \nabla \hat{v}) = 0 \quad \text{in } \Omega_i(t),\ i = 1, 2,\ t \in [0, T], \quad (2.17a)$$

$$[\![-\alpha \nabla \hat{v}]\!]_\Gamma \cdot \mathbf{n}_\Gamma = 0, \quad [\![\beta \hat{v}]\!]_\Gamma = 0 \quad \text{on } \Gamma, \quad (2.17b)$$

$$\hat{v}(\cdot, 0) = \hat{v}_T \quad \text{in } \Omega_i(0),\ i = 1, 2, \quad (2.17c)$$

$$\hat{v}(\cdot, t) = 0 \quad \text{on } \partial\Omega,\ t \in [0, T]. \quad (2.17d)$$

The analysis in [31] uses the assumption that the solution of this homogeneous backward problem has the regularity property $\|\hat{v}\|_{2,Q_1 \cup Q_2} \leq c\|\hat{v}_T\|_{L^2(\Omega)}$ where $\| \cdot \|_{2,Q_1 \cup Q_2}$ is the standard (isotropic) Sobolev norm. Based on this one can derive a second order error bound in the $L^2(\Omega)$ norm. In the analysis in [28] we replaced this assumption with the more realistic regularity assumption $\|\hat{v}\|_{2,Q_1 \cup Q_2} \leq c\|\hat{v}_T\|_{H^1(\Omega_1(T) \cup \Omega_2(T))}$ to arrive at the second order estimate given in the next theorem. The estimate gives a bound in a dual norm, the definition of which we repeat for

48   C. Lehrenfeld and A. Reusken

functions $v \in L^2(\Omega)$:

$$\|v\|_{-1,T} := \sup_{w \in H_0^1(\Omega_1(T) \cup \Omega_2(T))} \frac{(v,w)_{0,T}}{\|w\|_{H^1(\Omega_1(T) \cup \Omega_2(T))}}.$$

Theorem 2 *Assume that* (2.12) *has a solution* $u \in H_0^{2,1}(Q_1 \cup Q_2)$ *and* $u_h \in W^{\Gamma_*}$ *is the solution to* (2.16). *Under the assumption that the homogeneous backward problem* (2.17) *has a solution* $\hat{v} \in H_0^2(Q_1 \cup Q_2)$ *that has the regularity property* $\|\hat{v}\|_{2,Q_1 \cup Q_2} \le c\|\hat{v}_T\|_{H^1(\Omega_1(T) \cup \Omega_2(T))}$ *with a constant* c *independent of the initial data* $\hat{v}_T \in H^1(\Omega_1(T) \cup \Omega_2(T))$, *there holds the following bound for the discretization error* $u - u_h$:

$$\|(u - u_h)(\cdot, T)\|_{-1,T} \le c(h^2 + \Delta t^2)\|u\|_{2,Q_1 \cup Q_2}. \tag{2.18}$$

Hence, the discretization error of this method is of second order, both with respect to the spatial and time mesh size, under reasonable regularity assumptions. In numerical experiments, cf. Sect. 2.2.2.3 below, we observe that the second order convergence also holds in the L^2-norm. We do not assume any CFL-type conditions on the mesh sizes. To our knowledge there are no other Eulerian FE techniques which for this class of parabolic problems with an *evolving discontinuity* have a proven second order error bound.

2.2.2.3 Results of Numerical Experiments

We present results for a test problem also considered in [27]. In this problem a sphere is translated inside a cube with a time-dependent velocity. We give a summary of the setup and the results and refer to [27] for more details. The time interval is $[0, T]$ with $T = 0.5$ and the domain is the cube $\Omega = [-1, 1]^3$. The sphere is initially centered around $(-0.5, 0, 0)$, has a radius $R = \frac{1}{3}$ and is translated by the time-dependent velocity field $\mathbf{w} = (\frac{1}{2}\cos(2\pi t), 0, 0)$. Initial and boundary data are prescribed such that the solution takes the form

$$u_m(x, t) = \sin(\pi t) \cdot U^m(\|x - q(t)\|), \ m = 1, 2, \ U^1(y) = a + by^2, \ U^2(y) = \cos(\pi y).$$

where $q(t) = (\frac{1}{4\pi}\sin(2\pi t) - 0.5, 0, 0)$ describes the center of the sphere at time t and a and b are chosen such that the interface conditions hold. We chose $(\alpha_1, \alpha_2) = (10, 20)$ and $(\beta_1, \beta_2) = (2, 1)$.

In Fig. 2.6 we present results for the discretization error $\|u(\cdot, T) - u_h(\cdot, T)\|_{L^2(\Omega)}$. These results show a second order convergence of the error with respect to temporal and spatial refinements. We note that in further numerical studies we observed also third order convergence in time for many test cases as long as the quadrature error is sufficiently small. From the literature, e.g. [53], it is known that for *smooth* parabolic problems (i.e., without discontinuities) and with standard linear space-

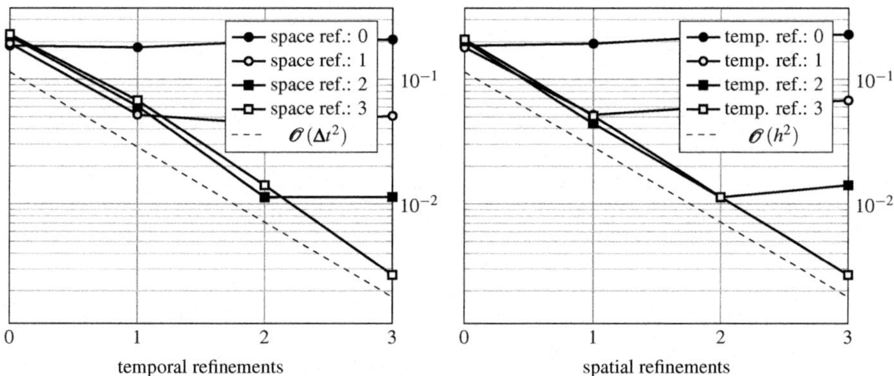

Fig. 2.6 Numerical results for the example in Sect. 2.2.2.3. L^2 error at final time after repeated (uniform) refinements in time (*left*) and space (*right*)

time FE spaces (no unfitted spaces, no Nitsche) the L^2 discretization error is of third order w.r.t. the time step.

2.3 Unfitted FEM for PDEs on Surfaces

Below, in Sect. 2.3.1 we first restrict to a model elliptic PDE, namely the Laplace-Beltrami equation on a *stationary* surface and then in Sect. 2.3.2 extend the problem class to parabolic PDEs on evolving interfaces. In the past decade several finite element techniques for the discretization of (elliptic and parabolic) PDEs on a smooth (evolving) surface have been developed. For a recent overview we refer to [11]. These methods can be classified as follows. Firstly, the (evolving) *surface finite element method* (SFEM), developed by Dziuk and Elliott in a series of papers (cf. [11]), is based on an explicit triangulation Γ_h of Γ. On this triangulation one uses a standard linear finite element space. In case of an evolving surface the vertices of the triangulation are transported with the surface velocity field. Thus this method is based on a Lagrangian approach. A second class of methods is based on an *extension of the PDE* (given on Γ) to a neighborhood of the surface. One then obtains a PDE in the volume, which can be discretized by standard FE techniques. The third class of methods consists of so-called *trace FEM* [39–41] in which one starts from a standard finite element space on an outer fixed volume mesh and then takes the trace on Γ of this space for the discretization of the surface PDE. This technique can also be applied to an evolving surface and results in a purely Eulerian approach. In this paper we restrict to the latter class of FE trace techniques, which can also be interpreted as an *unfitted* FEM: one starts from a standard finite element space on a (fixed) volume triangulation, which is not fitted to Γ, and differently from (2.5) or (2.14) one takes the restriction of this space on Γ instead of on the subdomains (separated by the surface Γ).

Below, both for the case of a stationary and an evolving surface, we explain these unfitted finite element techniques (or trace FEM), discuss optimal theoretical error bounds and present results of a few numerical experiments, which illustrate the behavior of the methods. For the case of a stationary interface we briefly address the issue of the conditioning of the resulting discrete problem (Remark 6). The results in this section are based on [17, 39–41].

2.3.1　A Model Elliptic PDE on a Stationary Surface

Let $\Omega \subseteq \mathbb{R}^3$ be a polygonal domain and $\Gamma \subset \Omega$ a smooth, closed, connected 2D surface. Given $f \in H^{-1}(\Gamma)$, with $f(1) = 0$ we consider the following Laplace–Beltrami equation: Find $u \in H^1_*(\Gamma) := \{ v \in H^1(\Gamma) \mid \int_\Gamma v \, ds = 0 \}$ such that

$$a(u, v) = f(v) \quad \text{for all } v \in H^1_*(\Gamma) \tag{2.19}$$

with

$$a(u, v) = \int_\Gamma \nabla_\Gamma u \cdot \nabla_\Gamma v \, ds.$$

2.3.1.1　An Unfitted FEM

We assume that the smooth interface Γ is the zero level of a smooth level set function ϕ, i.e., $\Gamma = \{ x \in \Omega \mid \phi(x) = 0 \}$. We will use *the same isoparametric mapping* Θ_h as in Sect. 2.2.1.1. Hence we assume that we have available $\phi_h \in V_h^k$ (degree k) and $\hat{\phi}_h = I_1 \phi_h$ (degree 1), which are finite element approximations of ϕ (in a neighborhood of Γ). These finite element functions are used as input for the isoparametric mapping Θ_h. Recall the local volume triangulation $\mathscr{T}^\Gamma := \{ T \in \mathscr{T}, T \cap \Gamma^{\mathrm{lin}} \neq \emptyset \}$. The standard affine polynomial finite element space V_h^k is restricted to $\mathscr{T}^\Gamma$, i.e., $(V_h^k)_{|\Omega^\Gamma}$. To this space we apply the transformation Θ_h resulting in the isoparametric space

$$V_{h,\Theta}^k := \{ v_h \circ \Theta_h^{-1} \mid v_h \in (V_h^k)_{|\Omega^\Gamma} \}. \tag{2.20}$$

The *unfitted* finite element space that we use is the trace of this space:

$$V_{h,\Theta}^\Gamma := \mathrm{tr}|_{\Gamma_h}(V_{h,\Theta}^k), \qquad V_{h,\Theta}^{\Gamma,0} := \{ v_h \in V_{h,\Theta}^\Gamma \mid \int_{\Gamma_h} v_h \, ds = 0 \}, \tag{2.21}$$

with $\Gamma_h := \Theta_h(\Gamma^{\mathrm{lin}})$. In the notation, we skip the polynomial degree k, and we use Γ to indicate that we *take the trace of the outer volume isoparametric space*. We introduce the bilinear form

$$a_h(u, v) := \int_{\Gamma_h} \nabla_{\Gamma_h} u \cdot \nabla_{\Gamma_h} v \, ds. \tag{2.22}$$

For the discrete problem we need a suitable extension of the data f to Γ_h, which is denoted by f_h. Specific choices for f_h are discussed in [47]. The discrete problem is as follows: Find $u_h \in V_{h,\Theta}^{\Gamma,0}$ such that

$$a_h(u_h, v_h) = \int_{\Gamma_h} f_h v_h \, dx \quad \text{for all } v_h \in V_{h,\Theta}^{\Gamma,0}. \tag{2.23}$$

Similar to the approach in Sect. 2.2.1.1, cf. (2.10), the implementation of the integrals in (2.23) is based on numerical integration rules with respect to Γ^{lin} and the transformation Θ_h. We illustrate this for the integral in (2.23). With $\tilde{u}_h = u_h \circ \Theta_h$, $\tilde{v}_h = v_h \circ \Theta_h \in V_h^{\Gamma}$, there holds

$$\int_{\Gamma_h} \nabla_{\Gamma_h} u_h \cdot \nabla_{\Gamma_h} v_h \, ds = \int_{\Gamma^{\mathrm{lin}}} \mathscr{J}_{\Gamma} \cdot \mathbf{P}_h D\Theta_h^{-T} \nabla \tilde{u}_h \cdot \mathbf{P}_h D\Theta_h^{-T} \nabla \tilde{v}_h \, ds,$$

with $\mathbf{P}_h = I - \mathbf{n}_h \mathbf{n}_h^T$ the tangential projection, $\mathbf{n}_h = \mathbf{N}/\|\mathbf{N}\|$ the unit-normal on Γ_h with $\mathbf{N} = (D\Theta_h)^{-T} \hat{\mathbf{n}}_h$ where $\hat{\mathbf{n}}_h = \nabla \hat{\phi}_h / \|\nabla \hat{\phi}_h\|$ is the normal on Γ^{lin}, and $\mathscr{J}_{\Gamma} = \det(D\Phi_h) \cdot \|\mathbf{N}\|$. This means that we only need an accurate integration with respect to the easy to construct piecewise linear geometry approximation Γ^{lin} and the explicitly available mesh transformation $\Theta_h \in (V_h^k)^d$.

2.3.1.2 Optimal Error Bound

The following optimal H^1-norm error bound is derived in [17]. One part of the discretization error stems from the approximation of the data f on the discrete surface Γ_h. Let $\mu_h : \Gamma_h \to \mathbb{R}$ describe the ratio in measures between integrals on Γ and Γ_h, so that there holds $\int_{\Gamma_h} \mu_h \, u \circ \Phi \, ds = \int_{\Gamma} u \, ds$ for $u \in L^2(\Gamma)$ and $\Phi : \Gamma_h \to \Gamma$ the closest point mapping. One can show that $\|1 - \mu_h\|_{L^\infty(\Gamma_h)} \lesssim h^{k+1}$. With $f^e(x) = f(\Phi(x))$ the constant extension in normal direction, we have for $f \in L^2(\Gamma)$ and $v, f_h \in L^2(\Gamma_h)$:

$$\int_{\Gamma} f \, v \circ \Phi^{-1} \, ds - \int_{\Gamma_h} f_h \, v \, ds = \int_{\Gamma_h} (\mu_h f \circ \Phi - f_h) v \, ds = \int_{\Gamma_h} (\mu_h f^e - f_h) v \, ds,$$

which characterizes one part of the consistency error stemming from the geometry approximation. We introduce the data error quantity $\delta_f := \mu_h f^e - f_h$ on Γ_h.

Theorem 3 *Let $u \in H^{k+1}(\Gamma)$ be the solution of (2.19) and $u_h \in V_{h,\Theta}^{\Gamma,0}$ the solution of (2.23). Assume that the data error satisfies $\|\delta_f\|_{L^2(\Gamma_h)} \lesssim h^{k+1}\|f\|_{L^2(\Gamma)}$. Then the following holds:*

$$\|u^e - u_h\|_{H^1(\Gamma_h)} \lesssim h^k \|u\|_{H^{k+1}(\Gamma)} + h^{k+1}\|f\|_{L^2(\Gamma)}. \tag{2.24}$$

Hence this method has the optimal h^k error bound in the H^1-norm, under optimal smoothness assumptions on u. An optimal L^2-norm error bound has not been derived, yet.

2.3.1.3 Results of Numerical Experiments

We consider an example taken from [15] and apply the discretization described above. The surface is a torus prescribed by the level set function ϕ, $\Gamma = \{x \in \Omega \,|\, \phi(x) = 0\}$ with

$$\phi(x) = \left(x_3^2 + \left((x_1^2 + x_2^2)^{\frac{1}{2}} - R \right)^2 \right)^{\frac{1}{2}} - r, \quad R = 1, r = 0.6.$$

The solution is given as $u(x) = \sin(3\varphi)\cos(3\theta + \varphi)$ where (φ, θ) are the angles describing a surface parametrization, cf. [15] for details. The function f is chosen accordingly. The functions u and f have mean value zero. We start from an initial mesh with mesh size $h \approx 0.1$ and repeatedly apply uniform refinements (at the interface).

In Fig. 2.7 we observe the convergence of the error in the $L^2(\Gamma_h)$ and the $H^1(\Gamma_h)$ semi-norm. The results clearly indicate optimal convergence rates both in the $H^1(\Gamma_h)$ semi-norm (as predicted by the theory) and in the $L^2(\Gamma_h)$ norm.

Remark 6 (Conditioning Stiffness Matrix) A disadvantage of the type of unfitted finite element method presented above is the fact that the stiffness matrix can be singular or extremely ill-conditioned. In particular this condition number depends not only on the mesh size h, but also on how the surface intersects the outer fixed triangulation $\mathcal{T}^\Gamma$. Recently, in [8, 16] a general (i.e., applicable also to higher order FEM) stabilization technique has been introduced. In this method one uses a stabilization term of the form

$$s_h(u_h, v_h) = \rho_s \int_{\Omega_\Theta^\Gamma} \nabla u_h \cdot \mathbf{n}_h \, \nabla v_h \cdot \mathbf{n}_h \, ds,$$

with $\rho_s > 0$ a stabilization parameter, $\Omega_\Theta^\Gamma := \Theta_h(\Omega^\Gamma)$ (recall that Ω^Γ is the domain formed by all simplices that are intersected by Γ^{lin}) and $\mathbf{n}_h$ an approximation of the unit normal on $\Gamma_h = \Theta_h(\Gamma^{\mathrm{lin}})$ (cf. [17] for details). This stabilization term is added on the left-hand side in (2.23). It can be shown that with appropriately chosen ρ_s

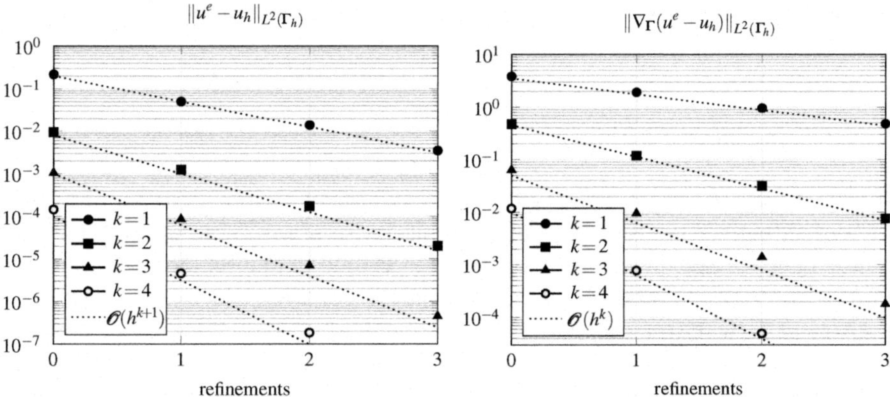

Fig. 2.7 Convergence history of the $L^2(\Gamma_h)$ norm and the $H^1(\Gamma_h)$ semi-norm of the error

an optimal order error bound as in Theorem 3 still holds and the resulting stiffness matrix has a condition number bounded by ch^{-2}, with a constant c independent of how the surface intersects the outer fixed triangulation.

2.3.2   Model Parabolic PDE on an Evolving Surface

Consider a surface $\Gamma(t)$ passively advected by a *given* smooth velocity field $\mathbf{w} = \mathbf{w}(x,t)$, i.e. the normal velocity of $\Gamma(t)$ is given by $\mathbf{w} \cdot \mathbf{n}$, with $\mathbf{n}$ the unit normal on $\Gamma(t)$. We assume that for all $t \in [0,T]$, $\Gamma(t)$ is a hypersurface that is closed ($\partial\Gamma = \emptyset$), connected, oriented, and contained in a fixed domain $\Omega \subset \mathbb{R}^d$, $d = 2, 3$. The convection-diffusion equation on the surface that we consider is given by:

$$\dot{u} + (\operatorname{div}_\Gamma \mathbf{w})u - \alpha_d \Delta_\Gamma u = f \qquad \text{on } \Gamma(t), \ \ t \in (0,T], \tag{2.25}$$

with a prescribed source term $f = f(x,t)$ and homogeneous initial condition $u(x,0) = u_0(x) = 0$ for $x \in \Gamma_0 := \Gamma(0)$. Here $\dot{u} = \frac{\partial u}{\partial t} + \mathbf{w} \cdot \nabla u$ denotes the material derivative, $\operatorname{div}_\Gamma := \operatorname{tr}\left((I - \mathbf{n}\mathbf{n}^T)\nabla\right)$ is the surface divergence and Δ_Γ is the Laplace-Beltrami operator, $\alpha_d > 0$ is the constant diffusion coefficient. If we take $f = 0$ and an initial condition $u_0 \neq 0$, this surface PDE is obtained from mass conservation of the scalar quantity u with a diffusive flux on $\Gamma(t)$ (cf. [20, 23]). A standard transformation to a homogeneous initial condition, which is convenient for a theoretical analysis, leads to (2.25). Several weak formulations of (2.25) are known in the literature, see [10, 20]. Most appropriate for our purposes is a space-time formulation on the space-time manifold $\Gamma_* = \{(x,t) \in \mathbb{R}^{d+1}, x \in \Gamma(t)\} \subset \mathbb{R}^{d+1}$ proposed in [41]. This well-posed space-time weak formulation, which we do not present here, forms the basis for the unfitted finite element presented in the next section.

2.3.2.1 An Unfitted FEM

We use the same setting as in Sect. 2.2.2.1. In particular, in the unfitted FEM that we present below we restrict ourselves to piecewise *bilinear* space-time functions (linear in space and linear in time). The space-time domain is denoted by $Q = \Omega \times (0, T] \subset \mathbb{R}^{d+1}$. A partitioning of the time interval is given by $0 = t_0 < t_1 < \ldots < t_N = T$, with (for simplicity) a uniform time step $\Delta t = T/N$. Corresponding to each time interval $I_n := (t_{n-1}, t_n]$ we assume a given shape regular simplicial triangulation $\mathcal{T}_n$ of the spatial domain Ω. In general this triangulation is *not fitted* to the interface $\Gamma(t)$. In the method we only need the local triangulation $\mathcal{T}_n^\Gamma$ (elements intersected by $\Gamma(t)$, $t \in I_n$). In this section we assume that the (implicit) geometry $\Gamma(t)$ can be handled without introducing additional errors. In practice we need a geometry approximation, for which we use the same second order accurate approximation as briefly addressed in Remark 4. We use the same "outer" space-time finite element spaces W_n, W as in (2.13) and for the unfitted finite element space we take the *trace on* Γ_*:

$$W^{\Gamma_*} := W|_{\Gamma_*}. \tag{2.26}$$

For the definition of the finite element method we need some further notation. On $L^2(\Gamma_*)$ we use the scalar product $(v, w)_0 = \int_0^T \int_{\Gamma(t)} vw \, ds \, dt$. For $u \in W_n$, the one-sided limits $u_+^n = u_+(\cdot, t_n)$ (i.e., $t \downarrow t_n$) and $u_-^n = u_-(\cdot, t_n)$ (i.e., $t \uparrow t_n$) are well-defined. At t_0 and t_N only u_+^0 and u_-^N are defined. For $v \in W^{\Gamma_*}$, a jump operator is defined by $[v]^n = v_+^n - v_-^n$, $n = 1, \ldots, N - 1$. For $n = 0$, we define $[v]^0 = v_+^0$. On the cross sections $\Gamma(t_n)$, $0 \leq n \leq N$, of Γ_* the L^2 scalar product is denoted by

$$(\psi, \phi)_{t_n} := \int_{\Gamma(t_n)} \psi\phi \, ds.$$

For the finite element discretization (which is based on a space-time weak formulation of (2.25)) we introduce the following bilinear forms:

$$a(u, v) = (\alpha_d \nabla_\Gamma u, \nabla_\Gamma v)_0 + (\mathrm{div}_\Gamma \mathbf{w}\, u, v)_0, \tag{2.27a}$$

$$d^n(u, v) = ([u]^{n-1}, v_+^{n-1})_{t_{n-1}}, \quad d(u, v) = \sum_{n=1}^N d^n(u, v), \tag{2.27b}$$

$$\langle \dot{u}, v \rangle_b = \sum_{n=1}^N \int_{t_{n-1}}^{t_n} \int_{\Gamma(t)} \left(\frac{\partial u}{\partial t} + \mathbf{w} \cdot \nabla u\right) v \, ds \, dt. \tag{2.27c}$$

The unfitted finite element discretization of (2.25) is as follows: Find $u_h \in W^{\Gamma_*}$ such that

$$\langle \dot{u}_h, v_h \rangle_b + a(u_h, v_h) + d(u_h, v_h) = (f, v_h)_0 \quad \text{for all } v_h \in W^{\Gamma_*}. \tag{2.28}$$

Note that we use the same trial and test space W^{Γ_*} and that this method allows a time stepping procedure. One easily checks that this discretization is consistent in the sense that a solution of (2.25) satisfies the variational equation (2.28). Finally note that this method is a Eulerian method, based on a fixed (per time slab) outer triangulation. For the implementation of the method one has to approximate the geometry and construct quadrature rules. So far, this has been done only for piecewise planar geometry approximations, cf. Remark 4.

2.3.2.2 Second Order Error Bound

An error analysis of this method is presented in [39]. We outline a main result. Define $H = \{ v \in L^2(\Gamma_*) \mid \|\nabla_\Gamma v\|_{L^2(\Gamma_*)} < \infty \}$ endowed with the scalar product $(u, v)_H = (u, v)_0 + (\nabla_\Gamma u, \nabla_\Gamma v)_0$. Define the total concentration $\bar{u}(t) := \int_{\Gamma(t)} u \, ds$. In the discretization error analysis we use a *consistent* stabilizing term involving the quantity $\bar{u}_h(t)$. More precisely, define

$$a_\sigma(u, v) := a(u, v) + \sigma \int_0^T \bar{u}(t) \bar{v}(t) \, dt, \quad \sigma \geq 0. \tag{2.29}$$

Instead of (2.28) we consider the stabilized version: Find $u_h \in W^{\Gamma_*}$ such that

$$\langle \dot{u}_h, v_h \rangle_b + a_\sigma(u_h, v_h) + d(u_h, v_h) = (f, v_h)_0 \quad \text{for all } v_h \in W^{\Gamma_*}. \tag{2.30}$$

Taking $\sigma > 0$ results in both a stabilizing effect and an improved discrete mass conservation property. Ellipticity estimates and error bounds are derived in the mesh-dependent norm:

$$\|u\|_h := \left(\|u_-^N\|_T^2 + \sum_{n=1}^N \|[u]^{n-1}\|_{t_{n-1}}^2 + \|u\|_H^2 \right)^{\frac{1}{2}}.$$

In the error analysis we need a condition which plays a similar role as the condition "$c - \frac{1}{2} \operatorname{div} b > 0$" used in standard analyses of variational formulations of the convection-diffusion equation $-\Delta u + b \cdot \nabla u + cu = f$ in an Euclidean domain $\Omega \subset \mathbb{R}^n$, cf. [49]. This condition is as follows: there exists a $c_0 > 0$ such that

$$\operatorname{div}_\Gamma \mathbf{w}(x, t) + \alpha_d c_F(t) \geq c_0 \quad \text{for all } x \in \Gamma(t), \ t \in [0, T]. \tag{2.31}$$

Here $c_F(t) > 0$ results from the Poincare inequality

$$\int_{\Gamma(t)} |\nabla_\Gamma u|^2 \, ds \geq c_F(t) \int_{\Gamma(t)} \left(u - \frac{1}{|\Gamma(t)|}\bar{u}\right)^2 ds \quad \forall \, t \in [0, T], \quad \forall \, u \in H. \tag{2.32}$$

A main result derived in [39] is given in the following theorem. To simplify the presentation, we assume that the time step Δt and the spatial mesh size parameter h have comparable size: $\Delta t \sim h$.

Theorem 4 *Assume* (2.31) *and take* $\sigma \geq \frac{\alpha_d}{2} \max_{t \in [0,T]} \frac{c_F(t)}{|\Gamma(t)|}$, *where* $c_F(t)$ *is defined in* (2.32). *Then the ellipticity estimate*

$$\langle \dot{u}_h, u_h \rangle_b + a_\sigma(u_h, u_h) + d(u_h, u_h) \geq c_s \|u_h\|_h^2 \quad \text{for all } u \in W^{\Gamma_*} \tag{2.33}$$

holds, with $c_s = \frac{1}{2}\min\{1, \alpha_d, c_0\}$ *and* c_0 *from* (2.31). *Let* $u \in H^2(\Gamma_*)$ *be the solution of* (2.25). *For the solution* $u_h \in W^{\Gamma_*}$ *of the discrete problem* (2.30) *the following error bound holds:*

$$\|u - u_h\|_h \leq ch\|u\|_{H^2(\Gamma_*)}.$$

A further main result derived in [39] is related to second order convergence. Denote by $\|\cdot\|_{-1}$ the norm dual to the $H_0^1(\Gamma_*)$ norm with respect to the L^2-duality. Under the conditions given in Theorem 4 and some further mild assumptions the error bound

$$\|u - u_h\|_{-1} \leq ch^2\|u\|_{H^2(\Gamma_*)}$$

holds. This second order convergence is derived in a norm weaker than the commonly considered $L^2(\Gamma_*)$ norm. The reason is that our arguments use isotropic polynomial interpolation error bounds on 4D space-time elements. Naturally, such bounds call for isotropic space-time H^2-regularity bounds for the solution. For our problem class such regularity is more restrictive than in an elliptic case, since the solution is generally less regular in time than in space. We can overcome this by measuring the error in the weaker $\|\cdot\|_{-1}$-norm.

2.3.2.3 Results of Numerical Experiments

For the results of experiments with the space-time unfitted FEM presented in Sect. 2.3.2.1 we refer to [16, 41]. In [41] results for the transport equation (2.25) on a smoothly evolving surface (e.g., shrinking sphere) are presented which clearly show a second order convergence (w.r.t. L^2-norm) both with respect to the space and the time mesh size. Furthermore, this convergence behavior occurs already on relatively coarse meshes. In [16] an example with a topological singularity is considered. For this example the assumptions we need in the error analysis are not satisfied. The evolving surface that we consider essentially consists of two disjoint spheres which approach each other and then merge. On this surface the transport equation (2.25) is considered. The space-time unfitted FEM presented in Sect. 2.3.2.1, cf. (2.28), is applied *without any modifications*. It turns out that despite the topological singularity the method yields satisfactory results, which indicates that this unfitted space-time FEM has very good robustness properties.

2.4 Applications to Two-Phase Flows

In this section we present two examples of more advanced applications of the unfitted finite element techniques presented above. These applications are directly related to the main topics of the Priority Program 1506.

2.4.1 Mass Transport in Two-Phase Flow

We present numerical simulations of a dissolution process of oxygen from a rising (and deforming) air bubble into a water-glycerol solution. This example is treated in detail in [28]. We take a setting that has also been considered in [4, 26], [25, Chap. 9.8.3] and [42, Chap. 4.3.2]. Close to the bottom of a container filled with a homogeneous water-glycerol mixture a 4 mm spherical air bubble is placed. Due to buoyancy the bubble rises and deforms. After some time the rising of the bubble reaches a quasi-stationary state with an ellipsoidal shape. Initially the concentration of oxygen inside the fluid is assumed to be zero and a constant concentration u_0 is prescribed inside the bubble. During the rise of the bubble oxygen dissolves from the bubble into the fluid and a wake of oxygen follows the path of the bubble. Contour lines of the concentration for different times are depicted in Fig. 2.8. The material parameters to this system are given in Table 2.1. While these parameters are realistic for the fluid dynamics, the diffusion coefficients are artificial in order to be able to prescribe a value for the Schmidt number $Sc = \frac{\mu}{\rho\alpha}$ in the liquid phase, $Sc = 10$. The

Table 2.1 Material parameters for the considered setting. The setting is the same as in [4, 25, 26]

	Liquid phase ($\Omega_2 = \Omega_L$)	Disperse phase ($\Omega_1 = \Omega_B$)
Density ρ (kg/m^3)	1205	1.122
Dynamic viscosity μ (Pa s)	0.075	1.824×10^{-5}
Henry weight β [1]	1	33
Diffusion coeff. α (m^2/s)	$6.224 \times 10^{-5} \cdot Sc^{-1}$	1.916×10^{-5}
Surface tension τ (N/m)	0.063	
Init. bubble diameter d (m)	0.004	
Gravity g (m/s^2)	9.81	

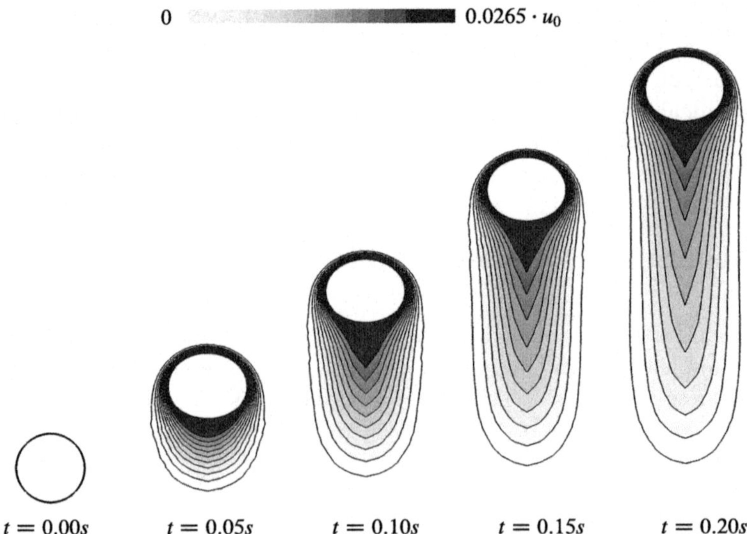

Fig. 2.8 Concentration contours in the fluid phase at several time for the dissolution process of oxygen from a rising air bubble in a water-glycerol mixture for Schmidt number $Sc = 10$

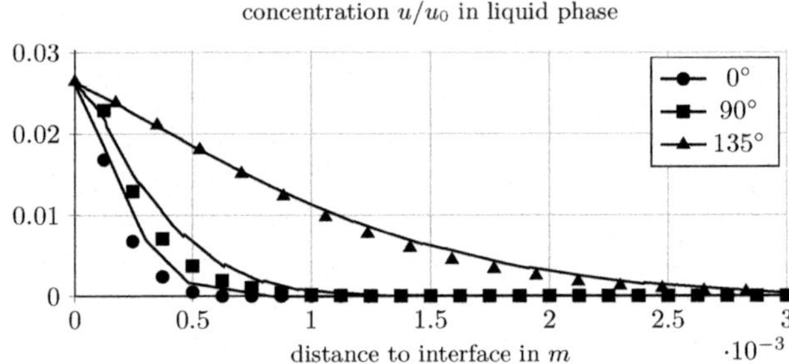

Fig. 2.9 Concentration layer profile in the liquid phase along lines of angles $0°$, $90°$ and $135°$ computed with the space-time unfitted finite element method (*lines*) and comparison data (*dots*)

simulation of the fluid dynamics is realized with the finite element software DROPS, cf. [19]. The software is based on a level set technique for interface capturing and (unfitted) modified $P_2 - P_1$ finite elements. We use adaptively refined tetrahedral grids close to the (evolving) interface and the wake of the bubble. We refer to [20] for details on the sharp interface fluid dynamics model and its numerical treatment. The results obtained for the fluid dynamics have been validated against experimental data from [54]. For details we refer to [28, Sect. 5.4.1.1].

For the discretization of the mass transport problem we consider the unfitted space-time finite element method discussed in Sect. 2.2.2 with bilinear functions (linear in space and linear in time) and a Nitsche stabilization parameter of $\lambda = 20$. At time $T = 0.2$ of the simulation we compare the concentration along straight lines which are crossing the bubble center in Fig. 2.9. On the lines through the tip ($0°$), the equator ($90°$) and close to the wake ($135°$) we compare the concentration with simulated data from [42, Fig. 9.35]. The results are in good agreement.

We can conclude that these numerical methods for transport equations on evolving domains are robust and accurate even in this challenging realistic configuration.

2.4.2   *Droplet Breakup with Surfactants*

We present results of a numerical experiment for a two-phase flow problem with surfactants. This example is treated in detail in [55]. We briefly describe the model of this flow problem, the important model parameters and discuss a simulation result. For more information we refer to Sect. 7.4 in [55]. The experiment is very similar to the one considered in [52]. In that paper, however, a *diffusive interface* model is used, whereas we consider a *sharp interface* model. A spherical droplet with radius $r = 1$ is put in a rectangular box of dimensions 12(length)×4(width)×4(height), cf. Fig. 2.10, and exposed to a shear flow in the length direction with shear rate $\dot{\gamma} = 1$. The two fluids are Newtonian, modeled by the incompressible Navier-Stokes equations, and there is a surface tension force at the sharp interface. Dirichlet shear flow boundary conditions are imposed on the upper and lower boundaries. Periodic boundary conditions are imposed on all other boundaries. For the densities and viscosities we take $\rho_1 = \rho_2 = \mu_1 = \mu_2 = 1$. The Reynold's and Capillary numbers have values $Re = 0.4$, $Ca = 0.42$ (as in [52]). The surface tension coefficient for a clean interface has value $\tau_0 = 2.38$. We assume that there are

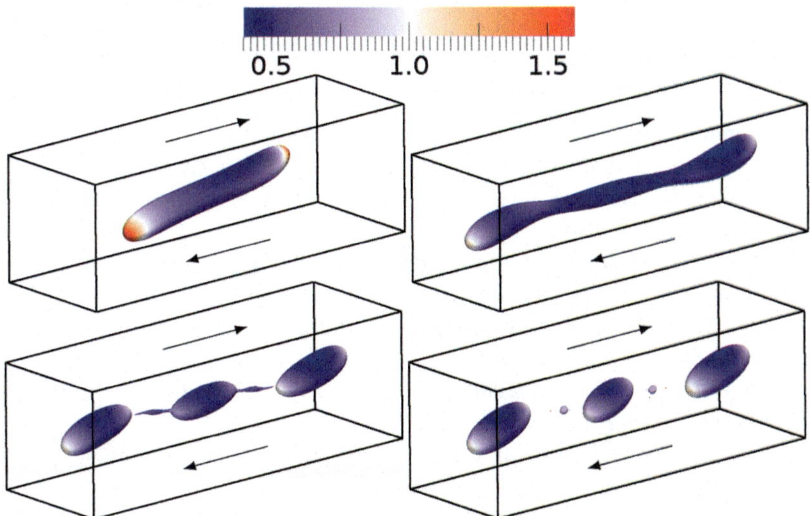

Fig. 2.10 Droplet in shear flow and surfactant concentration for $t = 10,\ 30,\ 37.5,\ 40\,$s

insoluble surfactants (only) on the interface. The transport of these surfactants is modeled by the equation (2.25) with source term $f = 0$ and $\mathbf{w}$ resulting from the Navier-Stokes fluid dynamics model. For the effect of the surfactant concentration S on the surface tension coefficient the Langmuir model is used:

$$\frac{\tau(S)}{\tau_0} = 1 + c_0 \ln(1 - \chi \frac{S}{S^*}),$$

with parameter values $c_0 = 0.2$, $\chi = 0.1$, $S^* = 1.91 \times 10^{-5}$. The surfactant diffusion coefficient has value $\alpha_d = 0.1$.

We do not further explain the components of the numerical solver used for the simulation of the fluid-dynamics, cf. the brief description in Sect. 2.4.1. For more information we refer to [20]. The surfactant equation on the evolving interface is treated with the unfitted space-time FEM discussed in Sect. 2.3.2.

In Fig. 2.10 we show a simulation result, which shows the interface evolution and surfactant concentration distribution. These (and further results in [55]) show a good agreement with simulation results from [52]. A comparison with a simulation for the clean droplet case shows that (as expected) due to the surfactants the droplet deforms more and breaks up earlier. The level set technique allows a robust handling of the droplet breakup. The unfitted space-time FEM is a robust discretization method for the surfactant transport equation. This method is exactly the same as the one presented in Sect. 2.3.2, without any "tricky" modifications to deal with the droplet breakup.

2.5 Summary and Outlook

We summarize the main new contributions obtained based on our research in the Priority Program 1506:

- We extended the combination of unfitted FEM with the Nitsche technique, already introduced in the literature for *stationary* interface problems, to problems with *evolving* interfaces. For this we developed and analyzed *space-time variants of unfitted FE and Nitsche methods* (Sect. 2.2.2).
- We developed and analyzed a new method for obtaining *higher order accurate* approximations of implicitly described geometries. We applied this method to stationary interface problems (Sect. 2.2.1) and to PDEs on stationary surfaces (Sect. 2.3.1).
- The space-time unfitted FE technique is further developed for the discretization of PDEs on *evolving* surfaces (Sect. 2.3.2).
- We developed a stabilization technique for convection dominated interface problems (Remark 2).
- We developed a new optimal preconditioner for the efficient solution of a discretized stationary interface problem (Remark 3).

- We applied these techniques not only to model problems, but also to more difficult applications in two-phase flow problems (Sect. 2.4).

We briefly address a few topics, which we consider to be relevant for further research:

- So far the higher order unfitted FEM has been studied only for a *stationary* interface/surface. The extension of this approach to problems with *evolving* interfaces/surfaces is a subject of current research.
- For the unfitted FEMs treated in the Sects. 2.2.2 and 2.3.1 optimal order L^2-error bounds are not available, yet.
- Concerning stabilization techniques and preconditioners for the resulting discrete problems only very few results are known. More research is required to improve the efficiency of (iterative) solvers for these systems.
- The higher order methods presented in Sects. 2.2.1 and 2.3.1 use different meshes, the original triangulation $\mathscr{T}$ and the curved mesh $\Theta_h(\mathscr{T})$, respectively, for the definition of the level set function ϕ_h and a scalar field u (solution of the PDE). In cases where these quantities are coupled, i.e. where the level set (evolution) may depend on u the transfer of information from one mesh to the other has to be provided. Realizing this in a proper way is another topic of current research.
- The performance of these relatively new (higher order) unfitted space-time FE techniques to challenging applications from e.g. two-phase flow problems should be further investigated.
- We need a better understanding of the unfitted space-time method for PDEs on evolving surfaces with topological singularities.

Acknowledgements The authors gratefully acknowledge funding by the German Science Foundation (DFG) within the Priority Program (SPP) 1506 "Transport Processes at Fluidic Interfaces". Furthermore, we thank Jörg Grande for his contributions related to the research on surface PDEs and Yuanjun Zhang for making the results in Sect. 2.4.2 available to us.

References

1. Bastian, P., Engwer, C.: An unfitted finite element method using discontinuous Galerkin. Int. J. Numer. Methods Eng. **79**, 1557–1576 (2009)
2. Becker, R., Burman, E., Hansbo, P.: A Nitsche extended finite element method for incompressible elasticity with discontinuous modulus of elasticity. Comput. Methods Appl. Mech. Eng. **198**, 3352–3360 (2009)
3. Bothe, D., Koebe, M., Wielage, K., Warnecke, H.-J.: VOF-simulations of mass transfer from single bubbles and bubble chains rising in aqueous solutions. In: Proceedings 2003 ASME Joint U.S.-European Fluids Engineering Conference, Honolulu (2003). ASME. FEDSM2003-45155
4. Bothe, D., Koebe, M., Warnecke, H.-J.: VOF-simulations of the rise behavior of single air bubbles with oxygen transfer to the ambient liquid. In: Schindler, F.-P. (ed.) Transport Phenomena with Moving Boundaries, pp. 134–146. VDI-Verlag, Düsseldorf (2004)

5. Bothe, D., Koebe, M., Wielage, K., Prüss, J., Warnecke, H.-J.: Direct numerical simulation of mass transfer between rising gas bubbles and water. In: Sommerfeld, M. (ed.) Bubbly Flows: Analysis, Modelling and Calculation. Heat and Mass Transfer. Springer, Berlin (2004)
6. Burman, E., Hansbo, P.: Fictitious domain finite element methods using cut elements: II. A stabilized Nitsche method. Appl. Numer. Math. **62**, 328–341 (2012)
7. Burman, E., Claus, S., Hansbo, P., Larson, M.G., Massing, A.: CutFEM: discretizing geometry and partial differential equations. Int. J. Numer. Methods Eng. (2014). doi:10.1002/nme.4823
8. Burman, E., Hansbo, P., Larson, M.G., Massing A.: Cut finite element methods for partial differential equations on embedded manifolds of arbitrary codimensions (2016). Preprint, arXiv:1610.01660
9. Carraro, T., Wetterauer, S.: On the implementation of the eXtended finite element method (XFEM) for interface problems. Archive Numer. Softw. **4**(2) (2016)
10. Dziuk, G., Elliott, C.: Finite elements on evolving surfaces. IMA J. Numer. Anal. **27**, 262–292 (2007)
11. Dziuk, G., Elliott, C.: Finite element methods for surface PDEs. Acta Numer. **22**, 289–396 (2013)
12. Engwer, C., Heimann, F.: Dune-UDG: a cut-cell framework for unfitted discontinuous Galerkin methods. In: Advances in DUNE. Springer, Berlin (2012), pp. 89–100
13. Guermond, J.-L., Ern, A.: Finite element quasi-interpolation and best approximation. ESAIM: Math. Model. Numer. Anal. (2015)
14. Fries, T.-P., Belytschko, T.: The extended/generalized finite element method: an overview of the method and its applications. Int. J. Numer. Methods Eng. **84**, 253–304 (2010)
15. Grande, J., Reusken, A.: A higher order finite element method for partial differential equations on surfaces. SIAM J. Numer. Anal. **54**, 388–414 (2016)
16. Grande, J., Olshanskii, M., Reusken, A.: A space-time FEM for PDEs on evolving surfaces. In: Onate, E., Oliver, J., Huerta, A. (eds.) Proceedings of the 11th World Congress on Computational Mechanics 2014 (2014)
17. Grande, J., Lehrenfeld, C., Reusken, A.: Analysis of a high order trace finite element method for PDEs on level set surfaces. arXiv preprint arXiv:1611.01100 (2016)
18. Groß, S., Reusken, A.: An extended pressure finite element space for two-phase incompressible flows. J. Comput. Phys. **224**, 40–58 (2007)
19. Groß, S., Peters, J., Reichelt, V., Reusken, A.: The DROPS package for numerical simulations of incompressible flows using parallel adaptive multigrid techniques, no. 227 (IGPM, RWTH Aachen, 2002). Preprint
20. Gross, S., Reusken, A.: Numerical Methods for Two-Phase Incompressible Flows. Springer, Heidelberg (2011)
21. Hansbo, A., Hansbo, P.: An unfitted finite element method, based on Nitsche's method, for elliptic interface problems. Comput. Methods Appl. Mech. Eng. **191**, 5537–5552 (2002)
22. Ishii, M.: Thermo-Fluid Dynamic Theory of Two-Phase Flow. Eyrolles, Paris (1975)
23. James, A., Lowengrub, J.: A surfactant-conserving volume-of-fluid method for interfacial flows with insoluble surfactant. J. Comput. Phys. **201**, 685–722 (2004)
24. Johansson, A., Larson, M.G.: A high order discontinuous Galerkin Nitsche method for elliptic problems with fictitious boundary. Numer. Math. **123**, 607–628 (2013)
25. Koebe, M.: Numerische Simulation aufsteigender Blasen mit und ohne Stoffaustausch mittels der Volume of Fluid (VOF) Methode. PhD thesis, Universität Paderborn (2004)
26. Koebe, M., Bothe, D., Warnecke, H.-J.: Direct numerical simulation of air bubbles in water/glycerol mixtures: shapes and velocity fields. In: Proceedings 2003 ASME joint U.S.-European Fluids Engineering Conference, Honolulu (2003) ASME. FEDSM2003-451354
27. Lehrenfeld, C.: The Nitsche XFEM-DG space-time method and its implementation in three space dimensions. SIAM J. Sci. Comput. **37**, A245–A270 (2015)
28. Lehrenfeld, C.: On a space-time extended finite element method for the solution of a class of two-phase mass transport problems. PhD thesis, RWTH Aachen, Feb (2015)
29. Lehrenfeld, C.: High order unfitted finite element methods on level set domains using isoparametric mappings. Comput. Methods Appl. Mech. Eng. **300**, 716–733 (2016)

30. Lehrenfeld, C., Reusken, A.: Nitsche-XFEM with streamline diffusion stabilization for a two-phase mass transport problem. SIAM J. Sci. Comput. **34**, 2740–2759 (2012)
31. Lehrenfeld, C., Reusken, A.: Analysis of a Nitsche-XFEM-DG discretization for a class of two-phase mass transport problems. SIAM J. Numer. Anal. **51**, 958–983 (2013)
32. Lehrenfeld, C., Reusken, A.: Finite element techniques for the numerical simulation of two-phase flows with mass transport. Comput. Methods Complex Liquid Fluid Interfaces **37**, 353–372 (2015)
33. Lehrenfeld, C., Reusken, A.: Analysis of a high order unfitted finite element method for elliptic interface problems (2017). Preprint, arXiv:1602.02970v2
34. Lehrenfeld, C., Reusken, A.: L^2-estimates for a high order unfitted finite element method for elliptic interface problems. arXiv preprint arXiv:1604.04529 (2016)
35. Lehrenfeld, C., Reusken, A.: Optimal preconditioners for Nitsche-XFEM discretizations of interface problems. Numer. Math. 1–20 First online (2016). doi:10.1007/s00211-016-0801-6
36. Massjung, R.: An unfitted discontinuous Galerkin method applied to elliptic interface problems. SIAM J. Numer. Anal. **50**, 3134–3162 (2012)
37. Mayer, U.M., Gerstenberger, A., Wall, W.A.: Interface handling for three-dimensional higher-order XFEM-computations in fluid–structure interaction. Int. J. Numer. Methods Eng. **79**, 846–869 (2009)
38. Nærland, T.A.: Geometry decomposition algorithms for the Nitsche method on unfitted geometries. Master's thesis, University of Oslo (2014)
39. Olshanskii, M.A., Reusken, A.: Error analysis of a space-time finite element method for solving PDEs on evolving surfaces. SIAM J. Numer. Anal. **52**, 2092–2120 (2014)
40. Olshanskii, M.A., Reusken, A., Grande, J.: A finite element method for elliptic equations on surfaces. SIAM J. Numer. Anal. **47**, 3339–3358 (2009)
41. Olshanskii, M.A., Reusken, A., Xu, X.: An Eulerian space-time finite element method for diffusion problems on evolving surfaces. SIAM J. Numer. Anal. **52**, 1354–1377 (2014)
42. Onea, A.: Numerical simulation of mass transfer with and without first order chemical reaction in two-fluid flows. PhD thesis, Forschungszentrum Karlsruhe, Sept (2007)
43. Oswald, P.: On a BPX-preconditioner for $\mathbb{P}_1$ elements. Computing **51**, 125–133 (1993)
44. Parvizian, J., Düster, A., Rank, E.: Finite cell method. Comput. Mech. **41**, 121–133 (2007)
45. Renard, Y., Pommier, J.: GetFEM++, an open-source finite element library (2014)
46. Reusken, A.: Analysis of an extended pressure finite element space for two-phase incompressible flows. Comput. Vis. Sci. **11**, 293–305 (2008)
47. Reusken, A.: Analysis of trace finite element methods for surface partial differential equations. IMA J. Numer. Anal. **11**, 1568–1590 (2015)
48. Reusken, A., Nguyen, T.: Nitsche's method for a transport problem in two-phase incompressible flows. J. Fourier Anal. Appl. **15**, 663–683 (2009)
49. Roos, H.-G., Stynes, M., Tobiska, L.: Numerical Methods for Singularly Perturbed Differential Equations. Springer Series in Computational Mathematics, vol. 24. Springer, Berlin (1996)
50. Sadhal, S., Ayyaswamy, P., Chung, J.: Transport Phenomena with Droplets and Bubbles. Springer, New York (1997)
51. Slattery, J., Sagis, L., Oh, E.-S.: Interfacial Transport Phenomena, 2nd edn. Springer, New York (2007)
52. Teigen, K., Song, P., Lowengrub, J., Voigt, A.: A diffuse-interface method for two- phase flows with soluble surfactants. J. Comput. Phys. **230**, 375–393 (2011)
53. Thomee, V.: Galerkin Finite Element Methods for Parabolic Problems. Springer, Berlin (1997)
54. Wegener, M.: Der Einfluss der konzentrationsinduzierten Marangonikonvektion auf den instationären Impuls- und Stofftransport an Einzeltropfen. PhD thesis, TU Berlin, Sept (2009)
55. Zhang, Y.: Numerical simulation of two-phase flows with complex interfaces. PhD thesis, RWTH Aachen (2015)

Chapter 3
An Extended Discontinuous Galerkin Framework for Multiphase Flows

Thomas Utz, Christina Kallendorf, Florian Kummer, Björn Müller, and Martin Oberlack

Abstract We present a framework for handling cut cells in a high order discontinuous Galerkin (DG) context. To describe the boundary between fluid phases, we use a level-set formulation. When the interface cuts a computational cell, we discretize the resulting sub-cells with the same DG method as used on standard cells. This requires a suitable quadrature procedure. Within this framework, we present a solver for the two-phase Navier-Stokes equation, a reinitialization procedure for the level-set and a solver for transport-processes on the surface.

3.1 Introduction

Over the past decades, increasing computational power has led to the widespread acceptance of numerical methods to solve engineering problems in fluid mechanics. While low-order methods for single phase flows are well established for research and industrial applications, the accuracy for the established 2nd-order methods scales with the gridsize as h^2, while the computational effort in 3D scales with h^3. Thus increasing accuracy requires a larger than linear increase in computing time. To circumvent this, one must resort to higher-order methods. If we choose a method based on a representation of the unknowns by a polynomial of degree k, the L^2-error ideally scales with h^{k+1}. The discontinuous Galerkin method (DG) allows a rather straight-forward use of higher-order polynomials, while maintaining a small stencil, which allows easy parallelization, h-p-adaptivity and meshing of complex geometries [12]. In addition, the method is locally conservative and stable for convection dominated problems.

Multiphase flows present another challenge: The interface between the fluid phases introduces an internal discontinuity in the material parameters and a singular force due to interface tension. We describe the interface implicitly using a level-set.

T. Utz • C. Kallendorf • F. Kummer • B. Müller • M. Oberlack (✉)
Institute of Fluid Dynamics, Technische Universität Darmstadt, Otto-Bernd-Straße 2, 64287 Darmstadt, Germany
e-mail: utz@fdy.tu-darmstadt.de; oberlack@fdy.tu-darmstadt.de

© Springer International Publishing AG 2017
D. Bothe, A. Reusken (eds.), *Transport Processes at Fluidic Interfaces*,
Advances in Mathematical Fluid Mechanics, DOI 10.1007/978-3-319-56602-3_3

In the context of classical low order numerical schemes, the level-set method is well established and available for a multitude of physical applications, see e.g. Osher and Fedkiw [53] for a review. Here we explicitly focus on the high-order discontinuous Galerkin (DG) method. Combining the discontinuous Galerkin method with the level set method is popular to simulate multiphase flows, since the low numerical dissipation allows highly accurate computation of the level-set movement [24, 43, 54, 55, 62].

With this level-set we define a sharp interface method, which directly implements the interface force and the jump in the material parameters. Thus we can accurately represent the resulting jump in the pressure field and the kink in the velocity field. Within the past decade, sharp-interface methods that strive to resolve local effects with sub-cell accuracy have gained more and more interest. Examples include the eXtended finite element method (XFEM) [9, 37, 46], the Finite Cell Method [13] and the discontinuous Galerkin method [20, 21, 35]. In [35] we developed a sharp interface DG method, which we call extended DG method due to its similarity to XFEM.

This work presents a solver framework, which is capable of calculating multiphase flows including surfactant transport with high order accuracy. Here, we present several building blocks for this solver:

1. A quadrature method for cut cells
2. Discretization of the flow using our method
3. A smoothing technique for the curvature computed from the level-set
4. A cell agglomeration technique to avoid ill-conditioned cut cells
5. A high-order reinitialization procedure for the level-set
6. A conservative form for the surfactant transport equation on the interface
7. Exact solutions of the surfactant transport equation for verification purposes
8. A numerical scheme for solving the surfactant transport

3.2   The Continuous Setting

We introduce the level-set φ, to divide the domain Ω into three parts: a phase $\mathfrak{A}$ where the level-set is positive, a second phase $\mathfrak{B}$, where it is negative, and an interface $\mathfrak{I}$, which is the zero iso-contour of the level-set.

$$\mathfrak{A}(\varphi) = \{\mathbf{x} \in \Omega : \varphi(\mathbf{x}) < 0\} \tag{3.1a}$$

$$\mathfrak{B}(\varphi) = \{\mathbf{x} \in \Omega : \varphi(\mathbf{x}) > 0\} \tag{3.1b}$$

$$\mathfrak{I}(\varphi) = \{\mathbf{x} \in \Omega : \varphi(\mathbf{x}) = 0\} \tag{3.1c}$$

Typically, the level-set is chosen as the signed distance to the interface $\mathfrak{I}$:

$$\varphi(\mathbf{x}) = \begin{cases} -\ \text{dist}(\mathbf{x}, \mathfrak{I}(\varphi)) & \text{in } \mathfrak{A}(\varphi) \\ \ \ \text{dist}(\mathbf{x}, \mathfrak{I}(\varphi)) & \text{in } \mathfrak{B}(\varphi) \end{cases} \tag{3.2}$$

For a given interface $\tilde{\mathfrak{I}}$, the signed distance formulation can be computed as the solution of the Eikonal equation

$$|\nabla\varphi| - 1 = 0 \quad \text{in } \Omega \tag{3.3a}$$

with the boundary condition

$$\varphi = 0 \quad \text{on } \tilde{\mathfrak{I}}. \tag{3.3b}$$

The interface normal can be obtained from the gradient of φ and the curvature can be calculated by Bonnet's formula:

$$\mathbf{n}_{\mathfrak{I}} = \frac{\nabla\varphi}{|\nabla\varphi|} \quad \text{and} \quad \kappa = \nabla \cdot \frac{\nabla\varphi}{|\nabla\varphi|}. \tag{3.4}$$

The level set moves due to a velocity field $\mathbf{u}$, according to the scalar advection equation

$$\frac{\partial\varphi}{\partial t} + \mathbf{u} \cdot \nabla\varphi = 0. \tag{3.5}$$

Separating the domain by the level set, we introduce two different fluid phases with piece-wise constant density ρ and viscosity μ:

$$\rho(\mathbf{x}) = \begin{cases} \rho_{\mathfrak{A}} & \text{for } \mathbf{x} \in \mathfrak{A} \\ \rho_{\mathfrak{B}} & \text{for } \mathbf{x} \in \mathfrak{B} \end{cases} \quad \text{and} \quad \mu(\mathbf{x}) = \begin{cases} \mu_{\mathfrak{A}} & \text{for } \mathbf{x} \in \mathfrak{A} \\ \mu_{\mathfrak{B}} & \text{for } \mathbf{x} \in \mathfrak{B} \end{cases}. \tag{3.6}$$

The two-fluid formulation of the incompressible Navier-Stokes equation is given by the equations in the bulk phases $\Omega_h \setminus \mathfrak{I}$,

$$\partial_t\mathbf{u} + \nabla \cdot (\rho\mathbf{u} \otimes \mathbf{u}) + \nabla p = \nabla \cdot \mu(\nabla\mathbf{u} + (\nabla\mathbf{u})^T) - \mathbf{F}, \tag{3.7}$$

$$\nabla \cdot \mathbf{u} = 0, \tag{3.8}$$

and the jump conditions at the interface $\mathfrak{I}$,

$$[\![\mathbf{u}]\!] = 0, \tag{3.9}$$

$$[\![p\mathbf{n}_{\mathfrak{I}} - \mu(\nabla\mathbf{u} + (\nabla\mathbf{u})^T)\mathbf{n}_{\mathfrak{I}}]\!] = \sigma\kappa\mathbf{n}_{\mathfrak{I}}, \tag{3.10}$$

where σ denotes the surface tension, κ the mean curvature of $\mathfrak{I}$ and $\mathbf{n}_{\mathfrak{I}}$ the normal of the interface $\mathfrak{I}$, with an orientation that "points from $\mathfrak{A}$ to $\mathfrak{B}$". With the jump operator being defined as $[\![\mathbf{y}]\!] = (\mathbf{y}_{\mathfrak{A}} - \mathbf{y}_{\mathfrak{B}})$, see Eq. (3.23).

Furthermore, one has to specify boundary conditions

$$\mathbf{u} = \mathbf{u}_\mathrm{D} \text{ on } \partial\Omega_\mathrm{D} \text{ and } \mu\left(\nabla\mathbf{u} + \nabla\mathbf{u}^T\right) \cdot \mathbf{n}_\Omega - p\mathbf{n}_\Omega = 0 \text{ on } \partial\Omega_\mathrm{N}, \tag{3.11}$$

at the Dirichlet $\partial\Omega_\mathrm{D}$ and Neumann boundary $\partial\Omega_\mathrm{N}$.

In many applications, one is interested in the transport process of a solute c whose concentration is small and whose influence on the flow field is negligible. We assume that no chemical reactions take place within or outside of the domain or the interface, i.e. no further solute is created within Ω.

Denote by c_i, $i = \mathfrak{A}, \mathfrak{B}$, the concentration of a surfactant in each of the bulks and by $c^{\mathfrak{J}}$ the surface excess density of surfactant and $D^{(\mathfrak{J})}$ the diffusion coefficient on the surface. The mass balance equation of the solute c on the interface $\mathfrak{J}$ results in the interfacial transport equation of the surfactant

$$\frac{dc^{\mathfrak{J}}}{dt} + c^{\mathfrak{J}}\nabla^{(\mathfrak{J})} \cdot \mathbf{u}^{(\mathfrak{J})} - D^{(\mathfrak{J})}\nabla^{(\mathfrak{J})} \cdot \nabla^{(\mathfrak{J})}c^{\mathfrak{J}} = S \text{ on } \mathfrak{J} \tag{3.12}$$

where the flow of mass from and to the bulks is summarized by the source term S,

$$S = D^{\mathfrak{A}}\nabla c_{\mathfrak{A}}|_{\mathfrak{J}} \cdot \mathbf{n}_{\mathfrak{J}} - D^{\mathfrak{B}}\nabla c_{\mathfrak{B}}|_{\mathfrak{J}} \cdot \mathbf{n}_{\mathfrak{J}}. \tag{3.13}$$

Here $\nabla^{(\mathfrak{J})} = \mathbf{P}^{\mathfrak{J}}\nabla$ denotes the projection of the ∇-operator on the interface with the projection-operator being $\mathbf{P}^{\mathfrak{J}} = \mathbf{I} - \mathbf{n}_{\mathfrak{J}} \otimes \mathbf{n}_{\mathfrak{J}}$.

3.3 Quadrature Using the Hierarchical Moment Fitting Method

When employing a sharp interface method to discretize a multiphysics problem, the burden of discretization lies in the numerical integration over complicated, typically curved domains where conventional quadrature rules are hard if not impossible to construct. As a consequence, the viability of the extension of these methods to higher approximation orders is directly linked to the affordable integration accuracy.

A prominent class of methods for this integration relies on an (adaptive) subdivision of cells intersected by the interface. They predominantly rely on a linear reconstruction of the interface [20, 44, 45], even though this is not strictly necessary [13, 50]. In [44] it has been shown, that this approach can lead to a robust, second-order convergent scheme. However, the scheme is limited to 5 to 7 subdivisions due to runtime restrictions, which does not give the accuracy needed for higher order methods.

Recently, a method based on a non-linear optimization procedure for the construction of (approximately) Gaussian quadrature rules on very general domains has been developed [8, 68], which has been applied successfully to the computation of integrals in the eXtended Finite Element Method [47]. In [48] and [61], the authors have shown how the non-linear optimization process can be reduced to a linear one in the context of piecewise linear interfaces. In [51] we introduced a general setting for the construction of efficient quadrature rules based on the so-called *moment-fitting* equations [8].

3.3.1 General Approach for Moment Fitting

A generic strategy for the construction of quadrature rules for a given set of integrands $\mathfrak{B} = \{f_j\}_{j=1,\dots,M}$ over a domain Ω is given by the solution of the moment-fitting equations

$$\underbrace{\begin{pmatrix} f_1(\mathbf{x}_1) & \dots & f_1(\mathbf{x}_N) \\ \vdots & \ddots & \vdots \\ f_M(\mathbf{x}_1) & \dots & f_M(\mathbf{x}_N) \end{pmatrix}}_{=:A} \begin{pmatrix} w_1 \\ \vdots \\ w_N \end{pmatrix} = \underbrace{\begin{pmatrix} \int_\Omega f_1 \, dV \\ \vdots \\ \int_\Omega f_M \, dV \end{pmatrix}}_{=:b} \tag{3.14}$$

with nodes $\mathfrak{X} = \{\mathbf{x}_i\}_{i=1,\dots,N}$ and weights $\mathfrak{W} = \{w_i\}_{i=1,\dots,N}$ [8]. For the purposes of this work, it is assumed that $\mathfrak{B}$ consists of a polynomial basis of $\mathbb{R}^D$ of degree k. Consequently, the system is linear in the weights and non-linear in the nodes.

In order to obtain an optimal quadrature rule, system (3.14) can be solved several times while eliminating nodes with marginal significance [68]. This approach is very well suited for the construction of quadrature rules on *fixed* domains where the right-hand side of system (3.14) can be computed a priori. On the other hand, this will be extremely costly in general applications. It is thus only reasonable to follow this approach if the resulting quadrature rules can be reused a large number of times.

Fortunately, we can greatly simplify the solution of Eq. (3.14) if:

1. The set of nodes $\mathfrak{X}$ can be predefined.
2. The right-hand side can be evaluated cheaply.

In such a situation, (3.14) reduces to the *linear* system

$$A_{ji} w_i = \int_\Omega f_j \, dV \tag{3.15}$$

which can be solved efficiently. While the obtained solution will usually be sub-optimal in the global sense, the gain in efficiency renders the approach attractive under very general conditions. We present a method to construct such a system for a polytope $K \subset \mathbb{R}$, which is divided by a level-set into two subdomains $\mathfrak{A}$ and $\mathfrak{B}$.

3.3.2 Hierarchy of Quadrature Rules

We are interested in the construction of quadrature rules for $\mathfrak{A}$ as well as for $\partial\mathfrak{A}$. The latter can be decomposed in the actual interface $\mathfrak{I}$ and the surface $\partial\mathfrak{A} \setminus \mathfrak{I} = \{\mathbf{x} \in \partial K : \varphi(\mathbf{x}) < 0\}$. Subsequently, we will use this decomposition to define a *hierarchy* of quadrature rules.

3.3.2.1 One-Dimensional Case

If K is one-dimensional (i.e., a line element), we construct a quadrature rule of order k by finding the roots of φ on K with the aid of a root-finding algorithm such as the safe-guarded Newton method described in [56]. We then employ a standard Gaussian quadrature rule of degree k on each of the induced sub-sections. Given a sufficient accuracy of the root-finding procedure, the composite quadrature rule will be exact up to numerical precision.

3.3.2.2 Two-Dimensional Case

If K is two-dimensional (i.e., a polygon such as a triangle or a quadrilateral), we construct quadrature rules for $\mathfrak{I}$ by reducing special integrals to integrals over $\partial\mathfrak{A} \setminus \mathfrak{I}$ (see Sect. 3.3.4). Sub-domain $\partial\mathfrak{A} \setminus \mathfrak{I}$, in turn, is composed of several line elements which can be treated as described in Sect. 3.3.2.1. Having constructed quadrature rules over $\mathfrak{I}$ and $\partial\mathfrak{A} \setminus \mathfrak{I}$, the methods presented in Sect. 3.3.4 can be applied to construct quadrature rules for $\mathfrak{A}$.

3.3.2.3 Three-Dimensional Case

If K is three-dimensional (i.e., a polyhedron such as a hexahedron), we construct quadrature rules for $\mathfrak{I}$ by reducing special integrals to integrals over $\partial\mathfrak{A} \setminus \mathfrak{I}$ (see Sect. 3.3.4). Sub-domain $\partial\mathfrak{A} \setminus \mathfrak{I}$ is composed of a set of implicit, two-dimensional domains. For each of these surfaces, we construct volume quadrature rules using the methods presented in Sect. 3.3.2.2. During this process, the algorithm will generate a volume quadrature rule on $\partial K \cap \mathfrak{I}$ (see Sect. 3.3.4) by creating a surface quadrature rule on $\partial(\partial\mathfrak{A}) \setminus \mathfrak{I}$ (see Sect. 3.3.2.1).

3.3.3 Choice of the Quadrature Nodes

In Sect. 3.3.1, we have formulated the requirement that the set $\mathfrak{X}$ should be kept constant during the optimization process. We emphasize that this does *not* imply that the quadrature nodes have to be located on $\mathfrak{I}$ in the surface case or solely in $\mathfrak{A}$ in the volume case, provided that $\mathfrak{B}$ consists of smooth functions. Our tests indicate that using the same points associated with traditional quadrature rules generally leads to the best results [51].

3.3.4 Construction of Surface and Volume Quadrature Rules

In general, the evaluation of

$$\int_{\mathfrak{I}} f_j \, \mathrm{d}S \tag{3.16}$$

is difficult since $\mathfrak{I}$ is only given implicitly as $\mathfrak{I}(\varphi)$ and an accurate (i.e., non-linear) interface reconstruction is cumbersome and computationally expensive. However, we can define a *divergence-free* basis $\mathfrak{B}' = \{\mathbf{f}'_k\}_{k=1,\dots,K}$ ($\mathbf{f}'_k : \mathbb{R}^D \to \mathbb{R}^D$) of degree k associated to $\mathfrak{B}$. It can be used for the definition of a set of integrands for which the evaluation is simpler, namely $\{\mathbf{f}_k \cdot \mathbf{n}_\mathfrak{I}\}_{k=1,\dots,K}$ where $\mathbf{n}_\mathfrak{I} = \nabla\varphi/||\nabla\varphi||$ is the level set normal. In this particular case, Gauss' theorem allows us to write

$$\int_{\mathfrak{I}} \mathbf{f}'_k \cdot \mathbf{n}_\mathfrak{I} \, \mathrm{d}S = \int_{\partial\mathfrak{A}} \mathbf{f}'_k \cdot \mathbf{n}_K \, \mathrm{d}S - \int_{\partial\mathfrak{A}\backslash\mathfrak{I}} \mathbf{f}'_k \cdot \mathbf{n}_K \, \mathrm{d}S \tag{3.17}$$

$$= \int_{\mathfrak{A}} \nabla \cdot \mathbf{f}'_k \, \mathrm{d}V - \int_{\partial\mathfrak{A}\backslash\mathfrak{I}} \mathbf{f}'_k \cdot \mathbf{n}_K \, \mathrm{d}S \tag{3.18}$$

$$= 0 - \int_{\partial\mathfrak{A}\backslash\mathfrak{I}} \mathbf{f}'_k \cdot \mathbf{n}_K \, \mathrm{d}S \tag{3.19}$$

where $\mathbf{n}_K$ is the outward unit normal vector on the boundary of K. A method for the construction of a suitable divergence-free basis is given in [50]. In $2D$ $\partial\mathfrak{A} \setminus \mathfrak{I}$ only consists of straight lines, for which exact quadrature rules are available as described previously in Sect. 3.3.2.1.

Once, a quadrature rule for the interface is available, we use a different basis $\mathfrak{B}'' = \{\mathbf{f}''_k\}_{k=1,\dots,K}$ ($\mathbf{f}''_k : \mathbb{R}^D \to \mathbb{R}^D$) with nonzero divergence and rewrite (3.17) and (3.18) as

$$\int_{\mathfrak{A}} \nabla \cdot \mathbf{f}''_k \, \mathrm{d}V = \int_{\mathfrak{I}} \mathbf{f}''_k \cdot \mathbf{n}_\mathfrak{I} \, \mathrm{d}S + \int_{\partial\mathfrak{A}\backslash\mathfrak{I}} \mathbf{f}''_k \cdot \mathbf{n}_k \, \mathrm{d}S \tag{3.20}$$

Since we have constructed a quadrature rule for the boundary of the phase $\mathfrak{A}$, we can evaluate the volume integral on the left hand side of Eq. (3.20). In [51] we present numerical test cases for the quadrature rules generated by this process. They show optimal h-convergence up to a polynomial order of 6 for two-dimensional and three dimensional test cases. The procedure was extended in [34] to preserve the Gauss and Stokes theorem on a discrete level, which further improves the methods accuracy.

3.4 Discretization of the Two-Phase Flow Problem

For the discontinuous Galerkin method, the computational domain $\Omega \subset \mathbb{R}$ is discretized into non-overlapping cells K_i with a characteristic mesh size h. Together, they form the grid $\mathfrak{K} = \{K_1, \ldots, K_J\}$ which defines the computational domain $\Omega_h = \bigcup_{K \in \mathfrak{K}} K$. The outward pointing normal of each cell is denoted by $\mathbf{n}$. The skeleton of the grid is defined as the union of all interior cell edges

$$\Gamma = \bigcup_{K \in \mathfrak{K}} \partial K \cap \partial \Omega. \tag{3.21}$$

On this grid the weak form of a differential equation is interpreted in a cell local way. To do so, the test function v and the trial functions u are defined on each cell separately. Thus, they are discontinuous across the cell interfaces. Typically, the functions are chosen from the polynomial space $\mathbb{P}$ of degree k. This means, we choose the test and trial functions

$$u, v \in \mathbb{V}_{\mathrm{DG}}(\Omega_h) \tag{3.22a}$$

from the broken polynomial space

$$\mathbb{V}_{\mathrm{DG}, k}(\Omega_h) := \{f \in L^2(\Omega_h); \ \forall K \in \mathfrak{K} : f|_K \in \mathbb{P}_k(K)\}. \tag{3.22b}$$

Across an edge shared by two cells, functions $u \in \mathbb{V}_{\mathrm{DG}}(\mathfrak{K})$ are discontinuous. We denote the value at the inner side of an element by the superscript "$-$", the outside by the superscript "$+$". Then, we can define the jump and mean operators:

$$\{\mathbf{u}\} = \frac{1}{2}\left(\mathbf{u}^+ + \mathbf{u}^-\right) \tag{3.23a}$$

$$[\![u]\!] = u^- - u^+ \tag{3.23b}$$

In order to represent the phases $\mathfrak{A}$ and $\mathfrak{B}$ as well as the interface $\mathfrak{I}$ numerically, we employ a level-set formulation.

Formally, a cell $K \in \mathfrak{K}$ is called to be cut (by the zero-level-set $\mathfrak{I}$) if $\oint_{K \cap \mathfrak{I}} 1 \, \mathrm{d}S \neq 0$. On a grid with cells cut by the interface we define the extended

DG space, or XDG-space as

$$\mathbb{V}_k^X(\varphi, \Omega_h) := \left\{ f \in L^2(\Omega_h);\ \forall\ K \in \mathfrak{K} : f|_{K \cap \mathfrak{A}}, f|_{K \cap \mathfrak{B}} \in \mathbb{P}_k(K) \right\}. \tag{3.24}$$

This functions space is identical to the standard DG space, when viewing each side of the cut cell as an individual subcell.

We define the narrow band subgrid $\mathfrak{N}^w$ which consists of the cut cells and their neighbors up to a width w. Formally the set of all cut-cells is

$$\mathfrak{N}^0 := \{ K \in \mathfrak{K};\ K \text{ is cut} \} \tag{3.25}$$

and the narrow Band of width $w > 0$ is

$$\mathfrak{N}^w := \{ K \in \mathfrak{K};\ K \in \mathfrak{N}^{w-1} \vee L^K \in \mathfrak{N}^{w-1} \}.$$

Later, we will need additional information about the mesh topology.

A neighbor cell $L^K \in \mathfrak{K}$ is called to be a neighbor of cell $K \in \mathfrak{K}$, if they share at least one point, i.e. if $\overline{L^K} \cap \overline{K} \neq \{\}$. An edge neighbor cell $L^{e,K} \in \mathfrak{K}$ is called to be an edge neighbor of cell $K \in \mathfrak{K}$, if they share a common edge, i.e. if $\oint \left(L^{e,K} \cap \overline{K} \right) dS \neq 0$.

3.4.1 The Spatial Discretization of the Two-Phase Problem

We discretize velocity and pressure in XDG spaces of order k and $k' := k - 1$, respectively, i.e.

$$(\mathbf{u}, p) \in \mathbb{V}_k^X(\varphi, \Omega_h)^2 \times \mathbb{V}_{k-1}^X(\varphi, \Omega_h) =: V_k. \tag{3.26}$$

To the best of our knowledge, there is no rigorous proof for the inf-sup stability of this velocity/pressure pair which would directly apply to the XDG setting. There is, however, some experimental evidence for the stability of this velocity/pressure pair, see e.g. [33].

In [34] we discretize the multiphase Navier-Stokes problem (3.7), (3.8) with jump conditions (3.9), (3.10) and boundary conditions (3.11) in the extended DG space: find $(\mathbf{u}, p) \in V_k$, such that for all $(\mathbf{v}, \tau) \in V_k$

$$Ns(\mathbf{u}, (\mathbf{u}, p), (\mathbf{v}, \tau)) = \underbrace{q(\mathbf{v}) + s(\mathbf{v}) + r(\tau)}_{=:rhsNs((\mathbf{v}, \tau))}, \tag{3.27}$$

where the Navier-Stokes form $Ns(-, -, -)$ is given as the sum

$$Ns(\mathbf{u}, (\mathbf{u}, p), (\mathbf{v}, \tau)) = b(p, \mathbf{v}) - a(\mathbf{u}, \mathbf{v}) - b(\tau, \mathbf{u}) + t(\mathbf{u}, \mathbf{u}, \mathbf{v}). \tag{3.28}$$

The convective terms $t(-, -, -)$ are discretized by a local Lax-Friedrichs flux,

$$t(\mathbf{w}, \mathbf{u}, \mathbf{v}) = -\int_{\Omega} \rho(\mathbf{u} \otimes \mathbf{w}) : \nabla \mathbf{v}\, dV - \oint_{\Gamma_{\text{int}} \cup \partial\Omega_{\text{N}} \cup face} (\{\mathbf{u} \otimes \mathbf{w}\}\, \mathbf{n}_{\mathfrak{I},\Gamma} + (\lambda/2)\, [\![\mathbf{u}]\!]) \cdot [\![\rho\mathbf{v}]\!]\, dS, \tag{3.29}$$

as used in the work of Shahbazi et al. [60]. Details on the stabilization parameter λ are given in [34].

Pressure gradient and velocity divergence $b(-, -)$ are discretized as

$$b(p, \mathbf{v}) = -\int_{\Omega} p\, \nabla \cdot \mathbf{v}\, dV - \oint_{(\Gamma \setminus \partial\Omega_{\text{N}}) \cup \mathfrak{I}} [\![\mathbf{v}]\!] \cdot \mathbf{n}_{\mathfrak{I},\Gamma}\, \{p\}\, dS. \tag{3.30}$$

To discretize the viscous terms $a(-, -)$, we employ an almost-standard symmetric interior penalty method (SIP), first introduced by Arnold [4] (for an extensive analysis, see e.g. [5]):

$$a(\mathbf{u}, \mathbf{v}) = -\int_{\Omega} \mu(\nabla\mathbf{u} : \nabla\mathbf{v} + (\nabla\mathbf{u})^T : \nabla\mathbf{v})\, dV$$

$$+ \oint_{(\Gamma \setminus \partial\Omega_{\text{N}}) \cup \mathfrak{I}} (\{\mu(\nabla\mathbf{u} + \nabla\mathbf{u}^T)\}\, \mathbf{n}_{\mathfrak{I},\Gamma}) \cdot [\![\mathbf{v}]\!] + (\{\mu(\nabla\mathbf{v} + \nabla\mathbf{v}^T)\}\, \mathbf{n}_{\mathfrak{I},\Gamma}) \cdot [\![\mathbf{u}]\!]\, dS$$

$$- \oint_{(\Gamma \setminus \partial\Omega_{\text{N}}) \cup \mathfrak{I}} \eta\, [\![\mathbf{u}]\!] \cdot [\![\mathbf{v}]\!]\, dS. \tag{3.31}$$

This is an extension of the classical SIP-form to the operator $\nabla \cdot \left(\mu \left(\nabla\mathbf{u} + \nabla\mathbf{u}^T\right)\right)$. Since $\nabla\mathbf{u}^T : \nabla\mathbf{v} = \nabla\mathbf{v}^T : \nabla\mathbf{u}$, the bilinear form $a(\mathbf{u}, \mathbf{v})$ is symmetric in $\mathbf{u}$ and $\mathbf{v}$. Details on the choice of the penalty parameter η are given in [34].

Finally, we specify the sources of the Stokes problem $s(-)$,

$$s(\mathbf{v}) = -\int_{\Omega} \mathbf{F} \cdot \mathbf{v}\, dV + \oint_{\mathfrak{I}} \kappa\, \sigma\, (\mathbf{n}_{\mathfrak{I}} \cdot [\![\mathbf{v}]\!])\, dS$$

$$- \oint_{\partial\Omega_{\text{D}}} \mathbf{u}_{\text{D}} \cdot \left(\nabla\mathbf{v}\mathbf{n}_{\Omega} + \nabla\mathbf{v}^T\mathbf{n}_{\Omega} - \eta\mathbf{v}\right)\, dS, \tag{3.32}$$

where the first term represents a volume force, the second term surface tension, and the third terms represents Dirichlet boundary conditions for the viscous part. The Dirichlet boundary conditions $r(-)$ for the continuity equation are given by

$$r(\tau) = \oint_{\partial\Omega_{\text{D}}} \tau\, \mathbf{u}_{\text{D}} \cdot \mathbf{n}_{\Omega}\, dS, \tag{3.33}$$

and for the convective terms $q(-)$ by

$$q(\mathbf{v}) = -\oint_{\partial\Omega_D} (\mathbf{u}_D \otimes \mathbf{u}_D)\,\mathbf{n}_{\mathfrak{J}} \cdot [\![\rho\mathbf{v}]\!]\; dS. \tag{3.34}$$

3.5 Stabilization Procedures for the Two-Phase Flow Solver: Curvature Recovery

The calculation of the interface curvature κ by Bonnet's formula (3.4) is highly sensitive to disturbances in the level-set function φ. In [36] we present a technique to reduce this issue.

The expression for the curvature $\kappa = \nabla \cdot \mathbf{n}_{\mathfrak{J}}$ can be reformulated using the level-set gradient $\nabla\varphi$ and the level-set Hessian $\mathbf{H} = \partial^2\varphi$ as

$$\nabla \cdot \frac{\nabla\varphi}{|\nabla\varphi|_2} = \frac{\mathrm{tr}(\mathbf{H})}{|\nabla\varphi|_2} - \frac{(\nabla\varphi)^T\,\mathbf{H}\,(\nabla\varphi)}{|\nabla\varphi|_2^3} . =: \kappa(\nabla\varphi, \mathbf{H}) \tag{3.35}$$

Note that by the introduction of φ, the properties $\mathbf{n}_{\mathfrak{J}}$ and κ are smoothly extended to the whole domain Ω.

For our method, the interface $\mathfrak{J}$ has to be a continuous surface, therefore it is required that also the numerical representation of the level-set field is at least continuous. To ensure this, the level-set function φ is projected onto the continuous piecewise polynomial space $\mathbb{V}_{\mathrm{FEM},2}(\Omega_h) = \mathbb{V}_{\mathrm{DG},2}(\Omega_h) \cap \mathscr{C}^0(\Omega_h)$. Here $\mathrm{Proj}_{\mathbb{V}}\,()$ denotes the projector onto sub-space $\mathbb{V}$ of $L^2(\Omega_h)$ in the L^2-norm.

As an example, we choose the interface $\mathfrak{J}$ to be a circle with radius r around the origin. Obviously, there are infinitely many choices for φ. We compare the two representations

$$\begin{aligned}
&\text{case (a) :}\ \varphi = r^2 - x^2 - y^2 &&\text{quadratic, } \varphi = \varphi_{\mathrm{ex}}\\
&\text{case (b) :}\ \varphi = \mathrm{Proj}_{\mathbb{V}_{\mathrm{FEM},2}}\left(r - \sqrt{x^2 - y^2}\right) &&\text{signed-distance, } \varphi \neq \varphi_{\mathrm{ex}}.
\end{aligned}$$

Obviously, the exact curvature at the interface is the radius r. Note that in case (a) the circular interface is represented *exactly* in the polynomial approximation space, which it is not in case (b).

After computing the curvature, it is projected back onto a broken polynomial space:

$$\kappa := \mathrm{Proj}_{\mathbb{V}_{\mathrm{DG},12}(K)}\left(\mathrm{curv}(\nabla\varphi, \partial^2\varphi)\right). \tag{3.36}$$

Note that we choose a rather high degree of 12 for the representation of the curvature κ, in order to diminish the projection error.

Table 3.1 Shape and curvature error for different circle representations, data from [36]

Error type		Quadratic	Signed-distance
Radius	$\forall \mathbf{x} \in \mathfrak{J} : \ \big| \ |\mathbf{x}|_2 - 8/10 \ \big|$	$\leq 10^{-11}$	$\leq 1.2 \cdot 10^{-4}$
Curvature	$\forall \mathbf{x} \in \mathfrak{J} : \ \big| \ |\kappa(\mathbf{x})| - (-10/8) \ \big|$	$\leq 2.2 \cdot 10^{-9}$	≤ 0.14

We discretize the domain $\Omega = (-3/2, 3/2)^2$ by 18×18 equidistant quadrilateral cells. Table 3.1 shows the error in the position of the interface and the curvature for both formulae. The quadratic formula is exactly contained in the DG-space $\mathbb{V}_{\mathrm{DG}}$, thus the radius and curvature are correct up to round-off errors. In contrast, the signed distance formulation gives a reasonable good representation of the shape, but the error in the curvature is unacceptably high. Such high errors in the curvature will result in spurious pressures and velocities, when employing a multiphase flow solver.

This example demonstrates how the result of a level-set based two-phase computation depends on the choice of the level-set field. In [36] we present a filtering procedure to minimize this dependence.

Originally, patch recovery was introduced by Zienkiewicz and Zhu [72] and used as a post-processing technique in order to obtain super-convergence. It can be applied to continuous as well as discontinuous Galerkin methods. In contrast to [72], where a nodal projection is used for the patch-recovery operator, we employ an L^2-projection. This allows to use patches of arbitrary shapes, which is beneficial if we only want to consider cells which are cut by the interface $\mathfrak{J}$.

We introduce the patch-recovery operator to filter the level-set field φ and its derivatives:

$$\mathrm{prc}_w : \mathbb{V}_{\mathrm{DG}, q}(K) \to \mathbb{V}_{\mathrm{DG}, r}(K). \tag{3.37}$$

Here, q and r, satisfying $r \geq q$, denote the polynomial degree of domain resp. codomain and w is the width of the patch.

The patch-recovery operator with width $w \in \{0, 1\}$ is defined as the L^2-projection onto the broken polynomial space on the composite cell Q_K, which is formed from all neighbor cells L^K of some cell K, which share at least one point with K.

For each cell K in the sub-grid $\mathfrak{N}^w$ we perform a L^2-projection of a broken polynomial function on the cells $\{K, L_1^K, \ldots, L_l^K\}$ onto polynomials which are continuous the composite cell Q_K. Then, the result of the patch-recovery operation in cell K, denoted as $v|_K$, is obtained by restricting the polynomial on Q_K to cell K.

For the cell K, the patch-recovery operation $\mathrm{prc}_w(u) =: v$ is then defined as the L^2-projection of u onto $\mathbb{V}_{\mathrm{DG}, r}(\{Q_K\})$, i.e.

$$v|_K := \begin{cases} \mathrm{Proj}_{\mathbb{V}_{\mathrm{DG}, r}(\{Q_K\})} \left(u|_{Q_K}\right)\big|_K & \text{if } K \in \mathfrak{N}^w \\ 0 & \text{if } K \notin \mathfrak{N}^w \end{cases}. \tag{3.38}$$

This procedure may be applied multiple times to the level-set and its derivatives, thus iteratively smoothing these fields.

When computing the curvature by Eq. (3.35), the patch recovery procedure may be applied to the level-set φ itself, but also to the derivatives of the level-set $\nabla\varphi$ and $\mathbf{H}$. In [36] we compare filtering of different derivatives of the level-set and the resulting curvature field, the width of the patch recovery domain w and the polynomial degrees r and q. We found that there is no need to increase the polynomial degree of the filtered properties by more than a factor of two times the polynomial degree of the original level-set field; second, that we found it sufficient to perform the patch recovery just on the layer of cut-cells themselves, and not considering any values outside of that band, which additionally keeps the run-time of the algorithm low. When projecting the level-set from a discontinuous Galerkin space into the continuous finite element space, filtering the gradient and the Hessian further reduces the error in curvature. This is not necessary, when computing the curvature based on the DG representation.

3.6 Stabilization Procedures for the Two-Phase Flow Solver: Cell Agglomeration

Up to this point, the discretization outlined in Sect. 3.4 admits arbitrarily small cut cells with arbitrary shapes. Obviously, this renders the method impractical due to the high condition number of the matrices resulting from cells with small volume fractions (cf. Fig. 3.1a). Moreover, the numerical experiments of [49] suggest that the presence of *sliver* elements with large aspect ratios have a strongly negative effect for $k \geq 2$. In Fig. 3.1b, we give an example for a corresponding cell K with a volume fraction of about 20%. While we did not encounter any problems for

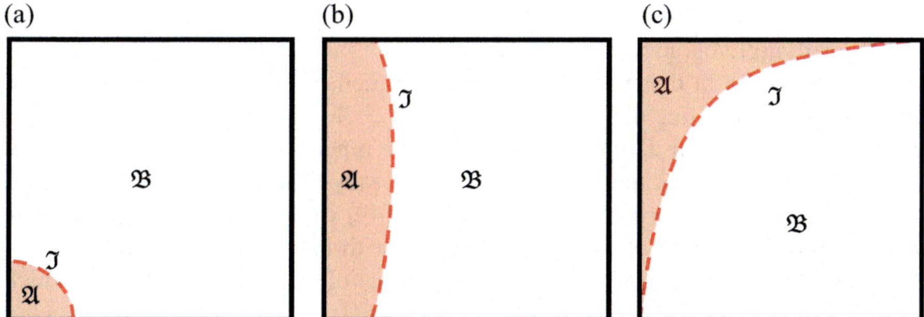

Fig. 3.1 Configurations where cell-agglomeration is required for higher order simulations. (**a**) Sub-cell with a small volume fraction [49]. (**b**) Sliver sub-cell. (**c**) Strongly curved sub-cell

low-order simulations ($k \leq 1$), using a higher-order basis $\mathfrak{B}$ defined on K results in strongly ill-conditioned mass matrices. This finding can be interpreted as a loss of linear independence of the basis functions within $\mathfrak{A}$ that has to be addressed. Unfortunately, redefining the basis on a bounding box, as proposed by Qin and Krivodonova [57] is not feasible for strongly curved interfaces as shown in Fig. 3.1c, where the bounding box is as large as the original cell. In order to overcome these issues, we employ the cell-agglomeration strategy proposed in [34].

3.6.1 Agglomeration Strategy

In the first step of the algorithm proposed in [34], we determine the list of agglomeration *source* cells $\{K_s^{\mathrm{src}}\} = \{K_i : \mathrm{frac}(K_{i,s}) \leq \delta\}$ for each phase $s = \mathfrak{A}, \mathfrak{B}$ based on the volume fraction $\mathrm{frac}(K_i, s)$ with a user-defined threshold $0 \leq \delta < 1$.

$$\mathrm{frac}(K_i, s) = \frac{\int_{K_i \cap s} 1 \, dV}{\int_{K_i} 1 \, dV} \tag{3.39}$$

Second, we determine the agglomeration *target* cell K^{tar} for each source cell K^{src} by finding the edge neighbor $L^{e,K^{\mathrm{src}}}$ with the largest volume fraction, i.e.

$$K^{\mathrm{tar}} = \underset{\{L^{e,K^{\mathrm{src}}}\}}{\mathrm{argmax}} \, \mathrm{frac}(L^{e,K^{\mathrm{src}}}, s). \tag{3.40}$$

In DG, the basis $\mathfrak{B}_i$ for each cell K_i can be chosen independently, which simply allows us to extend the basis of the target cell K^{tar} into the agglomeration source cell K^{src}. The source cell and the edge between the source and target cell is then formally eliminated from the discretization. Our actual implementation (cf.[34, 49]) exploits the locality of the involved operations and hence only requires few additional *cell-local* matrix-vector products that hardly affect the overall performance.

The cell-agglomeration process is illustrated in Fig. 3.2 using the example depicted in Fig. 3.2a that initially consists of 4 cells. Depending on the selected threshold δ, the resulting mesh consists of 3 (Fig. 3.2b) or 2 cells (Fig. 3.2c). In [34] we show, how the cell agglomeration technique renders the condition number of the DG discretization of the diffusion and the steady Stokes problem independent of the position f the interface in the cut cell. In [49] we use the cell agglomeration technique in the context of compressible flows with immersed boundaries. Again, the procedure greatly reduces the condition number of the involved matrices, which increases the accuracy of the method.

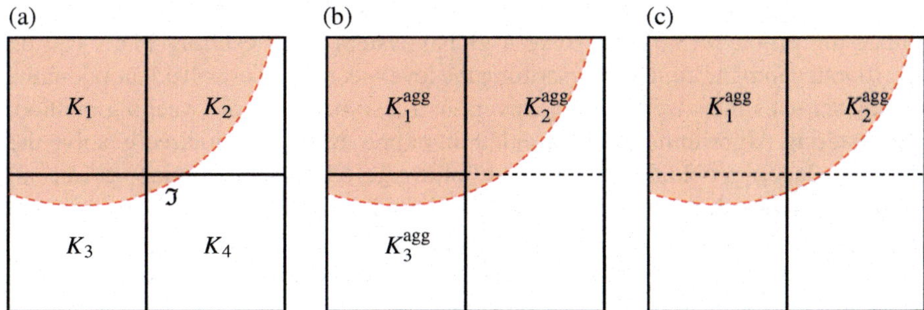

Fig. 3.2 Illustration of the cut-cell agglomeration strategy. For small values of δ (*middle*), only cell K_4 (*source*) will be agglomerated to the *direct* neighbor with the largest volume fraction, namely cell K_2 (*target*). If δ is increased (*right*), K_3 will additionally be agglomerated to cell K_1 [49]. (**a**) Initial mesh. (**b**) Agglomerated mesh, δ small. (**c**) Agglomerated mesh, δ large

3.7 Results for the Two-Phase Flow Solver

In [34] we use the discretization from Sect. 3.4.1 and evaluate the integrals using the hierarchical moment fitting from Sect. 3.3. We combine this with the patch recovery for the curvature from Sect. 3.5 and the cell agglomeration in Sect. 3.6 for a multiphase flow solver for the steady stokes problem using this approach. We present numerical test cases for h-convergence for polynomial orders of up to three. In [49] we present an immersed boundary solver with fixed boundaries for the compressible Navier-Stokes equations, where we present numerical examples for optimal h-convergence for polynomials up to order four.

3.8 Elliptic Reinitialization

Depending on the velocity field from the solution of the Navier-Stokes equations, this evolution of the level-set field due to the advection equation (3.5) may cause the level-set to become very steep or flat close to the interface. The goal of reinitialization is to restore the signed distance properties of the level-set, i.e. to find a solution of Eq. (3.3). Starting point is an initial function $\tilde{\varphi}$, which defines the interface $\tilde{\mathfrak{J}} = \mathfrak{J}(\tilde{\varphi})$ and does not have signed distance properties. In [64] we focus on multiphase flows in the context of high-order discontinuous Galerkin methods. There, reinitialization may suffer from instability and should provide the same accuracy as the flow solver.

The most popular algorithms for level-set reinitialization are geometry based algorithms such as direct approaches [11, 42, 43, 55, 59] or fast methods such as the fast marching method [18, 23, 32, 62] or the fast sweeping method [10, 40, 41, 67, 69, 70]. To the authors' knowledge, no such method is available, which allow higher than third-order accuracy on unstructured grids.

In contrast to these geometry based algorithms, PDE-based approaches reformulate the problem (3.3) and solve a global system. The resulting PDEs can be incorporated into the equation describing the level-set motion or solved individually, or may serve as the base for local solvers in a fast marching or sweeping method. The resulting Algorithms can be divided into approaches, which directly solve the Hamilton-Jacobi problem [10, 28], hyperbolic (e.g. [63]) and parabolic (e.g. Li et al. [39]) approaches. A recent addition to PDE based reinitialization techniques is the approach by Basting and Kuzmin [6], which is based on solving an elliptic problem. In [64], we extend their approach to the discontinuous Galerkin method and discuss the method based on multiple test cases.

3.8.1 An Elliptic PDE for Level-Set Reinitialization

The idea is to consider an energy functional R of the form

$$R(\varphi) = \int_\Omega \psi\left(|\nabla\varphi|\right) \, d\mathbf{x} \tag{3.41}$$

with the potential function $\psi(|\nabla\varphi|)$. The simplest choice for the potential is the single well $\psi(s) = 1/2\,(s-1)^2$:

$$R(\varphi) = \frac{1}{2} \int_\Omega \left(|\nabla\varphi| - 1\right)^2 \, d\mathbf{x}.$$

In [64], we will discuss the choice of this potential function. The function has an associated diffusion rate d, which is defined as

$$d(s) = \frac{d\psi(s)}{d(s)} \frac{1}{s}. \tag{3.42}$$

We follow the idea by Basting and Kuzmin [6] to directly solve the minimization problem

$$\min R(\varphi) \quad \text{s.t. } \varphi = 0 \text{ on } \Im(\tilde\varphi). \tag{3.43}$$

The constraint at the interface is enforced by a penalty term

$$P(\tilde\varphi, \varphi) = \alpha \oint_{\Im(\tilde\varphi)} \frac{\varphi^2}{2} \, dS, \tag{3.44}$$

where α is a large number.

The strong form of the minimization problem is

$$\frac{\partial R}{\partial \varphi} = \nabla \cdot (d\left(|\nabla\varphi|\right)\nabla\varphi) = 0 \quad \text{in } \Omega \tag{3.45a}$$

$$\varphi = 0 \quad \text{on } \Im(\tilde{\varphi}). \tag{3.45b}$$

Note, that Eq. (3.45a) can be split into a linear part $\Delta\varphi$ and a nonlinear part $f(\varphi)$

$$\Delta\varphi - \nabla \cdot (1 - d\left(|\nabla\varphi|\right)\nabla\varphi) = \Delta\varphi - f(\varphi) = 0 \tag{3.46}$$

3.8.2 Discontinuous Galerkin Discretization

We discretize the problem in the form of Eq. (3.46) with the standard symmetric interior penalty method for the linear terms $a_1 \ldots a_3$ (see, e.g. [12]) and the incomplete interior penalty method for the nonlinear terms b_1 and b_2. The penalty term a_4 enforces the boundary condition (see, e.g. [12]):

$$\text{find } \varphi \in \mathbb{V}_{\text{DG}} \text{ s.t.}$$

$$\underbrace{\int_{\Omega_h} \nabla\varphi \cdot \nabla v \, \mathrm{d}\mathbf{x}}_{a_1(\varphi,v),\ \text{SIP volume term}} \ \underbrace{- \oint_{\Gamma} (\{\nabla\varphi\}\,[\![v]\!]\cdot\mathbf{n} + \{\nabla v\}\,[\![\varphi]\!]\cdot\mathbf{n})\,\mathrm{d}S}_{a_2(\varphi,v),\ \text{SIP consistency \& symmetry term}} +$$

$$\underbrace{\oint_{\Gamma} \eta\,[\![\varphi]\!]\cdot[\![v]\!]\,\mathrm{d}S}_{a_3(\varphi,v),\ \text{SIP penalty term}} \ \underbrace{- \int_{\Omega_h} (1 - d\left(|\nabla\varphi|\right))\,\nabla\varphi\cdot\nabla v\,\mathrm{d}\mathbf{x}}_{b_1(\varphi,v),\ \text{nonlinear volume part}} +$$

$$\underbrace{\oint_{\Gamma} \{(1 - d\left(|\nabla\varphi|\right))\,\nabla\varphi\}\,[\![v]\!]\cdot\mathbf{n}\,\mathrm{d}S}_{b_2(\varphi,v),\ \text{nonlinear consistency part}} + \underbrace{\int_{\Gamma(\tilde{\varphi})} \alpha\varphi v\,\mathrm{d}S}_{a_4(\varphi,v),\ \text{B.C. at Interface}} = 0 \quad \forall v \in \mathbb{V}_{\text{DG}}(\mathfrak{K})$$

$$\tag{3.47}$$

In short, the above problem reads

$$\text{find } \varphi \in \mathbb{V}_{\text{DG}} \text{ s.t. } \sum_{i=1}^{4} a_i(\varphi,v) =: a(\varphi,v) = b(\varphi,v) := -\sum_{i=1}^{2} b_i(\varphi,v) \quad \forall v \in \mathbb{V}_{\text{DG}}(\mathfrak{K}).$$

The volume integrals a_1 and b_1 and the surface integrals on the cell edges a_2, a_3 and b_2 can be obtained using standard Gauß-quadrature, but the surface integral a_4 is only defined implicitly by the initial level-set field $\tilde{\varphi}$. For this integral we choose the quadrature method by Müller et al. [51], see Sect. 3.3. We solve the nonlinear problem by the iteration

$$a\left(\varphi^{n+1}, v\right) = b\left(\varphi^n, v\right) \tag{3.48}$$

$$\varphi^0 = \tilde{\varphi}. \tag{3.49}$$

In [64] we present numerical evidence, that the procedure for level-set reinitialization accurately preserves the initial position of the interface and converges in global norms of the level-set field. For the potential function ψ, we present two different variants, which exhibit individual drawbacks: While a single-well potential may lead to oscillations in the vicinity of singularities, which may cause divergence of the scheme, the double well potential gives better results. The method is stable even for interface features, which are about the same size as the grid cells and works on triangular and quadrilateral grids. However, applying the double-well potential in regions, where the initial conditions are overly flat, may lead to a constant solution, instead of the desired signed distance.

3.9   Surfactant Transport on a Surface

In [29] we develop a framework for the transport of unsoluble surfactants on an implicitly defined surface, as modeled by Eqs. (3.5) and (3.12). Simulating the transport of mass at fluidic interfaces a major numerical challenge, since these processes are modeled by equations defined on moving and deforming submanifolds of the original domain. Most commonly, the surface $\mathfrak{I}$ is approximated using a Langrangian approach, where the surface is explicitly reconstructed and discretized by some kind of surface grid, such as the Surface Finite Element Method (SFEM) [15] or the Evolving Surface Finite Element Methods (ESFEM) [14, 38]. In case of moving or deforming interfaces, these methods require an adaptation in every time step and re-meshing after certain periods of time [14, 19, 38]. Therefore we use the different approach of a fixed Eulerian grid. On this grid, the surface differential operators are given as the projection of the standard operators onto the tangential of the interface, which is given by the level-set function [16]. The surface equation can then be solved, by solving an extension in the whole domain [17].

3.9.1   A Conservative Form of the Surfactant Transport Equation

An Eulerian approach requires to re-write all equations in an extended form that embeds the interfacial quantities in the underlying three-dimensional domain. In order to employ a DG scheme properly, conserved forms of these extended surface transport equations are necessary. In [30] we develop a family of conservation laws using the direct construction method [1–3, 7]. Starting from the incompressibility constraint (3.8), the level-set transport equation (3.5) and the transport equation of the surfactant (3.12), we find an infinite family of conservation laws for the quantity $G = F(\varphi)|\nabla\varphi|$

$$\frac{\partial}{\partial t}(c^{3}G) + \nabla \cdot (\mathbf{A}G) = GS \quad \text{in } \Omega \tag{3.50}$$

with

$$\mathbf{A} = c^{3}\mathbf{u} - D^{(3)}\nabla^{(3)}c^{3} \tag{3.51}$$

where $F \neq 0$ is an arbitrary smooth function. Setting $F = 1$ yields the simplest form of this conservation law, which we use in the numerical discretization.

$$\frac{\partial}{\partial t}(c^{3}|\nabla\varphi|) + \nabla \cdot \left(c^{3}\mathbf{u}|\nabla\varphi| - D^{(3)}|\nabla\varphi|\nabla^{(3)}c^{3}\right) = |\nabla\varphi|S \quad \text{in } \Omega \tag{3.52}$$

3.9.2 *Exact Solutions to the Interfacial Surfactant Transport Equation on a Droplet in a Stokes Flow Regime*

In the literature, only few analytical solutions of the surfactant transport equation on an interface are available. A diffusion boundary-layer theory for the distribution of surfactant within the bulks around the stagnant cap bubble is presented by Harper [26, 27], where the surface excess is assumed to be related linearly to the bulk concentration and provides the boundary condition to the transport problem in the bulks. In [66], convection-diffusion in the bulk phases is investigated along with the convective transport at the interface as well as the Marangoni force created by the surfactants present. Furthermore, the effect of insoluble surfactants on small drop deformations in linear flows is theoretically derived by Vlahovska et al. [65]. All these analytical solutions are obtained with various simplifications and are available only for steady cases.

In [31], we derive exact analytical solution of the unsteady surfactant transport equation including the effects of both convection and diffusion for the interfacial transport of insoluble surfactants in a two-phase flow. We assume, that convection in the bulk phase can be neglected, i.e. $Re \ll 1$ and that the capillary number is small. This means, the flow-field around a bubble can be modeled by the Stokes equations and the bubble will keep a spherical shape. Under these assumptions, a solution to the flow field is the Hadamard-Rybzinksi [25, 58] solution, which models the flow around such a bubble with uniform rising velocity u_{∞}. In spherical coordinates the flow field in the outer phase (index o) and in the inner phase (index i) can be expressed by the explicit formula

$$u_\theta^s = u_\theta^i(r_0) = u_\theta^o(r_0) = -\frac{u_\infty}{2\left(1 + \frac{\mu^i}{\mu^o}\right)} \sin(\theta) \text{ and } u_\varphi^s = 0. \tag{3.53}$$

We further neglect the effect of the surface tension on the surfactant-concentration and vice-versa, i.e. we neglect the Marangoni effect. By introducing this simplifications, we can derive three different analytical solutions for surfactant concentration.

3.9.2.1　Unsteady Solution

We introduce the Peclet number $\text{Pe} = (u_\infty r_0) / \left(1 + \frac{\mu^i}{\mu^o}\right)$. Assuming that the solution can be written by separating the variables

$$c^s(\theta, \phi, t) = \Sigma_{k,l} S_{k,l}(\theta) T_k(t) W_l(\phi) \tag{3.54}$$

with $k, l \in \mathbb{N}$, we find the time depended part to be an exponential form

$$T_k(t) = \exp\left(-\frac{\lambda_k^2}{\text{Pe}\left(1 + \frac{\mu^i}{\mu^o}\right)} t\right), \tag{3.55}$$

the azimuthal coordinate to be periodic

$$W_l(\phi) = B_l^1 \sin(\mu_l \, \phi) + B_l^2 \cos(\mu_l \, \phi) \tag{3.56}$$

and the inclination coordinate to give a solution of the form

$$\hat{S}_{k,l}(s) = C_{k,l}^1 \, (s(s-1))^{\frac{l}{2}} \exp(\text{Pe}\, s) \, \text{HeunC}\left(\text{Pe}, l, l, -\text{Pe}, -\lambda_k^2 + \frac{l^2}{2} + \frac{\text{Pe}}{2}, s\right)$$

$$+ C_{k,l}^2 \, s^{-\frac{l}{2}} (s-1)^{\frac{l}{2}} \exp(\text{Pe}\, s) \, \text{HeunC}\left(\text{Pe}, -l, l, -\text{Pe}, -\lambda_k^2 + \frac{l^2}{2} + \frac{\text{Pe}}{2}, s\right) \tag{3.57}$$

where HeunC is the Heun's confluent function, which can be expressed by a series [31, 52] and $s = 1/2 \cos(\theta) + 1/2$. B_l^1, B_l^2 and λ_k are subject to the initial and boundary conditions.

For the steady case, it is shown that these solutions collapse to a simple exponential form. Furthermore, for the purely diffusive problem, exact solutions are constructed using Legendre polynomials. Such analytical solutions are very valuable as benchmark problems in numerical investigations. We use this analytical solutions in [29] to verify the solver for the diffusion-convection problem of a surfactant.

3.9.3 *An Extended DG Discretization for Surfactant Transport*

In [29] we present a narrow-band method to solve the embedded conservation law (3.52) on an arbitrary grid with an interface defined by a level-set. When employing an Eulerian approach to treat interfacial problems, it is first necessary to construct a representation of all relevant surface quantities on the underlying Eulerian grid. In each time step, a meaningful extension $\tilde{c}$ of the surfactant concentration $c^{\mathfrak{I}}$ to the updated narrow band $\mathfrak{N}$ has to be performed which, at the same time, conserves this quantity on the interface at a high level of accuracy. In particular, this method is important when the subgrid changes position and structure in accordance with the Level Set function. In order to reduce numerical errors, Greer and Bertozzi [22] have recommended to start off with variables that are constant in normal direction with respect to the interface. This means, an extension $\tilde{c}$ of a given variable c that solves the boundary value problem

$$\nabla \tilde{c} \cdot \nabla \Phi = 0 \quad \text{in } \mathfrak{N} \tag{3.58}$$

$$\tilde{c} = c^{\mathfrak{I}} \quad \text{on } \mathfrak{I}.$$

is employed.

In [29] we use a pseudo-timestepping method to calculate this extension by solving the equation

$$\frac{\partial \tilde{c}}{\partial \tau} + \text{sgn}(\varphi)\, \nabla \tilde{c} \cdot \nabla \varphi = 0 \quad \text{in } \Omega\, , \ \tau > 0 \tag{3.59}$$

$$\tilde{c}(0, \mathbf{x}) = c(\mathbf{x})$$

with respect to the pseudo-time τ until reaching a steady state using a local discontinuous Galerkin scheme as suggested by Greer et al. [22], which mimics a pseudo-timestepping scheme developed by Zhao for re-initializing the Level Set function in multiphase problems [71]:

$$\text{find } (\tilde{c}, \mathbf{p}) \in \mathbb{V}_{\text{DG}}(\mathfrak{N}) \times \mathbb{V}_{\text{DG}}^{D}(\mathfrak{N}) \text{ s.t.}$$

$$\int_{\Omega_h} \frac{\partial}{\partial \tau} \tilde{c} v \, dV + \int_{\Omega_h} \varphi \nabla \varphi \cdot \mathbf{p} v \, dV = 0 \tag{3.60}$$

$$\int_{\Omega_h} \mathbf{p} v \, dV - \int_{\Omega_h} \tilde{c}\, (\nabla v)\, dV + \oint_{\Gamma} F^*(\tilde{c}^+, \tilde{c}^-) \, [\![v]\!] \, \mathbf{n} \, dS = 0 \tag{3.61}$$

$$\forall v \in \mathbb{V}_{\text{DG}}(\mathfrak{N})$$

With the numerical flux being

$$F^*(c_h^+, c_h^-) = \begin{cases} c_h^+ & \text{if } \nabla \varphi \cdot \mathbf{n} \geq 0 \\ c_h^- & \text{if } \nabla \varphi \cdot \mathbf{n} < 0 \, . \end{cases} \tag{3.62}$$

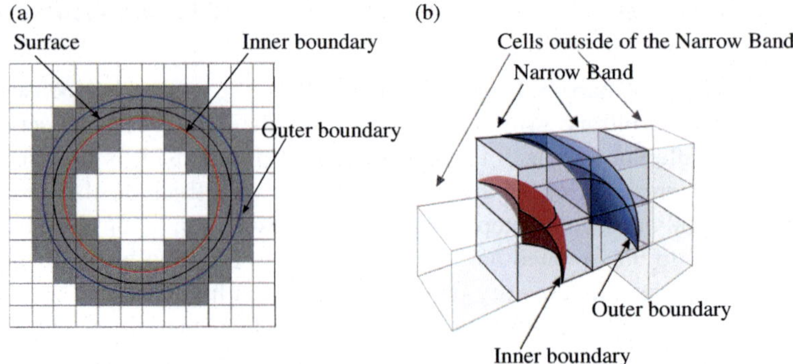

Fig. 3.3 Illustration of inner and outer boundaries $\partial\Omega_\gamma$ within a narrow band around a *circle* in 2D and a sphere in 3D, respectively. The cells of the narrow band $\mathfrak{N}^1$ are marked in *grey* [29]. (**a**) Eulerian grid in 2D. (**b**) Eulerian grid in 3D

After this extension, we can solve the extended surface transport equation in the embedded form (3.52). For the discretization, we split the operator into a convective part and a diffusive part. In [29] we discretize the convective operator on a narrowband grid $\mathfrak{N}$ using an explicit timestepping scheme and an upwind flux:

$$\text{find } \tilde{c} \in \mathbb{V}_{\text{DG}}(\mathfrak{N}) \text{ s.t.}$$

$$\int_{\mathfrak{N}} \frac{\partial}{\partial t}\tilde{c}v \, dV - \int_{\mathfrak{N}} \tilde{c}\mathbf{u} \cdot \nabla v \, dV + \oint_{\Gamma \cap \mathfrak{N}} \left((\mathbf{u} \cdot \mathbf{n})\{\tilde{c}\} + |\mathbf{u} \cdot \mathbf{n}|\frac{[\![\tilde{c}]\!]}{2} \right) [\![v]\!]\,\mathbf{n}\,dS$$

$$\tag{3.63}$$

$$\forall v \in \mathbb{V}_{\text{DG}}(\mathfrak{N})$$

On closed surfaces, the convection-diffusion problem does not have any boundary conditions. However, when extending the problem into the domain, we require additional boundary conditions. Since there is no physical boundary condition for the extended problem, we need to specify artificial ones. Following [17], we choose to generalize condition (3.58) on the whole narrow band, see Fig. 3.3. When employing a narrow band approach, the boundary of the domain includes cusps and corners, which restrict the order of convergence. Therefore, we introduce two curved boundaries within the narrowband, which coincide with isocontours of the level set. We restrict our domain to a curved narrow-band around the interface with width γ

$$\Omega_\gamma \subset \mathfrak{N} = \{\mathbf{x} : |\varphi(\mathbf{x})| < \gamma\} \tag{3.64}$$

and discretize the problem using the LDG form

$$\text{find } (\tilde{c}, \mathbf{p}) \in \mathbb{V}_{\text{DG}}(\Omega_\gamma) \times \mathbb{V}_{\text{DG}}^D(\Omega_\gamma) \text{ s.t.}$$

$$\int_{\Omega_\gamma} \frac{\partial}{\partial t}\tilde{c}v \, dV - D\int_{\Omega_\gamma} |\nabla\varphi|\mathbf{P}^3\mathbf{q} \cdot \nabla v \, dV - \oint_{\partial\Omega_\gamma} \eta\,[\![\mathbf{q}]\!] \cdot \mathbf{n}\,[\![v]\!]\,dS \tag{3.65}$$

$$+D \int_{\Gamma \cap \Omega_\gamma} \left(-\eta \, [\![\mathbf{q}]\!] - \frac{1}{2} \left\{ |\nabla \varphi| \mathbf{P}^3 \mathbf{q} \right\} \right) \cdot \mathbf{n} v \, \mathrm{d}S = 0 \qquad (3.66)$$

$$\int_{\Omega_\gamma} \mathbf{q} v \, \mathrm{d}V = - \int_{\Omega_\gamma} \frac{1}{|\nabla \varphi|} \tilde{c} \nabla \cdot \mathbf{v} \, \mathrm{d}V + \int_{\Gamma \cap \Omega_\gamma} \frac{1}{|\nabla \varphi|} \{\tilde{c}\} \, \mathbf{n} \cdot \mathbf{v} \, \mathrm{d}S \qquad (3.67)$$

$$\forall v \in \mathbb{V}_{\mathrm{DG}}(\Omega_\gamma), \mathbf{v} \in \mathbb{V}_{\mathrm{DG}}^D(\Omega_\gamma).$$

Since Ω_γ consists of cells, cut by the isocontours $\varphi = \pm\gamma$, we use the hierarchical moment fitting quadrature to evaluate the integrals and the cell agglomeration technique to deal with arbitrary cut positions.

Using the analytical solutions developed before, we show optimal order of convergence for the case of pure diffusion and pure convection up to polynomial order of two in [29]. In addition, we present a simulation of a convection-diffusion problem on an interface deforming due to a external flow field.

3.10 Conclusion

In this paper, we presented building blocks for solving multiphase flows using an extended discontinuous Galerkin discretization. The goal is, to use a DG method on a fixed grid for both the bulk and the interface problem. This is achieved by introducing a level-set, which implicitly defines cells cut by the interface. This allows a straightforward discretization [34], but requires a quadrature method for these cut-cells, which we developed in [50] and extended in [34]. The position of the cut within a cell may be at an arbitrary position, which spoils the condition number of the resulting matrices. A cell agglomeration technique [34, 49] deals with this issue. When the interface moves, we encounter additional issues: being a nonlinear second derivative of the level-set, the curvature may show oscillations, which result in spurious pressures and velocities due to surface tension. This requires additional smoothing of this field called patch recovery, which we present in [36]. When the level-set moves or even changes its topology, it might loose its signed-distance property. In [64], we present a novel technique for level-set reinitialization. For the discretization of a transport problem on the surface, we need a suitable form to discretize the problem on a fixed background grid. In [30], we derive a family of suitable conservation laws. In [29], we present a Discontinuous Galerkin solver for the simulation of such surface transport processes, which we verify using an analytical solution developed in [31]. To summarize, in this work we presented multiple building blocks, which in combination will allow the high-order accurate discontinuous Galerkin simulation of multiphase flows with the transport of a surfactant.

Acknowledgements This work is supported by the 'Excellence Initiative' of the German Federal and State Governments and the Graduate School of Computational Engineering at Technische Universität Darmstadt. The work of T. Utz and C. Kallendorf is supported by the German Science Foundation (DFG) within the Priority Program (SPP) 1506 "Transport Processes at Fluidic Interfaces". The work of F. Kummer is supported by the German DFG through Research Fellowship KU 2719/1-1. The work of B. Müller is supported by the German Research Foundation (DFG) through Research Grant WA 2610/2-1.

References

1. Anco, S., Bluman, G.: Direct construction method for conservation laws of partial differential equations. Part I: examples of conservation law classifications. Eur. J. Appl. Math. **13**, 545–566 (2002)
2. Anco, S., Bluman, G.: Direct construction method for conservation laws of partial differential equations. Part II: general treatment. Eur. J. Appl. Math. **13**, 567–585 (2002)
3. Anco, S.C., Bluman, G.W., Cheviakov, A.F.: Construction of conservation laws: how the direct method generalizes Noether's theorem. In: Proceedings of 4th Workshop "Group Analysis of Differential Equations & Integrability", vol. 1, pp. 1–23 (2009)
4. Arnold, D.N.: An interior penalty finite element method with discontinuous elements. SIAM J. Numer. Anal. **19**(4), 742 (1982)
5. Arnold, D.N., Brezzi, F., Cockburn, B., Marini, L.D.: Unified analysis of discontinuous Galerkin methods for elliptic problems. SIAM J. Numer. Anal. **39**(5), 1749–1779 (2002). doi:10.1137/S0036142901384162
6. Basting, C., Kuzmin, D.: A minimization-based finite element formulation for interface-preserving level set reinitialization. Computing **95**(1), 13–25 (2012). doi:10.1007/s00607-012-0259-z
7. Bluman, G., Cheviakov, A., Anco, S.: Applications of Symmetry Methods to Partial Differential Equations, vol. 168. Applied Mathematical Sciences. Springer, Berlin (2010)
8. Bremer, J., Gimbutas, Z., Rokhlin, V.: A nonlinear optimization procedure for generalized gaussian quadratures. SIAM J. Sci. Comput. **32**(4), 1761 (2010). doi:10.1137/080737046
9. Cheng, K.W., Fries, T.P.: Higher-order XFEM for curved strong and weak discontinuities. Int. J. Numer. Methods Eng. **82**(5), 564–590 (2010). doi:10.1002/nme.2768
10. Cheng, Y., Shu, C.W.: A discontinuous Galerkin finite element method for directly solving the Hamilton–Jacobi equations. J. Comput. Phys. **223**(1), 398–415 (2007). doi:10.1016/j.jcp.2006.09.012
11. Desjardins, O., Pitsch, H.: A spectrally refined interface approach for simulating multiphase flows. J. Comput. Phys. **228**(5), 1658–1677 (2009). doi:10.1016/j.jcp.2008.11.005
12. Di Pietro, D.A., Ern, A.: Mathematical Aspects of Discontinuous Galerkin Methods. Mathématiques et Applications, vol. 69. Springer, Berlin (2011). http://books.google.de/books?id=ak-qQvWGA5oC
13. Düster, A., Parvizian, J., Yang, Z., Rank, E.: The finite cell method for three-dimensional problems of solid mechanics. Comput. Methods Appl. Mech. Eng. **197**(45–48), 3768–3782 (2008). doi:10.1016/j.cma.2008.02.036
14. Dziuk, G., Elliott, M.C.: Finite elements on evolving surfaces. IMA J. Numer. Anal. **27**, 262–292 (2007)
15. Dziuk, G., Elliott, M.C.: Surface finite elements for parabolic equations. J. Comput. Math. **25**, 385–407 (2007)
16. Dziuk, G., Elliott, M.C.: Eulerian finite element method for parabolic PDEs on implicit surfaces. IMA J. Numer. Anal. **10**, 119–138 (2008)
17. Dziuk, G., Elliott, C.M.: An Eulerian approach to transport and diffusion on evolving implicit surfaces. Comput. Vis. Sci. **13**, 17–28 (2010)

18. Elias, R.N., Martins, M.A.D., Coutinho, A.L.G.A.: Simple finite element-based computation of distance functions in unstructured grids. Int. J. Numer. Methods Eng. **72**(9), 1095–1110 (2007). doi:10.1002/nme.2079
19. Elliott, M.C., Eilks, C.: Numerical simulation of dealloying by surface dissolution via the evolving surface finite element method. J. Comput. Phys. **227**, 9727–9741 (2008)
20. Engwer, C.: An unfitted discontinuous Galerkin scheme for micro-scale simulations and numerical upscaling. Ph.D. thesis, Heidelberg (2009)
21. Fröhlcke, A., Gjonaj, E., Weiland, T.: A boundary conformal DG approach for electro-quasistatics problems. In: Michielsen, B., Poirier, J.R. (eds.) Scientific Computing in Electrical Engineering SCEE 2010. Mathematics in Industry, vol. 16, pp. 153–161. Springer, Berlin (2012)
22. Greer, J.B., Bertozzi, A., Sapiro, G.: Fourth order partial differential equations on general geometries. J. Comput. Phys. **216**(1), 216–246 (2006)
23. Groß, S., Reichelt, V., Reusken, A.: A finite element based level set method for two-phase incompressible flows. Comput. Vis. Sci. **9**(4), 239–257 (2006). doi:10.1007/s00791-006-0024-y
24. Grooss, J., Hesthaven, J.S.: A level set discontinuous Galerkin method for free surface flows. Comput. Methods Appl. Mech. Eng. **195**(25–28), 3406–3429 (2006). doi:10.1016/j.cma.2005.06.020
25. Hadamard, J.: Mouvement permanent lent d' une sphere liquide et visqueuse dans un liquide visqueux. C. R. Acad. Sci. Paris **152**, 1735–1738 (1911)
26. Harper, J.F.: On spherical bubbles rising steadily in dilute surfactant solutions. Q. J. Mech. Appl. Math. **27**(1), 87–100 (1974). doi:10.1093/qjmam/27.1.87
27. Harper, J.F.: Stagnant-cap bubbles with both diffusion and adsorption rate-determining. J. Fluid Mech. **521**, 115–123 (2004). doi:10.1017/S0022112004001843
28. Hu, C., Shu, C.: A Discontinuous Galerkin finite element method for Hamilton–Jacobi equations. SIAM J. Sci. Comput. **21**(2), 666–690 (1999). doi:10.1137/S1064827598337282
29. Kallendorf, C.: An Eulerian discontinuous Galerkin method for the numerical simulation of interfacial transport. Ph.D. thesis, TU Darmstadt (2016)
30. Kallendorf, C., Cheviakov, A.F., Oberlack, M., Wang, Y.: Conservation laws of surfactant transport equations. Phys. Fluids **24**(10), 102105 (2012). doi:http://dx.doi.org/10.1063/1.4758184. http://scitation.aip.org/content/aip/journal/pof2/24/10/10.1063/1.4758184
31. Kallendorf, C., Fath, A., Oberlack, M., Wang, Y.: Exact solutions to the interfacial surfactant transport equation on a droplet in a stokes flow regime. Phys. Fluids **27**(8), 082104 (2015). doi:http://dx.doi.org/10.1063/1.4928547. http://scitation.aip.org/content/aip/journal/pof2/27/8/10.1063/1.4928547
32. Kimmel, R., Sethian, J.A.: Computing geodesic paths on manifolds. Proc. Natl. Acad. Sci. **95**(15), 8431–8435 (1998)
33. Klein, B., Kummer, F., Oberlack, M.: A SIMPLE based discontinuous Galerkin solver for steady incompressible flows. J. Comput. Phys. **237**, 235–250 (2013)
34. Kummer, F.: Extended discontinuous Galerkin methods for two-phase flows: the spatial discretization. Int. J. Numer. Methods Eng. **109**(2), 259–289 (2017)
35. Kummer, F., Oberlack, M.: An extension of the discontinuous Galerkin method for the singular Poisson equation. SIAM J. Sci. Comput. **35**(2), A603–A622 (2013)
36. Kummer, F., Warburton, T.: Patch-recovery filters for curvature in discontinuous Galerkin-based level-set methods. Commun. Comput. Phys. **19**(02), 329–353 (2016). http://tubiblio.ulb.tu-darmstadt.de/80852/
37. Legrain, G., Chevaugeon, N., Dréau, K.: High order x-FEM and levelsets for complex microstructures: uncoupling geometry and approximation. Comput. Methods Appl. Mech. Eng. **241–244**, 172–189 (2012). doi:10.1016/j.cma.2012.06.001
38. Lenz, M., Nemadjieu, S., Rumpf, M.: Finite volume method on moving surfaces. In: Eymard, R., Hérald, J.M. (eds.) Finite Volumes for Complex Applications V, pp. 561–576. Wiley, New York (2008)

39. Li, C., Xu, C., Gui, C., Fox, M.: Level set evolution without re-initialization: a new variational formulation. In: IEEE Computer Society Conference on Computer Vision and Pattern Recognition, 2005. CVPR 2005, vol. 1, pp. 430–436 (2005). doi:10.1109/CVPR.2005.213

40. Li, F., Shu, C.W., Zhang, Y.T., Zhao, H.: A second order discontinuous Galerkin fast sweeping method for Eikonal equations. J. Comput. Phys. **227**(17), 8191–8208 (2008). doi:10.1016/j.jcp.2008.05.018

41. Luo, S.: A uniformly second order fast sweeping method for Eikonal equations. J. Comput. Phys. **241**, 104–117 (2013). doi:10.1016/j.jcp.2013.01.042

42. Marchandise, E.: Simulation of three-dimensional two-phase flows: coupling of a stabilized finite element method with a discontinuous level set approach. Ph.D. thesis, Université Catholique de Louvain (2006)

43. Marchandise, E., Geuzaine, P., Chevaugeon, N., Remacle, J.F.: A stabilized finite element method using a discontinuous level set approach for the computation of bubble dynamics. J. Comput. Phys. **225**(1), 949–974 (2007). doi:10.1016/j.jcp.2007.01.005

44. Min, C., Gibou, F.: Geometric integration over irregular domains with application to level-set methods. J. Comput. Phys. **226**(2), 1432–1443 (2007). doi:16/j.jcp.2007.05.032

45. Min, C., Gibou, F.: Robust second-order accurate discretizations of the multi-dimensional heaviside and dirac delta functions. J. Comput. Phys. **227**(22), 9686–9695 (2008). doi:10.1016/j.jcp.2008.07.021

46. Moës, N., Dolbow, J., Belytschko, T.: A finite element method for crack growth without remeshing. Int. J. Numer. Methods Eng. **46**, 131–150 (1999)

47. Mousavi, S.E., Sukumar, N.: Generalized gaussian quadrature rules for discontinuities and crack singularities in the extended finite element method. Comput. Methods Appl. Mech. Eng. **199**(49–52), 3237–3249 (2010). doi:10.1016/j.cma.2010.06.031

48. Mousavi, S.E., Sukumar, N.: Numerical integration of polynomials and discontinuous functions on irregular convex polygons and polyhedrons. Comput. Mech. **47**(5), 535–554 (2011). doi:10.1007/s00466-010-0562-5

49. Müller, B., Krämer-Eis, S., Kummer, F., Oberlack, M.: A high-order discontinuous Galerkin method for compressible flows with immersed boundaries. Int. J. Numer. Methods Eng. pp. n/a–n/a (2016). doi:10.1002/nme.5343. http://dx.doi.org/10.1002/nme.5343. Nme.5343

50. Müller, B., Kummer, F., Oberlack, M., Wang, Y.: Simple multidimensional integration of discontinuous functions with application to level set methods. Int. J. Numer. Methods Eng. **92**(7), 637–651 (2012)

51. Müller, B., Kummer, F., Oberlack, M.: Highly accurate surface and volume integration on implicit domains by means of moment-fitting. Int. J. Numer. Methods Eng. **96**(8), 512–528 (2013). doi:10.1002/nme.4569

52. Olver, F.W., Lozier, D.W., Boisvert, R.F., Clark, C.W.: NIST Handbook of Mathematical Functions, 1st edn. Cambridge University Press, New York, NY (2010)

53. Osher, S., Fedkiw, R.P.: Level set methods: an overview and some recent results. J. Comput. Phys. **169**(2), 463–502 (2001). doi:10.1006/jcph.2000.6636

54. Owkes, M., Desjardins, O.: A discontinuous Galerkin conservative level set scheme for interface capturing in multiphase flows. J. Comput. Phys. **249**, 275–302 (2013). doi:10.1016/j.jcp.2013.04.036

55. Pochet, F., Hillewaert, K., Geuzaine, P., Remacle, J.F., Marchandise, M.: A 3d strongly coupled implicit discontinuous Galerkin level set-based method for modeling two-phase flows. Comput. Fluids **87**, 144–155 (2013). doi:10.1016/j.compfluid.2013.04.010

56. Press, W.H., Teukolsky, S.A., Vetterling, W.T., Flannery, B.P.: Numerical Recipes 3rd Edition: The Art of Scientific Computing, 3rd edn. Cambridge University Press, Cambridge (2007)

57. Qin, R., Krivodonova, L.: A discontinuous Galerkin method for solutions of the Euler equations on Cartesian grids with embedded geometries. J. Comput. Sci. **4**(1–2), 24–35 (2013). doi:10.1016/j.jocs.2012.03.008

58. Rybczynski, W.: Über die fortschreitende bewegung einer flüssigen kugel in einem zähen medium. Bull. Acad. Sci. de Cracovie A 40–46 (1911)

59. Saye, R.: High-order methods for computing distances to implicitly defined surfaces. Commun. Appl. Math. Comput. Sci. **9**(1), 107–141 (2014). doi:10.2140/camcos.2014.9.107
60. Shahbazi, K., Fischer, P., Ethier, R.C.: A high-order discontinuous Galerkin method for the unsteady incompressible Navier-Stokes equations. J. Comput. Phys. **222**(1), 391–407 (2007). doi:10.1016/j.jcp.2006.07.029
61. Sudhakar, Y., Wall, W.A.: Quadrature schemes for arbitrary convex/concave volumes and integration of weak form in enriched partition of unity methods. Comput. Methods Appl. Mech. Eng. (2013). doi:10.1016/j.cma.2013.01.007
62. Sussman, M., Hussaini, M.Y.: A discontinuous spectral element method for the level set equation. J. Sci. Comput. **19**(1–3), 479–500 (2003). doi:10.1023/A:1025328714359
63. Sussman, M., Smereka, P., Osher, S.: A level set approach for computing solutions to incompressible two-phase flow. J. Comput. Phys. **114**(1), 146–159 (1994). doi:10.1006/jcph.1994.1155
64. Utz, T., Kummer, F., Oberlack, M.: Interface-preserving level-set reinitialization for DG-FEM. Int. J. Numer. Meth. Fluids **84**(4), 183–198 (2017). doi:10.1002/fld.4344
65. Vlahovska, P.M., Blawzdziewicz, J., Loewenberg, M.: Small-deformation theory for a surfactant-covered drop in linear flows. J. Fluid Mech. **624** (2009). doi:10.1017/S0022112008005417
66. Wang, Y., Papageorgiu, D.T., Maldarelli, C.: Increased mobility of a surfactant-retarded bubble at high bulk concentrations. J. Fluid Mech. **390**, 251–270 (1999). doi:10.1017/S0022112099005157
67. Wu, L., Zhang, Y.T.: A Third order fast sweeping method with linear computational complexity for Eikonal equations. J. Sci. Comput. **62**(1), 198–229 (2014). doi:10.1007/s10915-014-9856-7
68. Xiao, H., Gimbutas, Z.: A numerical algorithm for the construction of efficient quadrature rules in two and higher dimensions. Comput. Math. Appl. **59**(2), 663–676 (2010). doi:10.1016/j.camwa.2009.10.027
69. Zhang, Y., Chen, S., Li, F., Zhao, H., Shu, C.: Uniformly accurate discontinuous Galerkin fast sweeping methods for Eikonal equations. SIAM J. Sci. Comput. **33**(4), 1873–1896 (2011). doi:10.1137/090770291
70. Zhao, H.: A fast sweeping method for Eikonal equations. Math. Comput. **74**(250), 603–627 (2005)
71. Zhao, H.K., Chan, T., Merriman, B., Osher, S.: A variational level set approach to multiphase motion. J. Comput. Phys. **127**(1), 179–195 (1996). doi:http://dx.doi.org/10.1006/jcph.1996.0167. http://www.sciencedirect.com/science/article/pii/S0021999196901679
72. Zienkiewicz, O., Zhu, J.: The superconvergent patch recovery (SPR) and adaptive finite element refinement. Comput. Methods Appl. Mech. Eng. **101**(1–3), 207–224 (1992). doi:10.1016/0045-7825(92)90023-D. http://linkinghub.elsevier.com/retrieve/pii/004578259290023D

Chapter 4
Building Blocks for a Strictly Conservative Generalized Finite Volume Projection Method for Zero Mach Number Two-Phase Flows

Matthias Waidmann, Stephan Gerber, Michael Oevermann, and Rupert Klein

Abstract Building blocks for a generalized fully conservative finite volume projection method for numerical simulation of immiscible zero Mach number two-phase flows on Cartesian grids are presented, focusing on the crucial issues of interface propagation, fluid phase conservation and discretization of the singular contribution due to surface tension, each in a discretely conservative fashion. Additionally, a solution approach for solving Poisson-type equations for two-phase flows at arbitrary ratio of coefficients is sketched. Further, (intermediate) results applying these building blocks are presented and open issues and future developments are proposed.

4.1 Introduction

Many important industrial chemical processes and naturally occurring phenomena can be described by the equations for variable density zero Mach number two-phase flow with heat and mass transfer across an interface separating the two fluids. Some of the main challenges in modeling and simulating such flows result from discontinuous distributions of pressure, density, and concentrations, from large ratios of fluid properties—such as viscosity, diffusivity, or conductivity—across the phase boundaries, from singular interface local contributions to the bulk flow and from the necessity of solving species transport equations on moving surfaces. This work focuses on features for the proper design of a discretely conservative

M. Waidmann (✉) • S. Gerber • R. Klein
Department of Mathematics and Computer Science, Freie Universität Berlin, Arnimallee 6, 14195 Berlin, Germany
e-mail: waidmann@math.fu-berlin.de; stephan.gerber@fu-berlin.de; rupert.klein@math.fu-berlin.de

M. Oevermann
Department for Applied Mechanics, Chalmers University of Technology, Gothenburg, Sweden
e-mail: michael.oevermann@chalmers.se

© Springer International Publishing AG 2017
D. Bothe, A. Reusken (eds.), *Transport Processes at Fluidic Interfaces*,
Advances in Mathematical Fluid Mechanics, DOI 10.1007/978-3-319-56602-3_4

method for immiscible zero Mach number two-phase flows, specifically addressing the design of a conservative sharp interface finite volume projection method on Cartesian grids with explicit representation of discontinuities and an asymptotics based efficient solution strategy for the variable coefficient Poisson projection equation for (stationary) interface problems with arbitrarily large ratio of the coefficients.

The computation of incompressible two-phase flows with large density ratio and surface tension remains a challenging problem and has attracted many researchers over the last decades. Overviews and detailed reference lists are, for example, given in [28] and [36]. There is now a huge variety of different flow solvers (finite elements, finite differences, finite volumes, boundary integral and more recently also discontinuous Galerkin methods) with different techniques for the representation of the interface (volume-of-fluid, level-set, front tracking, marker-and-cell and hybrid approaches) available and one can find almost any combination of these techniques in the literature. Due to the technological importance of variable density zero Mach number and incompressible two-phase flows for, e.g., the process industry, most commercial CFD codes offer options to compute such flows. While moving-grid techniques with a separate boundary fitted grid for each fluid phase potentially offer the highest accuracy at cost of difficulties in application to problems with changes in topology and the necessity of time consuming grid generation at every time step, fixed-grid methods as the present one solve the equations in the whole domain on a predefined fixed grid, which gives rise to grid cells intersected by the fluidic interface with different fluid properties on each side of the interface.

One major difficulty in computing two-phase flows results from the singular surface tension term in the Navier-Stokes equations. In many numerical methods on fixed grids the surface tension force is accounted for through variants of Peskin's immersed boundary method [25], in which singular forces are transferred to the fixed grid using discrete delta function approximations leading to a spreading of the discontinuity over several grid cells. In this method, which is often called continuous surface force (CSF) method as introduced in [6], singular forces appear as volume forces. In [27] surface tension is incorporated in a finite volume setting without smearing, applied solely to the finite volume cells containing the interface. The corresponding force appears as an integral along the intersecting line between the phase boundary and the boundaries of the control volume. To account for the pressure jump at the interface, a correction term is applied to the pressure Poisson equation based on the pressure of the old time step. In [26] an adaptive method for incompressible two-phase flows with volume-of-fluid based interface representation and improved height function based curvature computation is presented. A balanced-force CSF discretization of the surface tension term minimizes parasitic currents. Key ingredients are a mutually compatible discretization of pressure gradient and surface tension force—an idea that is in line with ideas of asymptotic well-balancing in [4]—and a good curvature estimate. This strategy is also followed in [9]. In [40] a volume constraint level-set interface representation is presented, improving the work from [33], and [41] presents a higher order sharp interface finite difference method.

Many of the available schemes, however, appear as a mix of different techniques without a consistent concept, especially once the solution of surface PDEs is included in a subsequent step. A consistent and rigorous application of the finite volume method with a sharp resolution of all interface discontinuities in combination with conservative discretizations only still seems to be missing and will be the main focus of this contribution based on—among others—the following preliminary work: Second-order projection methods for two-phase variable density low Mach-number flows have been presented in [35] in non-conservative form and in [31] in conservative form, the latter serving as basis for further developments such as [37], all of which can be traced back to [8] or [12]. Computation of updates for the entire grid cell only, also in case of intersected grid cells, is introduced in [32]. A mass conserving level-set method for incompressible two-phase flows has been presented in [30] for 2D problems. Mass conservation of the level-set is enforced by a combination of the level-set method with a volume-of-fluid algorithm. The method is applied to the computation of a rising bubble with a density ratio of 1:714. Surface tension is taken into account with a CSF method. The underlying flow solver implements the method proposed in [31]. In [21] and [24] Cartesian grid finite volume discretization schemes for elliptic equations with discontinuous solutions and normal derivatives across an interface are presented for two and three dimensional problems. The schemes incorporate the interfacial geometries through level-set functions. The presented results demonstrate a convergence rate of locally second order for ratios of the coefficients up to approx. 1:1000. In this line enhancements towards a generalized discretely conservative finite volume projection method for zero Mach number variable density two-phase flows are presented in Sect. 4.3, regarding building blocks for the approximation of the governing equations from the following Sect. 4.2.

4.2 Governing Equations

The leading order system of the governing equations for chemically reacting immiscible two-phase flows in the low Mach number limit for small length scales[1] and arbitrary equation of state in one-fluid formulation is obtained from the set of balances for mass, species masses, momentum and energy after an asymptotic expansion of the primitive quantities (density ρ, velocity $\mathbf{v}$, pressure p, species mass fractions Y_y) in the Mach number. It reads

$$\frac{\mathrm{d}}{\mathrm{d}t} \int_{\Omega} \tilde{\mathbf{q}} \, \mathrm{d}V + \oint_{\partial\Omega} \left(\hat{\mathbf{q}} \left(\mathbf{v} \cdot \mathbf{n} \right) - \mathbf{f} - \mathbf{f}_\Gamma \delta_\Gamma \right) \mathrm{d}A = \int_{\Omega} \mathbf{s} \, \mathrm{d}V \qquad (4.1)$$

---

[1] Small length scales compared to the atmospheric pressure scale height.

in conservative space-integral form for fixed control volumes Ω with boundary $\partial\Omega$ and outward pointing normal vector $\mathbf{n}$ and represents conservation of mass, species mass, momentum and both internal and potential energy. In particular, in the nondimensional form the expressions

$$\tilde{\mathbf{q}} := \mathrm{Sr}\,\rho\left(\mathbf{q}, 0\right) \qquad \hat{\mathbf{q}} := \rho\left(\mathbf{q}, \tfrac{1}{\rho}\right) \qquad \mathbf{q} := \left(1, Y_y, \mathbf{v}\right) \tag{4.2}$$

$$\mathbf{f} := \left(0,\ \tfrac{1}{\mathrm{Re\,Sc}}\left(\mathbf{j}_y \cdot \mathbf{n}\right),\ \left(\tfrac{1}{\mathrm{Re}}\mathbf{T} - \mathbf{I}p'\right)\cdot\mathbf{n},\ \tilde{f}\right) \tag{4.3}$$

$$\mathbf{f}_\Gamma := \left(0,\ 0,\ \frac{1}{\mathrm{We}}\mathbf{T}_\Gamma\cdot\mathbf{n},\ 0\right) \qquad \mathbf{s} := \rho\left(0,\ \mathrm{Da}\,\omega_y,\ \frac{1}{\mathrm{Fr}^2}\mathbf{g},\ \tilde{s}\right) \tag{4.4}$$

are generalized quantities with local values depending on the local fluid phase and the following characteristic numbers: Strouhal number $\mathrm{Sr} := 1$, Reynolds number Re, Schmidt number Sc, Weber number We, Damköhler number Da, and Froude number Fr. Further, t represents time, δ_Γ is the surface Dirac, $\mathbf{j}_y$ is the diffusive flux of species y (representing e.g. Fickian diffusion), $\mathbf{T} := 2\mu\mathbf{D} + \lambda\mathbf{I}(\nabla\cdot\mathbf{v})$ is the viscous stress tensor with strain rate tensor $\mathbf{D} := 0.5(\nabla\mathbf{v} + (\nabla\mathbf{v})^T)$, dynamic viscosity μ, second viscosity coefficient λ and identity tensor $\mathbf{I}$, and

$$\mathbf{T}_\Gamma := \sigma\left(\mathbf{I} - \mathbf{n}_\Gamma\,\mathbf{n}_\Gamma\right) \tag{4.5}$$

is the surface stress tensor with surface tension coefficient σ and interface normal vector $\mathbf{n}_\Gamma$, pointing from the fluid phase labeled $(+)$ to the one labeled $(-)$. The gravity vector is denoted by $\mathbf{g}$ and the species production rate due to chemical reactions is given by ω_y. Pressure

$$p = P + p' \tag{4.6}$$

decomposes into the perturbation pressure p' and the background pressure P. In general, P is an unknown which is determined from the global solution. For small (vertical) length scales P is homogeneous (which implies $\nabla P = \mathbf{0}$, $[\![P]\!] = 0$ and $[\![\nabla P]\!] = \mathbf{0}$ after definition of the interfacial jump $[\![\Phi]\!] := \Phi^{(+)} - \Phi^{(-)}$ of an arbitrary quantity Φ) and contributes to the energy source $\tilde{s}$ via its change over time as given in (4.8) from the following definitions for application to (4.3) and (4.4):

$$\tilde{f} := \frac{\eta}{\mathrm{Re\,Pr}}\left(\mathbf{j}_q \cdot \mathbf{n}\right) \tag{4.7}$$

$$\tilde{s} := -\eta\left(\mathrm{Sr}\,\Xi\,\frac{dP}{dt} + \mathrm{Nu}\,\dot{q} - \sum_y \dot{q}_y\right) \tag{4.8}$$

$$\eta := \frac{1}{\Xi\rho c^2} = \frac{\chi}{\Xi} \qquad \Xi := 1 - \rho\left(\frac{\partial h}{\partial p}\right)_{\rho, Y_y} \tag{4.9}$$

$$\dot{q}_y := \frac{1}{\mathrm{Re\,Sc}}\left(\mathbf{j}_y \cdot \nabla h_y\right) + \mathrm{Da}\left(h_y + \mathrm{Qr}\,(\Delta h)_y\right)\rho\omega_y \tag{4.10}$$

For (4.7)–(4.10) the following quantities remain to be defined: Prandtl number Pr, Nusselt number Nu, heat release parameter Qr, (frozen) speed of sound c, compressibility χ, enthalpy h, standard enthalpy of formation Δh, external heat sources $\dot{q}$ as well as heat flux $\mathbf{j}_q = k\nabla T$ with heat conductivity k and temperature T. In addition, the term $\dot{q}_y$ defined in (4.10), applied in (4.8), summarizes effects of chemical reactions and species diffusion on the energy budget. For (in particular isothermal) zero Mach number flow $\tilde{f} \equiv 0$ and $\tilde{s} \equiv 0$ holds. Additionally, $\omega_y = 0$ and $\mathbf{j}_y = \mathbf{0}$ is assumed in the following. The associated interface conditions read

$$[\![\mathbf{v}]\!] = \mathbf{0} \tag{4.11}$$

$$[\![p']\!] - \frac{2}{\mathrm{Re}}\mathbf{n}_\Gamma \cdot [\![\mu \mathbf{D}]\!] \cdot \mathbf{n}_\Gamma = \frac{1}{\mathrm{We}}\sigma\kappa \tag{4.12}$$

$$\frac{2}{\mathrm{Re}}\mathbf{t}_\Gamma \cdot [\![\mu \mathbf{D}]\!] \cdot \mathbf{n}_\Gamma = -\frac{1}{\mathrm{We}}\mathbf{t}_\Gamma \cdot \nabla_\Gamma\sigma \tag{4.13}$$

$$\left[\!\!\left[\frac{1}{\rho}\nabla p' \cdot \mathbf{n}_\Gamma \right]\!\!\right] = \left[\!\!\left[\frac{\mu}{\rho}(\nabla \cdot \nabla\mathbf{v}) \cdot \mathbf{n}_\Gamma \right]\!\!\right] \tag{4.14}$$

$$[\![Y_y]\!] = Y_y^{(+)}\left(1 - \frac{1}{H_y \sum_{\tilde{y}}\left(\frac{Y^{(+)}}{H}\right)_{\tilde{y}}} \right) \tag{4.15}$$

with definition of an interface tangential vector $\mathbf{t}_\Gamma$ and Henry's law constant H.

4.3   Numerical Scheme

The governing equations are solved on a Cartesian grid using finite volume discretizations by a generalized incremental pressure correction projection scheme, consisting of a predictor step, in which time dependence of p' is neglected within each fluid phase[2] and the divergence constraint (last vector component of (4.1)) is ignored, and a two-stage corrector step. As detailed in [38] and references therein, the divergence error is tracked in the quantity $\mathscr{P}$ (which is homogeneous and constant if P is). It satisfies

$$\nabla \cdot \mathbf{v} + \frac{1}{\rho c^2}(\mathbf{v} \cdot \nabla P) \overset{!}{=} \frac{1}{\mathscr{P}}\nabla \cdot (\mathscr{P}\mathbf{v}) \tag{4.16}$$

---

[2]Changes of fluid phase over time, however, are accounted for.

and accumulates divergence errors by solving an additional hyperbolic equation in the predictor auxiliary system

$$\frac{\mathrm{d}}{\mathrm{d}t} \int_\Omega \tilde{\mathbf{q}}^\star \, \mathrm{d}V + \oint_{\partial\Omega} \mathbf{F} \, \mathrm{d}A = \int_\Omega \mathbf{s} \, \mathrm{d}V \tag{4.17}$$

$$\mathbf{F} := \hat{\mathbf{q}}^\star \left(\mathscr{P} \mathbf{v} \cdot \mathbf{n}\right) - \mathbf{f} - \mathbf{f}_\Gamma \delta_\Gamma \tag{4.18}$$

with

$$\tilde{\mathbf{q}}^\star := \mathrm{Sr}\, \mathscr{P} \hat{\mathbf{q}}^\star \qquad \hat{\mathbf{q}}^\star := \frac{1}{\Theta}\left(\mathbf{q}, \Theta\right) \qquad \Theta := \frac{\mathscr{P}}{\rho} \tag{4.19}$$

for consistent correction of all fluxes based on the common carrier flux $(\mathscr{P}\mathbf{v} \cdot \mathbf{n})$ in the first of the two subsequent corrector steps, which requires solution of a cell-centered Poisson problem. The second corrector step is performed on a staggered grid by solving a node-centered Poisson problem on dual cells as done in [3], for example. The resulting approximate projection procedure both imposes the divergence constraint on the final cell-centered velocity field and incorporates time dependence of p' into the nodal pressure update, comparable to [1]. Both Poisson problems are solved in a discretely conservative integral flux based finite volume fashion as well.

4.3.1 Interface Representation

The sharp fluidic interface Γ, separating the two fluid phases, is modeled using a hybrid dual representation consisting of a level-set (LS) function and volume-of-fluid (VoF) information. The LS function G implicitly provides a continuous interface via its zero level $G = 0$ and serves as basis for computation of both

$$\mathbf{n}_\Gamma := -\frac{\nabla G}{|\nabla G|} \qquad \text{and} \qquad \kappa := -\nabla \cdot \mathbf{n}_\Gamma \tag{4.20}$$

with $\mathbf{n}_\Gamma$ as the interface normal vector and κ as the interface (mean) curvature. Additionally, the discrete explicit interface representation is obtained from interpolated nodal LS data, avoiding interface reconstruction procedures based on VoF information. The latter—containing the interface implicitly as well—is used in order to discretely conserve the mass of each of the fluid phases, which is crucial for both accuracy and stability of the numerical method. The evolution

$$G_t + \mathbf{v}_G \cdot \nabla G = 0 \tag{4.21}$$

of G with partial derivative G_t with respect to time t and $\mathbf{v}_G(\mathbf{x}_\Gamma) \equiv \mathbf{v}(\mathbf{x}_\Gamma)$ is approximated by

$$G_t = -\left[(\mathbf{v}_i + \mathbf{v}_i') \cdot \nabla G_i - Q_i((1 - |\nabla G_i|); \mathbf{D}, \Delta (\mathbf{v}_i \cdot \mathbf{n}_\Gamma))\right] \tag{4.22}$$

with index i indicating the discrete representation within a narrow band around Γ. Discretization of the LS gradient follows the spatially third order accurate upstream central finite difference approach UC3 according to [13], $\mathbf{v}_i$ indicates the discrete cell-centered velocity and $\mathbf{v}_i'$ is a discrete velocity increment to couple the interface to the flow, compensating for leading order discretization errors. The source term Q penalizes deviations of the magnitude of the LS gradient from unity, incorporating the local strain rate tensor $\mathbf{D}$ and the local discretization errors of the interface normal velocity $\Delta (\mathbf{v}_i \cdot \mathbf{n}_\Gamma)$ as sources of deterioration of the LS function via an hybrid approach borrowing from [20] for all narrow band cells and the non-iterative direct CR-2 re-initialization procedure from [14] in interface cells. VoF information, on the other hand, is obtained by means of one more additional hyperbolic conservation law in (4.17) and (4.18), yielding

$$\hat{\mathbf{q}}^\star := \frac{1}{\Theta} \left(\mathbf{q}, \Theta, \Theta\phi\right) \tag{4.23}$$

$$\mathbf{f} := \left(0, 0, \left(\tfrac{1}{\mathrm{Re}}\mathbf{T} - \mathbf{I}p'\right) \cdot \mathbf{n}, 0, 0\right) \tag{4.24}$$

$$\mathbf{f}_\Gamma := \left(0, 0, \frac{1}{\mathrm{We}}\mathbf{T}_\Gamma \cdot \mathbf{n}, 0, 0\right) \tag{4.25}$$

$$\mathbf{s} := \rho\left(0, 0, \frac{1}{\mathrm{Fr}^2}\mathbf{g}, 0, 0\right) \tag{4.26}$$

with fluid marker $\phi = 1 \ \forall \mathbf{x} \in \Omega^{(+)}$ and $\phi = 0$ otherwise. Both interface representations need to interact as detailed in Sect. 4.3.4 in order to compensate for the individual discretization errors in a physically reasonable way. During this two way coupling the integral averages $\overline{\mathscr{P}\phi}$, obtained after balancing of fluxes over $\partial\Omega$, are corrected based on LS information on the topology, while the LS correction velocity $\mathbf{v}'$ is determined by means of corrected VoF information. Consistent correction of all conserved quantities following the one for $\overline{\mathscr{P}\phi}$ guarantees for sharp conservation of each fluid phase.

4.3.2   Time Integration

Both the LS equation (4.22) and the set of conservation laws (4.17) with (4.23)–(4.26) are integrated in time using a low storage consuming second order two-stage strong stability preserving Runge-Kutta scheme, corresponding to the method of Heun, as given in [10]. The semi-discrete form of (4.17) with second order truncation errors reads

$$\frac{\mathrm{d}}{\mathrm{d}t}\overline{\tilde{\mathbf{q}}}^{\star}{}_i = \bar{\mathbf{s}}_i - \frac{1}{|\Omega_i|}\sum_{j=1}^{N_i}\left(|\partial\Omega_j|\sum_{\varphi=1}^{N_j^{(\varphi)}}\left[\beta_j^{\varphi}(t)\mathbf{F}_j^{\varphi}(\mathbf{x}_j^{c_\varphi}(t))\right]\right) \tag{4.27}$$

with time-dependent fraction

$$\beta_j^{\varphi}(t) := \frac{1}{|\partial\Omega_j|}\int_{\partial\Omega_j^{\varphi}(t)}\mathrm{d}A \tag{4.28}$$

of grid cell face j in fluid phase φ with $\varphi \in \{(+), (-)\}$, integral average

$$\overline{\Psi}_i := \frac{1}{|\Omega_i|}\int_{\Omega_i}\Psi\,\mathrm{d}V \tag{4.29}$$

and

$$\mathbf{x}_j^{c_\varphi}(t) := \frac{1}{|\partial\Omega_j^{\varphi}(t)|}\int_{\partial\Omega_j^{\varphi}(t)}\mathbf{x}\,\mathrm{d}A \tag{4.30}$$

as face fraction centroid. In (4.27) N_i represents the number of grid cell faces and $N_j^{(\varphi)}$ is the number of cell face fractions per cell face and therefore either equal to one or two. First the LS function is evolved and the time-dependent face fraction $\beta_j^{\varphi}(t)$ can be approximated by its temporal average

$$b_j^{\varphi} = \frac{1}{t^{n+1} - t^n}\int_{t^n}^{t^{n+1}}\beta_j^{\varphi}(t)\,\mathrm{d}t \tag{4.31}$$

within the respective time interval for time integration of the conserved quantities based on (4.27), determined from the nodal LS values at both time levels. Since fluxes are approximated at the old time level at the centroid of that time level in each stage of the method of lines approach, the method is only first order accurate in time in the L_∞ norm in grid cells, which contain Γ moving in time due to the loss of topological symmetry in space-time with respect to the temporal center of the time interval. The method is second order accurate in the L_∞ norm in grid cells which contain only one fluid phase throughout the entire time interval, and shows overall second order accuracy in the L_1 norm.

4.3.3 Flux Computation

Both on regular grid cell faces ($N_j^{(\varphi)} = 1$, $\beta_j^\varphi = 1$) and on grid cell faces which are intersected by Γ ($N_j^{(\varphi)} = 2$) a single effective numerical flux

$$\mathbf{F}_j \; := \; \sum_{\varphi=1}^{N_j^{(\varphi)}} \left[b_j^\varphi \mathbf{F}_j^\varphi \right] \tag{4.32}$$

is determined to yield an integral space-time average update for the entire grid cell according to (4.27), avoiding treatment of potentially arbitrarily small grid cell fractions as originated in [32]. In order to compute these numerical fluxes across entire grid cell faces, suitable data has to be available in the surrounding. While this is given by default sufficiently far away from Γ, in the vicinity of the latter suitable ghost data needs to be recovered for both fluid phases beyond Γ before numerical fluxes can be evaluated.

In order to apply standard stencils for flux computation in the vicinity of Γ as well, ghost data is recovered around Γ in a sufficiently large narrow band sub-set in both fluid phases layer by layer as shown in Fig. 4.1, starting with the cells intersected by Γ before progressing beyond Γ. Therefore available data is extrapolated linearly from all available one-dimensional directions per fluid phase into the ghost region and the average of all available extrapolations is determined per ghost fluid cell. If only one one-dimensional stencil is available for ghost data extrapolation due to the given topology, no data is recovered until sufficient information is available in the surrounding, including cells in which ghost data is already recovered in one of the previous recovery cycles.

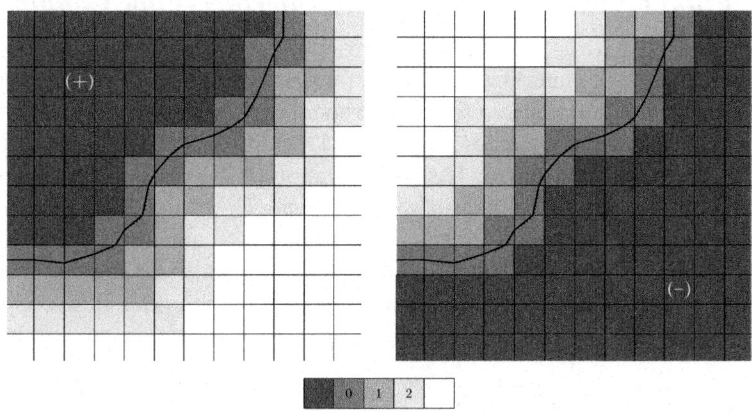

Fig. 4.1 Taken from [38]: Numbered ghost fluid layers for each fluid phase, $(+)$ and $(-)$, in the vicinity of Γ. Intersected grid cells are labeled 0. *White areas* represent regions in which only data for one fluid phase has to be available for proper flux evaluation

Once extrapolated average data is recovered in the entire ghost region, standard finite volume fluxes can be computed on each grid cell face for both fluid phases. Computation of numerical advective flux approximations requires a suitable state recovery on both sides of each grid cell face for each fluid phase, which is performed direction-by-direction using standard one-dimensional four point stencils for each grid cell face applying a van Leer slope limiter for determination of the grid cell face normal slopes in the adjacent grid cells. Due to the absence of shocks in the zero and low Mach number limit, no Riemann problems need to be solved and the average of the recovered grid cell face normal velocities on both sides of each grid cell face serves as basis for evaluation of the grid cell face normal carrier flux ($\mathscr{P}\mathbf{v}\cdot\mathbf{n}$), which the advective flux approximations of all conserved quantities are based on in turn, applying upwind values for $\mathscr{P}$ and $\hat{\mathbf{q}}^\star$ with respect to the obtained grid cell face normal velocity vector $\mathbf{v}_j$ at grid cell face j. The resulting fluxes at cut grid cell faces, evaluated in the center of the latter for both fluid phases, are interpolated to the centroid of each cell face fraction in the respective fluid phase considering cell face center fluxes in each ghost fluid region as well.

The final effective flux approximation

$$\mathbf{F} \;=\; \mathbf{F}^* + \left(\Delta\mathbf{F}^{(\mathrm{div.})} + \sum_k \Delta\mathbf{F}_k^{(\mathrm{cons.})} + \Delta\mathbf{F}^{(\mathrm{pres.})} \right) \tag{4.33}$$

of any conserved quantity consists of the predicted flux average $\mathbf{F}^*$, conservation related flux corrections $\Delta\mathbf{F}^{(\mathrm{cons.})}$, the correction due to divergence errors $\Delta\mathbf{F}^{(\mathrm{div.})}$ and the one, $\Delta\mathbf{F}^{(\mathrm{pres.})}$, incorporating time dependence of the pressure and adjusting the cell-centered velocity field to discretely satisfy the divergence constraint as well. The latter is only non-zero for momentum.

4.3.4   Fluid Phase Conservation and Interface-Flow Coupling

Both LS and VoF based interface representations lack discrete conservation[3] in their stand-alone fashion due to discretization errors during determination of (4.28) and its temporal evolution (or temporal average (4.31), respectively) on the one hand and the discrete treatment of the level-set equation (4.21) on the other hand. This results in smearing of conserved quantities around Γ, artificial mass changes of each fluid phase, unphysical discrete values of conserved quantities and eventually decoupling of level-set based interface representation and fluid flow. The applied procedure for discretely conservative fluid transport and interface-flow-coupling is described in detail in [38] and sketched in Fig. 4.2 for front transport from left to

---

[3]The VoF based interface representation itself is (discretely) conservative by default. However, discretization errors can produce VoF values which can not be assigned a physically reasonable interpretation anymore. Truncation of such over- and undershoots without suitable flux based redistribution leads to a lack of conservation on the discrete level.

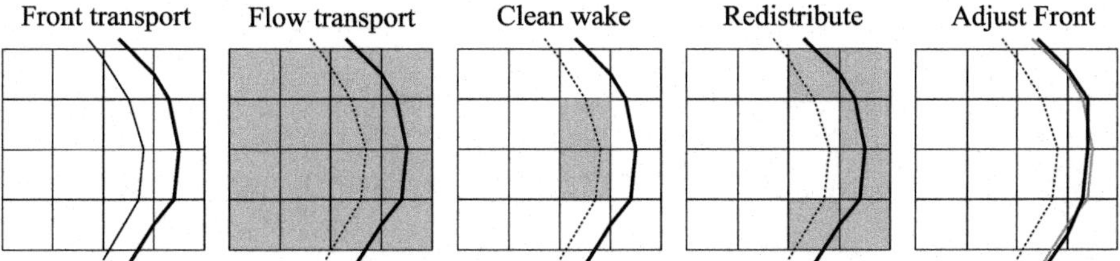

Fig. 4.2 Two-phase time stepping procedure from left to right: LS based front transport, transport of conserved quantities, removal of discretization errors around the front (mainly in its wake), redistribution of conserved quantities within intersected grid cells, adjustment of the LS based front to the conserved quantities of the flow field

right. In principle it follows the work done in [29] and [30] based on [5], adapted to the present generalized setting according to [15], in a modified version: After LS propagation, followed by advancing the conserved quantities of the flow field, in a first correction step (center sketch of Fig. 4.2) conserved quantities are corrected consistently in all grid cells, which have been cut during the time interval considered but end up regular at its end, based on known target values for the conserved VoF quantity $\overline{\mathscr{P}\phi}$: either $\overline{\mathscr{P}}$ or 0. Correction fluxes $\Delta\mathbf{F}^{(\mathrm{cons.})}$ point into cut neighbor cells only, in which, after an intermediate update, corrections to all conserved quantities need to be applied as well in the subsequent step by solving a coupled Poisson-type problem restricted to cut grid cells only, connecting the latter in order to solve for VoF based carrier correction fluxes $\Delta\mathbf{F}^{(\mathrm{cons.})}_{\overline{\mathscr{P}\phi}}$. As described in detail in [38], the required VoF target values are constructed conservatively and VoF-value-bounding, based on the information of the not yet corrected LS interface representation and the cut cell global sum of VoF values, which needs to be conserved within the set of cut grid cells. After adjusting all conserved flow quantities according to the VoF based carrier fluxes, the LS based interface representation is corrected applying local LS correction velocities that move the local segment of Γ towards the discontinuity of the conserved quantities. Figure 4.3, taken from [39], in which a more detailed overview on the present procedure is given, shows, that both the conserved quantities, represented by the volume fraction $\alpha := \overline{\phi}$, remain sharply separated (Fig. 4.3 *left* bottom) conserving the volume of each fluid phase up to machine accuracy, and Γ remains stably coupled to the flow (Fig. 4.3 *right*).

4.3.5   Corrector

In the first corrector step divergence errors in the predicted fluxes are corrected consistently for all conserved quantities by solving the cell-centered Poisson problem

$$\oint_{\partial\Omega_i}\left(\frac{\Delta t}{2}\Theta^{(n+\frac{1}{2}),*}\right)\nabla u^{(n+\frac{1}{2})}\cdot\mathbf{n}\,\mathrm{d}A = -\frac{|\Omega_i|}{\Delta t}\left(\overline{\mathscr{P}}_i^{(n+1),*} - \overline{\mathscr{P}}_i^{(n)}\right) \qquad (4.34)$$

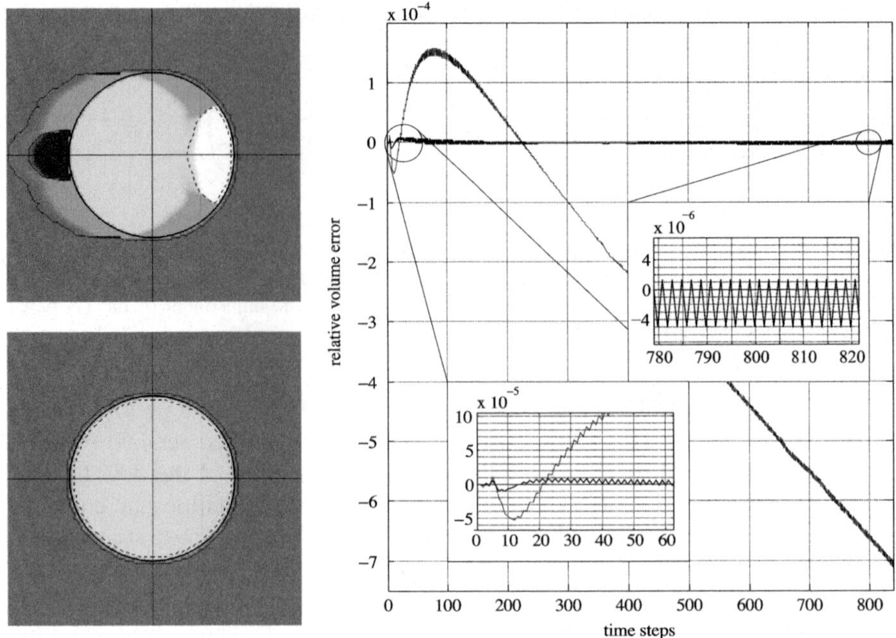

Fig. 4.3 Originating from [39], p. 464, with permission of Springer: *left:* Comparison of volume fractions $\alpha := \overline{\phi}$ without (*top*) and with (*bottom*) the described correction procedure, applied to a circular bubble in a homogeneous velocity field pointing from left to right applying periodic boundary conditions; *right:* Comparison of relative bubble volume error based on LS information w.r.t. initial data over time with (*solid line*) and without (*dotted line*) interface-flow-coupling. https://dx.doi.org/10.1007/978-3-319-05684-5_45

based on divergence errors tracked during the predictor step (superscript $*$) in $\mathscr{P}$. The resulting correction flux balance yields a corrective update per grid cell for all conserved quantities. As during the predictor step, the obtained common carrier correction flux $\Delta \mathbf{F}_{\mathscr{P}}^{(\mathrm{div.})} := \Delta (\mathscr{P} \mathbf{v} \cdot \mathbf{n}) = \left(\frac{\Delta t}{2} \Theta^{(n+\frac{1}{2}),*}\right)\left(\nabla u^{(n+\frac{1}{2})} \cdot \mathbf{n}\right)$ serves as basis for correction fluxes of any conserved quantity.

While the first correction step affects all conserved quantities (yielding values labeled by superscript $**$), the second corrector step addresses the momentum balance only. The respective correction flux $\Delta \mathbf{F}^{(\mathrm{pres.})} \equiv \Delta \mathbf{F}_{\rho \mathbf{v}}^{(\mathrm{pres.})} = \oint_{\partial \Omega} \partial p' \cdot \mathbf{n} \, dA$, obtained from nodal pressure increments $\partial p' = \mathscr{L}(u)$ with $\mathscr{L}(u) := u$ on the primal grid cell face after solution of the node-centered Poisson problem

$$\oint_{\partial \ddot{\Omega}_i} \left(\Delta t \, \Theta^{(n+1),**}\right) \nabla u^{(n+\frac{1}{2})} \cdot \mathbf{n} \, dA = \oint_{\partial \ddot{\Omega}_i} (\mathscr{P} \mathbf{v})^{(n+1),**} \cdot \mathbf{n} \, dA \qquad (4.35)$$

on a staggered grid with dual grid cells $\ddot{\Omega}_i$, guarantees for incorporation of time dependence of the perturbation pressure p' into the momentum balance such that the cell-centered velocity field satisfies the divergence constraint.

4.3.6 Surface Stress Approximation

A crucial feature of any two-phase flow method is suitable treatment of the singular contribution of surface tension to the momentum balance in order to guarantee for an accurate and stable method. The discretization of surface tension in two-phase flow solvers is notoriously difficult in nearly balanced quasi-static situations and in the limit of large surface tension since the discrete representation of a static state requires an exact balance of the discretized pressure gradient and surface tension. Due to the mathematically very different structure of these terms—a gradient on the one hand and a surface-distribution of the force on the other hand—it is hard to conceive discretizations that would realize this balance exactly. Nevertheless, such well-balancing schemes are attractive and aimed for as they balance these specific parts of the underlying equations on a discrete level and discretely mimic an analytical feature of the momentum equation.

4.3.6.1 Continuous Surface Stress Approach

An intermediate stage to a sharp conservative well-balanced discretization is obtained by a curvature-free continuous discretely conservative approximation of the surface stress tensor (4.5) within a narrow transition zone around Γ in order to discretize these contributions at the same location (the grid cell faces as done in [9]) the fluxes due to pressure and viscous forces are discretized at—a mandatory requirement for suitable local balances. Therefore a spatio-temporal average of the resulting fluxes over (portions of) grid cell faces is computed via analytical integration of ansatz functions for the respective grid cell face normal component of the surface stress tensor in space-time as described in detail in [38]. In the respective contribution

$$\frac{1}{\text{We}} \int_{t^n}^{t^{n+1}} \int_{\partial\Omega_j} \delta_\Gamma \mathbf{T}_\Gamma \cdot \mathbf{n} \, dA \, dt \tag{4.36}$$

to the space-time integral momentum equation with integrand

$$\delta_\Gamma \mathbf{T}_\Gamma \cdot \mathbf{n} = \delta_\Gamma \sigma \left(\mathbf{n} - \mathbf{n}_\Gamma \left(\mathbf{n}_\Gamma \cdot \mathbf{n} \right) \right) \tag{4.37}$$

due to (4.5) the surface Dirac δ_Γ is approximated using interface normal linear hat functions. Therefore—next to the distribution of the surface tension coefficient—(4.37) only depends on the LS function and its gradient due to (4.20). Both the LS function within the linear hat Dirac approximation and the contributions due to the surface stress tensor are approximated by spatio-temporal ansatz functions. First the resulting polynomial in space-time is integrated analytically over spatial bounds of spatio-temporal cell face fractions, bounded by the transition region boundary and Γ. The approximations of the latter depend linearly on time as shown

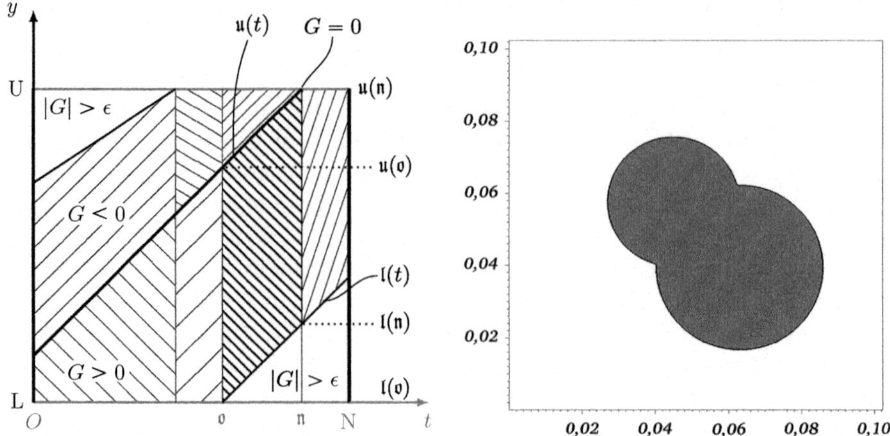

Fig. 4.4 With both pictures taken from [38]: *left*: Space-time scenario on a face of a two-dimensional grid cell: Integrals are evaluated analytically for each spatio-temporal cell face fraction (*shaded*) with time-dependent upper ($\mathfrak{u}$) and lower ($\mathfrak{l}$) bounds and time slice boundaries $\mathfrak{o}$ and $\mathfrak{n}$ as well as $O := t^n$ and $N := t^{n+1}$; Γ is represented by the *thick solid line* at $G = 0$; *right*: Initial geometric setting for a pair of oscillating soap bubbles according to [16]: $\rho^{(+)} = 1.2$, $\rho^{(-)}/\rho^{(+)} = 1$, $\mu^{(+)} = 1.71 \cdot 10^{-5}$, $\mu^{(-)}/\mu^{(+)} = 1$, $\sigma = 0.034$, $S_{\text{large}} = 0.0065$, $S_{\text{small}} = 0.0040$, $R_{\text{eq}} = 0.0252$ with S_i as the surfaces of the individual not merged bubbles and R_{eq} as the equilibrium radius of the resulting single bubble. The merged initial configuration is governed by the ratio of the radius of the resulting merging circle (half the distance between the two kinks) and the reference bubble radius (R_{large}). This ratio is chosen to be about 0.67778

on the left hand side of Fig. 4.4. The obtained polynomial in time is integrated analytically as well within the temporal bounds to finally yield the space-time average contribution due to the smoothed singular surface stress tensor within the time interval considered. Its difference to the space-time average contribution of the previous time interval is added to the right hand side of the nodal Poisson problem of the second corrector step, while—in analogy to the contribution due to the perturbation pressure p'—the space-time average contribution of the previous time interval is treated as passive momentum source in the predictor step.

Using this approach, experimental and theoretical results on the dynamics of a merged soap bubble configuration from [16] as shown in Fig. 4.4 are well reproduced on quite coarse grids at density ratio of $\rho^{(-)}/\rho^{(+)} \equiv 1$ and finite transition region thickness as shown in Table 4.1, presenting the first oscillation modes of the bubble surface.

Open issues in this context are matching approximations of surface stress tensor and perturbation pressure at the transition region boundary, robust transition to arbitrarily thin transition zones (although conceptually possible and straightforward) and the case of $\rho^{(-)}/\rho^{(+)} \neq 1$ for non-sharp transition regions due to density ratio dependent resulting spurious currents of different orders of magnitude in the different fluid phases using a conservative formulation of the governing equations.

Table 4.1 Numerical, experimental, and theoretical results for a configuration of merged soap bubbles of different size according to [16]

| | Frequencies | | |
| | Numerical (96 × 96) | Experimental | Theoretical |
Oscillation mode	[38]	[16]	[18]
2	19.996	19	20.75
3	39.992	35	39.21
4	59.988	57	59.90
5	82.841	72	82.76

4.3.6.2 Towards a Well-Balanced and Conservative Sharp Interface Approach for Surface Forces

A simplified version of the momentum equation—free of advection, friction and gravity—for balancing forces due to surface tension with the ones due to pressure is given through

$$\frac{\mathrm{d}}{\mathrm{d}t} \int_{\Omega} \rho \mathbf{v} \, \mathrm{d}V = -\oint_{\partial\Omega} p' \mathbf{n} \, \mathrm{d}A + \frac{1}{\mathrm{We}} \oint_{\partial\Omega \cap \Gamma(t)} \sigma \mathbf{t}_\Gamma \mathrm{d}\ell \tag{4.38}$$

considering the identity

$$\oint_{\partial\Omega \cap \Gamma(t)} \sigma \mathbf{t}_\Gamma \, \mathrm{d}\ell = \int_{\partial\Omega} \delta_\Gamma \sigma \mathbf{t}_\Gamma \, \mathrm{d}A = \int_{\partial\Omega} \delta_\Gamma \mathbf{T}_\Gamma \cdot \mathbf{n} \, \mathrm{d}A = \int_{\Omega} \delta_\Gamma \mathbf{k}_\sigma \, \mathrm{d}V = \int_{\Gamma(t)} \mathbf{k}_\sigma \, \mathrm{d}S$$

$$\tag{4.39}$$

with $\mathbf{k}_\sigma := \sigma \kappa \mathbf{n}_\Gamma + \nabla_\Gamma \sigma$ as surface force with surface gradient ∇_Γ, reducing to $\sigma \kappa \mathbf{n}_\Gamma$ for homogeneous surface tension coefficient $\sigma = \sigma_0$ along Γ. If the pressure integral with p' in Eq. (4.38) balances the one at Γ resulting from the surface internal stresses, the temporal change due to interaction of pressure and surface forces vanishes. In order to exactly balance these expressions, they have to yield the same result discretely up to machine accuracy (with opposing sign). While the pressure force term in Eq. (4.38) is given as a surface integral, the resulting surface force is given through a source term, which gives rise to rewriting these expressions in order to allow for similar calculation procedures for pressure and surface stress that offer the potential of exact matching in a stationary setting for a simple circle or sphere, in the following restricted to two space dimensions and divergence free velocities: For inviscid flow, constant surface tension σ_0 and zero mass flux through the surface the net effect of surface tension is to induce a pressure jump at the interface proportional to $\sigma\kappa$. The fluid mass, however, lies on either side of the surface, and therefore changes momentum exclusively due to the (smooth) pressure gradients found away from the surface. Therefore, fluid parcels never feel the singular surface force. This observation suggests a particularly simple discretization of the combined effect of pressure gradient and surface tension in a sharp interface finite volume method: Such a method provides an explicit, piecewise smooth representation of the pressure field including the pressure jump within the grid cells. Thus, the momentum change for the total cell due to the pressure gradient/surface tension combination can be determined by computing individual momentum changes for the partial cells based on the separate pressure gradients on either side of the interface and summing them up. This would give justice to the physics of the problem as described above. The surface force term therefore can be rewritten as

$$\int_{\partial\Omega\cap\partial\Gamma(t)} \sigma \mathbf{t}_\Gamma \, d\ell \;=\; \sigma_0 \int_{\ell_A}^{\ell_B} \kappa \mathbf{n}_\Gamma \, d\ell \;=\; \int_{\Omega\cap\partial\Gamma} [\![p']\!] \, dS \tag{4.40}$$

providing a surface integral on the very right hand side, which allows for a similar evaluation as for the pressure force term in Eq. (4.38). Evaluating a surface integral over $[\![p']\!]$ in a discrete cell intersected by Γ is possible due to [21, 24], which gives double bi/tri-linear solutions of the phase-wise pressure values in a two- or three-dimensional set-up, respectively. This allows for evaluation of $p'^{(+)}$ in $\Omega^{(-)}$ and vice versa.

In general it is aimed for the necessity of only small corrections in the underlying predictor-corrector scheme. Since the LS is pre-advected in the present explicit formulation of front transport for finite Weber numbers and, thus, known at both old and new time level, the above integrals can be evaluated at both time levels. Although the pressure is constant in time during the predictor step, interfacial movement and the involved phase change at the location of grid cell nodes is accounted for nevertheless. In nodes which swap the phase, the pressure is extrapolated based on the double bilinear ansatz function from [21]. Following this idea has the benefit, that the corrector only generates an incremental perturbation pressure update which corrects the spatial extrapolation error of the predictor step and the possible change of the perturbation pressure due to friction or curvature changes. Within the predictor step the curvature values are considered to remain the old ones also at the new interface position and therefore only an incremental curvature value is required in the corrector step to account for the change between old and new curvature at the new interface position.

This procedure shows fairly good results for both a resting circle as shown in Fig. 4.5 and a passively advected circle as shown in Fig. 4.6. In the former an initially resting, poorly resolved circle with a surface tension coefficient $\sigma = 1$ and a corresponding pressure jump at the interface is shown. The color bar, indicating

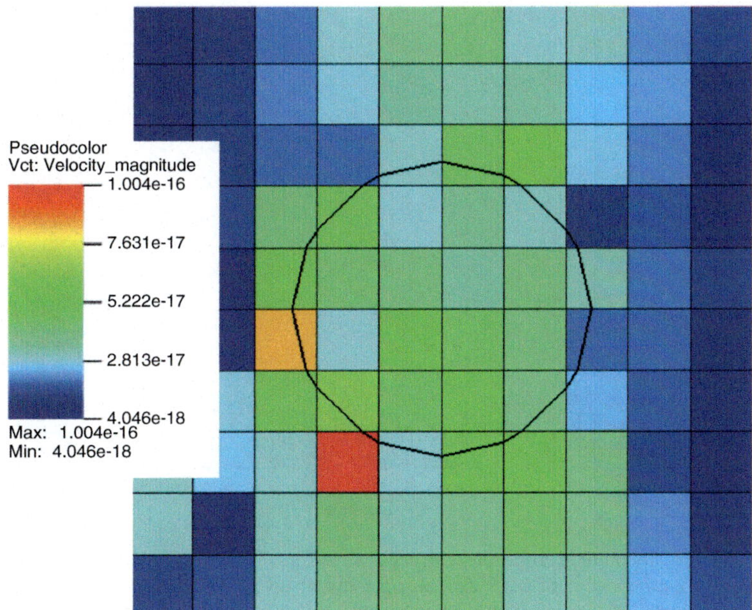

Fig. 4.5 Velocity magnitude after the corrector step for the resting circle

velocity after the correction step, clearly documents that the resulting velocity correction is in the range of machine precision, which is achieved via the discrete balancing of the perturbation pressure and the surface force within the predictor step. Therefore the discussed algorithm is able to preserve the resting circle after time integration by the two-stage Runge-Kutta time integrator. In each stage the forces are discretely balanced and the contributions of the corrector steps vanish, because all relevant forces are evaluated and balanced in the predictor step already.

A similar scenario is shown in Fig. 4.6, displaying a section of a poorly resolved circle moving from left to right. Again the velocity after the corrector step is shown, which again only differs from the background advection velocity by magnitudes of machine precision. The pressure force as well as the surface force integral account for the moving interface in both stages of the predictor step and balance each other in each stage. Thus, again the corrector does not contribute as long as the circle curvature does not change and the phase-wise pressure field is passively advected with the circle as well. This is the case because the pressure extrapolation within the predictor step is exact for nodes which swap phase due to interfacial movement for both homogeneous and spatially linearly changing pressure fields. A crucial basis for this approach to work even on very coarse grids is the following novel curvature discretization based on its geometrical definition.

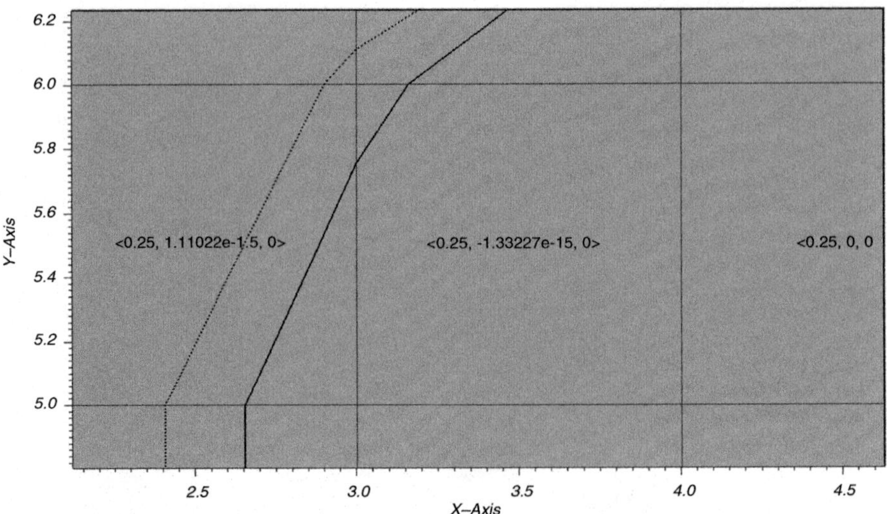

Fig. 4.6 Velocity magnitude after the corrector step for a passively advected circle with an initial velocity x-component of 0.25, *dashed line*: initial LS approximation, *solid line*: final LS approximation; only an excerpt of the entire poorly resolved circle, corresponding to Fig. 4.5, is shown in order to make the contributions to the displayed velocity vectors visible

Curvature Evaluation Using Fitting Tools

In order to evaluate the interface curvature κ from (4.20) numerically, typically finite difference formulae are used, yielding

$$\kappa = -\frac{\left[\frac{\partial^2 G}{\partial x_1^2}\left(\frac{\partial G}{\partial x_2}\right)^2 - 2\frac{\partial^2 G}{\partial x_1 \partial x_2}\frac{\partial G}{\partial x_1}\frac{\partial G}{\partial x_2} + \frac{\partial^2 G}{\partial x_1^2}\left(\frac{\partial G}{\partial x_1}\right)^2\right]}{\left(\sqrt{\left(\frac{\partial G}{\partial x_1}\right)^2 + \left(\frac{\partial G}{\partial x_2}\right)^2}\right)^3} \tag{4.41}$$

for the two-dimensional case with a second order approximation and x_1 and x_2 as the two-dimensional coordinate directions. However, there are several issues when calculating κ by finite difference formulae such as Eq. (4.41) in context of LS methods. Evaluating (4.41), for example, for a simple circle or sphere based on discrete values of an exact LS function subject to the signed distance property on a Cartesian grid already shows that there are significant fluctuations of κ along the arc length of the circle or sphere. Besides the general dependencies (such as grid size etc.) the quality of the approximation of κ depends on the position of the evaluation. This is caused by the discrete evaluation of κ and not by a possible error in G. Although discrete values of an *exact* LS function are used, the constraint $|\nabla G| = 1$ is not satisfied as well. Instead, $|\nabla G|$ shows a similar dependency on the point of evaluation along Γ and therefore influences the evaluation of κ through the first term of Eq. (4.41). The absolute error of the approximation of κ is in an order of magnitude that makes simulations of dispersed bubbly flows hardly possible without resolving Γ very finely, which involves excessive numerical effort.

In order to overcome these problems, evaluation of κ based on its geometrical definition

$$\kappa = \frac{1}{R} \tag{4.42}$$

as reciprocal of the radius R of a circle which touches the interface at the point in question is investigated rather than $\kappa = -\nabla \cdot \mathbf{n}_\Gamma$, focussing on the two-dimensional case.

Typically it can be evaluated whether or not there is an interface between two adjacent points by using the LS values at these points. In order to use the curvature value, e.g. in a finite volume scheme, its value is needed at the intersection of the interface and the connecting line between these two adjacent points. In a two-dimensional set-up there are six cells holding LS information, which can be used to describe the local curvature. One possible way of approximating κ is now to use all LS information closely surrounding the spatial point considered as part of an over-determined set of equations describing the radius and the centre ($\bullet$) of the osculating circle ($\circ$) by the following type of equation for the i'th point which holds LS information:

$$(x_{1,i} - x_{1,\circ,\bullet})^2 + (x_{2,i} - x_{2,\circ,\bullet})^2 = (G_i - R_\circ)^2 \tag{4.43}$$

Here $R_\circ$ is the radius of the osculating circle, $x_{j,\circ,\bullet}$ its coordinate in j direction and G_i and $x_{j,i}$ are the LS value at point i and the coordinate in j direction of point i. Solving this system, for example by a least squares solver, provides values for $x_{j,\circ,\bullet}$ and $R_\circ$. With the latter, curvature can be computed according to (4.42). A possible initial value for a least squares procedure is the radius of the osculating circle approximated through the standard finite difference formula of Eq. (4.41).

First results with this approach are encouraging, because the determined values of κ seem to be fairly accurate compared to Eq. (4.41). Nevertheless, the dependence of this procedure on grid size and disturbed LS values with possibly disturbed signed distance property is subject of further investigation. This aims for developing an accurate (also in absolute values) and robust method with sufficient accuracy to be applicable on coarse grids.

4.3.7 An Asymptotic Two-Step Poisson Solver with Discontinuous Coefficients

Sharp interface projection methods on fixed-grids require solution of Poisson-type equations

$$\nabla \cdot (\Theta^\pm(\mathbf{x})\nabla u(\mathbf{x})) = r^\pm(\mathbf{x}) \tag{4.44}$$

on domains Ω, which are separated by an interface Γ into two parts $\Omega^{(+)}$ and $\Omega^{(-)}$. The superscripts $\pm$ indicate that the coefficient $\Theta(x)$ as well as the function $r(\mathbf{x})$ may jump at Γ due to their dependence on the specific fluid phase.

Poisson-type equations arise from various problems in nature, among them species diffusion, heat conduction and fluid dynamics. Although there is a long tradition in solving this kind of equations involving multiple fluid phases (e.g. gas/solid), it still is a computational challenge, especially involving complex interface topologies. The major problems arise due to potentially large jumps of the coefficients like thermal conductivity, diffusion coefficient and density at Γ, usually differing by several orders of magnitude between the different fluid phases.

At Γ both u and the interface normal derivative of u are prescribed via

$$[\![u]\!]_\Gamma = u^{(+)}(\mathbf{x}) - u^{(-)}(\mathbf{x}) =: \psi(\mathbf{x}) \qquad \text{for } \mathbf{x} \in \Gamma \qquad (4.45)$$

$$[\![\Theta u_n]\!]_\Gamma = \Theta^{(+)}u_n^{(+)}(\mathbf{x}) - \Theta^{(-)}u_n^{(-)}(\mathbf{x}) =: \vartheta(\mathbf{x}) \qquad \text{for } \mathbf{x} \in \Gamma \qquad (4.46)$$

where $u_n := \nabla u \cdot \mathbf{n}_\Gamma$. There are various methods proposed in literature, [2, 11, 21, 23], that address this kind of problem while the present one, published in [22], particularly aims for ratios $1/\varepsilon := \Theta^{(+)}/\Theta^{(-)}$ of coefficients larger than 100 without showing the typically arising convergence problems and/or loss of efficiency. Furthermore, the present approach is designed to be applicable for standard black-box iterative solvers, aiming to fit problems arising in two-phase fluid dynamics involving the divergence-free momentum equation

$$\frac{\partial \mathbf{v}}{\partial t} + (\mathbf{v} \cdot \nabla)\mathbf{v} = -\frac{1}{\rho}\nabla p' + \nu \nabla \cdot \nabla \mathbf{v} \qquad (4.47)$$

with kinematic viscosity ν. Together with the coupling relation (4.11) for the velocity at Γ this results in the necessity to find pressure gradient terms in both phases which satisfy

$$\mathcal{O}\left(\frac{1}{\rho^{(+)}}\nabla p'^{(+)}\right) = \mathcal{O}\left(\frac{1}{\rho^{(-)}}\nabla p'^{(-)}\right) \qquad (4.48)$$

in order to derive velocity updates with an error in the same order of magnitude for both fluid phases.

Introducing an asymptotic expansion in terms of the (inverse) coefficient ratio ε for the solutions of u in both phases via

$$\begin{aligned} u^{(+)}(\mathbf{x}) &= u^{(0,+)}(\mathbf{x}) + \varepsilon\, u^{(1,+)}(\mathbf{x}) + \mathcal{O}(\varepsilon^2) \\ u^{(-)}(\mathbf{x}) &= u^{(0,-)}(\mathbf{x}) + \varepsilon\, u^{(1,-)}(\mathbf{x}) + \mathcal{O}(\varepsilon^2) \end{aligned} \qquad (4.49)$$

in Eq. (4.44) yields the following two decoupled problems:

1.

$$u^{(0,+)}(\mathbf{x}) = 0 \qquad \text{for} \quad \mathbf{x} \in \Omega^{(+)}$$
$$\nabla \cdot (\Theta^{(-)} \nabla u^{(0,-)}(\mathbf{x})) = r^{(-)}(\mathbf{x}) \qquad \text{for} \quad \mathbf{x} \in \Omega^{(-)}$$

(4.50)

2.

$$\nabla \cdot (\Theta^{(-)} \nabla u^{(1,+)}(\mathbf{x})) = r^{(+)}(\mathbf{x}) \qquad \text{for} \quad \mathbf{x} \in \Omega^{(+)}$$
$$\nabla \cdot (\Theta^{(-)} \nabla u^{(1,-)}(\mathbf{x})) = 0 \qquad \text{for} \quad \mathbf{x} \in \Omega^{(-)}$$

(4.51)

The leading order solution results from solving Eq. (4.50) with the trivial solution for $\mathbf{x} \in \Omega^{(+)}$. The correction solutions for both phases are found by solving Eq. (4.51) at cost of solving a second Poisson problem, however, without any jumps in the coefficients and with an almost constant condition number for arbitrary ratios of coefficients. In order to mimic a two-phase flow problem and underline the capability of the asymptotic solution procedure, a test problem is designed, which scales with $\mathcal{O}(\varepsilon)$ in $\Omega^{(+)}$ and $\mathcal{O}(1)$ in $\Omega^{(-)}$. Figure 4.7 shows the relative error of the standard one-step solution procedure and the derived asymptotic two-step solution procedure for a fairly small coefficient ratio of $\varepsilon = 10^{-6}$.

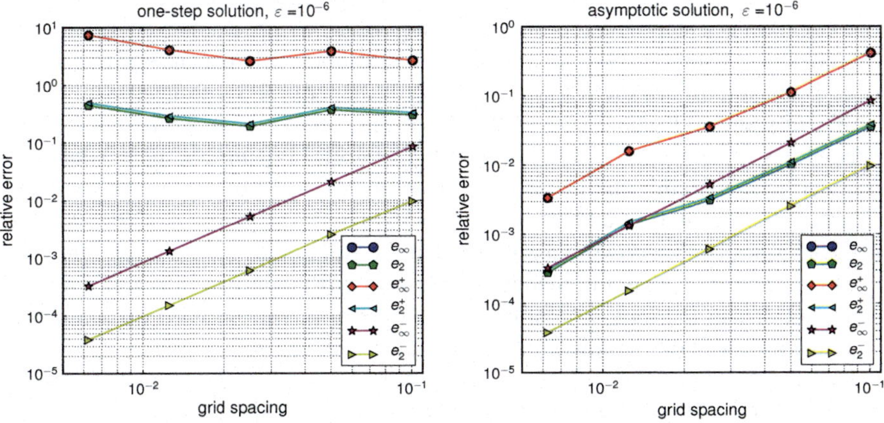

Fig. 4.7 Originating from [22], pp. 239 and 240, with permission from Elsevier: Relative error e of the one-step (*left*) and asymptotic (*right*) solution approach for a coefficient ratio of $\varepsilon = 10^{-6}$: With standard one-step approaches the solution in the light fluid phase (+) does not converge in both L_2 and L_∞ norm and, thus, the overall solution does not. The two-step asymptotic solution approach provides convergence in both fluid phases with coefficient ratio independent condition numbers. http://dx.doi.org/10.1016/j.jcp.2013.12.027

While the one-step procedure shows nice convergent results for $\Omega^{(-)}$ for both the L_2-norm as well as the L_∞-norm, it completely fails within $\Omega^{(+)}$. Since the overall error is dominated by the one of $\Omega^{(+)}$, the one-step solution procedure also fails to produce convergent results for the overall relative error. Considering the dependence of the ratio of the relative errors of both phases on ε yields an almost linear behavior, which foils the results of Eq. (4.48).

The two-step solution procedure, on the other hand, shows convergent solutions for both the L_2-norm and the L_∞-norm for both phases involved. The ratio of the relative errors of both phases is almost independent of the coefficient ratio ε for both the L_2-norm and the L_∞-norm. This feature ensures that the update of the phase-wise velocity through the result of the Poisson-type equation does not show any phase specific inaccuracies and follows the demands of the findings of Eq. (4.48).

4.3.8 A Poisson-Type Equation with Discontinuous Coefficients Including Interface Movement

While Poisson-type equations usually define purely spatial problems, time enters the system as additional variable in context of moving interfaces if the solution represents a temporal average. Within a finite volume discretization of a transient two-phase flow set-up one aims to find approximations of a specific numerical error. Using central differences to discretize second order derivatives of a Poisson-type equation is a standard approach resulting in second order errors. The temporal discretization is implicitly given through the mid-point-rule because of the assignment of the solution of the Poisson equation to the half time step level with proper coefficients and right hand side of the problem. Solving those equations for two-phase flows is more intricate because of the typically arising coefficient jumps at the interface separating the two phases (see e.g. [21]) and possible problems with accuracy of the solution and the efficiency of the solver (see e.g. [22]). Another problem arises concerning the assignment of the solution to a time level—the mid-point rule does not give automatically a second order approximation at half time step level due to the lack of topological symmetry in time caused by the locally changing presence of the individual fluid phases with respect to time.

In order to obtain second order results for this kind of problem the solution procedure from [21] is adopted and extended. A standard approximation to obtain a second order approximation in space and time is

$$\int_{t_0}^{t_1} \int_{x_0}^{x_1} \nabla p'(x,t)\, dt\, dx \;\approx\; \bar{p}'\,^{t=0.5(t_0+t_1)}_{x=0.5(x_0+x_1)} \tag{4.52}$$

where $\bar{p}'$ is the averaged solution quantity and the superscripts $t = 0.5(t_0 + t_1)$ and $x = 0.5(x_0 + x_1)$ indicate half time and half space approximations. A modification of this approximation reads

$$\int_{t_0}^{t_1}\int_{x_0}^{x_1}\nabla p'(x,t)dt\,dx \;=\; \int_{t_0}^{t_1}\int_{x_0}^{x_1}\left[\nabla p'^{\,(+)}(x,t)dx^{(+)} + \nabla p'^{\,(-)}(x,t)dx^{(-)}\right]dt$$

$$(4.53)$$

considering the two fluid phases, where $dx^{(\pm)}$ indicates the portions of dx belonging to the respective phase. Semi-discretization of the spatial problem yields

$$\int_{t_0}^{t_1}\int_{x_0}^{x_1}\nabla p'^{\,(\pm)}(x,t)dt\,dx \;\approx\; \int_{t_0}^{t_1}\left[\nabla p_m^{(+)}(t)\,\Delta x^{(+)}(t)\,t + \nabla p_m^{(-)}(t)\,\Delta x^{(-)}(t)\,t\right]dt$$

$$(4.54)$$

applying a mid-point-rule after introducing $p_m^{(+)}(t)$ and $p_m^{(-)}(t)$ as the time-dependent values of p at the mid of the corresponding time-dependent length $dx^{\pm}$. In order to evaluate the time-dependent functions $p_m'^{\,(\pm)}(t)$ and $dx^{(\pm)}(t)$ piecewise linearly, the temporal interval is further sub-divided between t_0 and t_1 at every intermediate point t_i, where either x_0 or x_1 changes the phase.

The extrapolation procedure for pressure nodes in the predictor step, changing the fluid phase affiliation within a time step, has an analogous part in the corrector step, which handles cells that end up without interface intersection at the end of the time step.

4.3.9 Pressure Boundary Condition

Accurate flow representation in the vicinity of a tracked interface supporting surface-bound physical processes is crucial, and the artificial pressure boundary layers known from classical projection methods is unacceptable in this context. Given a smooth solution $\mathbf{v}$ of the incompressible Navier-Stokes equations for constant density ρ and kinematic viscosity ν in a closed domain Ω with no-slip walls, the pressure p' satisfies a classical Poisson problem

$$\nabla\cdot\nabla p' \;=\; \rho\nabla\cdot(\mathbf{v}\cdot\nabla\mathbf{v}) \qquad\qquad (\mathbf{x}\in\Omega) \qquad\qquad (4.55)$$

$$\partial_n p' \;=\; \nu\,\mathbf{n}\cdot(\nabla\cdot\nabla\mathbf{v}) \qquad\qquad (\mathbf{x}\in\partial\Omega) \qquad\qquad (4.56)$$

with inhomogeneous Neumann boundary condition. When $\mathbf{v}$ is a Navier-Stokes solution as assumed, then according to [34] the pressure gradient and viscous forces will automatically cancel not only in the normal direction, as in (4.56), but also tangentially, such that $\nabla p = \nu\nabla\cdot\nabla\mathbf{v}$ for $\mathbf{x}\in\partial\Omega$. If, in contrast, $\mathbf{v}$ is an arbitrary smooth, divergence-free vector field that vanishes on $\partial\Omega$, then only the normal components of these fields will match in general, while the tangential ones will not.

Obviously, none of the approaches to consistent splitting schemes or higher-order accurate projection methods for the Navier-Stokes equations thus far ensures that—simultaneously—the flow divergence is exactly controlled and the pressure field satisfies its natural boundary condition requiring the balance of forces in the normal direction at rigid walls. If such simultaneous control could be achieved, the resulting scheme would combine the conceptual advantages of projection and consistent splitting schemes. A first investigation for constant density ($\rho = 1$) flow in a 2D slab Ω with rigid, straight no-slip walls indicates a possibility of devising such a scheme as follows:

Assuming the predictor step of a projection method or consistent splitting scheme [7, 19], to yield the predicted velocity $\mathbf{v}^*$, solving for a perturbation pressure increment π' is required such that $\mathbf{v}^* - \Delta t \nabla \pi'$ controls the divergence and achieves normal balance of forces near the walls. As in a consistent splitting scheme with

$$\Delta t \nabla \cdot \nabla \pi' = \nabla \cdot \left(\mathbf{v}^* - \mathbf{v}^{(n)}\right) \qquad\qquad (\mathbf{x} \in \Omega) \qquad\qquad (4.57)$$

near the container wall

$$\mathbf{n} \cdot \nabla \left(p'^{(n)} + \pi'\right) - \nu \nabla \cdot \nabla \left(\mathbf{v}^* - \Delta t \nabla \pi'\right) = 0 \qquad (\mathbf{x} \in \partial\Omega) \qquad\qquad (4.58)$$

would be required. At a first glance, this condition appears to impose a constraint on the third derivatives of the pressure at the boundary, while the pressure equation in the bulk of the domain is second-order only, and this would surely imply an ill-posed problem. Using the curl-curl-formulation of the viscous term, however, one can show that the constraint in (4.58) can be cast into the form

$$\left(-\frac{\nu \Delta t}{\rho} \partial_{\tau\tau}\right) \left(\partial_n \pi'\right) = \nu \partial_\tau \omega^* \qquad\qquad (4.59)$$

where ω is the vorticity of $\mathbf{v}$, and ∂_n and ∂_τ denote the normal and tangential derivatives along the (straight) container walls, respectively. Now, (4.59) is a Helmholtz-equation supported on the boundary of the domain only, and it can be solved for $\partial_n \pi'$. The effective boundary condition for the Poisson equation in (4.57) then becomes a classical inhomogeneous Neumann condition. For this particular case, the proposed formulation thus is well-posed contrary to immediate intuition as shown in [17]. Further investigation of this approach, its implementation and extension to more general variable density flows is subject of future work.

4.4   Conclusion

Several building blocks for a fully conservative generalized finite volume projection method for immiscible zero Mach number two-phase flows have been presented, extended and enhanced, focusing on the crucial issues of interface propagation, conservative fluid phase separation and discretizations of the singular contribution due to surface tension, the latter aiming for a scheme that balances with the perturbation pressure in physically stable situations. Additionally, an asymptotic two-step solution approach for Poisson-type equations with arbitrary ratio of coefficients in two-phase flow projection schemes is sketched and the main results concerning each building block are given. Future work has to address progress concerning the pressure boundary condition at solid and fluidic boundaries, moving interfaces in the Poisson solvers, discretization of surface PDEs including the related expressions in the species balance and accuracy of the time integrator in cut grid cells.

Acknowledgements This project was funded by German Research Foundation (DFG) Priority Program (SPP) 1506 on Transport Processes at Fluidic Interfaces from 2011 to 2016.

References

1. Almgren, A.S., Bell, J.B., Colella, P., Howell, L.H., Welcome, M.L.: A conservative adaptive projection method for the variable density incompressible Navier-Stokes equations. J. Comput. Phys. **142**, 1–46 (1998)
2. Bänsch, E.: Finite element discretization of the Navier-Stokes equations with free capillary surface. Numer. Math. **88**, 203–235 (2001)
3. Bell, J.B., Marcus, D.: A second-order projection method for variable-density flows. J. Comput. Phys. **101**, 334–348 (1992)
4. Botta, N., Klein, R., Langenberg, S., Lützenkirchen, S.: Well-balanced finite volume methods for nearly hydrostatic flows. J. Comput. Phys. **196**, 539–565 (2004)
5. Bourlioux, A.: A coupled level-set volume-of-fluid algorithm for tracking material interfaces. In: 3rd Annual Conference of the CFD Society, Banff (1995)
6. Brackbill, J.U., Kothe, D.B., Zemach, C.: A continuum method for modelling surface tension. J. Comput. Phys. **100**, 335–354 (1992)
7. Brown, D.L., Cortez, R., Minion, M.L.: Accurate projection methods for the incompressible Navier-Stokes equations. J. Comput. Phys. **168**, 464–499 (2001)
8. Chorin, A.J.: A numerical method for solving incompressible viscous flow problems. J. Comput. Phys. **2**, 12–26 (1967)
9. Francois, M.M., Cummins, S.J., Deny, E.D., Kothe, D.B., Sicilian, J.M., Williams, M.W.: A balanced-force algorithm for continuous and sharp interfacial surface tension models with a volume tracking framework. J. Comput. Phys. **213**, 141–173 (2006)
10. Gottlieb, S., Shu, C.-W., Tadmor, E.: Strong stability-preserving high-order time discretization methods. SIAM Rev. **43**(1), 89–112 (2001)
11. Hansbo, A., Hansbo, P.: An unfitted finite element method, based on Nitsche's method, for elliptic interface problems. Comput. Methods Appl. Mech. Eng. **191**, 5537–5552 (2002)
12. Harlow, F.H., Welch, J.E.: Numerical calculation of time-dependent viscous incompressible flow of fluid with a free interface. Phys. Fluids **8**, 2182–2189 (1965)

13. Hartmann, D.: A level-set based method for premixed combustion in compressible flow. Ph.D. thesis, Fakultät für Machinenwesen, Rheinisch-Westfälische Technische Hochschule Aachen, Aachen (2010)
14. Hartmann, D., Meinke, M., Schröder, W.: Differential equation based constrained reinitialization for level set methods. J. Comput. Phys. **227**, 6821–6845 (2008)
15. Klein, R.: Asymptotics, structure, and integration of sound-proof atmospheric flow equations. Theor. Comput. Fluid Dyn. **23**(3), 161–195 (2009)
16. Kornek, U., Müller, F., Harth, K., Hahn, A., Ganesan, S., Tobiska, L., Stannarius, R.: Oscillations of soap bubbles. New J. Phys. **12**, 073031 (2010)
17. Lakin, S.: Eine implizite Druckrandbedingung für die nichtstationäre Stokes-Gleichung. Master's thesis, Freie Universität Berlin (2012) [in German]
18. Lamb, H.: Hydrodynamics. Cambridge University Press, Cambridge (1932)
19. Liu, J.-G., Liu, J., Pego, R.L.: Stable and accurate pressure approximation for unsteady incompressible viscous flow. J. Comput. Phys. **229**, 3428–3453 (2010)
20. McCaslin, J.O., Desjardins, O.: A local re-initialization equation for the conservative level set method. J. Comput. Phys. **262**, 408–426 (2014)
21. Oevermann, M., Klein, R.: A Cartesian grid finite volume method for elliptic equations with variable coefficients and embedded interfaces. J. Comput. Phys. **219**, 749–769 (2006)
22. Oevermann, M., Klein, R.: An asymptotic solution approach for elliptic equations with discontinuous coefficients. J. Comput. Phys. **261**, 230–243 (2014)
23. Oevermann, M., Klein, R., Berger, M., Goodman, J.: A projection method for incompressible two-phase flow with surface tension. Technical Report 00-17, Konrad-Zuse-Zentrum, Berlin (2000)
24. Oevermann, M., Scharfenberg, C., Klein, R.: A sharp interface finite volume method for elliptic equations on Cartesian grids. J. Comput. Phys. **228**(14), 5184–5206 (2009)
25. Peskin, C.S.: Numerical analysis of blood flow in the heart. J. Comput. Phys. **25**, 220–252 (1977)
26. Popinet, S.: An accurate adaptive solver for surface-tension-driven interfacial flows. J. Comput. Phys. **228**, 5838–5866 (2009)
27. Popinet, S., Zaleski, S.: A front-tracking algorithm for accurate representation of surface tension. Int. J. Numer. Methods Fluids **30**(6), 775–793 (1999)
28. Prosperetti, A., Tryggvason, G. (eds.): Computational Methods for Multiphase Flow. Cambridge University Press, New York (2007)
29. Schneider, T.: Verfolgung von Flammenfronten und Phasengrenzen in schwachkompressiblen Strömungen. PhD thesis, Fakultät für Machinenwesen, Rheinisch-Westfälische Technische Hochschule Aachen, Germany (2000) [in German]
30. Schneider, T., Klein, R.: Overcoming mass losses in level-set-based interface tracking schemes. In: 2nd International Conference on Finite Volume for Complex Application (1999)
31. Schneider, T., Botta, N., Geratz, K.J., Klein, R.: Extension of finite volume compressible flow solvers to multi-dimensional, variable density zero Mach number flows. J. Comput. Phys. **155**, 248–286 (1999)
32. Smiljanovski, V., Moser, V., Klein, R.: A capturing-tracking hybrid scheme for deflagration discontinuities. Combust. Theor. Model. **1**(2), 183–215 (1997)
33. Sussman, M., Smith, K.M., Hussaini, M.Y., Ohta, M., Zhi-Wei, R.: A sharp interface method for incompressible two-phase flow. J. Comput. Phys. **221**, 469–505 (2007)
34. Temam, R.: Suitable initial conditions. J. Comput.Phys. **218**, 443–450 (2006)
35. Terhoeven, P.: Ein numerisches Verfahren zur Berechnung von Flammenfronten bei kleiner Mach-Zahl. PhD thesis, Rheinisch-Westfälische Technische Hochschule Aachen, Aachen (1998) [in German]
36. Tryggvason, G., Scardovelli, R., Zaleski, S.: Direct Numerical Simulations of Gas-Liquid Multiphase Flows. Cambridge University Press, Cambridge (2011)
37. Vater, S., Klein, R.: Stability of a Cartesian grid projection method for zero Froude number shallow water flows. Numer. Math. **113**(1), 123–161 (2009)

38. Waidmann, M.: Towards a strictly conservative hybrid level-set volume-of-fluid finite volume method for zero mach number two-phase flow. PhD thesis, Institut für Mathematik, Freie Universität Berlin, Berlin (2017)
39. Waidmann, M., Gerber, S., Oevermann, M., Klein, R.: A conservative coupling of level-set, volume-of-fluid and other conserved quantities. In: Fuhrmann, J., Ohlberger, M., Rohde, C. (eds.) Finite Volumes for Complex Applications VII - Methods and Theoretical Aspects. Springer Proceedings in Mathematics and Statistics, vol. 77, pp. 457–465. Springer, Berlin (2014)
40. Wang, Y., Simakhina, S., Sussman, M.: A hybrid levelset-volume constraint method for incompressible two-phase flow. J. Comput. Phys. **231**, 6438–6471 (2012)
41. Zhou, Y.C., Liu, J., Harry, D.L.: A matched interface and boundary method for solving multi-flow Navier-Stokes equations with applications to geodynamics. J. Comput. Phys. **231**, 223–242 (2012)

Chapter 5
Time Discretization for Capillary Flow: Beyond Backward Euler

Stephan Weller and Eberhard Bänsch

Abstract Development and analysis of numerical methods for two-phase flow has mainly focussed on spatial aspects of the discretization so far. In turn, most of the methods available are of first order in *time* at most and only conditionally stable. For many applications, however, these shortcomings may constitute the computational bottleneck, since both disadvantages dictate very small time step sizes in order to arrive at a decent approximation of the underlying problem. In this article we therefore focus on the time discretization of capillary free surface flows with the following features: higher order in time convergence, unconditional stability and low dissipativity. Applying the method of lines we use stiff integrators from the class of Rosenbrock and W-methods. As an alternative, a space-time Galerkin method is presented. These methods are compared computationally for examples of one- and two-phase capillary flows.

5.1 Introduction

Development and analysis of numerical methods for two-phase flow has mainly focused on spatial aspects of the discretization so far. In turn, most of the methods available are of first order in *time* at most and only conditionally stable in contrast to the case of one-phase flow, see [22, 23, 27, 30] for only a few references. For many applications, however, these shortcomings may constitute the computational bottleneck, since both disadvantages dictate very small time step sizes in order to arrive at a decent approximation of the underlying problem.

S. Weller
CD-adapco, Nürnberg Office, Nürnberg, Germany
e-mail: stephan.weller@cd-adapco.com

E. Bänsch (✉)
Department Mathematics, Universität Erlangen-Nürnberg, Cauerstr. 11, 91058 Erlangen, Germany
e-mail: baensch@math.fau.de

© Springer International Publishing AG 2017

D. Bothe, A. Reusken (eds.), *Transport Processes at Fluidic Interfaces*,
Advances in Mathematical Fluid Mechanics, DOI 10.1007/978-3-319-56602-3_5

Often the core of the problem lies in the coupling between the geometry and the fluid flow. This coupling is usually resolved explicitly in engineering applications, i.e. the fluid flow is solved for a fixed domain, then the domain is advected by the flow for one time step, then the flow is solved again etc. Such a strategy is only conditionally stable, typically a Courant-Friedrichs-Lewy (CFL) condition of the kind

$$\Delta t \leq \sqrt{\text{We}}\, h^{3/2},$$

where $\text{We} = \frac{\rho V^2 D}{\gamma_0}$ is the Weber number (with ρ, γ_0 density and surface tension, V, D characteristic velocity and length scales, respectively) and Δt describes the resolution in time and h the resolution in space, is encountered [24, 38]. It is therefore crucial to find other time discretization methods that resolve the coupling between fluid flow and geometry which do not suffer from this CFL condition but are rather unconditionally stable.

In this article we therefore focus on the time discretization of capillary free surface flows with the following features:

- higher order in time convergence (i.e. $k \geq 2$),
- unconditional stability (which rules out explicit methods),
- low dissipativity.

Since the focus will be on time discretization, we chose an ALE sharp interface, front tracking method for space discretization. Usually this kind of method is conceptually simpler and more accessible to analysis than interface capturing methods (like VOF, phase field or level set). We will not discuss the respective pros and cons of either class of methods here, since this is out of scope of the presentation.

To be more precise, we work with a *mesh moving* method, where the mesh moves with time and is aligned to the interface [5, 18]. The computations are then carried out in arbitrary Lagrangian Eulerian (ALE) coordinates. This methods is probably the conceptually easiest possibility. The drawback is that dealing with topological changes is, to say the least, very difficult. However, the amount of interesting numerical scenarios *without* topological changes is huge, and often a very precise resolution of interface quantities is required. Such a precise representation is the strong point of interface tracking methods using ALE coordinates.

This chapter is organized as follows. In Sect. 5.2 we introduce the problem and give a weak formulation on a *reference* domain. In Sect. 5.3 this weak formulation is used to discretize in space by finite elements. Section 5.4 discusses several time discretization methods based on the *method of lines* (MOL). A space-time Galerkin discretization is introduced in Sect. 5.5. The different discretizations are then compared by computational experiments in Sect. 5.6. Finally, in Sect. 5.7 these results are discussed and conclusions are drawn. More details can be found in [6–8, 42–44].

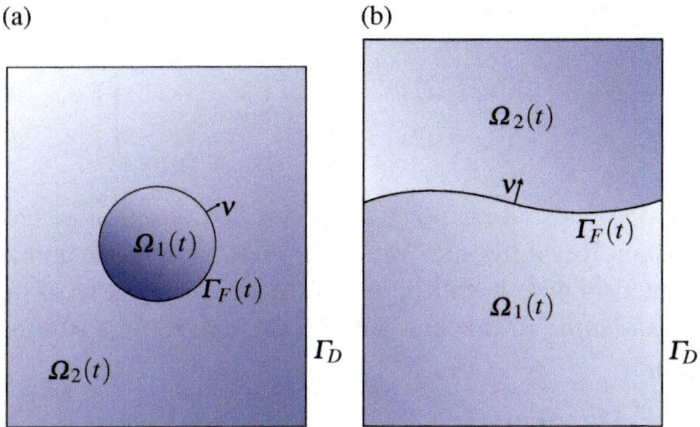

Fig. 5.1 Examples for two-phase capillary flow. (**a**) Rising drop. (**b**) Container with two fluids

5.2 Problem Setting

In what follows, we will consider one- or two-phase capillary flows. Examples will be given in Sect. 5.6. The setting for two-phase flows is typically a container with two immiscible fluids, see Fig. 5.1a, b.

The governing equations as well as the boundary conditions read in non-dimensional form: for a given sufficiently smooth right-hand side f, boundary values u_D, and initial data u_0, Ω_0, for all $t \in [0, T]$ find a domain $\Omega(t)$ with fixed Dirichlet boundary Γ_D, a velocity u, and pressure p, such that $u(0) = u_0$ and for all $t \in [0, T]$ fulfilling

$$\Lambda \frac{D}{Dt} u - \nabla \cdot \sigma(u, p) = \Lambda f \qquad \text{in } \Omega \qquad (5.1a)$$

$$\nabla \cdot u = 0 \qquad \text{in } \Omega \qquad (5.1b)$$

$$u = u_D \qquad \text{on } \Gamma_D \qquad (5.1c)$$

$$[\![\sigma(u, p)v]\!] = \frac{1}{\text{We}} \kappa v \qquad \text{on } \Gamma_F(t) \qquad (5.1d)$$

$$[\![u]\!] = 0 \qquad \text{on } \Gamma_F(t) \qquad (5.1e)$$

$$V_\Gamma(t) = u \cdot v \qquad \text{on } \Gamma_F(t) \qquad (5.1f)$$

$$\Omega(0) = \Omega_0, \quad u(0, \cdot) = u_0 \qquad (5.1g)$$

Here, $\sigma(u, p) = \frac{1}{\text{Re}} D[u] - pI$ denotes the stress tensor, $\frac{D}{Dt} := \partial_t u + u \cdot \nabla$ the material derivative, We the Weber number and κ the curvature of the interface. In the two-phase flow case, Re and Λ denote the phase-dependent Reynolds number and the nondimensional densities, respectively, defined by Λ, Re, where

$$\Lambda(x) := \begin{cases} \Lambda_1 := 1 & \text{if } x \in \Omega_1 \\ \Lambda_2 := \frac{\rho_1}{\rho_2} & \text{if } x \in \Omega_2 \end{cases}, \qquad \text{Re}(x) := \begin{cases} \text{Re}_1 & \text{if } x \in \Omega_1 \\ \text{Re}_2 & \text{if } x \in \Omega_2 \end{cases}$$

In the case of one-phase flow, no such piecewise definition is necessary, in this case assume constant Reynolds and Weber numbers Re and We and $\Lambda = 1$.

For the presentation here it will be useful to work with a weak formulation on a fixed *reference domain*. To this end we introduce the *ALE* (arbitrary Lagrangian Eulerian) mapping.

Definition 1 (ALE Mapping) Let $\hat{\Omega} \subset \mathbb{R}^d$ be an open domain that is fixed in time with a boundary part $\hat{\Gamma}_D \subset \partial\hat{\Omega}$ and a part $\hat{\Gamma}_F$.

Let $\Phi : [0,T] \times \hat{\Omega} \to \mathbb{R}^d)$ be a smooth mapping that is globally injective for fixed $t \in [0,T]$ and describes the movement of the domain, i.e. $\Phi(t, \hat{\Gamma}_D) = \Gamma_D$, $\Phi(t, \hat{\Gamma}_F) = \Gamma_F(t)$, $\Phi(t, \hat{\Omega}) = \Omega(t)$. Then, Φ is called an *ALE mapping*.

Denoting the time derivative in ALE coordinates by $\tilde{\partial}_t$, the Eulerian time derivative ∂_t is given by $\partial_t = \tilde{\partial}_t - w \cdot \nabla$, where $w = \partial_t\Phi$. Since from now on we are always working with the ALE derivative, the $\tilde{}$ will be dropped for the sake of better readability hereafter.

Transforming Eq. (5.1g) to the reference domain, multiplying by test functions and integrating by parts one arrives at the weak formulation on the reference domain $\hat{\Omega}$:

Problem 1 (Weak Formulation on the Reference Domain) Find a (smooth) ALE mapping and $\hat{u} \in \hat{u}_D + V_0 := \hat{u}_D + \{\hat{v} \in H^1(\hat{\Omega}) | \hat{v} = 0 \text{ on } \hat{\Gamma}_D, \underline{\nabla}v_{|\Gamma_F(t)} \in \mathscr{L}^2(\Gamma(t))^{d \times d}\}$, $\hat{p} \in \mathscr{L}^2(\hat{\Omega})$, such that the following equations hold for all $\hat{v} \in V_0, \hat{q} \in \mathscr{L}^2(\hat{\Omega})$ and for all $t \in [0,T]$:

$$\hat{m}(\Phi; \partial_t\hat{u}, \hat{v}) + \hat{k}(\Phi; \hat{u}, \hat{u}, \hat{v}) + \hat{s}(\Phi; \hat{u}, \hat{v}) + \hat{b}(\Phi; \hat{p}, \hat{v}) = \hat{f}(\Phi, \hat{v}) + e(\Phi; v) \tag{5.2a}$$

$$\hat{b}(\Phi; \hat{q}, \hat{u}) = 0 \tag{5.2b}$$

$$\hat{V}_{\Gamma_F} = \hat{u} \cdot \hat{v} \tag{5.2c}$$

$$\Phi(0, \hat{\Omega}) = \Omega_0, \quad \hat{u}(0, \cdot) = u_0 \circ \Phi(0, \cdot), \tag{5.2d}$$

where

$$\hat{m}(\Phi; \hat{u}, \hat{v}) := \int_{\hat{\Omega}} \hat{\Lambda}\hat{u} \cdot \hat{v} |\det D\Phi| \, dx \tag{5.3a}$$

$$\hat{k}(\Phi; \hat{u}, \hat{v}, \hat{w}) := \int_{\hat{\Omega}} \hat{\Lambda}\hat{w} \cdot (D\Phi^{-\top}\hat{\nabla}\hat{u})\hat{v} |\det D\Phi| \, dx \tag{5.3b}$$

$$\hat{s}(\varPhi;\hat{u},\hat{v}) := \int_{\hat{\Omega}} \frac{1}{2\mathrm{Re}} D\varPhi^{-\mathsf{T}}(\hat{\nabla}\hat{u} + \hat{\nabla}\hat{u}^{\mathsf{T}}) : D\varPhi^{-\mathsf{T}}(\hat{\nabla}v + \hat{\nabla}v^{\mathsf{T}})|\det D\varPhi|\,\mathrm{d}x \tag{5.3c}$$

$$\hat{b}(\varPhi;\hat{q},\hat{v}) := \int_{\hat{\Omega}} -\hat{q}D\varPhi^{-\mathsf{T}} : \hat{\nabla}\hat{v}|\det D\varPhi|\,\mathrm{d}x \tag{5.3d}$$

$$\hat{f}(t,\varPhi;\hat{v}) := \int_{\hat{\Omega}} \hat{\varLambda}(f(t) \circ \varPhi) \cdot \hat{v}|\det D\varPhi|\,\mathrm{d}x \tag{5.3e}$$

$$e(\varPhi;v) := \frac{1}{\mathrm{We}} \int_{\varPhi(t,\varGamma)} \underline{\nabla}\,\mathrm{id}_{\varPhi(t,\varGamma)} : \underline{\nabla}v \,\mathrm{d}S \tag{5.3f}$$

and $D\varPhi$ denotes the Jacobian of $\varPhi$.

Note 1 Note that in the above formulation the curvature term has been integrated by parts based on the identity

$$\kappa\nu = \underline{\Delta}\,\mathrm{id}\,.$$

This idea goes back to Dziuk [15] and has been applied to capillary flow for instance in [5].

5.3 Discretization in Space

In what follows we will make the following assumptions on the finite element approximation $\varPhi_h$ of $\varPhi$. Let $\hat{\Omega} \subset \mathbb{R}^n$ be a polygonally bounded reference domain. Let $\mathscr{T}$ be a conforming, regular triangulation of $\hat{\Omega}$ and let $S_h = S_h(\mathscr{T}) \subset C^0(\overline{\hat{\Omega}},\mathbb{R}^d) \cap H^1(\hat{\Omega},\mathbb{R}^d)$ be some finite element space on $\hat{\Omega}$.

Assumption 2

A) *For all $t \in [0,T]$ it holds $\varPhi_h(t) \in S_h$.*
B) *There is a $d-1$-dimensional conforming, regular sub-triangulation $\mathscr{S}$ of $\hat{\varGamma}_F \subset \hat{\Omega}$ such that the discrete free boundary fulfills $\varGamma_F(t) = \varPhi_h(t,\hat{\varGamma}_F)$ for all $t \in [0,T]$.*
C) *For any $t \in [0,T]$, $\varPhi_h$ is globally injective and $\inf_{\hat{x}\in\hat{\Omega}} \det D\varPhi_h(t,\hat{x}) > 0$.*

Sub-triangulation is meant in the following sense: we assume the existence of a finite element space $Z_h = Z_h(\mathscr{S}) \subset C^0(\hat{\varGamma}_F)$, such that for $\varLambda \in S_h$ we have

$$\varLambda_{|\hat{\varGamma}_F} = \lambda \in Z_h \tag{5.4}$$

and on the other hand there exists a continuous extension operator

$$E : Z_h \to S_h, \tag{5.5}$$

$$(E\lambda)_{|\hat{\Gamma}_F} = \lambda. \tag{5.6}$$

Now, if $\Phi_h = E\Psi_h$ for $\Psi_h \in Z_h, \Phi_h \in S_h$, all terms depending only on the phase boundary $\Gamma_F(t)$ can be parametrized by Ψ_h instead of Φ_h and are independent of the interior of Φ_h. In particular, we have $\hat{V}_\Gamma = \partial_t \Psi_h$ and also we can write $e(\Psi_h; v) = e(\Phi_h; v)$.

We will use *isoparametric* finite elements, i.e. the space used to represent the velocity will also be S_h (except for boundary conditions). W_h is chosen such that V_h, W_h is an inf-sup stable pair of elements.

It is now possible to define a discrete version of Problem (5.2). We postpone the precise definition of the domain boundary velocity and state:

Problem 2 (Semi-discrete Formulation of the Navier-Stokes Equations) For all $t \in [0, T]$, find $\Phi_h(t) \in S_h, \hat{u}_h(t, \cdot) \in V_h, \hat{p}_h(t, \cdot) \in W_h$, such that Φ_h is advected by the boundary velocity and the following equations hold for all $\hat{v}_h \in V_{h,0}, \hat{q}_h \in W_h$:

$$\hat{m}(\Phi_h; \partial_t \hat{u}_h, \hat{v}_h) + \hat{k}(\Phi_h; \hat{u}_h - \partial_t \Phi_h, \hat{u}_h, \hat{v}_h)$$

$$+\hat{s}(\Phi_h; \hat{u}_h, \hat{v}_h) + \hat{b}(\Phi_h; \hat{p}_h, \hat{v}_h) = \hat{f}(\Phi_h, \hat{v}_h) + \hat{e}(\Phi_h, v_h) \tag{5.7a}$$

$$\hat{b}(\Phi_h; \hat{q}_h, \hat{u}_h) = 0 \tag{5.7b}$$

$$\hat{u}_h(0, \cdot) = u_0 \circ \phi(0, \cdot) \tag{5.7c}$$

Note that the mapping Φ is fixed at $t = 0$ and at Γ_D. It is also coupled with the velocity on $\Gamma_F(t)$. However, nothing is prescribed in the interior of $\Omega(t)$. In other words, any continuous extension operator $E : Z_h \to S_h$ can be chosen in practice to prescribe a condition for Φ in the interior of the domain. This can be done in many different ways. The most classical is *harmonic extension* (used e.g. in [5]), i.e. after time discretization, $-\Delta\Phi = 0$ is solved in the interior with given boundary conditions for Φ. Another possibility is to use approaches based on linear [17] or nonlinear elasticity [35, 41]. While the choice of a suitable strategy for this *grid smoothing* is important for the numerical realization, it is not very relevant to time discretization and thus out of scope for this work.

The *kinematic* boundary condition (5.1f) is imposed by the so-called *full-update* strategy:

$$\partial_t \Phi(t, \hat{x}) = \hat{u}(t, \hat{x}) \; \forall t \in [0, T], \; \hat{x} \in \Gamma_F. \tag{5.8}$$

Upon expanding the unknowns in a given basis

$$\hat{u}_h = \sum_{i=1}^{n_\varphi} U_i \hat{\varphi}_i \qquad\qquad \hat{p}_h = \sum_{i=1}^{n_\psi} P_i \hat{\psi}_i$$

$$\Phi_h = \sum_{i=1}^{n_\varphi} X_i \hat{\varphi}_i \qquad\qquad \Psi_h = \sum_{i=1}^{n_\chi} X_{I_\Gamma(i)} \hat{\chi}_i$$

with $\hat{\varphi}_i, i = 1, \ldots, n_\varphi$ a basis of S_h, $\hat{\psi}_i, i = 1, \ldots, n_\psi$ a basis of W_h, and $\hat{\chi}_i, i = 1, \ldots, n_\chi$ a basis of Z_h and where $I_\Gamma(i)$ enumerates the coefficients $i \in \{1, \ldots, n_\varphi\}$ for which $\hat{\varphi}_{i|\hat{\Gamma}_F} \neq 0$, i.e. the same coefficients are used to represent Φ_h and Ψ_h one gets a matrix-vector form of Eq. (2) called MOL-DAE (*method-of-lines differential-algebraic equation*):

Problem 3 (MOL-DAE) For given u_0, Ω_0, and right-hand side F, and for all $t \in [0, T]$, find $\Phi_h(t) = \sum_{i=1}^{n_\varphi} X_i \hat{\varphi}_i \in S_h, \hat{u}_h(t, \cdot) = \sum_{i=1}^{n_\varphi} U_i \hat{\varphi}_i \in V_h, \hat{p}_h(t, \cdot) = \sum_{i=1}^{n_\psi} P_i \hat{\psi}_i \in W_h$, such that the following equations hold:

$$M[X]\partial_t U + (S[X] + K[X, U - \partial_t X])U + B[X]P = F[t, X] + E[X]X \tag{5.9a}$$

$$B[X]^\top U = 0 \tag{5.9b}$$

$$R\partial_t X = RU \tag{5.9c}$$

$$\Phi(0, \hat{\Omega}) = \Omega_0, \quad \hat{u}(0, \cdot) = u_0, \tag{5.9d}$$

where

$$M[X]_{ij} = \hat{m}(\Phi_h; \hat{\varphi}_j, \hat{\varphi}_i) \qquad\qquad i, j = 1, \ldots, n_\varphi$$

$$S[X]_{ij} = \hat{s}(\Phi_h; \hat{\varphi}_j, \hat{\varphi}_i) \qquad\qquad i, j = 1, \ldots, n_\varphi$$

$$K[X, U]_{ij} = \hat{k}\left(\Phi_h; \sum_{k=1}^{n_\varphi} U_k \hat{\varphi}_k, \hat{\varphi}_j, \hat{\varphi}_i\right) \qquad\qquad i, j = 1, \ldots, n_\varphi$$

$$B[X]_{ij} = \hat{b}(\Phi_h; \hat{\psi}_j, \hat{\varphi}_i) \qquad\qquad i = 1, \ldots, n_\varphi, \; j = 1, \ldots, n_\psi$$

$$F[t, X]_i = \hat{f}(t, \Phi_h; \hat{\varphi}_i) \qquad\qquad i = 1, \ldots, n_\varphi$$

$$E[X]_{ij} = e(\Phi_h; \chi_j, \chi_i) \qquad\qquad \text{if } i \in I_\Gamma, j = 1, \ldots, n_\chi$$

$$E[X]_{ij} = 0 \qquad\qquad \text{if } i \notin I_\Gamma$$

$$R_{ij} = 1 \qquad\qquad \text{if } i \in I_\Gamma, j = I_\Gamma(i)$$

$$R_{ij} = 0 \qquad\qquad \text{else}$$

and the vectors U, P, X comprise the coefficients for $u_h(t), p_h(t), \Phi_h(t)$, respectively:

$$U = (U_i)_i, \quad P = (P_i)_i, \quad X = (X_i)_i,$$

5.4 Time Discretization Based on the Method of Lines

Having discretized the continuous system in space, one ends up with a large system of ODEs that need to be further discretized in time. The arising systems of equations always fall into the class of *stiff* ordinary differential equations, so that for stability reasons (at least in part) implicit methods are mandatory. An exhaustive treatment of the stability of stiff ODEs can be found e.g. in [14, 20].

In this section we introduce several examples of time integrators, both from the class of one-step as well as multistep-method.

In order to fix notation, the basic time integrators will be defined for the following system of first order ODEs

$$\partial_t y = f(t, y), \qquad y(0) = y_0$$

with an arbitrary operator $f : [0, T] \times \mathbb{R}^k \to \mathbb{R}^k, k \in \mathbb{N}$.

A method that computes the solution y^{n+1} at time t_{n+1} from the solution y^n at the previous time step t_n with time step size $\tau = t_{n+1} - t_n$ by a rational function $R = P/R$ of the operator f, i.e.

$$Q(\tau f)y^{n+1} = P(\tau f)y^n.$$

with P, Q polynomials (that are prime to each other), is called a one-step method.

In contrast, a method that computes the solution y^{n+1} at time t_{n+1} from the solutions $y^n, \ldots, y^{n-s}$ at the previous time steps t_j by the following equations

$$\alpha_s y^n + \cdots + \alpha_0 y^{n-s} = \tau \left(\beta_s f(t^n, y^n) + \cdots + \beta_0 f(t^{n-s}, y^{n-s}) \right)$$

where $\alpha_0, \ldots, \alpha_s \in \mathbb{R}, \beta_0, \ldots, \beta_s \in \mathbb{R}, |\alpha_0| + |\beta_0| > 0, \alpha_s \neq 0$, is called an s-stage linear multistep method. For $\beta_s = 0$ the method is called *explicit*, else *implicit*.

5.4.1 Implicit Euler Method

We start our discussion with the most simple approach, the implicit Euler method:

Definition 3 (Implicit Euler Method) The linear one-step method with $R(\tau f) = \frac{\mathrm{id}}{\mathrm{id} - \tau f}$, i.e.

$$y^{n+1} = y^n + \tau f(t^{n+1}, y^{n+1})$$

is called the *implicit Euler method*.

The implicit Euler method is the simplest implicit one-step method, yet has the desirable property of L-stability. The drawbacks are high dissipativity and first order consistency only.

The properties of the implicit Euler method make it suitable for situations where stability is paramount and smoothing is required. One such case is that of using a time-discretization method to find a stationary or quasi-stationary solution. For example, for the solution of quasi-stationary *Taylor flow* (see Sect. 5.6) the implicit Euler method is well-suited.

Equations (5.7a)–(5.7c) of Problem 2 form the starting point for the fully discrete problem. First, divide the time interval $[0, T]$ in N subintervals $[t_i, t_{i+1}]$, $i = 0, \ldots, N - 1$, $t_j = \tau j$ with size $\tau = \frac{T}{N}$ (for the sake of simplicity variable time step size is not considered, although the generalization is straightforward).

Note that since Eq. (5.7a) is not an explicit ODE (because of the term $\partial_t \Phi_h$ in $\hat{k}$), the implicit Euler method is strictly speaking not applicable. However, the obvious idea to apply the implicit Euler method to the problem reads in matrix-vector notation:

For all $i = 1, \ldots, N$ find $\Phi_h^i = \sum_{j=1}^{n_\varphi} X_j^i \hat{\varphi}_j \in S_h$, $\hat{u}_h^i = \sum_{j=1}^{n_\varphi} U_j^i \hat{\varphi}_j \in V_h$, $\hat{p}_h^i = \sum_{j=1}^{n_\psi} P_j^i \hat{\psi}_i \in W_h$, such that the following equations hold:

$$M[X^i](\frac{U^i - U^{i-1}}{\tau}) + (S[X^i] + K[X^i, U^i - \frac{X^i - X^{i-1}}{\tau}])U^i$$

$$+ B[X^i]P^i = F[t^i, X^i] + E[X^i]X^i \tag{5.10a}$$

$$B[X^i]^\top U^i = 0 \tag{5.10b}$$

$$R\frac{X^i - X^{i-1}}{\tau} = RU^i \tag{5.10c}$$

$$\Phi^0(\hat{\Omega}) = \Omega_0, \quad \hat{u}_h^0 = u_0. \tag{5.10d}$$

Note 2 While the application of the implicit Euler method to the problem is relatively straightforward, its implementation is not. The dependence of all occurring terms, most importantly the curvature term e, on the discretization of the geometry X, is highly nonlinear.

5.4.2 Backward Differentiation Formulas

The most important drawback of the implicit Euler method is the low consistency order. There are two possibilities to extend the idea to a method of higher order: multistep methods and Runge-Kutta methods. We will first discuss multi-step methods. In particular we will concentrate on the class of backward differentiation formulas (BDF), especially of second order (BDF2).

Definition 4 (Backward Differentiation Formula) A backward differentiation formula (BDF) of order s is an s-stage linear multistep method with $\beta_s = 1, \beta_k = 0, k = 0, \ldots, s - 1$. The coefficients α_i are chosen by computing an interpolation polynomial of order s through the solution at the last s time steps such that the differential equation is fulfilled at t^n.

BDF2 The backward differentiation formula of order 2 (BDF2) is a 2-stage BDF method with $\alpha_2 = \frac{3}{2}, \alpha_1 = -2, \alpha_0 = \frac{1}{2}$, i.e. the iteration rule is the following:

$$\frac{3}{2}y^n - 2y^{n-1} + \frac{1}{2}y^{n-2} = \tau f(t^n, y^n). \tag{5.11}$$

BDF2 is a relatively simple computational extension to the implicit Euler method. The resulting system to solve is very similar to the implicit Euler one and thus the matrix-vector presentation is omitted here for brevity. Moreover, if an efficient solver for the implicit Euler time discretization is available, it can also be used for the BDF2-based discretization.

BDF2 is L-stable and of consistency order 2. Thus, among backward differentiation formulas, BDF2 is in a sense optimal, since a linear multistep method of order $p > 2$ cannot be A-stable, [20].

5.4.3 Linearly Implicit Approaches

A second possibility to derive a higher order method are one-step methods. Here, only information from time step t_{n-1} is used to compute the solution at time t_n. To this end, the operator is evaluated at several points in time and several intermediate solutions (*stages*) are used. If no other approximations are used, such methods are *Runge-Kutta* methods.

A special class of implicit Runge-Kutta methods, called DIRK (diagonally implicit Runge-Kutta) are those, where the stages form a lower triangular system and therefore only the the diagonal part has to be inverted.

Runge-Kutta methods could in principle be applied to capillary problems, however for second order methods no substantial advantage would be gained over BDF2, and for methods of order higher than 2 the costs of the solution of the implicit systems would be quite high. This is why we immediately skip to a cheaper variant

of such methods that require only the solution of *linear* systems even for nonlinear
ODEs.

5.4.3.1   Rosenbrock Methods

The basic idea of Rosenbrock methods is the following: Runge-Kutta methods lead
to nonlinear, implicit systems of equations (if the underlying ODE is nonlinear).
These systems could be solved by Newton's method. Now an alternative is to just
apply *one step* of Newton's method (with starting value 0) to the equation and then
immediately go to the next time step. Obviously for linear ODEs this is equivalent
to the original Runge-Kutta method. It turns out that for nonlinear problems stability
and convergence properties are almost as good as for the original Runge-Kutta
method. For an exhaustive treatment of Rosenbrock methods see [29] and [14].

Now a plethora of methods is available. The 3-stage, third order method ROS3P
by Lang and Verwer [28] is selected here, since it was designed as a solver for
parabolic problems and has been also applied successfully to the Navier-Stokes
equations [27].

Note 3 (Order Reduction)   When applying Rosenbrock (or Runge-Kutta-) methods
to parabolic partial differential equations discretized by the method of lines it is
important to use methods that are free of *order reduction*, see [16, 36, 40]. In the
following we use the so-called ROS3P and ROS34PW2 methods that are free of
order reduction, see [28, 33].

Let us mention that ROS3P is of third consistency order, A-stable, however not
L-stable, see [28].

The drawback of Rosenbrock methods is the need of the Jacobian matrix. In the
case of free surface flow, the Jacobian with respect to U can be readily computed,
whereas the Jacobian with respect to X is tedious to derive and very expensive to
assemble. The idea to use an approximation to the Jacobian matrix instead suggests
itself. We used two variants, called "semi-implicit" (where all terms regarding
linearization w.r.t the geometry are set to zero) and "inexact Jacobian" (where at
least the exact linearization of the curvature term w.r.t. the geometry is used).

5.4.3.2   W-Methods

W-methods are a straightforward extension of Rosenbrock methods in the sense that
an approximation W (in principle, any) is used instead of the true Jacobian.

It is out of scope of this chapter to go into details. We just mention that
indeed there exists a 4-stage W-method which is of third order *independent* of W,
namely ROS34PW2 [33]. If a consistent approximation to the Jacobian is used, i.e.
$W \overset{\tau \to 0}{\to} Df(t, y)$, then ROS34PW2 is also L-stable. ROS34PW2 was designed with

partial differential algebraic equations (PDAEs) in mind, thus it is well suited for our purposes.

Note that the implementation of such a method is much more elaborate than e.g. an implicit Euler time discretization. In one respect it is much simpler, however: no iterative method for any nonlinearity is necessary, as only linear equations are solved.

5.5 Discontinuous Galerkin Methods (dG)

As an alternative to the previous methods in the following we will consider (discontinuous) Galerkin methods in time. For this approach, we consider the prototype ODE $\partial_t y = f(t, y); y(0) = y_0$ in integral form instead:

$$y(t) = y_0 + \int_0^t \partial_t y(s) \, \mathrm{d}s = y_0 + \int_0^t f(s, y(s)) \, \mathrm{d}s. \tag{5.12}$$

$y(t)$ will now be replaced by an approximation $y_h(t)$ which is piecewise polynomial on each subinterval and globally discontinuous. This reduces the problem to a finite dimensional problem in the coefficients of the piecewise polynomial functions [37, 39]. More precisely, the discontinuous Galerkin method of order k is defined by:

Definition 5 (dG(k)) For each time interval $I_n := [t_{n-1}, t_n], n = 1, \ldots, N$, find $y_{h|I_n} \in \mathscr{P}_k(I_n)$ such that the following equation holds for all $v \in \mathscr{P}_k(I_n)$:

$$\int_{I_n} \partial_t y_h(t) v(t) \, \mathrm{d}t + [y_h](t_{n-1}) v(t_{n-1}) = \int_{I_n} f(t, y_h(t)) v(t) \, \mathrm{d}t, \tag{5.13}$$

where $[y_h](t_n)$ denotes the jump of y_h at t_n and $\lim_{t \nearrow 0} y_h(t) = y_0$ to account for the initial values.

Surprisingly, as is shown in [37], for a linear ODE, the approach can be classified as a Runge-Kutta method:

Theorem 1 *For a linear ODE dG(k) is equivalent to an implicit $k + 1$-stage Runge-Kutta method known as Radau IIA method. Moreover, dG(k) is L-stable and has consistency order $2k + 1$ (in nodal points).*

dG(k) methods have been successfully applied to the setting of parabolic PDEs (see e.g. [3, 37]). For $k \geq 1$, the efficient solution of the resulting dG(k) system of equations is subject to current research. In [9, 11], preconditioners are constructed for the case of parabolic PDEs (unfortunately this construction does not directly carry over to the case of free surface flows described in the next section).

5.5.1 Application to Free Surface Flows

The first step in deriving a full discretization based on dG in time is a weak formulation in space-time.

We will begin with a fundamental observation: from Problem 1 it follows that $\hat{V}_{\Gamma_F} = \hat{u} \cdot v$ holds at the boundary. For a Galerkin-type time discretization with polynomials of degree k for $\hat{u}$ this means that V_{Γ_F} is also polynomial in time of degree k. However, V_{Γ_F} is the *time derivative* of the ALE mapping Φ, implying that Φ has to be a polynomial of order $k + 1$ at the boundary. As the time-derivative of a continuous piecewise polynomial of order $k + 1$ is a discontinuous piecewise polynomial of order k, the discretization of $\hat{u}$ is based on dG(k) while the discretization of Φ is based on cGP($k + 1$). This will lead to $k + 1$ systems of equations of the size given by e.g. an implicit Euler discretization, where the number of equations and number of unknowns are equal.

For the space discretization, consider the finite element spaces S_h, W_h, Z_h as in Sect. 5.3. For the time discretization, for any $q \geq 0$ define the continuous and discontinuous piecewise polynomial functions on the time interval $[0, T]$:

$$P_q^{dc}([0, T]) = \{\Phi : [0, T] \to \mathbb{R} \mid \Phi_{|(t_n, t_{n+1}]} \in P^q((t_n, t_{n+1}]) \ \forall n = 0, \ldots, N\},$$

$$\tag{5.14}$$

$$P_q^c([0, T]) = C^0([0, T]) \cap P_q^{dc}([0, T]). \tag{5.15}$$

We are now looking for a $\Phi_h \in P_{k+1}^c([0, T]) \times S_h$ such that

$$\Phi_h(t, \hat{\Gamma}_D) = \Gamma_D, \quad \Omega(t) := \Phi_h(t, \hat{\Omega}), \quad \Gamma_F(t) := \Phi_h(t, \hat{\Gamma}_F), \ \forall t \in [0, T].$$

Define the space-time domain $G := \{0 < t < T, x \in \Phi(t, \hat{\Omega})\}$ and as in Sect. 5.3 we will use the hat symbol $(\hat{\ })$ for functions on the reference domain, i.e. for a function $v : G \to \mathbb{R}$ (or $v : G \to \mathbb{R}^d$), let $\hat{v}(t, \hat{x}) := v(t, \Phi_h(t, \hat{x}))$.

Again, we will use the space V_h for the space discretization of our velocity field, and W_h for the pressure. On the space-time domain G, the corresponding discrete spaces are the following:

$$\mathscr{V}_k := \{\hat{v} : [0, T] \times \hat{\Omega} \to \mathbb{R}^d : \hat{v} \in P_k^{dc} \times V_h\},$$

$$\mathscr{W}_k := \{\hat{q} : [0, T] \times \hat{\Omega} \to \mathbb{R}^d : \hat{q} \in P_k^{dc} \times W_h\},$$

$$\mathscr{Z}_k := \{\hat{z} : \Phi(\cdot, \hat{\Gamma}) \to \mathbb{R}^d : \hat{z} \in P_k^{dc} \times Z_h\}.$$

In what follows we write all terms with respect to a fixed triangulation $\mathscr{T}$ of the reference domain $\hat{\Omega}$. Again, we will assume that $\mathscr{S}$ is a triangulation of $\hat{\Gamma}_F$. The setting is very similar to Sect. 5.3. In addition to the bi-/trilinear forms defined in

Sects. 5.2 and 5.3, we will need the following definition:

$$\hat{m}_{\hat{F}}(\hat{u}, \hat{v})p := \int_{\hat{\Gamma}_F} \hat{u} \cdot \hat{v}. \tag{5.16}$$

On the reference domain $\hat{\Omega}$, the system to be solved then reads:

Problem 4 (Galerkin Space-Time Discretization) Find $\hat{u} \in \mathcal{V}_k, \hat{p} \in \mathcal{W}_k, \Phi_h \in P^c_{k+1}([0, T]) \times S_h$ such that for given initial data u_0, Ω_0 and right-hand side $f \in \mathcal{L}^2(G)$ the following equations hold for all $(\hat{v}, \hat{q}, \hat{w}) \in \mathcal{V}_k \times \mathcal{W}_k \times \mathcal{Z}_k$ and all n:

$$\hat{m}(\Phi_h(t_n), [\hat{u}](t_n), \hat{v}(t_{n-1})) + \int_{t_n}^{t_{n+1}} \hat{m}(\Phi_h, \partial_t \hat{u}^{n+1}, \hat{v})$$

$$+ \int_{t_n}^{t_{n+1}} \left(\hat{k}(\Phi_h; \hat{u} - \partial_t \Phi_h, \hat{u}, \hat{v}) + \hat{s}(\Phi_h; \hat{u}, \hat{v}) - \hat{b}(\Phi_h; \hat{p}, \hat{v}) + \hat{e}(\Phi_h; \Phi_h, \hat{v}) \right)$$

$$= \int_{t_n}^{t_{n+1}} \hat{f}(\Phi_h; \hat{v})$$

$$\int_{t_n}^{t_{n+1}} \hat{b}(\Phi_h; \hat{q}, \hat{u}) = 0$$

$$\int_{t_n}^{t_{n+1}} \hat{m}_{\hat{F}}(\partial_t \chi_h, \hat{w}) = \int_{t_n}^{t_{n+1}} \hat{m}_{\hat{F}}(\hat{u}, \hat{w}),$$

where $\chi_h = \mathrm{tr}_{\hat{F}}(\Phi_h)$.

As before, the above equation has to be complemented by a suitable extension operator for the interior.

DG methods in our context of free surface flow enjoy the desirable feature of a rigorous a priori estimate in terms of the natural energy norm of the problem. This implies unconditional stability of the method.

Theorem 2 *Let $\hat{u} \in \mathcal{V}_k, \hat{p} \in \mathcal{W}_k, \Phi_h \in P^c_{k+1}([0, T]) \times S_h$ solve Problem 4. Then the following estimate holds:*

$$\sup_{0 \le k \le N} \left(\frac{1}{2} \|\sqrt{\Lambda} u^k\|^2_{\Omega(t_k)} + \frac{1}{\mathrm{We}} |\Gamma(t_k)| \right) + \frac{1}{2} \sum_{n=0}^{N-1} \|\sqrt{\Lambda}(u^{n+0} - u^n)\|^2_{\Omega(t_n)} + \|\frac{1}{2\sqrt{\mathrm{Re}}} D[u]\|^2_G$$

$$\le C \|\sqrt{\mathrm{Re}\Lambda} f\|^2_G + \frac{1}{\mathrm{We}} |\Gamma_0| + \frac{1}{2} \|\sqrt{\Lambda} u^0\|^2_{\Omega_0}$$

Proof: see [44].

5.6   Experimental Investigation of Numerical Convergence and Dissipativity

In this section, computational results from three examples are presented and discussed. All examples were computed in 2D (the Taylor flow example in 2.5D, meaning two dimensional, rotationally symmetric) with an extended version of the code NAVIER [4] using the Taylor–Hood element, i.e. piecewise quadratics for velocity and piecewise lineares for pressure.

The first example is that of an **oscillating one-phase droplet**. The initial state is a droplet that is elongated around one axis and thus out of equilibrium. This results in a damped (because of the viscosity) oscillation until the drop eventually reaches steady state, which is a circle. This example exhibits almost all difficulties of capillary flow: the nonlinear coupling between fluid flow and the surface forces is distinctive, and the flow field is non-trivial.

Since no exact analytical solution is known, convergence rates are computed with respect to a numerical solution with very small time step size instead.

The physical parameters ($Re = 3 \times 10^2$, $We = 1$) are chosen according to [31] such that they correspond to the simulation of a 1 mm droplet of water.

The second example is the two-phase flow application probably most used for benchmarks, namely a **rising droplet**, see e.g. [25]. In this case several derived quantities can be used to measure deviation from a reference solution, e.g. the position of the barycenter of the drop.

The last example consists of an elongated bubble in a narrow channel. Such bubbles are called **Taylor bubbles** after having been first described by Taylor [13] and are of importance in many applications such as catalytic converters [45], multiphase monolith reactors [26, 34] and microfluidic channels [19, 32]. Given that the capillary is round, the flow field and bubble geometry will also be rotationally symmetric for the parameters used here and the simulations can be carried out in 2D. For the sketch of a rotationally symmetric computational domain of a single Taylor bubble, see Fig. 5.2.

Assuming a co-moving frame of reference, the flow can be considered in a steady state. Obviously here the order of the time discretization scheme is not of importance but it is paramount that it reaches the steady state quickly and is stable enough to allow for very big time step sizes. Thus, the experimental order of convergence was *not* determined for this example, but it is used instead to demonstrate how well the numerical code can reproduce a steady state known from experiments.

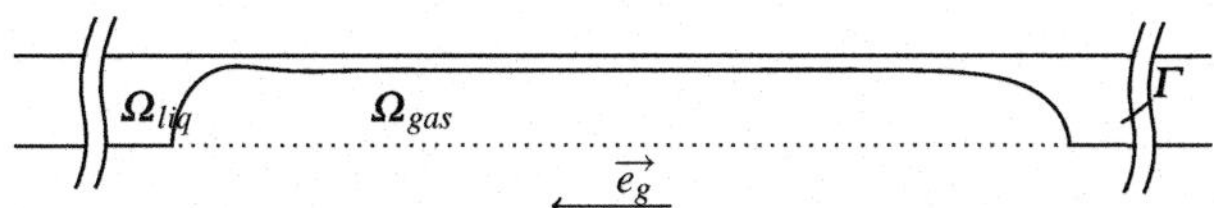

Fig. 5.2 Sketch of the computational domain for a rotationally symmetric Taylor bubble

Note 4 A verification of the *shape* of the Taylor bubble in the form of a comparison to other numerical codes and experimental data has been done using the same code as in this work [1, 2].

5.6.1 Error Measures

In order to investigate the experimental order of convergence (EOC), various error measure in space are used. These are combined with the discrete (in time nodal points) l_2 or l_∞ norm in time.

Numerical dissipativity is not that straightforward to define. As the Navier-Stokes equations contain a diffusive part, it is not enough to measure some energy balance, but rather the dependence of this balance on the time step size.

The example of an oscillating droplet was used to measure numerical dissipation in the following way: if the position $x(t)$ of the upper tip of the droplet is considered vs. time, it describes (approximately) a general damped oscillation, see Fig. 5.3 for an illustration. The dampened oscillation can be approximately described by the following ordinary differential equation:

$$\partial_t^2 x + \mu \partial_t x + \lambda x = 0 \tag{5.17}$$

with the solution

$$x(t) = \exp(-\mu t)(\lambda_1 \cos(t) + \lambda_2 \sin(t)). \tag{5.18}$$

The parameter μ then describes the dampening of the oscillation. For a simulated value $x_\tau(t)$, at time step size τ, the corresponding value μ_τ can be found by a simple

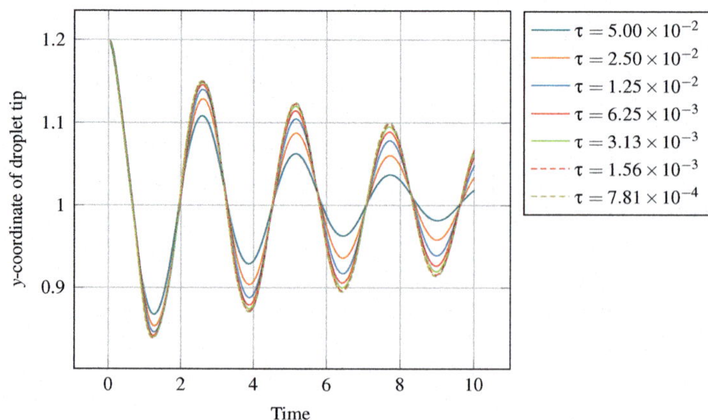

Fig. 5.3 *y*-Coordinate of oscillating bubble over time, using an implicit Euler time discretization

parameter fit. *Numerical dissipation* is now defined as $|\mu_\tau - \mu_\infty|$, where μ_τ is the dampening obtained with a certain method for time step size τ and μ_∞ the dampening for a very tiny time step size serving as a reference solution. Note that for a perfectly non-dissipative method we would expect $\mu_\tau = \mu_\infty$.

In addition to the time discretization methods described in the previous sections, the θ-splitting with a semi-implicit curvature treatment from [5] has also been included for comparison. In total eight different time discretization schemes were used in this numerical experiment:

1. The fully implicit Euler method ("Euler implicit")
2. The fully implicit BDF2 method ("BDF2 implicit")
3. The θ-splitting with semi-implicit curvature treatment as in [5].
4. ROS3P using the semi-implicit approximation to the Jacobian matrix
5. ROS3P using the inexact Jacobian matrix
6. ROS34PW2 using the semi-implicit approximation to the Jacobian matrix
7. ROS34PW2 using the inexact Jacobian matrix
8. The dG(1) based time discretization.

Note that by construction of W-methods, the two variants of ROS34PW2 should not differ greatly, as the approximation order is independent of the matrix used as a Jacobian approximation. Indeed the numerical dissipativity results are almost completely equal.

5.6.2 Numerical Dissipativity Results

The results for the numerical dissipativity for all considered methods are summarized in a single plot, Fig. 5.4.

5.6.3 Experimental Convergence Results

We first consider the one-phase oscillating drop. The numerical convergence rates are shown in Fig. 5.5. Again, the Euler method and the semi-implicit θ-splitting show first-order convergence. Unfortunately, the same convergence rates (with slightly better error constants, though) are achieved by both ROS3P variants. The other W-methods (ROS34PW2) perform somewhat better but still fall short of second order convergence. The only methods where second order convergence is clearly visible are the fully implicit BDF2 method and the space-time approach. The latter approach does not only have a very low error constant but its convergence rate is also even a little bit higher than 2.

It is noteworthy that for large time step sizes the ROS34PW2 methods in both variants show quite low error, i.e. while their convergence rate falls short of the

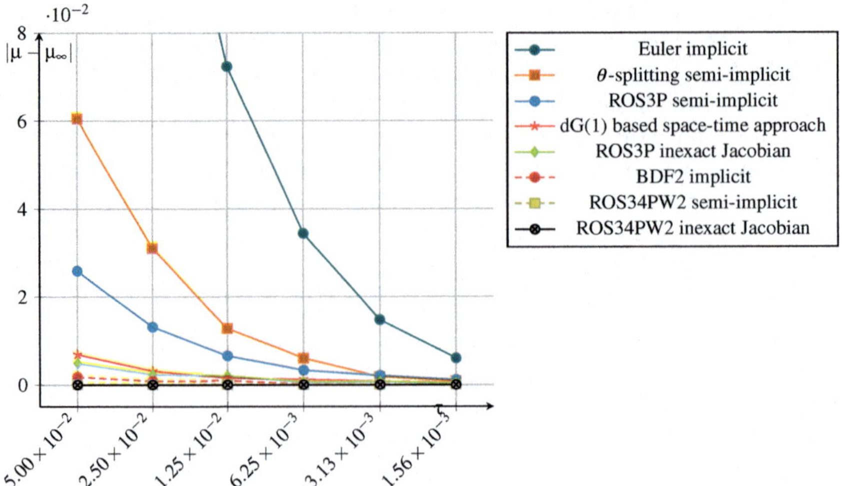

Fig. 5.4 Numerical dissipativity of all methods for different time step sizes. All results were computed using the oscillating drop scenario

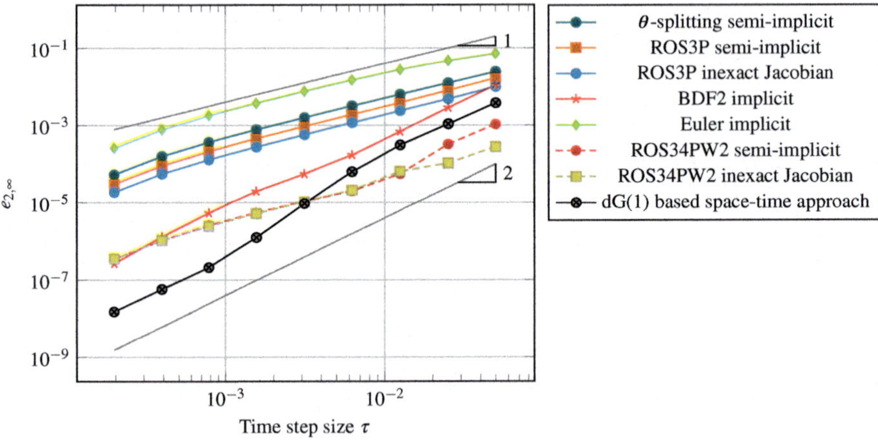

Fig. 5.5 Numerical convergence order for an oscillating one-phase drop

expectations, the error constants are very low and such methods may be competitive in scenarios with medium precision requirements.

The next convergence test is performed for the rising drop. For this numerical example, a variety of error measure exist. We chose the position of the barycenter with respect to time, since it gave a good discrimination among the different methods. The following parameters were used: $\Lambda_1 = 0.1818, \Lambda_2 = 1.818, \text{We} = 0.0275$.

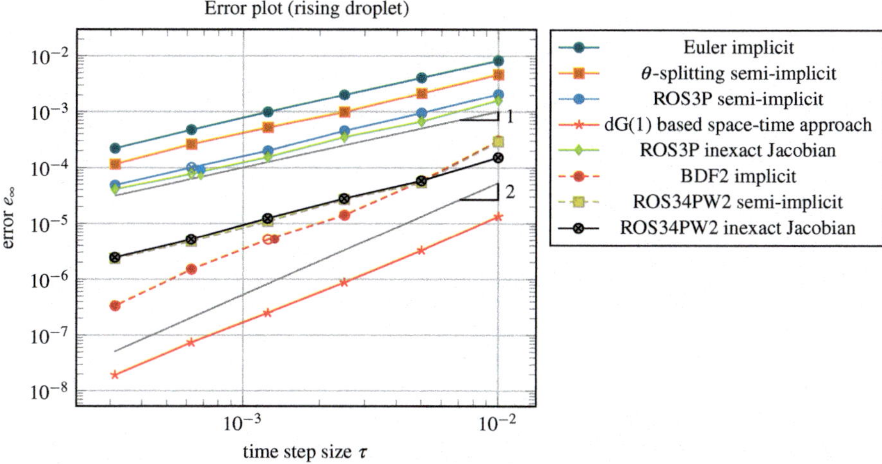

Fig. 5.6 Error in barycentric coordinate of different time discretization methods for the rising drop simulations

The results confirm the findings from the one-phase oscillating drop: implicit Euler, θ-splitting and the ROS3P variants are of first order, ROS34PW2 somewhere between first and second order, and BDF2 as well as dG(1) of second order (Fig. 5.6).

5.6.3.1 Results for Taylor flow

The Taylor flow simulation was carried out for four different capillary numbers (by varying the Weber number We at fixed Reynolds number Re $= 1 \times 10^{-1}$), Ca $\in \{2 \times 10^{-1}, 2 \times 10^{-2}, 2 \times 10^{-3}, 2 \times 10^{-4}\}$. The film width was then determined as an average over the side of the bubble, excluding the top and bottom 25%. This film width was compared to the predictions by the Bretherton and Halpern/Gaver models, see [12] and [21]. Figure 5.7 shows the results. As can be seen in the figure, the numerical results fit very well to the two models.

Note that the film width is extremely small for the lowest capillary number. However, because of the aligned mesh used for the simulation, a relatively coarse mesh suffices to simulate this scenario.

5.7 Discussion and Conclusions

Stability Computationally all the applied methods turned out to be stable. However, the only scheme that could rigorously be proved to be stable, is the dG method.

Numerical Dissipativity When considering the comparison of all implemented methods in Fig. 5.4, the low dissipativity of linearly implicit methods is striking.

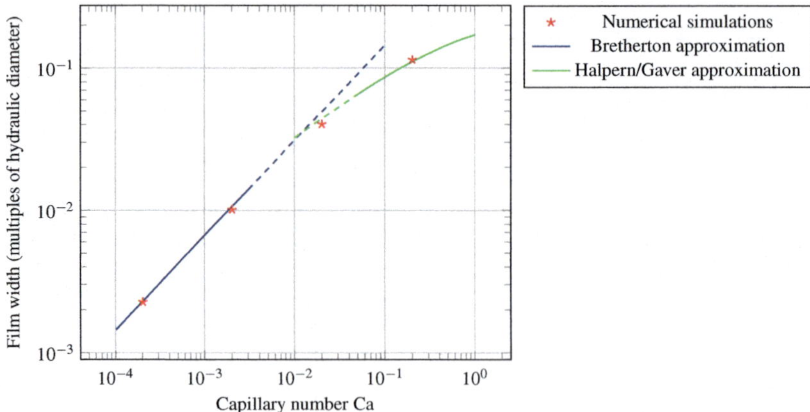

Fig. 5.7 Taylor bubble film width for different capillary numbers Ca

For ROS34PW2 with any of the two approximations, the dissipativity is almost zero. While both the space-time approach and the fully implicit BDF2 do offer low dissipativity as well, they are much more expensive in terms of numerical effort. Unsurprisingly, the Euler method exhibits the most numerical dissipativity.

Clearly, as dissipativity should tend to zero at the order of convergence, higher order methods are in principle less dissipative than the first order methods. For first order method the dissipativity is typically much more problematic.

Numerical Convergence For the numerical convergence, depending on the problem at hand, several of the methods considered might be best suited. Considering Fig. 5.5, which in point of view of the authors is best suited as an overview, the following can be said:

- The Euler method and semi-implicit θ-splitting are of first order as expected. The θ-splitting has a somewhat lower error constant and may therefore be preferred especially noting that it includes an efficient iterative solution strategy for the Navier-Stokes part of the equations.
- Both ROS3P variants clearly fall short of the expectations. Their convergence rate is only of first order and the error constant is only marginally smaller than for the θ-scheme. Apparently the approximation used for the Jacobian matrix is too crude, probably because these methods are no W-methods.
- For ROS34PW2, the situation is somewhat better. Both methods show a convergence rate that is somewhere between first and second order. The difference between both variants is only visible for large time step sizes so in practice the semi-implicit variant can be used as it is simpler to implement. All methods with better convergence rates are considerably more expensive in terms of numerical costs, and also the error constants for both ROS34PW2 variants are very low, so for moderate accuracy requirements, these methods are a viable alternative to lower-order methods.

- Finally, both the fully implicit BDF2 discretization and the dG(1) approach can be said to be of second order, the latter method exhibits an EOC that is even a little higher than 2. Unfortunately, both methods are relatively expensive to use. Because of the somewhat better convergence rate and the slightly lower error constant, the dG(1) method can be said to be superior, however this comes at the price of additional numerical costs and, not to be neglected, considerably higher implementational effort. While an existing e.g. implicit Euler method can be extended to an implicit BDF2 method with little implementational work, the implementation of a space-time method is much more intricate.

The present article summarizes results obtained within the funding period of SPP 1506. Because of lack of space, in the present chapter several topics could only be sketched or had to be omitted completely. For instance, the question of how to solve the arising non-linear systems is by far nontrivial. The development of efficient (nonlinear) solution strategies is a demanding task. In [44] an extensive discussion on different strategies can be found. However, this issue still needs further research. Moreover, computational cost may be a crucial argument when deciding for a particular time discretization. Thus we refer the reader to [1, 2, 6–8, 11, 42] for more details. See also [9, 10] for related work.

References

1. Aland, S., Boden, S., Hahn, A., Klingbeil, F., Weismann, M., Weller, S.: Quantitative comparison of Taylor flow simulations based on sharp-interface and diffuse-interface models. Int. J. Numer. Methods Fluids **73**(4), 344–361 (2013). doi:10.1002/fld.3802
2. Aland, S., Lehrenfeld, C., Marschall, H., Meyer, C., Weller, S.: Accuracy of two-phase flow simulations: the Taylor Flow benchmark. PAMM **13**(1), 595–598 (2013). doi:10.1002/pamm.201310278
3. Aziz, A., Monk, P.: Continuous finite elements in space and time for the heat equation. Math. Comput. **52**(186), 255–274 (1989). doi:10.1090/S0025-5718-1989-0983310-2
4. Bänsch, E.: Simulation of instationary, incompressible flows. Acta Math. Univ. Comenian. **67**(1), 101–114 (1998)
5. Bänsch, E.: Finite element discretization of the Navier-Stokes equations with a free capillary surface. Numer. Math. **88**, 203–235 (2001)
6. Bänsch, E., Weller, S.: Fully implicit time discretization for a free surface flow problem. PAMM **11**(1), 619–620 (2011). doi:10.1002/pamm.201110299
7. Bänsch, E., Weller, S.: A comparison of several time discretization methods for free surface flows. In: Handlovicova, A., Minarechova, Z., Sevcovic, D. (eds.) ALGORITMY 2012, 19th Conference on Scientific Computing, Vysoke Tatry – Podbanske, pp. 331–341 (2012). http://www.iam.fmph.uniba.sk/algoritmy2012/
8. Bänsch, E., Weller, S.: Linearly implicit time discretization for free surface problems. PAMM **12**(1), 525–526 (2012). doi:10.1002/pamm.201210251
9. Basting, S., Bänsch, E.: Preconditioners for the discontinuous Galerkin time-stepping method of arbitrary order. ESAIM: Math. Model. Numer. Anal. (2016). doi:10.1051/m2an/2016055
10. Basting, S., Weismann, M.: A hybrid level set-front tracking finite element approach for fluid-structure interaction and two-phase flow applications. J. Comput. Phys. **255**(0), 228–244 (2013). doi:http://dx.doi.org/10.1016/j.jcp.2013.08.018

11. Basting, S., Weller, S.: Efficient preconditioning of variational time discretization methods for parabolic partial differential equations. ESAIM: Math. Model. Numer. Anal. (2014). doi:10.1051/m2an/2014036
12. Bretherton, F.P.: The motion of long bubbles in tubes. J. Fluid Mech. **10**, 166 (1961)
13. Davies, R.M., Taylor, G.: The mechanics of large bubbles rising through extended liquids and through liquids in tubes. Proc. R. Soc. Lond. Ser. A Math. Phys. Sci. **200**(1062), 375–390 (1950)
14. Deuflhard, P., Bornemann, F.: Numerische Mathematik II. De Gruyter, Berlin (2002)
15. Dziuk, G.: An algorithm for evolutionary surfaces. Num. Math. **58**, 603–611 (1991)
16. Frank, R., Schneid, J., Ueberhuber, C.W.: Order results for implicit Runge-Kutta methods applied to stiff systems. SIAM J. Numer. Anal. **22**(3), 515–534 (1985). http://www.jstor.org/stable/2157080
17. Ganesan, S., Tobiska, L.: An accurate finite element scheme with moving meshes for computing 3D-axisymmetric interface flows. Int. J. Numer. Methods Fluids **57**(2), 119–138 (2008). doi:10.1002/fld.1624
18. Ganesan, S., Matthies, G., Tobiska, L.: On spurious velocities in incompressible flow problems with interfaces. Comput. Methods Appl. Mech. Eng. **196**(7), 1193–1202 (2007)
19. Günther, A., Jhunjhunwala, M., Thalmann, M., Schmidt, M.A., Jensen, K.F.: Micromixing of miscible liquids in segmented gas-liquid flow. Langmuir **21**(4), 1547–1555 (2005). doi:10.1021/la0482406
20. Hairer, E., Wanner, G.: Solving Ordinary Differential Equations II. Stiff and Differential-Algebraic Problems, 2nd revised edition. Springer, Berlin (1991)
21. Halpern, D., Gaver, P.: Boundary element analysis of the time-dependent motion of a semi-infinite bubble in a channel. J. Comput. Phys. **115**(366), 366–375 (1994)
22. Heywood, J.G., Rannacher, R.: Finite-element approximation of the nonstationary Navier-Stokes problem. IV. Error analysis for second-order time discretization. SIAM J. Numer. Anal. **27**(2), 353–384 (1990). doi:10.1137/0727022. http://dx.doi.org/10.1137/0727022
23. Hussain, S., Schieweck, F., Turek, S.: Higher order Galerkin time discretization for nonstationary incompressible flow. In: Numerical Mathematics and Advanced Applications 2011, pp. 509–517. Springer, Heidelberg (2013)
24. Hysing, S.: A new implicit surface tension implementation for interfacial flows. Int. J. Numer. Methods Fluids **51**(6), 659–672 (2006)
25. Hysing, S., Turek, S., Kuzmin, D., Parolini, N., Burman, E., Ganesan, S., Tobiska, L.: Quantitative benchmark computations of two-dimensional bubble dynamics. Int. J. Numer. Methods Fluids **60**(11), 1259–1288 (2009). doi:10.1002/fld.1934
26. Kreutzer, M.T., Kapteijn, F., Moulijn, J.A., Heiszwolf, J.J.: Multiphase monolith reactors: chemical reaction engineering of segmented flow in microchannels. Chem. Eng. Sci. **60**(22), 5895–5916 (2005). doi:10.1016/j.ces.2005.03.022
27. Lang, J., Teleaga, I.: Higher-order linearly implicit one-step methods for three-dimensional incompressible Navier-Stokes equations. Stud. Univ. "Babeş-Bolyai" Math. **LIII**(1), 109–121 (2008)
28. Lang, J., Verwer, J.: ROS3P – an accurate third-order Rosenbrock solver designed for parabolic problems. Bit Numer. Math. **41**(4), 731–738 (2001)
29. Lubich, C., Ostermann, A.: Linearly implicit time discretization of non-linear parabolic equations. IMA J. Numer. Anal. **15**(4), 555–583 (1995)
30. Lubich, C., Ostermann, A.: Runge-Kutta time discretization of reaction-diffusion and Navier-Stokes equations: nonsmooth-data error estimates and applications to long-time behaviour. Appl. Numer. Math. **22**(1–3), 279–292 (1996). doi:10.1016/S0168-9274(96)00038-4. http://dx.doi.org/10.1016/S0168-9274(96)00038-4. Special issue celebrating the centenary of Runge-Kutta methods
31. Matthies, G.: Finite Element Methods for Free Boundary Value Problems with Capillary Surfaces. Shaker, Aachen (2002)
32. Muradoglu, M., Günther, A., Stone, H.A.: A computational study of axial dispersion in segmented gas-liquid flow. Phys. Fluids **19**(7) (2007). doi:10.1063/1.2750295

33. Rang, J., Angermann, L.: New Rosenbrock W-methods of order 3 for partial differential algebraic equations of index 1. Bit Numer. Math. **45**, 761–787 (2005). doi:10.1007/s10543-005-0035-y
34. Roy, S., Bauer, T., Al-Dahhan, M., Lehner, P., Turek, T.: Monoliths as multiphase reactors: a review. AIChE J. **50**(11), 2918–2938 (2004). doi:10.1002/aic.10268
35. Rumpf, M.: A variational approach to optimal meshes. Numer. Math. **72**, 523–540 (1996)
36. Sanz-Serna, J., Verwer, J., Hundsdorfer, W.: Convergence and order reduction of Runge-Kutta schemes applied to evolutionary problems in partial differential equations. Numer. Math. **50**(4), 405–418 (1986). doi:10.1007/BF01396661
37. Schieweck, F., Matthies, G.: Higher order variational time discretizations for nonlinear systems of ordinary differential equations. Preprint 23/2011, Otto-von-Guericke-Universität Magdeburg (2011)
38. Sussman, M., Ohta, M.: A stable and efficient method for treating surface tension in incompressible two-phase flow. SIAM J. Sci. Comput. **31**(4), 2447–2471 (2009)
39. Thomée, V.: Galerkin Finite Element Methods for Parabolic Problems. Springer Lecture Notes in Mathematics, vol. 1054. Springer, Berlin (1984)
40. Verwer, J.: Convergence and order reduction of diagonally implicit Runge-Kutta schemes in the method of lines. Afdeling Numerieke Wiskunde: Band 8506 von Report NM, Centrum voor Wiskunde en Informatica Amsterdam (1985)
41. Weismann, M.: The hybrid level set – front tracking approach. Master Thesis. Department Mathematics, University of Erlangen (2012)
42. Weller, S.: Higher order time discretization for free surface flows. Diploma Thesis. Department Mathematics, University of Erlangen (2008)
43. Weller, S.: Higher order Galerkin methods in time for free surface flows. PAMM **14**(1), 855–856 (2014). doi:10.1002/pamm.201410408
44. Weller, S.: Time discretization for capillary problems (2015). https://opus4.kobv.de/opus4-fau/frontdoor/index/index/docId/6589
45. Williams, J.L.: Monolith structures, materials, properties and uses. ChemInform **33**(11) (2002). doi:10.1002/chin.200211225

Chapter 6
Upwind Schemes for Scalar Advection-Dominated Problems in the Discrete Exterior Calculus

Michael Griebel, Christian Rieger, and Alexander Schier

Abstract We present the *discrete exterior calculus* (DEC) to solve discrete partial differential equations on discrete objects such as cell complexes. To cope with advection-dominated problems, we introduce a novel stabilization technique to the DEC. To this end, we use the fact that the DEC coincides in special situations with known discretization schemes such as finite volumes or finite differences. Thus, we can carry over well-established upwind stabilization methods introduced for these classical schemes to the DEC. This leads in particular to a stable discretization of the Lie-derivative. We present the numerical features of this new discretization technique and study its numerical properties for simple model problems and for advection-diffusion processes on simple surfaces.

6.1 Introduction

The *discrete exterior calculus* provides a formalism to describe differential operators on discrete structures such as cell complexes. The application we have in mind to address in future work is that we are given a point cloud in the three-dimensional Euclidean space as sample from an unknown surface, i.e. a level-set function. This surface can for instance arise as a free boundary between the two phases in a flow problem which is modeled by the two-phase Navier-Stokes equation [7]. If this free surface stems from a numerical discretization of a continuous surface, then the number of points is naturally limited by the resolution of the underlying grid which is used for solving the given partial differential equation. This imposes the challenge to describe discrete differential operators and finally a discrete differential equation on the point cloud as surrogate for the continuous

M. Griebel (⊠) • C. Rieger • A. Schier
Institute for Numerical Simulation, University of Bonn, Wegelerstr. 6, 53115 Bonn, Germany
e-mail: griebel@ins.uni-bonn.de; rieger@ins.uni-bonn.de; schier@ins.uni-bonn.de

© Springer International Publishing AG 2017
D. Bothe, A. Reusken (eds.), *Transport Processes at Fluidic Interfaces*,
Advances in Mathematical Fluid Mechanics, DOI 10.1007/978-3-319-56602-3_6

differential equation on the surface. To this end, we use the fact that there are many algorithms which can produce a cell complex from a given point cloud, e.g. by using Delaunay triangulations or Voronoi tessellations. This cell complex then serves as a discretization of the free surface in each time step of the physical process.

The main emphasis of the *discrete exterior calculus* is to mimic continuous differential operators in a discrete setting, see [1] for a more abstract theory. As an example, we mention the discrete exterior derivative. As usual, a derivative assigns to each point on the surface a linear form on the tangent space attached to that point. This structure-preserving approach in the discrete exterior calculus is appealing since it does not rely on asymptotic (in the limit of finer and finer grid resolutions) arguments to ensure consistency with the continuous differential operators in the classical sense. The conventional DEC approach, however, suffers from stability issues if it is applied to advection-diffusion-type problems, see also [14]. This is similar to the standard finite element, finite volume or finite difference methods. There, stabilization techniques are necessary like SUPG or upwind, such that a stable discretization results for any advection strength albeit with a somewhat reduced convergence rate.

In this paper, we develop a modification of the DEC which leads to a stable discretization. To this end, we first show that the discrete partial differential equations resulting from the DEC are in simple cases (e.g. flat geometry, regular meshes) equivalent to classical numerical schemes such as finite differences or finite volumes discretizations, Here, finite volumes correspond to the use of dual two-forms whereas finite differences correspond to the use of primal zero-forms. This equivalency of the DEC to a standard scheme then allows us to mimic the well-established numerical upwind schemes from finite volumes and finite differences in the discrete exterior calculus. The resulting approach is new to the best of our knowledge and leads to a stable DEC also for larger advection. This is a desired prerequisite for any two-phase flow problem with an advection-diffusion-reaction process on the free surface.

The remainder of the manuscript is organized as follows: In Sect. 6.2, we recall the basics of a cell complex. We will explain how this concept of algebraic topology is related to the more common notions of Delaunay triangulation and Voronoi cells. In Sect. 6.3, we review the main objects of the discrete exterior calculus. Here, we do not aim at giving a complete overview but we just focus on the terms needed in the sequel. Our main results are derived in Sects. 6.4.1 and 6.4.2, namely the equivalence of the discrete exterior calculus with certain numerical schemes in special situations. Based on these observations, we then introduce novel upwind schemes for the discrete exterior calculus. Finally, we give numerical evidence for the convergence and stability properties of our new schemes in Sect. 6.5. We conclude this article with some comments in Sect. 6.6.

6.2 From Point Clouds to Cell Complexes

In our application, we need to generate a structure which allows to introduce differential operators in a meaningful way based on a given point cloud. Here, the construction of a cell complex from a point cloud is a well known task in topological data analysis, see for instance [10–12]. We describe the approach as outlined in [9]. The starting point is a point cloud $\mathbb{X}_M = \{p_1, \ldots, p_M\} \subset \Omega \subset \mathbb{R}^N$ and the associated (restricted) *Voronoi cells*

$$V_{p:\Omega} := \left\{ x \in \Omega \subset \mathbb{R}^N : \|x - p\|_2 \leq \|x - q\|_2 \ \text{ for all } q \in \mathbb{X}_M \right\}, \quad p \in \mathbb{X}_M.$$

To introduce an abstract simplicial complex, let $\mathscr{A} = \{A_1, \ldots, A_n\}$ be a finite collection of some sets. Then, the nerve of $\mathscr{A}$ is defined as

$$\mathrm{Nrv}\,\mathscr{A} := \left\{ \mathscr{B} \subseteq \mathscr{A} \ : \ \bigcap_{A_j \in \mathscr{B}} A_j \neq \emptyset \right\}.$$

Note here that we denote the initial sets of $\mathscr{A}$ by A_j and the collection of any of the subsets of $\mathscr{A}$ by calligraphic letters. Note also the important property that $\mathscr{C} \subseteq \mathscr{B}$ and $\mathscr{B} \in \mathrm{Nrv}\,\mathscr{A}$ implies $\mathscr{C} \in \mathrm{Nrv}\,\mathscr{A}$. This makes $\mathrm{Nrv}\,\mathscr{A}$ an abstract simplicial complex. We assume to have an injective function

$$\phi : \mathscr{A} \to \mathbb{R}^{n_{\mathscr{A}}}, \quad A_j \mapsto \phi(A_j),$$

where $n_{\mathscr{A}} \in \mathbb{N}$ is an arbitrary natural number. This map can be extended to arbitrary $\mathscr{B} \in \mathrm{Nrv}\,\mathscr{A}$ with $\mathscr{B} = \{A_{j_1}, \ldots, A_{j_n}\}$ via

$$\phi(\mathscr{B}) = \phi(\{A_{j_1}, \ldots, A_{j_n}\}) := \mathrm{conv}\left[\phi(A_{j_1}), \ldots, \phi(A_{j_n})\right]$$

$$:= \left\{ y \in \mathbb{R}^{n_{\mathscr{A}}} \ : \ y = \sum_{i=1}^{n} \alpha_i \phi(A_{j_i}),\ 0 \leq \alpha_i \text{ for all } 1 \leq i \leq n \text{ and } \sum_{i=1}^{n} \alpha_i = 1 \right\} \subset \mathbb{R}^{n_{\mathscr{A}}}$$

such that

$$\phi(\mathscr{B}) \cap \phi(\mathscr{C}) = \phi(\mathscr{B} \cap \mathscr{C}) \quad \text{for all } \mathscr{B}, \mathscr{C} \in \mathrm{Nrv}\,\mathscr{A}.$$

$$\mathscr{K}_{\mathscr{A}} := \left\{ \mathrm{conv}\left[\phi(\mathscr{B})\right] = \mathrm{conv}\left[\phi(B_{j_1}), \ldots, \phi(B_{j_n})\right] \ : \ \mathscr{B} \subset \mathscr{A} \right\} \tag{6.1}$$

a *simplicial complex* and the tuple $(\mathscr{K}_{\mathscr{A}}, \phi)$ a *geometric realization* of $\mathrm{Nrv}\,\mathscr{A}$. The underlying space is defined as $|\mathscr{K}_{\mathscr{A}}|_{\mathrm{sp}} := \bigcup \mathscr{K}_{\mathscr{A}}$. The (restricted) *Delaunay complex* of $\mathbb{X}_M$ is the geometric realization of $\mathrm{Nrv}\,V_{p:\Omega}$ and the map is $\phi : V_{p:\Omega} \mapsto p$.

This makes $n_{\mathscr{A}} = N$. In particular, we recover the standard definition of a simplex. From now on, we mainly follow [8] and [15].

Definition 1 A *k-simplex* $\sigma^{(k)}$ for $N \geq k \in \mathbb{N}$ in $\mathbb{R}^N$ is the convex hull of $k+1$ affinely independent vectors $v_i \in \mathbb{R}^N$, i.e.,

$$\sigma^{(k)} = [v_0, \ldots, v_k] = \left\{ \sum_{i=0}^{k} \alpha_i v_i, \,\middle|\, \alpha_i \geq 0, \text{ for } i = 0, \ldots, k \text{ and } \sum_{i=0}^{k} \alpha_i = 1 \right\}.$$

The points $v_0, \ldots, v_k$ are called the *vertices* of the simplex, and the number k is called the *dimension* of the simplex. A simplex $\sigma^{(\ell)}$ spanned by a (proper) subset of cardinality $0 \leq \ell \leq k$ of the vertices of $\sigma^{(k)}$ is a (proper) *ℓ-face* of $\sigma^{(k)}$. We write $\sigma^{(\ell)} \prec \sigma^{(k)}$, if $\sigma^{(\ell)}$ is a proper ℓ-face of $\sigma^{(k)}$ for some $0 \leq \ell \leq k$. We define the set of faces of a 0-simplex $\sigma^{(0)}$ as $\emptyset$.

Moreover, the Delaunay complex as simplicial complex has the following form:

Definition 2 A *simplicial complex* K of dimension $n \leq N$ is a collection of $0, \ldots, n$-simplices in $\mathbb{R}^N$, such that

- The faces of each simplex of K are in K.
- Two simplices have no intersection or they intersect at a common face.

We call a simplicial complex K of dimension n *manifold-like* if the polytope of K is a $\mathscr{C}^0$-manifold.

We can orient simplices using the order of their vertices.

Definition 3 Let $\sigma^{(k)}$ be a k-simplex, and fix the order of its vertices. Every even permutation of the vertices induces the same orientation, while an odd permutation reverses the sign of the orientation. For a 0-simplex we define the orientation as positive, for n-simplices we assume that the orientation is given.

From the orientation of a simplex, we can induce an orientation on its faces.

Definition 4 Let $\sigma^{(k)} = [v_0, \ldots, v_k]$ be an oriented simplex with $k \geq 1$. Consider the face $\sigma^{(k-1)}[v_0, \ldots, \hat{v}_i, \ldots, v_k]$, where the vertex $\hat{v}_i$ is omitted. This face has a positive orientation induced by $\sigma^{(k)}$ if i is even, otherwise its induced orientation is negative. Two k-simplices with $k > 0$ are *adjacent*, if they share a common $(k-1)$-face.

Finally, we will employ the following notion of a subcomplex.

Definition 5 Let K be a simplicial complex of dimension n, and let L be a subset of K that is a simplicial complex itself. Then L is called a subcomplex of K. For $k < n$, we call the subcomplex which contains all simplices of K of dimension at most k the *k-skeleton* of K.

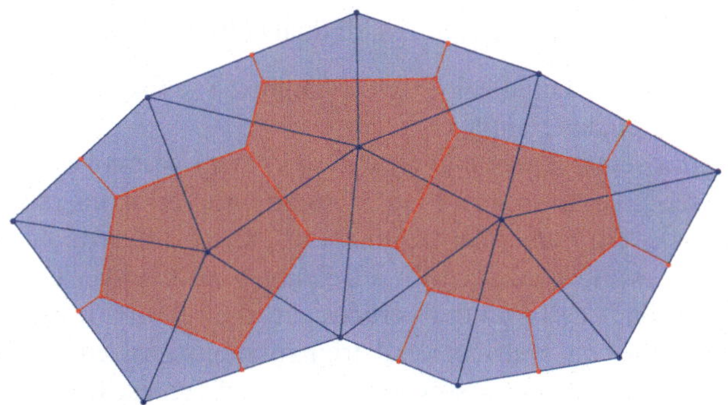

Fig. 6.1 A two-dimensional simplicial complex with its dual cells

6.2.1 Dual Complex

In this subsection, K always denotes an oriented manifold-like simplicial complex. We introduce the dual complex. Loosely speaking, a simplicial complex defines a dual complex $\star K$, by using the centers c_i of the n-simplices of the primal complex as vertex set and then connecting adjacent centers, where, on the boundary of the complex, we add the centers of the boundary faces as dual vertices. A simple example in the two-dimensional case is given in Fig. 6.1.
We will later on use a circumcentric dual complex.

Definition 6 The *circumcenter* of a k-simplex $\sigma^{(k)} = [v_0, \ldots, v_k]$ is given by the center of its k-circumsphere, which is the unique k-sphere that has the vertices $v_0, \ldots, v_k$ on its surface. Equivalently, the circumcenter is the unique point in the k-dimensional affine space that contains the k-simplex, which is equidistant to the $k + 1$ vertices of the simplex. We denote the circumcenter of a simplex $\sigma^{(k)}$ with $c_i = c(\sigma_i^{(k)})$.

We will need an additional assumption on the primal simplicial complex.

Definition 7 An n-simplex with $n \leq N$ is called *well-centered* if its circumcenter lies in its interior. A simplicial complex is called *well-centered* if each simplex in the complex is well-centered.

We are now in the position to define the dual cells.

Definition 8 Let K be a well-centered manifold-like simplicial complex of dimension n, and let $\sigma^{(k)}$ be one of the k-simplices of K with $0 \leq k \leq n \leq N$. We define the *dual cell* $\star\sigma^{(k)}$ as

$$\star\sigma^{(k)} := \bigcup_{r=0}^{n-k} \bigcup_{\sigma^{(k)} \prec \sigma_1 \prec \cdots \prec \sigma_r} [c(\sigma^{(k)}), c(\sigma_1), \ldots, c(\sigma_r)] \tag{6.2}$$

The *dual cell complex* $\star K$ is the union of all dual cells $\star \sigma^{(k)}, 0 \le k \le n \le N$.

We note that, in general, a dual cell is not a simplex. See Fig. 6.1 for an example of dual cells of the simplicial complex.

To conclude this section, note that we will work with Voronoi cells and Delaunay triangulations (see [3]). To this end, recall some basic relations (see [18]), which are common also in the numerical treatment of partial differential equations. The following facts are well established: In a Delaunay mesh with well-centered cells, the circumcentric dual cells of the vertices are their Voronoi regions. See also [24] for more details on well-centered triangulations. An edge connecting the circumcenters of two adjacent k-simplices of a well-centered mesh intersects the common $(k-1)$-face at its circumcenter and is perpendicular to the face.

6.3　The Discrete Exterior Calculus

We now want to employ the discrete exterior calculus (DEC) to solve PDEs. It is based on the representation of the domain of the partial differential equation as underlying space $|K|_{\mathrm{sp}}$ of a cell-complex K. We first introduce the basic operations of the discrete exterior calculus. Here, we follow [8] and [15]. We start with an oriented and manifold-like simplicial complex K of dimension $n \le N$ in $\mathbb{R}^N$ and denote its k-simplices by $\sigma^{(k)}$. The space of k-chains of k-simplices is given by the basic elements

$$C_k(K) := C_k(K; \mathbb{Z}) := \left\{ \sum_{j=1}^{N_k} \alpha_j \sigma_j^{(k)} \; : \; \alpha_j \in \mathbb{Z} \right\}, \tag{6.3}$$

where we use $N_k \in \mathbb{N}$ to denote the number of k-simplices. $C_k(K; \mathbb{Z})$ is naturally equipped with an Abelian group structure (with addition). The space of discrete k-forms (co-chains) is the dual space of the k-chains, i.e., the space of homomorphisms from the (group) of k-chains to the usual additive group $(\mathbb{R}, +)$. Precisely, we have

$$C^k(K) := \mathrm{Hom}(C_k(K; \mathbb{Z}), \mathbb{R}).$$

Naturally, a k-form is uniquely defined by its evaluation on the elements of $C_k(K)$. This leads to the idea that one simply has to store a list of all evaluations of a k-form

$$C^k(K) \ni \omega^k \leftrightarrow (\langle \omega^k, \sigma_1^k \rangle, \ldots, \langle \omega^k, \sigma_{N_k}^k \rangle)^\top \in \mathbb{R}^{N_k}, \tag{6.4}$$

where the evaluation of a k-form on a k-chain is written by the bilinear pairing $\langle \omega^k, \sigma^{(k)} \rangle \in \mathbb{R}$. The same constructions can be applied to the dual cell complex $\star K$.

We denote the Abelian group of k-chains on the dual complex by

$$C_k(\star K) := C_k(\star K; \mathbb{Z}) := \left\{ \sum_{j=1}^{N_k} \alpha_j \star \sigma_j^{(k)} \; : \; \alpha_j \in \mathbb{Z} \right\}.$$

Furthermore, we denote the space of k-forms on the dual complex as

$$C^k(\star K_d) := \mathrm{Hom}(C^k(\star K), \mathbb{R}).$$

There is a natural chain complex, i.e., a homomorphism $\partial_k : C_k(K) \to C_{k-1}(K)$, such that

$$0 \longrightarrow C_N(K) \xrightarrow{\partial_N} C_{N-1}(K) \xrightarrow{\partial_{N-1}} \cdots \xrightarrow{\partial_1} C_0(K) \longrightarrow 0,$$

with $\mathrm{Im}(\partial_{k+1}) \subset \ker(\partial_k)$, i.e., $\partial_k \circ \partial_{k+1} \equiv 0$ for all $0 \leq k \leq N - 1$. Furthermore, there is also an associated co-chain complex, i.e., a homomorphism $\delta_k : C^k(K) \to C^{k+1}(K)$ such that

$$0 \longleftarrow C^N(K) \xleftarrow{\delta_N} C^{N-1}(K) \xleftarrow{\delta_{N-1}} \cdots \xleftarrow{\delta_1} C^0(K) \longleftarrow 0.$$

We now identify a k-simplex σ^k with its ordered set of vertices $[v_0, \ldots, v_k]$. Then, we get a mapping from the simplex to a $(k-1)$-chain of its boundary simplices

$$\partial_k \sigma^k = \partial_k([v_0, \ldots, v_k]) = \sum_{i=0}^{k} (-1)^i [v_0, \ldots, \hat{v}_i, \ldots, v_k] \tag{6.5}$$

where $\hat{v}_i$ means that the vertex v_i is omitted.

Definition 9 The boundary operator $\partial_k : C_k(K) \to C_{k-1}(K)$ induces as formal adjoint with respect to the bilinear pairing $\langle \cdot, \cdot \rangle : C^k(K) \times C_k(K) \to \mathbb{R}$ the operator $\mathbf{d} : C^k(K) \to C^{k+1}(K)$. Precisely, we have

$$\langle \mathbf{d}\omega^k, \sigma^{k+1} \rangle := \langle \omega^k, \partial \sigma^{k+1} \rangle \tag{6.6}$$

for all $\sigma^{k+1} \in C_{k+1}(K)$. The operator $\mathbf{d}$ is called co-boundary operator and its definition can be seen as a variant of Stokes' theorem.

By construction, we have $\partial_k \circ \partial_{k+1} \equiv 0$. For our calculus, we need some more specific operations. They are defined in the following.

Definition 10 The discrete Hodge Star operator is a map $\star : C^k(K) \to C^k(\star K)$. For a k-cochain ω^k it is defined as

$$\frac{1}{|\sigma^k|} \langle \omega^k, \sigma^k \rangle = \frac{1}{|\star \sigma^k|} \langle \star \omega^k, \star \sigma^k \rangle \tag{6.7}$$

for all k-simplices σ^k.

We have the following identity

$$\star(\star\omega^k) = (-1)^{k(n-k)}\omega^k, \tag{6.8}$$

for all k-cochains ω^k. Furthermore, we also have the analogue relation for k-chains, i.e.,

$$\star(\star\sigma^{(k)}) = (-1)^{k(n-k)}\sigma^{(k)}. \tag{6.9}$$

A further necessary ingredient is an operation which allows to glue together k-forms. This operation is usually called *wedge product*.

Definition 11 Given a primal discrete k-form $\alpha^k \in C^k(K)$ and a primal discrete l-form $\beta^l \in C^l(K)$, the discrete *primal-primal wedge product* $\wedge : C^k(K) \times C^l(K) \to C^{k+l}(K)$ defined by the evaluation on a $(k+l)$-simplex $\sigma^{k+l} = [v_0, \ldots, v_{k+l}]$ is given by

$$\langle \alpha^k \wedge \beta^l, \sigma^{k+l} \rangle := \frac{1}{(k+l)!} \sum_{\tau \in S_{k+l+1}} \mathrm{sgn}(\tau) \frac{|\sigma^{k+l} \cap \star v_{\tau(k)}|}{|\sigma^{k+l}|} \langle \alpha^k \smile \beta^l, \tau(\sigma^{k+l}) \rangle$$

where S_{k+l+1} is the permutation group and its elements are thought of as permutations of the numbers $0, \ldots, (k+l+1)$. Here, we define $\star v_{\tau(k)}$ as the dual cell corresponding to the 0-simplex $[v_{\tau(k)}]$ and the notation $\tau(\sigma^{k+l})$ stands for the simplex $[v_{\tau(0)}, \ldots, v_{\tau(k+l)}]$. Moreover, the $\smile$ operator is defined by the evaluation of $\alpha \smile \beta$ on a $(k+l)$-simplex as

$$\langle \alpha^k \smile \beta^l, \tau(\sigma^{k+l}) \rangle := \langle \alpha^k, [v_{\tau(0)}, \ldots, v_{\tau(k)}] \rangle \langle \beta^l, [v_{\tau(k)}, \ldots, v_{\tau(k+l)}] \rangle.$$

For example, we get for the special case of a 1-form α^1 and a 0-form β^0:

$$\langle \alpha^1 \wedge \beta^0, [v_0, v_1] \rangle = \tag{6.10}$$

$$\left[\frac{|[v_0, v_1] \cap \star[v_1]|}{|[v_0, v_1]|} \langle \alpha^1, [v_0, v_1] \rangle \langle \beta^0, [v_1] \rangle \right] -$$

$$- \left[\frac{|[v_0, v_1] \cap \star[v_0]|}{|[v_0, v_1]|} \langle \alpha^1, [v_1, v_0] \rangle \langle \beta^0, [v_0] \rangle \right],$$

see also [20, Eq. (16)] for this expression. Now, in order to introduce an upwind-effect into the formalism of discrete exterior calculus, we have to modify the geometric weight factors $\frac{|\sigma^{k+l} \cap \star v_{\tau(k)}|}{|\sigma^{k+l}|}$. To this end, we replace $\frac{|[v_0, v_1] \cap \star[v_1]|}{|[v_0, v_1]|}$ by some

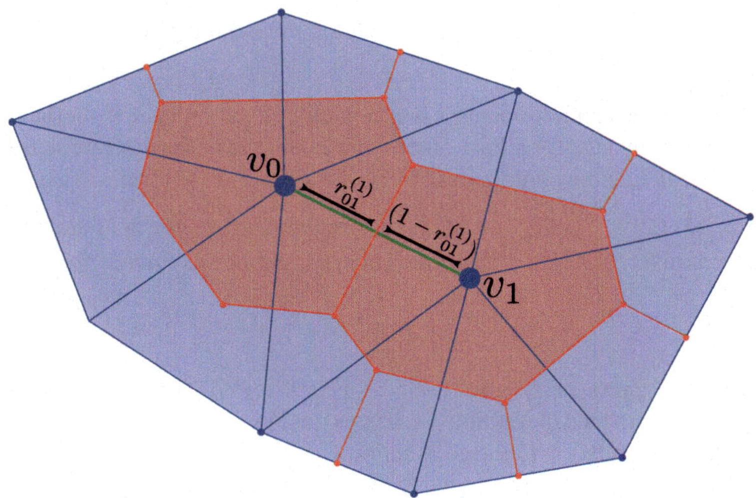

Fig. 6.2 Sketch of weight factor

$r_{0,1}^{(1)} \in [0, 1]$, and $\frac{|[v_0,v_1] \cap \star [v_0]|}{|[v_0,v_1]|}$ by $1 - r_{0,1}^{(1)}$, see Fig. 6.2 for an illustration. This gives

$$\langle \alpha^1 \wedge_{\mathrm{up}} \beta^0, [v_0, v_1] \rangle \tag{6.11}$$

$$:= \left[r_{0,1}^{(1)} \langle \alpha^1, [v_0, v_1] \rangle \langle \beta^0, [v_1] \rangle \right] - \left[(1 - r_{0,1}^{(1)}) \langle \alpha^1, [v_1, v_0] \rangle \langle \beta^0, [v_0] \rangle \right],$$

where we will provide some choices for the term $r_{0,1}^{(1)}$ in the sequel.

Furthermore, we need an operator which allows to assign a 1-form to a given vector field. Loosely speaking, a vector field on a dual cell complex $\star K$ is a smooth map $\mathbf{v} : |\star K|_{\mathrm{sp}} \to \mathbb{R}^N$. The operator which assigns a 1-form to $\mathbf{v}$ is called *flat operator* and the resulting form is denoted by $v^\flat$.

In the discrete setting, there are several choices of the flat operator depending on the discretization of the vector field, the areas over which we want to interpolate the field and if we want to evaluate $v^\flat$ on primal or dual edges. We refer to [15, Chap. 5] for a detailed discussion of the different flats. See also [22] for an other version of a discrete flat. In the following, we will use the flat operator $\flat_{dpp}$ for a dual vector field projected on primal edges with interpolation over primal volumes.

Definition 12 The *dual-primal-primal flat operator* $\flat_{dpp}$ is given by the evaluation of a vector field $\mathbf{v}$, which is assumed to be piecewise constant inside 2-simplices, on a primal 1-simplex $\sigma^{(1)} = [\sigma_l^{(0)}, \sigma_r^{(0)}]$. We have

$$\langle v^{\flat dpp}, \sigma^{(1)} \rangle := \sum_{\sigma^{(2)} \succ \sigma^{(1)}} \frac{|\star \sigma^{(1)} \cap \sigma^{(2)}|}{|\sigma^{(1)}|} \mathbf{v}(\sigma^{(2)}) \cdot (\sigma_r^{(0)} - \sigma_l^{(0)}), \tag{6.12}$$

where $\sigma_r^{(0)} - \sigma_l^{(0)} \in \mathbb{R}^2$ denotes the vector corresponding to the edge $\sigma^{(1)} = [\sigma_l^{(0)}, \sigma_r^{(0)}]$.

The dual-primal-primal flat operator is also well defined on curved meshes, which is not automatically true for all the other choices of flats in the discrete setting.

Finally, we define the contraction of a vector field and a k-form.

Definition 13 Let $\mathbf{v} : |\star K|_{\mathrm{sp}} \to \mathbb{R}^N$ be a given smooth vector field and let $\omega^{(k)} \in C^k(\star K)$ be a dual $0 \leq k \leq n \leq N$-form. Then, we define the *contraction*

$$\mathrm{i}_{\mathbf{v}}\omega^{(k)} := (-1)^{k(n-k)} \star \left(\star \omega^{(k)} \wedge v^{\flat dpp} \right) \in C^{k-1}(\star K).$$

This algebraic approach is also used in [20, Eq. (9)]. There is also a more geometrically motivated definition of the extrusion of a manifold with a flow. Theorem 3 gives a relation to the contraction.

Definition 14 Given a manifold M, a k-submanifold S and a vector field β on M, we call the manifold obtained by sweeping S along the flow of β for a time t the *extrusion* of S by β for a time t and denote it by $\mathrm{extr}_\beta(S, t)$. The manifold S carried by the flow for a time t will be denoted by $\varphi_\beta^t(S)$, see [8, Definition 10.1]. This extrusion $\mathrm{extr}_\beta(S, T)$ is the union of all manifolds $\varphi_\beta^t(S)$ with $t \in [0, T]$

$$\mathrm{extr}_\beta(S, T) := \bigcup_{t=0}^{T} \varphi_\beta^t(S).$$

We shall use this definition in Theorem 3.

The different operators introduced so far will be used in the following to define partial differential equations on simplicial complexes. For details on the implementation of such methods, see also [5].

6.4　Relation Between the DEC and Other Numerical Schemes

In this section, we consider an elliptic partial differential equation with $\alpha > 0$ and $f \in L^2(\Omega)$ of the following form

$$-\alpha \Delta u + \nabla \cdot (\mathbf{v}u) + cu = f \qquad \text{in } \Omega, \qquad (6.13)$$

$$u = 0 \qquad \text{on } \partial\Omega,$$

where $\mathbf{v} : \Omega \to \mathbb{R}^2$ is a given vector field. Note that this is the essential step in order to treat two-dimensional manifolds. The DEC has already proven to be appropriate for solving differential equations in many different situations. To name a few: For Darcy's equation, see [17]. For Navier-Stokes, see [20] and

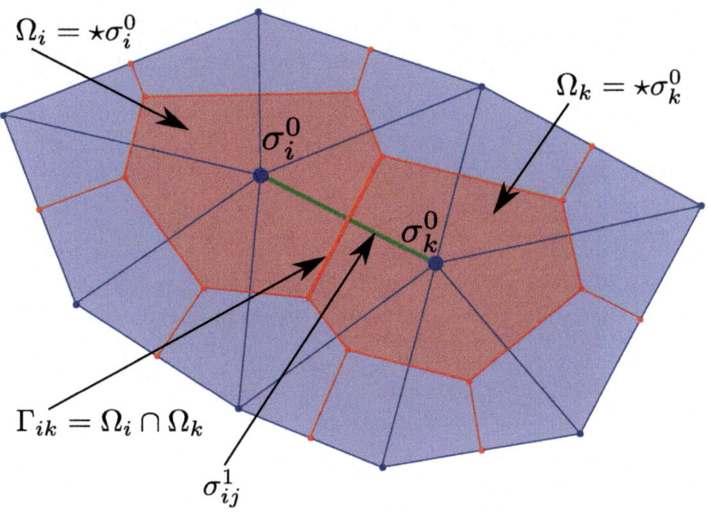

Fig. 6.3 Sketch of internal nodes and the corresponding primal and dual complexes

also [23]. For an energy minimization of Frank-Oseen-type, see [22]. In order to compare conventional numerical schemes for the discretization of this partial differential equation with schemes based on the discrete exterior calculus, we start with the points (primal vertices) $\mathbb{X}_{N_0}$ and form the restricted Voronoi cells $V_{p;\Omega}$. The corresponding Delaunay triangulation yields the primal 2-simplices. We recall the geometric setting in Fig. 6.3.

We first discuss finite volumes schemes but use the geometric terms from the cell complexes. Thus, we make the discussion in [21, Sect. 4.1] mathematically precise and provide an upwind scheme for the discrete exterior calculus.

6.4.1   Finite Volumes

As a classical discretization, we consider the finite volume method (FVM) in the two-dimensional case as described in [18, Chap. 6] and aim at an interpretation in the DEC formalism. To be precise, we use a dual grid with a *cell-centered* finite volume method approach as described in [6, Sect. 4.1.2] employing the perpendicular bisector method [6, Definition 4.1.4]. This is equivalent to the circumcentric dual grid of the DEC, as the intersection of the perpendicular bisectors inside a triangle is just the circumcenter. Using the notation from Sect. 6.2, we denote the Voronoi cell complex with $\star K$ and the primal simplicial complex with K. The numerical discretization with finite volumes uses piecewise linear polynomials on the Delaunay triangles as trial functions and piecewise constant functions on the

Voronoi cells as test functions. Given a (primal) vertex or 0-simplex $\sigma_i^{(0)}$, we denote its Voronoi cell by $\star\sigma_i^{(0)}$. Then, we have the finite volume discretization as: Find $u_h \in \mathbb{P}_1(K)$ such that

$$\int_{\star\sigma_i^{(0)}} \alpha \Delta u_h \mathrm{d}x - \int_{\star\sigma_i^{(0)}} \nabla \cdot (vu_h)\mathrm{d}x + \int_{\star\sigma_i^{(0)}} cu_h \mathrm{d}x = \int_{\star\sigma_i^{(0)}} f\mathrm{d}x \qquad (6.14)$$

for all internal control volumes. We now describe for each of the integrals in Eq. (6.14) how it is numerically evaluated. Following [18, p. 306], we approximate the reaction term and the right hand side with the midpoint rule, i.e.

$$\int_{\star\sigma_i^{(0)}} f\mathrm{d}x \approx |\star\sigma_i^{(0)}| f(\sigma_i^{(0)}) \quad \text{and} \quad \int_{\star\sigma_i^{(0)}} ru\mathrm{d}x \approx |\star\sigma_i^{(0)}| r(\sigma_i^{(0)})u(\sigma_i^{(0)}).$$

$$(6.15)$$

For the advection and diffusion terms, we use the divergence theorem

$$\int_{\star\sigma_i^{(0)}} \nabla \cdot (\alpha\nabla u - (vu))\mathrm{d}x = \int_{\partial\star\sigma_i^{(0)}} \alpha v \cdot \nabla u \mathrm{d}\sigma - \int_{\partial\star\sigma_i^{(0)}} v \cdot (vu)\mathrm{d}\sigma, \qquad (6.16)$$

where v denotes the outer unit normal to $\partial \star \sigma_i^{(0)}$. The boundary of a Voronoi cell can be decomposed into faces which are dual edges. We thus can write

$$\partial \star \sigma_i^{(0)} = \bigcup_{\sigma_i^{(0)} \prec \sigma^{(1)}} \star\left(\partial\sigma^{(1)} \setminus \{\sigma_i^{(0)}\}\right) \cap \star\sigma_i^{(0)} = \bigcup_{\partial\star\sigma_i^{(0)} \prec \star\sigma_k^{(0)}} \star\sigma_k^{(0)} \cap \star\sigma_i^{(0)}.$$

This is the reason why usually a dual edge $\Gamma_{i,k}$ is denoted by two indices indicating the two dual control volumes which are used to define this straight part of the boundary of the dual cell. We get

$$\int_{\partial\star\sigma_i^{(0)}} \alpha v \cdot \nabla u \mathrm{d}\sigma - \int_{\partial\star\sigma_i^{(0)}} v \cdot (vu)\mathrm{d}\sigma$$

$$= \sum_{\partial\star\sigma_i^{(0)} \prec \star\sigma_k^{(0)}} \left[\int_{\star\sigma_k^{(0)} \cap \star\sigma_i^{(0)}} \alpha v \cdot \nabla u \mathrm{d}\sigma - \int_{\star\sigma_k^{(0)} \cap \star\sigma_i^{(0)}} v \cdot (vu)\mathrm{d}\sigma \right].$$

Furthermore, we make the following approximations (using a constant $\alpha > 0$): We set

$$\alpha = \alpha|_{\sigma_k^{(0)} \cap \star\sigma_i^{(0)}} = \mu_{i,k}, \quad v \cdot v|_{\star\sigma_k^{(0)} \cap \star\sigma_i^{(0)}} \approx \gamma_{i,k},$$

which yields the following approximation to the boundary integrals

$$\sum_{\partial \star \sigma_i^{(0)} \prec \star \sigma_k^{(0)}} \left[\int_{\star \sigma_k^{(0)} \cap \star \sigma_i^{(0)}} \alpha v \cdot \nabla u \, d\sigma - \int_{\star \sigma_k^{(0)} \cap \star \sigma_i^{(0)}} v \cdot (vu) d\sigma \right]$$

$$\approx \sum_{\partial \star \sigma_i^{(0)} \prec \star \sigma_k^{(0)}} \left[\mu_{i,k} \int_{\star \sigma_k^{(0)} \cap \star \sigma_i^{(0)}} v \cdot \nabla u \, d\sigma - \gamma_{i,k} \int_{\star \sigma_k^{(0)} \cap \star \sigma_i^{(0)}} u \, d\sigma \right].$$

The gradient and hence the normal derivatives are discretized by standard difference quotients

$$(v \cdot \nabla u)\big|_{\star \sigma_k^{(0)} \cap \star \sigma_i^{(0)}} \approx \frac{u(\sigma_k^{(0)}) - u(\sigma_i^{(0)})}{\left| [\sigma_i^{(0)}, \sigma_i^{(0)}] \right|}.$$

The most delicate term is the advection since this is crucial for the stability of the resulting numerical scheme. We set

$$u\big|_{\star \sigma_k^{(0)} \cap \star \sigma_i^{(0)}} \approx r_{i,k} u(\sigma_k^{(0)}) + (1 - r_{i,k}) u(\sigma_i^{(0)}) \tag{6.17}$$

with $r_{i,k} \in [0, 1]$ as parameter, which then allows for upwind schemes. In summary, we obtain the following discretization of (6.14)

$$\sum_{\partial \star \sigma_i^{(0)} \prec \star \sigma_k^{(0)}} \left| \star \sigma_k^{(0)} \cap \star \sigma_i^{(0)} \right| \mu_{i,k} \frac{u(\sigma_k^{(0)}) - u(\sigma_i^{(0)})}{|v|}$$

$$- \sum_{\partial \star \sigma_i^{(0)} \prec \star \sigma_k^{(0)}} \left| \star \sigma_k^{(0)} \cap \star \sigma_i^{(0)} \right| \gamma_{i,k} \left[r_{i,k} u(\sigma_i^{(0)}) + (1 - r_{i,k}) u(\sigma_k^{(0)}) \right]$$

$$+ \left| \star \sigma_i^{(0)} \right| r(\sigma_i^{(0)}) u(\sigma_i^{(0)}) = \left| \star \sigma_i^{(0)} \right| f(\sigma_i^{(0)}). \tag{6.18}$$

Note again that this scheme is parametrized with $r_{i,k} \in [0, 1]$. These parameters can now be used to obtain a stable discretization provided that the $r_{i,k}$ are chosen appropriately.

6.4.1.1 Equivalence to the DEC Method

Now, we derive a DEC formulation of the advection-diffusion equation. Then, we will show that the discrete partial differential equation is indeed equivalent to the finite volume discretization. The Laplacian is a well-studied object in the discrete exterior calculus. We will therefore mainly focus on the advection term which is modeled by the Lie-advection in the discrete exterior calculus. We first follow [14] and get for a vector field $\boldsymbol{\beta} : \Omega \to \mathbb{R}^2$ the expression

$$L_{\boldsymbol{\beta}}\omega = i_{\boldsymbol{\beta}}d\omega + di_{\boldsymbol{\beta}}\omega$$

for the Lie advection of a differential k-form, where $i_{\boldsymbol{\beta}}$ is the contraction (see Definition 13) of the differential form and a vector field which yields a $(k-1)$-form. We will now show that the FVM is equivalent to solving the advection-diffusion-reaction PDE using dual 2-forms. For detailed definitions in the continuous context, see [14, Eqs. (2.23) and (2.24)]. We note here only the important equivalence

$$L_{\boldsymbol{\beta}}\omega \sim \begin{cases} \boldsymbol{\beta} \cdot \nabla u, & k = 0, \\ \nabla \cdot (\boldsymbol{\beta} u), & k = 2. \end{cases} \tag{6.19}$$

Hence, we expect to deal with differential 2-form if we study a discretization of (6.13). Since the testing is done on dual 2-cells, we have to use dual 2-forms. We use the subscript d to denote the fact that the forms are dual. We solve for the dual discrete differential 2-form ω_d^2, which corresponds to the integrated quantity $\int_{\star\sigma_i^{(0)}} u\mathrm{d}x\mathrm{d}y$ from the finite volume approach. The relation $\omega_d^2 = \star_0\omega^0$ defines the mapping between the integral and the piecewise constant function inside the control volume.

Theorem 1 *Let* $\omega_d^2, \eta_d^2 \in C^2(\star K)$ *be dual 2-forms,* $\alpha \geq 0$ *and* $c \in \mathbb{R}$ *given constants and* $\mathbf{v} : |\star K| \to \mathbb{R}^2$ *a vector field. We consider*

$$\alpha\langle \boldsymbol{d} \star \boldsymbol{d} \star \omega_d^{(2)}, \star[\sigma_i^{(0)}]\rangle + \langle \boldsymbol{d}\, i_{\mathbf{v}}\omega_d^{(2)}, \star[\sigma_i^{(0)}]\rangle + c\langle \omega_d^{(2)}, \star[\sigma_i^{(0)}]\rangle = \langle \eta_d^2, \star[\sigma_i^{(0)}]\rangle. \tag{6.20}$$

By setting $u := \star\omega_d^2$, $f = \star\eta_d^2$ *and* $r_{i,j} = r_{i,j}^{(1)}$ *according to (6.24), we obtain that (6.20) is equivalent to (6.14).*

Proof We directly compute

$$\langle \eta_d^{(2)}, \star[\sigma_i^{(0)}]\rangle = \langle \star\eta_d^{(2)}, [\sigma_i^{(0)}]\rangle = |\star[\sigma_i^{(0)}]|f(\sigma_i^{(0)}). \tag{6.21}$$

Also, by formula (6.8), we have

$$\langle \omega_d^2, \star[\sigma_i^{(0)}]\rangle = (-1)^{2\cdot(2-2)}\langle \star \star \omega_d^{(2)}, \star[\sigma_i^{(0)}]\rangle = \langle \star\omega_d^{(2)}, [\sigma_i^{(0)}]\rangle$$

$$= \left|\star[\sigma_i^{(0)}]\right| \langle \star\omega_d^{(2)}, \sigma_i^{(0)}\rangle \tag{6.22}$$

using (6.7). For the diffusion term, we get

$$\left\langle \mathbf{d} \star \mathbf{d} \star \omega_d^{(2)}, \star[\sigma_i^{(0)}]\right\rangle = \left\langle \star\mathbf{d} \star \omega_d^{(2)}, \partial \star [\sigma_i^{(0)}]\right\rangle$$

$$= \left\langle \star\mathbf{d} \star \omega_d^{(2)}, (-1)^{1(2-1)} \star \star\partial \star [\sigma_i^{(0)}]\right\rangle$$

$$= (-1) \sum_{[\sigma_i^{(0)}] \prec [\sigma_k^{(0)}, \sigma_i^{(0)}]} \frac{\left| \star [\sigma_i^{(0)}] \cap \star [\sigma_k^{(0)}] \right|}{\left| [\sigma_k^{(0)}, \sigma_i^{(0)}] \right|} \left\langle \mathbf{d} \star \omega_d^{(2)}, [\sigma_k^{(0)}, \sigma_i^{(0)}] \right\rangle$$

$$= (-1) \sum_{[\sigma_i^{(0)}] \prec [\sigma_k^{(0)}, \sigma_i^{(0)}]} \frac{\left| \star [\sigma_i^{(0)}] \cap \star [\sigma_k^{(0)}] \right|}{\left| [\sigma_k^{(0)}, \sigma_i^{(0)}] \right|} \left\langle \star \omega_d^{(2)}, \partial [\sigma_k^{(0)}, \sigma_i^{(0)}] \right\rangle$$

$$= (-1) \sum_{[\sigma_i^{(0)}] \prec [\sigma_k^{(0)}, \sigma_i^{(0)}]} \left| \star [\sigma_i^{(0)}] \cap \star [\sigma_k^{(0)}] \right| \frac{\langle \star \omega_d^{(2)}, [\sigma_k^{(0)}] \rangle - \langle \star \omega_d^{(2)}, [\sigma_i^{(0)}] \rangle}{\left| [\sigma_k^{(0)}, \sigma_i^{(0)}] \right|}.$$

Finally, we compute the advection term as

$$\left\langle \mathbf{d}\, i_v \omega_d^{(2)}, \star [\sigma_i^{(0)}] \right\rangle = \left\langle i_v \omega_d^{(2)}, \partial \star [\sigma_i^{(0)}] \right\rangle = \left\langle \star \left(\star \omega_d^{(2)} \wedge v^{\flat_{dpp}} \right), \partial \star [\sigma_i^{(0)}] \right\rangle$$

$$= \left\langle \star \left(\star \omega_d^{(2)} \wedge v^{\flat_{dpp}} \right), (-1) \star \star \partial \star [\sigma_i^{(0)}] \right\rangle$$

$$= (-1) \frac{\left| \partial \star \sigma_i^{(0)} \right|}{\left| \star \partial \star [\sigma_i^{(0)}] \right|} \left\langle \star \omega_d^{(2)} \wedge v^{\flat_{dpp}}, \star \partial \star [\sigma_i^{(0)}] \right\rangle$$

$$= (-1) \sum_{[\sigma_i^{(0)}] \prec [\sigma_k^{(0)}, \sigma_i^{(0)}]} \frac{\left| \star [\sigma_i^{(0)}] \cap \star [\sigma_k^{(0)}] \right|}{|[\sigma_k^{(0)}, \sigma_i^{(0)}]|} \left\langle \star \omega_d^{(2)} \wedge v^{\flat_{dpp}}, [\sigma_k^{(0)}, \sigma_i^{(0)}] \right\rangle.$$

Now, we can replace the usual wedge product $\wedge$ by $\wedge_{\mathrm{up}}$, use (6.11) and observe that, due to our geometric construction, we have

$$\frac{|[\sigma_i^{(0)}, \sigma_k^{(0)}] \cap \star [\sigma_k^{(0)}]|}{|[\sigma_i^{(0)}, \sigma_k^{(0)}]|} = \frac{1}{2}.$$

Thus, we derived

$$(-1) \left\langle \mathbf{d}\, i_v \omega_d^{(2)}, \star [\sigma_i^{(0)}] \right\rangle$$

$$= \sum_{[\sigma_i^{(0)}] \prec [\sigma_k^{(0)}, \sigma_i^{(0)}]} \frac{\left| \star [\sigma_i^{(0)}] \cap \star [\sigma_k^{(0)}] \right|}{|[\sigma_k^{(0)}, \sigma_i^{(0)}]|} \left\langle \star \omega_d^{(2)} \wedge_{\mathrm{up}} v^{\flat_{dpp}}, [\sigma_k^{(0)}, \sigma_i^{(0)}] \right\rangle$$

$$= \sum_{[\sigma_i^{(0)}] \prec [\sigma_k^{(0)}, \sigma_i^{(0)}]} \frac{\left| \star [\sigma_i^{(0)}] \cap \star [\sigma_k^{(0)}] \right|}{|[\sigma_k^{(0)}, \sigma_i^{(0)}]|} \left(\left[r_{k,i}^{(1)} \langle v^{\flat_{dpp}}, [\sigma_k^{(0)}, \sigma_i^{(0)}] \rangle \langle \star \omega_d^{(2)}, [\sigma_i^{(0)}] \rangle \right] \right.$$

$$\left. - \left[(1 - r_{k,i}^{(1)}) \langle v^{\flat_{dpp}}, [\sigma_i^{(0)}, \sigma_k^{(0)}] \rangle \langle \star \omega_d^{(2)}, [\sigma_k^{(0)}] \rangle \right] \right).$$

For the evaluation of the dual-primal-primal flat operator, we use definition (12). We then have

$$\langle v^{\flat dpp}, [\sigma_i^{(0)}, \sigma_k^{(0)}]\rangle = \sum_{\sigma^{(2)} \succ [\sigma_i^{(0)}, \sigma_k^{(0)}]} \frac{|\star [\sigma_i^{(0)}] \cap \star [\sigma_k^{(0)}] \cap \sigma^{(2)}|}{|\star [\sigma_i^{(0)}] \cap \star [\sigma_k^{(0)}]|} (\mathbf{v}(\sigma^{(2)}) \cdot (\sigma_k^{(0)} - \sigma_i^{(0)})),$$

$$\langle v^{\flat dpp}, [\sigma_k^{(0)}, \sigma_i^{(0)}]\rangle = \sum_{\sigma^{(2)} \succ [\sigma_k^{(0)}, \sigma_i^{(0)}]} \frac{|\star [\sigma_k^{(0)}] \cap \star [\sigma_i^{(0)}] \cap \sigma^{(2)}|}{|\star [\sigma_k^{(0)}] \cap \star [\sigma_i^{(0)}]|} (\mathbf{v}(\sigma^{(2)}) \cdot (\sigma_i^{(0)} - \sigma_k^{(0)}))$$

$$= -\langle v^{\flat dpp}, [\sigma_i^{(0)}, \sigma_k^{(0)}]\rangle,$$

where we understand $\mathbf{v}$ to be piecewise constant on each primal 2-simplex. Furthermore, we stress the similarity to the finite volume scheme by abbreviating

$$\gamma_{k,i} := \sum_{\sigma^{(2)} \succ [\sigma_k^{(0)}, \sigma_i^{(0)}]} \frac{|\star [\sigma_i^{(0)}] \cap \star [\sigma_k^{(0)}] \cap \sigma^{(2)}|}{|\star [\sigma_i^{(0)}] \cap \star [\sigma_k^{(0)}]|} \frac{1}{|\sigma_k^{(0)} - \sigma_i^{(0)}|} \mathbf{v}(\sigma^{(2)}) \cdot (\sigma_k^{(0)} - \sigma_i^{(0)}).$$

Now, we can define the function $r^{(1)} : \mathbb{R} \to [0, 1]$ by setting

$$r^{(1)}(z) := \begin{cases} \frac{1}{2}, & \text{central differences} \\ \frac{1}{2}(\operatorname{sgn}(z) + 1), & \text{full upwind.} \end{cases} \tag{6.23}$$

We set

$$r_{k,i}^{(1)} := r^{(1)}\left(\frac{\gamma_{k,i}\left|[\sigma_i^{(0)}, \sigma_k^{(0)}]\right|}{\alpha}\right). \tag{6.24}$$

The term $\frac{\gamma_{k,i}\left|[\sigma_i^{(0)}, \sigma_k^{(0)}]\right|}{\alpha}$ is called local Péclet number.

An other choice for the function $r^{(1)}(z)$ defined in [18, Sect. 8.2] is

$$r^{(1)}(z) := 1 - \frac{1}{z}\left(1 - \frac{z}{e^z - 1}\right) \qquad \text{exponential upwind.} \tag{6.25}$$

We obtain for the case of central differences, i.e. $r^{(1)} := \frac{1}{2}$

$$\langle \mathbf{d}\, i_{\mathbf{v}}\omega_d^{(2)}, \star[\sigma_i^{(0)}] \rangle$$

$$= (-1) \sum_{[\sigma_i^{(0)}] \prec [\sigma_k^{(0)}, \sigma_i^{(0)}]} \frac{1}{2} \frac{\left| \star[\sigma_i^{(0)}] \cap \star[\sigma_k^{(0)}] \right|}{|[\sigma_k^{(0)}, \sigma_i^{(0)}]|} \left((-1)\langle v^{b_{dpp}}, [\sigma_i^{(0)}, \sigma_k^{(0)}] \rangle \langle \star\omega_d^{(2)}, [\sigma_i^{(0)}] \rangle \right.$$

$$\left. - \langle v^{b_{dpp}}, [\sigma_i^{(0)}, \sigma_k^{(0)}] \rangle \langle \star\omega_d^{(2)}, [\sigma_k^{(0)}] \rangle \right)$$

$$= \sum_{[\sigma_i^{(0)}] \prec [\sigma_k^{(0)}, \sigma_i^{(0)}]} \frac{\left| \star[\sigma_i^{(0)}] \cap \star[\sigma_k^{(0)}] \right|}{2|[\sigma_k^{(0)}, \sigma_i^{(0)}]|} \langle v^{b_{dpp}}, [\sigma_i^{(0)}, \sigma_k^{(0)}] \rangle \left(\langle \star\omega_d^{(2)}, [\sigma_k^{(0)}] \rangle + \langle \star\omega_d^{(2)}, [\sigma_i^{(0)}] \rangle \right)$$

$$= \sum_{[\sigma_i^{(0)}] \prec [\sigma_k^{(0)}, \sigma_i^{(0)}]} \gamma_{k,i} \left| \star[\sigma_i^{(0)}] \cap \star[\sigma_k^{(0)}] \right| \left(\frac{1}{2}\langle \star\omega_d^{(2)}, [\sigma_k^{(0)}] \rangle + \frac{1}{2}\langle \star\omega_d^{(2)}, [\sigma_i^{(0)}] \rangle \right).$$

Finally, we obtain

$$\langle \mathbf{d}\, i_{\mathbf{v}}\omega_d^{(2)}, \star\sigma_i^{(0)} \rangle =$$

$$= \sum_{[\sigma_i^{(0)}] \prec [\sigma_k^{(0)}, \sigma_i^{(0)}]} \gamma_{k,i} \left| \star[\sigma_i^{(0)}] \cap \star[\sigma_k^{(0)}] \right| \left(\frac{1}{2}\langle \star\omega_d^{(2)}, [\sigma_k^{(0)}] \rangle + \frac{1}{2}\langle \star\omega_d^{(2)}, [\sigma_i^{(0)}] \rangle \right).$$

Collecting the terms yields the claim. $\square$

6.4.2 Finite Differences

Next, we consider the setting of the discrete Lie advection for differential 0-forms in the two-dimensional case.

We use the finite differences approximation as defined in [13, 3.2–3.5] to discretize (6.13). We employ a regular (square) grid for finite differences and get for a vertex $x_i \in \mathbb{R}^2$ on the grid with neighbors $x_i^N, x_i^S, x_i^W, x_i^E$ and edge length h

$$-\alpha_i \frac{u(x_i^N) + u(x_i^S) + u(x_i^W) + u(x_i^E) - 4u(x_i)}{h^2}$$

$$-v_x \frac{u(x_i^W) - u(x_i^E)}{2h} - v_y \frac{u(x_i^S) - u(x_i^N)}{2h}$$

$$+cu(x_i) = f(x_i), \tag{6.26}$$

where we refer for the notation to [13, 3.2–3.5]. Here, we applied central differences for the discretizations of the first derivatives.

6.4.2.1 Equivalence to the DEC Method

To compare this scheme with the DEC, we use a rectangular regular grid as primal grid, and calculate a well-centered dual rectangular regular grid which is again a regular grid shifted by $\frac{h}{2}$ in x and y direction. This is usually called staggered grid construction. The main geometric properties, e.g. that dual 1-simplices (edges) are perpendicular to the corresponding primal 1-simplices (edges) and that dual edges intersect themselves at the circumcenter of the primal 2-simplex, still hold on rectangular grids.

Theorem 2 *Let $\omega_p^{(0)}, \eta^{(0)} \in C^0(K)$ be primal 0-forms, $\alpha \geq 0$ and $c \in \mathbb{R}$ given constants and $\mathbf{v} : |\star K| \to \mathbb{R}^2$ a vector field. We consider*

$$\alpha \langle \star d \star d\omega_p^0, \sigma_i^{(0)} \rangle + \langle i_\mathbf{v} d\omega_p^{(0)}, \sigma_i^{(0)} \rangle + c\langle \omega_p^{(0)}, \sigma_i^{(0)} \rangle = \langle \eta_p^0, \sigma_i^{(0)} \rangle. \tag{6.27}$$

By setting $u := \omega_p^0, f = \eta_p^0$ and $r_{i,j} = \frac{1}{2}$, we get that (6.27) is equivalent to (6.14).

Proof We have by definition $\langle \omega^{(0)}, \sigma_i^0 \rangle = \omega^{(0)}(\sigma_i^{(0)})$. Analogue calculations yield $c\langle \omega^0, \sigma_i^{(0)} \rangle = c\omega^{(0)}(\sigma_i^{(0)})$ and $\langle \eta^0, \sigma_i^{(0)} \rangle = c\eta^{(0)}(\sigma_i^{(0)})$. For the diffusion term, we get

$$\langle \star\mathbf{d} \star \mathbf{d}\omega^0, [\sigma_i^{(0)}] \rangle = \frac{1}{\left|\star\sigma_i^{(0)}\right|} \langle \mathbf{d} \star \mathbf{d}\omega^0, \star[\sigma_i^{(0)}] \rangle = \frac{1}{\left|\star\sigma_i^{(0)}\right|} \langle \star\mathbf{d}\omega^0, \partial \star [\sigma_i^{(0)}] \rangle$$

$$= \frac{(-1)}{\left|\star\sigma_i^{(0)}\right|} \langle \star\mathbf{d}\omega^0, \star \star \partial \star [\sigma_i^{(0)}] \rangle$$

$$= \frac{(-1)}{\left|\star\sigma_i^{(0)}\right|} \sum_{[\sigma_i^{(0)}] \prec [\sigma_k^{(0)}, \sigma_i^{(0)}]} \frac{\left|\star[\sigma_k^{(0)}] \cap \star[\sigma_i^{(0)}]\right|}{\left|[\sigma_k^{(0)}, \sigma_i^{(0)}]\right|} \langle \mathbf{d}\omega^0, [\sigma_k^{(0)}, \sigma_i^{(0)}] \rangle$$

$$= \frac{(-1)}{\left|\star\sigma_i^{(0)}\right|} \sum_{[\sigma_i^{(0)}] \prec [\sigma_k^{(0)}, \sigma_i^{(0)}]} \frac{\left|\star[\sigma_k^{(0)}] \cap \star[\sigma_i^{(0)}]\right|}{\left|[\sigma_k^{(0)}, \sigma_i^{(0)}]\right|} \langle \omega^0, \partial[\sigma_k^{(0)}, \sigma_i^{(0)}] \rangle$$

$$= \frac{(-1)}{\left|\star\sigma_i^{(0)}\right|} \sum_{[\sigma_i^{(0)}] \prec [\sigma_k^{(0)}, \sigma_i^{(0)}]} \frac{\left|\star[\sigma_k^{(0)}] \cap \star[\sigma_i^{(0)}]\right|}{\left|[\sigma_k^{(0)}, \sigma_i^{(0)}]\right|} \left(\langle \omega^0, [\sigma_k^{(0)}] \rangle - \langle \omega^0, [\sigma_i^{(0)}] \rangle \right).$$

With a regular square grid, we have

$$\left|\star\sigma_i^{(0)}\right| = h^2 \quad \text{and} \quad \frac{\left|\star[\sigma_k^{(0)}] \cap \star[\sigma_i^{(0)}]\right|}{\left|[\sigma_k^{(0)}, \sigma_i^{(0)}]\right|} = \frac{h}{h} = 1,$$

where $h > 0$ denotes the mesh-width of the grid. For the advection term, we have to consider $\langle i_{\mathbf{v}} \mathbf{d}\omega_p^{(0)}, [\sigma_i^{(0)}]\rangle$. Here, we employ the more geometrically motived expression from [15, p. 93] (see also Theorem 3) for the contraction using the notion of extrusions. We first observe that we have (subscripts denote horizontal and vertical components of a vector)

$$\mathbf{v} := v_{\text{hor}}\mathbf{e}_1 + v_{\text{vert}}\mathbf{e}_2$$

$$= \frac{1}{2h}\left(v_{\text{hor}}(\sigma_{i,N}^{(0)} - \sigma_i^{(0)}) + v_{\text{hor}}(\sigma_i^{(0)} - \sigma_{i,S}^{(0)})\right) +$$

$$\frac{1}{2h}\left(v_{\text{vert}}(\sigma_{i,E}^{(0)} - \sigma_i^{(0)}) + v_{\text{vert}}(\sigma_i^{(0)} - \sigma_{i,W}^{(0)})\right).$$

Hence, we obtain

$$2h\langle i_{\mathbf{v}}\mathbf{d}\omega_p^{(0)}, [\sigma_i^{(0)}]\rangle =$$

$$= v_{\text{hor}}\left(\langle \mathbf{d}\omega_p^{(0)}, [\sigma_{i,N}^{(0)}, \sigma_i^{(0)}]\rangle + \langle \mathbf{d}\omega_p^{(0)}, [\sigma_i^{(0)}, \sigma_{i,S}^{(0)}]\rangle\right)$$

$$+ v_{\text{vert}}\left(\langle \mathbf{d}\omega_p^{(0)}, [\sigma_{i,E}^{(0)}, \sigma_i^{(0)}]\rangle + \langle \mathbf{d}\omega_p^{(0)}, [\sigma_i^{(0)}, \sigma_{i,W}^{(0)}]\rangle\right).$$

Collecting the terms yields the final assertion. $\qquad\square$

6.5 Numerical Experiments

Since we work in the DEC-setting, the objects we study are not functions but discrete differential forms. This implies that we work with vectors as in (6.4) for primal 0-forms and dual 2-forms. We identify a 0-form with a function $\omega_p^0 \leftrightarrow w^0 : |K|_{\text{sp}} \to \mathbb{R}$ and a dual 2-form with $\omega_d^2 \leftrightarrow w^2 dxdy$. Moreover, each σ_j^0 with $1 \le j \le N_0$ and N_0 defined in (6.3) is just a primal vertex for which we use the same notation. This yields

$$C^0(K) \ni \omega_p^0 \leftrightarrow (\langle \omega_p^0, \sigma_1^0\rangle, \ldots, \langle \omega_p^0, \sigma_{N_0}^0\rangle)^\top \in \mathbb{R}^{N_0}$$

$$= \Omega_p^0 := \left(w^0(\sigma_1^0), \ldots, w^0(\sigma_{N_0}^0)\right)^\top$$

$$C^2(\star K) \ni \omega_d^2 \leftrightarrow (\langle \omega_d^2, \star\sigma_1^0\rangle, \ldots, \langle \omega_d^2, \star\sigma_{N_0}^0\rangle)^\top \in \mathbb{R}^{N_0}$$

$$\approx \Omega_d^2 := \left(\left|\star\sigma_1^0\right| w^2(\sigma_1^0), \ldots, \left|\star\sigma_{N_0}^0\right| w^2(\sigma_{N_0}^0)\right)^\top,$$

where the last approximation corresponds to a mid-point quadrature rule. Note that on a square mesh with side-length h we have $\left|\star\sigma_j^0\right| = h^2$ for all $1 \le j \le N_0$. We measure the distances in a discrete L^2-norm, that is

$$\|\omega_p^0 - \tilde{\omega}_p^0\|_{L^2_{C^0(K)}}^2 := \frac{1}{N_0} \sum_{i=1}^{N_0} \left(w^0((\sigma_j^0) - \tilde{w}^0((\sigma_j^0)\right)^2$$

$$= \left(\Omega_p^0 - \tilde{\Omega}_p^0\right)^\top \mathrm{diag}(N_0^{-1}) \left(\Omega_p^0 - \tilde{\Omega}_p^0\right),$$

$$\|\omega_d^2 - \tilde{\omega}_d^2\|_{L^2_{C^2(\star K)}}^2 := \frac{1}{N_0} \sum_{i=1}^{N_0} \left(w^2(\sigma_j^0) - \tilde{w}^2(\sigma_j^0)\right)^2$$

$$= \left(\Omega_d^2 - \tilde{\Omega}_d^2\right)^\top \mathrm{diag}(|\star\sigma_1^0|^{-1} N_0^{-1}) \left(\Omega_d^2 - \tilde{\Omega}_d^2\right). \tag{6.28}$$

This norm is used to indicate the error in the numerical experiments.

6.5.1   *Pure Advection on a Flat Regular Mesh*

We work in the setting of Theorem 1, i.e., we consider the advection for a dual 2-form and discretize the spatial operators as outlined there. We set $u(\cdot, t) := w^2(t)$ in the notation of (6.28). The discrete 1-form corresponding to a vector field $\mathbf{v}$ reads $\langle \mathbf{v}^\flat, \mathbf{e} \rangle = \frac{1}{|\mathbf{e}|}$ for each edge $\mathbf{e}$, because all edges are axis-aligned either in positive x or y direction. As already mentioned, the rectangular grid has the same properties[1] as a well-centered triangulation and the dual grid is just a shifted rectangular grid. The flat operator mapping vectors to 1-forms can be implemented via splitting by dimension. This way, we can use rectangular grids for our experiments. We compare the standard DEC (central) scheme with our full upwind scheme, see (6.23). Hence, as a first example, we consider the PDE

$$\nabla \cdot (\mathbf{v}(x)u(\mathbf{x}, t)) = \frac{\partial u}{\partial t}(\mathbf{x}, t) \quad \text{for } \mathbf{x} = (x, y)^\top \in \Omega = [0, 1]^2, \tag{6.29}$$

i.e., a pure advection of the concentration u with the vector field $\mathbf{v}$. For the time-integration, we use an (explicit) Euler scheme with $\delta t = 0.5h^2$ such that the CFL condition holds. While the CFL condition would only require a time-step $\delta t < h/|v|$, we need to use here a time-step which is quadratic instead in h to show the second order rate of the central derivatives.

We consider as initial data a smooth Gaussian-like function (see Fig. 6.4), as defined by

$$u_0(x, y) = u(x, y, 0) = \exp(1) \exp\left(-\frac{1}{(1 - \hat{x}^2)(1 - \hat{y}^2)}\right), \tag{6.30}$$

---

[1]E.g. orthogonality of primal and dual edges.

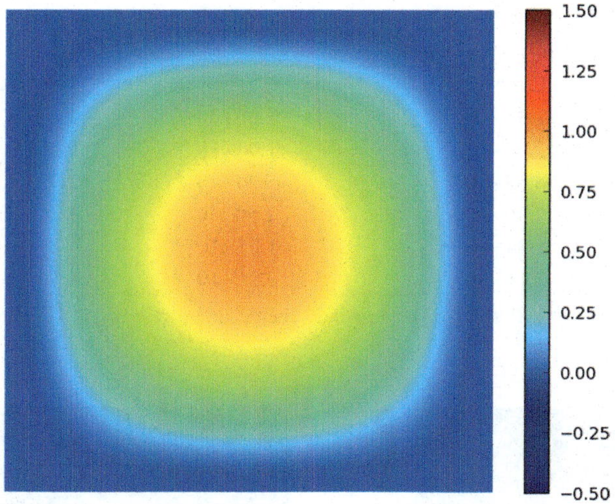

Fig. 6.4 Initial condition

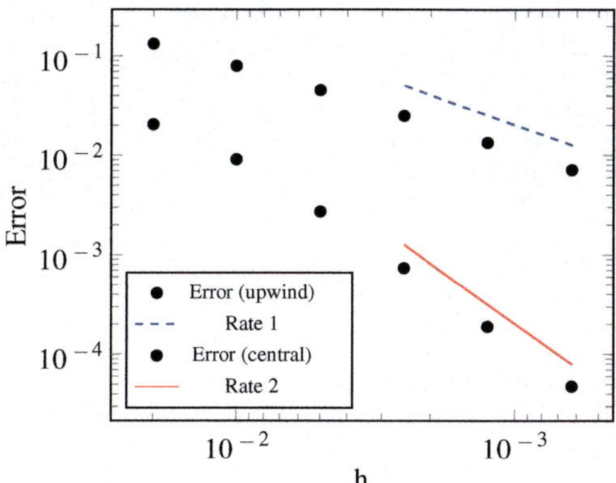

Fig. 6.5 Convergence rates of the advected concentration

where we set $\hat{x} = 2x - 1, \hat{y} = 2y - 1$ to center the function at $(0.5, 0.5)$. We set $\mathbf{v} = (1, 1)^T$ and $T = 1.0$ on a square grid on $\Omega = [0, 1]^2$ with cells of size $h \times h$ with periodic boundary conditions. The choices imply that the initial condition u_0 will be recovered at time $T = 1$, i.e., $u(x, y, 1) = u_0(x, y)$. As error measure we employ the difference between the analytic function and the result in the norm defined in (6.28). Precisely, we consider $u(\cdot, 0) = u_0 \leftrightarrow \omega_d^2$ and $u(\cdot, 0) \approx u(\cdot, 1) \leftrightarrow \tilde{\omega}_d^2$.

Figure 6.5 and Table 6.1 shows the error $\left\| \omega_d^2 - \tilde{\omega}_d^2 \right\|_{L^2_{C^2(\star K)}}$. We observe an asymptotic rate of one for our upwind approach and a rate of two for the standard DEC approach, when the central scheme is in a stable regime.

Table 6.1 Error versus mesh width h between the initial values and the advected function at time T

h	Error (upwind)	h	Error (central)
$2.00 \cdot 10^{-2}$	$1.33 \cdot 10^{-1}$	$2.00 \cdot 10^{-2}$	$2.03 \cdot 10^{-2}$
$1.00 \cdot 10^{-2}$	$7.98 \cdot 10^{-2}$	$1.00 \cdot 10^{-2}$	$9.10 \cdot 10^{-3}$
$5.00 \cdot 10^{-3}$	$4.53 \cdot 10^{-2}$	$5.00 \cdot 10^{-3}$	$2.72 \cdot 10^{-3}$
$2.50 \cdot 10^{-3}$	$2.49 \cdot 10^{-2}$	$2.50 \cdot 10^{-3}$	$7.36 \cdot 10^{-4}$
$1.25 \cdot 10^{-3}$	$1.34 \cdot 10^{-2}$	$1.25 \cdot 10^{-3}$	$1.89 \cdot 10^{-4}$
$6.25 \cdot 10^{-4}$	$7.11 \cdot 10^{-3}$	$6.25 \cdot 10^{-4}$	$4.75 \cdot 10^{-5}$

We achieve a rate of ≈ 0.915 for our upwind scheme and a rate of ≈ 1.993 with the central scheme

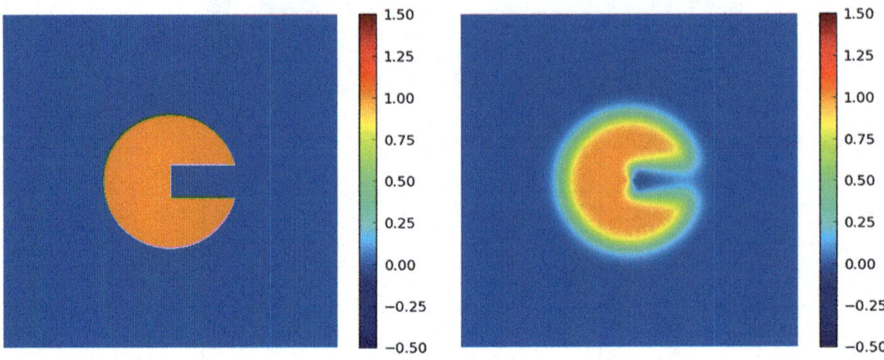

Fig. 6.6 *Left*: Initial condition, i.e., Zalesak's disc of radius 0.2 centered at $(0.5, 0.5)$. *Right*: After one full rotation on a mesh with 800×800 cells

6.5.2 Advection of Zalesak's Disc

Next, we consider an analogue problem to (6.29) but with non-smooth initial data u_0. We use Zalesak's disc (see [19, Sect. 5]) of radius 0.2 centered at $(0.5, 0.5)$ on the unit square (see Fig. 6.6) as initial data and employ a rotational vector field

$$\mathbf{v}(x, y) = \begin{pmatrix} 0.5 - y \\ x - 0.5 \end{pmatrix}.$$

The results are shown in Fig. 6.6. We compare the solution after one full rotation around the center of the square with the initial condition. We observe a slight smearing out of the boundary of the disc, which is due to the low order of polynomial exactness of (6.11). Moreover, we see in Fig. 6.7 a convergence rate of ≈ 0.259 for our upwind scheme and a rate of ≈ 0.302 for the central scheme as the grid is refined. The rates are worse than in the smooth case in Sect. 6.5.1 due to the non-smooth initial condition, but the observed rates match the results from the literature (see [19]) for the rotation of Zalesak's disc nevertheless quite well, see Table 6.2.

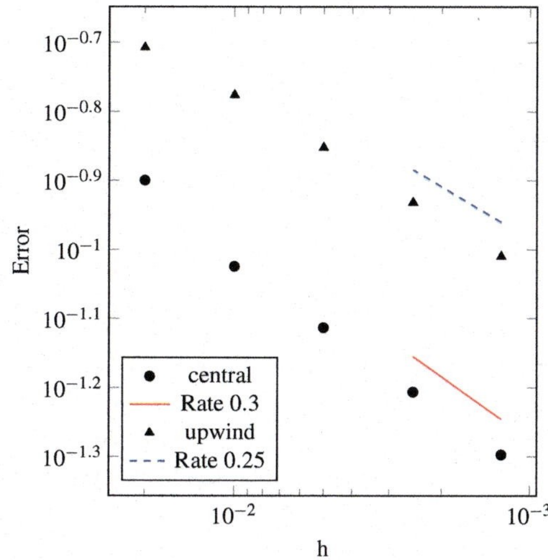

Fig. 6.7 The convergence rates after a full rotation of Zalesak's disc

Table 6.2 Errors after a full rotation of Zalesak's disc

h	Error (upwind)	h	Error (central)
$2.00 \cdot 10^{-2}$	$1.96 \cdot 10^{-1}$	$2.00 \cdot 10^{-2}$	$1.26 \cdot 10^{-1}$
$1.00 \cdot 10^{-2}$	$1.67 \cdot 10^{-1}$	$1.00 \cdot 10^{-2}$	$9.45 \cdot 10^{-2}$
$5.00 \cdot 10^{-3}$	$1.40 \cdot 10^{-1}$	$5.00 \cdot 10^{-3}$	$7.70 \cdot 10^{-2}$
$2.50 \cdot 10^{-3}$	$1.17 \cdot 10^{-1}$	$2.50 \cdot 10^{-3}$	$6.21 \cdot 10^{-2}$
$1.25 \cdot 10^{-3}$	$9.78 \cdot 10^{-2}$	$1.25 \cdot 10^{-3}$	$5.04 \cdot 10^{-2}$

6.5.3 *Advection-Diffusion Convergence Rates*

Now, we deal with an advection-diffusion problem. As in Sect. 6.5.1, we work in the setting of Theorem 1, i.e., we consider the advection for a dual 2-form and discretize the spatial operators as outlined there. Again, we set $u(\cdot, t) := w^2(t)$ and the discrete 1-form corresponding to a vector field $\mathbf{v}$ reads $\langle \mathbf{v}^\flat, \mathbf{e} \rangle = \frac{1}{|\mathbf{e}|}$ for each edge $\mathbf{e}$. We solve the equation

$$\frac{\partial u}{\partial t} - \alpha \Delta u + \nabla \cdot (\mathbf{v}u) = f$$

on $[0, 1]^2$ with periodic boundary conditions and $\mathbf{v} = (1, 1)^T$. As right hand side we choose

$$f(x, y, t) := -v_x 2\pi \cos(2\pi(x - tv_x) \sin(2\pi(y - tv_y)) \tag{6.31}$$

$$- v_y 2\pi \sin(2\pi(x - tv_x) \cos(2\pi(y - tv_y))$$

$$+ 2\pi \cos(2\pi(x - tv_x)) \sin(2\pi(y - tv_y)) +$$

$$+ 2\pi \sin(2\pi(x - tv_x)) \cos(2\pi(y - tv_y))$$

$$+ \alpha 8\pi^2 \sin(2\pi(x - tv_x)) \sin(2\pi(y - tv_y)).$$

This leads to an analytical solution in closed from

$$u(x, y, t) = \sin(2\pi(x - tv_x)) \sin(2\pi(y - tv_y)).$$

As time-integrator, we employ as in Sect. 6.5.1 the explicit Euler scheme with a time-step of $\delta t = \frac{h^2}{2}$. This allows for second order convergence although the time integrator is only of first order. Since we expect convergence rates ranging from first to second order, we use the exponential upwind coefficient depending on the local Péclet number as in (6.25). In the case of pure diffusion the coefficients simplify to the central differences scheme. Furthermore, one sided differences (full upwind) are recovered for $\alpha \to 0$.

For the convergence study, we make use of the fact that we have an analytical solution available. The error is measured after N time-steps, i.e, at time $T = N \cdot \delta_t$ for different diffusion coefficients $\alpha \geq 0$. The results are shown in Fig. 6.8 and

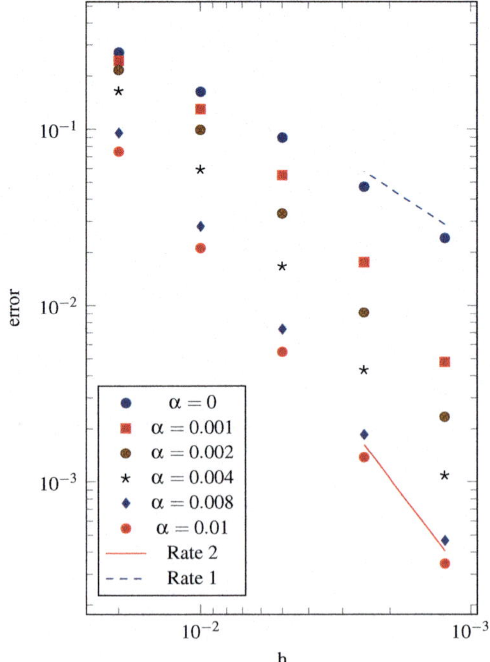

Fig. 6.8 Convergence rates for different diffusion coefficients α with the upwind scheme from (6.25) at time $t = 1.0$

Table 6.3 The errors for different diffusion coefficients α with the Péclet number depending upwind scheme from (6.25) at time $t = 1.0$

$\alpha = 0.0$		$\alpha = 0.001$	
h	Error	h	Error
$2.00 \cdot 10^{-2}$	$2.71 \cdot 10^{-1}$	$2.00 \cdot 10^{-2}$	$2.44 \cdot 10^{-1}$
$1.00 \cdot 10^{-2}$	$1.62 \cdot 10^{-1}$	$1.00 \cdot 10^{-2}$	$1.30 \cdot 10^{-1}$
$5.00 \cdot 10^{-3}$	$8.94 \cdot 10^{-2}$	$5.00 \cdot 10^{-3}$	$5.47 \cdot 10^{-2}$
$2.50 \cdot 10^{-3}$	$4.69 \cdot 10^{-2}$	$2.50 \cdot 10^{-3}$	$1.76 \cdot 10^{-2}$
$1.25 \cdot 10^{-3}$	$2.41 \cdot 10^{-2}$	$1.25 \cdot 10^{-3}$	$4.78 \cdot 10^{-3}$

$\alpha = 0.002$		$\alpha = 0.004$	
h	Error	h	Error
$2.00 \cdot 10^{-2}$	$2.16 \cdot 10^{-1}$	$2.00 \cdot 10^{-2}$	$1.64 \cdot 10^{-1}$
$1.00 \cdot 10^{-2}$	$9.91 \cdot 10^{-2}$	$1.00 \cdot 10^{-2}$	$5.90 \cdot 10^{-2}$
$5.00 \cdot 10^{-3}$	$3.31 \cdot 10^{-2}$	$5.00 \cdot 10^{-3}$	$1.67 \cdot 10^{-2}$
$2.50 \cdot 10^{-3}$	$9.13 \cdot 10^{-3}$	$2.50 \cdot 10^{-3}$	$4.31 \cdot 10^{-3}$
$1.25 \cdot 10^{-3}$	$2.34 \cdot 10^{-3}$	$1.25 \cdot 10^{-3}$	$1.09 \cdot 10^{-3}$

$\alpha = 0.008$		$\alpha = 0.01$	
h	Error	h	Error
$2.00 \cdot 10^{-2}$	$9.50 \cdot 10^{-2}$	$2.00 \cdot 10^{-2}$	$7.46 \cdot 10^{-1}$
$1.00 \cdot 10^{-2}$	$2.80 \cdot 10^{-2}$	$1.00 \cdot 10^{-2}$	$2.11 \cdot 10^{-2}$
$5.00 \cdot 10^{-3}$	$7.32 \cdot 10^{-2}$	$5.00 \cdot 10^{-3}$	$5.45 \cdot 10^{-3}$
$2.50 \cdot 10^{-3}$	$1.85 \cdot 10^{-3}$	$2.50 \cdot 10^{-3}$	$1.37 \cdot 10^{-3}$
$1.25 \cdot 10^{-3}$	$4.64 \cdot 10^{-3}$	$1.25 \cdot 10^{-3}$	$3.44 \cdot 10^{-4}$

Table 6.3. We clearly see the expected behavior, namely that we are able to smoothly vary the scheme (and the convergence rates) depending on the Péclet number.

6.5.4 Advection-Diffusion PDE on a Curved Mesh

As example of a smooth surface embedded into $\mathbb{R}^3$, we consider the unit sphere. We triangulate a sphere with well-centered meshes using the program Comsol. It provided us with eight different meshes with numbers of vertices ranging from $28, 64, 128, 234, 428, 656, 1408$ to 3288. To have finer meshes, we refine thereafter the meshes by adding vertices at all edge midpoints and projecting the new points on the sphere. The meshes, we obtain that way, are well-centered in our cases.[2] Contrary to Sect. 6.5.1, we work here in the setting of Theorem 2,

---

[2]In general this algorithm does not guarantee well-centered meshes.

since we study $\mathbf{v} \cdot \nabla_S u$ instead of $\nabla_S \cdot (\mathbf{v}u)$. The differential equation we study is

$$\frac{\partial u}{\partial t} - \alpha \Delta_S u + \mathbf{v} \cdot \nabla_S u = 0, \quad \alpha \geq 0,$$

where the subscript S denotes the spherical differential operators. We use the discretization

$$\frac{\langle \omega_d^{(2)[n+1]}, \sigma_d^2 \rangle - \langle \omega_d^{(2)[n]}, \sigma_d^2 \rangle}{\delta t} + \left\langle \alpha \mathbf{d} \star \mathbf{d} \star \omega_d^{(2)[n]} + \star i_v \mathbf{d} \star \omega_d^{(2)[n]}, \sigma_d^2 \right\rangle = 0.$$

$$(6.32)$$

With the choice $\sigma_d^2 = \star \sigma_i^{(0)}$ and $\omega_d^{(2)} = \star \omega_p^{(0)}$ this corresponds to Theorem 2. Here, we set $u(\cdot, t) := w^0(t)$ and the discrete 1-form corresponding to a vector field $\mathbf{v}$ reads $\langle \mathbf{v}^b, \mathbf{e} \rangle = \frac{1}{|\mathbf{e}|}$ for each edge $\mathbf{e}$. We first employed a tangential vector field

$$\mathbf{v}(x, y, t) := \begin{pmatrix} -r\sin(\varphi)\cos(\theta) \\ r\cos(\varphi)\cos(\theta) \\ 0 \end{pmatrix}$$

where (r, ϕ, θ) are the spherical coordinates. Recall that we work on the unit sphere, i.e., $r = 1$.

In this experiment, we use the central scheme and the full upwind scheme, see (6.23). As time-integrator we again use the explicit Euler scheme with a time step of $\delta t = \min(h)^2/2$, where $h = \min(\sigma_d^1) \propto \mathrm{diam}(T)$ for primal triangles T, such that the CFL condition holds. As initial condition, we use an exponential function

$$u_0\left(c(\sigma_d^2)\right) = \begin{cases} c_\phi & \text{for } d > r_a \\ c_\phi + e \cdot \exp\left(-\frac{1}{1-d^2}\right) & \text{else} \end{cases}$$

$$(6.33)$$

with $c(\sigma_d^2)$ as the center of the cell σ_d^2, $d = \arccos\left(c(\sigma_d^2) \cdot (0, 1, 0)^T\right)$ as the spherical distance of the cell center to the center of mass and with parameters c_ϕ, r_a.

We choose $c_\phi = 2.0$, $r_a = 1.2$ and $\alpha = 0$ for the advection experiment, using the standard scheme[3] and our upwind scheme.

The result for the standard scheme is depicted in Fig. 6.9. Note that we observe small oscillations due to the central difference scheme.

---

[3]Which is a weighted average of the finite differences scheme using adjacent vertices. This results on rectangular meshes indeed in the central difference scheme.

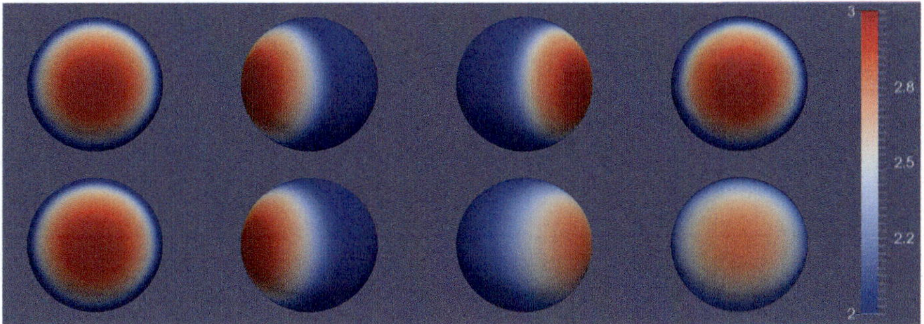

Fig. 6.9 Advection using the central scheme (*top*) and the novel upwind scheme (*bottom*) with a rotational vector field on the surface of a sphere with 13,146 points: Concentration at the initial condition, after $\frac{1}{8}$ rotation, after $\frac{7}{8}$ rotation and for the final state

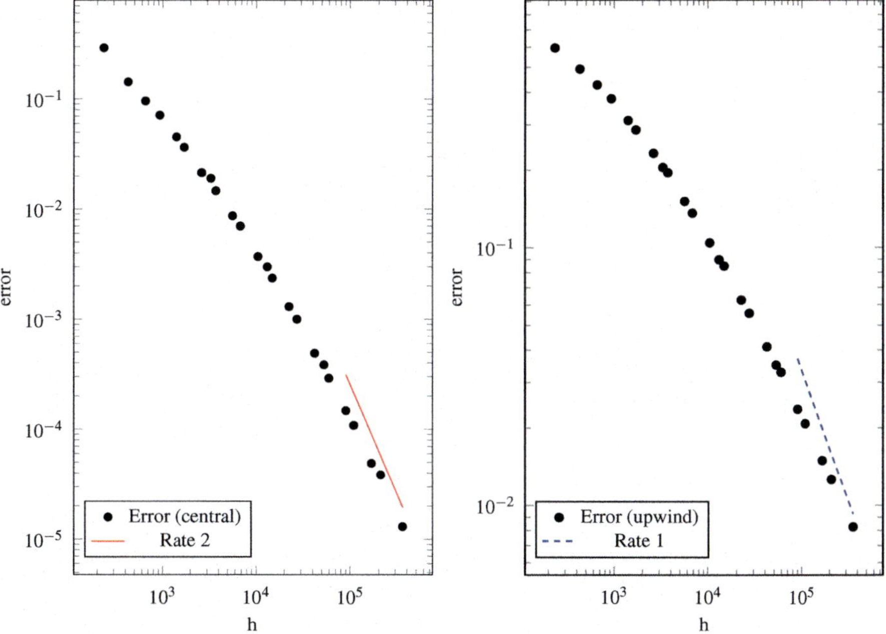

Fig. 6.10 Convergence rates after one rotation around the sphere. *Left*: central scheme, *Right*: upwind scheme

The convergence rate of the standard scheme and the upwind scheme for pure advection ($\alpha = 0$) is depicted in Fig. 6.10, see left panel for the central difference scheme and right panel for the upwind scheme. For the standard scheme, we observe from the left part of Table 6.4 a rate of ≈ 2.015 and for our upwind scheme, we observe a rate of ≈ 0.786 from the right part of Table 6.4. The slightly reduced convergence rate might be due to the curved geometry.

Table 6.4 Error of the concentration in the dual cells after one rotation around the sphere

Number of points	Error (central)	Number of points	Error (upwind)
$2.34 \cdot 10^2$	$2.92 \cdot 10^{-1}$	$2.34 \cdot 10^2$	$5.95 \cdot 10^{-1}$
$4.28 \cdot 10^2$	$1.42 \cdot 10^{-1}$	$4.28 \cdot 10^2$	$4.92 \cdot 10^{-1}$
$6.56 \cdot 10^2$	$9.55 \cdot 10^{-2}$	$6.56 \cdot 10^2$	$4.27 \cdot 10^{-1}$
$9.30 \cdot 10^2$	$7.10 \cdot 10^{-2}$	$9.30 \cdot 10^2$	$3.78 \cdot 10^{-1}$
$1.41 \cdot 10^3$	$4.51 \cdot 10^{-2}$	$1.41 \cdot 10^3$	$3.11 \cdot 10^{-1}$
$1.71 \cdot 10^3$	$3.64 \cdot 10^{-2}$	$1.71 \cdot 10^3$	$2.86 \cdot 10^{-1}$
$2.62 \cdot 10^3$	$2.14 \cdot 10^{-2}$	$2.62 \cdot 10^3$	$2.32 \cdot 10^{-1}$
$3.29 \cdot 10^3$	$1.90 \cdot 10^{-2}$	$3.29 \cdot 10^3$	$2.05 \cdot 10^{-1}$
$3.71 \cdot 10^3$	$1.47 \cdot 10^{-2}$	$3.71 \cdot 10^3$	$1.95 \cdot 10^{-1}$
$5.63 \cdot 10^3$	$8.63 \cdot 10^{-3}$	$5.63 \cdot 10^3$	$1.51 \cdot 10^{-1}$
$6.82 \cdot 10^3$	$6.97 \cdot 10^{-3}$	$6.82 \cdot 10^3$	$1.36 \cdot 10^{-1}$
$1.05 \cdot 10^4$	$3.69 \cdot 10^{-3}$	$1.05 \cdot 10^4$	$1.04 \cdot 10^{-1}$
$1.31 \cdot 10^4$	$2.98 \cdot 10^{-3}$	$1.31 \cdot 10^4$	$8.95 \cdot 10^{-2}$
$1.49 \cdot 10^4$	$2.36 \cdot 10^{-3}$	$1.49 \cdot 10^4$	$8.47 \cdot 10^{-2}$
$2.25 \cdot 10^4$	$1.30 \cdot 10^{-3}$	$2.25 \cdot 10^4$	$6.25 \cdot 10^{-2}$
$2.73 \cdot 10^4$	$9.98 \cdot 10^{-4}$	$2.73 \cdot 10^4$	$5.56 \cdot 10^{-2}$
$4.19 \cdot 10^4$	$4.87 \cdot 10^{-4}$	$4.19 \cdot 10^4$	$4.11 \cdot 10^{-2}$
$5.26 \cdot 10^4$	$3.82 \cdot 10^{-4}$	$5.26 \cdot 10^4$	$3.49 \cdot 10^{-2}$
$5.94 \cdot 10^4$	$2.89 \cdot 10^{-4}$	$5.94 \cdot 10^4$	$3.28 \cdot 10^{-2}$
$9.00 \cdot 10^4$	$1.47 \cdot 10^{-4}$	$9.00 \cdot 10^4$	$2.36 \cdot 10^{-2}$
$1.09 \cdot 10^5$	$1.08 \cdot 10^{-4}$	$1.09 \cdot 10^5$	$2.07 \cdot 10^{-2}$
$1.67 \cdot 10^5$	$4.88 \cdot 10^{-5}$	$1.67 \cdot 10^5$	$1.49 \cdot 10^{-2}$
$2.10 \cdot 10^5$	$3.83 \cdot 10^{-5}$	$2.10 \cdot 10^5$	$1.26 \cdot 10^{-2}$
$3.60 \cdot 10^5$	$1.30 \cdot 10^{-5}$	$3.60 \cdot 10^5$	$8.24 \cdot 10^{-3}$

Left: standard scheme, Right: upwind scheme

6.6 Concluding Remarks and Outlook

We have introduced an upwind scheme based on the observation that the discrete exterior calculus coincides with classical numerical techniques in certain special cases. We have shown the stabilization effect and the numerical convergence properties of the resulting modified DEC method in Sect. 6.5.

Altogether, this is a first important step towards treating two-phase flow problems with a free surface on which a time-dependent advection-diffusion equation for the surfactant concentration is taken into account. But this application is still current work. Due to our new stabilized DEC approach, we are now able to cope with the stability issues of the advection equation on the free surface for general velocities of the fluid. Without a stabilization method, the DEC is surely doomed to fail in such applications as soon as there are larger flow velocities involved.

Finally, we shall give a few comments on further possible modifications of our numerical scheme. A major prerequisite for our new stabilized DEC scheme is the

need of a well-centered mesh in the first place. Its construction in the general setting
of curved surfaces is an issue and a challenging task for complex geometries. It
is known since 1988 that there is a well-centered triangulation for every *planar*
polygonal area [4]. But the existence of well-centered triangulations for the curved
case is much less clear. The same holds for its efficient algorithmic construction.
Thus one may assume the weaker Delaunay property for which mesh generation is
well-known. We expect that for the geometries which arise for e.g. bubbles in two
phase flows, Delaunay triangulations are feasible and can be constructed without
much problems. Our DEC scheme then needs to be slightly modified to account for
negative volume of parts of the dual cell, when a dual vertex is the circumcenter
of a non-well-centered triangle. Furthermore, a special boundary treatment for non-
closed meshes [16] needs to be added. Another possibility is the use of barycentric
dual cells [2], though it is unclear if this is viable for the DEC since crucial features
like the orthogonality of primal and dual edges are lost.

Acknowledgements M. Griebel and A. Schier thank the *DFG* for the financial support through
the *Priority Programme 1506: Transport Processes at Fluidic Interfaces* (SPP 1506). The authors
would also like to thank B. Zwicknagl for reading parts of the manuscript and for inspiring
discussions.

Appendix: Derivation of the Discrete Contraction Formula

In this section, we give a formal derivation of the formula for the discrete
contraction. In contrast to [15, p. 93] we obtain slightly different normalization
factors. For the definitions see [15].

Theorem 3 *For piecewise constant differential forms, ω^{k+1} is the evaluation of i_{σ^1}
on a simplex σ^k with the algebraic definition from the geometric definition using the
extrusion of σ^k equivalent to the evaluation of the discrete differential form on σ^{k+1}.*

Proof We assume the differential form ω^{k+1} as piecewise constant on the $k + 1$-
simplices which yields

$$\langle \omega^{k+1}, \mathrm{extr}_{\sigma^1}(\sigma^k, t) \rangle = \frac{\left| \mathrm{extr}_{\sigma^1}(\sigma^k, t) \right|}{|\sigma^{k+1}|} \langle \omega^{k+1}, \sigma^{k+1} \rangle.$$

Hence, we obtain

$$\langle i_v \omega^{k+1}, \sigma^k \rangle := \left. \frac{\mathrm{d}}{\mathrm{d}t} \right|_{t=0} \langle \omega^{k+1}, \mathrm{extr}(\sigma^k, v_t) \rangle$$

$$= \left. \frac{\mathrm{d}}{\mathrm{d}t} \right|_{t=0} \left(\left| \mathrm{extr}_{\sigma^1}(\sigma^k, t) \right| \right) \frac{1}{|\sigma^{k+1}|} \langle \omega^{k+1}, \sigma^{k+1} \rangle.$$

This shows that we have to consider the map $t \mapsto \left| \mathrm{extr}_{\sigma^1}(\sigma^k, t) \right|$.

Following [15, Chap. 8.3], we get

$$\left|\text{extr}_{\sigma^1}(\sigma^k, t)\right| = \left|\sigma^{k+1}\right| \frac{|h(t)|}{|h(1)|},$$

where σ^k is the base side of a $(k+1)$-simplex and h its height. Here, we use the geometric definition and define the extrusion $\text{extr}_{\omega^1}(\sigma^k, t)$ of a face ω^k as the $(k+1)$-simplex spanned by the face and a vector $\dot{\mathbf{x}}(t) = \sigma^1(1-t), 0 \leq t \leq 1$. We assume without loss of generality that the common vertex σ^0 of the face and the edge lies at the origin. Because we have $\dot{\mathbf{x}}(0) = 0$, we get $\mathbf{x}(t) = t - \frac{t^2}{2}$. The height is proportional to the edge $\sigma^1 = \mathbf{x}(1)$, spanning the $(k+1)$-simplex together with the k-simplex σ^k. Thus, we have for $0 < t < 1$

$$\frac{h(t)}{h(1)} = \frac{|\mathbf{x}(t)|}{|\mathbf{x}(1)|} = \frac{|\sigma^1(t - \frac{t^2}{2})|}{|\sigma^1|} = t - \frac{t^2}{2}. \tag{6.34}$$

Hence, we obtain

$$\left.\frac{\mathrm{d}}{\mathrm{d}t}\right|_{t=0} \left(\left|\text{extr}_{\sigma^1}(\sigma^k, t)\right|\right) = 1 \left|\sigma^{k+1}\right|.$$

This yields the claim. $\square$

References

1. Arnold, D., Falk, R., Winther, R.: Finite element exterior calculus: from Hodge theory to numerical stability. Bull. Am. Math. Soc. **47**(2), 281–354 (2010)
2. Auchmann, B., Kurz, S.: A geometrically defined discrete Hodge operator on simplicial cells. IEEE Trans. Magn. **42**(4), 643 (2006)
3. Aurenhammer, F.: Voronoi diagrams – a survey of a fundamental geometric data structure. ACM Comput. Surv. **23**(3), 345–405 (1991)
4. Baker, B.S., Grosse, E., Rafferty, C.S.: Nonobtuse triangulation of polygons. Discret. Comput. Geom. **3**(2), 147–168 (1988)
5. Bell, N., Hirani, A.N.: PyDEC: software and algorithms for discretization of exterior calculus. ACM Trans. Math. Softw. (TOMS) **39**(1), (2012). doi:10.1145/2382585.2382588
6. Bey, J.: Finite–Volumen–und Mehrgitterverfahren für elliptische Randwertprobleme. Springer, New York (1998)
7. Croce, R., Griebel, M., Schweitzer, M.A.: Numerical simulation of bubble and droplet-deformation by a level set approach with surface tension in three dimensions. Int. J. Numer. Methods Fluids **62**(9), 963–993 (2009)
8. Desbrun, M., Hirani, A.N., Leok, M., Marsden, J.E.: Discrete exterior calculus. Preprint (2005). https://arxiv.org/abs/math/0508341v2
9. Edelsbrunner, H.: Shape reconstruction with Delaunay complex. In: Lucchesi, C.L., Moura, A.V. (eds.) LATIN'98: Theoretical Informatics: Third Latin American Symposium Campinas, Brazil, 20–24 Apr 1998 Proceedings, pp. 119–132. Springer, Berlin/Heidelberg (1998)

10. Edelsbrunner, H.: Geometry and Topology for Mesh Generation. Cambridge University Press, Cambridge (2001)
11. Edelsbrunner, H.: Roots of Geometry and Topology. Springer International Publishing, Cham (2014)
12. Edelsbrunner, H., Harer, J.: Computational Topology: An Introduction. American Mathematical Society, Providence, RI (2010)
13. Griebel, M., Dornseifer, T., Neunhoeffer, T.: Numerical Simulation in Fluid Dynamics: A Practical Introduction. Mathematical Modeling and Simulation, vol. 3. SIAM, Philadelphia, PA (1997)
14. Heumann, H.: Eulerian and semi-Lagrangian methods for advection-diffusion of differential forms, Ph.D. thesis, Dissertation, Eidgenössische Technische Hochschule ETH Zürich, Nr. 19608 (2011)
15. Hirani, A.N.: Discrete exterior calculus, Ph.D. thesis, California Institute of Technology (2003). http://thesis.library.caltech.edu/1885/
16. Hirani, A.N., Kalyanaraman, K., VanderZee, E.B.: Delaunay Hodge star. Comput. Aided Des. **45**(2), 540–544 (2013)
17. Hirani, A.N., Nakshatrala, K.B., Chaudhry, J.H.: Numerical method for Darcy flow derived using discrete exterior calculus. Int. J. Comput. Methods Eng. Sci. Mech. **16**(3), 151–169 (2015)
18. Knabner, P., Angermann, L.: Numerical Methods for Elliptic and Parabolic Partial Differential Equations. Texts in Applied Mathematics, vol. 44. Springer, New York (2003)
19. Laadhari, A., Saramito, P., Misbah, C.: Improving the mass conservation of the level set method in a finite element context. C.R. Math. **348**(9), 535–540 (2010)
20. Mohamed, M.S., Hirani, A.N., Ravi, S.: Discrete exterior calculus discretization of incompressible Navier–Stokes equations over surface simplicial meshes. J. Comput. Phys. **312**, 175–191 (2016)
21. Mullen, P., McKenzie, A., Pavlov, D., Durant, L., Tong, Y., Kanso, E., Marsden, J.E., Desbrun, M.: Discrete Lie advection of differential forms. Found. Comput. Math. **11**(2), 131–149 (2011)
22. Nestler, M., Nitschke, I., Praetorius, S., Voigt, A.: Orientational order on surfaces – the coupling of topology, geometry and dynamics. ArXiv e-prints (2016). https://www.arxiv.org/pdf/1608.01343v1.pdf
23. Nitschke, I., Reuther, S., Voigt, A.: Discrete exterior calculus (DEC) for the surface Navier-Stokes equation. ArXiv e-prints (2016). https://arxiv.org/abs/1611.04392
24. VanderZee, E., Hirani, A.N., Guoy, D., Zharnitsky, V., Ramos, E.: Geometric and combinatorial properties of well-centered triangulations in three and higher dimensions. Comput. Geom. Theory Appl. **46**(6), 700–724 (2013)

Chapter 7
Discrete Exterior Calculus (DEC) for the Surface Navier-Stokes Equation

Ingo Nitschke, Sebastian Reuther, and Axel Voigt

Abstract We consider a numerical approach for the incompressible surface Navier-Stokes equation. The approach is based on the covariant form and uses discrete exterior calculus (DEC) in space and a semi-implicit discretization in time. The discretization is described in detail and related to finite difference schemes on staggered grids in flat space for which we demonstrate second order convergence. We compare computational results with a vorticity-stream function approach for surfaces with genus $g(\mathcal{S}) = 0$ and demonstrate the interplay between topology, geometry and flow properties. Our discretization also allows to handle harmonic vector fields, which we demonstrate on a torus.

7.1 Introduction

We consider a compact smooth Riemannian surface $\mathcal{S}$ without boundary and an incompressible surface Navier-Stokes equation

$$\partial_t \mathbf{v} + \nabla_\mathbf{v} \mathbf{v} = -\operatorname{grad}_\mathcal{S} p + \frac{1}{\mathrm{Re}} \left(-\boldsymbol{\Delta}^{\mathrm{dR}} \mathbf{v} + 2\kappa \mathbf{v} \right) \tag{7.1}$$

$$\operatorname{div}_\mathcal{S} \mathbf{v} = 0 \tag{7.2}$$

in $\mathcal{S} \times (0, \infty)$ with initial condition $\mathbf{v}(\mathbf{x}, t = 0) = \mathbf{v}_0(\mathbf{x}) \in \mathsf{T}_\mathbf{x}\mathcal{S}$. Thereby $\mathbf{v}(t) \in \mathsf{T}\mathcal{S}$ denotes the tangential surface velocity, $p(\mathbf{x}, t) \in \mathbb{R}$ the surface pressure, Re the surface Reynolds number, κ the Gaussian curvature, $\mathsf{T}_\mathbf{x}\mathcal{S}$ the tangent space on $\mathbf{x} \in \mathcal{S}$, $\mathsf{T}\mathcal{S} = \cup_{\mathbf{x} \in \mathcal{S}} \mathsf{T}_\mathbf{x}\mathcal{S}$ the tangent bundle and $\nabla_\mathbf{v}, \operatorname{grad}_\mathcal{S}, \operatorname{div}_\mathcal{S}$ and $\boldsymbol{\Delta}^{\mathrm{dR}}$ the covariant directional derivative, surface gradient, surface divergence and surface Laplace-DeRham operator, respectively. As in flat space the equation results from conservation of mass and (tangential) linear momentum. However, differences are found in the appearing operators and the additional term including the Gaussian

I. Nitschke • S. Reuther • A. Voigt (✉)
Institute of Scientific Computing, Technische Universität Dresden, Zellescher Weg 12-14, 01069
Dresden, Germany
e-mail: ingo.nitschke@tu-dresden.de; sebastian.reuther@tu-dresden.de; axel.voigt@tu-dresden.de

© Springer International Publishing AG 2017
D. Bothe, A. Reusken (eds.), *Transport Processes at Fluidic Interfaces*,
Advances in Mathematical Fluid Mechanics, DOI 10.1007/978-3-319-56602-3_7

curvature. The Laplace-DeRham operator and the Gaussian curvature term thereby result from the divergence of the deformation tensor and the non-commutativity of the second covariant derivative in curved spaces, see e.g. [4, 23]. The unusual sign in front of the Laplacian results from the definition of the Laplace-DeRham operator [1], see Sect. 7.2. Alternatively, the equations can also be derived from the Rayleigh dissipation potential [9]. The equations are related to the Boussinesq-Scriven constitutive law for the surface viscosity in two-phase flow problems [6, 35, 36] and to fluidic biomembranes [4, 5, 15, 21]. Further applications can be found in computer graphics [14, 26, 40].

While a huge literature exists for the two-dimensional Navier-Stokes equation in flat space, results for its surface counterpart equations (7.1) and (7.2) are rare. For treatments in the mathematical literature we refer to [13, 23]. Numerical approaches are considered in [29, 32], where a surface vorticity-stream function formulation is introduced. This follows by considering the velocity $\mathbf{v}$ as the curl of a smooth scalar valued function ψ, i.e. $\mathbf{v} = \mathrm{rot}_S \, \psi$. For the correct definition of the curl operator $\mathrm{rot}_S(\cdot)$ we refer to [27]. On a compact, boundaryless, oriented Riemannian manifold of genus $g(S) = 0$, this representation is unique up to a constant by the Hodge decomposition theorem [1]. The resulting equations, after taking the curl and written as a system of two second order scalar surface partial differential equations, read

$$\partial_t \phi + J(\psi, \phi) = \frac{1}{\mathrm{Re}}(\Delta_S \phi + 2 \, \mathrm{div}_S(\kappa \, \mathrm{grad}_S \, \psi)) \tag{7.3}$$

$$\phi = \Delta_S \psi \tag{7.4}$$

in $S \times (0, \infty)$ with initial condition $\psi \, (\mathbf{x}, t = 0) = \psi_0(\mathbf{x}) \in \mathbb{R}$. Here ϕ is the surface vorticity, Δ_S the Laplace-Beltrami operator and $J(\psi, \phi) = \langle \mathrm{rot}_S \, \psi, \mathrm{grad}_S \, \phi \rangle$ the Jacobian. Equations (7.3) and (7.4) are either solved using the surface finite element approach [11, 12, 41], see [29, 32] for details, or the diffuse interface approach [31], see [33] for details. The equations, but without the Gaussian curvature term, has also been discretized using a discrete exterior calculus (DEC) approach [25]. We are not aware of any direct numerical approach for Eqs. (7.1) and (7.2), which will be the purpose of this paper. Such an approach will be desirable for surfaces with genus $g(S) \neq 0$, as it allows to also deal with harmonic vector fields. We will introduce a DEC approach and validate the results against a surface finite element discretization for the vorticity-stream function formulation in Eqs. (7.3) and (7.4) on surfaces with $g(S) = 0$ and show nontrivial solutions with $\mathrm{div}_S \, \mathbf{v} = 0$ and $\mathrm{rot}_S \, \mathbf{v} = 0$ on a torus.

The paper is organized as follows. In Sect. 7.2 we introduce the necessary notation and provide the formulation in covariant form. In Sect. 7.3 the DEC discretization is described in detail and compared with known discretizations in flat space. After some analytical results for the surface Navier-Stokes equation we use the properties of Killing vector fields to validate the approach on various surfaces and demonstrate the strong interplay of geometric and vortex interactions in Sect. 7.4. Conclusions are drawn in Sect. 7.5. In the Appendix we provide additional notation and prove second order convergence for the corresponding finite difference scheme in flat space.

7.2 Formulation in Covariant Form

For the readers convenience we here briefly review the basic notion. A more detailed description can be found in [27]. The key ingredient for a covariant formulation in local coordinates ϕ and θ is the positive definite metric tensor

$$\mathbf{g} = \begin{bmatrix} g_{\phi\phi} & g_{\phi\theta} \\ g_{\phi\theta} & g_{\theta\theta} \end{bmatrix} = g_{\phi\phi}\,d\phi^2 + 2g_{\phi\theta}\,d\phi\,d\theta + g_{\theta\theta}\,d\theta^2 . \tag{7.5}$$

$\mathbf{g}$ can be obtained from a surface parametrization $\mathbf{x} : \mathbb{R}^2 \supset U \to \mathbb{R}^3;\ (\phi,\theta) \mapsto \mathbf{x}(\phi,\theta)$, which maps local coordinates to the embedded $\mathbb{R}^3$ representation of the surface $\mathcal{S} = \mathbf{x}(U)$. The covariant components of the metric tensor are given by $\mathbb{R}^3$ inner products of partial derivatives of $\mathbf{x}$, i.e. $g_{ij} = \partial_i\mathbf{x} \cdot \partial_j\mathbf{x}$. The components of the inverse tensor $\mathbf{g}^{-1}$ are denoted by g^{ij} and the determinant of $\mathbf{g}$ by $|\mathbf{g}|$. We denote by $\{\partial_\phi\mathbf{x}, \partial_\theta\mathbf{x}\}$ the canonical basis to describe the contravariant (tangential) vector $\mathbf{v}(\mathbf{x}) \in \mathsf{T}_\mathbf{x}\mathcal{S}$, i.e. $\mathbf{v}(\mathbf{x}) = (u^\phi, u^\theta) = u^\phi\partial_\phi\mathbf{x} + u^\theta\partial_\theta\mathbf{x}$ at a point $\mathbf{x} \in \mathcal{S}$. Furthermore, with the arising dual basis $\{d\phi, d\theta\}$ we are able to write an arbitrary 1-form (covariant vector) $\mathbf{u}(\mathbf{x}) \in \mathsf{T}_\mathbf{x}^*\mathcal{S}$ as $\mathbf{u}(\mathbf{x}) = u_\phi d\phi + u_\theta d\theta$. This identifier choice of the covariant vector coordinates u_i in conjunction with representation of $\mathbf{v}$ as above implies that $\mathbf{u}$ and $\mathbf{v}$ are related by $\mathbf{u} = \mathbf{v}^\flat$ and $\mathbf{v} = \mathbf{u}^\sharp$, respectively. Explicitly lowering and rising the indices can be done using the metric tensor $\mathbf{g}$ by $u_i = g_{ij}u^j$ and $u^i = g^{ij}u_j$, respectively. The scalar $p(\mathbf{x})$ is also considered as a 0-form.

We now use exterior calculus (EC) to describe all present first order differential operators by the Hodge star $*$ and the exterior derivative $\mathbf{d}$, which arise algebraically (see [1] for details). In [1] the Laplace-deRham operator is defined for k-forms on a n-dimensional Riemannian manifold by $\mathbf{\Delta}^{\mathrm{dR}} := (-1)^{nk+1}\,(*\mathbf{d}*\mathbf{d} + \mathbf{d}*\mathbf{d}*)$. For vector fields the Laplace-deRham operator can thus be defined canonically as composition $(\sharp \circ \mathbf{\Delta}^{\mathrm{dR}} \circ \flat)$. This leads to $\mathbf{\Delta}^{\mathrm{dR}}\mathbf{v} = -\left(\mathbf{\Delta}^{\mathrm{RR}} + \mathbf{\Delta}^{\mathrm{GD}}\right)\mathbf{v}$, with the Rot-Rot-Laplace $\mathbf{\Delta}^{\mathrm{RR}}\mathbf{v} := \mathrm{rot}_\mathcal{S}\,\mathrm{rot}_\mathcal{S}\,\mathbf{v}$ and Grad-Div-Laplace $\mathbf{\Delta}^{\mathrm{GD}}\mathbf{v} := \mathrm{grad}_\mathcal{S}\,\mathrm{div}_\mathcal{S}\,\mathbf{v}$. Due to the incompressibility constraint $\mathrm{div}_\mathcal{S}\,\mathbf{v} = 0$ we thus have $\mathbf{\Delta}^{\mathrm{dR}}\mathbf{v} = -\mathbf{\Delta}^{\mathrm{RR}}\mathbf{v}$ and therefor in our case only $\mathbf{\Delta}^{\mathrm{dR}}\mathbf{u} = -(*\mathbf{d}*\mathbf{d})\mathbf{u}$. Equations (7.1) and (7.2) read in their covariant form

$$\partial_t\mathbf{u} + \nabla_\mathbf{u}\mathbf{u} = -\mathbf{d}p + \frac{1}{\mathrm{Re}}\left((*\mathbf{d}*\mathbf{d})\mathbf{u} + 2\kappa\mathbf{u}\right) \tag{7.6}$$

$$*\mathbf{d}*\mathbf{u} = 0 \tag{7.7}$$

in $\mathcal{S} \times (0,\infty)$ with $[\nabla_\mathbf{u}\mathbf{u}]_i = u^j u_{i|j}$ to be discussed below and initial conditions $\mathbf{u}(\mathbf{x}, t = 0) = \mathbf{u}_0(\mathbf{x}) \in \mathsf{T}_\mathbf{x}^*\mathcal{S}$.

7.3 DEC Discretization

The mathematical foundation of discrete exterior calculus (DEC) can be found in [8, 20, 22]. It follows by successively utilizing a discrete version of the Hodge star $*$ and the Stokes theorem for the exterior derivative $\mathbf{d}$. The approach has been successfully used in computer graphics, e.g. surface parametrization, see e.g. [17, 19, 38], and vector field decomposition and smoothing, see e.g. [16, 30, 37]. A rigorous treatment of the connection between discrete and continuous settings is given in [3]. We discuss the discretization for each term, introduce a time discretization and compare the resulting discrete system with known discretization schemes in flat space. However, we first introduce the degrees of freedom (DOFs) and rewrite the advection term to be suitable for the DEC discretization.

7.3.1 Degrees of Freedom (DOFs)

We consider a simplicial complex $\mathcal{K} = \mathcal{V} \sqcup \mathcal{E} \sqcup \mathcal{T}$ containing sets of vertices $\mathcal{V}$, edges $\mathcal{E}$ and (triangular) faces $\mathcal{T}$ which approximate $\mathcal{S}$. The quantities of interest in our DEC discretization are 0- and 1-forms, $p \in \Lambda^0(\mathcal{S})$ with $p(\mathbf{x}) \in \mathbb{R}$ and $\mathbf{u} \in \Lambda^1(\mathcal{S}) = T^*\mathcal{S}$, respectively. The discrete 0-forms are considered on $v \in \mathcal{V}$, $p_h(v) := p(\mathbf{x})_{|\mathbf{x}=v}$. For 1-forms we introduce DOFs as integral values on the edges $e \in \mathcal{E}$, i.e. $u_h(e) := \int_{\pi(e)} \mathbf{u}$, with the gluing map $\pi : \mathcal{E} \to \mathcal{S}$, which projects geometrically the edge e to the surface $\mathcal{S}$. The mapping $u_h \in \Lambda^1_h(\mathcal{K})$ is called the discrete 1-form of $\mathbf{u}$, since $u_h(e)$ approximates $\mathbf{u}(e) \equiv \mathbf{u}(e) = \langle \mathbf{v}, \mathbf{e} \rangle$ on an intermediate point $\xi \in \pi(e) \subset \mathcal{S}$, where the edge vector $\mathbf{e}$ exists in $T_\xi \mathcal{S}|_{\pi(e)}$ by the mean value theorem. Therefore, we approximate 1-forms on the restricted dual tangential space $T_\xi \mathcal{S}|_{\pi(e)}$, which is a one dimensional vector space in $\xi \in \mathcal{S}$ likewise the space of discrete 1-forms $\Lambda^1_h(\mathcal{K})|_e = \Lambda^1_h(\{e\})$ restricted to the edge e, see [27] for details. Furthermore, a discrete 1-form $u_h(e)$ can be approximated as $|e|\langle \mathbf{v}, \mathbf{e} \rangle (c(e))$ by the midpoint rule, with the midpoint $c(e) = \frac{v_1 + v_2}{2}$ of the edge $e = [v_1, v_2]$. If the mesh is considered to be flat and the faces are considered to be squares, we obtain the same DOF positions as for discretizations on a staggered grid, see Sect. 7.3.6.

7.3.2 Approximation of the Advection Term

The advection term in Eq. (7.6) is not yet written in an appropriate form for a DEC discretization. We linearize this term using a Taylor expansion with a known 1-form $\tilde{\mathbf{u}}$ and obtain

$$[\nabla_{\mathbf{u}}\mathbf{u}]_i = u^j u_{i|j} \approx \tilde{u}^j \tilde{u}_{i|j} + \tilde{u}_{i|j}\left(u^j - \tilde{u}^j\right) + \tilde{u}^j\left(u_{i|j} - \tilde{u}_{i|j}\right)$$

$$= u^j \tilde{u}_{i|j} + \tilde{u}^j u_{i|j} - \tilde{u}^j \tilde{u}_{i|j} = [\nabla_{\tilde{\mathbf{u}}}\mathbf{u} + \nabla_{\mathbf{u}}\tilde{\mathbf{u}} - \nabla_{\tilde{\mathbf{u}}}\tilde{\mathbf{u}}]_i .$$

With the Levi-Cevita-tensor $\mathbf{E}$ (defined by the volumetric form $\mathbf{E}(\mathbf{u}, \tilde{\mathbf{u}}) = \mu(\mathbf{u}, \tilde{\mathbf{u}})$) we obtain for $\mathbf{u}, \tilde{\mathbf{u}} \in T^*\mathcal{S}$

$$(\mathrm{rot}_\mathcal{S}\,\mathbf{u})\,[*\tilde{\mathbf{u}}]_i = E_{il}E_{jk}\tilde{u}^l u^{j|k} = \left(g_{ij}g_{lk} - g_{ik}g_{lj}\right)\tilde{u}^l u^{j|k} = \tilde{u}^l\left(u_{i|l} - u_{l|i}\right)$$

$$= [\nabla_{\tilde{\mathbf{u}}}\mathbf{u}]_i - \left[\tilde{\mathbf{u}}^\sharp \cdot \mathrm{grad}_\mathcal{S}\,\mathbf{u}\right]_i.$$

We further have

$$\left[\tilde{\mathbf{u}}^\sharp \cdot \mathrm{grad}_\mathcal{S}\,\mathbf{u} + \mathbf{u}^\sharp \cdot \mathrm{grad}_\mathcal{S}\,\tilde{\mathbf{u}}\right]_i = \tilde{u}^l u_{l|i} + u_l\tilde{u}^l_{|i} = \left(\tilde{u}^l u_l\right)_{|i} = \partial_i\,\langle\tilde{\mathbf{u}}, \mathbf{u}\rangle$$

and thus also $2\left[\mathbf{u}^\sharp \cdot \mathrm{grad}_\mathcal{S}\,\mathbf{u}\right]_i = \partial_i\,\|\mathbf{u}\|^2$. Putting everything together and using $\mathrm{rot}_\mathcal{S}\,\tilde{\mathbf{u}} = -\,\mathrm{div}_\mathcal{S}(*\tilde{\mathbf{u}})$ we thus obtain

$$\nabla_\mathbf{u}\mathbf{u} \approx \mathbf{d}\left(\langle\mathbf{u}, \tilde{\mathbf{u}}\rangle - \frac{1}{2}\,\|\tilde{\mathbf{u}}\|^2\right) + (\mathrm{rot}_\mathcal{S}\,\mathbf{u} - \mathrm{rot}_\mathcal{S}\,\tilde{\mathbf{u}})\,(*\tilde{\mathbf{u}}) - \mathrm{div}_\mathcal{S}(*\tilde{\mathbf{u}})(*\mathbf{u})$$

which provides a suitable form for a DEC approach. By using $\mathrm{rot}_\mathcal{S}\,\mathbf{u} = *\mathbf{du}$ and $\mathrm{div}_\mathcal{S}\,\mathbf{u} = *\mathbf{d}*\mathbf{u}$ we obtain

$$\nabla_\mathbf{u}\mathbf{u} \approx \mathbf{d}\left(\langle\mathbf{u}, \tilde{\mathbf{u}}\rangle - \frac{1}{2}\,\|\tilde{\mathbf{u}}\|^2\right) + (*\mathbf{du} - *\mathbf{d}\tilde{\mathbf{u}})\,(*\tilde{\mathbf{u}}) - (*\mathbf{d}*)(*\tilde{\mathbf{u}})(*\mathbf{u}) \qquad (7.8)$$

which will be used for discretization.

7.3.3 Time-Discrete Equations

We consider a semi-implicit Euler discretization and use the approximation of the advection term with $\tilde{\mathbf{u}} = \mathbf{u}_k$, the solution at time t_k. For $\tau_k := t_{k+1} - t_k$ and initial condition $\mathbf{u}_0$ we get a sequence of linear systems for $k = 0, 1, 2, \ldots$. We introduce the generalized pressure $q_{k+1} = p_{k+1} + \langle\mathbf{u}_{k+1}, \mathbf{u}_k\rangle - \frac{1}{2}\,\|\mathbf{u}_k\|^2$ and solve for $\mathbf{u}_{k+1}$, q_{k+1} and p_{k+1}

$$\frac{1}{\tau_k}\mathbf{u}_{k+1} + \mathbf{d}q_{k+1} + (*\mathbf{du}_{k+1})(*\mathbf{u}_k) - (*\mathbf{d}*)(*\mathbf{u}_k)(*\mathbf{u}_{k+1})$$

$$-\frac{1}{\mathrm{Re}}\left((*\mathbf{d}*\mathbf{d})\mathbf{u}_{k+1} + 2\kappa\mathbf{u}_{k+1}\right) = \frac{1}{\tau_k}\mathbf{u}_k + (*\mathbf{du}_k)(*\mathbf{u}_k) \qquad (7.9)$$

$$\langle\mathbf{u}_{k+1}, \mathbf{u}_k\rangle + p_{k+1} - q_{k+1} = \frac{1}{2}\,\|\mathbf{u}_k\|^2 \qquad (7.10)$$

$$*\mathbf{d}*\mathbf{u}_{k+1} = 0 \qquad (7.11)$$

on $\mathcal{S}$.

7.3.4 Fully-Discrete Equations

The used notation follows [27], see also Appendix 1. For the discrete 0-forms $p_h, q_h \in \Lambda_h^0(\mathcal{K})$, 1-forms $u_h \in \Lambda_h^1(\mathcal{K})$, sign mappings $s_{o,o} \in \{-1, +1\}$, volumes $|\cdot|$, Voronoi cells $\star v$, Voronoi edges $\star e$ and the "belongs-to" relations $\succ$ and $\prec$ we obtain for $\mathbf{\Delta}^{\mathrm{RR}}$, $\mathrm{div}_{\mathcal{S}}$ and $\mathrm{rot}_{\mathcal{S}}$

$$(\ast\mathbf{d}\ast\mathbf{d})\mathbf{u}(e) \approx -\frac{|e|}{|\star e|} \sum_{T \succ e} \frac{s_{T,e}}{|T|} \sum_{\tilde{e} \prec T} s_{T,\tilde{e}}\, u_h(\tilde{e}) ,$$

$$(\ast\mathbf{d}\ast\mathbf{u})(v) \approx -\frac{1}{|\star v|} \sum_{\tilde{e} \succ v} s_{v,\tilde{e}} \frac{|\star \tilde{e}|}{|\tilde{e}|} \mathbf{u}(\tilde{e}) ,$$

$$(\ast\mathbf{du})(c(e)) \approx \frac{1}{\sum_{T \succ e}|T|} \sum_{T \succ e}(\mathbf{du})(T) = \frac{1}{\sum_{T \succ e}|T|} \sum_{T \succ e}\sum_{\tilde{e} \prec T} s_{T,\tilde{e}}\mathbf{u}(\tilde{e})$$

respectively. The last line above follows from a special Hodge dualism between midpoint $c(e)$ and face union $\bigcup_{T \succ e} T =: \hat{\star} c(e)$, such that function evaluations at $c(e)$ are integral mean values over $\hat{\star} c(e)$. This allows to approximate

$$(\ast\mathbf{du})(\ast\tilde{\mathbf{u}})(e) \approx \frac{(\ast\tilde{\mathbf{u}})(e)}{\sum_{T \succ e}|T|} \sum_{T \succ e}\sum_{\tilde{e} \prec T} s_{T,\tilde{e}}\mathbf{u}(\tilde{e}) ,$$

$$(\ast\mathbf{d}\ast\tilde{\mathbf{u}})(\ast\mathbf{u})(e) \approx -\frac{1}{2}\left(\sum_{v \prec e} \frac{1}{|\star v|} \sum_{\tilde{e} \succ v} s_{v,\tilde{e}} \frac{|\star \tilde{e}|}{|\tilde{e}|} \tilde{\mathbf{u}}(\tilde{e})\right)(\ast\mathbf{u})(e) .$$

With the Stokes theorem we further have $(\mathbf{d}q)(e) = q(v_j) - q(v_i)$ for $e = [v_i, v_j]$. What remains to define is a discrete Hodge operator and a discrete version of the inner product. We approximate

$$(\ast\mathbf{u})\,(e) \approx \circledast u_h(e)$$

$$= \frac{1}{4}\sum_{T \succ e}\sum_{\substack{\tilde{e} \prec T \\ \tilde{e} \neq e}} \frac{s_{e\tilde{e}}}{\sqrt{|e|^2 |\tilde{e}|^2 - (\mathbf{e}\cdot\tilde{\mathbf{e}})^2}}\left((\mathbf{e}\cdot\tilde{\mathbf{e}})\, u_h(e) - |e|^2\, u_h(\tilde{e})\right) .$$

For other possibilities we refer to [24]. For the inner product we follow [27] and define

$$\langle \tilde{\mathbf{u}}, \mathbf{u}\rangle\,(c(e)) \approx \frac{1}{|e|^2}\left(\tilde{\mathbf{u}}(e)\mathbf{u}(e) + (\ast\tilde{\mathbf{u}})\,(e)\,(\ast\mathbf{u})\,(e)\right) .$$

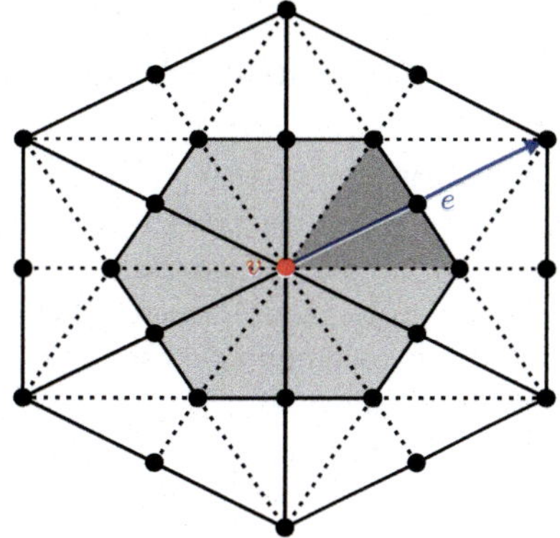

Fig. 7.1 Circumcentric subdivision of a simple simplicial complex around a vertex v. The Voronoi cell $\star v$ is marked *light gray* and the part A_{ve} *dark gray*

In order to approximate the inner product at a primal vertex we consider the decomposition of the Voronoi cell $\star v = \sum_{e \succ v} A_{ve}$, with $|A_{ve}| = (|e| \, |\star e|)/4$, see Fig. 7.1. We thus obtain

$$\langle \tilde{\mathbf{u}}, \mathbf{u} \rangle (v) \approx \frac{1}{|\star v|} \left(\ast \langle \tilde{\mathbf{u}}, \mathbf{u} \rangle \right) (\star v) = \frac{1}{|\star v|} \sum_{e \succ v} \left(\ast \langle \tilde{\mathbf{u}}, \mathbf{u} \rangle \right) (A_{ve})$$

$$\approx \frac{1}{|\star v|} \sum_{e \succ v} \frac{|e| \, |\star e|}{4} \langle \tilde{\mathbf{u}}, \mathbf{u} \rangle \left(c(e) \right)$$

$$\approx \frac{1}{4 \, |\star v|} \sum_{e \succ v} \frac{|\star e|}{|e|} \left(\tilde{\mathbf{u}}(e) \mathbf{u}(e) + (\ast \tilde{\mathbf{u}}) \, (e) \, (\ast \mathbf{u}) \, (e) \right) .$$

In case κ is not given analytically, a numerical approximation is required, which can effectively be done using a DEC approach for the Weingarten map [28].

7.3.5 Linear System

Putting everything together and using an additional equation

$$\circledast \mathbf{u}_{k+1}(e) - (\ast \mathbf{u})_{k+1}(e) = 0 \tag{7.12}$$

for all $e \in \mathcal{E}$ to determine the Hodge dual 1-form defines a linear system for $\mathbf{u}_{k+1}, (\ast \mathbf{u})_{k+1} \in \Lambda_h^1(\mathcal{K})$ and $q_{k+1}, p_{k+1} \in \Lambda_h^0(\mathcal{K})$. An appropriate assembly over $e \in \mathcal{E}$ and $v \in \mathcal{V}$ results in a sparse matrix $M_{k+1} \in \mathbb{R}^{2(|\mathcal{E}|+|\mathcal{V}|) \times 2(|\mathcal{E}|+|\mathcal{V}|)}$ and the right hand side vector $r_k \in \mathbb{R}^{2(|\mathcal{E}|+|\mathcal{V}|)}$. To determine the pressure we replace a row in M_{k+1} and r_{k+1} to ensure $p_{k+1}(v_0) = 0$ at $v_0 \in \mathcal{V}$. The linear system is solved using `umfpack`.

7.3.6 Comparison with Finite Difference Schemes on Uniform Rectangular Meshes in Two Dimensions

To compare the resulting scheme with known discretization schemes we consider the two-dimensional Navier-Stokes equation in flat space. The Gaussian curvature κ vanishes and the surface operators reduce to the classical two-dimensional operators grad, div, rot and Δ. Instead of the simplicial complex $\mathcal{K}$ we consider for simplicity a uniform rectangular mesh. The DEC discretization can then be considered as introduced above.

We identify the vector-components in the midpoints of the edges as the discrete 1-form u_h. We thus obtain with the grid spacing h and the notation in Fig. 7.2

$$u_{ij}^x := u^x(c(e_{i,j}^x)) = \frac{1}{h} u_h(e_{i,j}^x), \qquad u_{ij}^y := u^y(c(e_{i,j}^y)) = \frac{1}{h} u_h(e_{i,j}^y)$$

For the pressure we obtain with the discrete 0-form q_h

$$q_{i,j} := q(v_{i,j}) = q_h(v_{i,j}).$$

To analyze the scheme we here only consider the discretization of the Laplace operator, which is restricted to $\Delta = -\Delta^{\mathrm{RR}}$ in the present case, with

$$\left(\Delta^{\mathrm{RR}} u\right)^x = \partial_y^2 u^x - \partial_x \partial_y u^y, \qquad \left(\Delta^{\mathrm{RR}} u\right)^y = \partial_x^2 u^y - \partial_x \partial_y u^x \tag{7.13}$$

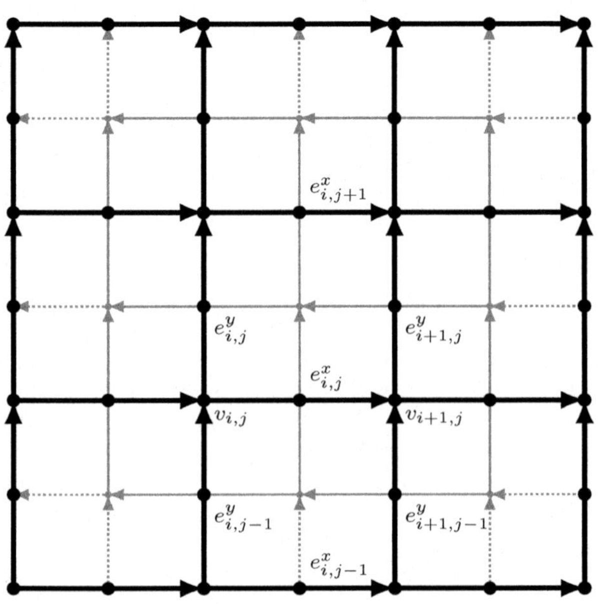

Fig. 7.2 Staggered grid with dual mesh and orientation. The components of the velocity u^x and u^y are defined on the midpoints of the edges, u^x on the horizontal e^x and u^y on the vertical e^y, and the pressure is defined on the vertices v. Such meshes are also known as Arakawa C-meshes [2]

and with our DEC discretization

$$(\mathbf{\Delta}^{\mathrm{RR}}u)^x_{i,j} =$$

$$\frac{1}{h^2}\left(u^x_{i,j+1} + u^x_{i,j-1} - 2u^x_{i,j} + u^y_{i,j} - u^y_{i+1,j} + u^y_{i+1,j-1} - u^y_{i,j-1}\right)$$

$$(\mathbf{\Delta}^{\mathrm{RR}}u)^y_{i,j} =$$

$$\frac{1}{h^2}\left(u^y_{i+1,j} + u^y_{i-1,j} - 2u^y_{i,j} - u^x_{i,j} + u^x_{i,j+1} - u^x_{i-1,j+1} + u^x_{i-1,j}\right).$$

$$(7.14)$$

This unusual stencil is visualized in Fig. 7.3. For the full Laplace operator $\mathbf{\Delta} = -\left(\mathbf{\Delta}^{\mathrm{RR}} + \mathbf{\Delta}^{\mathrm{GD}}\right)$, as considered in [27] and also typically used in flat space, we obtain

$$(\mathbf{\Delta}u)^{\{x,y\}}_{i,j} = \frac{1}{h^2}\left(u^{\{x,y\}}_{i+1,j} + u^{\{x,y\}}_{i-1,j} + u^{\{x,y\}}_{i,j+1} + u^{\{x,y\}}_{i,j-1} - 4u^{\{x,y\}}_{i,j}\right)$$

which is the usual five-point stencil, again visualized in Fig. 7.3. We thus have $(\mathbf{\Delta}u)^{\{x,y\}}_{i,j} \neq (\mathbf{\Delta}^{\mathrm{RR}}u)^{\{x,y\}}_{i,j}$, even if the identity holds in the continuous case under the incompressibility constraint. However, the order of consistency is $\mathcal{O}(h^2)$ for both stencils, which can be shown by a Taylor expansion for each component, see Appendix 2 for details.

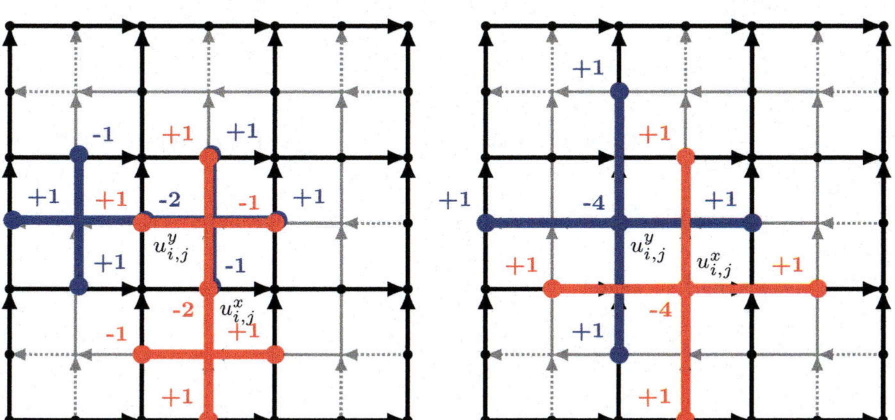

Fig. 7.3 *Left*: Illustration of the difference schemes for $\left(\mathbf{\Delta}^{\mathrm{RR}}u\right)^x$ (*red*) and $\left(\mathbf{\Delta}^{\mathrm{RR}}u\right)^y$ (*blue*). *Right*: Illustration of the schemes for $\left(\mathbf{\Delta}u\right)^x$ (*red*) and $\left(\mathbf{\Delta}u\right)^y$ (*blue*), which is the well known five-point stencil

The lower order terms can be compared in a similar way and lead to typical finite difference discretizations. However, a comparison of the full model strongly depends on the approximation of the advection term and will thus not be done. We conclude that the proposed DEC discretization, if considered on a uniform rectangular mesh in flat space, can be related to finite difference schemes with the same order of consistency as established approaches. Similar comparisons with finite difference schemes have also been considered for scalar valued problems in [18].

7.4 Results

7.4.1 Energy Dissipation

As in flat space we can show that $\dot{E} = \frac{d}{dt} \frac{1}{2} \int_{\mathcal{S}} \|\mathbf{u}\|^2 \mu = \int_{\mathcal{S}} \langle \mathbf{u}, \dot{\mathbf{u}} \rangle \mu \leq 0$. The only term which requires a remark is the viscous part $\frac{1}{\mathrm{Re}} ((*\mathbf{d} * \mathbf{d})\mathbf{u} + 2\kappa\mathbf{u})$. By using the Frobenius inner product for tensors we obtain

$$
\int_{\mathcal{S}} \left\langle \mathbf{u}, \frac{1}{\mathrm{Re}} ((*\mathbf{d} * \mathbf{d})\mathbf{u} + 2\kappa\mathbf{u}) \right\rangle \mu = \frac{1}{\mathrm{Re}} \int_{\mathcal{S}} \langle \mathbf{u}, \mathrm{div}_{\mathcal{S}} \, \mathcal{L}_{\mathbf{u}} \mathbf{g} \rangle \mu
$$

$$
= \frac{1}{\mathrm{Re}} \int_{\mathcal{S}} u^i [\mathcal{L}_{\mathbf{u}^\sharp} \mathbf{g}]_{ij}{}^{|j} \mu = -\frac{1}{\mathrm{Re}} \int_{\mathcal{S}} u^{i|j} \left(u_{i|j} + u_{j|i} \right) \mu
$$

$$
= -\frac{1}{\mathrm{Re}} \int_{\mathcal{S}} \langle \mathrm{grad}_{\mathcal{S}} \, \mathbf{u}, \mathrm{grad}_{\mathcal{S}} \, \mathbf{u} + (\mathrm{grad}_{\mathcal{S}} \, \mathbf{u})^T \rangle \mu
$$

$$
\overset{(*)}{=} -\frac{1}{2\mathrm{Re}} \int_{\mathcal{S}} \|\mathrm{grad}_{\mathcal{S}} \, \mathbf{u} + (\mathrm{grad}_{\mathcal{S}} \, \mathbf{u})^T\|^2 \mu = -\frac{1}{2\mathrm{Re}} \int_{\mathcal{S}} \|\mathcal{L}_{\mathbf{u}^\sharp} \mathbf{g}\|^2 \mu \leq 0,
$$

with the Lie-derivative $\mathcal{L}_{\mathbf{u}^\sharp}$ and $(*)$ following from the component wise computation

$$
u^{i|j} \left(u_{i|j} + u_{j|i} \right)
$$

$$
= \frac{1}{2} \left(u^{i|j} + u^{j|i} + u^{i|j} - u^{j|i} \right) \left(u_{i|j} + u_{j|i} \right)
$$

$$
= \frac{1}{2} \left[\left(u^{i|j} + u^{j|i} \right) \left(u_{i|j} + u_{j|i} \right) + \left(u^{i|j} u_{i|j} - u^{j|i} u_{j|i} \right) + \left(u^{i|j} u_{j|i} - u^{j|i} u_{i|j} \right) \right]
$$

$$
= \frac{1}{2} \left(u^{i|j} + u^{j|i} \right) \left(u_{i|j} + u_{j|i} \right) .
$$

As in flat space we obtain a non-dissipative system for the corresponding surface Euler equation (Re $\rightarrow \infty$). However, the system is also non-dissipative for $\mathcal{L}_{\mathbf{u}^\sharp} \mathbf{g} = 0$, so called Killing vector fields [1], which can be realized on rotational symmetric surfaces. We will use this property in various examples.

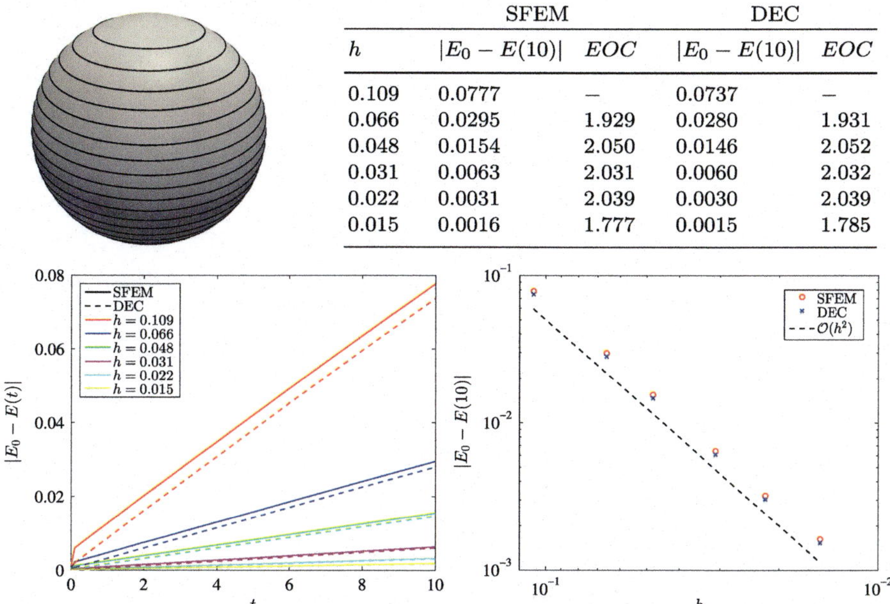

	SFEM		DEC	
h	$\|E_0 - E(10)\|$	EOC	$\|E_0 - E(10)\|$	EOC
0.109	0.0777	–	0.0737	–
0.066	0.0295	1.929	0.0280	1.931
0.048	0.0154	2.050	0.0146	2.052
0.031	0.0063	2.031	0.0060	2.032
0.022	0.0031	2.039	0.0030	2.039
0.015	0.0016	1.777	0.0015	1.785

Fig. 7.4 Streamlines of stationary solution on a sphere together with the error in the kinetic energy and the experimental order of convergence (EOC) for different mesh sizes h (maximum circumcircle diameter of all triangles) at time $t = 10$ for both numerical methods. E_0 denotes the exact kinetic energy. The timestep is $\tau = 0.1$ and Re $= 1$

7.4.2 Numerical Dissipation

We first consider a stationary solution on a sphere, with $\psi_0(\mathbf{x}) = z$ and $\mathbf{v}_0(\mathbf{x}) = (y, -x, 0)^T$ with coordinates $(x, y, z) \in \mathbb{R}^3$. Figure 7.4 shows the streamlines for the rotating flow together with the computed errors for the kinetic energy. The results essentially show second order convergence for both methods, the DEC approach for the surface Navier-Stokes equation and the surface finite element method (SFEM) for the vorticity-stream function equation.

7.4.3 Geometric Interaction

As already analyzed in detail in [32] the vortices in the flow, in the considered case two $+1$ defects, repel each other and are attracted by regions of high Gaussian curvature. We first consider an ellipsoid, represented by the level-set function $e(\mathbf{x}) = (x/a)^2 + (y/b)^2 + (z/c)^2$, with $(x, y, z) \in \mathbb{R}^3$, $a = b = 0.5$ and $c = 1.5$. We consider the initial solutions $\psi_0(\mathbf{x}) = y + 0.1z$ and $\mathbf{v}_0(\mathbf{x}) = \text{rot}_{\mathcal{S}}\, \psi_0(\mathbf{x})$ and use a timestep $\tau = 0.1$. Figure 7.5 shows the geometric properties, the streamlines at

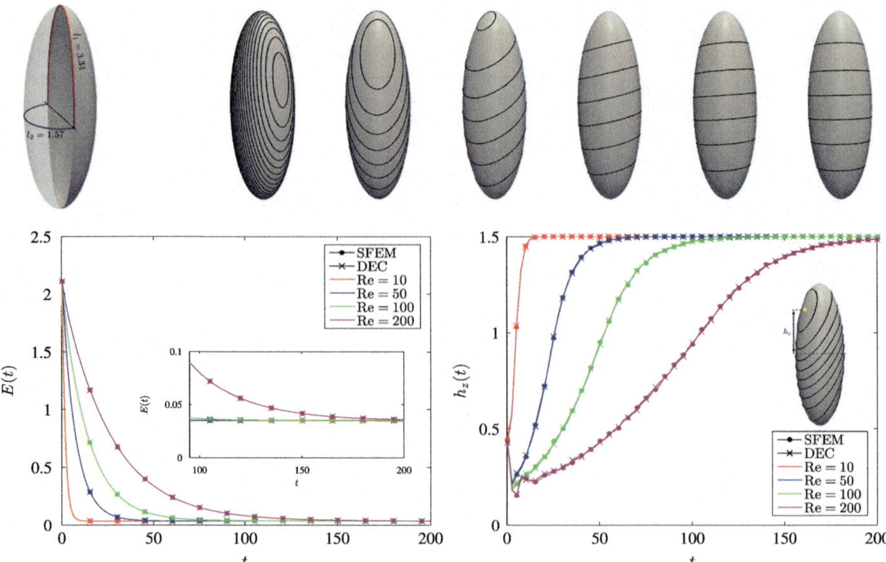

Fig. 7.5 *Top*: Distances on the ellipsoid together with the streamlines at $t = 0, 4, 8, 12, 16$ and 20 for the rotating flow. Results are shown for Re $= 10$. *Bottom*: Kinetic energy over time and the height for the upper vortex over time for both numerical approaches and various Re

various times for Re $= 10$ as well as the kinetic energy over time and the position of one vortex over time for both methods and various Re. The flow converges to a stationary solution with the vortices located at the high Gaussian curvature regions. However, these positions also favors the long range interaction between the vortices as they maximize their distance. We thus cannot argue on a geometric interaction. The time to reach the stationary solution strongly depends on Re, the lower Re the faster it is reached.

The second example considers a biconcave shape, represented by the level-set function $e(\mathbf{x}) = (a^2 + x^2 + y^2 + z^2)^3 - 4a^2(y^2 + z^2) - c^4$, with $(x, y, z) \in \mathbb{R}^3$, $a = 0.72$ and $c = 0.75$. We consider the initial solutions $\psi_0(\mathbf{x}) = y + z$ and $\mathbf{v}_0(\mathbf{x}) = \mathrm{rot}_{\mathcal{S}} \psi_0(\mathbf{x})$ and use a timestep $\tau = 0.1$. Figure 7.6 shows the geometric properties together with the trajectories of one vortex for different Re, the streamlines at various times, a plot of the Gaussian curvature and the kinetic energy over time. Again the flow converges to a stationary solution with the vortices located at the high Gaussian curvature regions. Here the location of the vortices clearly is a result of the geometric interaction, as their distance is not maximized. Again the time to reach the stationary solution strongly depends on Re, the lower Re the faster it is reached.

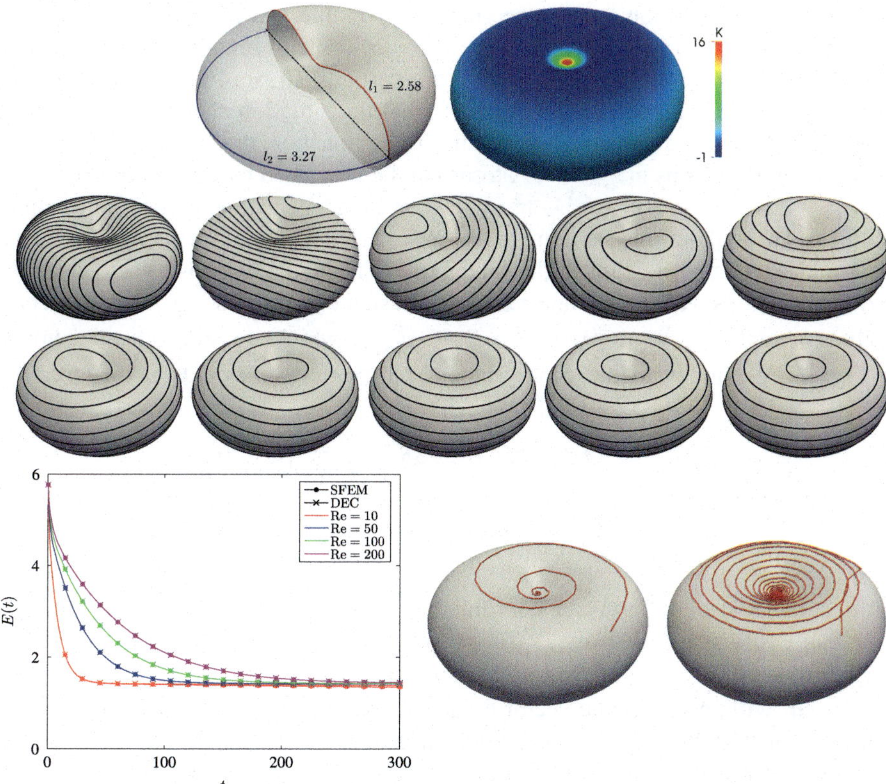

Fig. 7.6 *Top*: Distances on the biconcave shape together with the Gaussian curvature. *Middle*: Streamlines at $t = 0, 7, 14, 21, 28, 35, 42, 49, 56$ and 200 (*left* to *right*, *top* to *bottom*) for the rotating flow. Results are shown for Re $= 10$. *Bottom*: Kinetic energy over time for both numerical approaches and various Re together with two examples for the vortex trajectories for Re $= 10$ (*left*) and Re $= 100$ (*right*)

7.4.4   Surfaces with Genus $g(\mathcal{S}) \neq 0$

As mentioned above we will consider nontrivial solutions with $\mathrm{div}_{\mathcal{S}}\, \mathbf{v} = 0$ and $\mathrm{rot}_{\mathcal{S}}\, \mathbf{v} = 0$. The vorticity-stream function formulation in Eqs. (7.3) and (7.4) is based on the Hodge decomposition of the velocity field $\mathbf{v}$ which can be written as

$$\mathbf{v} = \mathbf{v}^{div} + \mathbf{v}^{rot} + \mathbf{v}^{harm} \tag{7.15}$$

on a general surface $\mathcal{S}$ with a divergence free vector field $\mathbf{v}^{div}$, a curl free vector field $\mathbf{v}^{rot}$ and a divergence as well as curl free vector field $\mathbf{v}^{harm}$. The first two parts are usually rewritten as $\mathbf{v}^{div} = \mathrm{rot}_{\mathcal{S}}\, \psi$ and $\mathbf{v}^{rot} = \mathrm{grad}_{\mathcal{S}}\, \Phi$ with scalar functions ψ and Φ. Since we require incompressibility of $\mathbf{v}$ one can easily verify that the curl free part $\mathbf{v}^{rot}$ vanishes identically. Furthermore, on spherical surfaces ($g(\mathcal{S}) = 0$)

we can drop the harmonic part since it is not possible to write a vector field that is divergence and curl free except of the zero vector field. Finally, this leads to the substitution $\mathbf{v} = \mathrm{rot}_{\mathcal{S}}\,\psi$ which is used in the vorticity-stream function approach in the prior sections. On surfaces with $g(\mathcal{S}) \neq 0$ the situation changes and the harmonic part $\mathbf{v}^{harm}$ does not vanish generally. To demonstrate this property we use the torus which has genus $g(\mathcal{S}) = 1$. A torus can be described by the levelset function $e(\mathbf{x}) = (\sqrt{x^2 + z^2} - R)^2 + y^2 - r^2$, with $(x, y, z) \in \mathbb{R}^3$, major radius R and minor radius r. Throughout this section we use $R = 2$ and $r = 0.5$. Let ϕ and θ denote the standard parametrization angles on the torus. Then, the two basis vectors can be written as $\partial_\phi \mathbf{x}$ as well as $\partial_\theta \mathbf{x}$ and read in Cartesian coordinates $\partial_\phi \mathbf{x} = (-z, 0, x)$ as well as $\partial_\theta \mathbf{x} = \left(-\frac{xy}{\sqrt{x^2+z^2}}, \sqrt{x^2 + z^2} - 2, -\frac{yz}{\sqrt{x^2+z^2}}\right)$ which are schematically shown in Fig. 7.7. We find two (linear independent) harmonic vector fields on the torus

$$\mathbf{v}_\phi^{harm} = (4 + \cos(\theta))^{-2}\partial_\phi \mathbf{x} = \frac{1}{4\left(x^2 + z^2\right)}\partial_\phi \mathbf{x}$$

$$\mathbf{v}_\theta^{harm} = (4 + \cos(\theta))^{-1}\partial_\theta \mathbf{x} = \frac{1}{2\sqrt{x^2 + z^2}}\partial_\theta \mathbf{x}$$

written in local and Cartesian coordinates, respectively, and shown in Fig. 7.8. One can easily verify that $\mathrm{div}_{\mathcal{S}}\,\mathbf{v}_\phi^{harm} = \mathrm{rot}_{\mathcal{S}}\,\mathbf{v}_\phi^{harm} = 0$ as well as $\mathrm{div}_{\mathcal{S}}\,\mathbf{v}_\theta^{harm} = \mathrm{rot}_{\mathcal{S}}\,\mathbf{v}_\theta^{harm} = 0$.

To start with, consider the vector field $\mathbf{v} = \partial_\phi \mathbf{x}$, which has zero divergence and non-zero curl. The Hodge decomposition equation (7.15) leads to $\mathbf{v}^{rot} = \mathbf{v}^{harm} = 0$. In that case the substitution $\mathbf{v} = \mathrm{rot}_{\mathcal{S}}(\psi)$ holds. The stream function ψ of the vector field $\partial_\phi \mathbf{x}$ can then be analytically written in local coordinates as $\psi = -\frac{1}{4}\sin(\theta) + \theta - \pi$. The linear contribution causes a discontinuity at $\theta = 2\pi$, which is shown in Fig. 7.7 together with the streamlines of $\partial_\phi \mathbf{x}$ (contour lines of ψ).

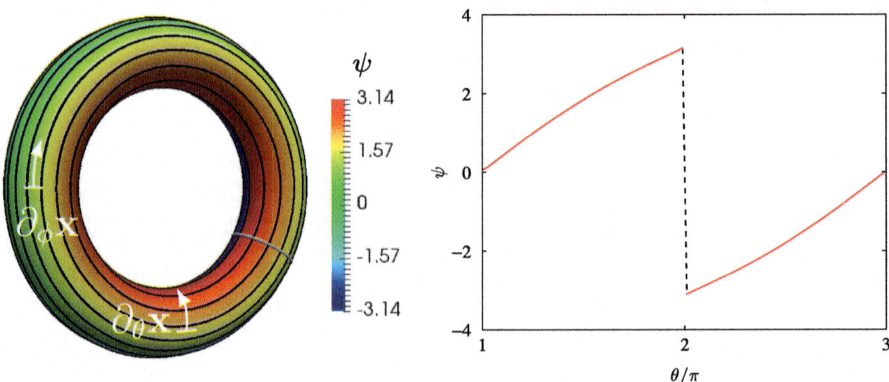

Fig. 7.7 *Left*: Streamlines and values of the discontinuous stream function ψ to represent the velocity field $\mathbf{v} = \partial_\phi \mathbf{x}$ on the torus and the two basis vectors $\partial_\phi \mathbf{x}$ and $\partial_\theta \mathbf{x}$. *Right*: Plot of the stream function values over the *gray contour line* in the *left figure*

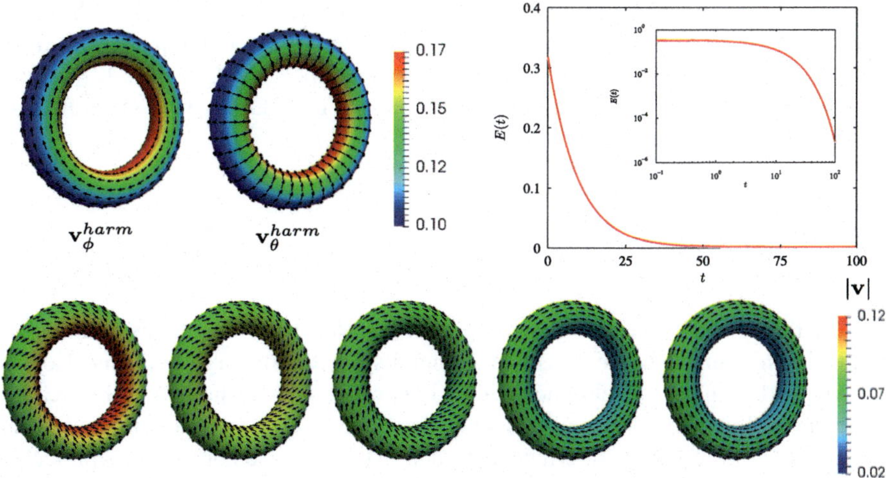

Fig. 7.8 *Top*: Harmonic vector fields $\mathbf{v}_\phi^{harm}$ and $\mathbf{v}_\theta^{harm}$ and kinetic energy E over time t for the simulation with $\mathbf{v}_\theta^{harm}$ as initial condition on both equally spaced and logarithmic scales (*right*). *Bottom*: Numerical solution of $\mathbf{v}$ for the simulation with $\frac{1}{2}(\mathbf{v}_\phi^{harm} + \mathbf{v}_\theta^{harm})$ as initial condition computed with the DEC algorithm at $t = 0, 2, 10, 30$ and 60 (*left* to *right*). The *arrows* are rescaled for better visualization

Solving the surface Navier-Stokes equations (7.1) and (7.2) directly circumvents the discontinuities.

In the next example we use the mean of the two harmonic vector fields as initial condition, i.e. $\mathbf{v}_0(\mathbf{x}) = \frac{1}{2}(\mathbf{v}_\phi^{harm} + \mathbf{v}_\theta^{harm})$. By considering the vorticity-stream function approach we have the initial conditions $\phi_0 = \psi_0 = 0$ and thus only the trivial solution. However, solving the surface Navier-Stokes equation directly covers also the harmonic parts. Figure 7.8 shows the numerical solution of $\mathbf{v}$ with the DEC algorithm in which we used the timestep $\tau = 0.1$ and Re $= 10$. In this case the reached steady state is again a Killing vector field and is proportional to the basis vector $\partial_\phi\mathbf{x}$. Interestingly, the curl of the vector field $\partial_\phi\mathbf{x}$ does not vanish.

Other linear combinations of the two harmonic vector fields $\mathbf{v}_\phi^{harm}$ and $\mathbf{v}_\theta^{harm}$ as initial condition leads to the same steady state solution (up to a proportionality constant) except of $\mathbf{v}_0(\mathbf{x}) = \mathbf{v}_\theta^{harm}$. In that case the vector field does not change its direction by symmetry and dissipates to zero. The results are shown in the energy plot in Fig. 7.8 which clearly shows the vanishing energy over time.

7.4.5   Comparison

All results for surfaces with genus $g(\mathcal{S}) = 0$ demonstrate the accuracy of the DEC discretization. The plotted vortex trajectories and kinetic energy values over time are almost indistinguishable from the SFEM results obtained by solving the vorticity-stream function formulation. The computational cost is larger for the DEC discretization, which however is also a consequence of the implementation. Both methods are implemented in the finite element toolbox AMDiS [41, 42], where the datastructures are optimized for SFEM, but not for DEC. A new general DEC toolbox is work in progress.

7.5   Conclusions

Even if the formulation of the incompressible surface Navier-Stokes equation is relatively old, numerical treatments on general surfaces are very rare. This also has not changed with the development of various numerical methods to solve scalar-valued partial differential equations on surfaces, such as the surface finite element method [12] or the diffuse interface approach [31]. They are not directly applicable to vector-valued partial differential equations on surfaces. One has to define what it means for a vector to be parallel on the discrete representation $\mathcal{K}$ of $\mathcal{S}$. The concept of discrete parallel transport can be easily realized using discrete exterior calculus (DEC), see [7] for details. DEC thus provides an ideal framework to solve vector-valued partial differential equations on surfaces. In [27] this is shown in detail for a surface Frank-Oseen model. We here use the approach to discretize the incompressible surface Navier-Stokes equation. The discretization is based on the covariant form and utilizing a discrete version of the Hodge star $*$ and the Stokes theorem for the exterior derivative $\mathbf{d}$. Non-standard in our discretization is the treatment of the discrete Hodge star and the discrete inner product. If considered in flat space the described discretization can be related to a finite difference schemes on a staggered grid. The resulting unusual stencil shows second order consistency.

Computationally we compare results of the DEC discretization with a vorticity-stream function approach for surfaces with genus $g(\mathcal{S}) = 0$. The examples use the properties of Killing vector fields and demonstrate the interplay between topology, geometry and flow properties. The numerical results are almost indistinguishable for all considered examples, varying the underlying surface $\mathcal{S}$ and the Reynolds number Re. We also demonstrate the possibility to deal with harmonic vector fields using the DEC approach. It would be interesting to compare the considered vortex trajectories for larger Re with results for point vortices on closed surfaces, as e.g. considered in [10] for ellipsoids or in [34] for toroidal surfaces.

Acknowledgements This work is partially supported by the German Research Foundation through grant Vo899/11. We further acknowledge computing resources provided at JSC under grant HDR06.

Appendix 1: Notation for DEC

We often use the strict order relation $\succ$ and $\prec$ on simplices, where $\succ$ is proverbial the "contains" relation, i.e. $e \succ v$ means: the edge e contains the vertex v. Correspondingly $\prec$ is the "part of" relation, i.e. $v \prec T$ means: the vertex v is part of the face T. Hence, we can use this notation also for sums, like $\sum_{f \succ e}$, i.e. the sum over all faces T containing the edge e, or $\sum_{v \prec e}$, i.e. the sum over all vertices v being part of the edge e. Sometimes we need to determine this relation for edges more precisely with respect to the orientation. Therefore, sign functions are introduced,

$$s_{T,e} := \begin{cases} +1 & \text{if } e \prec T \text{ and } T \text{ is on the left side of } e \\ -1 & \text{if } e \prec T \text{ and } T \text{ is on the right side of } e, \end{cases}$$

$$s_{e,\tilde{e}} := \begin{cases} +1 & \text{if } \angle(\mathbf{e}, \tilde{\mathbf{e}}) < \pi \\ -1 & \text{if } \angle(\mathbf{e}, \tilde{\mathbf{e}}) > \pi \end{cases}$$

$$s_{v,e} := \begin{cases} +1 & \text{if } v \prec e \text{ and } e \text{ points to } v \\ -1 & \text{if } v \prec e \text{ and } e \text{ points away from } v, \end{cases}$$

to describe such relations between faces and edges, edges and edges or vertices and edges, respectively. Figure 7.9 gives a schematic illustration.

The property of a primal mesh to be well-centered ensures the existence of a Voronoi mesh (dual mesh), which is also an orientable manifold-like simplicial complex, but not well-centered. The basis of the Voronoi mesh are not simplices, but chains of them. To identify these basic chains, we apply the (geometrical) star operator $\star$ on the primal simplices, i.e. $\star v$ is the Voronoi cell corresponding to the vertex v and inherits its orientation from the orientation of the polyhedron $|\mathcal{K}|$. From a geometric point of view, $\star v$ is the convex hull of circumcenters $c(T)$ of all triangles $T \succ v$. The Voronoi edge $\star e$ of an edge e is a connection of the right face $T_2 \succ e$ with the left face $T_1 \succ e$ over the midpoint $c(e)$. The Voronoi vertex $\star T$ of a

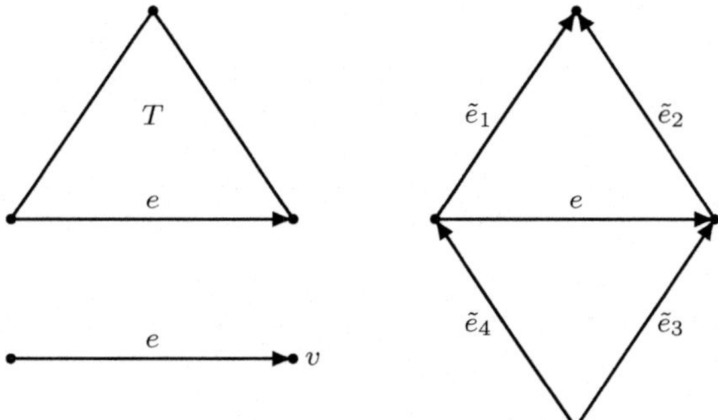

Fig. 7.9 These formations always yield positive signs $+1$ for $s_{T,e}$ (*top left*), $s_{v,e}$ (*bottom left*) and $s_{e,\tilde{e}_i}$ (*right*) for $i \in \{1, 2, 3, 4\}$, respectively. Every odd-numbered change in edge orientations results in a change of the sign $s_{.,.}$.

face T is simply its circumcenter $c(T)$, cf. Fig. 7.1. For a more detailed mathematical discussion see e.g. [20, 39].

The boundary operator ∂ maps simplices (or chains of them) to the chain of simplices that describes its boundary with respect to its orientation (see [20]), e.g. $\partial(\star v) = -\sum_{e \succ v} s_{v,e}(\star e)$ (formal sum for chains) and $\partial e = \sum_{v \prec e} s_{v,e} v$.

The expression $|\cdot|$ measures the volume of a simplex, i.e. $|T|$ the area of the face T, $|e|$ the length of the edge e and the 0-dimensional volume $|v|$ is set to be 1. Therefore, the volume is also defined for chains and the dual mesh, since the integral is a linear functional.

Appendix 2: Second Order Convergence

In this section we show that the discretization equation (7.14) of $\boldsymbol{\Delta}^{\mathrm{RR}}$, defined in Eq. (7.13) on a staggered grid, has a truncation error of order two. Without loss of generality, by a quarter turn of the difference scheme in Fig. 7.3 (left), we only elaborate on the discretization of $(\boldsymbol{\Delta}^{\mathrm{RR}} u)^x$ along the horizontal x-direction. The first three terms in Eq. (7.14) show the well-known second order central difference approximation in vertical direction of the first term in Eq. (7.13), i.e.

$$\frac{1}{h^2}\left(u^x_{i,j+1} + u^x_{i,j-1} - 2u^x_{i,j}\right) = \left(\partial^2_y u^x\right)^x_{i,j} + \mathcal{O}(h^2).$$

For the remaining terms, we first carry out a Taylor expansion on central vertices $v_{i+k,j} \in \mathcal{V}$ for $k \in \{0,1\}$ in the vertical edge columns, i.e.

$$u^y_{i+k,j-l} = \left(u^y + (-1)^l \frac{h}{2}\partial_y u^y + \frac{h^2}{8}\partial^2_y u^y + (-1)^l \frac{h^3}{48}\partial^3_y u^y + \frac{h^4}{384}\partial^4_y u^y\right)_{i+k,j}$$
$$+ \mathcal{O}(h^5)$$

for all $l \in \{0,1\}$. An additional horizontal expansion of sufficient order at the edge midpoint $c(e^x_{i,j})$ results in

$$u^y_{i+k,j-l} = \left(u^y + (-1)^{k+1}\frac{h}{2}\partial_x u^y + \frac{h^2}{8}\partial^2_x u^y + (-1)^{k+1}\frac{h^3}{48}\partial^3_x u^y + \frac{h^4}{384}\partial^4_x u^y\right.$$
$$+ (-1)^l \frac{h}{2}\partial_y u^y + (-1)^{l+k+1}\frac{h^2}{4}\partial_x\partial_y u^y + (-1)^l \frac{h^3}{16}\partial^2_x\partial_y u^y$$
$$+ (-1)^{l+k+1}\frac{h^4}{96}\partial^3_x\partial_y u^y + \frac{h^2}{8}\partial^2_y u^y + (-1)^{k+1}\frac{h^3}{16}\partial_x\partial^2_y u^y$$
$$+ \frac{h^4}{64}\partial^2_x\partial^2_y u^y + (-1)^l \frac{h^3}{48}\partial^3_y u^y + (-1)^{l+k+1}\frac{h^4}{96}\partial_x\partial^3_y u^y + \left.\frac{h^4}{384}\partial^4_y u^y\right)^x_{i,j}$$
$$+ \mathcal{O}(h^5)$$

for all $l, k \in \{0, 1\}$. Finally, we obtain

$$\frac{1}{h^2} \left(u^y_{i,j} - u^y_{i+1,j} + u^y_{i+1,j-1} - u^y_{i,j-1} \right)$$

$$= -\left(\partial_x \partial_y u^y + \frac{h^2}{96} \partial_x \partial_y \left(\partial_x^2 u^y + \partial_y^2 u^y \right) \right)^x_{i,j} + \mathcal{O}(h^3)$$

and thus a truncation error at most $\mathcal{O}(h^2)$ regarding $(\boldsymbol{\Delta}^{\mathrm{RR}} u)^x_{i,j}$ generally.

References

1. Abraham, R., Marsden, J., Ratiu, T.: Manifolds, Tensor Analysis, and Applications. Applied Mathematical Sciences, vol. 75. Springer, New York (1988)
2. Arakawa, A., Lamb, V.: Computational design of the basic dynamical processes of the UCLA general circulation model. In: General Circulation Models of the Atmosphere, pp. 173–265. Academic, New York (1977)
3. Arnold, D.N., Falk, R.S., Winther, R.: Finite element exterior calculus, homological techniques, and applications. Acta Numer. **15**, 1–155 (2006)
4. Arroyo, M., DeSimone, A.: Relaxation dynamics of fluid membranes. Phys. Rev. E **79**, 031915 (2009)
5. Barrett, J., Garcke, H., Nürnberg, R.: Numerical computations of the dynamics of fluidic membranes and vesicles. Phys. Rev. E **92**, 052704 (2015)
6. Bothe, D., Prüss, J.: On the two-phase Navier-Stokes equations with Boussinesq-Scriven surface. J. Math. Fluid Mech. **12**, 133–150 (2010)
7. Crane, K., de Goes, F., Desbrun, M., Schröder, P.: Digital geometry processing with discrete exterior calculus. In: ACM SIGGRAPH Courses, pp. 1–126 (2013)
8. Desbrun, M., Hirani, A., Leok, M., Marsden, J.: Discrete exterior calculus. arXiv:math/0508341 (2005)
9. Dörries, G., Foltin, G.: Energy dissipation of fluid membranes. Phys. Rev. E **53**, 2547–2550 (1996)
10. Dritschel, D.G., Boatto, S.: The motion of point vortices on closed surfaces. Proc. R. Soc. A **471**, 20140890 (2015)
11. Dziuk, G., Elliott, C.: Surface finite elements for parabolic equations. J. Comput. Math. **25**, 385–407 (2007)
12. Dziuk, G., Elliott, C.M.: Finite elements on evolving surfaces. IMA J. Numer. Anal. **27**, 262–292 (2007)
13. Ebin, D.G., Marsden, J.: Groups of diffeomorphisms and the motion of an incompressible fluid. Ann. Math. **92**, 102–163 (1970)
14. Elcott, S., Tong, Y., Kanso, E., Schröder, P., Desbrun, M.: Stable, circulation-preserving, simplicial fluids. ACM Trans. Graph. **26**, 4 (2007)
15. Fan, J., Han, T., Haataja, M.: Hydrodynamic effects on spinodal decomposition kinetics in planar lipid bilayer membranes. J. Chem. Phys. **133**, 235101 (2010)
16. Fisher, M., Springborn, B., Bobenko, A., Schröder, P.: An algorithm for the construction of intrinsic Delaunay triangulations with applications to digital geometry processing. In: ACM SIGGRAPH Courses, pp. 69–74 (2006)
17. Gortler, S., Gotsman, C., Thurston, D.: Discrete one-forms on meshes and applications to 3D mesh parameterization. Comput. Aided Geom. Des. **33**, 83–112 (2006)

18. Griebel, M., Rieger, C., Schier, A.: Discrete exterior calculus (DEC) for the surface Navier-Stokes equation. In: Bothe, D., Reusken, A. (eds.) Transport Processes at Fluidic Interfaces. Advances in Mathematical Fluid Mechanics. Springer, Cham (2017). doi 10.1007/978-3-319-56602-3_7
19. Gu, X., Yau, S.T.: Global conformal surface parameterization. In: ACM/EG Symposium on Geometry Processing, pp. 127–137 (2003)
20. Hirani, A.N.: Discrete exterior calculus. Ph.D. thesis, California Institute of Technology, Pasadena, CA (2003)
21. Hu, D., Zhang, P., E, W.: Continuum theory of a moving membrane. Phys. Rev. E **75**, 041605 (2007)
22. Mercat, C.: Discrete Riemann surfaces and the Ising model. Commun. Math. Phys. **218**, 177–216 (2001)
23. Mitrea, M., Taylor, M.: Navier-Stokes equations on Lipschitz domains in Riemannian manifolds. Math. Ann. **321**, 955–987 (2001)
24. Mohamed, M.S., Hirani, A.N., Samtaney, R.: Comparison of discrete hodge star operators for surfaces. Comput. Aided Des. (2016). doi:10.1016/j.cad.2016.05.002
25. Mohamed, M.S., Hirani, A.N., Samtaney, R.: Discrete exterior calculus discretization of incompressible Navier-Stokes equations over surface simplicial meshes. J. Comput. Phys. **312**, 175–191 (2016)
26. Mullen, P., Crane, K., Pavlov, D., Tong, Y., Desbrun, M.: Energy-preserving integrators for fluid animation. ACM Trans. Graph. **28**, 38 (2009)
27. Nestler, M., Nitschke, I., Praetorius, S., Voigt, A.: Orientational order on surfaces - the coupling of topology, geometry and dynamics. arXiv:1608.01343 (2016)
28. Nitschke, I., Voigt, A.: Curvature approximation of discrete surfaces - a discrete exterior calculus approach (in preparation)
29. Nitschke, I., Voigt, A., Wensch, J.: A finite element approach to incompressible two-phase flow on manifolds. J. Fluid Mech. **708**, 418–438 (2012)
30. Polthier, K., Preuß, E.: Identifying vector field singularities using a discrete Hodge decomposition. In: Hege, H., Polthier, K. (eds.) Visualization and Mathematics III, pp. 113–134. Springer, Heidelberg (2003)
31. Rätz, A., Voigt, A.: PDE's on surfaces: a diffuse interface approach. Commun. Math. Sci. **4**, 575–590 (2006)
32. Reuther, S., Voigt, A.: The interplay of curvature and vortices in flow on curved surfaces. Multiscale Model. Simul. **13**, 632–643 (2015)
33. Reuther, S., Voigt, A.: Incompressible two-phase flows with an inextensible Newtonian fluid interface. J. Comput. Phys. **322**, 850–858 (2016)
34. Sakajo, T., Shimizu, Y.: Point vortex interactions on a toroidal surface. Proc. R. Soc. A **472**, 20160271 (2016)
35. Scriven, L.E.: Dynamics of a fluid interface equation of motion for Newtonian surface fluids. Chem. Eng. Sci. **12**, 98–108 (1960)
36. Secomb, T.W., Skalak, R.: Surface flow of viscoelastic membranes in viscous fluids. Q. J. Mech. Appl. Math. **35**, 233–247 (1982)
37. Tong, Y., Lombeyda, S., Hirani, A.N., Desbrun, M.: Discrete multiscale vector field decomposition. ACM Trans. Graph. **22**, 445–452 (2003)
38. Tong, Y., Alliez, P., Cohen-Steiner, D., Desbrun, M.: Designing quadrangulations with discrete harmonic forms. In: ACM/EG Symposium on Geometry Processing, pp. 201–210 (2006)
39. VanderZee, E., Hirani, A.N., Guoy, D., Ramos, E.A.: Well-centered triangulation. SIAM J. Sci. Comput. **31**, 4497–4523 (2010)
40. Vaxman, A., Campen, M., Diamanti, O., Panozzo, D., Bommes, D., Hildebrandt, K., Ben-Chen, M.: Directional field synthesis, design and processing. In: EUROGRAPHICS - STAR, vol. 35, pp. 1–28 (2016)

41. Vey, S., Voigt, A.: AMDiS: adaptive multidimensional simulations. Comput. Vis. Sci. **10**, 57–67 (2007)
42. Witkowski, T., Ling, S., Praetorius, S., Voigt, A.: Software concepts and numerical algorithms for a scalable adaptive parallel finite element method. Adv. Comput. Math. **41**, 1145–1177 (2015)

Part II
Analysis and Simulation of Diffusive Interface Models

The second part reports on recent developments related to diffusive interface models for two-phase flows. Besides models for the pure fluid dynamics, where surface tension effects are described using the Cahn-Hilliard equation, additional transport processes such as mass transfer from the bulk to the interfacial region and the transport of surfactants are considered. Furthermore, the modeling of lipid bilayer membranes based on a diffuse interface approach is addressed and a diffuse interface model for a two-phase flow with noninteracting polymers is derived. Although the emphasis is on incompressible two-phase flows, also diffuse interface models with compressible fluids are treated. For the latter, recent results on existence and uniqueness of strong solutions for different classes of models are discussed. For diffusive interface models with *in*compressible fluids, theoretical analyses on thermodynamical consistency, existence of solutions and sharp interface limits are presented. A further aspect that is treated is the formulation and analysis of a class of optimal control problems for two-phase flows and the development of robust and efficient solution strategies for these.

In this part on analysis and simulation of diffuse interface models there are the following contributions:

Chapter 8. H. Abels, H. Garcke, G. Grün and S. Metzger, *Diffuse Interface Models for Incompressible Two-Phase Flows with Different Densities.*

Chapter 9. H. Abels, Y. Liu and A. Schöttl, *Sharp Interface Limits for Diffuse Interface Models for Two-Phase Flows of Viscous Incompressible Fluids.*

Chapter 10. H. Abels, H. Garcke, K. Fong Lam and J. Weber, *Two-Phase Flow with Surfactants: Diffuse Interface Models and Their Analysis.*

Chapter 11. S. Aland, *Phase Field Models for Two-Phase Flow with Surfactants and Biomembranes.*

Chapter 12. G. Grün and S. Metzger, *Micro-Macro-Models for Two-Phase Flow of Dilute Polymeric Solutions: Macroscopic Limit, Analysis, and Numerics.*

Chapter 13. M. Hintermüller, M. Hinze, C. Kahle and T. Keil, *Fully Adaptive and Integrated Numerical Methods for the Simulation and Control of Variable Density Multiphase Flows Governed by Diffuse Interface Models.*

Chapter 14.   H. Freistühler and M. Kotschote, *Dynamical Stability of Diffuse Phase Boundaries in Compressible Fluids.*

We outline the main topics of these contributions. In Chaps. 8–10, *theoretical analyses* of different aspects related to *diffusive interface models* are treated.

In Chap. 8, an overview of several important diffusive interface models is presented and the literature on this subject is reviewed. A recently developed *thermodynamically consistent model* with a *divergence free velocity* field for two-phase flows with *different densities* is treated in detail. A derivation of this model, based on the thermodynamical principles, is presented. In the recent literature several aspects of this new diffuse interface model have been studied. A few main results on (formal) sharp interface limits and existence of weak and strong solutions are discussed. It is explained how the transport of soluble species can be included in the model. A Galerkin finite element discretization of this model is briefly addressed. A strategy for iterative decoupling of the Cahn-Hilliard equation and the momentum equation is given and a convergence result for the discrete scheme is presented. Numerical simulation results for the new thermodynamically consistent model applied to a falling droplet flow problem are shown.

In Chap. 9, the topic of *sharp interface limits* for diffuse interface models is treated. Rigorous mathematical results regarding the limit of diffuse interface models (or of their solution) to sharp interface models, as the parameter $\epsilon > 0$ tends to zero, are discussed. The parameter ϵ is proportional to the "thickness" of the diffuse interface. It turns out that this is a subtle problem and the results depend significantly on the scaling of a mobility coefficient in the system as $\epsilon \to 0$. Formal sharp interface limits, based on the method of matched asymptotics, of the so-called model H (a Navier-Stokes/Cahn-Hilliard type system) are reviewed. Rigorous results on the convergence and non-convergence of the convective Cahn-Hilliard model for $\epsilon \to 0$, for a given velocity field, are presented. Results on convergence of weak solutions of the model H to so-called varifold solutions of the formal sharp interface limit model are discussed as well. Finally, a first rigorous convergence result with convergence rates in strong norms for the sharp interface limit of a two-phase diffuse interface model is explained. For this, a simplified model consisting of a two-phase Stokes flow problem coupled with the Allen-Cahn diffuse interface model is considered. It is shown that in this case the limit system is a Stokes system coupled to a mean curvature flow equation with an additional convection term.

In Chap. 10, new diffuse interface models are treated which include the *transport of soluble and insoluble surfactants*. The thermodynamically consistent model with a divergence free velocity field for two-phase flows with different densities, which is treated in Chap. 8, is extended with convection-diffusion type equations for the surfactant in the bulk phases and the interfacial region. The cases of instantaneous (i.e., diffusion controlled) and non-instantaneous (i.e., sorption controlled) adsorption are distinguished. Formal sharp interface limits, derived with the method of matched asymptotic expansions, of some of the diffuse interface models (including surfactants) are presented. For a particular diffuse interface model, including surfactant transport, recent results on the existence of weak solutions are discussed.

In Chap. 11, the diffusive interface (or phase field) model H, which is also discussed in Chap. 9, is considered. The issue of boundary conditions for dynamic and static contact angles (i.e., the fluid-fluid interface touches a wall) is briefly addressed. The two-phase flow model is extended with a diffuse interface model for soluble surfactants and different constitutive laws for adsorption isoterms (Henry and Langmuir law) are addressed. A main topic of this contribution is a *diffuse interface model* that is used for the simulation of the *behaviour of lipid bilayer membranes*. This model includes bending stiffness and membrane inextensibilty. It consists of the Navier-Stokes equations, with a bending stiffness force term, coupled with an equation that models this bending stiffness and a diffuse Willmore flow equation for the advection of the phase field. Different approaches for treating the membrane inextensibilty condition are reviewed. These different approaches are also compared in numerical experiments.

In Chap. 12, a diffuse interface model for a two-phase flow of incompressible fluids with dissolved noninteracting polymers is treated. The model consists of a Fokker-Planck type equation which describes the distribution and orientation of the polymer chains, coupled with Cahn-Hilliard and Navier-Stokes equations. The special stress tensor used in the momentum balance equation gives rise to non-Newtonian effects. The model that is presented allows flows with different densities and covers the case of one Newtonian fluid and one non-Newtonian fluid as well as the case of two non-Newtonian fluids. A main result of this contribution is *the derivation of this model*. Starting from general balance equations for mass, momentum and the configurational density of the polymer, Onsager's variational principle of minimum energy dissipation is applied and a thermodynamically consistent model for two-phase flow of polymeric solutions is derived. Moreover, a result on existence of weak solutions in the case of the same densities and constant mobility is discussed. Finally, results of a finite element based simulation are presented that give a qualitative validation of this model.

In Chap. 13, the Cahn-Hilliard-Navier-Stokes (CHNS) diffuse interface model treated in Chap. 8 is considered. Based on a weak formulation of this model an energy conserving adaptive finite element based discretization scheme is derived. Based on an energy inequality, an all-in-one *adaptive concept for the Cahn-Hilliard-Navier-Stokes system* with smooth free energy is introduced for which a-posteriori residual based error estimation techniques are developed. A second main contribution in this chapter is the formulation and analysis of *optimal control problems for two-phase flows* and the development of robust and reliable solution strategies for these. A semi-discrete, i.e., discretized in time, optimal control CHNS problem with a distributed force control is considered. For this optimization problem, first order optimality conditions are derived. In case of smooth free energies (used in the Cahn-Hilliard model) this results in classical KKT conditions, while in the case of non-smooth free energies (e.g., double obstacle potential) a result on so-called C(larke)-stationarity is derived. Also for the optimal control problem, the issue of (spatial) adaptivity is addressed. An adaptive finite element solver is developed which uses the fact that, for optimal control problems, one is usually interested in an accurate approximation of the target quantity, i.e., the objective functional.

In Chaps. 10–13 the two-phase flows that are treated assume *in*compressiblity of the fluids. In Chap. 14, two-phase flows with *compressible* fluids are studied. Three models are considered, namely the Navier-Stokes-Korteweg, the Navier-Stokes-Allen-Cahn and the Navier-Stokes-Cahn-Hilliard equations. In these models the profiles of diffuse phase boundaries are characterized as special instances of so-called heteroclinic traveling waves. *Existence and uniqueness results* of strong solutions, locally in time, for these models that have been presented in recent literature are discussed. For a certain class of Helmholtz energies, results on the existence of diffuse phase boundaries for Navier-Stokes-Allen-Cahn and the Navier-Stokes-Cahn-Hilliard equations are presented. Furthermore, a generic scenario for the fluid's constitutive law is identified to permit bifurcation of such waves at a critical value of a physical parameter (e.g., temperature). For the case of so-called no-flux diffuse boundaries spectral stability is treated.

Chapter 8
Diffuse Interface Models for Incompressible Two-Phase Flows with Different Densities

Helmut Abels, Harald Garcke, Günther Grün, and Stefan Metzger

Abstract Diffuse interface models have become an important analytical and numerical method to model two-phase flows. In this contribution we review the subject and discuss in detail a thermodynamically consistent model with a divergence free velocity field for two-phase flows with different densities. The model is derived using basic thermodynamical principles, its sharp interface limits are stated, existence results are given, different numerical approaches are discussed and computations showing features of the model are presented.

8.1 Introduction

A fundamental problem in fluid dynamics involves changes in topology of interfaces between immiscible or partially miscible fluids. Topological transitions such as pinch off and reconnection of fluid interfaces are important features of many systems and strongly affect the flow. Classical models based on sharp interface approaches typically fail to describe these phenomena. Although sometimes so called reconnection conditions can be formulated it is often not possible to justify them based on physical principles. An approach allowing for topological transitions is based on the level set method [41] which also allows for several computationally efficient discretizations. But also for the level set approach it is not clear whether topological transitions are described in a way which is based on physical principles. For example results often depend on discretization parameters and on the kind of "smoothing" applied.

Another so called implicit formulation is the volume of fluid method (VOF-method), see, e.g., [32]. Here, a so called color function is passively advected by

H. Abels • H. Garcke (✉)
Fakultät für Mathematik, Universität Regensburg, Universitätsstraße 31, 93053 Regensburg, Germany
e-mail: helmut.abels@ur.de; harald.garcke@ur.de

G. Grün • S. Metzger
Department Mathematik, Universität Erlangen-Nürnberg, Erlangen, Germany
e-mail: guenther.gruen@fau.de; stefan.metzger@fau.de

© Springer International Publishing AG 2017
D. Bothe, A. Reusken (eds.), *Transport Processes at Fluidic Interfaces*,
Advances in Mathematical Fluid Mechanics, DOI 10.1007/978-3-319-56602-3_8

the flow. This color function typically has no physical significance and dissipation inequalities are usually not known.

In recent years diffuse interface models turned out to be a promising approach to describe phenomena like droplet coalescence or droplet break-up. In diffuse interface (or phase field) methods an order parameter is introduced which allows for a mixing in an interfacial zone such that the sharp interface is replaced by a thin interfacial layer in which the order parameter, which can be a concentration, rapidly changes its value, see [2, 3, 14, 17, 18, 22, 24, 34–37, 39]. The diffuse interface approach turns out to be an attractive approach to model and to numerically simulate fluid interfaces.

In the literature, diffuse interface models are well established for two-phase flows of liquids with identical ("matched") densities. A first diffuse-interface model for two-phase flows of liquids with identical ("matched") densities was introduced by Hohenberg and Halperin [33]. In the general case, when the densities are different, several approaches have been discussed in the literature. Lowengrub and Truskinovsky [39] derived quasi-incompressible models, where the corresponding velocity field is not divergence free. On the other hand, Ding et al. [22] proposed a model with solenoidal fluid velocities which is not known to be thermodynamically consistent. Finally, we also mention the work of Aki et al. [12], who derived a diffuse-interface model for a two-phase flow with non-matched densities and a velocity field which is not divergence free.

The first three authors of this contribution proposed a thermodynamically consistent diffuse interface model which is based on a divergence free velocity, see [7]. The new model over the last years has been analyzed with respect to existence of weak and strong solutions, see [4, 8, 9, 42]. In order to show existence of weak solutions it is important that the model introduced in [7] is thermodynamically consistent and hence allows for a global energy inequality which can be used to obtain a priori estimates. The energy inequality is also important for the development of stable numerical schemes. In [27] fully discrete finite-element schemes for the model introduced in [7] have been developed and due to a priori estimates relying on the energy estimate Grün [26] was able to show convergence of discrete solutions. Later Garcke, Hinze, Kahle [25] and Grün, Guillén-González, Metzger [29] were able to derive simpler to solve stable and linear, and respectively, fully decoupled time discretizations for the model in [7].

In general, the evolution of fluid interfaces is influenced by many effects. Species or heat transport at the interfaces strongly affects the interface, e.g., because the surface tension depends on a surfactant concentration or on the temperature at the interfaces. The latter will lead to Marangoni effects at the interface. If in addition a surface active species diffuses on the interfaces, one has to consider convection and diffusion on the interface itself. All these additional effects can in principle be incorporated within a diffuse interface model for fluid interfaces. But so far in the literature only few results with respect to the modelling, analysis and numerics of these phenomena using a diffuse interface approach are known. We will discuss a simple case of transport at the interface in this contribution and refer to the companion contributions [10] and [28] for surfactant transport and micro-macro-models for transport across fluidic interfaces. Moreover, in [21] the model has been extended to ion transport—this way modeling dynamic electrowetting and other electrokinetic phenomena.

The outline of this paper is as follows. In the following section we derive the thermodynamically consistent phase field model for two-phase flows. Section 8.3 gives an overview about known analytic results for the model. In Sect. 8.4 stable and convergent fully discrete schemes are introduced, convergence results are stated, and a numerical computation is presented.

8.2 Diffuse Interface Models for Two-Phase Flows

8.2.1 Earlier Models

Before we discuss the diffuse interface models, let us recall the governing equations for the two-phase incompressible fluid flow in a classical sharp interface model. In the bulk, i.e., in open sets $\Omega_\pm$ which are separated by an interface Γ, one requires

$$\operatorname{div} \mathbf{v} = 0 \,,$$

$$\partial_t(\rho\mathbf{v}) + \operatorname{div}(\rho\mathbf{v} \otimes \mathbf{v}) + \nabla p = \operatorname{div}\mathbf{S} + \rho\mathbf{g} \,,$$

where $\mathbf{v}$ is the fluid velocity, ρ is the mass density, p is the pressure, $\mathbf{g}$ describes volume forces and $\mathbf{S}$ is the viscous stress tensor given as

$$\mathbf{S} = 2\eta\mathbf{Dv}, \quad \mathbf{Dv} = \frac{1}{2}(\nabla\mathbf{v} + (\nabla\mathbf{v})^T) \,.$$

At the interface Γ the jump conditions

$$[\mathbf{v}]_-^+ = 0, \ [-\mathbf{S}]_-^+\boldsymbol{v} + [p]_-^+\boldsymbol{v} = \sigma\kappa\boldsymbol{v} + \nabla_\Gamma\sigma$$

have to hold. Here $[.]_-^+$ is the jump across the interface, σ is the surface energy density, κ is the mean curvature (i.e., the sum of the principal curvatures) of the interface Γ, $\boldsymbol{v}$ is a unit normal to the interface Γ and ∇_Γ is the surface gradient. The term $\nabla_\Gamma\sigma$ takes effects due to a variable surface energy density into account and a non-zero $\nabla_\Gamma\sigma$ leads to Marangoni effects.

In a diffuse interface model, one allows for a partial mixing of the fluids on a small length scale. For systems where the two fluids have densities of a similar size such models are referred to as *model H* in the literature, see [33]. In the case of equal density, a derivation of the diffuse interface model was given by Gurtin, Polignone and Vinals [31] in the context of rational continuum mechanics.

In two-fluid flow, often the situation appears that both fluids are incompressible with a large density difference. This was the motivation of Lowengrub and Truskinovsky [39] to introduce so called quasi-incompressible diffuse interface models for two-fluid flow. The assumptions made are that both fluids are incompressible but in zones where the fluids mix a variable density is allowed and the velocity field is no longer divergence free. Lowengrub and Truskinovsky [39] introduce a mass concentration field c. The mass ρ is assumed to be independent of the pressure p (and vice versa), which is the assumption of *quasi-incompressibility*. The mass ρ is then postulated to be a function of c and in the case of a simple mixture the mixture relation is

$$\frac{1}{\rho(c)} = \frac{c}{\rho_-} + \frac{1-c}{\rho_+}$$

where ρ_- and ρ_+ are the mass densities of the two fluids involved. Then Lowengrub and Truskinovsky derived the system

$$D_t\rho + \rho \operatorname{div} \mathbf{v} = 0, \quad \rho D_t c + \nabla \cdot j = 0,$$

$$\rho D_t \mathbf{v} = \operatorname{div} \mathbf{T}$$

where D_t is the material derivative and

$$j = -m(c)\nabla\mu \quad \text{is a diffusive flux},$$

$$\mu = \frac{\sigma}{\varepsilon}\psi'(c) - \frac{\sigma\varepsilon}{\rho(c)}\nabla \cdot (\rho(c)\nabla c) - \rho'(c)\frac{p}{\rho^2} \quad \text{is the chemical potential},$$

$$\mathbf{T} = -p Id - \varepsilon\sigma\rho(c)\nabla c \otimes \nabla c + \eta(c)(\nabla\mathbf{v} + \nabla\mathbf{v}^T) \quad \text{is the stress tensor}$$

with ψ being a double well potential (e.g., of the form $\psi(c) = \frac{1}{2}(1-c^2)^2$), p being the pressure and ε being a small length scale parameter related to the thickness of the interface.

Specific features of the above system are that the pressure explicitly enters the chemical potential μ and that a dissipation inequality for the total energy

$$\rho\left(\frac{\mathbf{u}^2}{2} + \frac{\sigma\varepsilon}{2}|\nabla c|^2 + \frac{\sigma}{\varepsilon}\psi(c)\right) \tag{8.1}$$

holds, see [39] for details. We also remark that the model of Lowengrub and Truskinovsky reduces to the model H if the densities are equal. An alternative derivation based on the concept of microforces and similar to Gurtin et al. [31] is given in [1].

Some authors tried to modify the model H for the case of fluids with different densities keeping in particular a divergence free velocity field, see, e.g., Jacqmin [34] and Ding et al. [22]. We refer in particular to the paper of Ding et al. [22] who used a volume averaged velocity field to guarantee solenoidality of the velocity field. The first diffuse interface model which is based on a divergence free velocity field *and* at the same time strictly dissipates a physically sound energy has been introduced in [7] and we will now derive this model. The requirement that a dissipation inequality holds is not only necessary for thermodynamical consistency, it would also lead to a priori estimates which are essential for the analysis. They also allow, as we will see, for stable numerical discretizations.

There are several other generalizations of the model H for fluids with non-matched densities, cf. [15, 17, 38]. These models look simpler but they miss a physical sound derivation. Again, the validity of the second law of thermodynamics (in an appropriate mechanical version) is unclear and a global energy estimate is not known to hold. Under the assumption of small density variations, Boyer [17] was able to prove existence of strong solutions for the model derived therein.

8.2.2 A New Thermodynamically Consistent Model

In order to derive the model introduced by Abels, Garcke and Grün [7] we start with the basic balance equations. The two fluids, which are allowed to mix in a thin interfacial region are assumed to have mass densities ρ_- and ρ_+. The local mass balance equation for the two fluids is given by

$$\partial_t \rho_\pm + \operatorname{div} \widehat{\mathbf{J}}_\pm = 0 \tag{8.2}$$

with mass fluxes $\widehat{\mathbf{J}}_\pm$. With the help of the velocities

$$\mathbf{v}_\pm = \widehat{\mathbf{J}}_\pm / \rho_\pm$$

and the volume fractions

$$u_\pm = \rho_\pm / \tilde{\rho}_\pm ,$$

where $\tilde{\rho}_\pm$ are the constant specific densities of the unmixed states, we define the volume averaged velocity

$$\mathbf{v} = u_- \mathbf{v}_- + u_+ \mathbf{v}_+ = \frac{\rho_-}{\tilde{\rho}_-} \mathbf{v}_- + \frac{\rho_+}{\tilde{\rho}_+} \mathbf{v}_+ .$$

Assuming volume conservation during the mixing process we require

$$u_- + u_+ = 1$$

which is equivalent to the fact that the excess volume is zero. Weighting (8.2) with $1/\tilde{\rho}_\pm$ and adding leads to

$$0 = \partial_t(u_- + u_+) + \operatorname{div}\left(u_- \frac{\widehat{\mathbf{J}}_-}{\rho_-} + u_+ \frac{\widehat{\mathbf{J}}_+}{\rho_+} \right) = \partial_t(u_- + u_+) + \operatorname{div} \mathbf{v} = \operatorname{div} \mathbf{v} .$$

Introducing $\widetilde{\mathbf{J}}_{\pm} = \frac{1}{\tilde{\rho}_{\pm}}\widehat{\mathbf{J}}_{\pm} - u_{\pm}\mathbf{v}$ we obtain

$$\partial_t u_{\pm} + \operatorname{div}(u_{\pm}\mathbf{v}) + \operatorname{div}\widetilde{\mathbf{J}}_{\pm} = 0 .$$

It is now convenient to rephrase the governing equations with the help of the difference of the volume fractions and therefore we define

$$\varphi := u_+ - u_- .$$

It hence holds

$$\partial_t \varphi + \operatorname{div}(\varphi\mathbf{v}) + \operatorname{div}\mathbf{J}_{\varphi} = 0$$

where

$$\mathbf{J}_{\varphi} = \widetilde{\mathbf{J}}_+ - \widetilde{\mathbf{J}}_- .$$

The definitions of u_-, u_+, φ and the fact $u_- + u_+ = 1$ now imply that the total mass $\rho = \rho_+ + \rho_+$ is given as

$$\rho = \rho(\varphi) = \tilde{\rho}_+ \frac{1 + \varphi}{2} + \tilde{\rho}_- \frac{1 - \varphi}{2}$$

and hence due to $\partial_t \rho = \frac{\tilde{\rho}_+ - \tilde{\rho}_-}{2}\partial_t \varphi$ we obtain

$$\partial_t \rho + \operatorname{div}(\rho\mathbf{v}) + \operatorname{div}\tilde{\mathbf{J}} = 0, \tag{8.3}$$

where we define

$$\widetilde{\mathbf{J}} = \frac{\tilde{\rho}_+ - \tilde{\rho}_-}{2}\mathbf{J}_{\varphi} .$$

The classical conservation law for linear momentum in its pointwise formulation is given by

$$\partial_t(\rho\mathbf{v}) + \operatorname{div}(\rho\mathbf{v} \otimes \mathbf{v}) = \operatorname{div}\widetilde{\mathbf{T}}$$

with a stress tensor $\widetilde{\mathbf{T}}$. Rewriting this identity with the help of mass conservation (8.3) leads to

$$\rho(\partial_t\mathbf{v} + \mathbf{v} \cdot \nabla\mathbf{v}) = \operatorname{div}(\widetilde{\mathbf{T}} + \mathbf{v} \otimes \widetilde{\mathbf{J}}) - \widetilde{\mathbf{J}} \cdot \nabla\mathbf{v} .$$

It was discussed in [7], see also [13], that $\widetilde{\mathbf{T}}$ is not invariant under a change of observer. However, the tensor

$$\mathbf{T} = \widetilde{\mathbf{T}} + \mathbf{v} \otimes \widetilde{\mathbf{J}}$$

is an objective tensor and we obtain

$$\rho \partial_t \mathbf{v} + (\rho \mathbf{v} + \widetilde{\mathbf{J}}) \cdot \nabla \mathbf{v} = \operatorname{div} \mathbf{T},$$

see [13] and [7] for details.

To proceed further we need to introduce a total energy density which is given as the sum of kinetic energy and interfacial energy as follows

$$e(\mathbf{v}, \varphi, \nabla \varphi) = \frac{\rho}{2} |\mathbf{v}|^2 + \hat{\sigma} \left(\frac{\varepsilon}{2} |\nabla \varphi|^2 + \frac{1}{\varepsilon} \psi(\varphi) \right).$$

We now require the following energy inequality

$$\frac{d}{dt} \int_{V(t)} e(\mathbf{v}, \varphi, \nabla \varphi) dx + \int_{\partial V(t)} \mathbf{J}_e \cdot \nu d\mathcal{H}^{d-1} \leq 0 \tag{8.4}$$

where $\mathbf{J}_e$ is the energy flux, $\mathcal{H}^{d-1}$ is the $(d-1)$-dimensional surface measure and $V(t)$ is a time dependent volume transported with the velocity $\mathbf{v}$. The inequality (8.4) is the relevant formulation of the second law of thermodynamics in an isothermal situation, cf. [31]. With the help of a transport theorem and using the fact that $V(t)$ is arbitrary we obtain the pointwise inequality

$$-\mathscr{D} := \partial_t e + \operatorname{div}(\mathbf{v}e) + \operatorname{div} \mathbf{J}_e \leq 0. \tag{8.5}$$

Now every solution which fulfills the basic conservation laws and the energy inequality also fulfills

$$-\mathscr{D} = \partial_t e + \mathbf{v} \cdot \nabla \varphi + \operatorname{div} \mathbf{J}_e - \mu(\partial_t \varphi + \mathbf{v} \cdot \nabla \varphi + \operatorname{div} \mathbf{J}_\varphi) \leq 0$$

where μ is a Lagrange multiplier.

With the help of the mass balance and the balance of linear momentum we obtain

$$\partial_t \left(\frac{\rho}{2} |\mathbf{v}|^2 \right) + \operatorname{div} \left(\frac{\rho}{2} |\mathbf{v}|^2 \mathbf{v} \right) = -\frac{|\mathbf{v}|^2}{2} \operatorname{div} \widetilde{\mathbf{J}} + (\operatorname{div} \mathbf{T} - \widetilde{\mathbf{J}} \cdot \nabla \mathbf{v}) \cdot \mathbf{v}$$

$$= \operatorname{div} \left(-\frac{1}{2} |\mathbf{v}|^2 \widetilde{\mathbf{J}} + \mathbf{T}^T \mathbf{v} \right) - \mathbf{T} : \nabla \mathbf{v}.$$

Using the material derivative $D_t u = \partial_t u + \mathbf{v} \cdot \nabla u$ we obtain

$$D_t \nabla \varphi = \partial_t \nabla \varphi + \mathbf{v} \cdot \nabla(\nabla \varphi) = \nabla D_t \varphi - (\nabla \mathbf{v})^T \nabla \varphi\,.$$

The dissipation term $\mathscr{D}$ can hence be rewritten as

$$-\mathscr{D} = \operatorname{div}\left(\mathbf{J}_e - \widetilde{\mathbf{J}}\frac{|\mathbf{v}|^2}{2} + \mathbf{T}^T \mathbf{v} - \mu \mathbf{J}_\varphi + \hat\sigma\varepsilon\nabla\varphi D_t\varphi \right)$$

$$+ \left(\frac{\hat\sigma}{\varepsilon}\psi'(\varphi) - \mu - \hat\sigma\varepsilon\Delta\varphi \right) D_t\varphi$$

$$-(\mathbf{T} + \hat\sigma\varepsilon\nabla\varphi \otimes \nabla\varphi) : \nabla\mathbf{v} + \nabla\mu \cdot \mathbf{J}_\varphi = 0\,,$$

where the :-product for matrices $\mathbf{A}, \mathbf{B}$ is given as $\mathbf{A} : \mathbf{B} = tr(\mathbf{A}^T \mathbf{B})$. Defining the energy flux $\mathbf{J}_e$ as

$$\mathbf{J}_e = \widetilde{\mathbf{J}}\frac{|\mathbf{v}|^2}{2} - \mathbf{T}^T\mathbf{v} + \mu\mathbf{J}_\varphi - \hat\sigma\varepsilon\nabla\varphi D_t\varphi$$

and the chemical potential

$$\mu = -\hat\sigma\varepsilon\Delta\varphi + \frac{\hat\sigma}{\varepsilon}\psi'(\varphi)$$

the dissipation inequality is given as

$$\mathscr{D} = (\mathbf{T} + \hat\sigma\varepsilon\nabla\varphi \otimes \nabla\varphi) : \nabla\mathbf{v} - \nabla\mu \cdot \mathbf{J}_\varphi \geq 0\,. \tag{8.6}$$

As in [31] we introduce the pressure p and rewrite the above inequality as

$$\mathbf{S} : \nabla\mathbf{v} - \nabla\mu \cdot \mathbf{J}_\varphi \geq 0$$

with the viscous stress tensor

$$\mathbf{S} = \mathbf{T} + p\mathit{Id} + \hat\sigma\varepsilon\nabla\varphi \otimes \nabla\varphi\,.$$

The dissipation inequality is fulfilled by assuming viscous friction in the definition for $\mathbf{S}$ and Fick's law in the definition of $\mathbf{J}_\varphi$. We hence choose

$$\mathbf{S} = 2\eta(\varphi)D\mathbf{v}\,, \qquad \mathbf{J}_\varphi = -m(\varphi)\nabla\mu$$

with $\eta(\varphi), m(\varphi) \geq 0$.

The final diffuse interface model is given as follows

$$\rho \partial_t \mathbf{v} + (\rho \mathbf{v} + \widetilde{\mathbf{J}}) \cdot \nabla \mathbf{v} - \mathrm{div}(2\eta(\varphi)D\mathbf{v}) + \nabla p = -\hat{\sigma}\varepsilon\, \mathrm{div}(\nabla\varphi \otimes \nabla\varphi)\,, \quad (8.7)$$

$$\mathrm{div}\,\mathbf{v} = 0\,, \quad (8.8)$$

$$\partial_t \varphi + \mathbf{v} \cdot \nabla\varphi = \mathrm{div}(m(\varphi)\nabla\mu)\,, \quad (8.9)$$

$$\frac{\hat{\sigma}}{\varepsilon}\psi'(\varphi) - \hat{\sigma}\varepsilon\Delta\varphi = \mu\,, \quad (8.10)$$

where

$$\widetilde{\mathbf{J}} = \frac{\tilde{\rho}_+ - \tilde{\rho}_-}{2}\mathbf{J}_\varphi = -\frac{\tilde{\rho}_+ - \tilde{\rho}_-}{2}m(\varphi)\nabla\mu.$$

We notice that the term involving $\widetilde{\mathbf{J}}$ in the momentum equation vanishes for equal densities, i.e., if $\tilde{\rho}_+ = \tilde{\rho}_-$.

It is also possible to redefine the pressure as

$$\hat{p} = p - \hat{\sigma}\left(\frac{\varepsilon}{2}|\nabla\varphi|^2 + \frac{1}{\varepsilon}\psi(\varphi)\right) \quad (8.11)$$

and in this case the momentum equation becomes

$$\rho \partial_t \mathbf{v} + (\rho \mathbf{v} + \widetilde{\mathbf{J}}) \cdot \nabla \mathbf{v} - \mathrm{div}(2\eta(\varphi)D\mathbf{v}) + \nabla\hat{p} = \mu\nabla\varphi\,. \quad (8.12)$$

With even another pressure the momentum equation becomes

$$\rho \partial_t \mathbf{v} + (\rho \mathbf{v} + \widetilde{\mathbf{J}}) \cdot \nabla \mathbf{v} - \mathrm{div}(\eta(\varphi)D\mathbf{v}) + \nabla\tilde{p} \quad (8.13)$$

$$= \mathrm{div}\left(\hat{\sigma}\left(\frac{\varepsilon}{2}|\nabla\varphi|^2 + \frac{1}{\varepsilon}\psi(\varphi)\right)Id - \hat{\sigma}\varepsilon\nabla\varphi \otimes \nabla\varphi\right) \quad (8.14)$$

where $\hat{\sigma}\left(\frac{\varepsilon}{2}|\nabla\varphi|^2 + \frac{1}{\varepsilon}\psi(\varphi)\right)Id - \hat{\sigma}\varepsilon\nabla\varphi \otimes \nabla\varphi$ can be interpreted as an approximation of the surface stress tensor in a sharp interface model, see [13]. It is also sometimes convenient to write the acceleration part in a conservative form, see [25],

$$\partial_t(\rho\mathbf{v}) + \mathrm{div}(\mathbf{v} \otimes (\rho\mathbf{v} + \widetilde{\mathbf{J}})) - \mathrm{div}(2\eta(\varphi)D\mathbf{v}) + \nabla p = \mu\nabla\varphi\,. \quad (8.15)$$

8.2.3 Including Transport Effects of a Soluble Species

We now want to include the transport of a soluble species as an additional effect. We focus on a species that does not influence the surface tension at the interface and refer to [10] for the case of surface active agents (surfactants). Denoting the concentration of the soluble species by w we add to the conservation law

$$\partial_t w + \mathbf{v} \cdot \nabla w + \operatorname{div} \mathbf{J}_w = 0 \tag{8.16}$$

where $\mathbf{J}_w$ is the corresponding mass flux to the governing balance equations from the previous subsection. We also add the term

$$\int_\Omega \{g(w) + \beta(\varphi)w\}dx$$

to the total energy where g is an entropic term and β attains for $\varphi \leq -1$ or $\varphi \geq 1$ the values β_1 or β_2 respectively. These values will reappear in the Henry condition of the sharp interface limit in the following section. In [7] the authors derived a thermodynamically consistent model coupling (8.16) to the model of the previous subsection. In the model the momentum equation (8.7), the divergence equation (8.8) and the equation for the phase field (8.9) remain unchanged and the equation for the chemical potential now becomes

$$\mu = -\hat{\sigma}\varepsilon\Delta\varphi + \frac{\hat{\sigma}}{\varepsilon}\psi'(\varphi) + \beta'(\varphi)w.$$

All these equations are coupled to

$$\partial_t w + \mathbf{v} \cdot \nabla w - \operatorname{div}(K(\varphi)w\nabla(g'(w) + \beta(\varphi))) = 0,$$

where $K(\varphi)$ is related to the diffusion parameter in the phases. Under appropriate boundary condition the overall system decreases the total energy

$$\int_\Omega \frac{\rho}{2}|\mathbf{v}|^2 dx + \int_\Omega \hat{\sigma}\left(\frac{\varepsilon}{2}|\nabla\varphi|^2 + \frac{1}{\varepsilon}\psi(\varphi)\right)dx + \int_\Omega \{g(w) + \beta(\varphi)w\}dx.$$

8.3 Sharp Interface Limit

The parameter ε in the above models is related to the interfacial thickness of the diffuse interfacial layer. As ε tends to zero classical and new sharp interface models are attained. So far this has been shown by formally matched asymptotic expansions, see [7] and in certain situations a rigorous convergence result in a very weak varifold setting, see [5] or [11]. Which sharp interface limit we obtain depends on the mobility and we consider four cases

$$m_\varepsilon(\varphi) = \begin{cases} m_0 & \text{case I}, \\ \varepsilon m_0 & \text{case II}, \\ \frac{m_1}{\varepsilon}(1 - \varphi^2)_+ & \text{case III}, \\ m_1(1 - \varphi^2)_+ & \text{case IV}. \end{cases}$$

Here m_0 and m_1 are positive constants and $(1 - \varphi^2)_+ = \max(1 - \varphi^2, 0)$. We assume for simplicity that $g(w) = w(\log w - 1)$ and set $K_\pm = K(\pm 1)$, $\eta_\pm = \eta(\pm 1)$. In the cases II–IV we obtain in the bulk regions Ω_- and Ω_+ for $i \in \{-, +\}$

$$\rho_i(\partial_t \mathbf{v} + \mathrm{div}(\mathbf{v} \otimes \mathbf{v})) - \eta_i \Delta \mathbf{v} + \nabla p = 0,$$

$$\mathrm{div}\,\mathbf{v} = 0,$$

$$\partial_t w + \mathbf{v} \cdot \nabla w = K_i \Delta w$$

and in case I the momentum balance has to be replaced by

$$\rho_i \partial_t \mathbf{v} + \mathrm{div}\left(\mathbf{v} \otimes \left(\rho_i \mathbf{v} + \tfrac{\rho_- - \rho_+}{2} m_0 \nabla \mu\right)\right) - \eta_i \Delta \mathbf{v} + \nabla p = 0, \tag{8.17}$$

i.e., a non-classical term $\frac{\rho_- - \rho_+}{2} m_0 \nabla \mu$ appears for non-matched densities where μ solves the quasi-static diffusion equation

$$\Delta \mu = 0 \qquad \text{in} \qquad \Omega_\pm.$$

On the interface Γ we have in all four cases

$$[\mathbf{v}]_-^+ = 0,$$

$$\frac{w_+}{w_-} = \exp(\beta_- - \beta_+),$$

$$(\mathscr{V} - \mathbf{v} \cdot \boldsymbol{\nu})[w]_-^+ = -[K\nabla w]_-^+ \cdot \boldsymbol{\nu},$$

where $\mathscr{V}$ is the normal velocity of Γ. The remaining conditions depend on which of the cases we consider. In the cases II and IV we obtain

$$-[2\eta D\mathbf{v}]_-^+ \cdot \boldsymbol{\nu} + [p]_-^+ \boldsymbol{\nu} = \sigma \kappa \boldsymbol{\nu},$$

$$\mathscr{V} = \mathbf{v} \cdot \boldsymbol{\nu}$$

where the surface tension σ can be computed from $\hat{\sigma}$ and ψ, see [7] for details. We hence note that if one wants to approximate classical sharp interface models with the help of a diffuse interface model one hence has to choose

$$m_\varepsilon(\varphi) = \varepsilon m_0 \quad \text{or} \quad m_\varepsilon(\varphi) = m_1(1 - \varphi^2)_+.$$

In case I on the other hand we obtain

$$-[2\eta D\mathbf{v}]_-^+ \boldsymbol{\nu} + [p]_-^+ \boldsymbol{\nu} = \sigma \kappa \boldsymbol{\nu}, \tag{8.18}$$

$$2(\mathbf{v} \cdot \boldsymbol{\nu} - \mathscr{V}) = m_0[\nabla \mu]_-^+ \cdot \boldsymbol{\nu}, \tag{8.19}$$

$$2\mu = \sigma \kappa - [w]_-^+ \tag{8.20}$$

which is remarkable as the interface is not transported by the fluid velocity.

In case III we obtain

$$- \tfrac{1}{2}[\rho]_-^+ \hat{m}((\nabla_\Gamma \mu) \cdot \nabla_\Gamma)\mathbf{v} - 2[\eta D\mathbf{v}]_-^+ \boldsymbol{\nu} + [p]_-^+ \boldsymbol{\nu} = \sigma \kappa \boldsymbol{\nu}, \qquad (8.21)$$

$$2(\mathbf{v} \cdot \boldsymbol{\nu} - \mathcal{V}) = \hat{m} \Delta_\Gamma \mu, \qquad (8.22)$$

$$2\mu = \sigma \kappa - [w]_-^+ \qquad (8.23)$$

where Δ_Γ is the Laplace-Beltrami operator on the interface and $\hat{m}$ is a constant related to ψ. We hence obtain surface diffusion, cf. [23], in the equation for the normal velocity and we obtain the non-classical term $-\tfrac{1}{2}[\rho]_-^+ \hat{m}((\nabla_\Gamma \mu) \cdot \nabla_\Gamma)\mathbf{v}$ in the interfacial stress balance.

In [7] the following theorem has been shown in which in particular the different dissipation mechanisms can be read off.

Theorem 1 (Free Energy Inequality) *We assume $\eta, \rho, K, w, m_0, \hat{m} > 0$. Then a sufficiently smooth solution of the sharp interface problem defined in a spatial domain Ω with $\Gamma(t) \subset \Omega$, for all $t \geq 0$, fulfills*

$$\frac{d}{dt}\left[\int_\Omega \left(\frac{\rho}{2}|\mathbf{v}|^2 + (g(w) + \beta w) \right) dx + \int_{\Gamma(t)} \sigma \, d\mathcal{H}^{d-1} \right] = -\mathcal{D} \leq 0,$$

provided the integrals are finite. Here $\rho = \rho_-, \rho_+$ and $\beta = \beta_-, \beta_+$ holds in the two phases. The quantity $\mathcal{D}$ is given as

Case I $\quad : \quad \mathcal{D} = \displaystyle\int_\Omega 2\eta|D\mathbf{v}|^2 \, dx + \int_\Omega K\frac{1}{w}|\nabla w|^2 \, dx + \int_\Omega m_0|\nabla\mu|^2 \, dx,$

Cases II and IV $\quad : \quad \mathcal{D} = \displaystyle\int_\Omega 2\eta|D\mathbf{v}|^2 \, dx + \int_\Omega K\frac{1}{w}|\nabla w|^2 \, dx,$

Case III $\quad : \quad \mathcal{D} = \displaystyle\int_\Omega 2\eta|D\mathbf{v}|^2 \, dx + \int_\Omega K\frac{1}{w}|\nabla w|^2 \, dx + \int_{\Gamma(t)} \hat{m}|\nabla_\Gamma \mu|^2 d\mathcal{H}^{d-1}.$

8.4 Existence Results

In this subsection analytic results for the system (8.7)–(8.10) are discussed, i.e.,

$$\rho \partial_t \mathbf{v} + (\rho \mathbf{v} + \tilde{\mathbf{J}}) \cdot \nabla \mathbf{v} - \text{div}(2\eta(\varphi)D\mathbf{v}) + \nabla p = -\hat{\sigma}\varepsilon \, \text{div}(\nabla\varphi \otimes \nabla\varphi), \qquad (8.24)$$

$$\text{div}\,\mathbf{v} = 0, \qquad (8.25)$$

$$\partial_t \varphi + \mathbf{v} \cdot \nabla\varphi = \text{div}(m(\varphi)\nabla\mu), \qquad (8.26)$$

$$\mu = -\hat{\sigma}\varepsilon\Delta\varphi + \frac{\hat{\sigma}}{\varepsilon}\psi'(\varphi) \qquad (8.27)$$

in $\Omega \times (0, T)$, where $\Omega \subseteq \mathbb{R}^d$, $d = 2, 3$, is a bounded domain with smooth boundary and

$$\rho = \rho(\varphi) = \tilde{\rho}_+ \frac{1 + \varphi}{2} + \tilde{\rho}_- \frac{1 - \varphi}{2}, \tag{8.28}$$

$$\widetilde{\mathbf{J}} = \frac{\tilde{\rho}_+ - \tilde{\rho}_-}{2} \mathbf{J}_\varphi = -\frac{\tilde{\rho}_+ - \tilde{\rho}_-}{2} m(\varphi) \nabla \mu.$$

In order to obtain a well-posed problem we have to close the system with initial conditions

$$(\mathbf{v}, \varphi)|_{t=0} = (\mathbf{v}_0, \varphi_0) \tag{8.29}$$

and suitable boundary conditions. A standard choice, which is made for the following analytic results, is

$$\mathbf{v}|_{\partial\Omega} = \mathbf{0}, \, \nu \cdot \nabla\varphi|_{\partial\Omega} = \nu \cdot \nabla\mu|_{\partial\Omega} = 0. \tag{8.30}$$

For the following we set $\hat{\sigma} = 1$ for simplicity.

First of all smooth solutions of (8.24)–(8.27) together with (8.30) satisfy the energy dissipation identity

$$\frac{d}{dt} \int_\Omega \left(\frac{\rho|\mathbf{v}|^2}{2} + \frac{\varepsilon|\nabla\varphi|^2}{2} + \frac{\psi(\varphi)}{\varepsilon} \right) dx = -\int_\Omega 2\eta(\varphi)|D\mathbf{v}|^2 \, dx - \int_\Omega m(\varphi)|\nabla\mu|^2 \, dx \tag{8.31}$$

for all $t \in (0, T)$. This follows from integration of the dissipation relation (8.5) and (8.6). This identity yields a priori bounds for

$$\mathbf{v} \in L^\infty((0, \infty); L^2_\sigma(\Omega)) \cap L^2((0, \infty); H^1_0(\Omega)^d), \, \varphi \in L^\infty((0, \infty); H^1(\Omega)),$$

$$\nabla\mu \in L^2(\Omega \times (0, T))^d \text{ if } m(\varphi) \geq m_0 > 0$$

as long as $\rho(\varphi)$ is bounded below, which is the case if φ only attains values in the physical reasonable interval $[-1, 1]$. Here $L^p(\Omega)$, $1 \leq p \leq \infty$, is the standard Lebesgue space (with respect to the Lebesgue measure) consisting of measurable functions $f \colon \Omega \to \mathbb{R}$ such that $|f|^p$ is integrable if $p < \infty$, $|f|$ is essentially bounded if $p = \infty$, respectively. Moreover, $L^p(\Omega; X)$ is its X-valued analogue, $H^k(\Omega)$ is the L^2-Sobolev space of k-th order and $H^1_0(\Omega)$ is the closure of smooth, compactly supported functions in $H^1(\Omega)$. Furthermore, $L^2_\sigma(\Omega)$ denotes the closure of divergence-free, smooth, and compactly supported vector-fields.

For the mathematical analysis it is essential that ρ stays positive, which is true as long as $\varphi \in [-1, 1]$. In order to guarantee this property mathematically we worked with so-called singular free energy densities as e.g. given by

$$\psi(\varphi) = \frac{\theta}{2}\left((1+\varphi)\ln(1+\varphi) + (1-\varphi)\ln(1-\varphi)\right) - \frac{\theta_c}{2}\varphi^2 \quad \text{for } \varphi \in [-1,1],$$

$$(8.32)$$

where $0 < \theta < \theta_c$. This was used by Cahn and Hilliard [20] to describe physical relevant so-called regular solution models. Mathematically, the singularity in ψ' ensures that the order parameter φ stays in the physically reasonable interval $[-1,1]$. But it leads to additional difficulties due to singular terms in the equation for the chemical potential. For non-singular ψ (and non-degenerate $m(\varphi)$) this property is unknown.

Existence of weak solutions of (8.24)–(8.27) was shown by Abels, Depner, and Garcke in [8] and [9] in the case of singular free energies with non-degenerate and degenerate mobility, respectively. More precisely, in the non-degenerate case the following result was shown:

Theorem 2 (Existence of Weak Solutions [8, Theorem 3.4]) *Let* $m\colon \mathbb{R} \to \mathbb{R}$ *be continuously differentiable,* $\eta\colon \mathbb{R} \to \mathbb{R}$ *be continuous and assume that* $k \leq m(s), \eta(s) \leq K$ *for all* $s \in \mathbb{R}$ *and some* $k, K > 0$*. Moreover, let* ψ *be as in (8.32). Then for every* $v_0 \in L^2_\sigma(\Omega)$ *and* $\varphi_0 \in H^1(\Omega)$ *with* $|\varphi_0| \leq 1$ *almost everywhere and* $\frac{1}{|\Omega|} \int_\Omega \varphi_0\, dx \in (-1,1)$ *there exists a weak solution* (v, φ, μ) *of (8.24)–(8.27) together with (8.29)–(8.30) such that for any* $0 < T < \infty$

$$v \in L^\infty((0,\infty); L^2_\sigma(\Omega)) \cap L^2((0,\infty); H^1_0(\Omega)^d),$$

$$\varphi \in L^\infty((0,\infty); H^1(\Omega)) \cap L^2((0,T); H^2(\Omega)), \quad \psi'(\varphi) \in L^2(\Omega \times (0,T)),$$

$$\mu \in L^2((0,T); H^1(\Omega)) \quad \text{with} \quad \nabla\mu \in L^2(\Omega \times (0,\infty))^d.$$

Actually, the result holds true under more general assumptions. For the precise definition of weak solutions we refer to [8, Definition 3.3]. As usual the pressure p is not part of the weak formulation since (8.24) is tested with divergence free test functions.

The structure of the proof of Theorem 2 is as follows: System (8.24)–(8.27) is first approximated with the aid of a semi-implicit time discretization, which satisfies the same kind of energy identity as the continuous system. Hence one obtains a priori bounds for

$$\mathbf{v}^N \in L^\infty((0,\infty); L^2_\sigma(\Omega)) \cap L^2((0,\infty); H^1_0(\Omega)^d), \varphi^N \in L^\infty((0,\infty); H^1(\Omega)),$$

$$\nabla\mu^N \in L^2(\Omega \times (0,\infty)),$$

where $(\mathbf{v}^N, \varphi^N, \mu^N)$ are suitable interpolations of solutions of the time discretized system with discretization parameter $h = \frac{1}{N}$. In order pass to the limit $N \to \infty$ it is essential to obtain a bound for

$$\varphi^N \in L^2((0,T), H^2(\Omega)), \quad \psi'(\varphi^N) \in L^2(\Omega \times (0,T))$$

for any $T > 0$. To this end one uses that $\psi'(\varphi) = \psi_0'(\varphi) - \theta_c \varphi$, where ψ_0 is convex together with a priori estimates in $H^2(\Omega)$ for the non-linear elliptic equation

$$-\Delta\varphi + \psi_0'(\varphi) = g := \mu + \theta_c \varphi, \qquad \nu \cdot \nabla\varphi|_{\partial\Omega} = 0$$

due to Abels and Wilke [6, Theorem 4.3]. This theorem is also used to obtain existence of solutions for the time discretized system.

In the case of degenerate mobility it is assumed that

$$m(s) = \begin{cases} 1 - s^2 & \text{if } s \in [-1, 1], \\ 0 & \text{else} \end{cases}$$

and $\psi: \mathbb{R} \to \mathbb{R}$ is continuously differentiable. In this case one does not obtain an a priori bound for $\nabla\mu$ in $L^2((0,\infty) \times \Omega)$. Instead one obtains an a priori bound for $\widehat{\mathbf{J}} := \sqrt{m(\varphi)}\nabla\mu$ and $\mathbf{J} := m(\varphi)\nabla\mu$. Therefore one has to avoid $\nabla\mu$ in the weak formulation and has to formulate the equations in terms of $\mathbf{J}$. More precisely, the triple $(\mathbf{v}, \varphi, \mathbf{J})$ should satisfy the weak formulations:

$$-\int_0^T \int_\Omega \rho\mathbf{v} \cdot \partial_t\boldsymbol{\psi}\, dx\, dt + \int_0^T \int_\Omega \operatorname{div}(\rho\mathbf{v} \otimes \mathbf{v}) \cdot \boldsymbol{\psi}\, dx\, dt$$

$$+ \int_0^T \int_\Omega 2\eta(\varphi)D\mathbf{v} : D\boldsymbol{\psi}\, dx\, dt - \int_0^T \int_\Omega (\mathbf{v} \otimes \tfrac{\tilde{\rho}_+ - \tilde{\rho}_-}{2}\mathbf{J}) : \nabla\boldsymbol{\psi}\, dx\, dt \qquad (8.33)$$

$$= -\int_0^T \int_\Omega \varepsilon\Delta\varphi\, \nabla\varphi \cdot \boldsymbol{\psi}\, dx\, dt$$

for all $\boldsymbol{\psi} \in C_0^\infty(\Omega \times (0,T))^d$ with $\operatorname{div}\boldsymbol{\psi} = 0$,

$$-\int_0^T \int_\Omega \varphi\, \partial_t\zeta\, dx\, dt + \int_0^T \int_\Omega (\mathbf{v} \cdot \nabla\varphi)\, \zeta\, dx\, dt = \int_0^T \int_\Omega \mathbf{J} \cdot \nabla\zeta\, dx\, dt \qquad (8.34)$$

for all $\zeta \in C_0^\infty((0,T); C^1(\overline{\Omega}))$ and

$$\int_0^T \int_\Omega \mathbf{J} \cdot \boldsymbol{\eta}\, dx\, dt = -\int_0^T \int_\Omega \left(\tfrac{1}{\varepsilon}\psi'(\varphi)) - \varepsilon\Delta\varphi\right) \operatorname{div}(m(\varphi)\boldsymbol{\eta})\, dx\, dt \qquad (8.35)$$

for all $\boldsymbol{\eta} \in L^2(0,T; H^1(\Omega)^d) \cap L^\infty(\Omega \times (0,T))^d$ which fulfill $\nu \cdot \boldsymbol{\eta}|_{\partial\Omega} = 0$ on $\partial\Omega \times (0,T)$. Here $C_0^\infty(\Omega \times (0,T))$, $C_0^\infty((0,T); X)$ is the set of all smooth functions $\varphi: \Omega \times$

$(0, T) \to \mathbb{R}$, $\varphi \colon (0, T) \to X$ with compact support. We refer to [9, Definition 3.3] for the complete definition of weak solutions.

We remark that (8.35) is a weak formulation of

$$\mathbf{J} = -m(\varphi) \nabla \left(\frac{1}{\varepsilon} \psi'(\varphi) - \varepsilon \Delta \varphi \right).$$

Theorem 3 (Existence of Weak Solutions [9, Theorem 3.5]) *Let m, ψ be as before and η be as in Theorem 2, $v_0 \in L^2_\sigma(\Omega)$ and $\varphi_0 \in H^1(\Omega)$ with $|\varphi_0| \leq 1$ almost everywhere in Ω. Then there exists a weak solution $(v, \varphi, \mathbf{J})$ of (8.24)–(8.27) together with (8.29)–(8.30).*

The theorem is proved by approximating m by a sequence of strictly positive mobilities m_δ and ψ by

$$\psi_\delta(s) := \psi(s) + \delta(1 + s)\ln(1 + s) + \delta(1 - s)\ln(1 - s), \quad s \in [-1, 1],$$

where $\delta > 0$. Then existence of weak solutions $(v_\delta, \varphi_\delta, \mu_\delta)$ for $\delta > 0$ follows from Theorem 2. In order to pass to the limit one uses the energy identity (8.31). But this does not give a bound for $\varphi_\delta \in L^2(0, T; H^2(\Omega))$, which is essential to pass to the limit in the weak formulation of (8.24). In order to obtain this bound one tests the weak formulation of (8.26) with $G'_\delta(\varphi_\delta)$, where $G''(s) = \frac{1}{m_\delta(s)}$ for $s \in (-1, 1)$ and $G'_\delta(0) = G_\delta(0) = 0$, see [9, Proof of Lemma 3.7] for the details.

In the case of non-degenerate mobility existence of a unique strong solution was shown by Weber [42]:

Theorem 4 (Existence of Strong Solutions [42, Theorem 5.4]) *Let Ω, η, m, and ψ be sufficiently smooth, η, m be positive, and $4 < p < 6$. Moreover, let $v_0 \in H^1_0(\Omega)^d \cap L^2_\sigma(\Omega)$ and $\varphi_0 \in W^4_p(\Omega)$ with $v \cdot \nabla \varphi_0|_{\partial\Omega} = 0$ and $|\varphi_0| \leq 1$. Then there exists some $T > 0$ such that (8.24)–(8.27) in $\Omega \times (0, T)$ together with (8.29)–(8.30) has a unique strong solution*

$$\mathbf{v} \in W^1_2(0, T; L^2_\sigma(\Omega)) \cap L^2(0, T; H^2(\Omega)^d \cap H^1_0(\Omega)^d),$$

$$\varphi \in W^1_p(0, T; L^p(\Omega)) \cap L^p(0, T; W^4_p(\Omega)).$$

Finally, we mention that existence of weak solutions of (8.24)–(8.27) together with (8.29)–(8.30) was proven in the case of power-law type fluids of exponent $p > \frac{2d+2}{d+2}$, $d = 2, 3$, by Abels and Breit [4]. In this work the case of constant, positive mobility together with a suitable smooth free energy density ψ is considered. Unfortunately, in this case there is no mechanism, which enables to show that $\varphi \in [-1, 1]$. Hence one has to modify ρ, defined as in (8.28) for $\varphi \in [-1, 1]$, outside of $[-1, 1]$ suitably such that it stays positive. But then (8.3) is no longer valid and one obtains instead

$$\partial_t \rho + \mathrm{div}(\rho \mathbf{v} + \tilde{\mathbf{J}}) = R, \quad \text{where } R = -\nabla \rho'(\varphi) \cdot m\nabla\mu, \tilde{\mathbf{J}} = -\rho'(\varphi)m\nabla\mu. \qquad (8.36)$$

Here R is an additional source term, which vanishes in the interior of $\{\varphi \in [-1, 1]\}$. In order to obtain a local dissipation inequality and a global energy estimate the equation of linear momentum (8.24) has to be modified to

$$\varrho \partial_t \mathbf{v} + (\varrho \mathbf{v} + \tilde{\mathbf{J}})) \cdot \nabla \mathbf{v} + R \frac{\mathbf{v}}{2} - \mathrm{div} \mathbf{S}(\varphi, D\mathbf{v}) + \nabla p = -\varepsilon \mathrm{div}(\nabla \varphi \otimes \nabla \varphi).$$

Under these assumptions existence of weak solutions is shown with the aid of the so-called L^∞-truncation method, cf. [4] for the details.

8.5 Numerical Approximation

In this subsection, we discuss numerical aspects of the diffusive interface model (8.8)–(8.10) and (8.12). For the ease of presentation, we choose the parameters $\hat{\sigma}$ and ε equal to one.

Having a dissipative model consistent with thermodynamics at hand, it becomes realistic to look for stable numerical schemes and for convergence results—see [24] in the setting of Navier-Stokes-Cahn-Hilliard models with *identical* mass densities. Since system (8.8)–(8.10), (8.12) is formulated with solenoidal vector fields, one expects the classical discretization spaces for Navier-Stokes-Systems and Cahn-Hilliard equations to be applicable, too. For details of the discretization, e.g. the choice of spaces, initial data, discretization of nonlinearities etc., we refer to the Appendix.

To obtain a discrete formulation that permits to establish a discrete energy estimate just by testing with admissible functions, i.e. the momentum equation by the velocity field $\mathbf{v}$, the phase-field equation by the chemical potential μ and the weak formulation relating μ and φ by an appropriate discrete time derivative of φ, it is promising to replace the naive weak formulation of the momentum equation

$$\int \rho(\varphi) \partial_t \mathbf{v} \cdot \mathbf{w} + \int \rho \mathbf{v} \cdot (\nabla \mathbf{v})^T \mathbf{w} + \int \rho'(\varphi) \mathbf{J}_\varphi \cdot (\nabla \mathbf{v})^T \mathbf{w} + \int 2\eta(\varphi) D\mathbf{v} : D\mathbf{v}$$

$$= \int \mu \nabla \varphi \cdot \mathbf{w} \qquad \forall \mathbf{w} \in W^{1,2}_{0,\mathrm{div}}(\Omega)$$

by

$$\int \partial_t(\rho(\varphi)\mathbf{v})\mathbf{w} - \frac{1}{2} \int \partial_t \rho(\varphi) \langle \mathbf{v}, \mathbf{w} \rangle$$

$$- \frac{1}{2} \int \rho(\varphi) \langle \mathbf{v}, (\nabla \mathbf{w})^T \mathbf{v} \rangle + \frac{1}{2} \int \rho(\varphi) \langle \mathbf{v}, (\nabla \mathbf{v})^T \mathbf{w} \rangle$$

$$+ \frac{1}{2} \int \rho'(\varphi) \langle \mathbf{J}_\varphi, (\nabla \mathbf{v})^T \mathbf{w} \rangle - \frac{1}{2} \int \rho'(\varphi) \langle \mathbf{J}_\varphi, (\nabla \mathbf{w})^T \mathbf{v} \rangle$$

$$+ \int 2\eta(\varphi) D\mathbf{v} : D\mathbf{v} = - \int \varphi \langle \nabla \mu, \mathbf{w} \rangle \qquad \forall \mathbf{w} \in W^{1,2}_{0,\mathrm{div}}(\Omega)$$

which follows by adding an appropriate zero (see [27] for details).

Let $(\mathbf{W}_h, S_h)$ be an *admissible pair of discretization spaces for hydrodynamics* as defined in the Appendix, for instance the spaces associated with P_2P_1- or with P_2P_0-elements, and assume the hypotheses (H1)–(H6) and (T) in the Appendix to hold. In [27], the following discrete system has been suggested.

For given functions $(\varphi^0, \mathbf{v}^0) \in U_h \times \mathbf{W}_h$ and $k = 0, \ldots, N-1$ we have to find functions $(\varphi^{k+1}, \mu^{k+1}, \mathbf{v}^{k+1}, p^{k+1}) \in U_h \times U_h \times \mathbf{W}_h \times S_h$ such that

$$\left(\partial_\tau^- \varphi^{k+1}, \theta\right)_h - \int_\Omega \langle \mathbf{v}^{k+1}, \nabla\theta \rangle \varphi^k + \int_\Omega \langle \nabla\mu^{k+1}, \nabla\theta \rangle = 0 \quad \forall \theta \in U_h, \tag{8.37a}$$

$$\left(\mu^{k+1}, \theta\right)_h = \int_\Omega \langle \nabla\varphi^{k+1}, \nabla\theta \rangle + \int_\Omega \mathscr{I}_h\big(\psi'_{\tau h}(\varphi^{k+1}, \varphi^k)\theta\big) \quad \forall \theta \in U_h. \tag{8.37b}$$

$$\int_\Omega \langle \partial_\tau^-(\rho^{k+1}\mathbf{v}^{k+1}), \mathbf{w} \rangle - \tfrac{1}{2} \int_\Omega \partial_\tau^- \rho^{k+1} \langle \mathbf{v}^{k+1}, \mathbf{w} \rangle$$

$$- \tfrac{1}{2} \int_\Omega \rho^k \langle \mathbf{v}^k, (\nabla\mathbf{w})^T \mathbf{v}^{k+1} \rangle + \tfrac{1}{2} \int_\Omega \rho^k \langle \mathbf{v}^k, (\nabla\mathbf{v}^{k+1})^T \mathbf{w} \rangle$$

$$+ \tfrac{1}{2} \int_\Omega \tfrac{\delta\varrho}{\delta\varphi} \langle \mathbf{j}^{k+1}, (\nabla\mathbf{v}^{k+1})^T \mathbf{w} \rangle - \tfrac{1}{2} \int_\Omega \tfrac{\delta\varrho}{\delta\varphi} \langle \mathbf{j}^{k+1}, (\nabla\mathbf{w})^T \mathbf{v}^{k+1} \rangle$$

$$+ \int_\Omega 2\eta(\varphi^k)\mathbf{D}\mathbf{v}^{k+1} : \mathbf{D}\mathbf{w} - \int_\Omega p^{k+1} \operatorname{div}\mathbf{w}$$

$$= - \int_\Omega \varphi^k \langle \nabla\mu^{k+1}, \mathbf{w} \rangle \quad \forall \mathbf{w} \in \mathbf{X}_h, \tag{8.37c}$$

$$\int_\Omega \theta \operatorname{div}\mathbf{v}^{k+1} = 0 \quad \forall \theta \in S_h. \tag{8.37d}$$

Here, we use the abbreviations $\rho^{k+1} := \rho(\varphi^{k+1})$ and $\frac{\delta\varrho}{\delta\varphi} := \frac{\tilde{\varrho}_+ - \tilde{\varrho}_-}{2}$. Moreover, we define $\mathbf{j}^{k+1} := -\nabla\mu^{k+1}$. Note that $\psi'_{\tau h}(\varphi^{k+1}, \varphi^k)$ is an appropriate discretization of ψ'. Famous is the convex-concave splitting $\psi'_{\tau h}(\varphi^{k+1}, \varphi^k) = \psi'_+(\varphi^{k+1}) + \psi'_-(\varphi^k)$ where $\psi = \psi_+ + \psi_-$ with ψ_+ and $-\psi_-$ being convex functions.

Using the skew-symmetry of the convective term in (8.37c), an energy-estimate is readily established—with the free energy given by

$$\mathscr{E}(\mathbf{v}^k, \rho^k, \varphi^k) := \frac{1}{2} \int_\Omega \rho(\varphi^k)|\mathbf{v}^k|^2 + \frac{1}{2} \int_\Omega |\nabla\varphi^k|^2 + \int_\Omega \mathscr{I}_h\psi(\varphi^k).$$

Observe that for

$$\rho(\varphi) = \frac{\tilde{\rho}_+ - \tilde{\rho}_-}{2}\varphi + \frac{\tilde{\rho}_+ + \tilde{\rho}_-}{2}$$

we have positivity of $\varrho(\varphi)$ if $\varphi > -(\mathfrak{At})^{-1}$, where

$$\mathfrak{At} := \frac{\tilde{\rho}_+ - \tilde{\rho}_-}{\tilde{\rho}_+ + \tilde{\rho}_-}$$

for $\tilde{\rho}_+ \geq \tilde{\rho}_-$ is the Atwood number. In the engineering literature, it is used to measure density discrepancies of the two liquids involved—an Atwood number 0.998 corresponds to a ratio $\frac{\tilde{\rho}_+}{\tilde{\rho}_-} = 1000$.

Hence, the discrete energy estimate in its pure form guarantees stability only for values of $\varphi > -(\mathfrak{At})^{-1}$.

In the continuous setting (assuming in particular a mechanism which confines the values of the phase-field function to the interval $[-1, 1]$, for instance by choosing a degenerate mobility or a logarithmic potential ψ), ρ depends linearly on φ via (8.28) and is therefore bounded from below by a positive constant by definition. In the discrete setting, however, it is not possible to mimic singular or degenerate behaviour—regularization is indispensable. Hence, strict inclusions $\varphi \in [-1, 1]$ for discrete solutions φ cannot be expected in general. Bounds on solutions can only be obtained via integral estimates as the phase-field equation is fourth-order parabolic and therefore comparison principles do not hold. However, the energy of the system is not necessarily decreasing in time due to the work done by external forces. As a consequence, bounds on φ always will depend on the special choice of external forces. Therefore, we use the cut-off mechanism of (H6) to guarantee definiteness of ρ and hence definiteness of the density $\rho|\mathbf{v}|^2$ of the kinetic energy as well.

To proceed, observe that the discretizations of the momentum equation and of the Cahn-Hilliard equation in (8.37a)–(8.37d) would decouple if $\mathbf{v}^{k+1}$ was replaced on the left-hand-side of (8.37a) by a velocity-term not depending on the $(k + 1)$th time step in the momentum equation.

Following ideas in [16, 40], see [29], we update the velocity $\mathbf{v}^k$ by adding an additional momentum $\tau\varphi^k\nabla\mu^{k+1}$ divided by a mass density term for which we choose $\rho_{\min}^k$. Summing up, Eqs. (8.37a)–(8.37b) are replaced by equation

$$\left(\partial_\tau^-\varphi^{k+1}, \theta\right)_h - \int_\Omega \varphi^k\langle\mathbf{v}^k, \nabla\theta\rangle + \tau\int_\Omega \frac{|\varphi^k|^2}{\rho_{\min}^k}\langle\nabla\mu^{k+1}, \nabla\theta\rangle + \int_\Omega\langle\nabla\mu^{k+1}, \nabla\theta\rangle = 0 \tag{8.38}$$

for all $\theta \in U_h$, and

$$\left(\mu^{k+1}, \theta\right)_h = \int_\Omega\langle\nabla\varphi^{k+1}, \nabla\theta\rangle + \left(\psi'_{\tau h}(\varphi^{k+1}, \varphi^k), \theta\right)_h \tag{8.39}$$

for all $\theta \in U_h$. One readily obtains a discrete version of the energy estimate.

Lemma 1 *For every $1 \leq l \leq N$ we have*

$$
\frac{1}{2} \int_{\Omega} \rho^l \left| \mathbf{v}^l \right|^2 + \frac{1}{2} \int_{\Omega} \left| \nabla \varphi^l \right|^2 + \int_{\Omega} \mathscr{I}_h \big(\psi_{\tau h}(\varphi^l) \big)
$$

$$
+ \frac{\tau^2}{4} \sum_{m=0}^{l-1} \int_{\Omega} \rho^m \left| \partial_\tau^- \mathbf{v}^{m+1} \right|^2 + \frac{\tau^2}{2} \sum_{m=0}^{l-1} \int_{\Omega} \left| \partial_\tau^- \nabla \varphi^{m+1} \right|^2 + \frac{\tau^2}{3} \sum_{m=0}^{l-1} \int_{\Omega} \frac{|\varphi^m|^2}{\rho_{min}^m} |\mathbf{j}^{m+1}|^2
$$

$$
+ \tau \sum_{m=0}^{l-1} \int_{\Omega} 2\eta(\varphi^{m+1}) \left| \mathbf{D} \mathbf{v}^{m+1} \right|^2 + \tau \sum_{m=0}^{l-1} \int_{\Omega} \left| \mathbf{j}^{m+1} \right|^2
$$

$$
\leq \frac{1}{2} \int_{\Omega} \rho^0 \left| \mathbf{v}^0 \right|^2 + \frac{1}{2} \int_{\Omega} \left| \nabla \varphi^0 \right|^2 + \int_{\Omega} \mathscr{I}_h \big(\psi_{\tau h}(\varphi^0) \big) . \tag{8.40}
$$

For the resulting splitting scheme which consists of successively computing updates for $\varphi_{\tau h}^{k+1}$ and $\mu_{\tau h}^{k+1}$ and then in a second step for $\mathbf{v}_{\tau h}^{k+1}$, the following convergence results hold true.

Theorem 5 *Let $\Omega \subset \mathbb{R}^d$, $d \in \{2, 3\}$ be a convex polyhedral domain and let initial data Φ_0 and $\mathbf{V}_0$ be given. Let $I = (0, T)$ and suppose $(\mathbf{W}_h, S_h)$ to be an admissible pair of discretization spaces in hydrodynamics. Assume that hypotheses (T) and (H1)–(H6) are satisfied and that $(\varphi_{\tau h}, \mu_{\tau h}, \mathbf{v}_{\tau h})$ is a sequence of discrete solutions to the system (8.37). Then there is a subsequence which converges in $L^2(\Omega_T) \cap L^\infty_{weak}(I; H^1)) \cap H^1_{weak}(I; L^2(\Omega)) \times L^2_{weak}(I; H^1) \times L^2(\Omega_T) \cap L^2_{weak}(I; \mathbf{W}_0^{1,2}(\Omega))$ to functions $(\mathbf{v}, \varphi, \mu)$ which solve the system (8.8)–(8.10), (8.12) in the following generalized sense.*

$$
- \iint_{\Omega_T} \langle \rho \mathbf{v} - \rho(\Phi_0) \mathbf{V}_0, \partial_t \mathbf{w} \rangle - \frac{1}{2} \iint_{\Omega_T} \partial_t \rho \langle \mathbf{v}, \mathbf{w} \rangle - \frac{1}{2} \iint_{\Omega_T} \rho \langle \mathbf{v}, (\nabla \mathbf{w})^T \mathbf{v} \rangle
$$

$$
+ \frac{1}{2} \iint_{\Omega_T} \rho \langle \mathbf{v}, (\nabla \mathbf{v})^T \mathbf{w} \rangle + \frac{1}{2} \iint_{\Omega_T} \frac{\delta \rho}{\delta \varphi} \langle \mathbf{j}, (\nabla \mathbf{v})^T \mathbf{w} \rangle - \frac{1}{2} \iint_{\Omega_T} \frac{\delta \rho}{\delta \varphi} \langle \mathbf{j}, (\nabla \mathbf{w})^T \mathbf{v} \rangle
$$

$$
+ \iint_{\Omega_T} 2\eta(\varphi) d\mathbf{v} : d\mathbf{w} = \iint_{\Omega_T} \mu \langle \nabla \varphi, \mathbf{w} \rangle \tag{8.41}
$$

for all $\mathbf{w} \in C^1 \left(I; \mathbf{W}_{0,\mathrm{div}}^{1,2}(\Omega) \right)$ satisfying $\mathbf{w}(\cdot, T) = 0$,

$$
\iint_{\Omega_T} \partial_t \varphi \theta + \iint_{\Omega_T} \langle \nabla \varphi, \mathbf{v} \rangle \theta + \iint_{\Omega_T} \langle \nabla \mu, \nabla \theta \rangle = 0 \tag{8.42}
$$

for all $\theta \in L^2(I; H^1(\Omega))$, and

$$
\mu(\cdot, t) = -\Delta \varphi(\cdot, t) + \psi'(\varphi(\cdot, t)) \tag{8.43}
$$

for almost all $t \in I$.

Remark 1 If the solution φ only attains values in the linear regime of ϱ, then the weak formulation (8.41) of the momentum equation simplifies to become

$$- \iint_{\Omega_T} \langle \rho \mathbf{v} - \rho(\Phi_0) \mathbf{V}_0, \partial_t \mathbf{w} \rangle - \iint_{\Omega_T} \partial_t \rho \langle \mathbf{v}, \mathbf{w} \rangle + \iint_{\Omega_T} \rho \langle \mathbf{v}, (\nabla \mathbf{v})^T \mathbf{w} \rangle$$

$$+ \iint_{\Omega_T} \rho'(\varphi) \langle \mathbf{j}, (\nabla \mathbf{v})^T \mathbf{w} \rangle + \iint_{\Omega_T} 2\eta(\varphi) \mathbf{D}\mathbf{v} : \mathbf{D}\mathbf{w} = \iint_{\Omega_T} \mu \langle \nabla \varphi, \mathbf{w} \rangle \qquad (8.44)$$

for all $\mathbf{w} \in C^1 \left(I; W^{1,2}_{0,\mathrm{div}}(\Omega) \right)$ satisfying $\mathbf{w}(\cdot, T) = 0$.

Remark 2 In [25], a different splitting scheme has been suggested. It is a fully linear and stable, convergence, however, has not been studied, yet.

It is worth mentioning that this result and its predecessor—an analogous result for the scheme using the discrete Cahn-Hilliard equation (8.37a)—are—to the best of our knowledge—the only convergence results for numerical schemes for two-phase flows with different mass densities obtained so far.

Let us make some comments on the proof of Theorem 5. Naively, one would be tempted first to look for time compactness of the velocity field $\mathbf{v}_{\tau h}$. Obtaining formally only $(H^1)'$-regularity for time-derivatives of φ by the standard estimates of the Cahn-Hilliard equation, it seems necessary to control $\langle \mathbf{v}^{k+1}, \mathbf{w}_h \rangle$ in H^1. As $\mathbf{w}_h$ is the projection of an arbitrary element in $W^{1,2}_{0,\mathrm{div}}(\Omega)$ onto the space of discretely solenoidal vector fields, this is not possible. Hence, we look for higher regularity for φ. We have (see [26] and [29])

Lemma 2 *Let $(\varphi_{\tau h}, \mu_{\tau h}, \mathbf{v}_{\tau h}, p_{\tau h})$ be a discrete solution of (8.37) on $[0, T]$. Then the sum*

$$\|\varphi_{\tau h}\|_{L^4(0,T;L^\infty(\Omega))} + \|\Delta_h \varphi_{\tau h}\|_{L^\infty(0,T;L^2(\Omega))} + \|\partial_\tau^- \varphi_{\tau h}\|_{L^2(\Omega_T)}$$

is uniformly bounded in $(\tau, h) \to (0,0)$.

As a direct consequence, the discrete Laplacian of the chemical potential $\mu_{\tau h}$ can be bounded in $L^2(\Omega_T)$. Combining this result with a discrete Gagliardo-Nirenberg inequality, the fluxes $\mathbf{j}_{\tau h}$ are proven to converge—up to a subsequence—strongly in $L^2(0, T; L^2(\Omega))$. This allows us to pass to the limit in the $\mathbf{j}_{\tau h}(\nabla \mathbf{v}_{\tau h})^T$-term in the momentum equation.

Concerning the passage to the limit in the parabolic term of the momentum equation, it is sufficient to establish strong convergence of the discrete Helmholtz projection of $\rho_{\tau h} \mathbf{v}_{\tau h}$ towards the Helmholtz projection of $\rho \mathbf{v}$ which is achieved in Lemma 4.8 of [26].

Practical computations have been performed using the inhouse finite-element/finite-volume code EconDrop developed by G. Grün, F. Klingbeil, S. Metzger with contributions by H. Grillmeier and P. Weiss (see [27]). The scheme allows for adaptivity in space and time.

Let us discuss some aspects related to simulations of falling droplets. Due to numerical dissipation, the energy estimate (8.40) differs from its continuous counterpart in particular by showing an inequality sign. A practical consequence is evident: The higher the numerical dissipation, the lower the kinetic energy and the velocity of falling droplets. As a consequence, one is tempted to reduce numerical dissipation caused by the double-well potential ψ as much as possible—see [30]. For the classical double well potential $\psi_{class}(\varphi) := \frac{1}{4}(1 - \varphi^2)^2$, a reduction to zero is possible by discretizing ψ'_{class} by

$$
\begin{aligned}
\psi'_{dis}(\varphi^{k+1}, \varphi^k) &= \frac{\psi_{class}(\varphi^{k+1}) - \psi_{class}(\varphi^k)}{\varphi^{k+1} - \varphi^k} \\
&= \frac{1}{4}\left((\varphi^{k+1})^3 + (\varphi^{k+1})^2\varphi^k + \varphi^{k+1}(\varphi^k)^2 + (\varphi^k)^3\right) - \frac{1}{2}\left(\varphi^{k+1} - \varphi^k\right).
\end{aligned}
$$

In [29], this approach has been tested for two-phase flow with an Atwood number $\mathfrak{At} = 0.999$, i.e. corresponding to a density ratio above 1000. It turns out that φ^k does not stay in the admissible interval $(-\mathfrak{At}^{-1}, \mathfrak{At}^{-1})$ where ϱ is linear. As a consequence, modified energies have been studied in [29] as well. By adding a penalty term, we get

$$
\psi_{pen}(\varphi) := \psi_{class}(\varphi) + \tfrac{1}{\delta} \max\{|\varphi| - 1, 0\}^2
$$

to be a natural candidate. This leads to discretizations

$$
\psi'_{dis,pen}(a, b) := \psi'_{dis}(a, b) + \tfrac{1}{\delta}\tfrac{d}{d\varphi}|_{\varphi=a} \max\{|\varphi| - 1, 0\}^2
$$

and secondly

$$
\psi'_{conv,pen}(a, b) := \psi'_{pen}(a) - \tfrac{1}{2}(a + b).
$$

Decomposing $\psi_{pen}(\varphi) = \psi_+(\varphi) + \psi_-(\varphi)$ into a sum of a convex and concave function with the concave part given by $\psi_-(\varphi) := -\frac{1}{2}\varphi^2$, we note the discretization $\psi'_{dis,pen}$ to be close to the classical convex-concave splitting $\psi_{pen}(a, b) = \psi_+(a) + \psi_-(b)$. Numerical experiments on falling droplets presented in [29] indicate that already for parameters $\delta = 4 \cdot 10^{-3}$ φ-stability is achieved with φ staying bounded in $(-1.\overline{001}, 1.\overline{001})$ which corresponds to an Atwood number $\mathfrak{At} = 0.999$. Moreover, only negligible differences in falling velocities have been observed when comparing results corresponding to non-dissipative schemes with those corresponding to a discretization using ψ'_{dis}.

Finally, let us indicate that the scheme based on $\psi'_{dis,pen}$ has been implemented in rotational geometry, too. Figure 8.1 shows a characteristic sequence of a falling droplet entering a bath—an example for topological changes. The computational domain is given by $\Omega := \{\mathbf{x} \in \mathbb{R}^3 : x_1^2 + x_3^2 < 1,\, 0 < x_2 < 6\}$ with the gravitational acceleration $(0, -10, 0)$. As initial data for fluid phase 1, we take a ball around

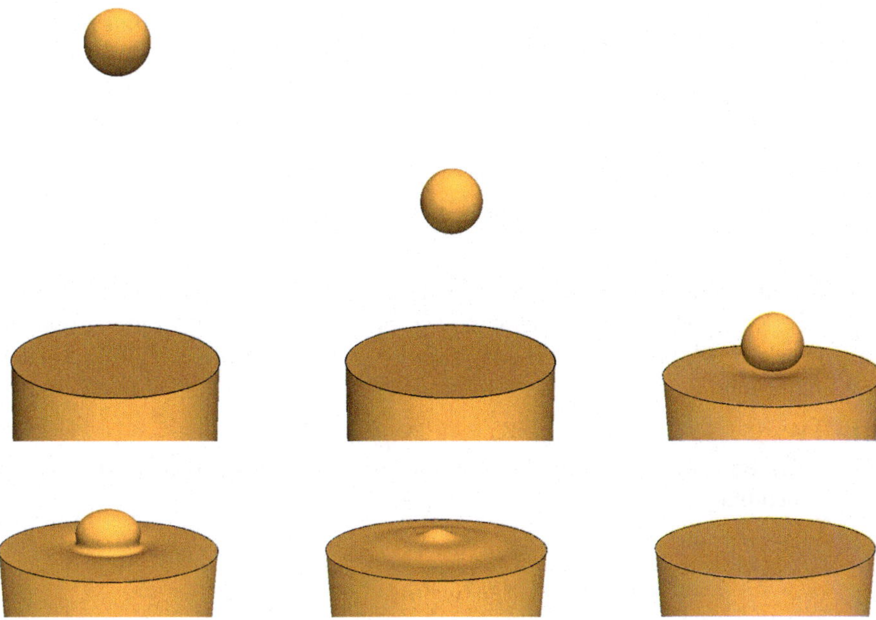

Fig. 8.1 Falling droplet at times $t = 0.0, 1.1, 1.75\,1.85, 1.95, 3.0$. Discretization based on $\psi_{dis,pen}$ with $\delta = 4 \cdot 10^{-3}$

$(0, 0, 5)$ of radius 0.3 and a bath of depth 2. The adaptive spatial grid attains grid parameters in $(0.0045, 0.0714)$, and the density ratio is 3.

Appendix

Discretization in Space and Time

We assume $\mathcal{T}_h$ to be a quasiuniform triangulation of Ω with simplicial elements in the sense of [19].

Concerning discretization with respect to time, we assume that

(T) the time interval $I := [0, T)$ is subdivided in intervals $I_k = [t_k, t_{k+1})$ with $t_{k+1} = t_k + \tau_k$ for time increments $\tau_k > 0$ and $k = 0, \cdots, N-1$. For simplicity, we take $\tau_k \equiv \tau$ for $k = 0, \cdots, N - 1$.

We write v^k for $v(\cdot, k\tau)$, $k \in \mathbb{N}$, and we denote step functions in time mapping $I = [0, T]$ onto one of the discrete function spaces $X_h, \ldots$ by an index τh.

For the approximation of both the phase-field φ and the chemical potential μ, we introduce the space U_h of continuous, piecewise linear finite element functions on

$\mathcal{T}_h$. The expression $\mathcal{I}_h$ stands for the nodal interpolation operator from $C^0(\Omega)$ to U_h defined by $\mathcal{I}_h u := \sum_{j=1}^{\dim U_h} u(x_j)\theta_j$, where the functions θ_j form a dual basis to the nodes x_j, i.e. $\theta_i(x_j) = \delta_{ij}$, $i,j = 1,\dots,\dim U_h$.

Let us furthermore introduce the well-known lumped masses scalar product corresponding to the integration formula

$$(\Theta, \Psi)_h := \int_\Omega \mathcal{I}_h(\Theta \Psi).$$

For the discretization of the velocity field $\mathbf{v}$ and the pressure p, we use function spaces $\mathbf{W}_h \subset \mathbf{X}_h \subset \mathbf{W}_0^{1,2}(\Omega)$ and $S_h \subset L_0^2(\Omega) := \{v \in L^2(\Omega)|\int_\Omega v = 0\}$ which form an *admissible pair of discretization spaces in hydrodynamics*. This means that besides the conditions

(S1) $\mathbf{W}_h := \{\mathbf{v}_h \in \mathbf{X}_h| \int_\Omega q_h \operatorname{div} \mathbf{v}_h = 0 \quad \forall q_h \in S_h\}$,

(S2) The Babuška-Brezzi condition is satisfied, i.e. a positive constant β exists such that

$$\sup_{\mathbf{v}_h \in \mathbf{X}_h} \frac{(q_h, \operatorname{div} \mathbf{v}_h)}{\|\mathbf{v}_h\|_{\mathbf{W}_0^{1,2}(\Omega)}} \geq \beta \, \|q_h\|_{L^2(\Omega)}$$

  for all $q_h \in S_h$

a number of additional conditions hold true which are specified in [26] and [29].

Taylor-Hood elements (i.e. $P_2 P_1$-elements) and $P_2 P_0$-elements are examples in agreement with these conditions. In both cases, $\mathbf{X}_h$ is given as

$$\mathbf{X}_h := \big\{\mathbf{w} \in (\mathbf{C}_0^0(\bar{\Omega})) : (\mathbf{w})_j|_K \in P_2(K), K \in \mathcal{T}_h, j = 1,\dots,d\big\}, \quad d = 2,3.$$

For Taylor-Hood elements, S_h is defined to be the subset of functions in U_h with vanishing mean value. In the case of $P_2 P_0$-elements, S_h is given by the set of elementwise constant functions with mean value zero. Let us specify the assumptions on initial data and the double-well potential.

Definition 1 Let $\psi \in C^1(\mathbb{R}; \mathbb{R}_0^+)$ be given such that ψ' is piecewise C^1 with at most quadratic growth of the derivatives for $|x| \to \infty$. We call $\psi'_{\tau h} \in C^0(\mathbb{R}^2; \mathbb{R})$ an admissible discretization of ψ' if the following is satisfied.

(H1) There is a positive constant C, such that

$$\big|\psi'_{\tau h}(a,b)\big| \leq C\big(1 + |a|^3 + |b|^3\big).$$

(H2) $\psi'_{\tau h}(a,b)(a - b) \geq F(a) - F(b)$ for all $a,b \in \mathbb{R}$.

(H3) $\psi'_{\tau h}(a,b) = F'(a)$ for all $a \in \mathbb{R}$.

(H4) There is a positive constant C such that

$$\big|\psi'_{\tau h}(a,b) - \psi'_{\tau h}(b,c)\big| \leq C\big(a^2 + b^2 + c^2\big)\big(|a - b| + |b - c|\big)$$

  for all $a,b,c \in \mathbb{R}$.

Moreover, we make the following assumptions on initial data and the regularized mass density $\rho(\varphi)$.

(H5)   Let initial data $\Phi_0 \in H^2(\Omega; [-1, 1])$ and $\mathbf{V}_0 \in \mathbf{W}^{1,2}_{0,\mathrm{div}}(\Omega)$ be given such that we have for discrete initial data $\varphi_h^0 := \mathscr{I}_h \Phi_0$ and $\mathbf{v}_h^0$—given by the orthogonal projection of $\mathbf{V}_0$ onto $\mathscr{W}_h$—uniformly in $h > 0$ that

$$\int_\Omega \left| \Delta_h \varphi_h^0 \right|^2 \leq C \left\| \Phi_0 \right\|^2_{H^2(\Omega)}$$

and that

$$\int_\Omega \rho(\varphi_h^0) \left| \mathbf{v}_h^0 \right|^2 + \frac{1}{2} \int_\Omega \left| \nabla \varphi_h^0 \right|^2 + \int_\Omega \mathscr{I}_h F(\varphi_h^0) \leq C \mathscr{E}(\mathbf{V}_0, \Phi_0).$$

Here, the discrete Laplacian $\Delta_h w \in U_h \cap H^1_*(\Omega)$ is defined by

$$(\Delta_h w, \Theta)_h = -\int_\Omega \langle \nabla w, \nabla \Theta \rangle \quad \forall \Theta \in U_h. \tag{8.45}$$

(H6)   Given mass densities $0 < \tilde{\rho}_- \leq \tilde{\rho}_+ \in \mathbb{R}$ of the fluids involved and an arbitrary, but fixed regularization parameter $\bar{\varphi} \in \left(\frac{\tilde{\rho}_-}{\tilde{\rho}_+ - \tilde{\rho}_-}, \frac{2\tilde{\rho}_-}{\tilde{\rho}_+ - \tilde{\rho}_-} \right)$, we define the regularized mass density of the two-phase fluid by a smooth, increasing, strictly positive function ρ of the phase-field φ which satisfies

$$\rho(\varphi)|_{(-1-\bar{\varphi}, 1+\bar{\varphi})} = \frac{\tilde{\rho}_+ - \tilde{\rho}_+}{2} \varphi + \frac{\tilde{\rho}_- + \tilde{\rho}_+}{2}, \tag{8.46}$$

$$\rho(\varphi)|_{(-\infty, -1-\frac{2\tilde{\rho}_-}{\tilde{\rho}_+ - \tilde{\rho}_-})} \equiv const., \tag{8.47}$$

$$\rho(\varphi)|_{(1+\frac{2\tilde{\rho}_-}{\tilde{\rho}_+ - \tilde{\rho}_-}, \infty)} \equiv const.. \tag{8.48}$$

Acknowledgements This work was supported within the DFG Priority Program 1506 "Transport Processes at Fluidic Interfaces".

References

1. Abels, H.: Diffuse interface models for two-phase flows of viscous incompressible fluids. Habilitation thesis, Leipzig (2007)
2. Abels, H.: Existence of weak solutions for a diffuse interface model for viscous, incompressible fluids with general densities. Commun. Math. Phys. **289**, 45–73 (2009)
3. Abels, H.: On a diffuse interface model for two-phase flows of viscous, incompressible fluids with matched densities. Arch. Ration. Mech. Anal. **194**, 463–506 (2009)
4. Abels, H., Breit, D.: Weak solutions for a non-Newtonian diffuse interface model with different densities. Nonlinearity **29**, 3426–3453 (2016)

5. Abels, H., Lengeler, D.: On sharp interface limits for diffuse interface models for two-phase flows. Interfaces Free Bound. **16**(3), 395–418 (2014)
6. Abels, H., Wilke, M.: Convergence to equilibrium for the Cahn-Hilliard equation with a logarithmic free energy. Nonlinear Anal. **67**, 3176–3193 (2007)
7. Abels, H., Garcke, H., Grün, G.: Thermodynamically consistent, frame indifferent diffuse interface models for incompressible two-phase flows with different densities. Math. Models Methods Appl. Sci. **22**(3), 1150013 (2012)
8. Abels, H., Depner, D., Garcke, H.: Existence of weak solutions for a diffuse interface model for two-phase flows of incompressible fluids with different densities. J. Math. Fluid Mech. **15**(3), 453–480 (2013)
9. Abels, H., Depner, D., Garcke, H.: On an incompressible Navier-Stokes/Cahn-Hilliard system with degenerate mobility. Ann. Inst. H. Poincaré Anal. Non Linéaire **30**(6), 1175–1190 (2013)
10. Abels, H., Garcke, H., Lam, Kei Fong, Weber, J.: Two-phase flow with surfactants: diffuse interface models and their analysis. In: Bothe, D., Reusken, A. (eds.) Transport Processes at Fluidic Interfaces. Springer, Cham (2017)
11. Abels, H., Liu, Y., Schöttl, A.: Sharp interface limits for diffuse interface models for two-phase flows of viscous incompressible fluids. In: Bothe, D., Reusken, A. (eds.) Transport Processes at Fluidic Interfaces. Springer, Cham (2017)
12. Aki, G., Dreyer, W., Giesselmann, J., Kraus, C.: A quasi-incompressible diffuse interface model with phase transition. Math. Models Methods Appl. Sci. **24**, 827–861 (2014)
13. Alt, H.W.: The entropy principle for interfaces. Fluids and solids. Adv. Math. Sci. Appl. **2**, 585–663 (2009)
14. Anderson, D.-M., McFadden, G.B., Wheeler, A.A.: Diffuse interface methods in fluid mechanics. Annu. Rev. Fluid Mech. **30**, 139–165 (1998)
15. Antanovskii, L.K.: A phase field model of capillarity. Phys. Fluids **7**(4), 747–753 (1995)
16. Armero, F., Simo, J.C.: Formulation of a new class of fractional-step methods for the incompressible mhd equations that retains the long-term dissipativity of the continuum dynamical system. Fields Inst. Commun. **10**, 1–24 (1996)
17. Boyer, F.: Nonhomogeneous Cahn-Hilliard fluids. Ann. Inst. H. Poincaré Anal. Non Linéaire **18**(2), 225–259 (2001)
18. Boyer, F.: A theoretical and numerical model for the study of incompressible mixture flows. Comput. Fluids **31**, 41–68 (2002)
19. Brenner, S.C., Scott, L.R.: The Mathematical Theory of Finite Element Methods. Springer, New York (2002)
20. Cahn, J.W., Hilliard, J.E.: Free energy of a nonuniform system. I. Interfacial energy. J. Chem. Phys. **28**(2), 258–267 (1958)
21. Campillo-Funollet, E., Grün, G., Klingbeil, F.: On modeling and simulation of electrokinetic phenomena in two-phase flow with general mass densities. SIAM J. Appl. Math. **72**(6), 825–854 (2012)
22. Ding, H., Spelt, P.D.M., Shu, C.: Diffuse interface model for incompressible two-phase flows with large density ratios. J. Comp. Phys. **22**, 2078–2095 (2007)
23. Elliott, C.M., Garcke, H.: Existence results for diffusive surface motion laws. Adv. Math. Sci. Appl. **7**(1), 467–490 (1997)
24. Feng, X.: Fully discrete finite element approximations of the Navier-Stokes-Cahn-Hilliard diffuse interface model for two-phase fluid flows. SIAM J. Numer. Anal. **44**, 1049–1072 (2006)
25. Garcke, H., Hinze, M., Kahle, C.: A stable and linear time discretization for a thermodynamically consistent model for two-phase incompressible flow. Appl. Numer. Math. **99**, 151–171 (2016)
26. Grün, G.: On convergent schemes for diffuse interface models for two-phase flow of incompressible fluids with general mass densities. SIAM J. Numer. Anal. **51**(6), 3036–3061 (2013)
27. Grün, G., Klingbeil, F.: Two-phase flow with mass density contrast: stable schemes for a thermodynamic consistent and frame-indifferent diffuse-interface model. J. Comput. Phys. **257**, Part A, 708–725 (2014)

28. Grün, G., Metzger, S.: On micro-macro models for two-phase flow of dilute polymeric solutions: modeling - analysis - simulation. In: Bothe, D., Reusken, A. (eds.) Transport Processes at Fluidic Interfaces. Springer, Cham (2017)
29. Grün, G., Guillén-González, F., Metzger, S.: On fully decoupled, convergent schemes for diffuse interface models for two-phase flow with general mass densities. Commun. Comput. Phys. **19**(5), 1473–1502 (2016)
30. Guillén-González, F., Tierra, G.: Splitting schemes for a Navier–Stokes–Cahn–Hilliard model for two fluids with different densities. J. Comput. Math. **32**(6), 643–664 (2014)
31. Gurtin, M.E., Polignone, D., Viñals, J.: Two-phase binary fluids and immiscible fluids described by an order parameter. Math. Models Methods Appl. Sci. **6**(6), 815–831 (1996)
32. Hirth, C.W., Nichols, B.D.: Volume of fluid (VOF) method for the dynamics of free boundaries. J. Comp. Phys. **39**, 201–225 (1981)
33. Hohenberg, P.C., Halperin, B.I.: Theory of dynamic critical phenomena. Rev. Mod. Phys. **49**, 435–479 (1977)
34. Jacqmin, D.: Calculation of two-phase Navier-Stokes flows using phase-field modeling. J. Comput. Phys. **155**, 96–127 (1999)
35. Kay, D., Styles, V., Welford, R.: Finite element approximation of a Cahn-Hilliard-Navier-Stokes system. Interfaces Free Bound. **10**, 15–43 (2008)
36. Kim, J., Kang, K., Lowengrub, J.: Conservative multigrid methods for Cahn-Hilliard fluids. J. Comput. Phys. **193**, 511–543 (2004)
37. Kotschote, M.: Strong solutions for a compressible fluid model of Korteweg type. Ann. Inst. H. Poincaré Anal. Non Linéaire **25**(4), 679–696 (2008)
38. Liu, C., Shen, J.: A phase field model for the mixture of two incompressible fluids and its approximation by a Fourier-spectral method. Physica D **179**(3–4), 211–228 (2003)
39. Lowengrub, J., Truskinovsky, L.: Quasi-incompressible Cahn-Hilliard fluids and topological transitions. R. Soc. Lond. Proc. Ser. A Math. Phys. Eng. Sci. **454**, 2617–2654 (1998)
40. Minjeaud, S.: An unconditionally stable uncoupled scheme for a triphasic Cahn–Hilliard/Navier-Stokes model. Numer. Methods Partial Differ. Equ. **29**(2), 584–618 (2013)
41. Sethian, J.A.: Level Set Methods. Evolving Interfaces in Geometry, Fluid Mechanics, Computer Vision, and Materials Science. Cambridge Monograph on Applied and Computational Mathematics. Cambridge University Press, Cambridge (1996)
42. Weber, J.: Analysis of diffuse interface models for two-phase flows with and without surfactants. Dissertation thesis, University of Regensburg (2016)

Chapter 9
Sharp Interface Limits for Diffuse Interface Models for Two-Phase Flows of Viscous Incompressible Fluids

Helmut Abels, YuNing Liu, and Andreas Schöttl

Abstract We consider the mathematical relation between diffuse interface and sharp interface models for the flow of two viscous, incompressible Newtonian fluids like oil and water. In diffuse interface models a partial mixing of the macroscopically immiscible fluids on a small length scale $\varepsilon > 0$ and diffusion of the mass particles are taken into account. These models are capable to describe such two-phase flows beyond the occurrence of topological singularities of the interface due to collision or droplet formation. Both for theoretical and numerical purposes a deeper understanding of the limit $\varepsilon \to 0$ in dependence of the scaling of the mobility coefficient m_ε is of interest. Here the mobility is the inverse of the Peclet number and controls the strength of the diffusion. We discuss several rigorous mathematical results on convergence and non-convergence of solutions of diffuse interface to sharp interface models in dependence of the scaling of the mobility.

9.1 Introduction

The dynamics of two immiscible, viscous fluids like oil and water or polymer blends is a fascinating and difficult topic because in general the interface, which separates both fluids, develops singularities in finite time, e.g. because of pinch off of droplets or collisions. In classical sharp interface models it is assumed that both fluids are separated by a (two-dimensional) surface. In so-called diffuse interface models a partial mixing of the fluids in a thin interfacial region is taken into account, cf. Fig. 9.1. Moreover, mixing and phase separation because of diffusion of molecules is modeled as well.

H. Abels (✉) • A. Schöttl
Faculty of Mathematics, University of Regensburg, Universitätsstraße 31, 93053 Regensburg, Germany
e-mail: helmut.abels@ur.de; andreas.schoettl@ur.de

YuNing Liu
NYU-ECNU Institute of Mathematical Sciences at NYU Shanghai, 3663 Zhongshan Road North, Shanghai 200061, China
e-mail: yl67@nyu.edu

© Springer International Publishing AG 2017

D. Bothe, A. Reusken (eds.), *Transport Processes at Fluidic Interfaces*,
Advances in Mathematical Fluid Mechanics, DOI 10.1007/978-3-319-56602-3_9

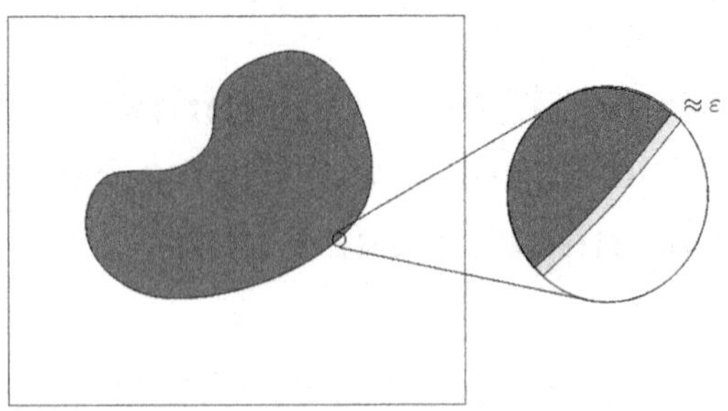

Fig. 9.1 Partial mixing in an interfacial region

The purpose of this contribution is to discuss recent rigorous mathematical results regarding the limit of diffuse interface models to sharp interface models as the parameter $\varepsilon > 0$, which is proportional to the "thickness" of the diffuse interface model, tends to zero. Actually, this is a subtle problem and the results depend significantly on the scaling of a mobility coefficient m_ε in the system as $\varepsilon \to 0$, as will be shown below.

A fundamental and well accepted diffuse interface model for the flow of two macroscopically immiscible viscous, incompressible Newtonian fluids with same densities is the so-called *model H*, cf. Hohenberg and Halperin [19] or Gurtin et al. [18]. This model leads to a system of Navier-Stokes/Cahn-Hilliard type:

$$\rho\partial_t \mathbf{v}_\varepsilon + \rho\mathbf{v}_\varepsilon \cdot \nabla\mathbf{v}_\varepsilon - \mathrm{div}(2\nu(c_\varepsilon)D\mathbf{v}_\varepsilon) + \nabla p_\varepsilon = -\varepsilon\,\mathrm{div}(\nabla c_\varepsilon \otimes \nabla c_\varepsilon) \quad \text{in } Q_T, \qquad (9.1)$$

$$\mathrm{div}\,\mathbf{v}_\varepsilon = 0 \qquad \text{in } Q_T, \qquad (9.2)$$

$$\partial_t c_\varepsilon + \mathbf{v}_\varepsilon \cdot \nabla c_\varepsilon = m_\varepsilon\Delta\mu_\varepsilon \qquad \text{in } Q_T, \qquad (9.3)$$

$$\mu_\varepsilon = -\varepsilon\Delta c_\varepsilon + \varepsilon^{-1}f'(c_\varepsilon) \qquad \text{in } Q_T, \qquad (9.4)$$

$$(\mathbf{v}_\varepsilon, c_\varepsilon)|_{t=0} = (\mathbf{v}_{0,\varepsilon}, c_{0,\varepsilon}) \qquad \text{in } \Omega. \qquad (9.5)$$

where $Q_T = \Omega \times (0, T)$ and $(a \otimes b)_{i,j} = a_i b_j$ for all $i, j = 1, \ldots, d$. Here $\mathbf{v}_\varepsilon$ and p_ε are the mean velocity and pressure of the fluid mixture, $D\mathbf{v}_\varepsilon = \frac{1}{2}(\nabla\mathbf{v}_\varepsilon + \nabla\mathbf{v}_\varepsilon^T)$, c_ε is an order parameter related to the concentrations of the fluids, which will be the concentration difference in the following. Furthermore, μ_ε is the chemical potential of the mixture, ν is the viscosity of the fluid mixture, ρ is the density of the fluids, which is assumed to be constant. Lastly, $f\colon \mathbb{R} \to \mathbb{R}$ is a suitable (homogeneous) free energy density of double well shape, which will be specified below, and $m_\varepsilon > 0$ is a (constant) mobility coefficient related to the strength of the diffusion in the mixture. In the following m_ε will depend on $\varepsilon > 0$ and it turns out that the choice of the scaling influences the results fundamentally. The system has to be closed by suitable boundary conditions if $\partial\Omega \neq \emptyset$. One standard choice are no-slip boundary conditions for $\mathbf{v}_\varepsilon$ and Neumann boundary conditions for c_ε and μ_ε, i.e.,

$$(\mathbf{v}_\varepsilon, \mathbf{n}_{\partial\Omega} \cdot \nabla c_\varepsilon, \mathbf{n}_{\partial\Omega} \cdot \nabla \mu_\varepsilon)|_{\partial\Omega} = 0 \qquad \text{on } \partial\Omega \times (0,T), \tag{9.6}$$

where $\mathbf{n}_{\partial\Omega}$ denotes the exterior normal of $\partial\Omega$. Moreover, $\Omega \subseteq \mathbb{R}^d$ will be a suitable domain, which will be specified in the following in dependence on the result we are discussing. Since the boundary conditions on $\partial\Omega$ will not play an important role, we will not specify it in the following for simplicity of the presentation. Finally, the results hold true for any $T \in (0, \infty)$ if not stated differently.

We note that sufficiently smooth solutions of (9.1)–(9.5) satisfy the *energy inequality*

$$E_\varepsilon(c_\varepsilon(t)) + \int_\Omega \frac{\rho |\mathbf{v}_\varepsilon(t)|^2}{2} \, dx$$

$$+ \int_0^t \int_\Omega \left(2\nu(c_\varepsilon)|D\mathbf{v}_\varepsilon|^2 + m_\varepsilon |\nabla\mu_\varepsilon|^2 \right) dx\, dt \leq E_\varepsilon(c_{0,\varepsilon}) + \int_\Omega \frac{\rho |\mathbf{v}_{0,\varepsilon}|^2}{2} \, dx \tag{9.7}$$

for all $t \in (0, T)$, where

$$E_\varepsilon(c) = \int_\Omega \left(\varepsilon \frac{|\nabla c|^2}{2} + \frac{f(c)}{\varepsilon} \right) dx$$

is the free energy of the fluid mixture. This energy inequality is fundamental for the analysis of weak and strong solutions for fixed $\varepsilon > 0$, cf. e.g. A. [2]. It also provides a limited control of $(\mathbf{v}_\varepsilon, c_\varepsilon, \mu_\varepsilon)$ as $\varepsilon \to 0$ as will be discussed in Sect. 9.3.

So far the sharp interface limits of diffuse interface models in fluid mechanics as (9.1)–(9.4) were mainly discussed with the method of formally matched asymptotics. There it is assumed that the quantities possess suitable power series expansions close to and away from the interface and suitable matching conditions between both expansions hold. With this method it was shown by A., Garcke and Grün [8] that there are (at least) two possible sharp interface limits of (9.1)–(9.5), which can both be formulated as the following system:

$$\rho\partial_t\mathbf{v} + \rho\mathbf{v} \cdot \nabla\mathbf{v} - \operatorname{div} \mathbf{T}(\mathbf{v},p) = 0 \qquad\qquad \text{in } \Omega^\pm(t), t \in (0,T), \tag{9.8}$$

$$\operatorname{div}\mathbf{v} = 0 \qquad\qquad \text{in } \Omega^\pm(t), t \in (0,T), \tag{9.9}$$

$$[\mathbf{v}] = 0 \qquad\qquad \text{on } \Gamma_t, t \in (0,T), \tag{9.10}$$

$$-[\mathbf{n}_{\Gamma_t} \cdot \mathbf{T}(\mathbf{v},p)] = \sigma H_{\Gamma_t}\mathbf{n}_{\Gamma_t} \qquad\qquad \text{on } \Gamma_t, t \in (0,T), \tag{9.11}$$

$$\mathbf{v}|_{t=0} = \mathbf{v}_0 \qquad\qquad \text{in } \Omega, \tag{9.12}$$

$$V_{\Gamma_t} - \mathbf{n}_{\Gamma_t} \cdot \mathbf{v}|_{\Gamma_t} = -\tfrac{\tilde{m}}{2}[\mathbf{n}_{\Gamma_t} \cdot \nabla\mu] \quad \text{on } \Gamma_t, t \in (0, T), \tag{9.13}$$

$$2\mu = \sigma H_{\Gamma_t} \quad \text{on } \Gamma_t, t \in (0, T), \tag{9.14}$$

$$\Delta\mu = 0 \quad \text{in } \Omega^{\pm}(t), t \in (0, T), \tag{9.15}$$

$$\Omega^{+}(0) = \Omega_0^{+}. \tag{9.16}$$

Here Ω is the disjoint union of $\Omega^{+}(t)$, $\Omega^{-}(t)$, and $\Gamma_t = \partial\Omega^{+}(t)$, $\mathbf{n}_{\Gamma_t}$ is the interior normal of Γ_t with respect to $\Omega^{+}(t)$, $V_{\Gamma_t}, H_{\Gamma_t}$ are the normal velocity, mean curvature of the interface, respectively, and $\mathbf{T}(\mathbf{v}, p) = 2v^{\pm}D\mathbf{v} - p\mathbf{I}$ in $\Omega^{\pm}(t)$ is the stress tensor in each fluid phase. Furthermore, $[u](x) = \lim_{h\to 0+} u(x + h\mathbf{n}_{\Gamma_t}(x)) - u(x - h\mathbf{n}_{\Gamma_t}(x))$, $x \in \Gamma_t$, denotes the jump of a quantity u at the interface Γ_t, $\sigma = \int_{-1}^{1} \sqrt{2f(s)}\, ds > 0$ is a (constant) surface tension coefficient, depending only on f, which will be specified below, $\tilde{m} = 0$ if $m_\varepsilon = m_0\varepsilon$ for some $m_0 > 0$ and $\tilde{m} = m_0$ if $m_\varepsilon = m_0$. Actually in the latter contribution a more general model for the case of fluids with different densities was considered.

We note that in the case $\tilde{m} = 0$ (9.14)–(9.15) and the chemical potential μ can be disregarded and solutions of (9.1)–(9.5) converge to solutions of the classical sharp interface model for a two-phase flow of viscous, incompressible fluids with pure transport of the interface Γ_t in time. That is

$$V_{\Gamma_t} = \mathbf{n} \cdot \mathbf{v}|_{\Gamma_t} \quad \text{on } \Gamma_t, t \in (0, T),$$

Moreover, we have the well-known Young-Laplace equation (9.11) relating the jump of the normal component of the stress tensor to surface tension forces. We note that this convergence was formally verified by Starovoitov [29] in the case $m_\varepsilon = m_0\varepsilon$ and Lowengrub and Truskinovsky [23] in the case $m_\varepsilon = m_0\varepsilon^2$. A more detailed analysis can be found in [8] and an overview in [17, Sect. 4].

In the case $\tilde{m} > 0$ the system is a new sharp interface model for a two-phase flow of viscous, incompressible fluids. In this case the dynamics of the interface Γ_t is given by a Mullins-Sekerka equation with additional convection term, cf. (9.13). Here m_0 is related to the strength of diffusion in the system and non-local effects like Ostwald ripening are present in the system as for the separate Mullins-Sekerka and Cahn-Hilliard system.

Here and in the following we will not consider the case that the interface Γ_t intersects the boundary $\partial\Omega$. Therefore we will not discuss situations where contact angle conditions at $\partial\Omega \cap \Gamma_t$ occur. Throughout this contribution we will not specify the precise (technical) assumptions on f for each result discussed. Typical assumptions are that f is sufficiently smooth, even, non-negative, $f(c) = 0$ if and only if $c = \pm 1$, and satisfies $f''(\pm 1) > 0$ together with suitable growth conditions. We refer to the corresponding publications for the precise assumptions for each result discussed in the following. A standard example is $f(c) = \tfrac{1}{8}(1 - c^2)^2$, $c \in \mathbb{R}$.

Moreover, the so-called *optimal profile* will play an important role for the following, which is the unique solution of

$$-\theta_0''(x) + f'(\theta_0(x)) = 0 \quad \text{for all } x \in \mathbb{R}, \tag{9.17}$$

$$\theta_0(0) = 0, \quad \theta_0(x) \to_{x \to \pm\infty} \pm 1.$$

If $f(c) = \frac{1}{8}(1-c^2)^2$, $c \in \mathbb{R}$, then $\theta_0(x) = \tanh(\frac{x}{2})$ for all $x \in \mathbb{R}$. In general θ_0 has the same qualitative behaviour as in the latter case. In particular, $\theta_0(x) \to_{x \to \pm\infty} \pm 1$ and $\theta_0'(x), \theta_0''(x) \to_{x \to \pm\infty} 0$ exponentially, cf. e.g. [28, Lemma 2.6.1]. In the following we will use that

$$\sigma = \int_{-\infty}^{\infty} \sqrt{2f(\theta_0(x))}\theta_0'(x)\, dx = \int_{-\infty}^{\infty} \theta_0'(x)^2\, dx$$

because of $f(\theta_0(x)) = \frac{\theta_0'(x)^2}{2}$, which follows from integration of (9.17) multiplied by $\theta_0'(x)$.

The structure of this contribution is as follows: In Sect. 9.2 we will discuss rigorous results on convergence and non-convergence of the convective Cahn-Hilliard equation (9.3)–(9.4) as $\varepsilon \to 0$ for a given velocity $\mathbf{v}$, which indicate what kind of results can be expected for the full coupled system (9.1)–(9.4). In Sect. 9.3 we discuss mathematical results on convergence of weak solutions of the model H to so-called varifold solutions of (9.8)–(9.16) for large times. A disadvantage of these results is that uniqueness of varifold solutions is unknown, convergence is only obtained for a suitable subsequence and no convergence rates can be shown. Therefore a goal is to prove convergence results of the full sequence together with convergence rates at least for sufficiently small times when the limit system possesses a smooth solution. In Sect. 9.4 we present a first convergence result of this kind, when the Navier-Stokes/Cahn-Hilliard system (9.1)–(9.4) is replaced by a Stokes/Allen-Cahn system. In this case the limit system is a Stokes system coupled to a mean curvature flow equation with an additional convection term. Finally, in Sect. 9.5 we discuss works in progress and further perspectives.

9.2 Sharp Interface Limit of the Convective Cahn-Hilliard Equation

In this section we discuss the sharp interface limit of the convective Cahn-Hilliard equation

$$\partial_t c_\varepsilon + \mathbf{v} \cdot \nabla c_\varepsilon = m_\varepsilon \Delta \mu_\varepsilon, \tag{9.18}$$

$$\mu_\varepsilon = -\varepsilon \Delta c_\varepsilon + \varepsilon^{-1} f'(c_\varepsilon). \tag{9.19}$$

Here $\mathbf{v}$ is assumed to be a given, sufficiently smooth velocity field such that div $\mathbf{v} = 0$ and $\mathbf{n} \cdot \mathbf{v}|_{\partial\Omega} = 0$. More precisely, we assume that $m_\varepsilon = m_0\varepsilon^k$ for some $k \geq 0$, $m_0 > 0$, and pose the following:

Question: *For which k do solutions c_ε of (9.18)–(9.19) (with suitably well prepared initial data) converge to a solution of*

$$2\partial_t\chi_{\Omega^+(t)} + 2\mathbf{v} \cdot \nabla\chi_{\Omega^+(t)} = \tilde{m}\Delta\mu \quad \text{in } \mathscr{D}'(\Omega \times (0, T)), \tag{9.20}$$

$$2\mu = \sigma H_{\Gamma_t} \quad \text{on } \partial\Omega^+(t) \tag{9.21}$$

for some $\tilde{m} \geq 0$ together with

$$\varepsilon\int_0^T \int_\Omega \nabla c_\varepsilon \otimes \nabla c_\varepsilon : \nabla\boldsymbol{\varphi}\, dx\, dt \to_{\varepsilon\to 0} \sigma\int_0^T \int_{\Gamma_t} \mathbf{n}_{\Gamma_t}\otimes\mathbf{n}_{\Gamma_t} : \nabla\boldsymbol{\varphi}\, d\mathscr{H}^{d-1}\, dt \tag{9.22}$$

for all $\boldsymbol{\varphi} \in C_0^\infty(\Omega \times (0, T))^d$ with div $\boldsymbol{\varphi} = 0$?

Here $\mathscr{H}^{d-1}$ denotes the $(d - 1)$-dimensional Hausdorff measure. We note that $\mathscr{H}^{d-1}(M)$ is the area of M if $d = 3$ and $M \subseteq \mathbb{R}^3$ is a smooth and compact surface. Moreover, we remark that

$$\int_{\Gamma_t} \mathbf{n}_{\Gamma_t}\otimes\mathbf{n}_{\Gamma_t} : \nabla\boldsymbol{\varphi}\, d\mathscr{H}^{d-1} = -\int_{\Gamma_t}(\mathbf{I}-\mathbf{n}_{\Gamma_t}\otimes\mathbf{n}_{\Gamma_t}) : \nabla\boldsymbol{\varphi}\, d\mathscr{H}^{d-1} = \int_{\Gamma_t} H_{\Gamma_t}\mathbf{n}_{\Gamma_t}\cdot\boldsymbol{\varphi}\, d\mathscr{H}^{d-1}$$

since div $\boldsymbol{\varphi} = 0$ *provided Γ_t is sufficiently regular. We will call*

$$\boldsymbol{\varphi} \mapsto -\int_{\Gamma_t}(\mathbf{I} - \mathbf{n}_{\Gamma_t} \otimes \mathbf{n}_{\Gamma_t}) : \nabla\boldsymbol{\varphi}\, d\mathscr{H}^{d-1} \tag{9.23}$$

the mean curvature functional *in the following. It can be used for weak formulations of the sharp interface models.*

The motivation for (9.22) comes from the final goal to pass to the limit in the coupled system (9.1)–(9.5), where the left-hand side of (9.22) is a weak formulation of the right-hand side of (9.1), which should converge to a weak formulation of the right-hand side of the Young-Lapace law (9.11).

Surprisingly, the answer to the question above is negative if $k > 3$, i.e., (9.22) does not hold in that case. But in the case $k = 0, 1$ it is possible to prove that (9.22) and (9.20)–(9.21) hold for $k = 0$ with $\tilde{m} = m_0$ and for $k = 1$ with $\tilde{m} = 0$. We will first explain the negative result.

In order to prove the negative result we first considered the case "$k = \infty$", which corresponds to choosing $m_\varepsilon = 0$ for all $\varepsilon \in (0, 1)$. Then (9.18) reduces to the transport equation, which can be solved by the method of characteristics. We will denote by c_ε^∞ its solution with initial value $c_{0,\varepsilon}$. Hence

$$c_\varepsilon^\infty(x, t) = c_{0,\varepsilon}\left(X_t^{-1}(x)\right) \qquad \text{for all } x \in \Omega, t \in (0, T),$$

where $X_t(\xi)$ is the solution of

$$\tfrac{d}{dt}X_t(\xi) = \mathbf{v}(X_t(\xi), t), \qquad X_t(\xi)|_{t=0} = \xi$$

for all $\xi \in \Omega$. Then one can show that

$$\varepsilon \int_0^T \int_\Omega \nabla c_\varepsilon^\infty \otimes \nabla c_\varepsilon^\infty : \nabla \boldsymbol{\varphi}\, dx\, dt \to_{\varepsilon\to 0} \sigma \int_0^T \int_{\Gamma_t} \kappa \mathbf{n}_{\Gamma_t} \otimes \mathbf{n}_{\Gamma_t} : \nabla \boldsymbol{\varphi}\, d\mathcal{H}^{d-1}\, dt$$

for all $\boldsymbol{\varphi} \in C_0^\infty(\Omega \times (0, T))^d$, where κ is determined explicitly by X_t and $\kappa \not\equiv 1$ in general, cf. [28, Theorem 5.2.2] or [6]. Hence (9.22) does not hold in general. To get a more precise description we assume for simplicity that $\Omega = \mathbb{R}^d$, $\Gamma_0 = \mathbb{R}^{d-1} \times \{0\}$, $v \equiv const.$, and $c_{0,\varepsilon}(x_d) = \theta_0(\tfrac{x_d}{\varepsilon})$. Then one has

$$
\begin{aligned}
\int_\Omega \nabla c_\varepsilon^\infty \otimes \nabla c_\varepsilon^\infty : \nabla \boldsymbol{\varphi}\, dx \;&=\; \tfrac{1}{\varepsilon} \int_{\mathbb{R}^d} \left(\theta_0'\big(\tfrac{x_d}{\varepsilon}\big)\right)^2 Ae_d \otimes Ae_d : \nabla\varphi \circ X_t(x)\, dx \\[2mm]
&\to_{\varepsilon\to 0} \sigma \int_{\mathbb{R}^{d-1}} Ae_d \otimes Ae_d : \nabla\varphi \circ X_t(x)\, dx \\[2mm]
&=\; \sigma \int_{\Gamma_t} \underbrace{|Ae_d| \circ X_t^{-1}}_{=\kappa} \mathbf{n}_{\Gamma_t} \otimes \mathbf{n}_{\Gamma_t} : \nabla\varphi\, d\mathcal{H}^{d-1}
\end{aligned}
$$

for every $t \in (0, T)$, where $\sigma = \int_{\mathbb{R}} |\theta_0'(s)|^2\, ds$,

$$A = (DX_t^{-1})^T = \mathrm{cof} DX_t, \qquad Ae_d = |Ae_d|\, \mathbf{n}_{\Gamma_t},$$

$\Gamma_t = X_t(\Gamma_0)$, $|Ae_d|$ is the surface element of Γ_t, and $\kappa = |Ae_d| \circ X_t^{-1}$. Choosing a velocity field $\mathbf{v}$ such that X_t increases or decreases the length locally, one can easily obtain that $\kappa \not\equiv 1$.

In the case $k > 3$ one gets the same result since one is able to prove that

$$\varepsilon \|\nabla(c_\varepsilon - c_\varepsilon^\infty)\|_{L^2(\Omega \times (0,T))}^2 \to_{\varepsilon\to 0} 0 \tag{9.24}$$

with the aid of the energy method applied to the equation for the difference $c_\varepsilon - c_\varepsilon^\infty$, cf. [28, Lemma 5.2.7]. Here c_ε is the solution of (9.18)–(9.19) and c_ε^∞ is the solution of the limit case "$k = \infty$" as above. Hence one obtains as before

$$\varepsilon \int_0^T \int_\Omega \nabla c_\varepsilon \otimes \nabla c_\varepsilon : \nabla \boldsymbol{\varphi}\, dx\, dt \to_{\varepsilon\to 0} \sigma \int_0^T \int_{\Gamma_t} \kappa \mathbf{n} \otimes \mathbf{n} : \nabla \boldsymbol{\varphi}\, d\mathcal{H}^{d-1}\, dt,$$

where $\kappa \not\equiv 1$ in general as in the case $k = \infty$. The reason that (9.22) does not hold in the case $k > 3$ in general is that the convection in the equation caused by $\mathbf{v} \cdot \nabla c$ dominates the diffusion related to the term $m_\varepsilon \Delta \mu_\varepsilon$. More precisely, in that case one does not have the relation

$$c_\varepsilon(x,t) = \theta_0\left(\frac{\mathrm{sdist}(x,\Gamma_t)}{\varepsilon}\right) + \mathcal{O}(\varepsilon) \quad \text{as } \varepsilon \to 0, \tag{9.25}$$

where $\mathrm{sdist}(x,\Gamma_t)$ is the signed distance of x to Γ_t, cf. (9.42) below. This relation holds true in the case $k = 0, 1$. If the initial values satisfy $c_{0,\varepsilon} \approx \theta_0(\frac{\mathrm{sdist}(x,\Gamma_0)}{\varepsilon})$, then

$$c_\varepsilon(x,t) \approx \theta_0\left(\frac{\mathrm{sdist}(X_t^{-1}(x),\Gamma_0)}{\varepsilon}\right) \neq \theta_0\left(\frac{\mathrm{sdist}(x,\Gamma_t)}{\varepsilon}\right) + \mathcal{O}(\varepsilon)$$

in general.

On the other-hand in the case $k = 0$ one can adapt the construction of suitable approximative solutions in [9] to include the convection term $\mathbf{v} \cdot \nabla c$. Then the approximative solution satisfies (9.25) and one can prove that the difference of approximative and exact solution of (9.18)–(9.19) converges to zero in a sufficiently strong norm provided that the limit system (9.20)–(9.21) possesses a sufficiently smooth solution, which is true at least for small times and sufficiently regular initial surfaces. Using (9.25) one is able to prove (9.22) in the case $k = 0$.

In the case $k = 1$ it is possible to modify the approach of [9] to prove that solutions of (9.18)–(9.19) converge to (9.20) with $\tilde{m} = 0$ and that (9.25) as well as (9.22) hold. We refer to [28, Chap. 6] for the details.

We note that the result is consistent with the scaling $m_\varepsilon = m_0\varepsilon^k$ with $k \in [1, 2)$ proposed and used by Jacqmin [20] for numerical simulations.

9.3 Sharp Interface Limit for the Navier-Stokes/Cahn-Hilliard System

9.3.1 A Counter Example for Too Fast Decreasing Mobility

As before we have considered a scaling of the mobility $m_\varepsilon = m_0\varepsilon^k$ for some $m_0 > 0$, $k \geq 0$ for the full system (9.1)–(9.5). Then the question arises: For which k do solutions of the system converge to solutions of (9.8)–(9.16)? Based on the experience for the convective Cahn-Hilliard system we were able to construct a radial symmetric solution of (9.1)–(9.5), which does not converge to a solution of (9.8)–(9.16) if $k > 3$. On the other hand, if $k \in [0, 1)$, we were able to prove convergence for the full system in the sense of varifold solutions due to Chen [13] for the Cahn-Hilliard equation.

More precisely, for the case $k > 3$ we consider (9.1)–(9.5) in the exterior domain $\Omega = \{x \in \mathbb{R}^d : |x| > 1\}$ with the "inflow" boundary conditions

$$\mathbf{v}_\varepsilon|_{\partial\Omega} = a\frac{x}{|x|} \equiv a\mathbf{e}_r \qquad \text{on } \partial\Omega \times (0, T),$$

$$c_\varepsilon|_{\partial\Omega} = 1 \qquad \text{on } \partial\Omega \times (0, T),$$

where $a > 0$. Moreover, we considered solutions of the form $\mathbf{v}_\varepsilon(x, t) = u_\varepsilon(r, t)\mathbf{e}_r$, $c_\varepsilon(x, t) = \tilde{c}_\varepsilon(r, t)$, where $r = |x|, \mathbf{e}_r = \frac{x}{|x|}$ such that $c_\varepsilon|_{t=0} = c_{0,\varepsilon} = \theta(\frac{r-r_0}{\varepsilon})$ with $r_0 > 1$ and

$$\theta(s) = \begin{cases} 1 & \text{if } s \leq -1, \\ -1 & \text{if } s \geq 1. \end{cases}$$

I.e., $c_{0,\varepsilon}$ describes a spherical droplet of radius $r_0 > 1$ with a diffuse interface of thickness 2ε, cf. Fig. 9.2.

Again we first considered the case $k = \infty$, i.e., $m_\varepsilon \equiv 0$, where the solution can be calculated explicitly. Since $\mathbf{v}_\varepsilon(x, t) = u_\varepsilon(r, t)\mathbf{e}_r$, (9.2) reduces to $\partial_r(r^{d-1}u_\varepsilon(r, t)) = 0$. Therefore

$$u_\varepsilon(r, t) = \frac{a}{r^{d-1}} \qquad \text{for all } r > 1, t > 0$$

because of the boundary condition for $\mathbf{v}_\varepsilon$. Hence (9.3) (with $m_\varepsilon \equiv 0$) reduces to

$$\partial_t c_\varepsilon(r, t) + \frac{a}{r^{d-1}}\partial_r c_\varepsilon(r, t) = 0.$$

Together with the initial value the solution is given by

$$c_\varepsilon^\infty(x, t) = \theta\left(\frac{\sqrt[d]{r^d - dat} - r_0}{\varepsilon}\right) \qquad \text{with } r = |x|.$$

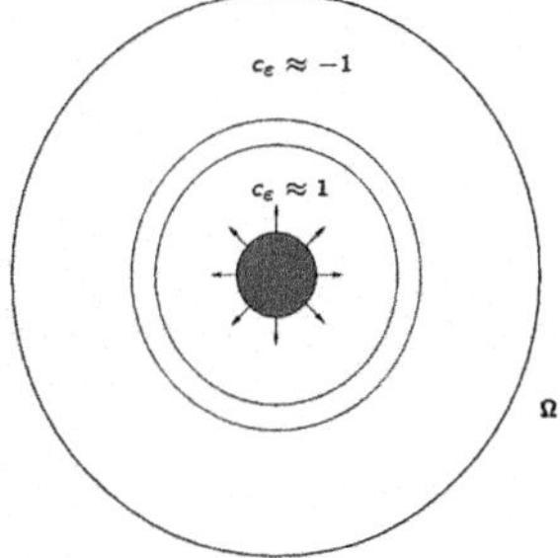

Fig. 9.2 Radial symmetric flow with inflow

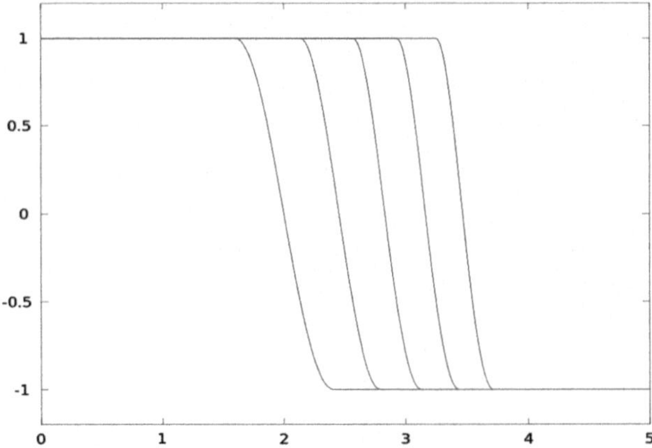

Fig. 9.3 Plot of the solution in radial direction for $t = 0, 1, 2, 3, 4$ (from *left* to *right*) with $a = 1, \varepsilon = 0.4, r_0 = 2, d = 2$

In Fig. 9.3 one can observe that the gradient in the diffuse interface regions increases as time increases and the diffuse interface thickness decreases. This results in an "increase in the surface tension coefficient" as follows: Using the explicit formulas for $\mathbf{v}_\varepsilon$ and c_ε^∞, one can determine p_ε^∞ uniquely (up to a constant) by (9.1). Then one verifies that

$$c_\varepsilon^\infty \to_{\varepsilon \to 0} 2\chi_{B_{R(t)}(0)} - 1 \qquad \text{for every } x \in \Omega \setminus \partial B_{R(t)}(0), t \in (0, T),$$

$$p_\varepsilon^\infty \to_{\varepsilon \to 0} p \qquad \text{for every } x \in \Omega \setminus \partial B_{R(t)}(0), t \in (0, T),$$

where $R(t) = \sqrt[d]{r_0^d + dat}$ and

$$[p] = \kappa(t)\sigma H \qquad \text{on } \Gamma_t = \partial B_{R(t)}(0) \tag{9.26}$$

with $1 < \kappa(t) \to_{t \to \infty} \infty$ and $\sigma = \int_{\mathbb{R}} |\theta_0'(s)|^2 \, ds$ as before. In particular $(\mathbf{v}, p)$ do not satisfy (9.11) since $[D\mathbf{v}] = 0$. As in the case of the convective Cahn-Hilliard equation (9.18)–(9.19), one can prove in the case $k > 3$ that (9.24) holds. This yields finally the same result as in the case $k = \infty$. In particular (9.26) holds with the same $\kappa(t)$. We refer to [3] for the details.

9.3.2 Convergence to Varifold Solutions

For the following we assume that $c_{0,\varepsilon}, \mathbf{v}_{0,\varepsilon}, \varepsilon \in (0, 1)$ are initial values such that

$$\sup_{\varepsilon \in (0,1)} \left(E_\varepsilon(c_{0,\varepsilon}) + \int_\Omega \frac{\rho |\mathbf{v}_{0,\varepsilon}|^2}{2} \, dx \right) < \infty.$$

Then (9.7) yields that

$$E_\varepsilon(c_\varepsilon(t)) + \int_\Omega \frac{\rho|\mathbf{v}_\varepsilon(t)|^2}{2}\, dx$$

is uniformly bounded with respect to $t \in (0, T)$ and $\varepsilon \in (0, 1)$ as well. Moreover, we note that by Modica and Mortola [26] or Modica [25], we have

$$E_\varepsilon \to_{\varepsilon \to 0} \mathscr{P} \qquad \text{w.r.t. } L^1\text{-}\Gamma\text{-convergence,} \tag{9.27}$$

where

$$\mathscr{P}(u) = \begin{cases} \sigma\, \mathscr{H}^{d-1}(\partial^* E) & \text{if } u = -1 + 2\chi_E \text{ and } E \text{ has finite perimeter,} \\ +\infty & \text{else.} \end{cases}$$

Here E is a set of finite perimeter (in Ω) if and only if the distribution $\nabla\chi_E$ coincides with a vector-valued Radon measure and $\partial^* E$ is the reduced boundary of a set of finite perimeter, cf. e.g. [16]. We remark that for sets of finite perimeter we have $|\nabla\chi_E| = \mathscr{H}^{d-1}\lfloor\partial^* E$, where $|\nabla\chi_E|$ denotes the total variation measure of $\nabla\chi_E$, and we have that the Radon-Nikodym derivative $\frac{\nabla\chi_E}{|\nabla\chi_E|}$ coincides with a measure theoretic interior normal of $\partial^* E$. For the following we denote by $\mathscr{M}(\Omega)$ the set of all signed Radon measures on Ω and recall that $\mathscr{M}(\Omega) = C_c(\Omega)'$ by the Riesz representation theorem. Here $C_c^k(\Omega)$, $k \in \mathbb{N}_0$, denotes the set of all k-times continuously differentiable functions $f\colon \Omega \to \mathbb{R}$ with compact support and $C_c(\Omega) = C_c^0(\Omega)$. Furthermore,

$$BV(\Omega) = \{f \in L^1(\Omega) : \nabla f \in \mathscr{M}(\Omega)^d\},$$

where $L^p(\Omega)$, $1 \le p \le \infty$, denotes the standard Lebesgue space with respect to the Lebesgue measure and $L^p(0, T; X)$ denotes its X-valued variant for $\Omega = (0, T)$, where X is a Banach space. $H^1(\Omega)$ denotes the L^2-Sobolev space of first order.

The proof of the following convergence result as $\varepsilon \to 0$ is based on ideas of the proof of the latter results by Modica and Mortola, similar to Chen [13]. First of all, using the uniform boundedness of $E_\varepsilon(c_\varepsilon(t))$, $f''(\pm 1) > 0$ and a suitable growth condition for f'' one obtains that there is some $C > 0$ such that

$$\int_\Omega (|c_\varepsilon(t)| - 1)^2\, dx \le C\varepsilon R \tag{9.28}$$

for all $t \in (0, T)$, $\varepsilon \in (0, 1)$. Hence there is a subsequence $(c_{\varepsilon_k})_{k \in \mathbb{N}}$, which will for simplicity be again denoted by $(c_\varepsilon)_{\varepsilon \in (0,1)}$, such that

$$c_\varepsilon(x, t) \to_{\varepsilon \to 0} \pm 1 \qquad \text{for almost all } x \in \Omega, t \in (0, T). \qquad (9.29)$$

Moreover, if we define

$$W(c) = \int_{-1}^{c} \sqrt{2\tilde{f}(s)}\, ds, \quad \text{where } \tilde{f}(s) = \min(f(s), 1 + |s|^2),$$

and

$$w_\varepsilon(x, t) = W(c_\varepsilon(x, t)),$$

then $(w_\varepsilon)_{\varepsilon \in (0,1)}$ are uniformly bounded in $L^\infty(0, T; BV(\Omega))$ since

$$\int_\Omega |\nabla w_\varepsilon(x, t)|\, dx = \int_\Omega \sqrt{2\tilde{f}(c_\varepsilon(x, t))}|\nabla c_\varepsilon(x, t)|\, dx \le E_\varepsilon(c_\varepsilon(t)) \qquad (9.30)$$

is uniformly bounded with respect to $t \in (0, T)$, $\varepsilon \in (0, 1)$. Hence there is some $w \in L^\infty_{w*}(0, T; BV(\Omega))$ such that

$$w_\varepsilon \overset{*}{\to}_{\varepsilon \to 0} w \qquad \text{in } L^\infty_{w*}(0, T; BV(\Omega))$$

for a suitable subsequence. Here $L^\infty_{w*}(0, T; X')$ is the space of all weakly-$*$ measurable and essentially bounded functions $u \colon (0, T) \to X'$ for a Banach space X. Moreover, using that $BV(\Omega)$ is compactly embedded into $L^1(\Omega)$, $w_\varepsilon \to_{\varepsilon \to 0} w$ in $L^1(\Omega)$ for every $t \in (0, T)$ (and a suitable subsequence). Because of (9.29), we have $w(x, t) \in \{0, \sigma\}$. Hence there is a measurable set E such that $w = \sigma \chi_E$ and we define $c = -1 + 2\chi_E$. Together with suitable estimates for differences in time one can show, cf. [3, Lemmas 3.4 and 3.6]:

Lemma 1 *There exists a subsequence and a measurable set $E \subset \Omega \times [0, T]$ such that, as $\varepsilon \to 0$,*

$$w_\varepsilon \to \sigma \chi_E \quad \text{a.e. in } \Omega \times (0, T) \text{ and in } C^{\frac{1}{9}}([0, T]; L^1(\Omega))$$

$$c_\varepsilon \to -1 + 2\chi_E \quad \text{a.e. in } \Omega \times (0, T) \text{ and in } C^{\frac{1}{9}}([0, T]; L^2(\Omega))$$

Moreover, $\chi_E \in L^\infty_{w*}(0, T; BV(\Omega)) \cap C^{\frac{1}{4}}([0, T]; L^1(\Omega))$ and for all $t \in [0, T]$ we have $|E_t| = |E_0|$.

Here $C^\alpha([0, T]; X)$, $\alpha \in (0, 1)$, is the set of all $f \colon [0, T] \to X$ that are Hölder continuous with exponent α.

With aid of the last considerations and a suitable one-sided estimate of the so-called discrepancy measure $\xi_\varepsilon := \varepsilon \frac{|\nabla c_\varepsilon|^2}{2} - \frac{f(c_\varepsilon)}{\varepsilon}$ due to [13, Theorem 3.6], the following result was obtained:

Theorem 1 *Let* $(\mathbf{v}_\varepsilon, c_\varepsilon, \mu_\varepsilon)_{0<\varepsilon\leq 1}$ *be weak solutions of* (9.1)–(9.4) *with* $m_\varepsilon \to m_0 \geq 0$ *such that* $\lim_{\varepsilon\to 0} \varepsilon m_\varepsilon^{-1} = 0$. *Then for a suitable subsequence*

$$(\mathbf{v}_\varepsilon, m(\varepsilon)\mu_\varepsilon) \rightharpoonup_{\varepsilon\to 0} (\mathbf{v}, m_0\mu) \quad \text{in } L^2(0,T; H^1(\Omega))$$

$$c_\varepsilon \to_{\varepsilon\to 0} -1 + 2\chi_E \quad \text{in } C^{\frac{1}{9}}([0,T]; L^2(\Omega)) \text{ and a.e.}$$

where $\chi_E \in L^\infty(0,T; BV(\Omega))$ *and*

$$\partial_t(\rho\mathbf{v}) + \operatorname{div}(\rho\mathbf{v}\otimes\mathbf{v}) - \operatorname{div}(2\nu(\chi_E)D\mathbf{v}) + \nabla q = \delta V$$

$$\partial_t\chi_E + v\cdot\nabla\chi_E = \tfrac{m_0}{2}\Delta\mu$$

$$\text{If } m_0 > 0 : -2\mu\cdot\nabla\chi_E = \sigma\delta V$$

in the sense of distributions on $\Omega \times (0,T)$, *where* $V\colon (0,T) \to \mathcal{M}(\Omega \times G_{d-1})$ *is bounded and weak-$*$ measurable,* G_{d-1} *is the space of* $(d-1)$-*dimensional linear subspaces of* $\mathbb{R}^d$, *and*

$$\langle \delta V(t), \boldsymbol{\psi} \rangle = \int_{\Omega \times G_{d-1}} (\mathbf{I} - \mathbf{s}\otimes\mathbf{s}) : \nabla\boldsymbol{\psi}(x)\, dV(t)(x,\mathbf{s})$$

for all $\boldsymbol{\psi} \in C_c^1(\Omega)^d$, *cf. Chen [13].*

Here elements in $\mathcal{M}(\Omega \times G_{d-1})$ are called (general) *varifolds*. Therefore (v, χ_E, μ, V) is called varifold solution. The result was proved by A. and Röger [5, Appendix] in the case $m_\varepsilon \equiv m_0 > 0$ and by A. and Lengeler [3] for the assumptions above even for a model with different densities.

Since every $(d-1)$-dimensional hyperplane $U \subseteq \mathbb{R}^d$ is uniquely determined by its normal (up to orientation), we can identify G_{d-1} with $\mathbb{S}^{d-1}/\sim$, where $\mathbf{p}_0 \sim \mathbf{p}_1$ for $\mathbf{p}_0, \mathbf{p}_1 \in \mathbb{S}^{d-1}$ if and only if $\mathbf{p}_0 = \pm\mathbf{p}_1$ and $\mathbb{S}^{d-1}$ is the unit sphere in $\mathbb{R}^d$. Then by disintegration $V(t)$ can be decomposed in a non-negative measure $|V_t| \in \mathcal{M}(\Omega)$ and a family of probability measures $V_{t,x} \in \mathcal{M}(\mathbb{S}^{d-1}/\{\pm\mathbf{p}\})$, $x \in \Omega$, $t \in (0,T)$ such that

$$\langle V(t), \psi \rangle = \int_\Omega \int_{\mathbb{S}^{d-1}} \psi(x, \mathbf{p})\, dV_{t,x}(\mathbf{p})\, d|V_t|(x) \quad \text{for all } \psi \in C_c(\Omega \times \mathbb{S}^{d-1}),$$

cf. [10, Theorem 2.28]. Here $|V_t|$ corresponds to the measure of "area of the interface" and $V_{t,x}$ can be considered as a probability for the "normal at the interface". If $\Gamma(t) \subseteq \Omega$, $t \in (0,T)$ is a smoothly evolving family of hypersurfaces, then the canonically associated varifold $V(t)$ associated to it is given by

$$\langle V(t), \psi \rangle = \int_{\Gamma_t} \psi(x, [\mathbf{n}_{\Gamma_t}(x)]) \, d\mathcal{H}^{d-1}(x) \qquad \text{for all } \psi \in C_c(\Omega \times \mathbb{S}^{d-1}/\{\pm \mathbf{p}\}).$$

Then

$$\langle \delta V(t), \boldsymbol{\psi} \rangle = \int_{\Gamma_t} \underbrace{(\mathbf{I} - \mathbf{n}_{\Gamma_t} \otimes \mathbf{n}_{\Gamma_t}) : \nabla \boldsymbol{\psi}}_{=\operatorname{div}_\tau \boldsymbol{\psi}} \, d\mathcal{H}^{d-1} = -\int_{\Gamma_t} H_{\Gamma_t} \mathbf{n}_{\Gamma_t} \cdot \boldsymbol{\psi} \, d\mathcal{H}^{d-1}$$

for all $\boldsymbol{\psi} \in C_c^1(\Omega)^d$ because of Gauß' theorem on hypersurfaces for non-tangential vectorfields. Hence $\delta V(t)$ generalizes the mean curvature functional (9.23).

Since the normal $\mathbf{n}_{\Gamma_t}$ is replaced by a probability distribution for the normal, varifold solutions are similar to measure-valued solutions, cf. e.g. [24]. We note that in the case $m_0 > 0$ existence of weak solutions to the Navier-Stokes/Mullins-Sekerka system (9.8)–(9.16) was proved by A. and Röger in [5]. In the definition of weak solutions a suitable formulation of (9.23) instead of the first variation of a (general) varifold $V(t)$ is used. These weak solutions are also varifold solutions as in Theorem 1. But it is unknown whether the converse statement is true and convergence to weak solutions in the sharp interface limit $\varepsilon \to 0$ is an open problem even in the case of the Cahn-Hilliard equation, i.e., (9.3)–(9.4) with $\mathbf{v}_\varepsilon \equiv 0$. The convergence to varifold solutions was shown by Chen [13]. Finally, we remark that in the case $m_0 = 0$ even existence of weak solutions to the limit system (9.8)–(9.16) is an open problem. We refer to [1] or [17, section on "Weak Solutions and Diffuse Interface Models for Incompressible Two-Phase Flows"] for a further discussion.

Finally, we remark that every sufficiently smooth solution of (9.8)–(9.16) satisfies

$$\frac{d}{dt} \frac{1}{2} \int_\Omega \rho \, |\mathbf{v}|^2 \, dx + \sigma \frac{d}{dt} \mathcal{H}^{d-1}(\Gamma_t) = -\int_\Omega 2\nu(c)|D\mathbf{v}|^2 \, dx - \int_\Omega m_0 |\nabla \mu|^2 \, dx,$$
$$\tag{9.31}$$

where $c(t,x) = -1 + 2\chi_{\Omega^+(t)}(x)$. In view of (9.27) this is consistent with (9.7).

9.4 Sharp Interface Limit of a Stokes/Allen-Cahn System

In this section we discuss a first rigorous convergence result with convergence rates in strong norms for the sharp interface limit $\varepsilon \to 0$ in the case of a two-phase flow in fluid mechanics, which is comparable to results known for single phase field models like the Allen-Cahn equation, which is due to De Mottoni and Schatzman [15], or to the Cahn-Hilliard equation, which is due to Alikakos et al. [9].

More precisely, we consider the asymptotic limit $\varepsilon \to 0$ of the following system:

$$-\Delta \mathbf{v}_\varepsilon + \nabla p_\varepsilon = -\varepsilon \operatorname{div}(\nabla c_\varepsilon \otimes \nabla c_\varepsilon) \quad \text{in } \Omega \times (0, T_0), \tag{9.32}$$

$$\operatorname{div} \mathbf{v}_\varepsilon = 0 \quad \text{in } \Omega \times (0, T_0), \tag{9.33}$$

$$\partial_t c_\varepsilon + \mathbf{v}_\varepsilon \cdot \nabla c_\varepsilon = \Delta c_\varepsilon - \frac{1}{\varepsilon^2} f'(c_\varepsilon) \quad \text{in } \Omega \times (0, T_0), \tag{9.34}$$

$$\mathbf{v}_\varepsilon|_{\partial\Omega} = 0, \quad c_\varepsilon|_{\partial\Omega} = -1, \quad \text{on } \partial\Omega \times (0, T_0), \tag{9.35}$$

$$c_\varepsilon|_{t=0} = c_{0,\varepsilon} \quad \text{in } \Omega, \tag{9.36}$$

where $\Omega \subseteq \mathbb{R}^2$ is bounded domain with smooth boundary, and for suitable "well-prepared" initial data $c_{0,\varepsilon}$.

Similar to (9.1)–(9.4), every sufficiently smooth solution of (9.32)–(9.36) satisfies the *energy identity*

$$E_\varepsilon(c_\varepsilon(t)) + \int_0^t \int_\Omega \left(|\nabla \mathbf{v}_\varepsilon|^2 + \frac{1}{\varepsilon}|\mu_\varepsilon|^2 \right) dx \, d\tau = E_\varepsilon(c_{0,\varepsilon}) \tag{9.37}$$

for all $t \in (0, T_0)$, where $\mu_\varepsilon = -\varepsilon \Delta c_\varepsilon + \frac{1}{\varepsilon} f'(c_\varepsilon)$ and

$$E_\varepsilon(c_\varepsilon(t)) = \int_\Omega \varepsilon \frac{|\nabla c_\varepsilon(x, t)|^2}{2} \, dx + \int_\Omega \frac{f(c_\varepsilon(x, t))}{\varepsilon} \, dx$$

as before. The sharp interface limit of (9.32)–(9.36) is the system

$$-\Delta \mathbf{v} + \nabla p = 0 \quad \text{in } \Omega^\pm(t), t \in [0, T_0], \tag{9.38}$$

$$\operatorname{div} \mathbf{v} = 0 \quad \text{in } \Omega^\pm(t), t \in [0, T_0], \tag{9.39}$$

$$-[2D\mathbf{v} - p\mathbf{I}]\mathbf{n}_{\Gamma_t} = \sigma H_{\Gamma_t}\mathbf{n}_{\Gamma_t} \quad \text{on } \Gamma_t, t \in [0, T_0], \tag{9.40}$$

$$V_{\Gamma_t} - \mathbf{n}_{\Gamma_t} \cdot \mathbf{v}|_{\Gamma_t} = H_{\Gamma_t} \quad \text{on } \Gamma_t, t \in [0, T_0], \tag{9.41}$$

where Ω is the disjoint union of $\Omega^+(t), \Omega^-(t), \Gamma_t$ for every $t \in [0, T_0]$, $\Omega^\pm(t)$ are smooth domains, $\Gamma_t = \partial\Omega^+(t)$ and $\sigma = \int_\mathbb{R} \theta_0'(x)^2 \, dx$ as before.

We note that, if the material time derivative $\partial_t \mathbf{v}_\varepsilon + \mathbf{v}_\varepsilon \cdot \nabla \mathbf{v}_\varepsilon$ is added to the left-hand side of (9.32) (i.e., the Navier-Stokes equations are considered), the system (9.32)–(9.36) was suggested by Liu and Shen in [21] as an alternative approximation of a classical sharp interface model for a two-phase flow of viscous, incompressible, Newtonian fluids. It has advantages for numerical simulations since the Allen-Cahn equation is of second order and not of fourth order as the Cahn-Hilliard equation. A disadvantage of it is that the total mass $\int_\Omega c_\varepsilon(x, t) \, dx$ is in general not preserved in time for solutions of (9.32)–(9.36). This property is desirable if the model is used to approximate a two-phase flow without phase transitions. But (9.32)–(9.36) can be considered as a simplified model for a two-phase flow with phase transitions.

For such flows one can obtain systems of Navier-Stokes/Allen-Cahn type, cf. e.g. Blesgen [11].

Existence of weak solutions of the sharp interface model (9.38)–(9.41) was proved by Liu et al. [22] if the Stokes equation on the right-hand side is replaced by a modified Navier-Stokes equation for a shear thickening non-Newtonian fluid of power-law type. To this end they used a Galerkin approximation by a corresponding Navier-Stokes/Allen-Cahn system. Then they pass to the limit in the Galerkin approximation and $\varepsilon \to 0$ simultaneously. But they do not perform a sharp interface limit separately.

In the following we assume that $(\mathbf{v}, p, \Gamma)$ is a smooth solution of (9.38)–(9.41) for some $T_0 > 0$, where $(\Gamma_t)_{t\in[0,T_0]}$ is a family of smoothly evolving compact, non-selfintersecting, closed curves in Ω. More precisely, we assume that

$$\Gamma := \bigcup_{t\in[0,T_0]} \Gamma_t \times \{t\}$$

is a smooth two-dimensional submanifold of $\Omega \times \mathbb{R}$ (with boundary), and $\mathbf{v}|_{\Omega^\pm} \in C^\infty(\overline{\Omega^\pm})^2, p|_{\Omega^\pm} \in C^\infty(\overline{\Omega^\pm})$, where

$$\Omega^\pm = \bigcup_{t\in[0,T_0]} \Omega^\pm(t) \times \{t\}.$$

Moreover, let $(\mathbf{v}_\varepsilon, p_\varepsilon, c_\varepsilon)$ be the (classical) solution of (9.32)–(9.36) with smooth initial values $c_{0,\varepsilon}: \Omega \to \mathbb{R}$, which will be specified in the main result below.

We note that the mean curvature and the Stokes equation in (9.38)–(9.41) are coupled by terms of lower order, which is also the case for the Navier-Stokes/Mullins-Sekerka system (9.8)–(9.16). Hence existence of a local strong solution of the system can be obtained by adapting the strategy in [7], where this result was proven for (9.8)–(9.16). This was carried out by Moser in [27] in the case that the Stokes system (9.38)–(9.39) is replaced by the instationary Navier-Stokes system. By standard arguments from the regularity theory of parabolic equations and the Stokes system, one can prove that the solution is indeed smooth for smooth initial values.

For the statement of our main result we need tubular neighborhoods of Γ_t. For $\delta > 0$ and $t \in [0, T_0]$ we define

$$\Gamma_t(\delta) := \{y \in \Omega : \mathrm{dist}(y, \Gamma_t) < \delta\}, \quad \Gamma(\delta) = \bigcup_{t\in[0,T_0]} \Gamma_t(\delta) \times \{t\}.$$

Moreover, we define the signed distance function

$$d_\Gamma(x, t) := \mathrm{sdist}(\Gamma_t, x) = \begin{cases} \mathrm{dist}(\Omega^-(t), x) & \text{if } x \notin \Omega^-(t) \\ -\mathrm{dist}(\Omega^+(t), x) & \text{if } x \in \Omega^-(t) \end{cases} \qquad (9.42)$$

for all $x \in \Omega, t \in [0, T_0]$. Since Γ is smooth and compact, there is some $\delta > 0$ sufficiently small such that $d_\Gamma \colon \Gamma(3\delta) \to \mathbb{R}$ is smooth.

The main result of A. and L. [4] is:

Theorem 2 *Let $(\mathbf{v}, \Gamma)$ be a smooth solution of (9.38)–(9.41) for some $T_0 \in (0, \infty)$ and let*

$$c_{A,0}^0(x) = \zeta(d_{\Gamma_0}(x))\theta_0\left(\frac{d_{\Gamma_0}(x)}{\varepsilon}\right) + (1 - \zeta(d_{\Gamma_0}(x)))\left(\chi_{\Omega^+(0)}(x) - \chi_{\Omega^-(0)}(x)\right)$$

for all $x \in \Omega$, where $d_{\Gamma_0} = d_\Gamma|_{t=0}$ is the signed distance function to Γ_0 and $\zeta \in C^\infty(\mathbb{R})$ such that

$$\zeta(s) = 1, \ \text{if } |s| \leq \delta; \ \zeta(s) = 0, \ \text{if } |s| \geq 2\delta; \ 0 \leq s\zeta'(s) \leq 4 \text{ if } \delta \leq |s| \leq 2\delta.$$
$$(9.43)$$

Moreover, let $c_{0,\varepsilon} \colon \Omega \to \mathbb{R}, \ 0 < \varepsilon \leq 1$, be smooth such that

$$\|c_{0,\varepsilon} - c_{A,0}^0\|_{L^2(\Omega)} \leq C\varepsilon^{2+\frac{1}{2}} \qquad \text{for all } \varepsilon \in (0, 1] \tag{9.44}$$

and some $C > 0$, $\sup_{0 < \varepsilon \leq 1} \|c_{0,\varepsilon}\|_{L^\infty(\Omega)} < \infty$ and $(\mathbf{v}_\varepsilon, c_\varepsilon)$ the corresponding solutions of (9.32)–(9.36). Then there are some $\varepsilon_0 \in (0, 1], \ R > 0, \ T \in (0, T_0]$, and $c_A \colon \Omega \times [0, T_0] \to \mathbb{R}, \ \mathbf{v}_A \colon \Omega \times [0, T_0] \to \mathbb{R}^2$ (depending on ε) such that

$$\sup_{0 \leq t \leq T} \|c_\varepsilon(t) - c_A(t)\|_{L^2(\Omega)} + \|\nabla(c_\varepsilon - c_A)\|_{L^2(\Omega \times (0,T)\setminus\Gamma(\delta))} \leq R\varepsilon^{2+\frac{1}{2}} \tag{9.45a}$$

$$\|(\nabla_\tau(c_\varepsilon - c_A), \varepsilon\partial_\mathbf{n}(c_\varepsilon - c_A))\|_{L^2(\Omega \times (0,T)\cap\Gamma(2\delta))} \leq R\varepsilon^{2+\frac{1}{2}} \tag{9.45b}$$

and

$$\|\mathbf{v}_\varepsilon - \mathbf{v}_A\|_{L^2(0,T;H^1(\Omega))} \leq R\varepsilon^2 \tag{9.46}$$

hold true for all $\varepsilon \in (0, \varepsilon_0]$. Moreover,

$$\lim_{\varepsilon \to 0} c_A = \pm 1 \qquad \text{uniformly on compact subsets in } \Omega^\pm.$$

and

$$\mathbf{v}_A = \mathbf{v} + O(\varepsilon) \qquad \text{in } L^\infty(\Omega \times (0, T)) \text{ as } \varepsilon \to 0.$$

Here c_A is constructed such that

$$c_A(x, t) = \underbrace{\theta_0\left(\frac{d_\Gamma(x, t)}{\varepsilon} - h_\varepsilon(S(x, t), t)\right)}_{=c_{A,0}} + O(\varepsilon^2), \qquad \text{where } h_\varepsilon = h_1 + \varepsilon h_{2,\varepsilon},$$

in $\Gamma(\delta)$ for some suitable $h_1, h_{2,\varepsilon} \colon \mathbb{T}^1 \times [0, T_0] \to \mathbb{R}$, where $S(\cdot, t) \colon \Gamma_t(3\delta) \to \mathbb{T}^1$ is the pull-back of a suitable parametrization of Γ_t for every $t \in [0, T_0]$ and $\mathbb{T}^1 \cong \mathbb{S}^1 = \{x \in \mathbb{R}^2 : x_1^2 + x_2^2 = 1\}$ is the one dimensional torus, which corresponds to $[0, 2\pi]$ if 0 and 2π are identified. Here h_ε is introduced in order to obtain that the zero-level set of $c_{A,0}$ approximates the zero-level set of c_ε sufficiently well. More precise information on c_A can be found in [4, Sect. 4].

Here 2 is the basic convergence order (w.r.t. the $L^\infty(\Omega)$-norm). The order of convergence differs by $\frac{1}{2}$ if the L^2-norm is considered in space and by $-\frac{1}{2}$ if L^2-norm of the gradient is considered. Here the additional $\frac{1}{2}$ in the power of ε is natural since e.g. a simple change of variables yields

$$\left\| \theta_0\!\left(\tfrac{\cdot}{\varepsilon}\right) - \left(\chi_{[0,\infty)} - \chi_{(-\infty,0)}\right) \right\|_{L^2(\mathbb{R})} = M\varepsilon^{\frac{1}{2}}.$$

Here $\theta_0(s) \to_{s \to \pm\infty} \pm 1$ exponentially is used. Moreover, from the L^2-estimates one can obtain an L^∞-estimate in normal direction with the aid of the interpolation inequality

$$\|u\|_{L^\infty(-2\delta,2\delta)} \leq C_\delta \|u\|_{L^2(-2\delta,2\delta)}^{\frac{1}{2}} \|u\|_{H^1(-2\delta,2\delta)}^{\frac{1}{2}} \qquad \text{for all } u \in H^1(-2\delta, 2\delta).$$

Hence (9.45) yields a control of $c_\varepsilon - c_A$ of the order ε^2 in the L^∞-norm in normal direction and L^2-norms in the tangential direction.

For the proof of our main results we will follow the same basic strategy, which was already successfully used in [15] for the Allen-Cahn equation, in [9] for the Cahn-Hilliard equation, and in [14] for the mass-preserving Allen-Cahn equation. Following this strategy the proof consists of two parts. In the first part a suitable approximative solution for (9.32)–(9.36) upto an error term of a certain order in ε is constructed, which will be denoted by $(c_A, \mathbf{v}_A)$ in the following. In the second step the error of the approximative $(c_A, \mathbf{v}_A)$ and the exact solutions $(c_\varepsilon, \mathbf{v}_\varepsilon)$ is estimated with the aid of a suitable estimate for the linearized Allen-Cahn operator $\mathscr{L}_\varepsilon$, where

$$\mathscr{L}_\varepsilon u = -\Delta u + \frac{1}{\varepsilon^2} f''(c_A) u \qquad \text{for all } u \in H^2(\Omega)$$

is used, cf. (9.52) below.

But in order to adapt this strategy to the present system several new difficulties, which are mainly related to the coupling of the Allen-Cahn and the Stokes system, have to be overcome. More precisely, in order to estimate the difference $u := c_\varepsilon - c_A$ a suitable estimate of the convection term $\mathbf{v}_\varepsilon \cdot \nabla c_\varepsilon$ is needed. To this end it is essential how this term is approximated in the equation of c_A. More precisely, c_A is constructed such that the following result holds true, cf. [4, Theorem 1.3]:

Theorem 3 *Let the assumption of Theorem 2 be satisfied and $R \geq 1$. Then for every $\varepsilon \in (0, 1)$ there are*

$$\mathbf{v}_A, \mathbf{w}_1, \mathbf{w}_2 \colon \Omega \times [0, T_0] \to \mathbb{R}^2, \quad c_A \colon \Omega \times [0, T_0] \to \mathbb{R}, \quad r_A \colon \Omega \times [0, T_0] \to \mathbb{R}$$

(depending on $\varepsilon \in (0,1]$) such that $\mathbf{v}_\varepsilon = \mathbf{v}_A + \varepsilon^2 \mathbf{w}_1 + \varepsilon^2 \mathbf{w}_2$ and

$$\partial_t c_A + (\mathbf{v}_A + \varepsilon^2 \mathbf{w}_2) \cdot \nabla c_A + \varepsilon^2 \mathbf{w}_1|_\Gamma \cdot \nabla c_A = \Delta c_A - \frac{f'(c_A)}{\varepsilon^2} + r_A \tag{9.47}$$

in $\Omega \times [0, T_0]$. Moreover, there are some $\varepsilon_0 > 0$, $T_1 > 0$ and $M_R\colon (0,1] \times (0, T_0] \to (0, \infty)$, which is increasing with respect to both variables, such that $M_R(\varepsilon, T) \to_{(\varepsilon, T) \to 0} 0$ and, if

$$\sup_{0 \le t \le T_\varepsilon} \|c_\varepsilon(t) - c_A(t)\|_{L^2(\Omega)} + \|\nabla(c_\varepsilon - c_A)\|_{L^2(\Omega \times (0, T_\varepsilon) \setminus \Gamma(\delta))} \le R\varepsilon^{2+\frac{1}{2}}, \tag{9.48a}$$

$$\|(\nabla_\tau(c_\varepsilon - c_A), \varepsilon \partial_\mathbf{n}(c_\varepsilon - c_A))\|_{L^2(\Omega \times (0, T_\varepsilon) \cap \Gamma(2\delta))} \le R\varepsilon^{2+\frac{1}{2}} \tag{9.48b}$$

holds true for some $T_\varepsilon \in (0, T_0]$, $\varepsilon_0 \in (0,1]$, and all $\varepsilon \in (0, \varepsilon_0]$, then

$$\int_0^T \left| \int_\Omega r_A(x,t)(c_\varepsilon(x,t) - c_A(x,t))\,dx \right| dt \le M_R(\varepsilon, T)\varepsilon^{2(2+\frac{1}{2})}, \tag{9.49}$$

for all $T \in (0, \min(T_\varepsilon, T_1))$, $\varepsilon \in (0, \varepsilon_0]$.

Here ∇_τ denotes the tangential gradient and $\mathbf{w}_1$ is the leading part of the error $\mathbf{w} = \frac{\mathbf{v}_\varepsilon - \mathbf{v}_A}{\varepsilon^2}$ and $\mathbf{w}_1|_\Gamma(x,t) = \mathbf{w}_1(P_{\Gamma_t}(x), t)$ for $x \in \Gamma_t(2\delta)$, where P_{Γ_t} denotes the orthogonal projection onto Γ_t. Moreover, $\mathbf{v}_A$ is a suitable approximation of the solution of

$$-\Delta \mathbf{v} + \nabla p = -\varepsilon \operatorname{div}(\nabla c_{A,0} \otimes \nabla c_{A,0}) \quad \text{in } \Omega, \text{ for almost all } t \in (0, T_0),$$

$$\operatorname{div} \mathbf{v} = 0 \qquad\qquad\qquad \text{in } \Omega, \text{ for almost all } t \in (0, T_0).$$

We note that in (9.47) $\varepsilon^2 \mathbf{w}_2 \cdot \nabla c_A$ could also be omitted since it is of the same order as r_A. But the presence of the term $\varepsilon^2 \mathbf{w}_1|_\Gamma \cdot \nabla c_A$ is essential for the error estimates. Theorem 3 is proved with the aid of finitely many terms of an expansion in $\varepsilon > 0$ close to Γ_t (the inner expansion) and away from Γ_t (the outer expansion), using the method of formally matched asymptotics. It is shown rigorously that all terms in the expansion are well-defined, which yields that c_A is well-defined, and that the estimates above hold true.

Sketch of the proof of Theorem 2 Because of the regularity of c_ε and c_A for every fixed ε and the assumption on the initial data, for every $\varepsilon > 0$ (9.48) holds true for some $T_\varepsilon \in (0, T_0]$. The essential step in the proof of Theorem 2 is to show that for sufficiently small $\varepsilon > 0$ there is some $T \in (0, T_0)$ independent of ε such that (9.48) hold for $T_\varepsilon \ge T$. To this end the equation for $u = c_\varepsilon - c_A$ is considered, which can be written as

$$\partial_t u + \mathbf{v}_\varepsilon \cdot \nabla u + \mathcal{L}_\varepsilon u = -r_\varepsilon(c_\varepsilon, c_A) - r_A + \mathcal{R} \tag{9.50}$$

where

$$r_\varepsilon(c_\varepsilon, c_A) = \frac{1}{\varepsilon^2}\left(f'(c_\varepsilon) - f'(c_A) - f''(c_A)(c_\varepsilon - c_A)\right),$$

$$\mathscr{R} = -\varepsilon^2 \mathbf{w}_1 \cdot \nabla c_A + \varepsilon^2 \mathbf{w}_1|_\Gamma \cdot \nabla c_A.$$

Here $r_\varepsilon(c_\varepsilon, c_A)$ is the error due to a linearization of $\frac{1}{\varepsilon^2}f'(c_\varepsilon)$, r_A is the error in the equation for the approximation c_A, cf. (9.47), and $\mathscr{R}$ is the error in the approximation of the convection term $\mathbf{v}_\varepsilon \cdot \nabla c_\varepsilon$. Multiplying (9.50) with u and integrating yield

$$\frac{1}{2}\|u(t)\|^2_{L^2(\Omega)} + \int_0^t \int_\Omega \left(|\nabla u|^2 + \frac{f''(c_A)}{\varepsilon^2}u^2\right) dx\, ds \tag{9.51}$$

$$\leq \int_0^t \int_\Omega |r_\varepsilon(c_\varepsilon, c_A)u|\, dx\, ds + \int_0^t \left|\int_\Omega r_A\, u\, dx\right| ds + \int_0^t \left|\int_\Omega \mathscr{R}\, u\, dx\right| ds + \frac{1}{2}\|u(0)\|^2_{L^2(\Omega)}$$

for all $t \in [0, T_\varepsilon]$. Here

$$\int_\Omega \left(|\nabla u|^2 + \frac{f''(c_A)}{\varepsilon^2}u^2\right) dx = \langle \mathscr{L}_\varepsilon u, u\rangle.$$

Next the following estimate is used: There are some $C, \varepsilon_1 > 0$ such that for all $\psi \in H^1(\Omega)$, $\varepsilon \in (0, \varepsilon_1]$, and $t \in [0, T_0]$

$$\int_\Omega \left(|\nabla\psi(x)|^2 + \varepsilon^{-2}f''(c_A(x,t))\psi^2(x)\right) dx$$

$$\geq -C\int_\Omega \psi^2\, dx + \int_{\Omega\setminus\Gamma_t(\delta)} |\nabla\psi|^2\, dx + \int_{\Gamma_t(2\delta)} |\nabla_\tau\psi|^2\, dx. \tag{9.52}$$

This is called a spectral estimate since it implies that the L^2-spectrum of $\mathscr{L}_\varepsilon$ is bounded below by $-\sqrt{C}$ independent of $\varepsilon \in (0, \varepsilon_1)$. The estimate holds true if c_A is of suitable form, which is in particular true for c_A in Theorem 3. The estimate is a refinement of [12, Theorem 2.3], cf. [4, Theorem 2.12]. Now using the latter spectral estimate, the assumption on the initial data (9.44), and (9.49) from Theorem 3, we obtain

$$\sup_{0\leq s\leq t} \|u(s)\|^2_{L^2(\Omega)} + \|\nabla_\tau u\|^2_{L^2(\Omega\times(0,t)\cap\Gamma(2\delta))} + \|\nabla u\|^2_{L^2(\Omega\times(0,t)\setminus\Gamma(\delta))}$$

$$\leq C\int_0^t \|u(s)\|^2_{L^2(\Omega)}\, ds + \int_0^t \int_\Omega |r_\varepsilon(c_\varepsilon, c_A)u|\, dx\, ds + \int_0^t \left|\int_\Omega \mathscr{R}\, u\, dx\right| ds$$

$$+ \left(\frac{R^2}{4} + M_R(\varepsilon, t)\right)\varepsilon^{2(2+\frac{1}{2})} \tag{9.53}$$

for all $t \in [0, T_\varepsilon]$ and some $C > 0$ independent of $\varepsilon, T_\varepsilon$. Using suitable interpolation inequalities in normal direction one can estimate the error due to linearization by

$$\int_0^T \int_\Omega |r_\varepsilon(c_\varepsilon, c_A) u| \, dx \, dt \leq C(R, T_0) \varepsilon^{3 \cdot 2 + 1}$$

as long as (9.48) is valid, cf. [4, Lemma 5.4] for the details. Moreover, using the structure of the leading part of c_A and $\operatorname{div} \mathbf{w}_1 = 0$ one can prove for the error in the convection term

$$\int_0^t \left| \int_\Omega \mathscr{R} \, u \, dx \right| ds \leq C(R, T_\varepsilon, \varepsilon) \varepsilon^{2(2 + \frac{1}{2})},$$

where $C(R, T, \varepsilon) \to 0$ as $(T, \varepsilon) \to 0$, cf. [4, Lemma 5.3]. Together with Gronwall's inequality we obtain

$$\sup_{0 \leq t \leq T_\varepsilon} \|u(t)\|^2_{L^2(\Omega)} + \|\nabla_\tau u\|^2_{L^2(\Omega \times (0,t) \cap \Gamma(2\delta))}$$

$$+ \|\nabla u\|^2_{L^2(\Omega \times (0,t) \setminus \Gamma(\delta))} \leq C(R, T_\varepsilon, \varepsilon) \varepsilon^{2(2 + \frac{1}{2})}$$

for all $t \in [0, T_\varepsilon]$, where $C(R, T, \varepsilon) \to 0$ as $(T, \varepsilon) \to 0$. Moreover, putting in (9.51) the integral of $\frac{1}{\varepsilon^2} f''(c_A)$ on the right-hand side and using the previous estimates, one can add $\varepsilon^2 \|\partial_\mathbf{n} u\|^2_{L^2(\Omega \times (0,t) \cap \Gamma(2\delta))}$ on the left-hand side of the latter estimate. Hence choosing T, ε_0 sufficiently small, one obtains that (9.48) remains true for some $T_\varepsilon \geq T$ and all $\varepsilon \in (0, \varepsilon_0]$, which shows (9.45). Afterwards the rest of the statements of Theorem 2 follow from the construction of c_A and $\mathbf{v}_A$ and estimates for $\mathbf{v}_\varepsilon - \mathbf{v}_A$ in dependence of (9.45). $\qquad \square$

9.5 Works in Progress and Perspectives

The convergence result in the last section should be considered as a first step since only a simplified model in two dimensions is treated. More research needs to be done in order to extend this result to more realistic models and situations as e.g. to the three dimensional case, to a Navier-Stokes/Cahn-Hilliard system, to the case of different viscosities and to the case of different densities. Moreover, we expect that the convergence result holds true as long as the limit system possesses a sufficiently smooth solution, as it is the case for the Cahn-Hilliard equation. But even in this simplified situation the proof is already quite long and involved. Extending it in one of the above mentioned directions will probably result in the need of approximative solutions $(\mathbf{v}_A, c_A)$ of higher order than in the last section. Moreover, additional perturbation terms might require new ideas and refinements.

The original goal of this research project was to obtain a convergence result for the model H (9.1)–(9.4) for short times and for a suitable scaling of m_ε. In the case $m_\varepsilon \equiv m_0 > 0$ this is work in progress and will be treated in the third author's PhD-thesis. So far (9.1)–(9.4) is considered in a two-dimensional torus with constant viscosity ν and neglected material time derivative of the velocity. In this case convergence has been proved provided a suitable approximative solution is constructed. The construction of the approximative solution and the inclusion of the convection term is work in progress. But even when this work is finished the same kind of extensions as for the Stokes/Allen-Cahn system remain open. Furthermore, other scalings of the mobility as e.g. $m_\varepsilon = m_0 \varepsilon$ remain to be discussed.

Since the rigorous analysis of such sharp interface limits is rather involved and consists of tedious and technical constructions of approximative solutions, it is very desirable to develop a more systematic calculus for such constructions, which could be applied to general classes of diffuse interface models. Finally, even if the approximative solutions are constructed, suitable estimates of the coupling terms and results for the linearized operators are needed. There might still be a great potential for improvements concerning these results and estimates.

References

1. Abels, H.: On the notion of generalized solutions of two-phase flows for viscous incompressible fluids. RIMS Kôkyûroku Bessatsu **B1**, 1–15 (2007)
2. Abels, H.: On a diffuse interface model for two-phase flows of viscous, incompressible fluids with matched densities. Arch. Rat. Mech. Anal. **194**(2), 463–506 (2009)
3. Abels, H., Lengeler, D.: On sharp interface limits for diffuse interface models for two-phase flows. Interfaces Free Bound. **16**(3), 395–418 (2014)
4. Abels, H., Liu, Y.: Sharp interface limit for a Stokes/Allen-Cahn system. Preprint arXiv:1611.04422 (2016)
5. Abels, H., Röger, M.: Existence of weak solutions for a non-classical sharp interface model for a two-phase flow of viscous, incompressible fluids. Ann. Inst. H. Poincaré Anal. Non Linéaire **26**, 2403–2424 (2009)
6. Abels, H., Schaubeck, S.: Nonconvergence of the capillary stress functional for solutions of the convective Cahn-Hilliard equation. In: Mathematical Fluid Dynamics, Present and Future. Springer Proceedings in Mathematics & Statistics, vol. 183. Springer, Cham (2016)
7. Abels, H., Wilke, M.: Well-posedness and qualitative behaviour of solutions for a two-phase Navier-Stokes-Mullins-Sekerka system. Interfaces Free Bound. **15**(1), 39–75 (2013)
8. Abels, H., Garcke, H., Grün, G.: Thermodynamically consistent, frame indifferent diffuse interface models for incompressible two-phase flows with different densities. Math. Models Methods Appl. Sci. **22**(3), 1150013, 40 pp. (2012)
9. Alikakos, N.D., Bates, P.W., Chen, X.: Convergence of the Cahn-Hilliard equation to the Hele-Shaw model. Arch. Ration. Mech. Anal. **128**(2), 165–205 (1994)
10. Ambrosio, L., Fusco, N., Pallara, D.: Functions of Bounded Variation and Free Discontinuity Problems. Oxford Mathematical Monographs, pp. xviii, 434. Clarendon Press, Oxford (2000)
11. Blesgen, T.: A generalization of the Navier-Stokes equations to two-phase flows. J. Phys. D (Appl. Phys.) **32**, 1119–1123 (1999)
12. Chen, X.: Spectrum for the Allen-Cahn, Cahn-Hilliard, and phase-field equations for generic interfaces. Commun. Partial Differ. Equ. **19**(7–8), 1371–1395 (1994)

13. Chen, X.: Global asymptotic limit of solutions of the Cahn-Hilliard equation. J. Differ. Geom. **44**(2), 262–311 (1996)
14. Chen, X., Hilhorst, D., Logak, E.: Mass conserving Allen-Cahn equation and volume preserving mean curvature flow. Interfaces Free Bound. **12**(4), 527–549 (2010)
15. De Mottoni, P., Schatzman, M.: Geometrical evolution of devoloped interfaces. Trans. Am. Math. Soc. **347**(5), 1533–1589 (1995)
16. Evans, L.C., Gariepy, R.F.: Measure Theory and Fine Properties of Functions. Studies in Advanced Mathematics. CRC Press, Boca Raton, FL (1992)
17. Giga, Y., Novotny, A. (ed.): Handbook of Mathematical Analysis in Mechanics of Viscous Fluids. Springer, Cham (2018)
18. Gurtin, M.E., Polignone, D., Viñals, J.: Two-phase binary fluids and immiscible fluids described by an order parameter. Math. Models Methods Appl. Sci. **6**(6), 815–831 (1996)
19. Hohenberg, P.C., Halperin, B.I.: Theory of dynamic critical phenomena. Rev. Mod. Phys. **49**, 435–479 (1977)
20. Jacqmin, D.: Calculation of two-phase Navier-Stokes flows using phase-field modeling. J. Comput. Phys. **155**(1), 96–127 (1999)
21. Liu, C., Shen, J.: A phase field model for the mixture of two incompressible fluids and its approximation by a Fourier-spectral method. Phys. D **179**(3–4), 211–228 (2003)
22. Liu, C., Sato, N., Tonegawa, Y.: Two-phase flow problem coupled with mean curvature flow. Interfaces Free Bound. **14**(2), 185–203 (2012)
23. Lowengrub, J., Truskinovsky, L.: Quasi-incompressible Cahn-Hilliard fluids and topological transitions. R. Soc. Lond. Proc. Ser. A: Math. Phys. Eng. Sci. **454**(1978), 2617–2654 (1998)
24. Málek, J., Nečas, J., Rokyta, M., Ružička, M.: Weak and Measure-Valued Solutions to Evolutionary PDEs. Applied Mathematics and Mathematical Computation, vol. 13, vii, 317 pp. Chapman & Hall, London (1996)
25. Modica, L.: The gradient theory of phase transitions and the minimal interface criterion. Arch. Ration. Mech. Anal. **98**(2), 123–142 (1987)
26. Modica, L., Mortola, S.: Un esempio di Γ^--convergenza. Boll. Un. Mat. Ital. B (5) **14**(1), 285–299 (1977)
27. Moser, M.: Lokale Wohlgestelltheit für ein Navier-Stokes/mittleren Krümmungsfluss-System. Master thesis, Regensburg (2016)
28. Schaubeck, S.: Sharp interface limits for diffuse interface models. PhD thesis, University Regensburg, urn:nbn:de:bvb:355-epub-294622 (2014)
29. Starovoĭtov, V.N.: A model of the motion of a two-component fluid taking into account capillary forces. Prikl. Mekh. Tekhn. Fiz. **35**(6), 85–92 (1994)

Chapter 10
Two-Phase Flow with Surfactants: Diffuse Interface Models and Their Analysis

Helmut Abels, Harald Garcke, Kei Fong Lam, and Josef Weber

Abstract New diffuse interface and sharp interface models for soluble and insoluble surfactants fulfilling energy inequalities are introduced. We discuss their relation with the help of asymptotic analysis and present an existence result for a particular diffuse interface model.

10.1 Introduction

Surface active agents (or commonly known as surfactants) are compounds that are able to lower the surface tension of fluid interfaces, and thus have found numerous applications in both biological systems and industrial processes.

For systems with two or more immiscible fluids, surfactants can be broadly classified into two types: insoluble and soluble. In the latter case, surfactants can exist in both the bulk fluid phases and also on the fluid interfaces, but in the former case, insoluble surfactants will only exist on the interfaces. When introduced to a multi-fluid system, the soluble surfactants may migrate towards the fluid interfaces and are incorporated to the interface by the process of adsorption.

One of the simplest model of adsorption dynamics is that studied by Ward and Tordai [27] and is defined on $(0, \infty)$ with the interface at the origin:

$$\partial_t c = D\partial_{xx}c \ \text{ for } \ x > 0, \ t > 0, \quad \partial_t c^\Gamma = D\partial_x c \ \text{ at } \ x = 0, \ t > 0,$$

$$\lim_{x \to \infty} c(x,t) = c_b \ \text{ for } \ t > 0, \quad c(x,0) = c_b, \quad c^\Gamma(0) = 0.$$

Here, c and c^Γ denote the concentration of the bulk and interfacial surfactants, c_b is the initial and far-field boundary condition, D denotes the diffusion coefficient, and the term $D\partial_x c$ is a source term for the surfactant concentration on the interface stemming from the surfactant flux in the bulk. In their work, Ward and Tordai derived an equation relating the interfacial surfactant concentration $c^\Gamma(t)$ and the

H. Abels • H. Garcke (✉) • K.F. Lam • J. Weber
Fakultät für Mathematik, Universität Regensburg, Universitätsstraße 31, 93053 Regensburg, Germany
e-mail: helmut.abels@ur.de; harald.garcke@ur.de; kei-fong.lam@ur.de; josef.weber@ur.de

© Springer International Publishing AG 2017
D. Bothe, A. Reusken (eds.), *Transport Processes at Fluidic Interfaces*,
Advances in Mathematical Fluid Mechanics, DOI 10.1007/978-3-319-56602-3_10

surfactant concentration of the sub-layer $c(0, t)$, where the sub-layer is defined as the bulk region immediately adjacent to the interface. To solve for $c^{\Gamma}(t)$, Ward and Tordai assumed that the sub-layer and the interface are in thermodynamical equilibrium, and thus postulates a relation between $c^{\Gamma}(t)$ and $c(0, t)$, which is given by

$$c^{\Gamma}(t) = g(c(0, t)) \tag{10.1}$$

for some function g. This functional relation is termed *adsorption isotherm* [13], which relates the interfacial concentration with the sub-layer concentration.

A key assumption in the work of Ward and Tordai is that the interface and the sub-layer are in equilibrium, that is, the process of adsorption is fast compared to the kinetics in the bulk regions. This case is called *instantaneous adsorption* or *diffusion controlled adsorption*. However, there are systems in which instantaneous adsorption is not valid, for example in the context of ionic surfactants [12], and in these situations a closure relation akin to (10.1) between $c(0, t)$ and $c^{\Gamma}(t)$ is not available. For such cases, which we denote as *non-instantaneous adsorption* or *dynamic adsorption*, we will have to postulate alternative closure relations.

Two-phase flow with surfactant is classically modelled with moving hypersurfaces describing the interfaces separating the two fluids. In [17], the following sharp interface model for a domain Ω containing two fluids of different mass densities in the presence of soluble surfactants is derived. We denote by $\Omega_-(t)$, $\Omega_+(t)$ the domains of the fluids which are separated by an interface $\Gamma(t)$:

$$\operatorname{div} \mathbf{v} = 0 \qquad \text{in } \Omega_{\pm}(t), \tag{10.2a}$$

$$\partial_t(\tilde{\rho}_{\pm}\mathbf{v}) + \operatorname{div}(\tilde{\rho}_{\pm}\mathbf{v} \otimes \mathbf{v}) = \operatorname{div}(-p\mathit{Id} + 2\eta_{\pm}\mathbf{Dv}) \quad \text{in } \Omega_{\pm}(t), \tag{10.2b}$$

$$\partial_t^{\bullet} c_{\pm} = \operatorname{div}(M_c^{\pm} \nabla G'_{\pm}(c_{\pm})) \quad \text{in } \Omega_{\pm}(t), \tag{10.2c}$$

$$[\mathbf{v}]_-^+ = 0, \quad \mathbf{v} \cdot \boldsymbol{\nu} = \mathcal{V} \qquad \text{on } \Gamma(t), \tag{10.2d}$$

$$[p\mathit{Id} - 2\eta\mathbf{Dv}]_-^+ \boldsymbol{\nu} = \sigma(c^{\Gamma})\kappa\boldsymbol{\nu} + \nabla_{\Gamma}\sigma(c^{\Gamma}) \quad \text{on } \Gamma(t), \tag{10.2e}$$

$$\partial_t^{\bullet} c^{\Gamma} + c^{\Gamma}\operatorname{div}_{\Gamma}\mathbf{v} - [M_c\nabla G'(c)]_-^+ \boldsymbol{\nu} = \operatorname{div}_{\Gamma}(M_{\Gamma}\nabla_{\Gamma}\gamma'(c^{\Gamma})) \quad \text{on } \Gamma(t), \tag{10.2f}$$

$$\mp\alpha_{\pm}M_c^{\pm}\nabla G'_{\pm}(c^{\pm}) \cdot \boldsymbol{\nu} = -(\gamma'(c^{\Gamma}) - G'_{\pm}(c_{\pm})) \quad \text{on } \Gamma(t). \tag{10.2g}$$

Here $\mathbf{v}$ denotes the fluid velocity, $\tilde{\rho}_{\pm}$ and $\eta_{\pm}$ are the constant mass densities and viscosities of the fluids, respectively, $\mathbf{Dv} = \frac{1}{2}(\nabla\mathbf{v} + (\nabla\mathbf{v})^{\top})$ is the rate of deformation tensor, p is the pressure, Id is the identity tensor, $\partial_t^{\bullet}(\cdot) = \partial_t(\cdot) + \mathbf{v} \cdot \nabla(\cdot)$ is the material derivative, $c_{\pm}$ are the bulk densities of the surfactants, $M_c^{\pm}$ are the bulk mobilities, and $G_{\pm}$ are the bulk free energy densities. We point out that the densities $c_\pm$ need not both correspond to same surfactant species, and it is also possible to consider different surfactant species in each bulk phases by introducing further density functions and additional conservation laws.

On the interface, $\mathcal{V}$ is the normal velocity, $\boldsymbol{v}$ is the unit normal on Γ pointing into Ω_+, c^Γ is the interfacial surfactant density, $\gamma(c^\Gamma)$ is the interfacial free energy density, $\sigma(c^\Gamma) := \gamma(c^\Gamma) - c^\Gamma \gamma'(c^\Gamma)$ is the density dependent surface tension, κ is the mean curvature of Γ, ∇_Γ is the surface gradient operator, div_Γ is the surface divergence, M_Γ is the interfacial mobility, and $\alpha_\pm \geq 0$ are kinetic factors which are related to the speed of adsorption.

Equations (10.2a) and (10.2b) are the classical incompressibility condition and momentum equation, respectively. The mass balance equation for bulk surfactants is given by (10.2c). Equation (10.2d) states that the interface is transported with the flow and that not only the normal components but also the tangential components of the velocity field match up (cf. [9, 23, 26] for a more general condition that takes into account the relative velocities between the bulk phases and the interface due to mass transfer). The force balance on the interface (10.2e) relates the jump in the stress tensor across the interface to the surface tension force and the Marangoni force at the interface. The mass balance of the interfacial surfactants is given by (10.2f), and the closure condition (10.2g) tells us whether adsorption is instantaneous ($\alpha = 0$, an isotherm is obtained) or dynamic ($\alpha > 0$, the mass flux into the interface is proportional to the difference in chemical potentials).

To see this, suppose the process of adsorption at the interface is instantaneous, i.e., fast compared to the timescale of convective and diffusive transport in the bulk regions. This local equilibrium corresponds to the case that the bulk chemical potential $G'_\pm(c)$ and the interface chemical potential $\gamma'(c^\Gamma)$ are equal, which is the case if we set $\alpha = 0$ in (10.2g) (we here only consider one of the bulk phases adjacent to the interface and, for simplicity, drop the subscript $\pm$). We obtain the following relation

$$\gamma'(c^\Gamma) = G'(c) \quad \Longleftrightarrow \quad c^\Gamma = g(c) := (\gamma')^{-1}(G'(c)), \tag{10.3}$$

where $g : \mathbb{R}_+ \to \mathbb{R}_+$ is strictly increasing. This function g plays the role of various adsorption isotherms which state the equilibrium relations between the two densities. The novelty of the model (10.2) is twofold; we can realise various adsorption isotherms by choosing appropriate functional forms for the free energy densities G and γ, see Table 10.1 below and also [17, Table 2.1], [18] for examples. Moreover, with positive values of α, we can include the effects of non-equilibrium adsorption dynamics, and thus (10.2) is a generalisation of the model studied in [10] to the case of dynamic adsorption. Furthermore a model involving insoluble surfactants easily arises by setting $c_\pm = M_c^\pm = 0$ and neglecting (10.2g) in (10.2), while the case where the surfactant is only soluble in Ω_+ can be obtained by setting $c_- = M_c^- = G_- = 0$ in (10.2) and neglecting the corresponding equation in (10.2g).

Table 10.1 Possible functional forms for γ and G to obtain the Henry and Langmuir adsorption isotherms and equations of state

Isotherm	Henry	Langmuir
Relation	$Kc = \dfrac{c^\Gamma}{c_M^\Gamma}$	$Kc = \dfrac{c^\Gamma}{c_M^\Gamma - c^\Gamma}$
$\gamma(c^\Gamma) - \sigma_0$	$Bc^\Gamma \left(\log \dfrac{c^\Gamma}{c_M^\Gamma} - 1\right)$	$B\left(c^\Gamma \log \dfrac{c^\Gamma}{c_M^\Gamma - c^\Gamma} + c_M^\Gamma \log(1 - \dfrac{c^\Gamma}{c_M^\Gamma})\right)$
$G(c)$	$Bc(\log(Kc) - 1)$	$Bc(\log(Kc) - 1)$
$\sigma - \sigma_0$	$-Bc^\Gamma$	$Bc_M^\Gamma \log\left(1 - \dfrac{c^\Gamma}{c_M^\Gamma}\right)$

See [17] for the details regarding the Freundlich, Volmer and Frumkin isotherms

In Table 10.1 the functional forms for γ and G for the Henry and Langmuir adsorption isotherms are stated. Here, c_M^Γ is the maximum interfacial surfactant density, K is a constant relating the surface density to the bulk density in equilibrium, σ_0 denotes the surface tension of a clean interface, and B is the sensitivity of the surface tension to surfactant. We point out that the model (10.2) satisfies the second law of thermodynamics in an isothermal situation in the form of an energy dissipation inequality (under suitable boundary conditions), cf. [15, 17],

$$0 = \frac{d}{dt}\left(\left[\sum \int_{\Omega_\pm} \left(\frac{\tilde{\rho}_\pm}{2} |\mathbf{v}|^2 + G_\pm(c_\pm)\right)\right] + \int_\Gamma \gamma(c^\Gamma)\right) + \int_\Gamma M_\Gamma |\nabla_\Gamma \gamma'(c^\Gamma)|^2$$

$$+ \sum \left(\int_{\Omega_\pm} \left(\eta_\pm |\mathbf{D}\mathbf{v}|^2 + M_c^\pm |\nabla G_\pm'(c_\pm)|^2\right) + \int_\Gamma \frac{1}{\alpha_\pm} |\gamma'(c^\Gamma) - G_\pm'(c_\pm)|^2\right). \tag{10.4}$$

The model (10.2) constitutes a free boundary problem, in which the interface $\Gamma(t)$ is unknown a priori and has to be computed as part of the solution. For numerical simulations of two-phase flow with surfactants based on the above models, we refer the reader to the work of [7, 8].

A second approach to model the dynamics is to relax the immiscibility assumption of the fluids, and assume that there is some microscopic mixing of the macroscopically immiscible fluids. This replaces the hypersurface description with an interfacial layer of small and finite width. Within this layer the fluids are assumed to be mixed and thus we have to account for the mixing energies. These models are commonly termed as *diffuse interface* or *phase field models*, and in [1] Abels, Garcke and Grün derived a diffuse interface model for non-matched densities with a solenoidal velocity field based on a volume-averaged velocity. We refer to the companion contribution [2] in this book for more details. The goal is to derive thermodynamically consistent diffuse interface models for two-phase flow with soluble surfactants, using the model of Abels et al. [1] as our basis.

10.2 Diffuse Interface Models

At the core of any diffuse interface model lies the Ginzburg–Landau functional

$$\mathscr{E}(\varphi) := \int_{\Omega} \left(\frac{\varepsilon}{2} |\nabla \varphi|^2 + \frac{1}{\varepsilon} \psi(\varphi) \right).$$

Here, $\varphi : \Omega \to \mathbb{R}$ denotes the order parameter used to distinguish the bulk fluid phases, $\varepsilon > 0$ is a parameter related to the thickness of the interfacial layer, and ψ is a potential with two equal minima (which we will take to be ± 1). Through the work of Modica and Mortola [21], it is well-known that the Ginzburg–Landau functional $\mathscr{E}(\varphi)$ converges to a multiple of the perimeter functional of the set $\{\varphi = 1\}$ in the sense of Γ-convergence. Hence $\mathscr{E}$ is often used to approximate the surface energy on the interface. Let us denote by δ_Γ the Hausdorff measure restricted to Γ, and by $\chi_{\Omega_\pm}$ the characteristic function of the set $\Omega_\pm$. With the help of [6, §2.7 and Theorem 2.8] (see also [17, §6] and [19, Appendix B]) the surfactant subsystem (10.2c), (10.2f), (10.2g) can be reformulated into an equivalent distributional form

$$\partial_t(\chi_{\Omega_\pm} c_\pm) + \operatorname{div}(\chi_{\Omega_\pm} c_\pm \mathbf{v} - \chi_{\Omega_\pm} M_c^\pm \nabla G'_\pm(c_\pm)) = \delta_\Gamma j_\pm, \tag{10.5a}$$

$$\partial_t(\delta_\Gamma c^\Gamma) + \operatorname{div}(\delta_\Gamma c^\Gamma \mathbf{v} - \delta_\Gamma M_\Gamma \nabla \gamma'(c^\Gamma)) = -\delta_\Gamma(j_- + j_+), \tag{10.5b}$$

$$\frac{1}{\alpha_\pm}(\gamma'(c^\Gamma) - G'_\pm(c_\pm)) = j_\pm. \tag{10.5c}$$

The idea of [17] is to replace the distributions δ_Γ and $\chi_{\Omega_\pm}$ with regularisations $\delta_\varepsilon(\varphi, \nabla \varphi)$ and $\xi_{\pm,\varepsilon}(\varphi)$ indexed by the width of the interfacial layer $\varepsilon > 0$. This is done in the spirit of the so-called diffuse domain approach [20, 24]. For a rigorous treatment of the diffuse domain approach in the limit $\varepsilon \to 0$ we refer the reader to [4, 11]. One example of a regularisation δ_ε is the Ginzburg–Landau density

$$\delta_\varepsilon(\varphi, \nabla \varphi) = \mathscr{W} \left(\frac{\varepsilon}{2} |\nabla \varphi|^2 + \frac{1}{\varepsilon} \psi(\varphi) \right), \quad \frac{1}{\mathscr{W}} := \int_{-1}^{1} \sqrt{2\psi(s)} ds.$$

In the following, we will rescale the potential ψ so that $\mathscr{W} = 1$. Meanwhile, we can take $\xi_{-,\varepsilon}(\varphi) = \frac{1-\varphi}{2}$ and $\xi_{+,\varepsilon}(\varphi) = 1 - \xi_{-,\varepsilon}$. Then, for $\alpha_\pm > 0$, the diffuse interface model for soluble surfactants of [17] (denoted as Model A) in the case of dynamic adsorption is given as (dropping the subscript ε from $\xi_{\pm,\varepsilon}$ and δ_ε)

$$\operatorname{div} \mathbf{v} = 0, \tag{10.6a}$$

$$\partial_t(\rho \mathbf{v}) + \operatorname{div}(\rho \mathbf{v} \otimes \mathbf{v}) = \operatorname{div}\left(-p Id + 2\eta(\varphi) \mathbf{D}\mathbf{v} + \mathbf{v} \otimes \frac{\tilde{\rho}_+ - \tilde{\rho}_-}{2} m(\varphi) \nabla \mu\right) \tag{10.6b}$$

$$+ \operatorname{div}\left(\sigma(c^\Gamma)(\delta(\varphi, \nabla \varphi) Id - \varepsilon \nabla \varphi \otimes \nabla \varphi)\right),$$

$$\partial_t^\bullet \varphi = \mathrm{div}\,(m(\varphi)\nabla\mu), \tag{10.6c}$$

$$\mu + \mathrm{div}\,(\varepsilon\sigma(c^\Gamma)\nabla\varphi) = \frac{\sigma(c^\Gamma)}{\varepsilon}\psi'(\varphi) + \sum \xi'_\pm(\varphi)(G_\pm(c_\pm) - G'_\pm(c_\pm)c_\pm), \tag{10.6d}$$

$$\partial_t^\bullet(\xi_\pm(\varphi)c_\pm) = \mathrm{div}\,(M_c^\pm(c_\pm)\xi_i(\varphi)\nabla G'_i(c_\pm)) \tag{10.6e}$$

$$+ \frac{1}{\alpha_\pm}\delta(\varphi,\nabla\varphi)(\gamma'(c^\Gamma) - G'_\pm(c_\pm)),$$

$$\partial_t^\bullet(\delta(\varphi,\nabla\varphi)c^\Gamma) = \mathrm{div}\,\big(M_\Gamma(c^\Gamma)\delta(\varphi,\nabla\varphi)\nabla\gamma'(c^\Gamma)\big) \tag{10.6f}$$

$$- \delta(\varphi,\nabla\varphi)\sum\frac{1}{\alpha_\pm}(\gamma'(c^\Gamma) - G'_\pm(c_\pm)),$$

where the density $\rho(\varphi)$ and viscosity $\eta(\varphi)$ are defined as

$$\rho(\varphi) := \frac{\tilde\rho_+ - \tilde\rho_-}{2}\varphi + \frac{\tilde\rho_+ + \tilde\rho_-}{2}, \quad \eta(\varphi) := \frac{\eta_+ - \eta_-}{2}\varphi + \frac{\eta_+ + \eta_-}{2}. \tag{10.7}$$

Equations (10.6a) and (10.6b) are the incompressibility condition and the phase field momentum equations, respectively. Equation (10.6c) together with (10.6d) forms a Cahn–Hilliard type equation which governs how the order parameter evolves and equations (10.6e) and (10.6f) are the bulk and interfacial surfactant equations, respectively. In (10.6d), the variable μ is often denoted as the chemical potential for the order parameter φ, and in (10.6c), $m \geq 0$ denotes a mobility for φ. The above model (10.6) is derived by modifying the approach of Teigen et al. [25] for the surfactant subsystem such that the following energy inequality is obtained (under suitable boundary conditions):

$$0 = \frac{d}{dt}\int_\Omega \left(\sum \xi_\pm(\varphi)G_\pm(c_\pm) + \delta(\varphi,\nabla\varphi)\gamma(c^\Gamma) + \frac{\rho(\varphi)}{2}|\mathbf{v}|^2\right)$$

$$+ \int_\Omega \left(m(\varphi)|\nabla\mu|^2 + 2\eta(\varphi)|\mathbf{Dv}|^2 + M_\Gamma\delta(\varphi,\nabla\varphi)|\nabla\gamma'(c^\Gamma)|^2\right)$$

$$+ \int_\Omega \sum \left(M_c^\pm\xi_\pm(\varphi)|\nabla G'_\pm(c_\pm)|^2 + \frac{\delta(\varphi,\nabla\varphi)}{\alpha_\pm}|\gamma'(c^\Gamma) - G'_\pm(c_\pm)|^2\right). \tag{10.8}$$

Here, we observe the similarities between (10.4) and (10.8). In particular, $\delta(\varphi,\nabla\varphi)\gamma(c^\Gamma)$ can be seen as an approximation of the interfacial surfactant energy density.

In the case of instantaneous adsorption for both fluid phases, that is, when the sub-layers in both bulk phases are in equilibrium with the interface, the ansatz is to assume that the chemical potentials $\gamma'(c^\Gamma)$ and $G'_\pm(c_\pm)$ are equal on the interface. We can introduce the chemical potential as a new continuous variable q and consider this as an unknown field and define the surfactant densities $c_\pm, c^\Gamma$ as functions of q:

$$c_\pm(q) := (G'_\pm)^{-1}(q), \quad c^\Gamma(q) := (\gamma')^{-1}(q), \tag{10.9}$$

for strictly convex free energies $G_\pm$ and γ. The surfactant densities are well-defined as the derivatives $G'_\pm$ and γ' are monotone and one-to-one. Then, summing (10.6e) and (10.6f) leads to one equation for q:

$$\partial_t^\bullet \left(\xi_- c_-(q) + \xi_+ c_+(q) + \delta c^\Gamma(q) \right) = \operatorname{div}\left(\left(M_c^- \xi_- + M_c^+ \xi_+ + M_\Gamma \delta \right) \nabla q \right). \tag{10.10}$$

We define the surface tension $\tilde\sigma(q)$ by

$$\tilde\sigma(q) := \sigma(c^\Gamma(q)) = \gamma(c^\Gamma(q)) - q c^\Gamma(q). \tag{10.11}$$

Then the diffuse interface model for soluble surfactants of [17] (denoted as Model C) in the case of instantaneous adsorption consists of (10.6a)–(10.6d) and (10.10) [with $\tilde\sigma$ replacing σ in (10.6b) and (10.6d), and q replacing $G'_\pm(c_\pm)$ in (10.6d)].

It is also possible to consider a model which has instantaneous adsorption in Ω_+ and dynamic adsorption in Ω_-. In this case we use (10.3) to express c^Γ as a function of c_+, and add (10.6f) to the equation (10.6e) for c_+. This yields a equation in c_+ that is coupled to the equation of c_- via a source term $\frac{1}{\alpha}\delta(G'_+(c_+) - G'_-(c_-))$. This is denoted as Model B in [17].

In [17], for the choice of a degenerate mobility $m(\varphi) = (1-\varphi^2)_+ = \max(0, 1 - \varphi^2)$, it has been shown via the method of formally matched asymptotic expansions that the sharp interface model (10.2) with $\alpha_\pm > 0$ is recovered from Model A in the limit $\varepsilon \to 0$, and analogously (10.2) with (10.3) instead of (10.2g) is recovered from both Model C and Model A with the particular scaling $\alpha_\pm = \varepsilon$. We point out that the same sharp interface models can be recovered from (10.6) if we consider the choice $m(\varphi) = \varepsilon m_0$ for some positive constant $m_0 > 0$.

In terms of the mathematical analysis of the aforementioned diffuse interface models, the main difficulty lies in getting a compactness result for the surfactant densities. Take for example Model C with constant mobilities $M_c^\pm = M_\Gamma = 1$ and equal bulk energy densities $G_- = G_+$ (and hence $c_-(q) = c_+(q) =: c(q)$). Then, Model C admits an energy identity of the form

$$0 = \frac{d}{dt} \int_\Omega \left(G(c(q)) + \delta(\varphi, \nabla\varphi)\gamma(c^\Gamma(q)) + \frac{\rho(\varphi)}{2} |\mathbf{v}|^2 \right)$$
$$+ \int_\Omega \left(m(\varphi) |\nabla\mu|^2 + 2\eta(\varphi) |\mathbf{D}\mathbf{v}|^2 + (1 + \delta(\varphi, \nabla\varphi)) |\nabla q|^2 \right), \tag{10.12}$$

where we used $\xi_- + \xi_+ = 1$. If γ is bounded from below by a positive constant, then one obtains spatial estimates for φ in $H^1(\Omega)$, and compactness with respect to time follows from standard arguments. However, any time compactness for q has to come from Eq. (10.10), which now reads as

$$\partial_t^\bullet \left(c(q) + \delta(\varphi, \nabla\varphi) c^\Gamma(q) \right) = \operatorname{div}\left((1 + \delta(\varphi, \nabla\varphi)) \nabla q \right).$$

This is not a trivial matter as $|\nabla\varphi|^2$ appears under the time derivative, and thus compactness with respect to the strong topologies for $\nabla\varphi$ has to be derived beforehand. The appearance of $|\nabla\varphi|^2$ under the time derivative comes from the fact that we used

$$\int_\Omega \gamma(c^\Gamma(q)) \left(\frac{\varepsilon}{2}|\nabla\varphi|^2 + \frac{1}{\varepsilon}\psi(\varphi)\right) \tag{10.13}$$

as an approximation to the interfacial surfactant energy. An alternative is to model the interfacial surfactant energy with the help of the functional

$$\int_\Omega \left(\frac{\varepsilon}{2}|\nabla\varphi|^2 + \frac{d(q)}{\varepsilon}\psi(\varphi)\right), \text{ with } d(q) := h(q) - h'(q)q, \ h(q) := (\tilde{\sigma}(q))^2, \tag{10.14}$$

i.e., h is the square of the surface tension $\tilde{\sigma}$ and d is the Legendre transform of the square of the surface tension. The difference between the original approximation (10.13) and the alternate approximation (10.14) is that there are no functions involving q that are multiplied with $|\nabla\varphi|^2$. Heuristically, we have transferred the interfacial surfactant energy from the gradient part all onto the potential part. It turns out that the correct prefactor in front of the potential part is the Legendre transform of the square of the surface tension if we want to recover the appropriate sharp interface model. Consequently, following the derivation in [1, 17], we obtain the model (denoted as Model D hereafter)

$$\operatorname{div} \mathbf{v} = 0, \tag{10.15a}$$

$$\partial_t(\rho\mathbf{v}) + \operatorname{div}(\rho\mathbf{v}\otimes\mathbf{v}) = \operatorname{div}\left(-pId + 2\eta\mathbf{Dv} + \mathbf{v}\otimes\frac{\tilde{\rho}_+ - \tilde{\rho}_-}{2}m(\varphi)\nabla\mu\right) \tag{10.15b}$$

$$+ \operatorname{div}\left(\left(\frac{\varepsilon}{2}|\nabla\varphi|^2 + \frac{(\tilde{\sigma}(q))^2}{\varepsilon}\psi(\varphi)\right)Id - \varepsilon\nabla\varphi\otimes\nabla\varphi\right),$$

$$\partial_t^\bullet\varphi = \operatorname{div}(m(\varphi)\nabla\mu), \tag{10.15c}$$

$$\mu - \frac{(\tilde{\sigma}(q))^2}{\varepsilon}\psi'(\varphi) = -\varepsilon\Delta\varphi + \sum \xi'_\pm(\varphi)(G_\pm(c_\pm(q)) - qc_\pm(q)), \tag{10.15d}$$

$$\partial_t^\bullet\left(\frac{2}{\varepsilon}\psi(\varphi)\tilde{\sigma}(q)c^\Gamma(q) + \xi_-(\varphi)c_-(q) + \xi_+(\varphi)c_+(q)\right) \tag{10.15e}$$

$$= \operatorname{div}\left(\left(M_c^-\xi_-(\varphi) + M_c^+\xi_+(\varphi) + \frac{2}{\varepsilon}M_\Gamma\tilde{\sigma}(q)\psi(\varphi)\right)\nabla q\right).$$

Let us point out the main differences between (10.15) and Model C. For the equation involving the chemical potential μ, the surface tension is only paired with $\psi'(\varphi)$. This is also the case in the momentum equation. Meanwhile, in the surfactant equation, the prefactor $\delta(\varphi, \nabla\varphi) = \frac{\varepsilon}{2}|\nabla\varphi|^2 + \frac{1}{\varepsilon}\psi(\varphi)$ is replaced by $\frac{2}{\varepsilon}\psi(\varphi)\tilde{\sigma}(q)$.

Unlike in the previous models where the surface tension appears as a common factor in both the gradient term and the potential term, in this new model, we have transferred the prefactors all onto the potential term. As we will discuss in Sect. 10.3, this causes the interfacial thickness to depend on the chemical potential q [specifically see (10.17)]. We point out that a similar situation also occurs when the interfacial energy depends on the orientation of the interface, see Garcke, Nestler and Stoth [16] for example.

Under suitable boundary conditions, Model D (10.15) admits the following energy identity

$$0 = \frac{d}{dt} \int_{\Omega} \left(\frac{\rho(\varphi)}{2} |\mathbf{v}|^2 + \frac{\varepsilon}{2} |\nabla \varphi|^2 + \frac{d(q)}{\varepsilon} \psi(\varphi) + \sum \xi_{\pm}(\varphi) G_{\pm}(c_{\pm}(q)) \right)$$

$$+ \int_{\Omega} \left(2\eta(\varphi) |\mathbf{D}\mathbf{v}|^2 + m(\varphi) |\nabla \mu|^2 + \left(\sum M_c^{\pm} \xi_{\pm}(\varphi) + \frac{2}{\varepsilon} M_{\Gamma} \tilde{\sigma}(q) \psi(\varphi) \right) |\nabla q|^2 \right),$$

where $d(q)$ is defined in (10.14). The above energy identity will be useful to show the existence of weak solutions to a simplified version of (10.15) in Sect. 10.4 below. For numerical computations based on the models discussed in this section, we refer to [5, 17].

As we have introduced several variants of diffuse interface models for surfactant dynamics in this section, we now summarise the distinction between the models. All of the models mentioned above allow the surfactants to be soluble in both fluid phases. The cases of insoluble surfactants and solubility in only one phase can be obtained by setting certain variables to zero and neglecting the appropriate equations. The distinction between Models A, B, C of [17] and Model D presented above is whether the adsorption dynamics is instantaneous or non-instantaneous. Model A deals with the case where the dynamics are non-instantaneous in both phases, and Model B corresponds to the case where the dynamics is instantaneous in one phase and is non-instantaneous in the other. Meanwhile both Model C and Model D are used for the case where the dynamics are instantaneous in both phases.

10.3   Sharp Interface Limit

The sharp interface limit of diffuse interface models can be derived with the method of formally matched asymptotic expansions, which is described in detail in [1, 14, 17]. In this section, we derive the sharp interface limit for Model D. The procedure is similar to the one performed for Model C in [17], and so, in the following we only give a brief overview of the analysis for the surfactant equation (10.15e). We make the following assumptions:

$$\tilde{\sigma}(s) > 0 \quad \forall s \in \mathbb{R}, \quad \psi(\pm 1) = \psi'(\pm 1) = 0, \quad \psi(s) > 0 \quad \forall s \neq \pm 1,$$

$$\xi_{-}(1) = 0, \ \xi_{-}(-1) = 1, \ \xi_{+} = 1 - \xi_{-}, \quad m(\varphi) = \varepsilon m_0,$$

where $m_0 > 0$ is a fixed constant. The idea of the method is as follows: We assume that for small ε, the domain Ω can be divided into two open subdomains $\Omega_\pm(t; \varepsilon)$ at each time, separated by an interface $\Gamma(t; \varepsilon)$. There exists a family of solutions $(\varphi^\varepsilon, \mu^\varepsilon, \mathbf{v}^\varepsilon, p^\varepsilon, q^\varepsilon)$ to (10.15), sufficiently smooth and indexed by ε such that the solutions have asymptotic expansions in ε in the bulk regions (away from $\Gamma(t; \varepsilon)$ which are denoted as outer expansions) and another set of expansions in the interfacial regions (close to $\Gamma(t; \varepsilon)$ which are denoted as inner expansions). The idea is to analyse these expansions order by order in suitable transition regions where they should match up.

Heuristically, the appearance of $\frac{1}{\varepsilon}\psi(\varphi^\varepsilon)$ in (10.14), along with the fact that $\psi(s) = 0$ only when $s = \pm 1$ forces the order parameter φ^ε to converge to a function φ_0 that take values $\{\pm 1\}$ almost everywhere in the limit $\varepsilon \to 0$. This allows us to define the domains $\Omega_- := \{\varphi_0(x) = -1\}$ and $\Omega_+ := \{\varphi_0(x) = 1\}$ as the bulk fluid regions. Using the properties of $\xi_\pm$, we obtain by substituting $\varphi = \pm 1$ into (10.15e):

$$\partial_t^\bullet c_\pm(q) = \mathrm{div}\left(M_c^\pm \nabla q\right) \text{ in } \Omega_\pm.$$

The most difficult part is to derive the interfacial equation (10.2f) from (10.15e). We assume that the zero level sets of the order parameter converge to some limiting hypersurface Γ as $\varepsilon \to 0$. Then, we introduce a new variable $z = d(x,t)/\varepsilon$, where $d(x,t)$ denotes the signed distance function to Γ, and consider a change of coordinates. For points (x,t) close to Γ, functions $u(x,t)$ will be expressed as $U(t,s,z)$ in the new coordinate system, where s denotes the tangential spatial coordinates on Γ. Then, we assume that each variable in the model (10.15) has a power series expansion of the form:

$$u^\varepsilon(x,t) = U_0(t,s,z) + \varepsilon U_1(t,s,z) + \varepsilon^2 U_2(t,s,z) + \dots,$$

for $u \in \{\varphi, \mu, \mathbf{v}, p, q\}$ and $U \in \{\Phi, \varXi, \mathbf{V}, P, Q\}$. Employing a similar expansion for the flux

$$\mathbf{J} := \left(M_c^+ \xi_+(\varphi) + M_c^- \xi_-(\varphi) + \frac{2}{\varepsilon} M_\Gamma \tilde{\sigma}(q)\psi(q)\right)\nabla q$$

as considered in [17], we obtain from (10.15e) to leading order $\partial_z Q_0 = 0$ and to first order $\partial_z Q_1 = 0$. Meanwhile, to leading order, Eq. (10.15d) yields

$$\partial_{zz}\Phi_0 - (\tilde{\sigma}(Q_0))^2 \psi'(\Phi_0) = 0. \tag{10.16}$$

We consider the function $\phi(z)$ satisfying

$$\phi''(z) = \psi'(\phi(z)), \quad \lim_{z \to \pm\infty} \phi(z) = \pm 1, \quad \phi(0) = 0.$$

For the double-well potential $\psi(s) = \frac{1}{4}(1-s^2)^2$, the solution is $\phi(z) = \tanh(z/\sqrt{2})$. We now set

$$\Phi_0(t,s,z) = \phi(\tilde{\sigma}(Q_0(t,s))z). \tag{10.17}$$

A short calculation shows that Φ_0 indeed solves (10.16) with $\Phi(t,s,0) = 0$. Here we point out that, in the asymptotic analysis of Model A and Model C, Φ_0 is a function depending only on z, and so (10.16) is a new feature of Model D, which states that the interfacial thickness depends on q. Multiplying (10.16) with $\partial_z \Phi_0$, integrating over z and applying the matching conditions [14] leads to the equipartition of energy:

$$\frac{1}{2}|\partial_z\Phi_0|^2(t,s,z) = (\tilde{\sigma}(Q_0))^2(t,s)\psi(\Phi_0)(t,s,z),$$

$$\text{with } \int_{\mathbb{R}} |\partial_z\Phi_0|^2(t,s,z)dz = (\tilde{\sigma}(q_0))^2 \int_{\mathbb{R}} 2\psi(\phi(\tilde{\sigma}(q_0)z)dz = \tilde{\sigma}(q_0),$$

where for the last equality a change of variables $t \mapsto \tilde{\sigma}(q_0)z$ and the fact that ψ is rescaled so that $\int_{\mathbb{R}} 2\psi(\phi(t))dt = \int_{-1}^{1} \sqrt{2\psi(s)}ds = 1$ are used. Using (10.9), (10.11) and (10.14), we obtain the relations:

$$h'(q) = 2\tilde{\sigma}(q)\tilde{\sigma}'(q), \quad \tilde{\sigma}'(q) = \gamma'(c^\Gamma(q))(c^\Gamma)'(q) - c^\Gamma(q) - q(c^\Gamma)'(q) = -c^\Gamma(q).$$

Using (10.14), this leads to the correct formula for the total energy across the interface:

$$\int_{\mathbb{R}} \frac{1}{2}|\partial_z\Phi_0|^2 + d(q_0)\psi(\phi(\tilde{\sigma}(q_0)z))dz = \tilde{\sigma}(q_0) - \tilde{\sigma}'(q_0)q_0 = \gamma(c^\Gamma(q_0)).$$

To second order, we obtain from (10.15e) the equation

$$\partial_t^\bullet c^\Gamma(q) + c^\Gamma(q)\mathrm{div}_\Gamma\,\mathbf{v} = \mathrm{div}_\Gamma\,(M_\Gamma\nabla_\Gamma q) + [M_c\nabla q]_-^+\mathbf{v}.$$

For more details, we refer the reader to [17, §4.5.3]. Hence, the sharp interface model from Model D (10.15) is (10.2a), (10.2b), (10.2d), (10.2e) with $\tilde{\sigma}(q)$ replacing $\sigma(c^\Gamma)$, together with the surfactant system

$$\partial_t^\bullet c_\pm(q) = \mathrm{div}\,(M_c^\pm\nabla q) \quad \text{in } \Omega_\pm(t),$$

$$\partial_t^\bullet c^\Gamma(q) + c^\Gamma(q)\mathrm{div}_\Gamma\,\mathbf{v} - \mathrm{div}_\Gamma\,(M_\Gamma\nabla_\Gamma q) = [M_c\nabla q]_-^+\mathbf{v} \quad \text{on } \Gamma(t).$$

10.4   Existence Result

Setting $G_- = G_+ =: G$, $M_c^- = M_c^+ =: M_c$, and $c_- = c_+ =: c$, the surfactant equation (10.15e) can be expressed as

$$\partial_t^\bullet \left(\frac{f(q)}{\varepsilon} \psi(\varphi) + c(q) \right) = \operatorname{div} \left(M(\varphi, q) \nabla q \right), \qquad (10.18)$$

where

$$f(q) := -h'(q) = 2\tilde\sigma(q) c^\Gamma(q), \quad M(\varphi, q) := M_c + \frac{2}{\varepsilon} M_\Gamma \tilde\sigma(q) \psi(\varphi).$$

In this section, let $T > 0$ be fixed and $\Omega \subset \mathbb{R}^d$, $d = 2, 3$, be a bounded domain with sufficiently smooth boundary $\partial\Omega$. Setting $Q_T := \Omega \times (0, T)$, we provide an existence result to the following model:

$$\operatorname{div} \mathbf{v} = 0 \qquad\qquad\qquad \text{in } Q_T, \qquad (10.19a)$$

$$\partial_t(\rho\mathbf{v}) + \operatorname{div}\left(\mathbf{v} \otimes (\rho\mathbf{v} + \tilde{\mathbf{J}}) \right) = -\nabla p + \operatorname{div}\left(2\eta \mathbf{D}\mathbf{v} \right)$$
$$+ \left(\mu - \frac{h(q)}{\varepsilon} \psi'(\varphi) \right) \nabla\varphi + \frac{1}{2} R\mathbf{v} \text{ in } Q_T, \qquad (10.19b)$$

$$\partial_t^\bullet \varphi = \operatorname{div}\left(m(\varphi) \nabla\mu \right) \qquad\qquad \text{in } Q_T, \qquad (10.19c)$$

$$\mu = -\varepsilon \Delta\varphi + \frac{h(q)}{\varepsilon} \psi'(\varphi) \qquad\qquad \text{in } Q_T, \qquad (10.19d)$$

$$\partial_t^\bullet \left(\frac{f(q)}{\varepsilon} \psi(\varphi) + c(q) \right) = \operatorname{div}\left(M(\varphi, q) \nabla q \right) \qquad \text{in } Q_T, \qquad (10.19e)$$

where

$$\tilde{\mathbf{J}} := -\rho'(\varphi) m(\varphi) \nabla\mu, \quad R := -m(\varphi) \nabla\rho'(\varphi) \cdot \nabla\mu,$$

together with the initial conditions

$$\varphi(0) = \varphi_0, \quad \mathbf{v}(0) = \mathbf{v}_0,$$
$$\frac{f(q(0))}{\varepsilon} \psi(\varphi(0)) + c(q(0)) = \frac{f(q_0)}{\varepsilon} \psi(\varphi_0) + c(q_0) \quad \text{in } \Omega, \qquad (10.20)$$

and the boundary conditions

$$\mathbf{v} = \mathbf{0}, \quad \partial_n\varphi = \partial_n\mu = \partial_n q = 0 \text{ on } \partial\Omega \times (0, T), \qquad (10.21)$$

where $\partial_n f$ denotes the normal derivative of f on $\partial\Omega$.

Note that (10.19e) is exactly (10.18), and by choosing the free energy G such that $G(c(q)) = qc(q)$, the second term on the right-hand side of (10.15d) vanishes, leading to (10.19d). Furthermore, in (10.15b) we have replaced $p - \frac{h(q)}{\varepsilon}\psi(\varphi)$ by a rescaled pressure, which we call p again, and used the relation

$$\mathrm{div}\left(\frac{\varepsilon}{2}|\nabla\varphi|^2\, Id - \varepsilon\nabla\varphi \otimes \nabla\varphi\right) = -\varepsilon\Delta\varphi\nabla\varphi = \left(\mu - \frac{h(q)}{\varepsilon}\psi'(\varphi)\right)\nabla\varphi.$$

As the density is a physical quantity that is positively valued, the explicit form (10.7) for the density $\rho(\varphi)$ may become negative for certain values of φ, and in general, it is not guaranteed that the values of the order parameter will not deviate from the physical interval $[-1, 1]$. Hence, for the mathematical analysis of the models, the expression (10.7) has to be modified in such a way that $\rho(s) > 0$ for all $s \in \mathbb{R}$, and this modification leads to the appearance of the terms $\tilde{\mathbf{J}}$ and $R\mathbf{v}$ in the momentum equation (10.19b). Furthermore, in the physical interval $\varphi \in [-1, 1]$, it holds that $\rho'(\varphi) = \frac{1}{2}(\tilde{\rho}_+ - \tilde{\rho}_-)$, and thus $\tilde{\mathbf{J}} = -\frac{1}{2}(\tilde{\rho}_+ - \tilde{\rho}_-)m(\varphi)\nabla\mu$, while $R = 0$. Then, the corresponding momentum equation (10.19b) is identical to (10.15b) (with a rescaled pressure).

To state the existence results, we introduce some notation and function spaces. For $\mathbf{a}, \mathbf{b} \in \mathbb{R}^d$, the tensor product $\mathbf{a} \otimes \mathbf{b}$ is defined as $\mathbf{a} \otimes \mathbf{b} := (a_i b_j)_{i,j=1}^d$. If $\mathbf{A}, \mathbf{B} \in \mathbb{R}^{d\times d}$, then we set $\mathbf{A} : \mathbf{B} := \sum_{i,j=1}^d A_{ij}B_{ij}$. Let $\Omega \subset \mathbb{R}^d$ be a bounded domain with C^2-boundary $\partial\Omega$. For $1 \le p \le \infty$ and $k \in \mathbb{N} \cup \{0\}$, we denote by $L^p(\Omega)$ and $W^{k,p}(\Omega)$ the usual Lebesgue and Sobolev spaces equipped with the norms $||\cdot||_{L^p}$ and $||\cdot||_{W^{k,p}}$, respectively. In the case $p = 2$, we use the notation $H^k(\Omega) := W^{k,2}(\Omega)$ for $k \ge 1$, along with the norm $||\cdot||_{H^k} := ||\cdot||_{W^{k,2}}$. We denoted $C^\infty_{0,\sigma}(\Omega) := \{\mathbf{u} \in C^\infty_0(\Omega)^d : \mathrm{div}\,\mathbf{u} = 0\}$, and define $L^2_\sigma(\Omega)$ as the completion of $C^\infty_{0,\sigma}(\Omega)$ with respect to the $||\cdot||_{L^2}$ norm. Furthermore, we define the space $H^1_0(\Omega)$ as the completion of $C^\infty_0(\Omega)$ with respect to the $||\cdot||_{H^1}$ norm, and use the notation $H^2_n(\Omega) := \{f \in H^2(\Omega) : \partial_n f = 0 \text{ on } \partial\Omega\}$.

Definition 10.4.1 (Weak Solution)   Let $T \in (0, \infty)$, $\mathbf{v}_0 \in L^2_\sigma(\Omega)$, $\varphi_0 \in H^2_n(\Omega)$, and $q_0 \in L^2(\Omega)$ be given. We call $(\mathbf{v}, \varphi, \mu, q)$ a weak solution of (10.19)–(10.21) if

$$\mathbf{v} \in L^2(0, T; H^1_0(\Omega)^d) \cap L^\infty(0, T; L^2_\sigma(\Omega)), \quad q \in L^2(0, T; H^1(\Omega)) \cap L^\infty(0, T; L^2(\Omega)),$$

$$\varphi \in L^\infty(0, T; H^1(\Omega)) \cap L^2(0, T; H^2(\Omega)), \quad \mu \in L^2(0, T; H^1(\Omega)),$$

and the following equations are satisfied:

$$\int_{Q_T} -\rho(\varphi)\mathbf{v} \cdot \partial_t \mathbf{w} - (\rho(\varphi)\mathbf{v} \otimes \mathbf{v} + \mathbf{v} \otimes \tilde{\mathbf{J}}) : \nabla\mathbf{w} + 2\eta(\varphi)\mathbf{Dv} : \mathbf{Dw}\, dx\, dt \qquad (10.22)$$

$$+ \int_{Q_T} \frac{1}{2}m(\varphi)\left(\nabla\rho'(\varphi) \cdot \nabla\mu\right)\mathbf{v} \cdot \mathbf{w} - \left(\mu - \frac{h(q)}{\varepsilon}\psi'(\varphi)\right)\nabla\varphi \cdot \mathbf{w}\, dx\, dt = 0$$

for all $\mathbf{w} \in C_0^\infty(0, T; C_{0,\sigma}^\infty(\Omega))$ and

$$\int_{Q_T} M(\varphi, q)\nabla q \cdot \nabla \xi - \left(\frac{1}{\varepsilon} f(q)\psi(\varphi) + c(q)\right) \partial_t^\bullet \xi \, dx \, dt = 0, \tag{10.23}$$

$$\int_{Q_T} m(\varphi)\nabla \mu \cdot \nabla \xi - \varphi \partial_t \xi + \nabla \varphi \cdot \mathbf{v} \xi \, dx \, dt = 0, \tag{10.24}$$

$$\int_{Q_T} \mu \xi - \varepsilon \nabla \varphi \cdot \nabla \xi - \frac{h(q)}{\varepsilon} \psi'(\varphi)\xi \, dx \, dt = 0 \tag{10.25}$$

for all $\xi \in C_0^\infty(0, T; C^1(\overline{\Omega}))$. Moreover, the energy inequality

$$E(t) + \int_s^t \int_\Omega \left(M(\varphi, q)|\nabla q|^2 + m(\varphi)|\nabla \mu|^2 + 2\eta(\varphi)|\mathbf{Dv}|^2\right) dx \, d\tau \leq E(s) \tag{10.26}$$

has to hold for all $t \in [s, T)$ and almost all $s \in [0, T)$, where E is defined as

$$E(t) := \int_\Omega \frac{\rho(\varphi(t))}{2}|\mathbf{v}(t)|^2 + \frac{\varepsilon}{2}|\nabla \varphi(t)|^2 + \frac{d(q(t))}{\varepsilon}\psi(\varphi(t)) + G(c(q(t))) \, dx. \tag{10.27}$$

To obtain weak solutions to (10.19)–(10.21), we make the following assumptions:

Assumption 4.1 *We assume that $\Omega \subset \mathbb{R}^d$, $d = 2, 3$, is a bounded domain with C^2-boundary $\partial\Omega$. The assumptions on the initial data $(\mathbf{v}_0, \varphi_0, q_0)$ are as stated in Definition 10.4.1. Furthermore, we assume that*

1. *ψ, ρ, η, M, and m are smooth functions, and there exist positive constants c_0, c_1, c_2, c_3, c_4, c_5, c_6, c_7 such that, for all $s, t \in \mathbb{R}$,*

$$c_0 < \rho(s), \eta(s) < c_1, \quad |\rho'(s)| + |\rho''(s)| \leq c_2, \quad c_3 \leq M(s, t), m(s) \leq c_4,$$

$$\psi(s) \geq 0, \quad |\psi(s)| \leq c_5(|s|^3 + 1), \quad |\psi'(s)| \leq c_5(|s|^2 + 1), \quad \psi(s) \geq c_6|s| - c_7,$$

with $\rho(s) = \frac{\tilde{\rho}_+ - \tilde{\rho}_-}{2}s + \frac{\tilde{\rho}_+ + \tilde{\rho}_-}{2}$ if $s \in [-1, 1]$. If it holds that $\rho'(\varphi)$ is not constant, then there exists a positive constant c_8 and $p \in (0, 1)$ such that

$$|\psi'(s)| \leq c_8(|s|^p + 1) \quad \forall s \in \mathbb{R}.$$

2. *h is a smooth concave function and d, f are smooth functions that satisfy the relations*

$$d(s) = h(s) + f(s)s, \quad h'(s) = -f(s),$$

and there exist constants $q_{\min}, q_{\max} \in \mathbb{R}$ with $q_{\min} < q_{\max}$ such that $d(s)$ is constant for $s \notin [q_{\min}, q_{\max}]$.

3. *The function $c \in C^2(\mathbb{R})$ is strongly monotone, i.e., for some positive constant K,*

$$(c(a) - c(b))(a - b) \geq K|a - b|^2 \quad \forall a, b \in \mathbb{R}.$$

The composite function $\hat{G} := G \circ c \in C^2(\mathbb{R})$ is strictly convex and there exist positive constants c_9, c_{10} such that

$$\hat{G}'(0) = 0, \quad \hat{G}'(r) < c_9 r, \quad \hat{G}'(t) > c_9 t,$$

$$\hat{G}'(s) = sc'(s), \quad |\hat{G}(s)| \leq c_{10}(|s|^2 + 1), \quad |\hat{G}'(s)| \leq c_{10}(|s| + 1),$$

for all $s \in \mathbb{R}, r < 0, t > 0$.

We now state the existence result:

Theorem 4.1 (Existence of Weak Solutions) *Under Assumption 4.1, for any $0 < T < \infty$, there exists a weak solution $(\mathbf{v}, \varphi, \mu, q)$ to (10.19)–(10.21) in the sense of Definition 10.4.1.*

The idea of the proof is to first show the existence a weak solution $(\mathbf{v}^\delta, \varphi^\delta, \mu^\delta, q^\delta)$ to a regularized version of (10.19) with an additional $\delta \partial_t \varphi$ on the right-hand side of (10.19d), and an additional $\delta \Delta^2 \mathbf{v}$ on the left-hand side of (10.19b) for $\delta > 0$. This is achieved with an semi-implicit time discretization, where the existence of time-discrete solutions $(\mathbf{v}_N^\delta, \varphi_N^\delta, \mu_N^\delta, q_N^\delta)_{N \in \mathbb{N}}$ are established with the aid of the Leray–Schauder principle. A crucial step is to show the compactness of $\{q_N^\delta\}_{N \in \mathbb{N}}$ in $L^2(Q_T)$, which is obtained with the aid of a compactness result due to Simon [22] and (10.19e). Then, by passing to the limit $\delta \to 0$, we obtain a weak solution to (10.19)–(10.21). For more details we refer the reader to [3, 28].

References

1. Abels, H., Garcke, H., Grün, G.: Thermodynamically consistent, frame indifferent diffuse interface models for incompressible two-phase flow with different densities. Math. Models Methods Appl. Sci. **22**(3), 1150013, 40 pp. (2012)
2. Abels, H., Garcke, H., Grün, G., Metzger, S.: Diffuse interface models for incompressible two-phase flows with different densities. In: Bothe, D., Reusken, A. (eds.) Transport Processes at Fludic Interfaces. Advances in Mathematical Fluid Mechanics. Springer International Publishing AG, Cham (to appear)
3. Abels, H., Garcke, H., Weber, J.: Existence of weak solutions for a diffuse interface model for two-phase flow with surfactants (in preparation)
4. Abels, H., Lam, K.F., Stinner, B.: Analysis of the diffuse domain approach for a bulk-surface coupled PDE system. SIAM J. Math. Anal. **47**(5), 3687–3725 (2015)
5. Aland, S., Hahn, A., Kahle, C., Nürnberg, R.: Comparative simulations of Taylor-flow with surfactants based on sharp- and diffuse-interface methods. In: Bothe, D., Reusken, A. (eds.)

Transport Processes at Fludic Interfaces. Advances in Mathematical Fluid Mechanics. Springer International Publishing AG, Cham (to appear)

6. Alt, H.W.: The entropy principle for interfaces. Fluids and solids. Adv. Math. Sci. Appl. **19**(2), 585–663 (2009)

7. Barrett, J.W., Garcke, H., Nürnberg, R.: On the stable numerical approximation of two-phase flow with insoluble surfactant. ESAIM: M2AN **49**(2), 421–458 (2015)

8. Barrett, J.W., Garcke, H., Nürnberg, R.: Stable finite element approximations of two-phase flow with soluble surfactant. J. Comput. Phys. **297**, 530–564 (2015)

9. Bedeaux, D.: Nonequilibrium thermodynamics and statistical physics of surfaces. In: Prigogine, I., Rice, S.A. (eds.) Advances in Chemical Physics, vol. 64, pp. 47–109. Wiley, Hoboken, NJ (1986)

10. Bothe, D., Prüss, J., Simonett, G.: Well-posedness of a two-phase flow with soluble surfactant. In: Brezis, H., Chipot, M., Escher, J. (eds.) Nonlinear Elliptic and Parabolic problems, Progress in Nonlinear Differential Equations and Their Applications, vol. 64, pp. 37–61. Springer, New York (2005)

11. Burger, M., Elvetun, O.L., Schlottbom, M.: Analysis of the diffuse domain method for second order elliptic boundary value problems. Found. Comput. Math. 1–48 (2015). doi:10.1007/s10208-015-9292-6

12. Diamant, H., Andelman, D.: Kinetics of surfactant adsorption at fluid–fluid interfaces. J. Phys. Chem. **100**, 13732–13742 (1996)

13. Eastoe, J., Dalton, J.S.: Dynamic surface tension and adsorption mechanisms of surfactants at the air–water interface. Adv. Colloid Interface Sci. **85**, 103–144 (2000)

14. Garcke, H., Stinner, B.: Second order phase field asymptotics for multi-component systems. Interfaces Free Bound. **8**(2), 131–157 (2006)

15. Garcke, H., Wieland, S.: Surfactant spreading on thin viscous films: Nonnegative solutions of a coupled degenerate system. SIAM J. Math. Anal. **37**(6), 2025–2048 (2006)

16. Garcke, H., Nestler, B., Stoth, B.: On anisotropic order parameter models for multi-phase systems and their sharp interface limits. Phys. D **115**(1–2), 87–108 (1998)

17. Garcke, H., Lam, K.F., Stinner, B.: Diffuse interface modelling of soluble surfactants in two-phase flow. Commun. Math. Sci. **12**(8), 1475–1522 (2014)

18. Kralchevsky, P.A., Danov, K.D., Denkov, N.D.: Chemical physics of colloid systems and interfaces. In: Birdi, K.S. (ed.) Handbook of Surface and Colloid Chemistry, 3rd edn., pp. 199–355. CRC Press, Boca Raton, FL (2008)

19. Lam, K.F.: Diffuse interface models of soluble surfactants in two-phase fluid flows. PhD thesis, University of Warwick (2014)

20. Li, X., Lowengrub, J., Rätz, A., Voigt, A.: Solving PDEs in complex geometries: a diffuse domain approach. Commun. Math. Sci. **7**(1), 81–107 (2009)

21. Modica, L.: The gradient theory of phase transitions and the minimal interface criterion. Arch. Ration. Mech. Anal. **98**(2), 123–142 (1987)

22. Simon, J.: Compact sets in space $L^p(0, T; B)$. Ann. Mat. Pura Appl. **146**(1), 65–96 (1986)

23. Slattery, J.C., Sagis, L., Oh, E.-S.: Interfacial Transport Phenomena, 2nd edn. Springer, New York (2007)

24. Teigen, K.E., Li, X., Lowengrub, J., Wang, F., Voigt, A.: A diffuse-interface approach for modeling transport, diffusion and adsorption/desorption of material quantities on a deformable interface. Commun. Math. Sci. **7**(4), 1009–1037 (2009)

25. Teigen, K.E., Song, P., Lowengrub, J., Voigt, A.: A diffuse-interface method for two-phase flows with soluble surfactants. J. Comput. Phys. **230**, 375–393 (2011)

26. Wang, Y., Oberlack, M.: A thermodynamic model of multiphase flows with moving interfaces and contact line. Continuum Mech. Thermodyn. **23**, 409–433 (2011)

27. Ward, A.F.H., Tordai, L.: Time dependence of boundary tensions of solutions I. The role of diffusion in time effects. J. Chem. Phys. **14**(7), 453–461 (1946)

28. Weber, J.T.: Analysis of diffuse interface models for two-phase flows with and without surfactants. PhD thesis, Universität Regensburg (2016)

Chapter 11
Phase Field Models for Two-Phase Flow with Surfactants and Biomembranes

Sebastian Aland

Abstract We give an overview on recent developments of phase field models for two-phase flows with surfactants and lipid bilayer membranes. Starting from the two-phase flow model of a clean fluid-fluid interface we discuss the time discretization and boundary conditions for dynamic and static contact angles. Using the adsorption models of Henry and Langmuir, soluble surfactants are included in the diffuse interface formulation. To consider lipid bilayer membranes the model is extended by membrane bending stiffness and membrane inextensibility. We present phase field models to include these elastic effects, with a particular focus on the inextensibility constraint for which we discuss different phase field variants from the literature and present numerical tests.

11.1 Introduction

A large collection of fluid problems involve moving interfaces. The simplest examples include everyday phenomena, such as when pouring water in an empty glass filled with ambient air. While the dynamics at the interface between water and air are mostly governed by surface tension, things become more subtle if molecules of an additional chemical species are present at the interface. Such species can range from single surfactant molecules that locally lower the surface tension, to dense structures of elastic lipid molecules that completely isolate the two fluids from each other.

Many numerical techniques are available to handle such two-phase flow problems with additional interfacial particles [48, 49, 59, 60]. One of the simplest and most flexible among these is the phase field (or diffuse interface) method. The method has a sound physical background and can easily handle topological changes of the fluid phases, as well as contact of the fluids with a solid body (or boundary).

In this chapter, we give an overview on recent developments of phase field models for two-phase flow with interfacial particles, be it surfactants or lipid

S. Aland (✉)
Institute of Scientific Computing, TU Dresden, 01062 Dresden, Germany
e-mail: sebastian.aland@tu-dresden.de

© Springer International Publishing AG 2017
D. Bothe, A. Reusken (eds.), *Transport Processes at Fluidic Interfaces*,
Advances in Mathematical Fluid Mechanics, DOI 10.1007/978-3-319-56602-3_11

molecules. Both cases can be handled quite similar by additional surface and bulk equations and their coupling to the flow dynamics. We start in Sect. 11.2 with the two phase flow model of a clean fluid-fluid interface and discuss the time discretization and boundary conditions for dynamic and static contact angles. Then, we include soluble surfactants and focus on the adsorption models of Henry and Langmuir. In Sect. 11.3 we switch to fluidic membranes where surfactants are replaced by a dense layer of lipid molecules. These molecules give rise to elastic properties, in particular bending stiffness and membrane inextensibility. We present phase field models to include these elastic effects, with a particular focus on the inextensibility constraint. We present different modeling approaches to include this constraint in phase field models and compare them in numerical tests.

11.2 The Diffuse Interface Model for Two Phase Flow with Surfactants

We consider isothermal, incompressible flow of two immiscible fluids. The interface between both fluids is usually considered as a free boundary that evolves in time. Instead of explicitly tracking the fluid-fluid interface, the phase field method uses an auxiliary function (the phase field) and the interface position is implicitly described by a level-set of this function. As opposed to the level-set method, the phase field ϕ takes distinct values in each of the fluid phases (e.g. $\phi = -1$ and $\phi = 1$) with a smooth transition in between, around the interface. Hence, the interface is diffuse with a finite width and an intermediate level set of the phase field (e.g. $\phi = 0$) may be used to get a discrete interface location. Figure 11.1 illustrates the corresponding sharp and diffuse interface settings.

A fine computational grid is needed to numerically resolve the transition layer of thickness ε, which makes phase field computations often more expensive than other interface capturing methods. On the other hand, phase field methods offer

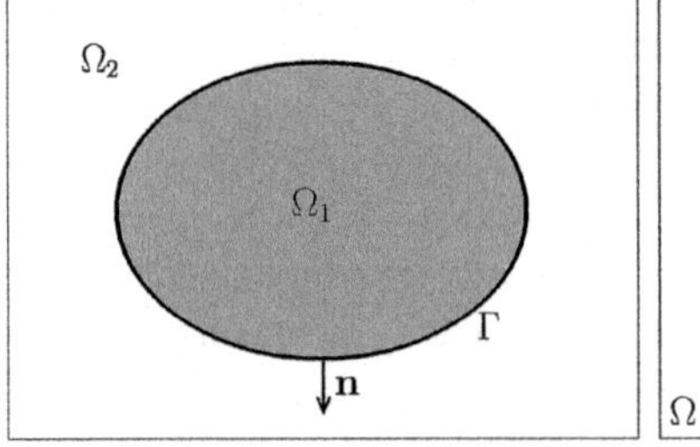

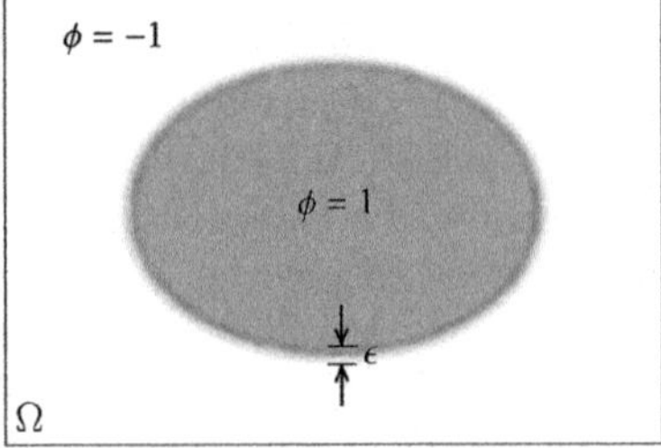

Fig. 11.1 Comparison of a sharp interface model (*left*) with a diffuse interface model (*right*). In the diffuse interface model the domains Ω_1, Ω_2 are implicitly defined by regions where $\phi \approx 1$, $\phi \approx -1$, respectively. The interface Γ is smeared out to an interface region of finite thickness ε, where $-1 < \phi < 1$

many advantages in the mathematical modeling. One distinct advantage is their simplicity. As opposed to Level-Set methods, mass is conserved and reinitialization or convection stabilization are not necessary for the interface advection. Phase field methods naturally admit physical energy laws which makes it possible to develop energy stable discretization schemes. Also topological changes, such as during vesicle fission and fusion, are naturally captured and additional physical processes can be easily included within the phase field formulation.

11.2.1 Phase Field Equations

Given a sharp interface Γ, the surface energy of a fluid-fluid interface can be written as

$$E_s = \int_\Gamma \sigma \mathrm{dA}, \tag{11.1}$$

where σ is the surface tension. Van der Waals was the first who realized that real interfaces are not sharp but diffuse with a finite thickness [53]. Based on physical arguments, Cahn and Hilliard [16] formulated the diffuse interface version of a fluid-fluid interface,

$$\tilde{E}_s = \int_\Omega \tilde{\sigma} \left(\frac{\varepsilon}{2} |\nabla \phi|^2 + \frac{1}{\varepsilon} W(\phi) \right) \mathrm{dV}, \tag{11.2}$$

where $\tilde{\sigma} = 3\sigma/2\sqrt{2}$ is a scaled surface tension. The scaling factor depends on the chosen double-well potential W, which is here $W(\phi) = \frac{1}{4}(\phi^2 - 1)^2$.

Hohenberg and Halperin [36] were the first to couple the idea of a diffuse interface to flow dynamics. The simplest form of the resulting Navier-Stokes-Cahn-Hilliard equations is,

$$\rho \partial_t^\bullet \mathbf{v} + \nabla p - \nabla \cdot \left(\eta (\nabla \mathbf{v} + (\nabla \mathbf{v})^T) \right) = -\varepsilon \nabla \cdot (\tilde{\sigma} \nabla \phi \otimes \nabla \phi) + \mathbf{F}, \tag{11.3}$$

$$\nabla \cdot \mathbf{v} = 0, \tag{11.4}$$

$$d_t^\bullet \phi = \nabla \cdot (M(\phi) \nabla \mu), \tag{11.5}$$

$$\mu = \frac{1}{\varepsilon} W'(\phi) - \varepsilon \Delta \phi. \tag{11.6}$$

Here $\mathbf{v}, p, \mu$ and $\mathbf{F}$ are the (volume-averaged) velocity, pressure, chemical potential and body force, respectively. The parameter ε defines the interface thickness, $\partial_t^\bullet = \partial_t + \mathbf{v} \cdot \nabla$ is the material derivative. The term $-\varepsilon \nabla \cdot (\tilde{\sigma} \nabla \phi \otimes \nabla \phi)$ provides the surface tension force, many other forms of this term are possible, see, e.g., [40].

The density $\rho(\phi)$ and viscosity $\eta(\phi)$ are interpolations of the corresponding values in the two fluids, e.g.

$$\rho(\phi) = \rho_1(1 + \phi)/2 + \rho_2(1 - \phi)/2, \qquad \eta(\phi) = \eta_1(1 + \phi)/2 + \eta_2(1 - \phi)/2 \tag{11.7}$$

The Mobility $M(\phi)$ The function $M(\phi)$ is a mobility. In practice, M is often either a constant or a double well potential similar to W. In general for the diffuse interface fluid method, it is desirable to keep M small such that the interface position is primarily advected. At the same time the mobility needs to be big enough to ensure that the interface profile stays accurately modeled and the interface thickness is approximately constant. Asymptotic analysis [1] and numerical benchmarks [6] show that this can be achieved by taking $M = \mathcal{O}(\varepsilon)$.

Non-matched Densities For matched densities, i.e., $\rho_1 = \rho_2$, the above system is thermodynamically consistent, i.e., there exist physical energy estimates. This property is not maintained for non-matched densities ($\rho_1 \neq \rho_2$) which has led to different extensions of the model [1, 15, 45]. The model described in [1] involves an additional term in the momentum equation. The model in [45], defines $\mathbf{v}$ as the mass-average of the two fluid velocities, which leads to a relaxed incompressibility condition around the interface. However, in many practical applications, the physical interface thickness is very small compared to the domain size. In this case the diffuse interface model aims to approximate the sharp interface solution and the interface thickness ε needs to be chosen very small. For small ε the error introduced by the diffuse interface itself becomes larger than the differences between the various diffuse interface models [6]. In this case, the different diffuse-interface models provide equally good approximations to the sharp-interface solution.

11.2.2 Boundary Conditions

The boundary conditions to the above defined system are particularly interesting when $\partial\Omega$ is a solid boundary. In this case the fluid-fluid interface may touch the solid wall which leads to a moving contact line problem [5]. The Generalized-Navier-Boundary-Condition (GNBC) holds for the fluid velocity [50]

$$\left[l(\phi)(\mathbf{v} - \mathbf{v}_{\text{wall}}) + \eta(\phi)(\nabla\mathbf{v} + \nabla\mathbf{v}^T) \cdot \mathbf{n}_\Omega - L(\phi)\nabla\phi \right] \times \mathbf{n}_\Omega = 0. \tag{11.8}$$

Here $l(\phi)$ is the phase-dependent inverse of the slip length, $\mathbf{v}_{\text{wall}}$ is the wall velocity, $\mathbf{n}_\Omega$ is the outer normal to $\partial\Omega$. The function $L(\phi)$ is defined as

$$L(\phi) = \frac{3\sigma}{2\sqrt{2}}\varepsilon\,\mathbf{n}_\Omega \cdot \nabla\phi - \gamma'(\phi), \qquad (11.9)$$

where $\gamma'(\phi)$ is the first derivative of the fluid-wall energy potential

$$\gamma(\phi) = \frac{\sigma}{2}\cos(\theta)\sin(\frac{\pi}{2}\phi), \qquad (11.10)$$

with the equilibrium contact angle θ. Note, that γ describes the energy density associated with a fluid touching the wall. In particular, the difference in γ between the two phases is $\gamma(1) - \gamma(-1) = \sigma\cos(\theta)$, which is the difference in the wall energies of fluid 1 and fluid 2, respectively, due to Young's law, $\sigma\cos(\theta) = \sigma_1 - \sigma_2$. The different wall potential $\gamma(\phi) = \sigma\,\cos(\theta)\,(3\phi - \phi^3)/4$ has been used in [7, 56], which leads to equal static contact angles for all level-sets of the phase field [2].

Additionally, the movement of the contact line is determined by the dynamic contact line boundary condition [37, 50]

$$\partial_t\phi + \mathbf{v}\cdot\nabla\phi = -\beta L(\phi) \qquad \text{(dynamic angle BC)}, \qquad (11.11)$$

for a relaxation parameter β. Hence, the actual contact angle is dynamic and relaxes to the equilibrium angle θ with a speed controlled by β. If this relaxation is very fast, i.e. $\beta \to \infty$, Eq. (11.11) reduces to $L(\phi) = 0$ which implies the following two boundary conditions from Eqs. (11.8), (11.9):

$$\left[l(\phi)(\mathbf{v} - \mathbf{v}_{\text{wall}}) + \eta(\phi)(\nabla\mathbf{v} + \nabla\mathbf{v}^T)\cdot\mathbf{n}_\Omega\right] \times \mathbf{n}_\Omega = 0 \qquad \text{(Navier BC)}, $$

$$(11.12)$$

$$\frac{3\sigma}{2\sqrt{2}}\varepsilon\mathbf{n}_\Omega \cdot \nabla\phi = \gamma'(\phi) \qquad \text{(static angle BC)}. $$

$$(11.13)$$

If additionally the slip length is close to zero, i.e., $l(\phi) \to \infty$, the Navier condition reduces to the common no slip condition

$$\mathbf{v} = \mathbf{v}_{\text{wall}} \qquad \text{(no slip BC)}. \qquad (11.14)$$

To close the system and to ensure conservation of fluid masses, a no flux condition is specified for the chemical potential,

$$\mathbf{n}_\Omega \cdot \nabla\mu = 0. \qquad (11.15)$$

11.2.3 Numerical Discretization

Most numerical methods for the Navier-Stokes-Cahn-Hilliard system (11.3)–(11.6) focus on the no slip boundary condition with no contact lines or a static 90° contact angle, i.e., $\mathbf{n}_\Omega \cdot \nabla\phi = 0$. In this case a lot of different discretization techniques are available for the individual subsystems, see, for example, [38, 51] for Navier-Stokes and [14, 31, 58] for the Cahn-Hilliard system.

Coupling Between Navier-Stokes and Cahn-Hilliard Only in recent years, efficient discretization techniques have been proposed for the coupled Navier-Stokes-Cahn-Hillard system [26, 39]. For small capillary numbers, the surface tension force introduces a strong coupling between the Navier-Stokes equation providing the flow field and the Cahn-Hilliard equation evolving the phase field. In [3] it was shown by numerical tests, that an explicit coupling of these two problems can result in a time step restriction of the form

$$dt_{\max} \lesssim \varepsilon\rho^{2/3}M^{1/3}\sigma^{-1/3}, \tag{11.16}$$

where dt is the time step and M and ρ are assumed to be constant. In contrast to other numerical models for two-phase flow, this time step restriction is independent of the grid size but strongly dependent on ε. It is suspected that this dependence is a consequence of the smallest capillary wavelength that can be resolved in a diffuse interface model. Given the grid resolution is high enough to resolve the diffuse interface properly, this wavelength is proportional to the interface thickness and independent of the grid size. Special techniques have been introduced in [3] to lift the time step constraint, including additional stabilizing terms as well as monolithic coupling of both subproblems. Apart from increasing the computational performance, such improved time integration schemes allow to choose lower mobility, which reduces the non-physical interface movement due to the Cahn-Hilliard dynamics.

Energy-Stable Schemes One advantage of phase field models is that they admit physical energy laws which makes it possible to develop energy stable discretization schemes. Energy stable schemes for the fully discrete Navier-Stokes-Cahn-Hilliard system have been proposed, both in linear form [28] and nonlinear form [29, 30]. Liu et al. [44] also presented an energy stable scheme where the Navier-Stokes and Cahn-Hilliard system are completely decoupled during the solution process. For a modified energy law, even second order convergence in time could be established [32].

Energy stable schemes for the moving contact line problem with the Generalized-Navier-Boundary-Condition have been proposed in [33]. A linear scheme was developed in [5] which enables a very robust unconditionally energy-stable approximation of the Navier-Stokes-Cahn-Hilliard system.

11.2.4 Inclusion of Surfactants

Surfactants are amphiphilic molecules that tend to adsorb at fluid-fluid interfaces. Their presence lowers the surface tension locally and thus has a significant effect on the flow dynamics. A relation between the surfactant concentrations and the local surface tension is given by Gibbs adsorption equation. Let c_1, c_2, c_Γ the surfactant concentrations in phase 1, phase 2 and on the surface, respectively. Gibbs adsorption equation states that

$$\frac{\partial \sigma}{\partial c_1} = -RT \frac{c_\Gamma}{c_1}, \tag{11.17}$$

and analogously for c_2, where RT is gas constant times temperature. The functional dependence between c_Γ and c_1, c_2 is also called isotherm and depends on the underlying assumptions of the considered physical system. The two most important isotherms are the laws of Henry and Langmuir, discussed in the next sections.

The evolution of surfactant concentrations c_1, c_2 and c_Γ can be determined by additional equations in the bulk phases and on the surface. Phase field modeling provides an easy way to account for such equations by use of the characteristic functions χ_1, χ_2 and δ indicating phase 1, phase 2 and the surface, respectively. A popular choice to approximate these functions in the diffuse interface context is

$$\chi_1 = \frac{1+\phi}{2}, \qquad \chi_2 = \frac{1-\phi}{2}, \qquad \delta = \frac{1}{2}|\nabla\phi|. \tag{11.18}$$

By use of these characteristic functions, equations given in one of the bulk domains or on the surface, can be extended to the full computational domain Ω which greatly simplifies their discretization. To illustrate this, let us consider a surface advection equation

$$\partial_t c_\Gamma + \nabla_\Gamma \cdot (\mathbf{v} c_\Gamma) = 0 \qquad \text{on } \Gamma, \tag{11.19}$$

which describes mass conservation of the concentration c_Γ on the surface. Now multiply with a test function ψ and integrate over Γ to obtain the weak form of this equation. Subsequently, we can conclude

$$\int_\Gamma \partial_t c_\Gamma \psi + \nabla_\Gamma \cdot (\mathbf{v} c_\Gamma)\psi = d_t \int_\Gamma c_\Gamma \psi - \int_\Gamma c_\Gamma \partial_t^\bullet \psi \qquad \text{(Leibnitz rule)}$$

$$= d_t \int_\Omega c_\Gamma \delta \psi - \int_\Omega c_\Gamma \delta \partial_t^\bullet \psi \qquad \text{(delta function)}$$

$$= \int_\Omega \partial_t (c_\Gamma \delta)\psi - \int_\Omega c_\Gamma \delta \mathbf{v} \cdot \nabla\psi \qquad \text{(since } \partial_t \psi = 0\text{)}$$

$$= \int_\Omega \partial_t (c_\Gamma \delta)\psi + \nabla \cdot (c_\Gamma \delta \mathbf{v})\psi.$$

The strong form of the latter expression is,

$$\partial_t(\delta c) + \nabla \cdot (\delta c \mathbf{v}) = 0 \qquad \text{in } \Omega. \qquad (11.20)$$

Hence, Eq. (11.20) provides an extended form of the surface advection equation (11.19) that is now valid in the whole domain Ω.

Similar reformulations can be done for a large variety of bulk and surface equations, see [42, 52] for examples and justification by matched asymptotic analysis. In the following, we will focus on such diffuse interface formulations for the mass balance equations for the surfactants. The detailed form of these equations depends on the chosen isotherm. We refer to [27] for an overview of different adsorption isotherms and the corresponding diffuse interface formulations for two-phase flow with soluble surfactants.

11.2.4.1 The Henry Model

Based on experimental findings the Henry model assumes that the amount of molecules that are adsorbed at an interface is proportional to the concentration of these molecules in the ambient bulk phase [35]. This condition, also called Henry's law, means mathematically that the concentrations in the bulk phases, c_1, c_2, and on the interface, c_Γ, are related by the Henry constants H, H_Γ as follows:

$$c_2 = Hc_1, \qquad c_\Gamma = H_\Gamma c_1 \qquad \text{on } \Gamma. \qquad (11.21)$$

This condition allows to define a single continuous field $c : \Omega \to \mathbb{R}_{\geq 0}$ describing the surfactant concentration in the whole domain by

$$c_1 = c_{|\bar{\Omega}_1}, \qquad c_2 = Hc_{|\bar{\Omega}_2}, \qquad c_\Gamma = H_\Gamma c_{|\Gamma}, \qquad (11.22)$$

Consequently, the total surfactant mass can be expressed as $\int_\Omega [\chi_1 + \chi_2 H + \delta H_\Gamma]\, c\, dV$ and one can derive the mass balance equation,

$$\partial_t^\bullet ([\chi_1 + \chi_2 H + \delta H_\Gamma] c) - \nabla \cdot ([\chi_1 D_1 + \chi_2 H D_2 + \delta H_\Gamma D_\Gamma] \nabla c) = 0, \qquad (11.23)$$

where D_1, D_2, D_Γ are the diffusion constants in the corresponding phases. Approximating the characteristic functions by Eq. (11.18) allows to handle the surfactant concentrations in the two phases and on the interface by solving a single equation.

Henry's law together with Gibb's isotherm (11.17) yields the surfactant-dependent surface tension

$$\sigma = \sigma_0 - RT\, H_\Gamma c. \qquad (11.24)$$

Hence, Eqs. (11.3)–(11.6), (11.23)–(11.24) form a coupled system for two-phase flow with soluble surfactants.

The underlying assumptions restrict the use of Henry's law to situations of very low surface concentrations. However, due to its simplicity the model is still widely used and, even if further simplified, can explain many complicated flow phenomena. For example, in [54] it is assumed that surfactants do not dissolve in phase 2 ($H = 0$) and that interfacial adsorption is very low ($H_\Gamma \approx 0$). Still, the model was able to explain and qualitatively describe the surprising relaxation oscillations that occur at droplet surfaces exposed to an ambient fluid with a surfactant gradient.

11.2.4.2 Langmuir Model

The Langmuir model is probably the most popular adsorption isotherm. It can be derived from statistical physics and provides not only a stationary isotherm but also a rate of exchange between surfactant molecules in the bulk and on the surface. Accordingly, the flux from phase i to the surface is defined by

$$j_i := a_i c_i (1 - c_\Gamma / c_\infty) - d_i c_\Gamma \qquad\qquad i = 1, 2. \qquad\qquad (11.25)$$

Here, a_i, d_i are adsorption and desorption rates and the saturation constant c_∞ defines the concentration of maximal interface coverage. The mass balance equations for the phase-dependent surfactant concentrations become,

$$\partial_t^\bullet (\chi_i c_i) - D_i \nabla \cdot (\chi_i \nabla c_i) = -\delta j_i \qquad\qquad i = 1, 2, \qquad\qquad (11.26)$$

$$\partial_t^\bullet (\delta c_\Gamma) - D_\Gamma \nabla \cdot (\delta \nabla c_\Gamma) = \delta (j_1 + j_2). \qquad\qquad (11.27)$$

Evaluating the Gibbs isotherm (11.17) at equilibrium ($j_1 = j_2 = 0$) yields the surfactant-dependent surface tension

$$\sigma = \sigma_0 + RT c_\infty \ln(1 - c_\Gamma / c_\infty). \qquad\qquad (11.28)$$

Hence, Eqs. (11.3)–(11.6), (11.25)–(11.28) compose a diffuse interface model for two-phase flows with soluble surfactants based on the Langmuir isotherm.

In reality, also the Langmuir isotherm is limited by some underlying assumptions. In particular it accounts only for a monolayer of interfacial surfactants which is often doubtful as frequently more molecules adsorb to the monolayer. However, despite its limitations, Langmuir's isotherm is often the first choice for many practical applications.

The model has been used to describe soluble surfactants on droplets [55] and for Taylor bubbles, see the corresponding Chapter of this book [10]. An improved thermodynamically consistent form has been proposed and tested in [27]. In [8] the model was adapted for elastic interfacial nano-particles. There, c defines a particle number density instead of a concentration and elastic particle interactions give rise to additional surface forces.

11.3   Phase Field Models for Fluidic Elastic Membranes

Fluid membranes are one of the essential building blocks of biological cells. They separate the interior of all cells from the outside environment and divide intracellular regions into different functional compartments. The membrane itself consists of a dense bilayer of lipid molecules (and other proteins) and is mostly impermeable for fluids, see Fig. 11.2.

Although, the mathematical description by a phase field is identical to previous case of two-phase flow with surfactants, the dense packing of the lipids along the membrane gives now rise to elastic properties. Assuming the membrane to behave like a thin elastic shell leads to two essential elastic contributions: the membranes tendency to assume a preferred curvature (bending stiffness) and its tendency to locally conserve its surface area (inextensibility). In the following we will discuss how these constraints can be included in phase field models.

11.3.1   Bending Stiffness

In separated pioneering work, Helfrich [34] and Canham [17] assumed a Hookean response of the membrane to bending and derived the bending energy, also called Helfrich or Canham-Helfrich energy,

$$E_B = \int_\Gamma \frac{1}{2} k_B (H - H_0)^2 \, \mathrm{dA}. \qquad (11.29)$$

Here, k_B is the bending stiffness, H the total curvature and H_0 the preferred curvature of the membrane. For a homogeneous lipid bilayer $H_0 = 0$.

Given a phase field that describes the membrane position, curvature as well as the interface delta function can be expressed in terms of ϕ, which allows to define the a

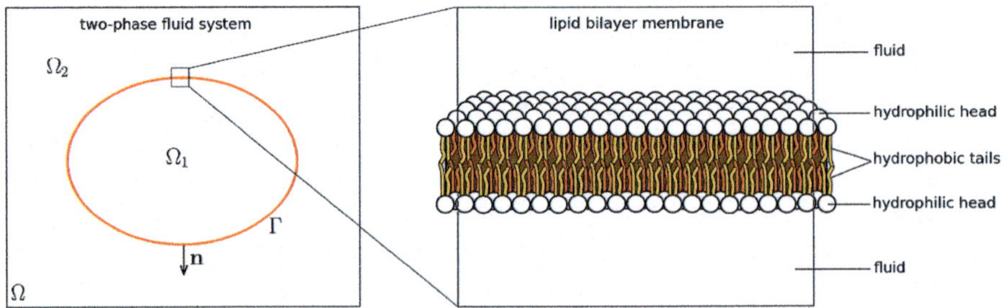

Fig. 11.2 Illustration of a closed fluid membrane. *Left*: Sharp interface setting, the membrane surface Γ separates two fluid phases Ω_1 and Ω_2. *Right*: A close-up shows the physical form of the membrane as a lipid bilayer. Image adapted from [4]

phase field variant of the bending energy. Du et al. proposed the following energy,

$$\tilde{E}_B = \int_\Omega \frac{3k_B}{8\sqrt{2}\varepsilon} \left(\varepsilon \Delta\phi - \frac{1}{\varepsilon} W_0'(\phi) \right)^2 dV, \tag{11.30}$$

where $W_0'(\phi) = (\phi^2 - 1)(\phi + \sqrt{2}\varepsilon H_0)$ is the first derivative of a modified double well potential [22]. In [21] and more rigorous in [57] convergence of $\tilde{E}_B$ toward the sharp interface bending energy E_B was shown as $\varepsilon \to 0$.

Early phase field methods for biomembranes were used without the coupling to hydrodynamics of the surrounding fluids. Such models were subject of numerical studies to describe equilibrium shapes of vesicles [20, 23] as well as analytical studies on existence and convergence of the proposed equations [19, 22]. Topological considerations including the calculation of the Euler number are addressed in [23, 25]. Reviews of existing phase field models for the minimization of the Helfrich energy can be found in [18, 41].

The coupling of the membrane dynamics with the fluid flow is typically derived by an energy variation approach. The resulting Navier-Stokes equations with additional bending stiffness force are [9, 24],

$$\rho \partial_t^\bullet \mathbf{v} + \nabla p - \nabla \cdot \left(\eta(\nabla\mathbf{v} + \nabla\mathbf{v}^T) \right) = -\varepsilon \nabla \cdot (\sigma \nabla\phi \otimes \nabla\phi) + \frac{\delta \tilde{E}_B}{\delta\phi} \nabla\phi, \tag{11.31}$$

$$\nabla \cdot \mathbf{v} = 0, \tag{11.32}$$

where the surface tension σ acts now as a Lagrange multiplier to enforce inextensibility, see Sect. 11.3.2. Further

$$\frac{\delta \tilde{E}_B}{\delta\phi} = \frac{3k_B}{4\sqrt{2}} \left[\frac{1}{\varepsilon^2} \mu W_0''(\phi) - \Delta\mu \right], \tag{11.33}$$

$$\mu = \frac{1}{\varepsilon} W_0'(\phi) - \varepsilon \Delta\phi. \tag{11.34}$$

Additionally, the phase field needs to be advected with the flow which is usually realized by an advected Willmore flow equation with or without volume conservation,

$$d_t\phi + \mathbf{v} \cdot \nabla\phi = -M \frac{\delta \tilde{E}_B}{\delta\phi} \qquad \text{(non-conserved)}, \tag{11.35}$$

$$d_t\phi + \mathbf{v} \cdot \nabla\phi = \nabla \cdot \left(M\nabla \frac{\delta \tilde{E}_B}{\delta\phi} \right) \qquad \text{(conserved)}. \tag{11.36}$$

Both choices for the advection of ϕ ensure thermodynamic consistency in the case of matched densities. Other approaches to advect the phase field include the advected field approach [11, 12] where a modified Allen-Cahn equation is used.

For the system of equations (11.31)–(11.35) without surface tension and inextensibility, existence of global weak solutions and uniqueness under extra regularity were proven [24]. Local time existence and uniqueness of strong solutions is shown in Liu et al. [43].

11.3.2 Inextensibility

The presence of the lipid molecules at the interface introduces a local inextensibility constraint, since the lipids resist interfacial compression and stretching. The local form of the inextensibility constraint reads

$$\nabla_\Gamma \cdot \mathbf{v} = 0, \tag{11.37}$$

which describes a surface incompressibility, similar to the bulk incompressibility condition in the fluid. The necessary Lagrange multiplier to enforce this condition, σ, enters Eq. (11.31) as a surface tension force, similar to the pressure in the Navier-Stokes equation to enforce incompressibility. Different approaches have been presented to realize the inextensibility constraint in phase field models. In the following we discuss and compare the approaches given in [4, 9, 12, 13, 24].

Global Inextensibility (Model A) In [24] the local inextensibility constraint is approximated by a weaker global constraint, i.e., a global conservation of surface area. To realize this, a penalty method is proposed and the total energy of the system is augmented by the penalty term

$$E_p = \frac{k_p}{2}(\mathcal{A}(\phi) - \mathcal{A}_0)^2, \tag{11.38}$$

$$\mathcal{A}(\phi) = \int_\Omega \frac{\varepsilon}{2}|\nabla\phi|^2 + \frac{1}{\varepsilon}W(\phi)dV, \tag{11.39}$$

where $\mathcal{A}_0$ is the initial value of $\mathcal{A}(\phi)$. The corresponding global surface tension is

$$\sigma = k_p(\mathcal{A}(\phi) - \mathcal{A}_0), \tag{11.40}$$

which is positive if the surface area $\mathcal{A}(\phi)$ is too big and negative if $\mathcal{A}(\phi)$ is too small. The approach ensures conservation of the surface area given the factor k_p is large enough. A similar approach with penalty terms coupled to Lagrange multipliers has been presented in [23] and gives identical results for large k_p. In [9] it has been shown that the global and local inextensibility constraints may lead to significantly different membrane dynamics. In particular for situations of multiple approaching membranes the global constraint is not a good approximation and may lead to incorrect flow dynamics.

Local Inextensibility (Model B) A phase field model to account for the full local inextensibility, has been proposed in [9]. Therefore, the following regularized extension of the inextensibility constraint over the whole domain was introduced,

$$\xi\varepsilon^2\nabla\cdot\left(\phi^2\nabla\sigma\right) + |\nabla\phi|\tilde{\nabla}_\Gamma\cdot\mathbf{v} = 0, \tag{11.41}$$

where the diffuse inextensibility constraint is defined as

$$\tilde{\nabla}_\Gamma\cdot\mathbf{v} = -\mathbf{n}\cdot\nabla\mathbf{v}\cdot\mathbf{n}. \tag{11.42}$$

Here, $\mathbf{n} = \nabla\phi/|\nabla\phi|$ represents the membrane normal, ξ is a regularization parameter independent of ε. Eq. (11.41) becomes $\tilde{\nabla}_\Gamma\cdot\mathbf{v} = \nabla_\Gamma\cdot\mathbf{v} = 0$ at the interface where $\phi \approx 0$, and reduces to $\Delta\sigma = 0$ away from Γ since $\phi^2 \approx 1$ and $|\nabla\phi| \approx 0$. This provides a harmonic extension of σ off Γ, while maintaining the local inextensibility constraint near Γ. Eq. (11.41) and the Navier-Stokes equation are coupled implicitly to determine velocity and surface tension simultaneously. The formulation of diffuse inextensibility is validated by asymptotic analysis and numerical tests in [9]. In [47] the model has been used to simulate the interaction between red and white blood cells within a blood vessel.

Local Inextensibility with Relaxation (Model C) The use of the regularization term in Eq. (11.41) can introduce small errors that may accumulate over time and thus lead to spurious local stretching and compression of the membrane. To correct these errors and drive a slightly stretched or compressed surface back to equilibrium, a relaxation mechanism has been proposed in [9]. A field c is introduced to measure local stretching of the interface. Setting the initial value $c(\mathbf{x}, 0) = 1$ and taking c to evolve by a surface mass conservation equation (11.19), locations where c deviates from 1 represent regions of compression ($c > 1$) or stretching ($c < 1$). As seen earlier the surface mass conservation equation can be approximated in the diffuse interface context by use of the characteristic delta function $\delta = |\nabla\phi|$. Additional normal diffusion ensures that the concentration is constantly extended off the interface and leads to the evolution equation

$$\partial_t(|\nabla\phi|c) + \nabla\cdot(|\nabla\phi|c\mathbf{v}) - D_n\nabla\cdot(|\nabla\phi|\mathbf{n}\,\mathbf{n}\cdot\nabla c) = 0. \tag{11.43}$$

Here, D_n is the normal diffusion constant and the use of $|\nabla\phi|$ in the diffusion operator ensures that the bulk concentration does not influence the surface concentration.

Hooke's law can be used to relax the local changes in interfacial area, which amounts in replacing the inextensibility constraint by $\nabla_\Gamma\cdot\mathbf{v} = \zeta(c-1)/c$ for a relaxation constant ζ. Within the diffuse interface formulation this means replacing Eq. (11.41) by

$$\xi\varepsilon^2\nabla\cdot\left(\phi^2\nabla\sigma\right) + |\nabla\phi|\tilde{\nabla}_\Gamma\cdot\mathbf{v} = \zeta|\nabla\phi|(c-1)/c, \tag{11.44}$$

The inextensibility with relaxation has been proven very effective with almost zero stretching and compression of the membrane [9]. Recently, the model has been applied to simulate the formation of a membrane vesicle from a larger membrane during endocytosis [46]. There, curvature-inducing molecules sit on a part of the membrane and induce strong tangential forces which makes the process strongly dependent on the accuracy of the inextensibility constraint. Due to the size difference between the forming vesicle and the large membrane, only a part of the latter is considered and a special boundary condition for c is applied to regulate the influx of membrane area from the domain boundary.

Inextensibility from Membrane Stretch Elasticity (Model D) An alternative approach can be constructed from the physical origin of the membrane inextensibility, namely the membrane stretch energy. Assuming a Hookean response of lipid molecules against compression and stretching leads to the additional stretch energy

$$E_{\text{stretch}} = \int_\Gamma \frac{1}{2} k_s (J-1)^2 \mathrm{dA}, \tag{11.45}$$

where J is the local area stretch of the membrane and k_s the stretching modulus. The local area stretch J is defined as the current area of a surface element divided by the initial area of the same material part of the surface: $J = A/A_0$, hence, $J = 1$ corresponds to the reference state [4].

The above energy was already present in the famous paper of Helfrich [34] and leads, in first order, to a surface tension

$$\sigma = k_s (J-1), \tag{11.46}$$

see [4]. The common inextensibility constraint and corresponding surface tension can actually be seen as approximations to this stretching tension for very large k_s.

The evolution of the local area stretch is determined by the surface evolution equation

$$d_t J + \mathbf{v} \cdot \nabla J = J \nabla_\Gamma \cdot \mathbf{v} \qquad \text{on } \Gamma. \tag{11.47}$$

However it might be more favorable to introduce the inverse of the stretch, $c = 1/J$, which leads to a conservative evolution

$$d_t c + \nabla_\Gamma \cdot (\mathbf{v} c) = 0 \qquad \text{on } \Gamma. \tag{11.48}$$

Approximating this with a diffuse interface model results in Eq. (11.43) again, but contrary to model C this equation is now used to provide the surface tension directly [without involving Eq. (11.44)]. Accordingly, from Eq. (11.46) we obtain $\sigma = k_s (1-c)/c$.

The model is quite similar to the approach described in [12, 13], where a non-conserved evolution equation for c was used. However, the conservation property of Eq. (11.43) is quite desirable since it can be carried over to the discretized model and hence ensures highly accurate conservation of the local membrane area for all times.

11.3.3 Model Comparison

In the following, we perform numerical simulations to test models A, B, C and D in two dimensions. Therefore, the conserved Willmore flow equation (11.36) is coupled to the Navier-Stokes equation (11.31)–(11.32) and the respective inextensibility constraint. We simulate shear flow around a single elliptical vesicle and compare the four models in terms of vesicle dynamics and accuracy of the inextensibility constraint.

Discretization and Parameters We use a Finite-Element discretization with semi-implicit Euler timestepping. P2 elements are used to discretize the velocity, phase field, chemical potential and concentration while the pressure is discretized with P1 elements. Details on the discretization of the Willmore and Navier-Stokes problem can be found in [9], the handling of the surface equation (11.43) for c is described in [46].

A vesicle with major axis of length $45\,\mu$m and minor axis of length $15\,\mu$m is placed in the center of a domain $\Omega = [0, 60\,\mu\text{m}]^2$. We prescribe a shear rate of 0.2s^{-1} at the domain boundary. Typical experimental parameters are used for density, $\rho_2 = 10^3\,\text{kg/m}^3$, $\rho_1 = 1.125\rho_2$, viscosity $\eta_2 = 10^{-3}\text{Pa}\,\text{s}$, $\eta_1 = 20\eta_2$, spontaneous curvature $H_0 = 0$ and bending stiffness $k_B = 10^{-19}\,\text{Nm}$.

The mobility $M = 5 \cdot 10^{-12}\,\text{m}^3\text{s/kg}$ is chosen small enough that the interface is primarily advected. The parameters for model B and C are discussed further in [9], here we use $\xi = 1.33 \cdot 10^4\,\text{m}^{-1}\text{s}^{-1}$, $\zeta = 20\text{s}^{-1}$, $D_n = 3 \cdot 10^{-5}\,\text{m/s}$. The penalty coefficient, $k_p = 3 \cdot 10^4\,\text{N/m}^3$, and stretching coefficient, $k_s = 6.8 \cdot 10^{-6}\,\text{N/m}$, are chosen large enough to guarantee proper global and local inextensibility, respectively. Finally, the time step size is $\tau = 5\,\text{ms}$, the interface thickness is $\varepsilon = 0.45\,\mu$m and the spatial mesh is adaptive with a minimum grid size of $h = 0.5\,\mu$m around the interface. Note that this ensures a resolution of approximately 7 degrees of freedom across the interface, since the corresponding thickness of the interface region (where $-0.9 < \phi < 0.9$) is $1.87\,\mu$m and due to the use of P2 elements.

The use of the conserved Willmore equation (11.36) already leads to a very good conservation of vesicle volume. To ensure even perfect volume conservation we add a growth relaxation term. To this end, we specify the vesicle volume as $\mathcal{V} = \int_\Omega (\phi + 1)/2\,\text{dV}$ and denote the initial value of $\mathcal{V}$ by $\mathcal{V}_0$. The term $|\nabla\phi|(\mathcal{V}_0 - \mathcal{V})$ is added to the left-hand side of Eq. (11.36) aiming to compensate small changes in vesicle volume by growing or shrinking the vesicle around the interface. Note that the restriction of this growth to the interface can be derived from the corresponding sharp interface volume constraint. This restriction is an essential difference to other models which simply add constant values to the phase field to conserve the integral of the phase field, e.g. in [9, 23, 24, 47].

11.3.3.1 Inextensibility Test

As seen in Fig. 11.3, the vesicle volume and the total interface area are conserved very well for all four models. Model B exhibits the largest errors of up to 0.16%, which is still very small. Note, that the area conservation in model B is a pure result of a diffusely inextensible velocity field since there is no relaxation mechanism. To reduce the small error further, a global area conservation mechanism as in model A can be added to model B, like proposed in [9, 47]. For the vesicle volume, we find extremely good conservation due to the above growth relaxation term. Without this term a slight volume loss of up to 0.1% is observed.

The accumulated stretching is presented in Fig. 11.4. To this end, the value of the concentration c is plotted over the arc length of the interface at time $t = 35$ s,

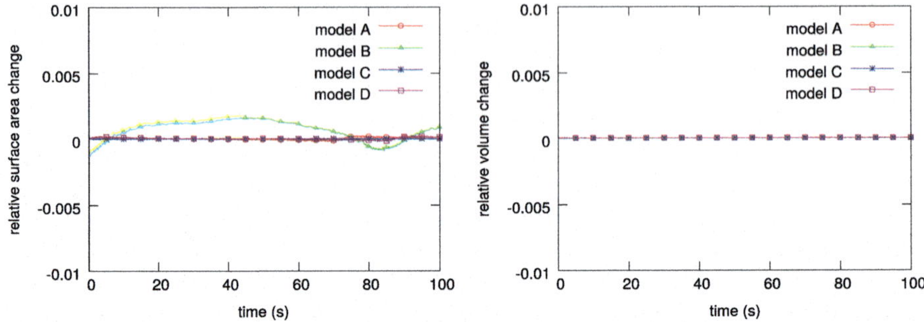

Fig. 11.3 Relative change in vesicle surface area (*left*) and volume (*right*) over time. The relative change in surface area is measured by $(\mathscr{A}(\phi) - \mathscr{A}_0)/\mathscr{A}_0$, the relative change in vesicle volume is $(\mathscr{V}(\phi) - \mathscr{V}_0)/\mathscr{V}_0$. All models show very good global conservation. Errors in surface area are less than 0.2%, errors in volume are less than 0.01%

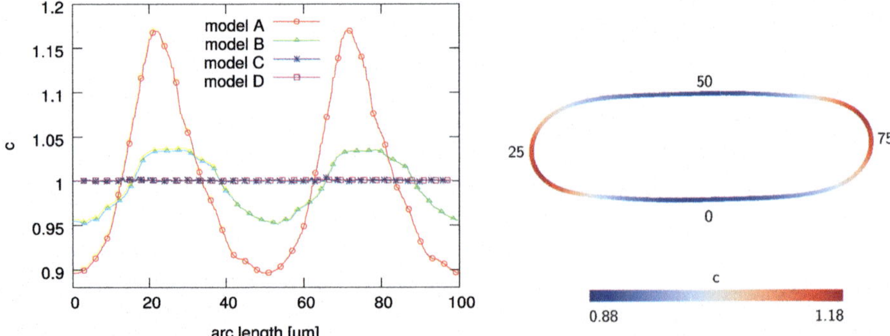

Fig. 11.4 The accumulated local stretching at $t = 35$ s. *Left*: The concentration c over the arc length of the interface. Model A exhibits interfacial stretching of up to 17%, model B reduces local stretching/compression significantly, models C and D show perfect inextensibility ($c = 1$). *Right*: Spatial distribution of c along the interface for model A showing compression at the tips and stretching along the sides. Numbers around the vesicle indicate the arc length

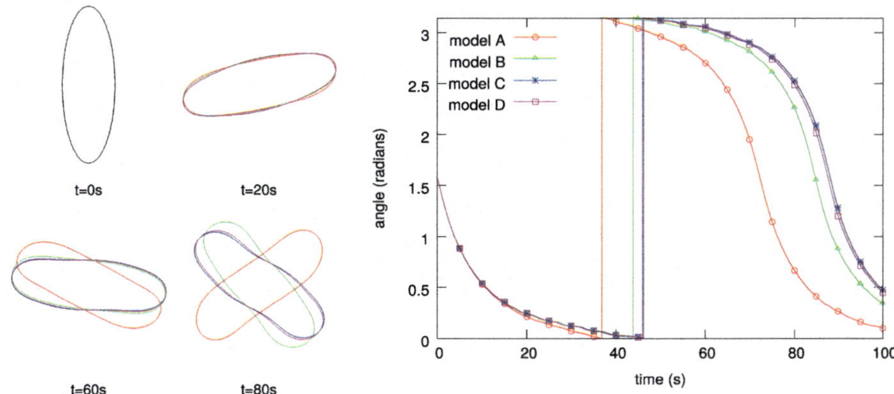

Fig. 11.5 Vesicle shapes (*left*) and inclination angles (*right*) over time. The local inextensibility constraints in models B, C, D significantly delays the tumbling time as compared to model A. Models C and D give almost identical results

when the vesicle is in all models in an almost horizontal state. Note, that the total length of the initial elliptical vesicle is almost exactly $100\,\mu$m. For Model A we find strong interface compression (up to 17%) around the tips of the vesicle while the sides of the vesicle are stretched (up to 11%). This behavior was also observed in [9]. The local inextensibility constraint in models B,C,D suppresses these compression/stretching errors. In particular models C and D show absolutely no local stretching or compression. The relatively large values for model B are actually inherited from the first time steps and might result from the slight interfacial stretching and compression during the initial equilibration of the interface.

11.3.3.2   Model Comparison

The time evolution of the vesicle shapes and the corresponding inclination angles are depicted in Fig. 11.5. For the given parameter set, the vesicle is in the tumbling regime. We find that the vesicle rotates significantly faster in model A than in the locally inextensible models. While models C and D lead to almost identical evolutions, the vesicle dynamics is slightly accelerated in model B.

We conclude that the accurate incorporation of local inextensibility is important to obtain reliable vesicle dynamics in numerical simulations. The global inextensibility constraint in model A is not sufficient and leads to accelerated vesicle dynamics in shear flow. The inextensible models B, C and D provide much more accurate solutions, in particular models C and D seem to completely eliminate local stretching or compression. From a numerical point of view, models B and C are typically coupled monolithically to the Navier-Stokes equation, which makes these models very stable. For model B also energy laws can be shown, see [9] for the matched density case. Model D is explicitly coupled to the Navier-Stokes equations which compromises the stability, hence, for large k_s time step constraints will appear. However, for moderate k_s, model D provides a simple and yet very effective numerical model for highly accurate local inextensibility. Since model D is based on the stretching energy (11.45), it might be possible to derive energy stable numerical schemes for this model as well.

Acknowledgements The author acknowledges support from the German Science Foundation through grant SPP-1506 (AL 1705/1).

References

1. Abels, H., Garcke, H., Grün, G.: Thermodynamically consistent, frame indifferent diffuse interface models for incompressible two-phase flows with different densities. Math. Models Methods Appl. Sci. **22**(3) (2012). doi:10.1142/S0218202511500138
2. Aland, S.: Modelling of two-phase flow with surface-active particles. Dissertation, TU Dresden (2012)
3. Aland, S.: Time integration for diffuse interface models for two-phase flow. J. Comput. Phys. **262**, 58–71 (2014). doi:10.1016/j.jcp.2013.12.055
4. Aland, S.: Phase field modeling of inhomogeneous biomembranes in flow. In: Becker, S. (ed.) Microscale Transport Modelling in Biological Processes, chap. 9. Elsevier, Amsterdam (2016)
5. Aland, S., Chen, F.: An efficient and energy stable scheme for a phase-field model for the moving contact line problem. Int. J. Numer. Methods Fluids **81**(11), 657–671 (2015). doi:10.1002/fld.4200
6. Aland, S., Voigt, A.: Benchmark computations of diffuse interface models for two-dimensional bubble dynamics. Int. J. Num. Meth Fluids **69**, 747–761 (2012). doi:10.1002/fld.2611
7. Aland, S., Lowengrub, J.S., Voigt, A.: Two-phase flow in complex geometries: a diffuse domain approach. CMES **57**(1), 77–106 (2010)
8. Aland, S., Lowengrub, J., Voigt, A.: A continuum model of colloid-stabilized interfaces. Phys. Fluids **23**(6), 062103 (2011). doi:10.1063/1.3584815
9. Aland, S., Egerer, S., Lowengrub, J., Voigt, A.: Diffuse interface models of locally inextensible vesicles in a viscous fluid. J. Comput. Phys. **277**, 32–47 (2014). doi:10.1016/j.jcp.2014.08.016
10. Aland, S., Hahn, A., Kahle, C., Nürnberg, R.: Comparative simulations of Taylor-flow with surfactants based on sharp- and diffuse-interface methods. In: Reusken, A., Bothe, D. (eds.) Advances in Mathematical Fluid Mechanics. Springer, New York (2017)
11. Beaucourt, J., Rioual, F., Séon, T., Biben, T., Misbah, C.: Steady to unsteady dynamics of a vesicle in a flow. Phys. Rev. E **69**(1), 011906 (2004)
12. Biben, T., Misbah, C.: Tumbling of vesicles under shear flow within an advected-field approach. Phys. Rev. E **67**(3), 31908 (2003). doi:10.1103/PhysRevE.67.031908
13. Biben, T., Kassner, K., Misbah, C.: Phase-field approach to three-dimensional vesicle dynamics. Phys. Rev. E **72**(4), 41921 (2005). doi:10.1103/PhysRevE.72.041921
14. Boyanova, P., Do-Quang, M., Neytcheva, M.: Efficient preconditioners for large scale binary Cahn-Hilliard models. Comput. Methods Appl. Math. **12**(1), 1–22 (2012). doi:10.2478/cmam-2012-0001
15. Boyer, F.: A theoretical and numerical model for the study of incompressible mixture flows. Comput. Fluids **31**(1), 41–68 (2002)
16. Cahn, J.W., Hilliard, J.E.: Free energy of a nonuniform system. I. Interfacial free energy. J. Chem. Phys. **28**(2), 258–267 (1958)
17. Canham, P.B.: The minimum energy of bending as a possible explanation of the biconcave shape of the human red blood cell. J. Theor. Biol. **26**(1), 61–81 (1970)

18. Du, Q.: Phase field calculus, curvature-dependent energies, and vesicle membranes. Philos. Mag. **91**(1), 165–181 (2011)
19. Du, Q., Zhu, L.: Analysis of a mixed finite element method for a phase field bending elasticity model of vesicle membrane deformation. J. Comput. Math. **24**(3), 265–280 (2006)
20. Du, Q., Liu, C., Wang, X.: A phase field approach in the numerical study of the elastic bending energy for vesicle membranes. J. Comput. Phys. **198**(2), 450–468 (2004)
21. Du, Q., Liu, C., Ryham, R., Wang, X.: Modeling the spontaneous curvature effects in static cell membrane deformations by a phase field formulation. Commun. Pure Appl. Anal. **4**(3), 537–548 (2005)
22. Du, Q., Liu, C., Ryham, R., Wang, X.: A phase field formulation of the Willmore problem. Nonlinearity **18**, 1249–1267 (2005). doi:10.1088/0951-7715/18/3/016
23. Du, Q., Liu, C., Wang, X.: Simulating the deformation of vesicle membranes under elastic bending energy in three dimensions. J. Comput. Phys. **212**(2), 757–777 (2006)
24. Du, Q., Li, M., Liu, C.: Analysis of a phase field Navier-Stokes vesicle-fluid interaction model. Discrete Continuous Dyn. Syst. **8**(3), 539–556 (2007)
25. Esedoglu, S., Rätz, A., Röger, M.: Colliding interfaces in old and new diffuse-interface approximations of willmore-flow. Commun. Math. Sci. **12**(1), 125–147 (2013). doi:10.4310/CMS.2014.v12.n1.a6
26. Feng, X.: Fully discrete finite element approximations of the Navier–Stokes–Cahn-Hilliard diffuse interface model for two-phase fluid flows. SIAM J. Numer. Anal. **44**, 1049–1072 (2006). doi:10.1137/050638333
27. Garcke, H., Lam, K., Stinner, B.: Diffuse interface modelling of soluble surfactants in two-phase flow. Commun. Math. Sci. **12**(8), 1475–1522 (2014)
28. Garcke, H., Hinze, M., Kahle, C.: A stable and linear time discretization for a thermodynamically consistent model for two-phase incompressible flow. Appl. Numer. Math. **99**, 151–171 (2016). doi:10.1016/j.apnum.2015.09.002
29. Grün, G.: On convergent schemes for diffuse interface models for two-phase flow of incompressible fluids with general mass densities. {SIAM} J. Numer. Anal. **51**(6), 3036–3061 (2013). doi:10.1137/130908208
30. Grün, G., Klingbeil, F.: Two-phase flow with mass density contrast: stable schemes for a thermodynamic consistent and frame-indifferent diffuse-interface model. J. Comput. Phys. **257**, 708–725 (2014). doi:10.1016/j.jcp.2013.10.028
31. Guillén-González, F., Tierra, G.: On linear schemes for a Cahn–Hilliard diffuse interface model. J. Comput. Phys. **234**, 140–171 (2013). doi:10.1016/j.jcp.2012.09.020
32. Han, D., Wang, X.: A second order in time, uniquely solvable, unconditionally stable numerical scheme for Cahn–Hilliard–Navier–Stokes equation. J. Comput. Phys. **290**, 139–156 (2015). doi:10.1016/j.jcp.2015.02.046
33. He, Q., Glowinski, R., Wang, X.P.P.: A least-squares/finite element method for the numerical solution of the {Navier-Stokes-Cahn-Hilliard} system modeling the motion of the contact line. J. Comput. Phys. **230**(12), 4991–5009 (2011). doi:10.1016/j.jcp.2011.03.022
34. Helfrich, W.: Elastic properties of lipid bilayers: theory and possible experiments. Zeitschrift fur Naturforschung. Teil C: Biochemie, Biophysik, Biologie, Virologie **28**(11), 693–703 (1973). doi;10.1002/mus.88004021l
35. Henry, W.: Experiments on the quantity of gases absorbed by water, at different temperatures, and under different pressures. Philos. Trans. R. Soc. Lond. **93**, 29–276 (1803)
36. Hohenberg, P.C., Halperin, B.I.: Theory of dynamic critical phenomena. Rev. Mod. Phys. **49**(3), 435 (1977)
37. Jaqmin, D.: Calculation of two-phase Navier-Stokes flows using phase-field modelling. J. Comput. Phys. **155**, 96–127 (1999). doi:10.1006/jcph.1999.6332
38. Kay, D., Loghin, D., Wathen, A.: A preconditioner for the steady-state navier–stokes equations. SIAM J. Sci. Comput. **24**(1), 237–256 (2002)
39. Kay, D., Welford, R.: Efficient numerical solution of Cahn-Hilliard-Navier-Stokes fluids in 2D. SIAM J. Sci. Comput. **29**, 15–43 (2007). doi:10.1137/050648110

40. Kim, J.: A continuous surface tension force formulation for diffuse-interface models. J. Comput. Phys. **204**(2), 784–804 (2005). doi:10.1016/j.jcp.2004.10.032. http://linkinghub.elsevier.com/retrieve/pii/S0021999104004383
41. Lázaro, G.R., Pagonabarraga, I., Hernández-Machado, A.: Phase-field theories for mathematical modeling of biological membranes. Chem. Phys. Lipids **185**, 46–60 (2015)
42. Li, X., Lowengrub, J., Rätz, A., Voigt, A.: Solving PDEs in complex geometries: a diffuse domain approach. Commun. Math. Sci. **7**(1), 81–107 (2009)
43. Liu, Y., Takahashi, T., Tucsnak, M.: Strong solutions for a phase field Navier–Stokes vesicle–fluid interaction model. J. Math. Fluid Mech. **14**(1), 177–195 (2012)
44. Liu, C., Shen, J., Yang, X.: Decoupled energy stable schemes for a phase-field model of two-phase incompressible flows with variable density. J. Sci. Comput. **62**(2), 601–622 (2014). doi:10.1007/s10915-014-9867-4
45. Lowengrub, J., Truskinovsky, L.: Quasi-incompressible Cahn–Hilliard fluids and topological transitions. In: Proceedings of the Royal Society of London A: Mathematical, Physical and Engineering Sciences, vol. 454, pp. 2617–2654. The Royal Society, London (1998)
46. Lowengrub, J., Allard, J., Aland, S.: Numerical simulation of endocytosis: viscous flow driven by membranes with non-uniformly distributed curvature-inducing molecules. J. Comput. Phys. **309**, 112–128 (2016). doi:10.1016/j.jcp.2015.12.055
47. Marth, W., Aland, S., Voigt, A.: Margination of white blood cells - a computational approach by a hydrodynamic phase field model. J. Fluid Mech. **790**, 389–406 (2016)
48. Muradoglu, M., Tryggvason, G.: A front-tracking method for computation of interfacial flows with soluble surfactants. J. Comput. Phys. **227**(4), 2238–2262 (2008)
49. Pozrikidis, C.: Boundary Integral and Singularity Methods for Linearized Viscous Flow. Cambridge University Press, Cambridge (1992)
50. Qian, T., Wang, X.P., Sheng, P.: Molecular scale contact line hydrodynamics of immiscible flows. Phys. Rev. E **68**(1), 016306 (2003)
51. Rannacher, R.: Finite element methods for the incompressible Navier-Stokes equations. Ph.D. thesis (2000)
52. Rätz, A., Voigt, A.: PDE's on surfaces—a diffuse interface approach. Commun. Math. Sci. **4**(3), 575–590 (2006)
53. Rowlinson, J.: Translation of J.D. van der Waals' "The thermodynamik theory of capillarity under the hypothesis of a continuous variation of density". J. Stat. Phys. **20**(2), 197–200 (1979)
54. Schwarzenberger, K., Aland, S., Domnick, H., Odenbach, S., Eckert, K.: Relaxation oscillations of solutal Marangoni convection at curved interfaces. Colloids Surf. A **481**, 633–643 (2015)
55. Teigen, K.E., Song, P., Lowengrub, J., Voigt, A.: A diffuse-interface method for two-phase flows with soluble surfactants. J. Comput. Phys. **230**, 375 (2011)
56. Villanueva, W., Amberg, G.: Some generic capillary-driven flows. Int. J. Multiphase Flow **32**(9), 1072–1086 (2006). doi:10.1016/j.ijmultiphaseflow.2006.05.003
57. Wang, X.: Asymptotic analysis of phase field formulations of bending elasticity models. SIAM J. Math. Anal. **39**(5), 1367–1401 (2008). doi:10.1137/060663519
58. Wise, S., Kim, J., Lowengrub, J.: Solving the regularized, strongly anisotropic cahn–hilliard equation by an adaptive nonlinear multigrid method. J. Comput. Phys. **226**(1), 414–446 (2007)
59. Xu, J.J., Li, Z., Lowengrub, J., Zhao, H.: A level-set method for interfacial flows with surfactant. J. Comput. Phys. **212**(2), 590–616 (2006)
60. Zhang, J., Johnson, P.C., Popel, A.S.: An immersed boundary lattice Boltzmann approach to simulate deformable liquid capsules and its application to microscopic blood flows. Phys. Biol. **4**(4), 285 (2007)

Chapter 12
Micro-Macro-Models for Two-Phase Flow of Dilute Polymeric Solutions: Macroscopic Limit, Analysis, and Numerics

Günther Grün and Stefan Metzger

Abstract We derive a diffuse-interface model for two-phase flow of incompressible fluids with dissolved noninteracting polymers. Describing the polymers as bead chains governed by general elastic spring potentials, including in particular Hookean and finitely extensible, nonlinear elastic (FENE) potentials, it couples a Fokker-Planck type equation describing distribution and orientation of the polymer chains with Cahn–Hilliard and Navier–Stokes type equations describing the balance of mass and momentum. Allowing for different solubility properties which are modelled by Henry type energy functionals, the presented model covers the case of one Newtonian fluid and one non-Newtonian fluid as well as the case of two non-Newtonian fluids. In the case of Hookean spring potentials, we derive a macroscopic diffuse-interface model for two-phase flow of Oldroyd-B-type liquids.

In the case of dumbbell models, we show existence of solutions and present numerical simulations in two space dimensions on oscillating polymeric droplets.

12.1 Introduction

We are concerned with modelling and analysis of two-phase flow of dilute polymeric solutions. For the ease of presentation, we focus initially on dumbbells to describe polymer chains. These dumbbells consist of two beads which are connected by an elastic spring with some spring potential U. We are not only interested in the spatial distribution of the polymers, but also in their configuration, i.e. in their elongation and orientation. Therefore, we have to distinguish between the spatial coordinate $\mathbf{x} \in \Omega$ referring to the barycenter of the dumbbell and the configurational coordinate $\mathbf{q} \in D$ describing its configuration. Consequently, we will use $\nabla_{\mathbf{x}}$, $\mathrm{div}_{\mathbf{x}}$, $\nabla_{\mathbf{q}}$ and $\mathrm{div}_{\mathbf{q}}$ to denote the differential operators with respect to $\mathbf{x}$ and $\mathbf{q}$, respectively.

In recent years, many authors have contributed to a mathematical theory for multi-scale modeling of dilute polymeric flows (see [3–5, 9, 11, 15, 18–20, 22,

G. Grün (✉) • S. Metzger
Department Mathematik, Friedrich-Alexander Universität Erlangen-Nürnberg, Cauerstrasse 11, 91058 Erlangen, Germany
e-mail: gruen@am.uni-erlangen.de; stefan.metzger@fau.de

© Springer International Publishing AG 2017

291

D. Bothe, A. Reusken (eds.), *Transport Processes at Fluidic Interfaces*, Advances in Mathematical Fluid Mechanics, DOI 10.1007/978-3-319-56602-3_12

23, 27, 29, 31] and the references therein). To describe dilute polymeric solutions, different approaches are at hand. Without claiming to be exhaustive, we list the concepts of the stochastic Brownian configuration field approaches coupling stochastic differential equations describing the evolution of the polymer chains with macroscopic equations describing the balance of momentum, deterministic Fokker–Planck approaches, and purely macroscopic approaches like Oldroyd-B type models. For the theoretical background, we refer to [17] and the references therein. Information on the numerical treatment of these approaches can be found e.g. in [13]. Focussing on Fokker–Planck–Navier–Stokes approaches, a generic single phase model reads as

$$\partial_t \mathbf{u} + (\mathbf{u} \cdot \nabla_\mathbf{x})\,\mathbf{u} - \mathrm{div}_\mathbf{x}\,\{2\eta \mathbf{D}\mathbf{u}\} + \nabla_\mathbf{x} p = \mathrm{div}_\mathbf{x} \left\{ \int_D \psi\, U'(\tfrac{1}{2}|\mathbf{q}|^2)\mathbf{q} \otimes \mathbf{q} - \int_D \psi\, \mathbb{1} \right\},$$
$$\tag{12.1a}$$

$$\mathrm{div}_\mathbf{x}\,\mathbf{u} = 0 \qquad\qquad \text{in } \Omega \times (0, T)\ ,$$
$$\tag{12.1b}$$

$$\partial_t \psi + \mathbf{u} \cdot \nabla_\mathbf{x} \psi + \mathrm{div}_\mathbf{q}\,\{\psi \nabla_\mathbf{x}\mathbf{u} \cdot \mathbf{q}\} = c_\mathbf{q}\,\mathrm{div}_\mathbf{q}\,\{\psi \nabla_\mathbf{q}\mu_\psi\} + c_\mathbf{x}\,\mathrm{div}_\mathbf{x}\,\{\psi \nabla_\mathbf{x}\mu_\psi\}\ ,$$
$$\tag{12.1c}$$

$$\mu_\psi = \log\left(\tfrac{\psi}{M}\right) \qquad\qquad \text{in } \Omega \times D \times (0, T)$$
$$\tag{12.1d}$$

with $\eta, c_\mathbf{q} > 0$, $c_\mathbf{x} \geq 0$ and the symmetrized gradient $\mathbf{D} := \tfrac{1}{2}\left(\nabla_\mathbf{x} + \nabla_\mathbf{x}^T\right)$. The function $\psi(\mathbf{x}, t)\ :\ D \to \mathbb{R}_0^+$ describes a distribution on the configuration space $D \subset \mathbb{R}^d$ of admissible elongations/orientations of the polymers. In particular, the marginal

$$\omega(\mathbf{x}, t) := \int_D \psi(\mathbf{x}, t, \mathbf{q}) d\mathbf{q}$$

gives the number density of the polymers. The function

$$M(\mathbf{q}) := \left(\int_D \exp\left(-U(\tfrac{1}{2}\,|\overline{\mathbf{q}}|^2)\right) d\overline{\mathbf{q}}\right)^{-1} \exp\left(-U(\tfrac{1}{2}\,|\mathbf{q}|^2)\right) \tag{12.2}$$

denotes the Maxwellian associated with the spring potential U. Velocity and pressure field are denoted by $(\mathbf{u}, p)$. The tensor on the right-hand side of Eq. (12.1a) is the Kramers stress tensor (see [7, 16] and the references therein) which models effects exerted by the polymers on the solvent flow and gives rise to non-Newtonian effects. Equation (12.1c) is the usual evolution equation for the probability density on the configuration space. Note that $\mathrm{div}_\mathbf{x}\,\{\psi \nabla_\mathbf{x}\mu_\psi\}$ describes the center-of-mass diffusion of the dumbbells. Although the parameter $c_\mathbf{x}$ is magnitudes smaller than one, it seems reasonable to keep the corresponding diffusion term for mathematical reasons: It guarantees parabolicity of (12.1c). For approaches setting $c_{\mathbf{x}} \equiv 0$, see [22, 23] and the references therein.

In [14], we extend (12.1) to the case of two-phase flows with different mass densities. Allowing for different solubility properties, the resulting model covers the case of one Newtonian fluid and one non-Newtonian fluid as well as the case of two non-Newtonian fluids.

12.2 The Governing Equations

12.2.1 Derivation of the Model: A Novel Approach Based on Onsager's Variational Principle

To derive a micro-macro model describing two-phase flow of dilute polymeric solutions, we apply Onsager's variational principle of minimum energy dissipation (cf. [26]), which turned out to be an efficient workhorse in the derivation of thermodynamically consistent models. There are various examples where it has successfully been applied in fluid dynamics. First, Qian et al. [28] introduced it to model contact line motion for two-phase flow with wall contact. Other applications range from different models for two-phase flow with general mass densities (cf. [21] and [2]) to models for species transport (cf. [2]), electrowetting and other electrokinetic phenomena [8]. Onsager's principle is supposed to provide the most probable evolution of a dissipative process by postulating a linear relation between the rates of displacement from the thermodynamic equilibrium and the applied forces, i.e. the rate of energy (cf. (12.8) below).

In general, there are many examples of models having an underlying energetic structure which may be derived by the Onsager formalism. It is worth mentioning that the micro-macro model obtained in [4] to describe single-phase flows of dilute polymeric solutions, can also be derived by the Onsager method.

Recapitulating the approach of [14], we start with general balance equations and—based on the above observation—apply Onsager's variational principle of minimum energy dissipation to obtain a thermodynamically consistent model for two-phase flow of polymeric solutions. Concerning the balance of mass and momentum, we apply the diffuse-interface method which was used e.g. in [10, 21], and [2] and which allows for a smooth transition between the fluid phases. Following in particular the approach used in [2], we define volume fractions $\phi_i := \frac{\rho_i}{\tilde{\rho}_i}$ ($i = 1, 2$), where $\tilde{\rho}_i$ denotes the mass density of the ith fluid in a pure phase and ρ_i denotes the actual mass density in the two-phase flow. Assuming $\phi_1 + \phi_2 \equiv 1$, the conservation of the phase-field parameter $\phi := \phi_2 - \phi_1$ implies also conservation of the individual mass densities. Defining the momentum as $(\rho \mathbf{u})$, where $\mathbf{u}$ is the solenoidal, volume averaged velocity field, and imposing that the momentum is transported with the same velocity as the mass—i.e. relative velocities are taken into account—yields

$$\partial_t \phi + \mathbf{u} \cdot \nabla_{\mathbf{x}} \phi + \mathrm{div}_{\mathbf{x}} \, \mathbf{J}_\phi = 0 \,, \tag{12.3}$$

$$\rho(\phi) \, \partial_t \mathbf{u} + \left(\left(\rho(\phi) \mathbf{u} + \frac{\partial \rho}{\partial \phi} \mathbf{J}_\phi \right) \cdot \nabla_{\mathbf{x}} \right) \mathbf{u} = \mathrm{div}_{\mathbf{x}} \, \mathbf{S} - \nabla_{\mathbf{x}} p + \mathbf{k} \,, \tag{12.4}$$

with the correction flux $\mathbf{J}_\phi$, the symmetric stress tensor $\mathbf{S}$ and the force term $\mathbf{k}$ which still are to be determined. For further details, we refer to [2].

In the same spirit we state a balance equation for the configurational density ψ on $\Omega \times D$,

$$\partial_t \psi + \mathrm{div}_{\mathbf{x}} \{ \tilde{\mathbf{v}}_{\mathbf{x}} \psi \} + \mathrm{div}_{\mathbf{q}} \{ \tilde{\mathbf{v}}_{\mathbf{q}} \psi \} = 0 \,. \tag{12.5}$$

The next step is to decompose $\tilde{\mathbf{v}}_{\mathbf{x}} \psi$ into $\mathbf{u}\psi + \mathbf{J}_{\psi,\mathbf{x}}$ and $\tilde{\mathbf{v}}_{\mathbf{q}} \psi$ into $\mathbf{u}_{\mathbf{q}} \psi + \mathbf{J}_{\psi,\mathbf{q}}$ with correction fluxes $\mathbf{J}_{\psi,\mathbf{x}}$ and $\mathbf{J}_{\psi,\mathbf{q}}$, respectively, which are unknowns at this stage of the derivation. Note that there are two velocities to be distinguished—on the physical space the hydrodynamic velocity $\mathbf{u}$, on the configurational space a drift velocity $\mathbf{u}_{\mathbf{q}}$ which is intended to model the drift caused by the different velocities experienced by the dumbbell at head and tail. It is given by

$$\mathbf{u}_{\mathbf{q}}(\mathbf{x}, \mathbf{q}, t) := \frac{\mathbf{u}(\mathbf{x} + \frac{\varepsilon}{2}\mathbf{q}, t) - \mathbf{u}(\mathbf{x} - \frac{\varepsilon}{2}\mathbf{q}, t)}{\varepsilon} = \nabla_{\mathbf{x}} \left\{ \frac{1}{\varepsilon} \int_{-\frac{\varepsilon}{2}}^{\frac{\varepsilon}{2}} \mathbf{u}(\mathbf{x} + \theta\mathbf{q}, t) \, \mathrm{d}\theta \right\} \cdot \mathbf{q} \,, \tag{12.6}$$

where the parameter ε reflects the different length scales in Ω and D.
Following the approach in [4], we replace the directional mollifier on the right-hand side of (12.6) by an isotropic one which we denote by $\mathscr{J}_\varepsilon$ in the scalar case and J_ε or $\mathfrak{J}_\varepsilon$, if applied to vector-valued or matrix-valued quantities. Therefore, the last balance equation reads

$$\partial_t \psi + \mathbf{u} \cdot \nabla_{\mathbf{x}} \psi + \mathrm{div}_{\mathbf{q}} \{ \nabla_{\mathbf{x}} J_\varepsilon \{ \mathbf{u} \} \cdot \mathbf{q} \psi \} = - \mathrm{div}_{\mathbf{x}} \mathbf{J}_{\psi,\mathbf{x}} - \mathrm{div}_{\mathbf{q}} \mathbf{J}_{\psi,\mathbf{q}} \,. \tag{12.7}$$

Imposing suitable, mass conserving boundary conditions, the unknown quantities are determined using Onsager's variational principle which is supposed to model the most probable behavior of an irreversible process and postulates the relation

$$\delta_{(\mathbf{J}_\phi, \mathbf{J}_{\psi,\mathbf{x}}, \mathbf{J}_{\psi,\mathbf{q}}, \mathbf{S})} \left(\frac{\mathrm{d}\mathscr{E}}{\mathrm{d}t} + \Phi \right) \stackrel{!}{=} 0 \tag{12.8}$$

between thermodynamical forces (given by the rate of energy in the system) and an appropriate dissipation function Φ which will be defined in (12.9). In (12.8), $\delta_{(\mathbf{J}_\phi, \mathbf{J}_{\psi,\mathbf{x}}, \mathbf{J}_{\psi,\mathbf{q}}, \mathbf{S})}$ denotes the first variation with respect to the quantities $\mathbf{J}_\phi$, $\mathbf{J}_{\psi,\mathbf{x}}$, $\mathbf{J}_{\psi,\mathbf{q}}$, and $\mathbf{S}$.

The energy of the system is assumed to be the sum of the kinetic energy $\frac{1}{2} \int_\Omega |\mathbf{u}|^2$, a Cahn-Hilliard energy $\frac{\delta}{2} \int_\Omega |\nabla_{\mathbf{x}} \phi|^2 + \frac{1}{\delta} \int_\Omega W(\phi)$, where $W(\phi) = \frac{1}{4}\left(1 - \phi^2\right)^2$ is a standard double-well potential, an entropic component $\int_{\Omega \times D} \psi \left(\log \left(\frac{\psi}{M} \right) - 1 \right)$, and the so called Henry energy $\int_{\Omega} \beta \left(\phi \right) \mathscr{J}_\varepsilon \left\{ \int_D \psi \, d\mathbf{q} \right\} d\mathbf{x}$. By choosing an appropriate function β, the latter energy component will allow to constrain the polymers to one dedicated phase, and therefore allows for one Newtonian phase. Assuming a dissipation functional of the form

$$\Phi \left(\mathbf{J}_\phi, \mathbf{J}_{\psi,\mathbf{x}}, \mathbf{J}_{\psi,\mathbf{q}}, \mathbf{S} \right) := \int_{\Omega} \frac{|\mathbf{J}_\phi|^2}{2m(\phi)} + \int_{\Omega \times D} \frac{|\mathbf{J}_{\psi,\mathbf{x}}|^2}{2c_\mathbf{x}\psi} + \int_{\Omega \times D} \frac{|\mathbf{J}_{\psi,\mathbf{q}}|^2}{2c_\mathbf{q}\psi} + \int_{\Omega} \frac{|\mathbf{S}|^2}{4\eta(\phi)}$$

(12.9)

with $m, c_\mathbf{x}, c_\mathbf{q}, \eta > 0$ leads to the following set of equations.

$$\rho \left(\phi \right) \partial_t \mathbf{u} + \left(\left(\rho \left(\phi \right) \mathbf{u} - m \left(\phi \right) \frac{\partial \rho}{\partial \phi} \nabla_\mathbf{x} \mu_\phi \right) \cdot \nabla_\mathbf{x} \right) \mathbf{u} - \mathrm{div}_\mathbf{x} \left\{ 2\eta \left(\phi \right) \mathbf{Du} \right\} + \nabla_\mathbf{x} p$$

$$= \mu_\phi \nabla_\mathbf{x} \phi + \int_D \mu_\psi \nabla_\mathbf{x} \psi + \mathrm{div}_\mathbf{x} \left\{ \mathfrak{J}_\varepsilon \left\{ \int_D M \nabla_\mathbf{q} \frac{\psi}{M} \otimes \mathbf{q} \right\} \right\} , \qquad (12.10\text{a})$$

$$\mathrm{div}_\mathbf{x} \mathbf{u} = 0 , \qquad (12.10\text{b})$$

$$\partial_t \phi + \mathbf{u} \cdot \nabla_\mathbf{x} \phi - \mathrm{div}_\mathbf{x} \left\{ m \left(\phi \right) \nabla_\mathbf{x} \mu_\phi \right\} = 0 , \qquad (12.10\text{c})$$

$$\mu_\phi = -\delta \Delta_\mathbf{x} \phi + \tfrac{1}{\delta} W' \left(\phi \right) + \beta' \left(\phi \right) \mathscr{J}_\varepsilon \left\{ \int_D \psi \right\} , \qquad (12.10\text{d})$$

$$\partial_t \psi + \mathbf{u} \cdot \nabla_\mathbf{x} \psi + \mathrm{div}_\mathbf{q} \left\{ \psi \nabla_\mathbf{x} J_\varepsilon \left\{ \mathbf{u} \right\} \cdot \mathbf{q} \right\} = \mathrm{div}_\mathbf{q} \left\{ c_\mathbf{q} \psi \nabla_\mathbf{q} \mu_\psi \right\} + \mathrm{div}_\mathbf{x} \left\{ c_\mathbf{x} \psi \nabla_\mathbf{x} \mu_\psi \right\} , \qquad (12.10\text{e})$$

$$\mu_\psi = \log \left(\frac{\psi}{M} \right) + \mathscr{J}_\varepsilon \left\{ \beta \left(\phi \right) \right\} , \qquad (12.10\text{f})$$

on $\Omega \times \mathbb{R}^+$ or $\Omega \times D \times \mathbb{R}^+$, respectively, with the boundary conditions

$$\mathbf{u} = 0 \qquad \text{on } \partial\Omega \times \mathbb{R}^+, \qquad (12.11\text{a})$$

$$\nabla_\mathbf{x} \phi \cdot \mathbf{n}_\mathbf{x} = 0 \qquad \text{on } \partial\Omega \times \mathbb{R}^+, \qquad (12.11\text{b})$$

$$\nabla_\mathbf{x} \mu_\phi \cdot \mathbf{n}_\mathbf{x} = 0 \qquad \text{on } \partial\Omega \times \mathbb{R}^+, \qquad (12.11\text{c})$$

$$\psi \nabla_\mathbf{x} \mu_\psi \cdot \mathbf{n}_\mathbf{x} = 0 \qquad \text{on } \partial\Omega \times D \times \mathbb{R}^+, \qquad (12.11\text{d})$$

$$\psi \left(\nabla_\mathbf{x} J_\varepsilon \left\{ \mathbf{u} \right\} \cdot \mathbf{q} - c_\mathbf{q} \nabla_\mathbf{q} \mu_\psi \right) \cdot \mathbf{n}_\mathbf{q} = 0 \qquad \text{on } \Omega \times \partial D \times \mathbb{R}^+. \qquad (12.11\text{e})$$

Restricting (12.10) to the case of a single-phase flow, i.e. ϕ, ρ, and η constant, allows to recover the set of equations derived in [4]. In practical applications, the constant $c_\mathbf{x}$ is magnitudes smaller than $c_\mathbf{q}$. Therefore, some authors decided to neglect diffusion in $\mathbf{x}$-direction (cf. [3], [22, 23]). As the $\mathbf{x}$-diffusion guarantees parabolicity of the Fokker–Planck type equation which is convenient for the analytical treatment, we follow the ansatz used e.g. in [4] and keep the $c_{\mathbf{x}} > 0$.

It is straight forward to adapt this modeling approach to polymer models consisting of $K + 1$ beads. In this case, we have to consider K elongation vectors $(\mathbf{q}_i)_{i=1,\ldots,K} \in \bigotimes_{i=1}^{K} D_i =: D \subset \mathbb{R}^{Kd}$. Each of these vectors has its own entropic potential U_i associated with a Maxwellian M_i. The Maxwellian of the complete chain is defined as $M(\mathbf{q}) := \Pi_{i=1}^{K} M_i(\mathbf{q}_i)$ (cf. [5]). Consequently, the momentum and Fokker–Planck equations are changed to become

$$\rho\left(\phi\right)\partial_t\mathbf{u} + \left(\left(\rho\left(\phi\right)\mathbf{u} - m\left(\phi\right)\frac{\partial\rho}{\partial\phi}\nabla_{\mathbf{x}}\mu_\phi\right)\cdot\nabla_{\mathbf{x}}\right)\mathbf{u} - \mathrm{div}_{\mathbf{x}}\left\{2\eta\left(\phi\right)\mathbf{Du}\right\} + \nabla_{\mathbf{x}}p$$

$$= \mu_\phi\nabla_{\mathbf{x}}\phi + \int_D \mu_\psi\nabla_{\mathbf{x}}\psi + \mathrm{div}_{\mathbf{x}}\left\{\mathfrak{J}_\varepsilon\left\{\int_D M\sum_{i=1}^{K}\nabla_{\mathbf{q}_i}\frac{\psi}{M}\otimes\mathbf{q}_i\right\}\right\}, \qquad (12.12a)$$

$$\partial_t\psi + \mathbf{u}\cdot\nabla_{\mathbf{x}}\psi + \sum_{i=1}^{K}\mathrm{div}_{\mathbf{q}_i}\left\{\psi\nabla_{\mathbf{x}}\mathbf{J}_\varepsilon\left\{\mathbf{u}\right\}\cdot\mathbf{q}_i\right\}$$

$$= \sum_{i=1}^{K}\sum_{j=1}^{K}\mathrm{div}_{\mathbf{q}_i}\left\{A_{ij}\psi\nabla_{\mathbf{q}_j}\mu_\psi\right\} + \mathrm{div}_{\mathbf{x}}\left\{c_{\mathbf{x}}\psi\nabla_{\mathbf{x}}\mu_\psi\right\}, \qquad (12.12b)$$

$$\mu_\psi = \log\left(\frac{\psi}{M}\right) + \mathscr{J}_\varepsilon\left\{\beta\left(\phi\right)\right\}, \qquad (12.12c)$$

on $\Omega\times\mathbb{R}^+$ or $\Omega\times D\times\mathbb{R}^+$, respectively. The matrix $A = (A_{ij})_{i,j=1}^{K}$ is the symmetric positive definite Rouse matrix, or connectivity matrix. For further information we refer to [25] and the references therein. Moreover, the boundary conditions need to be adapted to guarantee conservation of particle number.

12.2.2 A Two-Phase Oldroyd-B Model

It is well known that micro-macro models may be used to derive deterministic viscoelastic models by taking the expected value in the Fokker–Planck equation. If the elastic potential is of Hookean type, i.e. $U(s) = s$ and $D = \mathbb{R}^d$, particular versions of the Oldroyd-B model are obtained (see [4] and the references therein). As the elastic potential U and therefore the admissible configurational space D was not specified in the derivation of our model, we may assume that our potential is of Hookean type and formulate a Oldroyd-B type model for two-phase flow by deriving an evolution equation for the additional stress tensor $\mathbf{C} := \int_D M\nabla_{\mathbf{q}}\frac{\psi}{M}\otimes\mathbf{q}$. In [14] the authors derived the following set of equations.

$$\rho\left(\phi\right)\partial_t\mathbf{u} + \left(\left(\rho\left(\phi\right)\mathbf{u} - m\left(\phi\right)\tfrac{\partial\rho}{\partial\phi}\nabla_{\mathbf{x}}\mu_\phi\right)\cdot\nabla_{\mathbf{x}}\right)\mathbf{u} - \mathrm{div}_{\mathbf{x}}\left\{2\eta\left(\phi\right)\mathbf{Du}\right\} + \nabla_{\mathbf{x}}p$$

$$= \mu_\phi\nabla_{\mathbf{x}}\phi + \mathscr{J}_\varepsilon\left\{\beta\left(\phi\right)\right\}\nabla_{\mathbf{x}}\omega + \mathrm{div}_{\mathbf{x}}\left\{\mathfrak{J}_\varepsilon\left\{\mathbf{C} - \omega\mathbb{1}\right\}\right\}, \qquad (12.13\mathrm{a})$$

$$\mathrm{div}_{\mathbf{x}}\mathbf{u} = 0, \qquad (12.13\mathrm{b})$$

$$\partial_t\phi + \mathbf{u}\cdot\nabla_{\mathbf{x}}\phi - \mathrm{div}_{\mathbf{x}}\left\{m\left(\phi\right)\nabla_{\mathbf{x}}\mu_\phi\right\} = 0, \qquad (12.13\mathrm{c})$$

$$\mu_\phi = -\delta\Delta_{\mathbf{x}}\phi + \tfrac{1}{\delta}W'\left(\phi\right) + \beta'\left(\phi\right)\mathscr{J}_\varepsilon\left\{\omega\right\}, \qquad (12.13\mathrm{d})$$

$$\partial_t\mathbf{C} + \left(\mathbf{u}\cdot\nabla_{\mathbf{x}}\right)\mathbf{C} - \nabla_{\mathbf{x}}\mathbf{J}_\varepsilon\left\{\mathbf{u}\right\}\mathbf{C} - \mathbf{C}\left(\nabla_{\mathbf{x}}\mathbf{J}_\varepsilon\left\{\mathbf{u}\right\}\right)^T$$
$$= 2c_{\mathbf{q}}\omega\mathbb{1} - 2c_{\mathbf{q}}\mathbf{C} + c_{\mathbf{x}}\Delta_{\mathbf{x}}\mathbf{C} + c_{\mathbf{x}}\mathrm{div}_{\mathbf{x}}\left(\mathbf{C}\otimes\nabla_{\mathbf{x}}\mathscr{J}_\varepsilon\left\{\beta\left(\phi\right)\right\}\right), \qquad (12.13\mathrm{e})$$

$$\partial_t\omega + \left(\mathbf{u}\cdot\nabla_{\mathbf{x}}\right)\omega - c_{\mathbf{x}}\Delta_{\mathbf{x}}\omega = c_{\mathbf{x}}\mathrm{div}_{\mathbf{x}}\left\{\omega\nabla_{\mathbf{x}}\mathscr{J}_\varepsilon\left\{\beta(\phi)\right\}\right\}, \qquad (12.13\mathrm{f})$$

on $\Omega\times\mathbb{R}^+$.

Model (12.13) seems to be the first two-phase model for visco-elastic polymeric flows which takes the density of the polymers in the different fluid phases and their impact on the rheology into account. This way, it may serve to model two-phase flow with Newtonian and visco-elastic phases, too. Conceptually, this approach is different from other methods found in the literature (see e.g. [30]).

To point out this difference, we rewrite (12.13e) in terms of $\mathbf{K} := \mathbf{C} - \omega\mathbb{1}$.

$$\partial_t\mathbf{K} + \left(\mathbf{u}\cdot\nabla_{\mathbf{x}}\right)\mathbf{K} - \nabla_{\mathbf{x}}\mathbf{J}_\varepsilon\left\{\mathbf{u}\right\}\cdot\mathbf{K} - \mathbf{K}\cdot\left(\nabla_{\mathbf{x}}\mathbf{J}_\varepsilon\left\{\mathbf{u}\right\}\right)^T - 2\omega\mathbf{D}\mathbf{J}_\varepsilon\left\{\mathbf{u}\right\}$$

$$= -2c_{\mathbf{q}}\mathbf{K} + c_{\mathbf{x}}\Delta_{\mathbf{x}}\mathbf{K} + c_{\mathbf{x}}\mathrm{div}_{\mathbf{x}}\left\{\mathbf{K}\otimes\nabla_{\mathbf{x}}\mathscr{J}_\varepsilon\left\{\beta(\phi)\right\}\right\}. \qquad (12.14)$$

Denoting the upper convected time derivative, i.e. the first four terms in (12.14), by $\mathbf{K}^*$ (cf. [4]), we obtain

$$\mathbf{K} + \tfrac{1}{2c_{\mathbf{q}}}\mathbf{K}^* = \tfrac{\omega}{cq}\mathbf{D}\mathbf{J}_\varepsilon\left\{\mathbf{u}\right\}, \qquad (12.15)$$

as evolution equation for $\mathbf{K}$ after setting $c_{\mathbf{x}} = 0$. I.e. the velocity field $\mathbf{u}$ depends on an additional stress tensor $\mathbf{K}$ in both fluids, but the back coupling of the velocity field depends on the number density of the fluids which may be chosen fluid-dependent. A different approach was used in [30]. Yue et al. used

$$\mathbf{K} + \lambda_H\mathbf{K}^* = 2\mu_P\mathbf{Du} \qquad (12.16)$$

with two constants λ_H and μ_P as an evolution equation for the additional stress tensor. As (12.16) is independent of the phase-field and the species concentration, they had to distinguish between the Newtonian and the non-Newtonian fluid by using $\tfrac{1}{2}(1 + \phi)\mathbf{K}$ on the right-hand side of (12.13a), i.e. they considered the additional stress tensor only in the non-Newtonian fluid indicated by $\phi = 1$. In particular, the model in [30] does not allow for gradual changes in the rheology.

12.3 Existence of Solutions

One disadvantage of Hookean elasticity is well known—it permits arbitrarily large polymer elongations. Instead, one may use the FENE (*f*initely *e*xtensible, *n*onlinear *e*lastic) spring potential which reads

$$U(s) := -\frac{Q_{\max}^2}{2} \log \left(1 - \frac{2s}{Q_{\max}^2}\right) . \tag{12.17}$$

In this case the critical polymer length of $Q_{\max}$ may not be exceeded, i.e. $D = B(0, Q_{\max})$. Making appropriate assumptions, the existence of weak solutions to (12.10) was established in [14] for the case of equal mass densities and a constant mobility m.

To describe the regularity of these solutions, we introduce the weighted Lebesgue and Sobolev spaces

$$L^2(\Omega \times D; M^{-1}) := \left\{\theta \in L_{\text{loc}}^1(\Omega \times D) : \int_{\Omega \times D} M^{-1} |\theta|^2 < \infty\right\} , \tag{12.18}$$

$$X := \left\{\theta \in L^2(\Omega \times D; M^{-1}) : \int_{\Omega \times D} M \left|\nabla_{\mathbf{x}} \tfrac{\theta}{M}\right|^2 + \int_{\Omega \times D} M \left|\nabla_{\mathbf{q}} \tfrac{\theta}{M}\right|^2 < \infty\right\} , \tag{12.19}$$

$$X_+ := \{\theta \in X : \theta(\mathbf{x}, \mathbf{q}) \geq 0 \text{ for a.e. } (\mathbf{x}, \mathbf{q}) \in \Omega \times D\} . \tag{12.20}$$

Furthermore, we denote the dual space of $H_{0,\text{div}}^1(\Omega) := \{\mathbf{w} \in H_0^1(\Omega) : \text{div}_{\mathbf{x}} \mathbf{w} = 0\}$ by $(H_{0,\text{div}}^1(\Omega))'$ and the associated dual pairing by $\langle ., . \rangle$. The following result was deduced in [14] by proving existence of solutions to a time discrete version of (12.10) and passing to the limit. For a detailed list of assumptions, we refer to [14].

Theorem 1 *Let $d \in \{2, 3\}$. Given initial data $(\phi_0, \mathbf{u}_0, \psi_0) \in H^2(\Omega) \times H_{0,\text{div}}^1(\Omega) \times L^2(\Omega \times D; M^{-1})$, there is a quadruple $(\phi, \mu, \mathbf{u}, \psi)$ which solves the equal-density version of system (12.10) combined with the boundary conditions (12.11) in the following weak sense.*

$$\int_{\Omega_T} (\phi_0 - \phi) \, \partial_t \theta - \int_{\Omega_T} \phi \mathbf{u} \cdot \nabla_{\mathbf{x}} \theta + \int_{\Omega_T} \nabla_{\mathbf{x}} \mu_\phi \cdot \nabla_{\mathbf{x}} \theta = 0$$

$$\forall \theta \in C^1\left([0, T]; H^1(\Omega)\right) \text{ with } \theta(., T) \equiv 0, \tag{12.21a}$$

$$\int_{\Omega_T} \mu_\phi \theta = \int_{\Omega_T} \delta \nabla_x \phi \cdot \nabla_x \theta + \int_{\Omega_T} \tfrac{1}{\delta} W'(\phi) \theta + \int_{\Omega_T} \beta'(\phi) \mathscr{J}_\varepsilon \left\{ \int_D \psi \right\} \theta$$

$$\forall \theta \in L^2\left(0,T;H^1(\Omega)\right), \qquad (12.21\mathrm{b})$$

$$\int_0^T \langle \partial_t \mathbf{u}, \mathbf{w} \rangle + \int_{\Omega_T} (\mathbf{u} \cdot \nabla_x)\,\mathbf{u} \cdot \mathbf{w} + \int_{\Omega_T} 2\eta(\phi)\,\mathbf{Du} : \mathbf{Dw}$$

$$= \int_{\Omega_T} \mathrm{div}_x \left\{ \mathfrak{J}_\varepsilon \left\{ \int_D \psi U' \mathbf{q} \otimes \mathbf{q} \right\} \right\} \cdot \mathbf{w} + \int_{\Omega_T} \mu_\phi \nabla_x \phi \cdot \mathbf{w}$$

$$+ \int_0^T \int_{\Omega \times D} \mathscr{J}_\varepsilon \{\beta(\phi)\} \nabla_x \psi \cdot \mathbf{w}$$

$$\forall \mathbf{w} \in L^{4/(4-d)}\left(0,T;H^1_{0,\mathrm{div}}(\Omega)\right),$$
$$(12.21\mathrm{c})$$

$$\int_0^T \int_{\Omega \times D} (\psi_0 - \psi)\,\partial_t \tfrac{\theta}{M} + \int_0^T \int_{\Omega \times D} M \tfrac{\theta}{M} \mathbf{u} \cdot \nabla_x \tfrac{\psi}{M} - \int_0^T \int_{\Omega \times D} \psi \left(\nabla_x J_\varepsilon\{\mathbf{u}\} \cdot \mathbf{q}\right) \cdot \nabla_q \tfrac{\theta}{M}$$

$$+ c_\mathbf{q} \int_0^T \int_{\Omega \times D} M \nabla_\mathbf{q} \tfrac{\psi}{M} \cdot \nabla_\mathbf{q} \tfrac{\theta}{M} + c_\mathbf{x} \int_0^T \int_{\Omega \times D} M \nabla_\mathbf{x} \tfrac{\psi}{M} \cdot \nabla_\mathbf{x} \tfrac{\theta}{M}$$

$$+ c_\mathbf{x} \int_0^T \int_{\Omega \times D} \psi \nabla_\mathbf{x} \mathscr{J}_\varepsilon \{\beta(\phi)\} \cdot \nabla_\mathbf{x} \tfrac{\theta}{M} = 0$$

$$(12.21\mathrm{d})$$

for all θ such that $\frac{\theta}{M} \in C^1\left([0,T];C^\infty\left(\overline{\Omega \times D}\right)\right)$ and $\theta(\cdot,T) \equiv 0$. Moreover, the solution has the following regularity properties.

$$\mathbf{u} \in L^\infty\left(0,T;L^2(\Omega)\right) \cap L^2\left(0,T;H^1_{0,\mathrm{div}}(\Omega)\right) \cap W^{1,4/d}\left(0,T;\left(H^1_{0,\mathrm{div}}(\Omega)\right)'\right),$$
$$(12.22\mathrm{a})$$

$$\phi \in L^\infty\left(0,T;H^1(\Omega)\right) \cap L^2\left(0,T;H^2(\Omega)\right), \qquad (12.22\mathrm{b})$$

$$\mu_\phi \in L^2\left(0,T;H^1(\Omega)\right), \qquad (12.22\mathrm{c})$$

$$\psi \in L^2\left(0,T;X_+\right) \cap L^\infty\left(0,T;L^2(\Omega \times D;M^{-1})\right), \qquad (12.22\mathrm{d})$$

$$\omega := \int_D \psi(\cdot,\mathbf{q})\mathrm{d}\mathbf{q} \in L^\infty\left(0,T;L^2(\Omega)\right) \cap L^2\left(0,T;H^1(\Omega)\right), \qquad (12.22\mathrm{e})$$

with $\omega \geq 0$ a.e. in Ω_T.

The above mentioned result relies on the mollification of certain terms to compensate for the little regularity for the configurational density ψ provided by the first stability result. For single-phase flow, this lack of regularity may also be overcome by using a cut-off operator in the $\mathbf{q}$-convective term, although revoking the cut-off when passing to the limit is not an easy task and requires an L^∞-bound on the marginal $\int_D \psi$ in space and time (cf. [5]). With a view to the numerical treatment of the system, the application of the mollifier also seems to be preferable, as by now the cut-off operator may not be revoked when passing to the limit simultaneously in space and time (cf. [6]).

Similar considerations apply to the convolution in the Henry energy. The strong nonlinear coupling between (12.10c) and (12.10e) requires some regularization. Without mollification, a s-Dirichlet integral might be introduced in the Cahn–Hilliard energy density (cf. [1, 12])—leading, however, to additional severe nonlinearities in the chemical potential.

12.4 Simulations Based on a Convergent Finite Element Scheme

A stable, fully discrete finite element scheme suitable for the numerical treatment of model (12.10) is proposed in [24]. Simulations based on this scheme are used for a qualitative validation of the presented model. Placing an elliptical shaped, non-Newtonian droplet ($\phi = 1$) with axes of length 1.3 and 0.7 and barycenter at $(0,0)$ in a domain ($\Omega = (-1, 1)^2$) filled with a Newtonian fluid ($\phi = -1$), we observe its oscillatory behaviour and compute the stresses induced by the additional stress tensor (see Fig. 12.1). As the polymer chains are initially assumed to be in equilibrium, i.e. their distribution on the configurational space $D = B(0, 10)$ is prescribed by the Maxwellian (i.e. $\psi_0(\mathbf{x}, \mathbf{q}) = \max\{0, \phi_0(\mathbf{x})\}3M(\mathbf{q})$), there are no additional stresses at the beginning of the simulation. The remaining physical parameters are listed below, where σ represents the surface tension and weights the Cahn–Hilliard energy.

σ	δ	m	$\rho(\pm 1)$	$\eta(\pm 1)$	$c_\mathbf{x}$	$c_\mathbf{q}$	$\beta(-1)$	$\beta(+1)$	ε
10	0.01	10^{-4}	2	0.005	0.01	0.1	5	1	0.01

While the triangulation of Ω is adaptive and consists of simplices with diameters between approximately 0.0667 and 0.0083, the approximation of D is adapted to the Maxwellian and consists of simplices with diameters between approximately 3.5355 and 0.3115. Concerning the discretization in time, we decided for a fixed time increment $\tau = 10^{-4}$. For the discrete scheme and discretization related regularization parameters, we refer to [24] (in particular Sect. 4.3.1).

In the course of the simulation, the droplet tries to attain an energetically more preferable circular shape and thereby gives rise to velocity fields. As these velocity fields induce deviations in the configurational distribution of the polymers, additional stresses arise. To visualize those stresses, we computed the eigenvalues and eigenvectors of the stress tensor on every simplex of the triangulation and depicted the eigenvector to the largest positive, real eigenvalue as a yellow line in Fig. 12.1. In comparison to its initial state, the droplet is stretched vertically and therefore the eigenvectors also point mainly in vertical direction.

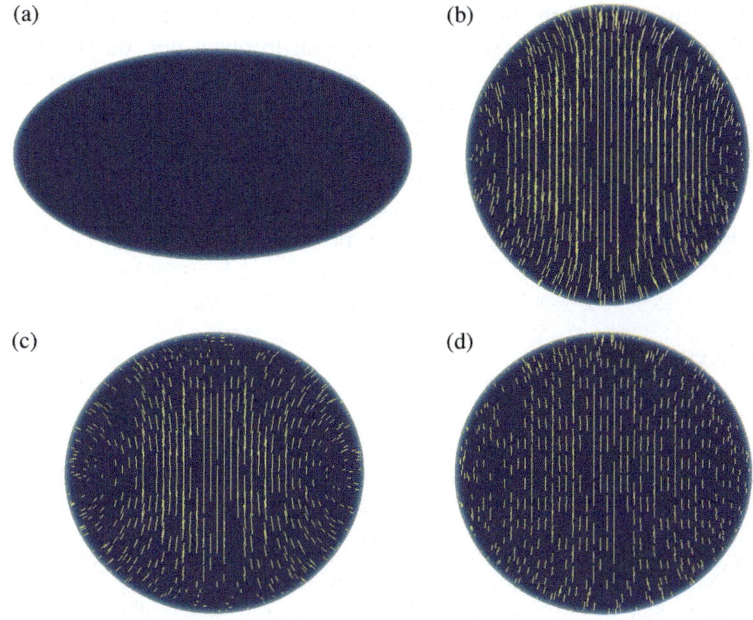

Fig. 12.1 Evolution of a non-Newtonian droplet. (**a**) $t = 0.0$. (**b**) $t = 0.2$. (**c**) $t = 1.0$. (**d**) $t = 2.0$

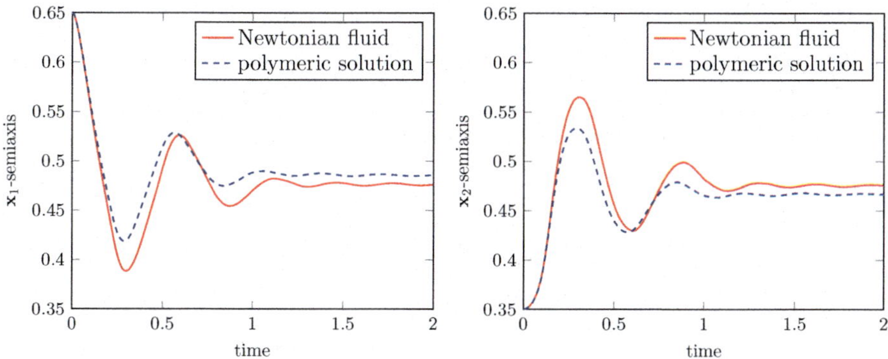

Fig. 12.2 Length of the semiaxes of oscillating droplets

To compare the oscillatory behaviour of a Newtonian and a non-Newtonian droplet, we measured and plotted the length of the $\mathbf{x}_1$- and $\mathbf{x}_2$-semiaxes of the droplets (see Fig. 12.2). When comparing the evolution of the axes' length, two phenomena are noticeable. First, the damping of the oscillation is asymmetric for the non-Newtonian droplet. In comparison to the axes of the Newtonian droplet, the maximal elongation of the vertical axis is smaller, while the horizontal axis reaches almost the same length in the ensuing oscillation as the one of the Newtonian droplet. Secondly, the shape of the non-Newtonian droplet is not completely circular at the end of the simulation, as the stresses induced by the deformation of the polymer chains are not completely released (cf. Fig. 12.1d).

Details on the scheme and the implementation, as well as the proof of stability can be found in [24]. Given fluids with equal mass densities, the convergence of discrete solutions towards solutions of a continuous weak formulation, which is comparable to the one in Theorem 1, is also proven in [24].

References

1. Abels, H.: Existence of weak solutions for a diffuse interface model for viscous incompressible fluids with general densities. Commun. Math. Phys. **289**(1), 45–73 (2009)
2. Abels, H., Garcke, H., Grün, G.: Thermodynamically consistent, frame indifferent diffuse interface models for incompressible two-phase flows with different densities. Math. Models Methods Appl. Sci. **22**(3), 1150013 (2012)
3. Barrett, J.W., Schwab, C., Süli, E.: Existence of global weak solutions for some polymeric flow models. Math. Models Methods Appl. Sci. **15**, 939–983 (2005)
4. Barrett, J.W., Süli, E.: Existence of global weak solutions to some regularized kinetic models for dilute polymers. Multiscale Model. Simul. **6**, 506–546 (2007)
5. Barrett, J.W., Süli, E.: Existence and equilibration of global weak solutions to kinetic models for dilute polymers I: Finitely extensible nonlinear bead-spring chains. Math. Models Methods Appl. Sci. **21**, 1211–1289 (2011)
6. Barrett, J.W., Süli, E.: Finite element approximation of finitely extensible nonlinear elastic dumbbell models for dilute polymers. ESAIM: Math. Modell. Numer. Anal. **46**, 949–978 (2012)
7. Bird, R.B., Curtiss, C.F.: Molecular theory expressions for the stress tensor in flowing polymeric liquids. J. Polymer Sci.: Polym. Symp. **73**, 187–199 (1985)
8. Campillo-Funollet, E., Grün, G., Klingbeil, F.: On modeling and simulation of electrokinetic phenomena in two-phase flow with general mass densities. SIAM J. Appl. Math. **72**(6), 1899–1925 (2012)
9. Constantin, P.: Nonlinear Fokker–Planck Navier–Stokes systems. Commun. Math. Sci. **3**(4), 531–544 (2005)
10. Ding, H., Spelt, P.D.M., Shu, C.: Diffuse interface model for incompressible two-phase flows with large density ratios. J. Comput. Phys. **226**, 2078–2095 (2007)
11. E, W., Li, T., Zhang, P.-W.: Well-posedness for the dumbbell model of polymeric fluids. Commun. Math. Phys. **248**(2), 409–427 (2004). MR2073140 (2005d:35210)
12. Fontelos, M.A., Grün, G., Jörres, S.: On a phase-field model for electrowetting and other electrokinetic phenomena. SIAM J. Math. Anal. **43**(1), 527–563 (2011)
13. Griebel, M., Rüttgers, A.: Multiscale simulations of three-dimensional viscoelastic flows in a square-square contraction. J. Non-Newtonian Fluid Mech. **205**, 41–63 (2014)
14. Grün, G., Metzger, S.: On micro-macro-models for two-phase flow with dilute polymeric solutions – modeling and analysis. Math. Models Methods Appl. Sci. **26**(05), 823–866 (2016)
15. Jourdain, B., Leliévre, T., Le Bris, C.: Existence of solution for a micro–macro model of polymeric fluid: the FENE model. J. Funct. Anal. **209**(1), 162–193 (2004)
16. Kramers, H.A.: Het gedrag van macromoleculen in een stroomende vloeistof. Physica **11**, 1–19 (1944)

17. Le Bris, C., Lelièvre, T.: Multiscale modelling of complex fluids: a mathematical initiation. In: Engquist, B., Lötstedt, P., Runborg, O. (eds.) Multiscale Modeling and Simulation in Science. Lecture Notes in Computational Science and Engineering, vol. 66, pp. 49–137. Springer, Berlin (2009)
18. Le Bris, C., Lelièvre, T.: Micro-macro models for viscoelastic fluids: modelling, mathematics and numerics. Sci. China Math. **55**(2), 353–384 (2012)
19. Li, T., Zhang, H., Zhang, P.-W.: Local existence for the dumbbell model of polymeric fluids. Commun. Partial Differ. Equ. **29**(5–6), 903–923 (2004)
20. Lions, P.L., Masmoudi, N.: Global existence of weak solutions to some micro-macro models. C. R. Math. Acad. Sci. Paris **345**(1), 15–20 (2007)
21. Lowengrub, J., Truskinovsky, L.: Quasi-incompressible Cahn–Hilliard fluids and topological transitions. Proc. R. Soc. Lond. Ser. A Math. Phys. Eng. Sci. **454**(1978), 2617–2654 (1998)
22. Masmoudi, N.: Well-posedness for the fene dumbbell model of polymeric flows. Commun. Pure Appl. Math. **61**(12), 1685–1714 (2008)
23. Masmoudi, N.: Global existence of weak solutions to the FENE dumbbell model of polymeric flows. Invent. Math. **191**(2), 427–500 (2013)
24. Metzger, S.: Diffuse interface models for complex flow scenarios: modeling, analysis, and simulation. Ph.D. thesis, Friedrich-Alexander-Universität Erlangen-Nürnberg (2017)
25. Nitta, K.: A graph-theoretical approach to statistics and dynamics of tree-like molecules. J. Math. Chem. **25**(2), 133–143 (1999)
26. Onsager, L.: Reciprocal relations in irreversible processes. II. Phys. Rev. **38**(12), 2265–2279 (1931)
27. Otto, F., Tzavaras, A.: Continuity of velocity gradients in suspensions of rod-like molecules. Commun. Math. Phys. **277**(3), 729–758 (2008) (English)
28. Qian, T., Wang, X.-P., Sheng, P.: A variational approach to the moving contact line hydrodynamics. J. Fluid Mech. **564**, 333–360 (2006)
29. Renardy, M.: An existence theorem for model equations resulting from kinetic theories of polymer solutions. SIAM J. Math. Anal. **22**(2), 313–327 (1991)
30. Yue, P., Feng, J.J., Liu, C., Shen, J.: A diffuse-interface method for simulating two-phase flows of complex fluids. J. Fluid Mech **515**, 293–317 (2004)
31. Zhang, H., Zhang, P.-W.: Local existence for the fene-dumbbell model of polymeric fluids. Arch. Ration. Mech. Anal. **181**(2), 373–400 (2006)

Chapter 13
Fully Adaptive and Integrated Numerical Methods for the Simulation and Control of Variable Density Multiphase Flows Governed by Diffuse Interface Models

Michael Hintermüller, Michael Hinze, Christian Kahle, and Tobias Keil

Abstract The present work is concerned with the simulation and optimal control of two-phase flows. We provide stable time discretization schemes for the simulation based on both, smooth and non-smooth free energy densities, which we combine with a practical, reliable and efficient adaptive mesh refinement concept for the spatial variables. Furthermore, we consider optimal control problems for two-phase flows and, among other things, derive first order optimality conditions. In the presence of smooth free energies we encounter classical Karush-Kuhn-Tucker (KKT) conditions, while in the case of non-smooth free energies we can prove C(larke)-stationarity. Moreover, we propose a dual weighted residual concept for spatial mesh adaptivity which is based on the newly derived stationarity conditions. We also address future research directions, including closed-loop control concepts and model order reduction techniques for simulation and control of variable density multiphase flows.

M. Hintermüller
Weierstraß-Institut, Mohrenstrasse 39, 10117 Berlin, Germany

Institut für Mathematik, Humboldt-Universität zu Berlin, Unter den Linden 6, 10099 Berlin, Germany
e-mail: hint@math.hu-berlin.de

M. Hinze (✉)
Fachbereich Mathematik, Universität Hamburg, Bundesstrasse 55, 20146 Hamburg, Germany
e-mail: michael.hinze@uni-hamburg.de

C. Kahle
Center for Mathematical Sciences, Technical University of Munich, Chair of Optimal Control, M17, Boltzmannstraße 3, 85748 Garching bei Muenchen, Germany
e-mail: christian.kahle@ma.tum.de

T. Keil
Institut für Mathematik, Humboldt-Universität zu Berlin, Unter den Linden 6, 10099 Berlin, Germany
e-mail: tkeil@math.hu-berlin.de

© Springer International Publishing AG 2017

D. Bothe, A. Reusken (eds.), *Transport Processes at Fluidic Interfaces*, Advances in Mathematical Fluid Mechanics, DOI 10.1007/978-3-319-56602-3_13

13.1 Introduction

In the present work we

(a) develop and analyze numerical discretization concepts for the simulation of two-phase flow problems with variable fluid densities that guarantee a locally refined resolution of the local processes at the interface and preserve the thermodynamically consistency of the underlying models;
(b) formulate and analyze optimal control problems for two-phase flows and develop robust and reliable solution strategies for optimal control of two-phase flows governed by diffuse interface models.

Concerning (a) we extend the work [65] to the thermodynamically consistent model for two-phase flow with different densities proposed in [2]. This allows to accurately resolve the transition region between the fluid phases, thus yielding quantifiable simulation of the physical processes located in the interfacial region. Special care is taken to preserve the thermodynamical consistency on the time discrete and fully (i.e., space and time) discrete level. In addition, residual-based a posteriori error estimation of the flow is incorporated into our approach. The benchmark for rising bubble dynamics from [79] is used as a validation.

In many applications, one is interested in steering the underlying multiphase flow towards a desired phase pattern, e.g., at a specific (final) time and/or to a desired flow profile, e.g., yielding rotation-free flows. Particular applications can be found in polymer science, where membrane formation or blending are of importance. In the former case, the porosity pattern of the membrane determines the membrane's use and quality, whereas in the latter context particular material properties of the new blends can be obtained. Moreover, for instance, in the case of immiscible polymer blends, on a macroscopic scale pure phases are present, but close to the interface certain diffusion processes take place. The latter are modeled by diffusive interfaces with certain partial differential equations (PDEs) posed within the narrow band of the diffuse interface. In this context, it appears that the double-obstacle potential is well suited for modeling the interface (indeed, the double-well potential allows unphysical violations of the constraints for the concentration, whereas a logarithmic potential would not allow to reach a pure phase). We also mention that further applications of surface active agents obeying PDEs on an (diffuse) interface include drug delivery, industrial emulsification, or liquid/liquid extraction and hydrodesulfurization of crude oil. These topics are analyzed within objective (b).

Mathematically, the underlying state system, i.e. the multi-phase flow model, is given by a coupled Cahn-Hilliard-Navier-Stokes (CHNS) system, where the Cahn-Hilliard part models the phase separation and the Navier-Stokes system captures the dynamics of the fluid. Task (b), which is mentioned above, hence requires to establish existence of solutions to the underlying control problems, stability and sensitivity of the CHNS system subject to perturbations, and the derivation of first-order optimality or stationarity conditions. In this context, the semi-discretization (in time) of the forward model (CHNS) has to additionally guarantee consistency properties of the resulting adjoint system in order to enable the derivation of certain energy-type estimates which yield existence of a solution of the semi-discrete forward problem and the associated adjoint. Such a discretization technique is proposed in [45, 71], where the regularization approach presented in the first reference yields the desired stationarity conditions of ε-almost C-stationary type and facilitates the implementation of a solution algorithm. Concerning the control action in the context of optimal control problems, we use Dirichlet boundary control as in [62] and control of the amplitudes of given distributed control actions [72].

Objective (b) is further concerned with the design and analysis of numerical solvers for the underlying control problems. Due to the extreme computational cost caused by solving the Navier-Stokes system coupled to a non-linear and non-smooth parabolic system for the phase separation for each time step, this especially requires an efficient mesh refinement technique for the underlying finite elements method. Therefore it is our goal to incorporate suitable error estimates which take the special structure of the optimal control problem into account by estimating the discretization error with respect to the objective functional, i.e. the quantity of interest.

13.1.1 Related Work

We commence with an overview of the state-of-the-art for the numerical treatment and analysis for optimal control problems for variable density two-phase flows.

We consider the diffuse interface approach for the simulation of two-phase flows. In contrast to so called sharp interface approaches, the interface between immiscible fluids in this approach is assumed to have a positive but small thickness. In sharp interface approaches this interface is assumed to be a lower dimensional manifold that is represented during numerical simulation, either explicitly or implicitly. Here we only refer to [47, 79] and [104] and the references therein. We further stress that several projects inside Special Priority Programme (SPP) 1506 of the German Research Foundation (DFG) worked on numerical realizations for sharp interface models and we refer to the corresponding proceedings for further readings.

In the following we restrict ourselves to diffuse-interface approaches.

13.1.1.1 Work Related to the Simulation of Two-Phase Flows Using the Diffuse Interface Approach

Since the pioneering work [30] and the famous model 'H' in [78] many authors have dealt with the investigation of two-phase flows using diffuse interface models with equal density fluids. In parallel, several attempts where made to generalize the model 'H' to the case of different densities.

For stable discretization schemes for the Cahn–Hilliard equation we refer to [53, 55] and for the Cahn–Hilliard Navier–Stokes system to [85]. Multigrid solvers for the Cahn–Hilliard equation are proposed e.g. in [83, 87], and residual based error estimation is proposed in [65, 67]. For a fully coupled solver for model 'H' we refer to [40].

Focusing now on models with different densities, one notes that one of the main limitations of model 'H' is that it is only thermodynamically consistent in situations where both fluids (roughly) have the same density. Indeed, in [37, 105] it is shown that the model is also consistent in the situation of different densities if the kinetic energy of the fluid is defined by using $\sqrt{\rho}|v|^2$ instead of $\rho|v|^2$, where ρ is the distributed density of the fluid and v is the velocity field. The notion of a distributed density is based on φ and by using the densities of the individual fluid components, a global density field is defined by attaching to every point of the computational domain the density of the fluid.

In [9], solutions to the model in [37] using a discretization that sequentially couples the Cahn–Hilliard and the Navier–Stokes equations are compared to results for a rising bubble benchmark proposed in [79] using sharp interface numerics. The authors obtained good agreement of the numerical results. A critical point in numerical treatment of these equations is the discretization of interfacial forces that appear from surface tension. In [8], different stabilization schemes are proposed for the discretization of this terms and a CFL-like condition for admissible time step sizes is derived. In [86] a completely new model for this forces is proposed.

A first thermodynamically consistent model for two-phase flow with different densities is considered in [92]. Here the velocity is not solenoidal so that analytical and numerical investigations of this scheme explicitly have to consider the pressure. Further the pressure enters the equation for the two-phase structure leading to a strong coupling of the resulting equations. In [57], a time discretization scheme is proposed that preserves the consistency with thermodynamics and numerical examples are provided.

Another diffuse interface model for fluids of different densities is proposed in [25]. Here the velocity is solenoidal, but the model is not consistent with thermodynamics. In [9], results for an implementation of this model are compared with results for sharp interface models for a benchmark of rising bubble simulations as proposed in [79].

In [2], a thermodynamically consistent model for two-phase flows is proposed. It contains a solenoidal velocity field and can be regarded as an extension of model 'H', as it resembles its structure and only differs from it by using variable densities and by an additional term in the convective term in the Navier–Stokes equation. The latter term vanishes in the case of equal densities. In [2], three variants of this model are proposed that can also handle, e.g., additional particles that are transported across the interface but do not interact with it. The existence of a weak solution for the case of constant mobility is shown in [3] for the logarithmic free energy that guarantees a-priori bounds on the range of the phase field. In [4], the existence of weak solutions is established for general smooth free energies together with a degenerate mobility that also guarantees these bounds. The existence of a weak solution for non-Newtonian fluids is discussed in [1] for a polynomially bounded free energy and constant mobility. In the latter work, also an extension of the model of [2] is proposed that allows to use nonlinear but smooth relations between the phase field φ and the density field $\rho(\varphi)$ for the case where $|\varphi|$ is not bounded by one, which appears due to a smooth free energy. We note here that by convention

$\varphi \equiv -1$ if a pure fluid phase is reached and $\varphi \equiv 1$ if the respective other pure fluid phase is present.

In [48], the existence of generalized solutions is shown for the case of a polynomially bounded free energy and constant mobility. Depending on the densities of the individual fluids, expressed using the Atwood number, these generalized solutions are weak solutions for some time horizon. The analysis is based on proceeding to the limit in a numerical scheme.

In [9], a discretization of this model is used to simulate the rising bubble benchmark from [79]. These authors obtained good agreements with results from sharp interface numeric. The free energy density is chosen as the smooth polynomially free energy and is linearized using Taylor expansion. The Navier–Stokes equation and the Cahn–Hilliard equation are used sequentially coupled by using the velocity field from the old time instance in the Cahn–Hilliard equation and solving the Navier–Stokes equation afterwards using the phase field and chemical potential obtained from the Cahn–Hilliard equation. We note that the important property of thermodynamically consistency is not preserved in the numerical realization by this approach.

Throughout the following publications, thermodynamically consistent time discretization schemes for the model from [2] are proposed.

A first thermodynamically consistent scheme is proposed in [49] that strongly couples the Chan–Hilliard and the Navier–Stokes equation. It is implemented in a splitting scheme, where the Navier–Stokes and the Cahn–Hilliard system as solved on each time instance subsequently until convergence. This allows the treatment of the (typically dominant) convection in the Cahn–Hilliard equation using a higher order finite volume scheme.

In [54] and [50], a thermodynamically consistent splitting scheme is proposed. Here the time discretization is used to sequentially couple the Cahn–Hilliard and Navier–Stokes equation, such that the two systems can be solved one after the other. In [50], additional various discretization methods for the polynomially bounded free energy are proposed and convergence of the scheme for vanishing discretization parameter to the model of [2] is shown.

In [71], for the purpose of optimal control of two-phase flow a stable time discretization is proposed. A scheme that also preserves the thermodynamical consistency in the fully discrete setting is proposed in [45]. The scheme is linear except for the usual non-linearity resulting from the free energy that is polynomially bounded. The mobility is not degenerate. The authors provide rigorous residual based error estimation to formulate an adaptive scheme. The consistency in the fully discrete setting is obtained by a suitable post processing step after the marking of cells for refinement and coarsening.

Based on the model of [2] in [42], a thermodynamically consistent model for two-phase flow with different densities is proposed that can also handle additional surface active agents, so called surfactants. These particles adhere to the interface, following some advection-diffusion equation and some sorption laws. On the interface they lower locally the surface tension of the interface. Thus, this model especially contains a locally varying surface tension and a partial differential equation on a diffuse interface. This work also contains numerical results, where especially the results of [39] on the simulation of partial differential equations on evolving interfaces that are given by a diffuse interface approach are used.

For a model that allows phase transition we refer to [7].

In the situation of multi phase flows with more then two fluid components a vector-valued phase field equation is used. Here we only refer to [23, 26].

Finally we stress that results concerning the simulation of one-phase flows stay valid or are a good starting point for two-phase flow and we only refer to the book [47] for the former.

For further reading on diffuse interface models, we refer to the contributions of H. Abels (University of Regensburg), S. Aland (TU Dresden), and G. Grün (University of Erlangen-Nuremberg) and H. Garcke (University of Regensburg) in this book.

13.1.1.2 Work Related to Adaptive Concepts for Two-Phase Flows Using the Diffuse Interface Approach

The special structure of the phase field that models the spatial distribution makes it necessary to use an appropriately adapted spatial discretization. This is commonly achieved by heuristic mesh refinement. As the interfacial region is known to be characterized by $|\varphi| < 1$ typically local refinement based on the modulus of φ is used, see e.g. [9, 22, 85]. On the other hand, as at the center of the interfacial region we have $|\nabla\varphi| \approx \frac{1}{\pi\varepsilon}$, the value $|\nabla\varphi|$ is used as an indicator for the interface in [49]. The first variant leads to a homogeneously refined mesh across the interface, while in the second case most refinement takes place around the zero level line of φ where $|\nabla\varphi|$ takes its maximum. We refer to [67], for a comparison of different refinement and marking strategies.

In [65], a reliable and efficient residual based error estimation is proposed for the Cahn–Hilliard system with a relaxed non-smooth double-obstacle free energy. In [67], the former work is extended to the simulation of two-phase flow based on model 'H' and it is further extended to the simulation of variable density two-phase flow based on the model of [2] in [45], where additionally arbitrary polynomially bounded free energies are used. We note that based on results of [31] in [43] for a Cahn–Hilliard type model it is argued that an estimator based on the jumps of normal derivatives in general will result in well adapted meshes.

A-posteriori error estimation for the Cahn–Hilliard systems with non-smooth double obstacle free energy is proposed in [11, 12]. There, also residual based error estimation is proposed and reliability of the derived estimator is shown.

13.1.1.3 Work Related to MPECs

The presence of a non-smooth homogeneous free energy density associated with the underlying Ginzburg-Landau energy in the Cahn-Hilliard system gives rise to an optimal control problem governed by a variational inequality. Hence the problem falls into the realm of so-called mathematical programs with equilibrium constraints (MPECs) in function spaces, cf. [62, 71]. The main difficulty in dealing with MPECs is that the feasible set which can be characterized by the solution operator of a variational inequality is usually non smooth and non convex and therefore violates the typical constraint qualifications known in classical optimization theory. As a result, stationarity conditions for this problem class are no longer unique. In finite dimensions, MPECs and the associated difficulties are already fairly well understood, see, e.g., the monographs [93, 98, 101] and the references therein.

In contrast, the literature on infinite dimensional MPECs is comparatively scarce. In [20, 94, 95], the authors use the conical derivative of the solution operator of the variational inequality to derive a stationarity system for the control problem, which one would classify now as strong stationarity. A different approach is introduced in [14], where the variational inequalities are approximated by variational equations and optimality conditions are derived by a passage to the limit in the approximation process. This technique typically yields a weaker stationarity system only. Further contributions to the topic include [15, 21, 41, 80, 106] most of which use regularization-penalization methods.

A first step towards the systematization and completion of stationarity concepts in function space was undertaken in [60], where the concept of ε-almost C-stationarity is introduced, paving the way for various contributions in the recent past. Here, we mention [69] where an abstract first-order optimality system is derived by means of variational analysis. In [109], the MPEC is approximated by a sequence of non smooth problems similar to the virtual control approach from [88].

13.1.1.4 Work Related to Adaptive Concepts for Optimal Control Problems

In Sect. 13.1.1.2, we already discussed the importance of an adequate mesh refinement technique for solving the Cahn–Hilliard Navier–Stokes system numerically. In this subsection, we briefly comment on the available literature on adaptive finite element methods (AFEMs) for optimal control problems. Whereas AFEMs for partial differential equations have been studied in great detail over the last decades, see, e.g., [13, 99] and the references therein, the research on AFEMs for variational inequalities and optimal control problems is comparatively recent. It started in the beginning of the century with the works by Becker et al. [16, 17] which pioneered a new approach to error control and mesh adaptivity in the numerical solution of unconstrained optimal control problems governed by elliptic differential equations. Here the mesh adaptation is driven by weighted residual-based a posteriori error estimates which are derived by global duality arguments and include the error in the state, the adjoint state and the control. This general approach facilitated the control of the error with respect to any quantity of physical interest such as, e.g., the given objective functional. In the following years the approach was successfully transferred to optimal control problems with control constraints, see, e.g., [58, 64, 108], as well as state constraints, see, e.g., [18] and [59] where additional error terms coming from data oscillations have been considered.

Here we also mention the works by Deckelnick et al. [36], Günther and Hinze [56], and Hintermüller et al. [66] where reliable a-posteriori error bounds for optimal control problems governed by point wise gradient constraints on the state were derived and an adaptive solution algorithm was presented.

In [90, 91] and [100], some other approaches are depicted which directly utilize the residuals of the associated first order optimality systems of the optimal control problems. The first paper presents a-posteriori error estimates for finite element approximation of distributed optimal control problems with convex constraint sets, whereas the second one provides a reliable a-posteriori error estimator for optimal control problems with state and control constraints containing the L^∞-error of the state, the H^{-2}-residual of the adjoint equation as well as the L^2-residual of the variational inequality if the Slater condition is satisfied. Very recently, [103] presented a-posteriori error estimates for control-constrained, linear-quadratic optimal control problems using an altered norm motivated by the objective functional in order to measure the error.

In contrast to PDE-constrained optimal control problems, the literature on goal-oriented mesh adaptivity methods appears rather scarce with respect to MPECs in function spaces. However, in [29, 68] the method was successfully applied to the optimal control of elliptic variational inequalities.

13.2 An Energy Conserving Adaptive Discretization Scheme for Variable Density Two-Phase Flows

The subsequent investigations are based on the following diffuse interface model for two-phase flows as proposed in [2]:

$$\rho\partial_t v + ((\rho v + J) \cdot \nabla)\, v - \operatorname{div}(2\eta Dv) + \nabla p = \mu\nabla\varphi + \rho g \quad \forall x \in \Omega,\ \forall t \in I, \tag{13.1}$$

$$-\operatorname{div}(v) = 0 \qquad \forall x \in \Omega,\ \forall t \in I, \tag{13.2}$$

$$\partial_t\varphi + v \cdot \nabla\varphi - \operatorname{div}(m\nabla\mu) = 0 \qquad \forall x \in \Omega,\ \forall t \in I, \tag{13.3}$$

$$-\sigma\varepsilon\Delta\varphi + \Psi'(\varphi) - \mu = 0 \qquad \forall x \in \Omega,\ \forall t \in I, \tag{13.4}$$

$$v(0,x) = v_0(x) \qquad \forall x \in \Omega, \tag{13.5}$$

$$\varphi(0,x) = \varphi_0(x) \qquad \forall x \in \Omega, \tag{13.6}$$

$$v(t,x) = 0 \qquad \forall x \in \partial\Omega,\ \forall t \in I, \tag{13.7}$$

$$\nabla \mu(t,x) \cdot \vec{v}_\Omega = \nabla \varphi(t,x) \cdot \vec{v}_\Omega = 0 \qquad\qquad \forall x \in \partial\Omega, \ \forall t \in I,$$

$$(13.8)$$

where $J = -\frac{d\rho}{d\varphi} m \nabla \mu$. Here $\Omega \subset \mathbb{R}^n$, $n \in \{2,3\}$, denotes an open and bounded domain with outer normal $\vec{v}_\Omega$, $I = (0,T]$ with $0 < T < \infty$ a time interval, φ denotes the phase field, μ the chemical potential, v the volume averaged velocity, p the pressure, and $\rho = \rho(\varphi) = \frac{1}{2}((\rho_2 - \rho_1)\varphi + (\rho_1 + \rho_2))$ the mean density, where $0 < \rho_1 \le \rho_2$ denote the densities of the involved fluids. The viscosity is denoted by η and can be chosen arbitrarily positive, fulfilling $\eta(-1) = \tilde{\eta}_1$ and $\eta(1) = \tilde{\eta}_2$, with individual fluid viscosity $\tilde{\eta}_1, \tilde{\eta}_2$. Here we restrict to Newtonian fluids, but non-Newtonian fluids are covered by the model as well. The mobility is denoted by $m = m(\varphi)$. The gravitational force is denoted by g. By $Dv = \frac{1}{2}(\nabla v + (\nabla v)^t)$ we denote the symmetrized gradient. The scaled surface tension is denoted by σ and the interfacial width is proportional to ε. The scaling of the physical surface tension is required due to the diffuse interface approach, see [2, Sect. 4.3.4]. The free energy density is denoted by Ψ and fulfills $\mathrm{argmin}(\Psi) = \pm 1$. The initial data is given by $(\varphi_0, v_0) = (\varphi_a, v_a)$.

The above model couples the Navier–Stokes equations (13.1)–(13.2) to the Cahn–Hilliard model (13.3)–(13.5) in a thermodynamically consistent way, i.e. the energy inequality from Theorem 1 holds for the total energy of the system, which is the sum of a Ginzburg–Landau energy for the interface between the two fluids and the kinetic energy of the fluids.

Theorem 1 *Let v, φ, μ be a sufficiently smooth solution to* (13.1)–(13.8). *Then it holds that*

$$\frac{d}{dt}\left(\int_\Omega \frac{\rho}{2}|v|^2 + \frac{\sigma\varepsilon}{2}|\nabla\varphi|^2 + \frac{\sigma}{\varepsilon}\Psi(\varphi)\,dx\right) = -\int_\Omega 2\eta|Dv|^2 + m|\nabla\mu|^2\,dx + \int_\Omega \rho g v\,dx.$$

Furthermore, the existence of weak solutions to system (13.1)–(13.8) for specific choices of data, i.e. Ψ, m, η, is shown in [1, 3, 4, 48].

As mentioned above, it is our goal to extend the existing theory by a more application driven perspective focusing on the development of efficient numerical solvers for the problem. For this purpose, we subsequently present some advanced numerical solution techniques for the simulation of the problem itself, as well as for generic optimal control problems which contain the system (13.1)–(13.8) as constraints. This will be accompanied by a rigorous analysis of the underlying problems concerning, e.g., the thermodynamical consistency of the discretization scheme, mesh stability, existence of solutions, characterization of stationarity conditions, and a-priori error estimation.

The rest of this section is organized as follows. In Sect. 13.2.1 we propose a discretization scheme for the numerical treatment of (13.1)–(13.8) that preserves the consistency with thermodynamics in the fully discrete setting and that is nearly linear. This scheme is proposed and analytically investigated for polynomially bounded free energies. Residual based error estimation is proposed for the spatial discretization. Further optimal control and instantaneous control of two-phase fluids are proposed.

In Sect. 13.3, we analyse a general optimal control problem associated to a semi-discrete version of (13.1)–(13.8) where the Ginzburg-Landau energy is characterized by the double-obstacle potential. This includes the existence of feasible points, as well as global optimal solutions, and culminates in the derivation of ε-almost C-stationarity conditions via a Yosida regularization technique with a subsequent passage to the limit with the Yosida parameter.

Section 13.4 provides a rigorous derivation of a goal-oriented dual-weighted error estimator for the problem considered in Sect. 13.3 and presents some numerical results along with the details of the numerical implementation of the solution algorithm for the optimal control problem.

Notation

Let $\Omega \subset \mathbb{R}^n$, $n \in \{2,3\}$ denote a bounded domain with sufficiently smooth boundary $\partial\Omega$ and outer normal $\vec{\nu}_\Omega$. Let $I = (0,T]$ denote a time interval.

We use the conventional notation for Sobolev and Hilbert Spaces, see e.g. [5]. By $L^p(\Omega)$, $1 \leq p \leq \infty$, we denote the space of measurable functions on Ω, whose modulus to the power p is Lebesgue-integrable. $L^\infty(\Omega)$ denotes the space of measurable functions on Ω, which are essentially bounded. For $p = 2$ we denote by $L^2(\Omega)$ the Hilbert space of square integrable functions on Ω with inner product $(\cdot,\cdot)$ and norm $\|\cdot\|$. For a subset $D \subset \Omega$ and functions $f, g \in L^2(\Omega)$ we denote by $(f,g)_D$ the inner product of f and g restricted to D, and by $\|f\|_D$ the respective norm. By $W^{k,p}(\Omega)$, $k \geq 1$, $1 \leq p \leq \infty$, we denote the Sobolev space of functions admitting weak derivatives up to order k in $L^p(\Omega)$. If $p = 2$ we write $H^k(\Omega)$ to acknowledge the Hilbertian structure of the space. The subset $H_0^1(\Omega)$ denotes $H^1(\Omega)$ functions with vanishing boundary trace. For $k \in \mathbb{N}$, we further set

$$H_{0,\sigma}^k(\Omega;\mathbb{R}^n) := \left\{ f \in H^k(\Omega;\mathbb{R}^n) \cap H_0^1(\Omega;\mathbb{R}^n) : \mathrm{div} f = 0, \text{ a.e. on } \Omega \right\};$$

$$\overline{H}^k(\Omega) := H_{(0)}^k(\Omega) := \left\{ f \in H^k(\Omega) : \int_\Omega f dx = 0 \right\};$$

$$\overline{H}_{\partial_n}^k(\Omega) := \left\{ f \in \overline{H}^k(\Omega) : \partial_n f_{|\partial\Omega} = 0 \text{ on } \partial\Omega \right\}, \quad k \geq 2;$$

where 'a.e.' stands for 'almost everywhere' and the boundary condition is supposed to hold true in the trace sense. We stress that the subscript σ used here is not related to the surface tension, but using σ here is common notation.

Unless otherwise noted, $\langle\cdot,\cdot\rangle := \langle\cdot,\cdot\rangle_{\overline{H}^{-1},\overline{H}^1}$ represents the duality pairing between $\overline{H}^1(\Omega)$ and $\overline{H}^{-1}(\Omega)$. For $u \in L^q(\Omega)^n$, $q > n$, and $v, w \in H^1(\Omega)^n$ we introduce the trilinear form

$$a(u,v,w) = \frac{1}{2}\int_\Omega ((u \cdot \nabla) v) w \, dx - \frac{1}{2}\int_\Omega ((u \cdot \nabla) w) v \, dx. \tag{13.9}$$

Note that it holds that $a(u,v,w) = -a(u,w,v)$, and especially $a(u,v,v) = 0$.

13.2.1 A Stable Time Discretization for Smooth Free Energies

In this section we summarize the fully practical scheme for the numerical treatment of (13.1)–(13.8) as proposed in [45]. Here we assume that the free energy density Ψ is smooth, see Assumption A2. To emphasize this, in the following we denote the free energy density by W, and we introduce the splitting $W = W_+ + W_-$, where W_+ denotes the convex part of W and W_- denotes the concave part.

Assumption 1 *Concerning the data of our problem we invoke the following assumptions:*

A1 *There exist constants $\overline{\rho} \geq \underline{\rho} > 0$, $\overline{\eta} \geq \underline{\eta} > 0$, and $\overline{m} \geq \underline{m} > 0$ such that the following relations are satisfied:*

- *$\overline{\rho} \geq \rho(\varphi) \geq \underline{\rho} > 0$,*
- *$\overline{\eta} \geq \eta(\varphi) \geq \underline{\eta} > 0$,*
- *$\overline{m} \geq m(\varphi) \geq \underline{m} > 0$.*

  Especially we assume that the mobility is non degenerate. In addition, we assume that ρ, η, and m are continuous.

A2 *$W : \mathbb{R} \to \mathbb{R}$ is continuously differentiable.*

A3 *W and the derivatives W'_+ and W'_- are polynomially bounded, i.e. there exists $C > 0$ such that $|W(x)| \leq C(1 + |x|^q)$, $|W'_+(x)| \leq C(1 + |x|^{q-1})$ and $|W'_-(x)| \leq C(1 + |x|^{q-1})$ holds for some $q \in [1, 4]$ if $n = 3$ and $q \in [1, \infty)$ if $n = 2$,*

A4 *W'_+ is Newton (sometimes called slantly) differentiable (see e.g. [63]) regarded as nonlinear operator $W'_+ : H^1(\Omega) \to \left(H^1(\Omega)\right)^*$ with Newton derivative G satisfying*

$$(G(\varphi)\delta\varphi, \delta\varphi) \geq 0$$

for each $\varphi \in H^1(\Omega)$ and $\delta\varphi \in H^1(\Omega)$.

To ensure Assumption A1 we introduce a cut-off mechanism to guarantee the bounds on ρ defined in Assumption A1 independently of φ. Note that $\eta(\varphi)$ and $m(\varphi)$ can be chosen arbitrarily fulfilling the stated bounds. We define the mass density as a smooth, monotone and strictly positive function $\rho(\varphi)$ fulfilling

$$\rho(\varphi) = \begin{cases} \frac{\rho_2-\rho_1}{2}\varphi + \frac{\rho_1+\rho_2}{2} & \text{if } -1 - \frac{\rho_1}{\rho_2-\rho_1} < \varphi < 1 + \frac{\rho_1}{\rho_2-\rho_1}, \\ \text{const} & \text{if } \varphi > 1 + \frac{2\rho_1}{\rho_2-\rho_1}, \\ \text{const} & \text{if } \varphi < -1 - \frac{2\rho_1}{\rho_2-\rho_1}. \end{cases}$$

For a discussion we refer to [48, Remark 2.1].

Remark 1 We stress that in the following we require the affine linearity of $\rho(\varphi) := \frac{1}{2}\left((\rho_2 - \rho_1)\varphi + (\rho_1 + \rho_2)\right)$ to derive our numerical scheme. Also, we note that it is sufficient to have the affine-linearity of ρ for all values of φ that appear in a simulation. This is the case in all of our numerical examples. Concerning the necessity of the cut-off procedure we also note Remark 5. For the case of a nonlinear dependence between φ and ρ we also refer to [1], where an extension of model (13.1)–(13.8) to non-linear functions $\rho(\varphi)$ is proposed. Extending our results to this case is subject to future work.

Remark 2 Assumptions A2–A4 are for example fulfilled by the polynomial free energy

$$W^{poly}(\varphi) = \frac{1}{4}\left(1 - \varphi^2\right)^2.$$

Another free energy fulfilling these assumptions is the relaxed double-obstacle free energy given by

$$W^{rel}(\varphi) = \frac{1}{2}\left(1 - \varphi^2 + s\lambda^2(\varphi)\right),$$

with

$$\lambda(\varphi) := \max(0, \varphi - 1) + \min(0, \varphi + 1),$$

where $s \gg 0$ denotes the relaxation parameter. W^{rel} is introduced in [65] as Moreau–Yosida relaxation of the double-obstacle free energy

$$W^{obst}(\varphi) = \begin{cases} \frac{1}{2}\left(1 - \varphi^2\right) & \text{if } |\varphi| \le 1, \\ 0 & \text{else,} \end{cases}$$

which is proposed in [96] to model phase separation and is first analytically investigated in [24].

In the numerical examples of this section we use the free energy $W \equiv W^{rel}$. For this choice the splitting into convex and concave part reads

$$W_+(\varphi) = s\frac{1}{2}\lambda^2(\varphi), \qquad W_-(\varphi) = \frac{1}{2}(1 - \varphi^2).$$

We note that the minima of W^{rel} are at $\pm\frac{s}{s-1} \neq 1$. One might rescale the argument of W^{rel} to obtain $\mathrm{argmin}(W) = \pm 1$. As s is typically large and thus $\frac{s}{s-1} \approx 1$ we refrain from this rescaling.

With the above assumptions we introduce a weak formulation of (13.1)–(13.8) that we use to derive our discrete scheme. For this purpose, we restrict ourselves to solenoidal velocity fields in what follows, i.e., we neglect the pressure p.

For a sufficiently smooth solution (φ, μ, v) of (13.1)–(13.8) and using the linearity of ρ it holds that

$$\partial_t \rho + \mathrm{div}\left(\rho v + J\right) = 0,$$

see [2, p. 14]. Using this mass balance we can rewrite (13.1) as

$$\partial_t(\rho v) + \text{div}(\rho v \otimes v) + \text{div}(v \otimes J) - \text{div}(2\eta Dv) + \nabla p = \mu \nabla \varphi + \rho g. \tag{13.10}$$

We also observe that the term $\rho v + J$ in (13.1) is not solenoidal (which might lead to difficulties both in the analytical and the numerical treatment) and that the trilinear form $(((\rho v + J) \cdot \nabla)u, w)$ thus is not anti-symmetric. To obtain a weak formulation yielding an anti-symmetric convection term we use a convex combination of (13.1) and (13.10) to define a weak formulation.

This leads to the following weak formulation.

Definition 1 We call v, p, φ, μ a weak solution to (13.1)–(13.8) if $v(0) = v_0$, $\varphi(0) = \varphi_0$, $v(t) \in H^1_{0,\sigma}(\Omega; \mathbb{R}^n)$ for a.e. t, and

$$\frac{1}{2} \int_\Omega (\partial_t(\rho v) + \rho \partial_t v) \, w \, dx + \int_\Omega 2\eta Dv : Dw \, dx$$

$$+a(\rho v + J, v, w) - (p, \text{div}w) = \int_\Omega \mu \nabla \varphi w + \rho g w \, dx \quad \forall w \in H^1_0(\Omega)^n, \tag{13.11}$$

$$-(\text{div}v, q) = 0 \quad \forall q \in L^2_{(0)}(\Omega) \tag{13.12}$$

$$\int_\Omega (\partial_t \varphi + v \cdot \nabla \varphi) \, \Phi \, dx + \int_\Omega m(\varphi) \nabla \mu \cdot \nabla \Phi \, dx = 0 \quad \forall \Phi \in H^1(\Omega), \tag{13.13}$$

$$\sigma \varepsilon \int_\Omega \nabla \varphi \cdot \nabla \Psi \, dx + \frac{\sigma}{\varepsilon} \int_\Omega W'(\varphi) \Psi \, dx - \int_\Omega \mu \Psi \, dx = 0 \quad \forall \Psi \in H^1(\Omega), \tag{13.14}$$

is satisfied for almost all $t \in I$.

Using this weak formulation, the energy equality in Theorem 1 can be derived by using $w \equiv v$, $q \equiv p$, $\Phi \equiv \mu$, and $\Psi \equiv \varphi$ and summation over the resulting equations.

13.2.1.1 The Energy Stable Scheme

For a numerical treatment we next discretize the weak formulation (13.11)–(13.14) in time and space. We aim at an adaptive discretization of the domain Ω, and thus obtain a different spatial discretization in every time step.

For the discretization in time, let $0 = t_0 < t_1 < \ldots < t_{k-2} < t_{k-1} < t_k < \ldots < t_K = T$ denote an equidistant subdivision of the interval $\bar{I} = [0, T]$ with $\tau_{k+1} - \tau_k \equiv \tau$. From here onwards the superscript k denotes the corresponding variables at time instance t_k, and we abbreviate $\rho^k := \rho(\varphi^k)$, $\eta^k := \eta(\varphi^k)$, and $m^k := m(\varphi^k)$.

For the discretization in space let $\mathscr{T}^k := \bigcup_{i=1}^{N_T} T_i$ denote a conforming triangulation of $\overline{\Omega}$ with closed simplices $T_i, i = 1, \ldots, N_T$ and edges $E_i, i = 1, \ldots, N_E$, $\mathscr{E}^k := \bigcup_{i=1}^{N_E} E_i$. Here, k refers to the time instance t_k. On $\mathscr{T}^k$ we define the following finite element spaces:

$$V_1^k = \{v \in C(\mathscr{T}^k) \mid v|_T \in P^1(T) \; \forall T \in \mathscr{T}^k\},$$

$$V_2^k = \{v \in C(\mathscr{T}^k)^n \mid v|_T \in (P^2(T))^n \; \forall T \in \mathscr{T}^k, v|_{\partial\Omega} = 0\},$$

where $P^l(S)$ denotes the space of polynomials up to order l defined on S.

We further introduce an H^1-stable projection operator $\mathscr{P}^k : H^1(\Omega) \rightarrow V_1^k$ satisfying

$$\|\mathscr{P}^k v\|_{L^p(\Omega)} \le \|v\|_{L^p(\Omega)} \text{ and } \|\nabla \mathscr{P}^k v\|_{L^r(\Omega)} \le \|\nabla v\|_{L^r(\Omega)}$$

for $v \in H^1(\Omega)$ with $r \in [1, 2]$ and $p \in [1, 6]$ if $n = 3$, and $p \in [1, \infty)$ if $n = 2$. Possible choices are the H^1-projection, the Clément operator or, by restricting the preimage to $C(\overline{\Omega}) \cap H^1(\Omega)$, the Lagrangian interpolation operator, see e.g. [28]. For triangulations with specific properties, also the L^2-projection fulfills these assumption, see [27].

Using these spaces we state the fully discrete counterpart of (13.11)–(13.14): Let $k \ge 2$, given $\varphi^{k-2} \in V_1^{k-2}$, $\varphi^{k-1} \in V_1^{k-1}$, $\mu^{k-1} \in V_1^{k-1}$, $v^{k-1} \in V_2^k$, find $v_h^k \in V_1^k$, $p_h^k \in V_k^1 \cap L_{(0)}^2(\Omega)$, $\varphi_h^k \in V_1^k$, $\mu_h^k \in V_1^k$ such that for all $w \in V_2^k$, $q \in V_1^k \cap L_{(0)}^2(\Omega)$, $\Phi \in V_1^k$, $\Psi \in V_1^k$ it holds that:

$$\frac{1}{2\tau}(\rho^{k-1}v_h^k - \rho^{k-2}v^{k-1} + \rho^{k-2}(v_h^k - v^{k-1}), w) + a(\rho^{k-1}v^{k-1} + J^{k-1}, v_h^k, w)$$

$$+(2\eta^{k-1}Dv_h^k, Dw) - (p_h^k, \text{div} w) - (\mu_h^k \nabla \varphi^{k-1} + \rho^{k-1} g, w) = 0, \tag{13.15}$$

$$-(\text{div} v_h^k, q) = 0, \tag{13.16}$$

$$\frac{1}{\tau}(\varphi_h^k - \mathscr{P}^k \varphi^{k-1}, \Phi) + (m^{k-1}\nabla \mu_h^k, \nabla \Phi) + (v_h^k \nabla \varphi^{k-1}, \Phi) = 0, \tag{13.17}$$

$$\sigma\varepsilon(\nabla \varphi_h^k, \nabla \Psi) + \frac{\sigma}{\varepsilon}(W_+'(\varphi_h^k) + W_-'(\mathscr{P}^k \varphi^{k-1}), \Psi) - (\mu_h^k, \Psi) = 0, \tag{13.18}$$

where $J^{k-1} := -\frac{d\rho}{d\varphi} m^{k-1} \nabla \mu^{k-1}$, $\varphi^0 = P\varphi_0$ denotes the L^2 projection of φ_0 onto V_1^0, $v^0 = P^L v_0$ denotes the Leray projection of v_0 onto H_σ^0, see [34]. The functions

$\varphi_h^1 \in V_1^1$, $\mu_h^1 \in V_1^1$ and $v_h^1 \in V_2^1$ are obtained by a one-step-scheme for equation (13.1)–(13.8), see [45].

Remark 3 Due to the appearance of ρ^{k-2} in (13.15) the scheme (13.15)–(13.18) is a two-step scheme and thus needs additional initialization. By using ρ^{k-2} we obtain a scheme that is only nonlinear in the discretization of the convex part of the free energy density. However we can replace (13.15) by

$$\frac{1}{2\tau} \int_\Omega \left(\rho_h^k v_h^k - \rho^{k-1} v^{k-1} \right) w + \rho^{k-1}(v_h^k - v^{k-1}) w \, dx$$

$$+ a(\rho^{k-1} v^{k-1} + J^{k-1}, v_h^k, w) + \int_\Omega 2\eta^{k-1} Dv_h^k : Dw \, dx \tag{13.19}$$

$$- (p_h^k, \mathrm{div} w) - \int_\Omega \mu_h^k \nabla \varphi^{k-1} w + \rho^{k-1} gw \, dx = 0 \quad \forall w \in H_\sigma^k,$$

which leads to a one-step-scheme that then also is nonlinear in the Navier–Stokes equation. For the scheme including (13.15) the unique solution can be found by Newton's method, while—up to now—this has not been proven for the scheme including (13.19).

We further argue that if we have initial data φ_{-1} at some virtual time instance $t_{-1} := -\tau$ together with v_0, we can solve (13.17)–(13.18) to obtain φ_h^0 and μ_h^0 and then proceed with the scheme (13.15)–(13.18).

To guarantee mass conservation for φ_h^k we implement the term $(v_h^k \nabla \varphi^{k-1}, \Phi)$ in (13.17) by using integration by parts as $-(v_h^k \varphi^{k-1}, \nabla \Phi)$.

The following is proven in [45].

Theorem 2 *There exist unique* $v_h^k \in V_2^k$, $p_h^k \in V_1^k \cap L_{(0)}^2(\Omega)$, $\varphi_h^k \in V_1^k$, $\mu_h^k \in V_1^k$ *solving* (13.15)–(13.18).

Remark 4 As (13.15)–(13.18) is linear in all terms but the free energy density W_+ which is Newton differentiable and convex by Assumption A4 using the same arguments as for Theorem 2 one can show that Newton's method can be used to find the unique solution from Theorem 2.

The time discretization is chosen such that the following fully discrete correspondence to the energy equality in Theorem 1 holds.

Theorem 3 *Let* $(\varphi_h^k, \mu_h^k, v_h^k, p_h^k)$ *be a solution to* (13.15)–(13.18). *Then for* $k \geq 2$:

$$\frac{1}{2} \int_\Omega \rho^{k-1} \left| v_h^k \right|^2 dx + \frac{\sigma \varepsilon}{2} \int_\Omega |\nabla \varphi_h^k|^2 \, dx + \frac{\sigma}{\varepsilon} \int_\Omega W(\varphi_h^k) \, dx$$

$$+ \frac{1}{2} \int_\Omega \rho^{k-2} |v_h^k - v^{k-1}|^2 \, dx + \frac{\sigma \varepsilon}{2} \int_\Omega |\nabla \varphi_h^k - \nabla \mathscr{P}^k \varphi^{k-1}|^2 \, dx$$

$$+\tau \int_\Omega 2\eta^{k-1}|Dv_h^k|^2\,dx + \tau \int_\Omega m^{k-1}|\nabla\mu_h^k|^2\,dx$$

$$\leq \frac{1}{2}\int_\Omega \rho^{k-2}\left|v^{k-1}\right|^2\,dx + \frac{\sigma\varepsilon}{2}\int_\Omega |\nabla\mathscr{P}^k\varphi^{k-1}|^2\,dx + \frac{\sigma}{\varepsilon}\int_\Omega W(\mathscr{P}^k\varphi^{k-1})\,dx$$

$$+\tau\int_\Omega \rho^{k-1}gv_h^k.$$

$$(13.20)$$

So we observe that the energy

$$E(\varphi_h^k, v_h^k) := \frac{1}{2}\int_\Omega \rho^{k-1}|v_h^k|^2\,dx + \sigma\int_\Omega \frac{\varepsilon}{2}|\nabla\varphi^k|^2 + \frac{1}{\varepsilon}W(\varphi^k)\,dx$$

decays as in the continuous case due to diffusion and increases due to outer forces, like the gravitational force. We further have some addition loss of energy due to numerical dissipation. We further observe that $E(\varphi_h^k, v_h^k)$ is in fact bounded by $E(\mathscr{P}^k\varphi_h^{k-1}, v_h^{k-1})$ which is not necessarily bounded by the energy at time t_{k-1}, i.e. $E(\varphi_h^{k-1}, v_h^{k-1})$. This will be dealt with during our adaptive scheme to obtain a guaranteed decrease of the total energy. We formulate this in the following assumption.

Assumption 2 *For every solution $v_h^{k-1}, p_h^{k-1}, \varphi_h^{k-1}, \mu_h^{k-1}$ to (13.15)–(13.18) at time t_{k-1}, we assume that the following inequality is satisfied*

$$E(\mathscr{P}^k\varphi_h^{k-1}, v_h^{k-1}) \leq E(\varphi_h^{k-1}, v_h^{k-1}),$$

which is equivalent to

$$\sigma\int_\Omega \frac{\varepsilon}{2}|\nabla\mathscr{P}^k\varphi^{k-1}|^2 + \frac{1}{\varepsilon}W(\mathscr{P}^k\varphi^{k-1})\,dx \leq \sigma\int_\Omega \frac{\varepsilon}{2}|\nabla\varphi^{k-1}|^2 + \frac{1}{\varepsilon}W(\varphi^{k-1})\,dx.$$

We can use the unique solution from Theorem 2 in a Galerkin approach to investigate the limit $h \to 0$ in the scheme (13.15)–(13.18). This leads to the following time discrete scheme:
Given $\varphi^{k-2} \in H^1(\Omega)\cap L^\infty(\Omega)$, $\varphi^{k-1} \in H^1(\Omega)\cap L^\infty(\Omega)$, $\mu^{k-1} \in W^{1,q}(\Omega)$, $q > n$, $v^{k-1} \in H_0^1(\Omega)^n$,
find $v^k \in H_0^1(\Omega)^n$, $p^k \in L_{(0)}^2(\Omega)$, $\varphi^k \in H^1(\Omega)$, $\mu^k \in H^1(\Omega)$ satisfying

$$\frac{1}{2\tau}\int_\Omega \left(\rho^{k-1}v^k - \rho^{k-2}v^{k-1}\right)w + \rho^{k-2}(v^k - v^{k-1})w\,dx$$

$$+a(\rho^{k-1}v^{k-1} + J^{k-1}, v^k, w) + \int_\Omega 2\eta^{k-1}Dv^k : Dw\,dx$$

$$-(p^k, \mathrm{div}w) - \int_\Omega \mu^k \nabla\varphi^{k-1} w - \rho^{k-1} g w \, dx = 0 \quad \forall w \in H_0^1(\Omega)^n,$$

$$(13.21)$$

$$-(\mathrm{div}v^k, q) = 0 \qquad \forall q \in L_{(0)}^2(\Omega),$$

$$(13.22)$$

$$\frac{1}{\tau} \int_\Omega (\varphi^k - \varphi^{k-1})\Phi \, dx + \int_\Omega (v^k \cdot \nabla\varphi^{k-1})\Phi \, dx$$

$$+ \int_\Omega m^{k-1} \nabla\mu^k \cdot \nabla\Phi \, dx = 0 \quad \forall \Phi \in H^1(\Omega),$$

$$(13.23)$$

$$\sigma\varepsilon \int_\Omega \nabla\varphi^k \cdot \nabla\Psi \, dx - \int_\Omega \mu^k \Psi \, dx$$

$$+ \frac{\sigma}{\varepsilon} \int_\Omega ((W_+)'(\varphi^k) + (W_-)'(\varphi^{k-1}))\Psi \, dx = 0 \quad \forall\Psi \in H^1(\Omega),$$

$$(13.24)$$

where $J^{k-1} := -\frac{d\rho}{d\varphi} m^{k-1} \nabla\mu^{k-1}$.

Theorem 4 *Let* $\varphi^{k-2} \in H^1(\Omega) \cap L^\infty(\Omega)$, $\varphi^{k-1} \in H^1(\Omega) \cap L^\infty(\Omega)$, $\mu^{k-1} \in W^{1,q}(\Omega)$, $q > n$, $v^{k-1} \in H_0^1(\Omega)^n$, *be given. Then there exists a unique solution* $(v^k, p^k, \varphi^k, \mu^k) \in H_0^1(\Omega)^n \times L_{(0)}^2(\Omega) \times H^1(\Omega) \times H^1(\Omega)$ *to* (13.21)–(13.24). *In fact it holds* $\varphi^k \in H^2(\Omega) \subset H^1(\Omega) \cap L^\infty(\Omega)$ *and* $\mu^k \in W^{1,q}(\Omega), q > n$.

Also the energy inequality stays valid in the time discrete setting.

Theorem 5 *Let* $(\varphi^k, \mu^k, v^k, p^k)$ *be a solution to* (13.21)–(13.24). *Then for* $k \geq 2$:

$$\frac{1}{2} \int_\Omega \rho^{k-1} |v^k|^2 \, dx + \frac{\sigma\varepsilon}{2} \int_\Omega |\nabla\varphi^k|^2 \, dx + \frac{\sigma}{\varepsilon} \int_\Omega W(\varphi^k) \, dx$$

$$+ \frac{1}{2} \int_\Omega \rho^{k-2} |v^k - v^{k-1}|^2 \, dx + \frac{\sigma\varepsilon}{2} \int_\Omega |\nabla\varphi^k - \nabla\varphi^{k-1}|^2 \, dx$$

$$+ \tau \int_\Omega 2\eta^{k-1} |Dv^k|^2 \, dx + \tau \int_\Omega m^{k-1} |\nabla\mu^k|^2 \, dx \qquad (13.25)$$

$$\leq \frac{1}{2} \int_\Omega \rho^{k-2} |v^{k-1}|^2 \, dx + \frac{\sigma\varepsilon}{2} \int_\Omega |\nabla\varphi^{k-1}|^2 \, dx + \frac{\sigma}{\varepsilon} \int_\Omega W(\varphi^{k-1}) \, dx$$

$$+ \tau \int_\Omega \rho^{k-1} g v^k.$$

Remark 5 Let W denote the relaxed double-obstacle free energy W^{rel} introduced in Remark 2 with relaxation parameter s. Let $(v_s, p_s, \varphi_s, \mu_s)_{s\in\mathbb{R}}$ denote the sequence of solutions of (13.21)–(13.24) for a sequence $(s_l)_{l\in\mathbb{N}}$ with $s_l \to \infty$. From the linearity of (13.21) and [65, Prop. 4.2] it follows that there exists a subsequence, still denoted by $(v_s, p_s, \varphi_s, \mu_s)_{s \in \mathbb{R}}$, such that

$$(v_s, p_s, \varphi_s, \mu_s)_{s \in \mathbb{R}} \rightarrow (v^\star, p^\star, \varphi^\star, \mu^\star) \quad \text{in } H^1(\Omega),$$

where $(v^\star, p^\star, \varphi^\star, \mu^\star)$ denotes the solution of (13.21)–(13.24), where W^{obst}, denoted in Remark 2, is chosen as free energy. Especially $|\varphi^*| \leq 1$ holds true.

In the following argumentation we concentrate on the phase field only. >From the regularity $\varphi_s \in H^2(\Omega)$ together with a-priori estimates on the solution of the Poisson problem and the energy inequality of Theorem 5, we obtain the existence of a strongly convergent subsequence $\varphi_{s'} \rightarrow \varphi^*$ in $C^{0,\alpha}(\overline{\Omega})$, where we use the compact embedding $H^2(\Omega) \hookrightarrow C^{0,\alpha}(\overline{\Omega})$ for $2\alpha < 4 - n$.

Thus, for s large enough we have $|\varphi_s| \leq 1 + \theta$ with θ arbitrarily small. In fact, using arguments from [70] in [82] a rate

$$\| \max(|\varphi_s| - 1, 0) \|_{L^\infty(\Omega)} \leq Cs^{-1}$$

is argued for the pure Cahn–Hilliard system. This is also observed in the numerical tests of these works.

13.2.2 A Posteriori Error Estimation

For an efficient solution of (13.15)–(13.18) we next describe an a-posteriori error estimator based mesh refinement scheme that is reliable and efficient up to terms of higher order and errors introduced by the projection. We propose an all-in-one adaptation concept for the fully coupled Cahn–Hilliard Navier–Stokes system, where we exploit the energy inequality of Theorem 3. Further we describe how we can guarantee that the total energy can not increase in absence of outer forces, i.e. how we can guarantee the validity of Assumption 2.

We define the following error terms:

$$e_v := v_h^k - v^k, \qquad\qquad e_p := p_h^k - p^k, \qquad\qquad (13.26)$$

$$e_\varphi := \varphi_h^k - \varphi^k, \qquad\qquad e_\mu := \mu_h^k - \mu^k, \qquad\qquad (13.27)$$

as well as the discrete element residuals

$$r_h^{(1)} := \frac{\rho^{k-1} + \rho^{k-2}}{2} v_h^k - \rho^{k-2} v^{k-1} + \tau(b^{k-1}\nabla)v_h^k + \frac{1}{2}\tau \mathrm{div}(b^{k-1})v_h^k$$
$$- 2\tau \mathrm{div}\left(\eta^{k-1} D v_h^k\right) + \tau \nabla p_h^k - \tau \mu_h^k \nabla \varphi^{k-1} - \rho^{k-1} g,$$

$$r_h^{(2)} := \varphi_h^k - \mathscr{P}^k \varphi^{k-1} + \tau v_h^k \nabla \varphi^{k-1} - \tau \mathrm{div}(m^{k-1} \nabla \mu_h^k),$$

$$r_h^{(3)} := \frac{\sigma}{\varepsilon}\left(W_+'(\varphi_h^k) + W_-'(\mathscr{P}^k \varphi^{k-1})\right) - \mu_h^k,$$

where $b^{k-1} := \rho^{k-1} v^{k-1} + J^{k-1}$. Furthermore we define the error indicators

$$\eta_T^{(1)} := h_T \|r_h^{(1)}\|_T, \quad \eta_E^{(1)} := h_E^{1/2} \|2\eta^{k-1}\left[Dv_h^k\right]_{\vec{v}_E} \|_E,$$

$$\eta_T^{(2)} := h_T \|r_h^{(2)}\|_T, \quad \eta_E^{(2)} := h_E^{1/2} \|m^{k-1}\left[\nabla \mu_h^k\right]_{\vec{v}_E} \|_E, \qquad (13.28)$$

$$\eta_T^{(3)} := h_T \|r_h^{(3)}\|_T, \quad \eta_E^{(3)} := h_E^{1/2} \|\left[\nabla \varphi_h^k\right]_{\vec{v}_E} \|_E.$$

Here $[\cdot]_{\vec{v}_E}$ denotes the jump of a discontinuous function across an edge E of $\mathscr{T}^k$ in normal direction $\vec{v}_E$ pointing from the triangle with lower global number to the triangle with higher global number. Thus $\eta_E^{(j)}, j = 1, 2, 3$, measures the jump of the corresponding variable across the edge E, while $\eta_T^{(j)}, j = 1, 2, 3$, measures the triangle-wise residuals.

Theorem 6 *There exists a constant $C > 0$ only depending on the domain Ω and the regularity of the mesh $\mathscr{T}^k$ such that*

$$\underline{\rho}\|e_v\|^2 + \tau \underline{\eta}\|\nabla e_v\|^2 + \tau \underline{m}\|\nabla e_\mu\|^2 + \sigma \varepsilon \|\nabla e_\varphi\|^2 + \frac{\sigma}{\varepsilon}(W_+'(\varphi_h^k) - W_+'(\varphi^k), e_\varphi)$$

$$\leq C\left(\eta_\Omega^2 + \eta_{h.o.t} + \eta_C\right),$$

holds with

$$\eta_\Omega^2 = \frac{1}{\tau \underline{\eta}} \sum_{T \in \mathscr{T}^k} \left(\eta_T^{(1)}\right)^2 + \frac{\tau}{\underline{\eta}} \sum_{E \in \mathscr{E}^k} \left(\eta_E^{(1)}\right)^2$$

$$\frac{1}{\tau \underline{m}} \sum_{T \in \mathscr{T}^k} \left(\eta_T^{(2)}\right)^2 + \frac{\tau}{\underline{m}} \sum_{E \in \mathscr{E}^k} \left(\eta_E^{(2)}\right)^2$$

$$\frac{1}{\sigma \varepsilon} \sum_{T \in \mathscr{T}^k} \left(\eta_T^{(3)}\right)^2 + \sigma \varepsilon \sum_{E \in \mathscr{E}^k} \left(\eta_E^{(3)}\right)^2,$$

$$\eta_{h.o.t.} = \tau (\mathrm{div}(e_v), e_p),$$

and $\eta_C = (\mathscr{P}^k \varphi^{k-1} - \varphi^{k-1}, e_\mu) - \frac{\sigma}{\varepsilon}(W_-'(\mathscr{P}^k \varphi^{k-1}) - W_-'(\varphi^{k-1}), e_\varphi).$

Remark 6 • The term $\eta_{h.o.t.}$ is of higher order. By approximation results it can be estimated in terms of h_T to a higher order then the orders included in $\eta_T^{(i)}, \eta_E^{(i)}$, $i = 1, 2, 3$. Thus it is neglected in the numerics.

- The term η_C arises due to the transfer of φ^{k-1} from the old grid $\mathcal{T}^{k-1}$ to the new grid $\mathcal{T}^k$ through the projection $\mathcal{P}^k$. In our numerics we use Lagrangian interpolation $\mathcal{I}^k$ as projection operator. We note that $\mathcal{I}^k \varphi^{k-1}$ and φ^{k-1} do only differ in regions of the domain where coarsening in the last time step took place, if bisection is used as refinement strategy. Since it seems unlikely that elements being coarsened in the last time step are refined again in the present time step, this term is neglected in the numerics. We note that this term might be further estimated to obtain powers of h_T by approximation results for the Lagrange interpolation, see e.g. [28].
- Due to these two terms involved the estimator is not fully reliable.
- Neglecting these two terms the estimator can be shown to be efficient by the standard bubble technique, see e.g. [6, 65].
- An adaptation of the time step size is not considered in the present work, since it would conflict with the time discretization over three time instances. In our numerics we have to choose time steps small enough to sufficiently well resolve the interfacial force $\mu_h^k \nabla \varphi^{k-1}$.

In the numerical part, this error estimator is used together with the mesh adaptation cycle described in [65]. The overall adaptation cycle

$$\text{SOLVE} \to \text{ESTIMATE} \to \text{MARK} \to \text{ADAPT}$$

is performed once per time step. For convenience of the reader we state the marking strategy here.

Algorithm 1 (Marking Strategy)

- *Fix $a_{\min} > 0$ and $a_{\max} > 0$, and set $\mathcal{A} = \{T \in \mathcal{T}^{k+1} \mid a_{\min} \leq |T| \leq a_{\max}\}$.*
- *Define indicators:*

  1. $\eta_T = \frac{1}{\tau \underline{\eta}} \left(\eta_T^{(1)}\right)^2 + \frac{1}{\tau \underline{m}} \left(\eta_T^{(2)}\right)^2 + \frac{1}{\sigma \varepsilon} \left(\eta_T^{(3)}\right)^2$,

  2. $\eta_{TE} = \sum_{E \subset T} \left[\frac{\tau}{\underline{\eta}} \left(\eta_{TE}^{(1)}\right)^2 + \frac{\tau}{\underline{m}} \left(\eta_{TE}^{(2)}\right)^2 + \sigma \varepsilon \left(\eta_{TE}^{(3)}\right)^2 \right]$.

- *Refinement: Choose $\theta^r \in (0, 1)$,*

  1. Find a set $R^T \subset \mathcal{T}^{k+1}$ with $\theta^r \sum_{T \in \mathcal{T}^{k+1}} \eta_T \leq \sum_{T \in R^T} \eta_T$,
  2. Find a set $R^{TE} \subset \mathcal{T}^{k+1}$ with $\theta^r \sum_{T \in \mathcal{T}^{k+1}} \eta_{TE} \leq \sum_{T \in R^{TE}} \eta_{TE}$.

- *Coarsening: Choose $\theta^c \in (0, 1)$,*

  1. Find the set $C^T \subset \mathcal{T}^{k+1}$ with $\eta_T \leq \frac{\theta^c}{N} \sum_{T \in \mathcal{T}^{k+1}} \eta_T \; \forall T \in C^T$,
  2. Find the set $C^{TE} \subset \mathcal{T}^{k+1}$ with $\eta_{TE} \leq \frac{\theta^c}{N} \sum_{T \in \mathcal{T}^{k+1}} \eta_{TE} \; \forall T \in C^{TE}$.

- *Mark all triangles of $\mathcal{A} \cap (R^T \cup R^{TE})$ for refining.*
- *Mark all triangles of $\mathcal{A} \cap (C^T \cup C^{TE})$ for coarsening.*

13.2.2.1 Ensuring the Validity of the Energy Estimate

To ensure the validity of the energy estimate during the numerical computations
we ensure that the total energy does not increase triangle-wise. For the following
considerations we restrict to bisection as refinement strategy combined with a
hierarchically organised mesh, such that coarsening inverses prior refinement steps,
see e.g. [102]. Note that in such a situation local coarsening, e.g. substituting
four triangles by two triangles, only appears if all four triangles are marked for
coarsening. We call such a set of four (resp. two cells at $\partial\Omega$), triangles a nodeStar,
following [32].

By using this strategy, we do not harm the Assumption 2 on triangles that are
refined. We note that this assumption can only be violated on patches of triangles
where coarsening appears.

After marking triangles for refinement and coarsening and before applying
refinement and coarsening to $\mathscr{T}^k$ we make a post-processing of all triangles that
are marked for coarsening.

Let M^C denote the set of triangles marked for coarsening obtained by the marking
strategy described in Algorithm 1. To ensure the validity of the energy estimate
(13.20) we perform the following post processing steps:

Algorithm 2 (Post Processing)

1. For each triangle $T \in M^C$:
   if T is not part of a nodeStar
   then set $M^C := M^C \setminus T$.
2. For each nodeStar $S \in M^C$:
   if Assumption 2 is not fulfilled on S
   then set $M^C := M^C \setminus S$.

The resulting set M^C only contains triangles yielding nodeStars on which
Assumption 2 is fulfilled.

13.2.2.2 A Numerical Example

The proposed scheme is implemented in C++ using a mesh based on [32] and
using the package [35]. We note that the linear systems that are solved during
Newton's method have a saddle point structure, where the diagonal blocks again
have saddle point structure. For saddle point problems efficient preconditioning
techniques are available, we refer to [19, 84] for details and [45] for the actual
realization in the present setting. In the following we use $\mathscr{P}^k \equiv \mathscr{I}^k$, where $\mathscr{I}^k$
denotes the Lagrangian interpolation, associated with the space V_1^k. This leads to a
small deviation in the total mass, see [45].

To test the validity of the energy inequality in the fully discrete setting, we use
the classic example of spinodal decomposition [30] as test case. The parameters are

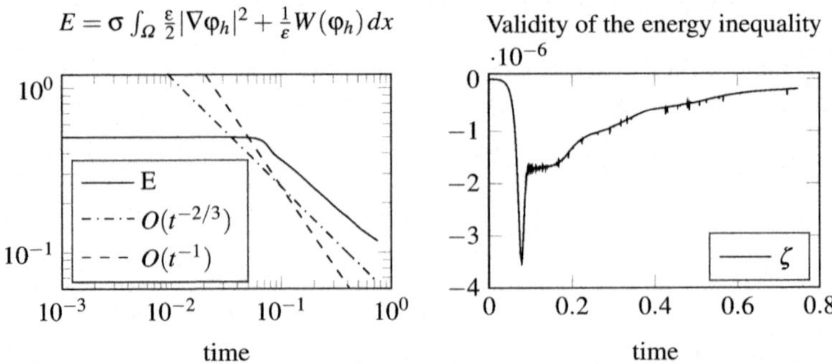

Fig. 13.1 Time evolution of the Ginzburg–Landau energy (*left*), and validity of the energy inequality (*right*)

chosen as: $\rho_1 = \rho_2 = 1$, $\eta_1 = \eta_2 = 0.004$, $g \equiv 0$, and $m \equiv \varepsilon$, $\varepsilon = 0.004$, $\sigma = 0.004$ and $\tau = 10^{-4}$.

In Fig. 13.1 we show the time evolution of the monotonically decreasing Ginzburg–Landau energy (left plot). We obtain a time span where $E \sim t^{-1}$ holds, as predicted e.g. in [97], and a later stage where $E \sim t^{-2/3}$ holds.

Next we investigate the validity of the energy inequality; see Fig. 13.1 (right plot). The plot depicts the time evolution of the term

$$\zeta = \frac{1}{2} \int_\Omega \rho^{k-1} \left| v_h^k \right|^2 dx + \frac{\sigma \varepsilon}{2} \int_\Omega |\nabla \varphi_h^k|^2 dx + \frac{\sigma}{\varepsilon} \int_\Omega W(\varphi_h^k) dx$$

$$+ \frac{1}{2} \int_\Omega \rho^{k-2} (v_h^{k-1} - v^k)^2 dx + \frac{\sigma \varepsilon}{2} \int_\Omega |\nabla \varphi_h^k - \nabla \mathscr{I}^k \varphi^{k-1}|^2 dx$$

$$+ \tau \int_\Omega 2\eta^{k-1} |Dv_h^k|^2 dx + \tau \int_\Omega m^{k-1} |\nabla \mu_h^k|^2 dx$$

$$- \left(\frac{1}{2} \int_\Omega \rho^{k-2} \left| v^{k-1} \right|^2 dx + \frac{\sigma \varepsilon}{2} \int_\Omega |\nabla \mathscr{I}^k \varphi^{k-1}|^2 dx \right.$$

$$\left. + \frac{\sigma}{\varepsilon} \int_\Omega W(\mathscr{I}^k \varphi^{k-1}) dx + \int_\Omega \rho^{k-1} g v_h^k \right).$$

The post processing of Algorithm 2 guarantees that this term is always negative as we observe in the plot.

We also simulated the benchmark for rising bubble dynamics from [79]: Here we observed results that are closer to sharp interface numeric when we use the relaxed double obstacle free energy W^{rel} then when using the polynomially free energy W^{poly}, see Remark 2, which clearly shows the benefits of using the relaxed double-obstacle free energy.

13.3 Optimal Control of Variable Density Two-Phase Flow

In this section, we focus on a double-obstacle potential type free energy density which yields an optimal control problem for a family of coupled systems in each time instant of a variational inequality of fourth order and the Navier–Stokes equation. By proposing a suitable time discretization, we establish the existence of solutions to the primal system and of optimal controls for the original problem as well as for a family of regularized problems which is introduced to handle the constraint degeneracy. The latter correspond to Moreau–Yosida type approximations of the double-obstacle potential. We further show the consistency of these approximations and derive first-order optimality conditions for the regularized problems. Through a limit process with respect to the regularization parameter, we obtain a stationarity system for the original problem which corresponds to a function space version of C-stationarity. The following results have been derived in [71].

13.3.1 The Semi-discrete CHNS-System and the Optimal Control Problem

The presence of a non smooth free energy density gives rise to a degenerate constraint system with the overall problem falling into the realm of mathematical programs with equilibrium constraints (MPECs). This evokes a variety of new challenges when it comes to the analytical treatment of the problem, cf. Sect. 13.1.1.3, and demands for a slightly different set of assumptions and definitions as used in previous section which is provided in the following.

First note that, assuming integrability in time, from (13.2), (13.3), (13.7), and (13.8), it follows that

$$\int_\Omega \partial_t \varphi dx = -\int_\Omega v \nabla \varphi dx + \int_\Omega \operatorname{div}(m(\varphi)\nabla\mu)dx = 0.$$

Hence, utilizing (13.6) the integral mean of φ satisfies

$$\frac{1}{|\Omega|}\int_\Omega \varphi dx \equiv \frac{1}{|\Omega|}\int_\Omega \varphi_a dx =: \overline{\varphi_a},$$

i.e., it is constant in time. By assuming $\overline{\varphi_a} \in (-1, 1)$, we exclude the uninteresting case $|\overline{\varphi_a}| = 1$. This can be achieved by considering the shifted system (13.1)–(13.8), where φ is replaced by its projection onto $\overline{L}^2(\Omega)$. Consequently, we need to work with shifted variables such as, e.g. $m(y + \overline{\varphi_a})$, which we again denote by $m(y)$ in a slight misuse of notation.

As in the previous section, we assume throughout that the mobility and viscosity coefficients are strictly positive as specified in Assumption 3 below. Furthermore, we extend the connection between φ and ρ to all of $\mathbb{R}$, as our studies include certain double-well type potentials which allow for values of φ outside the physically relevant interval $[-1, 1]$.

Assumption 3

1. *The coefficient functions $m, \eta \in C^2(\mathbb{R})$ in (13.1) and (13.3) as well as their derivatives up to second order are bounded, i.e. there exist constants $0 < b_1 \le b_2$ such that for every $x \in \mathbb{R}$, it holds that $b_1 \le \min\{m(x), \eta(x)\}$ and*

$$\max\{m(x), \eta(x), |m'(x)|, |\eta'(x)|, |m''(x)|, |\eta''(x)|\} \le b_2.$$

2. *The initial state satisfies $(v_a, \varphi_a) \in H^2_{0,\sigma}(\Omega; \mathbb{R}^n) \times \left(\overline{H}^2_{\partial_n}(\Omega) \cap \mathbb{K}\right)$ where*

$$\mathbb{K} := \left\{ v \in \overline{H}^1(\Omega) : \psi_1 \le v \le \psi_2 \text{ a.e. in } \Omega \right\},$$

with $-1 - \overline{\varphi_a} =: \psi_1 < 0 < \psi_2 := 1 - \overline{\varphi_a}$.
3. *The density ρ depends on the order parameter φ via*

$$\rho(\varphi) = \max\left\{ \frac{\rho_1 + \rho_2}{2} + \frac{\rho_2 - \rho_1}{2}(\varphi + \overline{\varphi_a}), 0 \right\} \ge 0.$$

We note that the pure phases are attained at x when $\varphi(x) = \psi_1$ or $\varphi(x) = \psi_2$. The *max*-operator in Assumption 3.3 ensures that the density remains always non-negative and maintains the affine connection of ρ and φ if φ is contained in the interval $[\psi_1, \psi_2]$. This is necessary to derive appropriate energy estimates.

With these assumptions we now state the semi-discrete Cahn–Hilliard Navier–Stokes system. For the sake of generality, we additionally introduce a distributed force on the right-hand side of the Navier-Stokes equation, which will later serve the purpose of a distributed control. As before, $\tau > 0$ denotes the time step-size and $K \in \mathbb{N}$ the total number of time instants in the semi-discrete setting. We further set $\sigma := \frac{1}{\varepsilon}$ and $\kappa := \frac{\tilde{\kappa}}{\varepsilon^2}$ in order to keep the notation as short as possible.

Definition 2 (Semi-discrete CHNS-System) Let $\Psi_0 : \overline{H}^1(\Omega) \to \mathbb{R}$ be a convex functional with subdifferential $\partial \Psi_0$. Fixing $(\varphi_{-1}, v_0) = (\varphi_a, v_a)$ we say that a triple

$$(\varphi, \mu, v) = ((\varphi_i)_{i=0}^{K-1}, (\mu_i)_{i=0}^{K-1}, (v_i)_{i=1}^{K-1})$$

in $\overline{H}^2_{\partial_n}(\Omega)^K \times \overline{H}^2_{\partial_n}(\Omega)^K \times H^1_{0,\sigma}(\Omega; \mathbb{R}^n)^{K-1}$ solves the semi-discrete CHNS system with respect to a given control $u = (u_i)_{i=1}^{K-1} \in L^2(\Omega; \mathbb{R}^n)^{K-1}$, denoted as $(\varphi, \mu, v) \in$

$S_\psi(u)$, if it holds for all $\phi \in \overline{H}^1(\Omega)$ and $\psi \in H^1_{0,\sigma}(\Omega;\mathbb{R}^n)$ that

$$\left\langle \frac{\varphi_{i+1} - \varphi_i}{\tau}, \phi \right\rangle + \langle v_{i+1}\nabla\varphi_i, \phi \rangle + (m(\varphi_i)\nabla\mu_{i+1}, \nabla\phi) = 0, \tag{13.29}$$

$$(\nabla\varphi_{i+1}, \nabla\phi) + \langle \partial\Psi_0(\varphi_{i+1}), \phi \rangle - \langle \mu_{i+1}, \phi \rangle - \langle \kappa\varphi_i, \phi \rangle = 0, \tag{13.30}$$

$$\left\langle \frac{\rho(\varphi_i)v_{i+1} - \rho(\varphi_{i-1})v_i}{\tau}, \psi \right\rangle_{H^{-1}_{0,\sigma}, H^1_{0,\sigma}} - (v_{i+1} \otimes \rho(\varphi_{i-1})v_i, \nabla\psi)$$

$$+ \left(v_{i+1} \otimes \frac{\rho_2 - \rho_1}{2} m(\varphi_{i-1})\nabla\mu_i, \nabla\psi \right) + (2\eta(\varphi_i)\varepsilon(v_{i+1}), \varepsilon(\psi))$$

$$- \langle \mu_{i+1}\nabla\varphi_i, \psi \rangle_{H^{-1}_{0,\sigma}, H^1_{0,\sigma}} = \langle u_{i+1}, \psi \rangle_{H^{-1}_{0,\sigma}, H^1_{0,\sigma}}. \tag{13.31}$$

The first two equations are supposed to hold for every $0 \le i + 1 \le K - 1$ and the last equation holds for every $1 \le i + 1 \le K - 1$.

Remark 7 In general, the subdifferential of a convex function Ψ_0 can be a multivalued, see, e.g., [38]. In this case, by Eq. (13.30) there exists $\beta \in \partial\Psi_0(\varphi_{i+1})$ such that

$$(\nabla\varphi_{i+1}, \nabla\phi) + \langle \beta, \phi \rangle - \langle \mu_{i+1}, \phi \rangle - \langle \kappa\varphi_i, \phi \rangle = 0, \ \forall\phi \in \overline{H}^1(\Omega).$$

Remark 8 We note that in the above system the boundary conditions specified in (13.7) and (13.8) are incorporated in the respective function spaces. Furthermore, the definition already includes the inherent regularity properties of φ and μ which anticipates the results obtained in Lemma 2 below.

Due to the choice of solenoidal test functions, the term $\langle \nabla p_i, \psi \rangle = 0$ vanishes and is not included in the semi-discrete Navier Stokes equation (13.31). As a consequence, the pressure can be disregarded in our subsequent analysis.

We point out that, as also noted in Remark 3, this semi-discretization of (13.1)–(13.8) in time involves three time instants $(i-1, i, i+1)$ and (φ_0, μ_0) is characterized in an initialization step by the (decoupled) Cahn-Hilliard system only. At the subsequent time instants, however, the strong coupling of the Cahn-Hilliard and Navier-Stokes system which, in the case of non-matched densities, is additionally enforced through the presence of the relative flux J is preserved. As a result, well-posedness of the time discrete scheme can be guaranteed and energy estimates mirroring the physical fact of decreasing energies can be argued as seen below.

Finally, we present the optimal control problem for the semi-discrete CHNS system. For its formulation, let $U_{ad} \subset L^2(\Omega;\mathbb{R}^n)^{K-1}$ and $\mathscr{J} : \mathscr{X} \to \mathbb{R}$ be a Fréchet differentiable function, with

$$\mathscr{X} := \overline{H}^1(\Omega)^K \times \overline{H}^1(\Omega)^K \times H^1_{0,\sigma}(\Omega;\mathbb{R}^n)^{K-1} \times L^2(\Omega;\mathbb{R}^n)^{K-1}.$$

Further requirements on U_{ad} and $\mathscr{J}$ are made explicit in connection with the existence result, Theorem 11, below.

Definition 3 The optimal control problem is given by

$$\min \ \mathscr{J}(\varphi, \mu, v, u) \ \text{over} \ (\varphi, \mu, v, u) \in \mathscr{X}$$
$$\text{s.t.} \ u \in U_{ad}, \ (\varphi, \mu, v) \in S_\psi(u). \qquad (P_\psi)$$

In many applications, $\mathscr{J}$ is given by a tracking-type functional and U_{ad} by unilateral or bilateral box constraints.

13.3.2 Existence of Feasible Points

In this subsection, we investigate the existence of feasible points for the optimization problem (P_ψ). For this purpose, we first study the solvability of the semi-discrete Cahn–Hilliard Navier–Stokes system. Since we will later on approximate the double-obstacle potential by a sequence of smooth potentials of double-well type, we consider here the following two types of free energy densities.

Assumption 4 *The functional* $\Psi_0 : \overline{H}^1(\Omega) \to \mathbb{R}$ *is convex, proper and lower-semicontinuous. It has one of the two subsequent properties:*

1. Either it is given by $\Psi_0(\varphi) := \int_\Omega \psi_0(\varphi(x))dx$ *where* $\psi_0 : \mathbb{R} \to \overline{\mathbb{R}} := \mathbb{R} \cup \{+\infty\}$ *represents the double-obstacle potential,*

$$\psi_0(z) := i_{[\psi_1;\psi_2]} := \begin{cases} +\infty \ \textit{if} \ z < \psi_1, \\ 0 \quad \ \textit{if} \ \psi_1 \leq z \leq \psi_2, \\ +\infty \ \textit{if} \ z > \psi_2. \end{cases}$$

2. Or it satisfies:

 a. Ψ_0 *is Fréchet differentiable with* $\{\Psi_0'(\varphi)\} = \partial\Psi_0(\varphi) \subset L^2(\Omega)$ *for every* $\varphi \in \overline{H}^1(\Omega)$;
 b. There exists $B_u \in \mathbb{R}$ *such that* $\Psi_0(\varphi) \leq B_u$ *for every* $\varphi \in \mathbb{K}$.

Note that these conditions are satisfied for double-well type potential.

Additionally, we assume that the functional $\Psi(\varphi) := \Psi_0(\varphi) - \int_\Omega \frac{\kappa}{2}\varphi(x)^2 dx$, $\kappa > 0$, *is bounded from below by a constant* $B_l \in \mathbb{R}$.

As a first observation, we state that the chosen time discretization does not break the thermodynamical consistency of the system. In the subsequent lemma, $(\varphi_i, \varphi_{i-1}, \mu_i, v_i)$ characterizes the state of the system at a given time step i. Then the total energy of the next time step is non-increasing if the external force u_{i+1} is set to zero.

Lemma 1 (Energy Estimate for a Single Time Step) *Let $\varphi_i, \varphi_{i-1}, \mu_i \in \overline{H}^1(\Omega)$, $v_i \in H_{0,\sigma}^1(\Omega; \mathbb{R}^n)$, $u_{i+1} \in (H_{0,\sigma}^1(\Omega; \mathbb{R}^n))^*$ be given such that*

$$\rho(\varphi_i), \rho(\varphi_{i-1}) > 0. \tag{13.32}$$

In case of the double-obstacle potential suppose additionally that $\varphi_i, \varphi_{i-1} \in \mathbb{K}$.

Then, if $(\varphi_{i+1}, \mu_{i+1}, v_{i+1}) \in \overline{H}^1(\Omega) \times \overline{H}^1(\Omega) \times H_{0,\sigma}^1(\Omega; \mathbb{R}^n)$ solves the system (13.29)–(13.31) for one time step, the following energy estimate holds true:

$$\int_\Omega \frac{\rho(\varphi_i) |v_{i+1}|^2}{2} dx + \int_\Omega \frac{|\nabla \varphi_{i+1}|^2}{2} dx + \Psi(\varphi_{i+1})$$

$$+ \int_\Omega \rho(\varphi_{i-1}) \frac{|v_{i+1} - v_i|^2}{2} dx + \int_\Omega \frac{|\nabla \varphi_{i+1} - \nabla \varphi_i|^2}{2} dx$$

$$+ \tau \int_\Omega 2\eta(\varphi_i) |\varepsilon(v_{i+1})|^2 dx + \tau \int_\Omega m(\varphi_i) |\nabla \mu_{i+1}|^2 dx + \int_\Omega \kappa \frac{(\varphi_{i+1} - \varphi_i)^2}{2}$$

$$\leq \int_\Omega \frac{\rho(\varphi_{i-1}) |v_i|^2}{2} dx + \int_\Omega \frac{|\nabla \varphi_i|^2}{2} dx + \Psi(\varphi_i) + \langle u_{i+1}, v_{i+1} \rangle_{H_{0,\sigma}^{-1}, H_{0,\sigma}^1},$$

$$\tag{13.33}$$

Remark 9 Note that in case of the double-obstacle potential the positivity of the density in (13.32) is always satisfied , since $\rho(\varphi_i) \geq \rho(\psi_1) > 0$. For double-well type potentials, however, the assumption is necessary, since φ may attain arbitrary values in $\mathbb{R}$. Nevertheless, it can be argued that the order parameter attains values in a neighborhood of the interval $[\psi_1, \psi_2]$, if the double-well type potential approximates the double-obstacle potential in a certain sense, cf. Theorem 9.

Besides reflecting an important physical property of the Cahn-Hilliard-Navier-Stokes system, the energy estimate also constitutes a valuable ingredient in the proof of existence of solutions to the system (13.29)–(13.31). As it serves to verify the boundedness constraint of Schaefer's fixed point theorem, also called the Leray-Schauder principle, which, in combination with arguments from monotone operator theory, yields the following result concerning the solvability of the semi-discrete system (13.29)–(13.31) for single time steps, cf. [71].

Theorem 7 (Existence of Solutions to the CHNS System for a Single Time Step)
Let the assumptions of Lemma 1 be satisfied. Then the system (13.29)–(13.31) admits a solution $(\varphi_{i+1}, \mu_{i+1}, v_{i+1}) \in \overline{H}^1(\Omega) \times \overline{H}^1(\Omega) \times H_{0,\sigma}^1(\Omega; \mathbb{R}^n)$ for one time step.

In our setting, the control force u_{i+1} is contained in $L^2(\Omega; \mathbb{R}^n)$. As a result the corresponding solution possesses higher regularity properties which can be shown via a bootstrap argument and well-known regularity results for the stationary Stokes equation.

Lemma 2 (Regularity of Solutions) *Let the assumptions of Lemma 1 be satisfied, and suppose additionally that $\varphi_i \in H^2(\Omega)$.*

Then it holds that $\varphi_{i+1}, \mu_{i+1} \in \overline{H}^2_{\partial_n}(\Omega)$ and $v_{i+1} \in H^2(\Omega; \mathbb{R}^n)$, provided that $(\varphi_{i+1}, \mu_{i+1}, v_{i+1}) \in \overline{H}^1(\Omega) \times \overline{H}^1(\Omega) \times H^1_{0,\sigma}(\Omega; \mathbb{R}^n)$ satisfies the system (13.29)–(13.31). Moreover, there exists a constant $C = C(N, \Omega, b_1, b_2, \tau, \kappa) > 0$ such that

$$\|\varphi_{i+1}\|_{H^2} + \|\mu_{i+1}\|_{H^2} + \|v_{i+1}\|_{H^2}$$

$$\leq C(\|\varphi_{i+1}\| + \|\mu_{i+1}\| + \|\varphi_i\| + \|v_{i+1}\|_{H^1} \|\varphi_i\|_{H^2} + \|\Psi_0'(\varphi_{i+1})\|).$$
$$(13.34)$$

In case of the double-obstacle potential, it also holds that $\varphi_{i+1} \in \mathbb{K}$ and the term $\|\Psi_0'(\varphi)\|$ in the above inequality is dropped.

Repeated applications of these statements for each time step $i = 0, .., K - 2$ directly verifies the existence of feasible points in the case of a double-obstacle potential.

Theorem 8 (Existence of Feasible Points) *Let $u \in L^2(\Omega; \mathbb{R}^n)^{K-1}$. Let $\overline{\Psi}_0$ be the double-obstacle potential defined in Assumption 4.1.*

Then the system (13.29)–(13.31) admits a solution $(\varphi, \mu, v) \in \overline{H}^2_{\partial_n}(\Omega)^K \times \overline{H}^2_{\partial_n}(\Omega)^K \times H^1_{0,\sigma}(\Omega; \mathbb{R}^n)^{K-1}$.

In order to approach the case of the double-well type potentials, we need to ensure the positivity of the density as explained in Remark 9. Using a technique from [70], the subsequent lemma guarantees that the order parameter of a solution to the system (13.29)–(13.31) for the double-well type potentials under consideration is always greater than $\psi_1 - \varepsilon$ for some small $\varepsilon > 0$.

Theorem 9 *Let $u \in L^2(\Omega; \mathbb{R}^n)^{K-1}$ be given and $\left\{\Psi_0^{(k)}\right\}_{k \in \mathbb{N}}$ a sequence of functions which satisfies the following two conditions:*

1. *For every $k \in \mathbb{N}$ $\Psi_0^{(k)}$ fulfills Assumption 4.*
2. *If $\left\{\hat{\varphi}^{(k)}\right\}_{k \in \mathbb{N}}$ is a sequence in $\overline{H}^1(\Omega)$ such that there exists a constant $C > 0$ with $\Psi_0^{(k)}(\hat{\varphi}^{(k)}) \leq C$ for $k \in \mathbb{N}$, then*

$$\left\|\max(-\hat{\varphi}^{(k)} + \psi_1, 0)\right\|_{L^1} \to 0, \text{ as } k \to \infty.$$

Furthermore, let $\left\{(\varphi^{(k)}, \mu^{(k)}, v^{(k)})\right\}_{k \in \mathbb{N}}$ be a sequence of solutions to the systems (13.29)–(13.31) with $\Psi_0 = \Psi_0^{(k)}$. Then

$$\left\|\max(-\varphi^{(k)} + \psi_1, 0)\right\|_{L^\infty} \to 0, \text{ as } k \to \infty.$$

Employing the previous theorem, we can verify that the semi-discrete CHNS system (13.29)–(13.31) has a solution if the double-well type potential under consideration is close enough to the double-obstacle potential.

Theorem 10 (Existence of Feasible Points) *Let* $u \in L^2(\Omega; \mathbb{R}^n)^{K-1}$. *Let* $\left\{\Psi_0^{(k)}\right\}_{k \in \mathbb{N}}$ *be a sequence which satisfies the conditions of Theorem 9.*

Then there exists $k^* \in \mathbb{N}$ *such that the system (13.29)–(13.31) admits a solution* (φ, μ, v) *for every* $\Psi_0 \in \left\{\Psi_0^{(k)}\right\}_{k \geq k^*}$.

For more details on the proof of the above theorem, we refer to [71]. In Definition 4 below, we propose a specific regularization which satisfies the conditions of Theorems 9 and 10, respectively.

13.3.3 Existence of Globally Optimal Points and Convergence of Minimizers

In the previous section, the existence of feasible points for the optimal control problem (P_Ψ) is verified. Our next goal is to investigate the existence of an optimal solution to (P_Ψ). We commence by presenting the following lemma which states some important properties of the corresponding control-to-state operator.

Lemma 3 (Regularity and Boundedness of the State) *There exists a positive constant* $C = C(N, \Omega, b_1, b_2, \tau, \kappa, v_a, \varphi_a, u) > 0$ *such that for every solution* $(\varphi, \mu, v) \in \overline{H}_{\partial_n}^2(\Omega)^K \times \overline{H}_{\partial_n}^2(\Omega)^K \times H_{0,\sigma}^1(\Omega; \mathbb{R}^n)^{K-1}$ *of Theorems 8 and 10 it holds that*

$$\|v\|_{(H^2)^K}^2 + \|\mu\|_{(H^2)^K}^2 + \|\varphi\|_{(H^2)^{K+1}}^2 \leq C. \tag{13.35}$$

Furthermore, the operator $L^2(\Omega, \mathbb{R}^n)^{K-1} \ni u \longmapsto C(N, \Omega, b_1, b_2, \tau, \kappa, v_a, \varphi_a, u) \in \mathbb{R}$ *is bounded.*

With the help of Lemma 3 it is possible to verify the existence of globally optimal points via standard arguments from optimization theory if some classical assumptions on the objective functional and the constraint set U_{ad} are imposed.

Theorem 11 (Existence of Global Solutions) *Suppose that the objective functional* $\mathscr{J} : \overline{H}_{\partial_n}^2(\Omega)^K \times \overline{H}_{\partial_n}^2(\Omega)^K \times H_{0,\sigma}^1(\Omega; \mathbb{R}^n)^{K-1} \times L^2(\Omega; \mathbb{R}^n)^{K-1} \to \mathbb{R}$ *is convex and weakly lower-semi-continuous and* U_{ad} *is non-empty, closed and convex. Assume that either* U_{ad} *is bounded or* $\mathscr{J}$ *is partially coercive, i.e. for every sequence* $\left\{(\varphi^{(k)}, \mu^{(k)}, v^{(k)}, u^{(k)})\right\}_{k \in \mathbb{N}}$ *with* $\lim_{k \to \infty} \|u^{(k)}\| = +\infty$ *it holds that* $\lim_{k \to \infty} \mathscr{J}(\varphi^{(k)}, \mu^{(k)}, v^{(k)}, u^{(k)}) = +\infty$.

Then the optimization problem (P_Ψ) *admits a global solution.*

Next, we turn our focus to the consistency of the regularization, i.e. the convergence of a sequence of solutions to $(P_{\Psi^{(k)}})$ with $\Psi^{(k)}$ a double-well potential approaching the double-obstacle potential in the limit as $k \to \infty$, to a solution of (P_Ψ) with Ψ the double-obstacle potential. For this purpose, we consider a sequence of functionals $\{\Psi^{(k)}\}_{k\in\mathbb{N}}$ satisfying Assumption 4.2 and a corresponding limit functional $\overline{\Psi}$.

The following theorem provides conditions under which a sequence of globally optimal solutions to $(P_{\Psi^{(k)}})$ converge to a global solution of $(P_{\overline{\Psi}})$, as $k \to \infty$.

Theorem 12 (Consistency of the Regularization) *Let the assumptions of Theorem 11 be fulfilled. The objective* $\mathscr{J} : \overline{H}^1(\Omega)^K \times \overline{H}^1(\Omega)^K \times H^1_{0,\sigma}(\Omega;\mathbb{R}^n)^{K-1} \times L^2(\Omega;\mathbb{R}^n)^{K-1} \to \mathbb{R}$ *is supposed to be upper-semicontinuous, and let* $\{\Psi^{(k)}\}_{k\in\mathbb{N}}$ *be a sequence of potentials satisfying Assumption 4.2. Assume further that* $\overline{\Psi}$ *is given such that for every sequence* $\{(x^{(k)}, y^{(k)})\}_{k\in\mathbb{N}} \subset \overline{H}^1(\Omega) \times \overline{H}^{-1}(\Omega)$ *with* $y^{(k)} = \Psi^{(k)'}(x^{(k)})$ *and* $(x^{(k)}, y^{(k)}) \to (x^{(\infty)}, y^{(\infty)})$ *strongly in* $\overline{H}^1(\Omega) \times \overline{H}^{-1}(\Omega)$ *it holds that* $y^{(\infty)} \in \partial\overline{\Psi}(x^{(\infty)})$.

Then a sequence $\{(\varphi^{(k)}, \mu^{(k)}, v^{(k)}, u^{(k)})\}_{k\in\mathbb{N}}$ *of global solutions to* $(P_{\Psi^{(k)}})$ *in* $\overline{H}^2(\Omega)^K \times \overline{H}^2(\Omega)^K \times H^1_{0,\sigma}(\Omega;\mathbb{R}^n)^{K-1} \times U_{ad}$ *converges to a global solution of* $(P_{\overline{\Psi}})$, *provided that* $\{\mathscr{J}(\varphi^{(k)}, \mu^{(k)}, v^{(k)}, u^{(k)})\}_{k\in\mathbb{N}}$ *is assumed bounded, whenever* U_{ad} *is unbounded.*

In summary, the optimal control problems under consideration are well-posed and admit globally optimal solutions. Furthermore, the chosen regularization approach is consistent in the sense of Theorem 12.

13.3.4 Stationarity Conditions

At this point, we turn our attention to the derivation of stationarity conditions for the optimal control problem. In this case of smooth potential functions which satisfy Assumption 4.2, stationarity or first-order optimality conditions for the problem (P_Ψ) can be derived by applying classical results from Zowe and Kurcyusz concerning the existence of Lagrange multipliers. The latter approach is employed in the following theorem.

Theorem 13 (First-Order Optimality Conditions for Smooth Potentials) *Let* $\mathscr{J} : \overline{H}^1(\Omega)^K \times \overline{H}^1(\Omega)^K \times H^1_{0,\sigma}(\Omega;\mathbb{R}^n)^{K-1} \times L^2(\Omega;\mathbb{R}^n)^{K-1} \to \mathbb{R}$ *be Fréchet differentiable and let* Ψ_0 *satisfy Assumption 4.2 such that* Ψ'_0 *maps* $\overline{H}^2_{\partial_n}(\Omega)$ *continuously Frèchet-differentiably into* $L^2(\Omega)$. *Further, let* $\bar{z} := (\bar{\varphi}, \bar{\mu}, \bar{v}, \bar{u})$ *be a minimizer of* (P_Ψ).

Then there exist $(\chi, r, q) \in \overline{H}^1(\Omega)^K \times \overline{H}^1(\Omega)^{K-1} \times H^1_{0,\sigma}(\Omega;\mathbb{R}^n)^{K-1}$, *with* $\chi = (\chi_{-1}, \dots \chi_{K-2})$, $r = (r_{-1}, \dots r_{K-2})$, $q = (q_0, \dots q_{K-2})$, *such that*

$$-\frac{1}{\tau}(\chi_i - \chi_{i-1}) + m'(\varphi_i)\nabla\mu_{i+1} \cdot \nabla\chi_i - \operatorname{div}(\chi_i v_{i+1}) - \Delta r_{i-1}$$

$$+\Psi_0''(\varphi_i)^* r_{i-1} - \kappa r_{i+1} - \frac{1}{\tau}\rho'(\varphi_i)v_{i+1} \cdot (q_{i+1} - q_i)$$

$$-(\rho'(\varphi_i)v_{i+1} - \frac{\rho_2 - \rho_1}{2}m'(\varphi_i)\nabla\mu_{i+1})(Dq_{i+1})^{\mathsf{T}}v_{i+2}$$

$$+2\eta'(\varphi_i)\varepsilon(v_{i+1}) : Dq_i + \mathrm{div}(\mu_{i+1}q_i) = \frac{\partial\mathscr{J}}{\partial\varphi_i}(\bar{z}),$$

$$(13.36)$$

$$-r_{i-1} - \mathrm{div}(m(\varphi_{i-1})\nabla\chi_{i-1}) - \mathrm{div}(\frac{\rho_2 - \rho_1}{2}m(\varphi_{i-1})(Dq_i)^{\mathsf{T}}v_{i+1})$$

$$-q_{i-1}\cdot\nabla\varphi_{i-1} = \frac{\partial\mathscr{J}}{\partial\mu_i}(\bar{z}),$$

$$(13.37)$$

$$-\frac{1}{\tau}\rho(\varphi_{j-1})(q_j - q_{j-1}) - \rho(\varphi_{j-1})(Dq_j)^{\mathsf{T}}v_{j+1}$$

$$-(Dq_{j-1})(\rho(\varphi_{j-2})v_{j-1} - \frac{\rho_2 - \rho_1}{2}m(\varphi_{j-2})\nabla\mu_{j-1})$$

$$- \mathrm{div}(2\eta(\varphi_{j-1})\varepsilon(q_{j-1})) + \chi_{j-1}\nabla\varphi_{j-1} = \frac{\partial\mathscr{J}}{\partial v_j}(\bar{z}),$$

$$(13.38)$$

$$\left(\frac{\partial\mathscr{J}}{\partial u_k}(\bar{z}) - q_{k-1}\right)_{k=1}^{K-1} \in \left[\mathbb{R}_+(U_{ad} - \bar{u})\right]^+,$$

$$(13.39)$$

for all $i = 0,\ldots,K-1$ and $j = 1,\ldots,K-1$. Here, $\left[\mathbb{R}_+(U_{ad} - \bar{u})\right]^+$ denotes the polar cone of the set $\{r(w - u)|w \in U_{ad} \wedge r \in \mathbb{R}^+\}$. Furthermore, we use the convention that χ_i, r_i, q_i are equal to 0 for $i \geq K-1$ along with q_{-1} and φ_i, μ_i, v_i for $i \geq K$.

In [71], it was further shown that the adjoint state (χ, r, q) is bounded independently of the regularization parameter. This enables the derivation of a slightly weaker form of stationarity for certain non-smooth potentials via a limiting process which is given in the following theorem.

Theorem 14 (Stationarity Conditions) *Suppose that the following assumptions are satisfied.*

1. *$\mathscr{J}'$ is a bounded mapping from $\overline{H}^1(\Omega)^K \times \overline{H}^1(\Omega)^K \times H_{0,\sigma}^1(\Omega;\mathbb{R}^n)^{K-1} \times U_{ad}$ into the space $(\overline{H}^1(\Omega)^K \times \overline{H}^1(\Omega)^K \times H_{0,\sigma}^1(\Omega;\mathbb{R}^n)^{K-1} \times L^2(\Omega;\mathbb{R}^n)^{K-1})^*$ and $\frac{\partial\mathscr{J}}{\partial u}$ satisfies the following weak lower-semicontinuity property*

$$\left\langle\frac{\partial\mathscr{J}}{\partial u}(\hat{z}), \hat{u}\right\rangle \leq \liminf_{n\to\infty}\left\langle\frac{\partial\mathscr{J}}{\partial u}(\hat{z}^{(n)}), \hat{u}^{(n)}\right\rangle,$$

for $\hat{z}^{(n)} = (\hat{\varphi}^{(n)}, \hat{\mu}^{(n)}, \hat{v}^{(n)}, \hat{u}^{(n)})$ converging weakly in $\overline{H}_{\partial_n}^2(\Omega)^K \times \overline{H}_{\partial_n}^2(\Omega)^K \times H_{0,\sigma}^1(\Omega;\mathbb{R}^n)^{K-1} \times U_{ad}$ to $\hat{z} = (\hat{\varphi}, \hat{\mu}, \hat{v}, \hat{u})$.

2. *For every $n \in \mathbb{N}$ let $\Psi_0^{(n)} : \overline{H}_{\partial_n}^2(\Omega) \to \overline{\mathbb{R}}$ be a convex, lower-semicontinuous and proper functional satisfying the assumptions of Theorem 13.*

3. *Let $(\varphi^{(n)}, \mu^{(n)}, v^{(n)}, u^{(n)}) \in \overline{H}_{\partial_n}^2(\Omega)^K \times \overline{H}_{\partial_n}^2(\Omega)^K \times H_{0,\sigma}^1(\Omega; \mathbb{R}^n)^{K-1} \times U_{ad}$ be a minimizer for $(P_{\Psi^{(n)}})$ and let further $(\chi^{(n)}, r^{(n)}, q^{(n)}) \in \overline{H}^1(\Omega)^K \times \overline{H}^1(\Omega)^K \times H_{0,\sigma}^1(\Omega; \mathbb{R}^n)^{K-1}$ be given as in Theorem 13.*

Then there exists an element $(\varphi, \mu, v, u, \chi, r, q)$ and a subsequence denoted by $\{(\varphi^{(m)}, \mu^{(m)}, v^{(m)}, u^{(m)}, \chi^{(m)}, r^{(m)}, q^{(m)})\}_{m \in \mathbb{N}}$ with

$$\varphi^{(m)} \to \varphi \text{ weakly in } \overline{H}_{\partial_n}^2(\Omega)^K, \ \mu^{(m)} \to \mu \text{ weakly in } \overline{H}_{\partial_n}^2(\Omega)^{K-1},$$

$$v^{(m)} \to v \text{ weakly in } H^2(\Omega; \mathbb{R}^n)^{K-1}, \ u^{(m)} \to u \text{ weakly in } L^2(\Omega; \mathbb{R}^n)^{K-1},$$

$$\chi^{(m)} \to \chi \text{ weakly in } \overline{H}^1(\Omega)^K, \ r^{(m)} \to r \text{ weakly in } \overline{H}^1(\Omega)^{K-1},$$

$$q^{(m)} \to q \text{ weakly in } H_{0,\sigma}^1(\Omega; \mathbb{R}^n)^{K-1}, \ \Psi_0^{(m)''}(\varphi_{i+1}^{(m)})^* r_i^{(n)} \to \lambda_i \text{ weakly in } \overline{H}^1(\Omega)^*,$$

for all $i = -1, \ldots, K - 2$ such that for $z = (\varphi, \mu, v, u)$ and $\tilde{q}_k := q_{k-1}$ it holds that

$$-\frac{1}{\tau}(\chi_i - \chi_{i-1}) + m'(\varphi_i)\nabla\mu_{i+1} \cdot \chi_i - \mathrm{div}(\chi_i v_{i+1}) - \Delta r_{i-1}$$

$$+\lambda_{i-1} - \kappa r_{i+1} - \frac{1}{\tau}\rho'(\varphi_i)v_{i+1} \cdot (q_{i+1} - q_i)$$

$$-(\rho'(\varphi_i)v_{i+1} - \frac{\rho_2 - \rho_1}{2}m'(\varphi_i)\nabla\mu_{i+1})(Dq_{i+1})^\top v_{i+2}$$

$$+2\eta'(\varphi_i)\varepsilon(v_{i+1}) : Dq_i + \mathrm{div}(\mu_{i+1}q_i) = \frac{\partial \mathscr{J}}{\partial \varphi_i}(z),$$

$$\tag{13.40}$$

$$-r_{i-1} - \mathrm{div}(m(\varphi_{i-1})\nabla\chi_{i-1}) - \mathrm{div}(\frac{\rho_2 - \rho_1}{2}m(\varphi_{i-1})(Dq_i)^\top v_{i+1})$$

$$-q_{i-1} \cdot \nabla\varphi_{i-1} = \frac{\partial \mathscr{J}}{\partial \mu_i}(z),$$

$$\tag{13.41}$$

$$-\frac{1}{\tau}\rho(\varphi_{j-1})(q_j - q_{j-1}) - \rho(\varphi_{j-1})(Dq_j)^\top v_{j+1}$$

$$-(Dq_{j-1})(\rho(\varphi_{j-2})v_{j-1} - \frac{\rho_2 - \rho_1}{2}m(\varphi_{j-2})\nabla\mu_{j-1})$$

$$-\mathrm{div}(2\eta(\varphi_{j-1})\varepsilon(q_{j-1})) + \chi_{j-1}\nabla\varphi_{j-1} = \frac{\partial \mathscr{J}}{\partial v_j}(z),$$

$$\tag{13.42}$$

$$\frac{\partial \mathscr{J}}{\partial u}(z) - \tilde{q} \in [\mathbb{R}_+(U_{ad} - u)]^+.$$

$$\tag{13.43}$$

We point out that a tracking-type functional, like, e.g.,

$$\mathscr{J}(\varphi,\mu,v,u) := \sum_{i=0}^{K-1} \frac{1}{2} \left\| \varphi_i - \varphi_d^i \right\|^2 + \frac{\xi}{2} \|u\|_{(L^2)^{(K-1)}}^2 \,, \quad \xi > 0, \tag{13.44}$$

with desired states $\varphi_d^i \in L^2(\Omega)$, satisfies the assumptions of Theorem 14.

If the set U_{ad} is bounded, Theorem 14 holds also true for a sequence of stationary points for $(P_{\psi^{(n)}})$. If it is unbounded, then the result can still be transferred to sequences of stationary points by assuming that the sequence $\{u^{(n)}\}_{n\in\mathbb{N}}$ is bounded in $L^2(\Omega;\mathbb{R}^n)^{K-1}$.

In order to apply the developed theory, in particular Theorem 14, to the initially stated optimal control problem associated to the double-obstacle potential, we provide the following definition which characterizes the sequence of approximating double-well type potentials. For this purpose, let ψ_0 be defined as in Assumption 4.1 and set $\gamma := \partial\psi_0 \subset \mathbb{R}\times\mathbb{R}$.

Definition 4 Let a mollifier $\zeta \in C^1(\mathbb{R})$ with $\operatorname{supp}\zeta \subset [-1,1]$, $\int_{\mathbb{R}}\zeta = 1$ and $0 \leq \zeta \leq 1$ a.e. on $\mathbb{R}$, and a function $\theta : \mathbb{R}^+ \to \mathbb{R}^+$, with $\theta(\alpha) > 0$ and $\frac{\theta(\alpha)}{\alpha} \to 0$ as $\alpha \to 0$, be given. For the Yosida approximation γ_α with parameter $\alpha > 0$ of γ define

$$\zeta_\alpha(s) := \tfrac{1}{\alpha}\zeta\left(\tfrac{s}{\alpha}\right), \quad \widetilde{\gamma}_\alpha := \gamma_\alpha * \zeta_{\theta(\alpha)}, \quad \psi_{0\alpha}(s) := \int_0^s \widetilde{\gamma}_\alpha(t)\,dt,$$

$$\Psi_{0\alpha}(c) := \int_\Omega (\psi_{0\alpha} \circ c)(t)\,dt.$$

Moreover, we set $\alpha_n := n^{-1}$, $\Psi_0^{(n)} := \Psi_{0\alpha_n}$.

Utilizing Theorem 14 with respect to the approximating sequence from Definition 4 yields a stationarity system for the optimal control problem of the semi-discrete CHNS system with the double-obstacle potential. Through a careful limiting analysis the system can be extended by additional complementarity conditions which are presented in the subsequent theorem which can be found in [71].

Theorem 15 (Limiting ε-Almost C-stationarity) *Let* $\Psi_0^{(n)}$, $n \in \mathbb{N}$ *be the functionals of Definition 4, and let the tuples* $(\varphi^{(m)}, \mu^{(m)}, v^{(m)}, u^{(m)}, \chi^{(m)}, r^{(m)}, q^{(m)})$, $(\varphi,\mu,v,u,\chi,r,q)$ *and* $\mathscr{J}$ *be as in Theorem 14. Moreover, let* $\Lambda : \mathbb{R} \to \mathbb{R}$ *be a Lipschitz function with* $\Lambda(\psi_1) = \Lambda(\psi_2) = 0$. *For*

$$a_i^{(m)} := \Psi_0^{(m)'}(\varphi_i^{(m)}), \quad \lambda_i^{(m)} := \Psi_0^{(m)''}(\varphi_i^{(m)}) * r_{i-1}^{(m)}$$

for $i = 0,\ldots,K$, *and for* a_i *denoting the limit of* $a_i^{(m)}$, *it holds that*

$$(a_i, \Lambda(\varphi_i))_{L^2} = 0, \qquad\qquad \langle \lambda_i, \Lambda(\varphi_i)\rangle = 0, \tag{13.45}$$

$$(a_i, r_{i-1})_{L^2} = 0, \qquad\qquad \liminf(\lambda_i^{(m)}, r_{i-1}^{(m)})_{L^2} \geq 0. \tag{13.46}$$

Moreover, for every $\varepsilon > 0$ there exist a measurable subset M_i^ε of $M_i := \{x \in \Omega :$ $\psi_1 < \varphi_i(x) < \psi_2\}$ with $|M_i \setminus M_i^\varepsilon| < \varepsilon$ and

$$\langle \lambda_i, v \rangle = 0 \qquad \forall v \in \overline{H}^1(\Omega), \ v|_{\Omega \setminus M_i^\varepsilon} = 0.$$

In combination with the results from Theorem 14, the last theorem states stationarity conditions corresponding to a function space version of C-stationarity for MPECs, cf. [60, 69]. More precisely, the resulting stationarity system is of limiting ε-almost C-stationarity type. For the underlying problem class, this is currently the most (and, to the best of our knowledge, only) selective stationarity system available.

13.4 Goal Oriented Adaptivity for Optimal Control of Two-Phase Flow

The specific semi-discretization in time for the coupled CHNS system with non-matched fluid densities of the previous section represents a first step towards a numerical investigation/realization of the problem. Furthermore, the constructive nature of our derivation of the stationarity conditions facilitates the implementation of a solution algorithm for the problem which solves each approximating problem by a Newton method applied to a suitable finite element discretization in space. For this purpose, it is necessary to solve a sequence of large-scale nonlinear optimization problems. As already mentioned in Sect. 13.2.2 for the primal system, this might cause an immense numerical expense. Hence, we aim to reduce the computational effort by developing a beneficial adaptation process for the underlying space mesh which incorporates the fact that, for optimal control problems, one is usually interested in an accurate estimation of the target quantity, i.e., the objective functional.

For this purpose, we present an adaptive finite elements solver for the optimal control problem of the Cahn–Hilliard Navier–Stokes-system. This includes an adequate error estimator which consists of dual-weighted primal residuals, primal-weighted dual residuals and complementarity errors. It is based on the notion of a modified Lagrangian associated with the MPEC and uses the associated saddle-point condition for optimal points to characterize the error in the objective function between the continuous solution and a fully discretized problem.

The next subsection is devoted to the derivation of the error estimator, whereas Sect. 13.4.2 deals with the numerical details and showcases some of the obtained results.

13.4.1 Goal-Oriented Error Estimator

In order to treat the problem numerically and to derive the aforementioned error estimates, it is necessary to establish a fully discretized version of the problem. Hereby, we follow the so called first optimize, then discretize approach in that we directly discretize the optimality conditions given in Sect. 13.3.4. The spacial discretization uses Taylor-Hood finite elements which are known to be LBB-stable in case of the Navier-Stokes equation, cf., e.g., [46, 107]. More precisely, the phase field and the chemical potential are discretized via piecewise linear and continuous finite elements, whereas the discretization of the velocity field utilizes piecewise quadratic and continuous finite elements. For more details on the chosen discretization approach we refer the reader to [72]. Furthermore, we consider the concrete objective functional given in (13.44).

The subsequent definition characterizes the MPCC-Lagrangian of the optimal control problem (P_ψ), which is defined on the product function space

$$\mathscr{Y} := \overline{H}^1(\Omega)^K \times \overline{H}^1(\Omega)^K \times H^1_{0,\sigma}(\Omega;\mathbb{R}^n)^{K-1} \times \overline{L}^2(\Omega)^K \times L^2(\Omega;\mathbb{R}^n)^{K-1} \times \overline{H}^1(\Omega)^K$$

$$\times \overline{H}^1(\Omega)^K \times H^1_{0,\sigma}(\Omega;\mathbb{R}^n)^{K-1} \times \overline{H}^1(\Omega)^K \times \left(\overline{H}^1(\Omega)^*\right)^K \times \left(\overline{H}^1(\Omega)^*\right)^K.$$

In contrast to the classical Lagrange function, the MPCC-Lagrangian does not include a multiplier for the complementarity condition. It rather corresponds to the Lagrange function of certain tightend nonlinear problems associated to the MPEC, cf., e.g., [93, 101].

Definition 5 The MPCC-Lagrangian $L : \mathscr{Y} \to \mathbb{R}$ corresponding to (P_ψ) (see Definition 3) is given by

$$L(\varphi,\mu,v,a,u,\chi,r,q,\pi,\lambda^+,\lambda^-) := \mathscr{J}(\varphi,\mu,v,u)$$

$$+ \sum_{i=-1}^{K-2} \left[\left\langle \frac{\varphi^{i+1}-\varphi^i}{\tau}, \chi^{i+1} \right\rangle + \left\langle v^{i+1}\nabla\varphi^i, \chi^{i+1} \right\rangle - \left\langle \mathrm{div}(m(\varphi^i)\nabla\mu^{i+1}), \chi^{i+1} \right\rangle \right]$$

$$+ \sum_{i=-1}^{K-2} \left[\left\langle -\Delta\varphi^{i+1}, r^{i+1} \right\rangle + \left\langle a^{i+1}, r^{i+1} \right\rangle - \left\langle \mu^{i+1}, r^{i+1} \right\rangle - \left\langle \kappa\varphi^i, r^{i+1} \right\rangle \right]$$

$$+ \sum_{i=0}^{K-2} \left[\left\langle \frac{\rho(\varphi^i)v^{i+1}-\rho(\varphi^{i+1})v^i}{\tau}, q^{i+1} \right\rangle_{H^{-1},H^1_0} + \left\langle \mathrm{div}(v^{i+1}\otimes\rho(\varphi^{i+1})v^i), q^{i+1} \right\rangle_{H^{-1},H^1_0} \right.$$

$$- \left\langle \mathrm{div}\left(v^{i+1}\otimes\frac{\rho_2-\rho_1}{2}m(\varphi^{i+1})\nabla\mu^i\right), q^{i+1} \right\rangle_{H^{-1},H^1_0} + (2\eta(\varphi^i)\varepsilon(v^{i+1}), \varepsilon(q^{i+1}))$$

$$\left. - \left\langle \mu^{i+1}\nabla\varphi^i, q^{i+1} \right\rangle_{H^{-1},H^1_0} - \left\langle u^{i+1}, q^{i+1} \right\rangle_{H^{-1},H^1_0} \right]$$

$$- \sum_{i=0}^{K-1} \left\langle a^i, \pi^i \right\rangle - \sum_{i=0}^{K-1} \left\langle (\lambda^i)^+, \varphi^i - \psi_2 \right\rangle - \sum_{i=0}^{K-1} \left\langle (\lambda^i)^-, \varphi^i - \psi_1 \right\rangle.$$

For the sake of readability, we subsequently collect the primal variables in $y :=(\varphi, \mu, a, v)$ which describes the state of the optimal control problem and the adjoint variables in $\Phi := (\chi, r, q)$. Furthermore, $\mathscr{Y}_h$ denotes the discrete equivalent to $\mathscr{Y}$.

Remark 10 Note that if (y, u) is an ε-almost C-stationary point of (P_ψ) with adjoints $(\Phi, \pi, \lambda^+, \lambda^-)$ then

$$L(y, u, \Phi, \pi, \lambda^+, \lambda^-) = \mathscr{J}(\varphi, \mu, v, u). \tag{13.47}$$

Based on the MPCC-Lagrangian we provide a first characterization of the difference of the objective values at stationary points of the semi-discrete and the fully discretized problem. Subsequently, the index δ denotes the difference of the discrete and the continuous variables, e.g. $(y_\delta, u_\delta, \Phi_\delta) := (y_h, u_h, \Phi_h) - (y, u, \Phi)$.

Theorem 16 *Let* $(y, u, \Phi, \pi, \lambda^+, \lambda^-)$ *be a stationary point of the optimal control problem* (P_ψ) *and assume that* $(y_h, u_h, \Phi_h, \pi_h, \lambda_h^+, \lambda_h^-) \in \mathscr{Y}_h$ *satisfy the discretized stationarity system. Then it holds that*

$$\mathscr{J}(\varphi_h, \mu_h, v_h, u_h) - \mathscr{J}(\varphi, \mu, v, u) = \frac{1}{2}\left(\sum_{i=0}^{K-1}\langle a_h^i, \pi^i\rangle - \sum_{i=0}^{K-1}\langle a^i, \pi_h^i\rangle\right)$$

$$+ \frac{1}{2}\left(\sum_{i=0}^{K-1}\langle(\lambda^i)^+, \varphi_h^i - \psi_2\rangle - \sum_{i=0}^{K-1}\langle(\lambda_h^i)^+, \varphi^i - \psi_2\rangle\right)$$

$$+ \frac{1}{2}\left(\sum_{i=0}^{K-1}\langle(\lambda^i)^-, \varphi_h^i - \psi_1\rangle - \sum_{i=0}^{K-1}\langle(\lambda_h^i)^-, \varphi^i - \psi_1\rangle\right)$$

$$+ \frac{1}{2}\nabla_x L(y_h, u_h, \Phi_h, \pi_h, \lambda_h^+, \lambda_h^-)((y_h, u_h, \Phi_h) - (y, u, \Phi)). \tag{13.48}$$

The last term on the right-hand side of equation (13.48) assembles the weighted dual and primal residuals. Whereas the other terms display the mismatch in the complementarity between the discretized solution and the original one.

For each time step $i \in \{0, .., K-1\}$, the latter can be split into the following four parts

$$\eta_{CM1,i} := \frac{1}{2}\langle a_h^i, \pi^i - \pi_h^i\rangle, \quad \eta_{CM2,i} := \frac{1}{2}\langle(\lambda_h^i), \varphi^i - \varphi_h^i\rangle,$$

$$\eta_{CM3,i} := \frac{1}{2}\langle a^i, \pi_h^i - \pi^i\rangle, \quad \eta_{CM4,i} := \frac{1}{2}\langle(\lambda^i)^+, \varphi_h^i - \psi_2\rangle + \langle(\lambda^i)^-, \varphi_h^i - \psi_1\rangle.$$

The so-called dual-weighted primal residual $\eta_{CHNS,i} := \eta_{CH1,i} + \eta_{CH2,i} + \eta_{NS,i}$ consists of the three parts coming from the respective primal equations (for $i = -1, .., K - 2$)

$$\eta_{CH1,i+1} := \left\langle \frac{\varphi_h^{i+1} - \varphi_h^i}{\tau}, \chi_\delta^{i+1} \right\rangle + \left\langle v_h^{i+1} \nabla \varphi_h^i, \chi_\delta^{i+1} \right\rangle - \left\langle \operatorname{div}(m(\varphi_h^i) \nabla \mu_h^{i+1}), \chi_\delta^{i+1} \right\rangle,$$

$$\eta_{CH2,i+1} := \left\langle -\Delta \varphi_h^{i+1}, r_\delta^{i+1} \right\rangle + \left\langle a_h^{i+1}, r_\delta^{i+1} \right\rangle - \left\langle \mu_h^{i+1}, r_\delta^{i+1} \right\rangle - \left\langle \kappa \varphi_h^i, r_\delta^{i+1} \right\rangle,$$

$$\eta_{NS,i+1} := \left\langle \frac{\rho(\varphi_h^i) v_h^{i+1} - \rho(\varphi_h^{i-1}) v_h^i}{\tau}, q_\delta^{i+1} \right\rangle_{H^{-1}, H_0^1}$$
$$+ \left\langle \operatorname{div}(v_h^{i+1} \otimes \rho(\varphi_h^{i-1}) v_h^i), q_\delta^{i+1} \right\rangle_{H^{-1}, H_0^1}$$
$$- \left\langle \operatorname{div}(v_h^{i+1} \otimes \frac{\rho_2 - \rho_1}{2} m(\varphi_h^{i-1}) \nabla \mu_h^i), q_\delta^{i+1} \right\rangle_{H^{-1}, H_0^1}$$
$$+ (2\eta(\varphi_h^i) \varepsilon(v_h^{i+1}), \varepsilon(q_\delta^{i+1}))$$
$$- \left\langle \mu_h^{i+1} \nabla \varphi_h^i, q_\delta^{i+1} \right\rangle_{H^{-1}, H_0^1} - \left\langle u_h^{i+1}, q_\delta^{i+1} \right\rangle_{H^{-1}, H_0^1}.$$

Finally, the primal-weighted dual residuals can be defined for each $i \in \{0, .., K - 1\}$ (with $\eta_{ADv,0} := 0$) in three steps by

$$\eta_{AD\varphi,i} := \left[\varphi_h^i - \varphi_d^i - \frac{1}{\tau}(\chi_h^{i+1} - \chi_h^i) + m'(\varphi_h^i) \nabla \mu_h^{i+1} \cdot \nabla \chi_h^{i+1} - \operatorname{div}(\chi_h^{i+1} v_h^{i+1}) - \Delta r_h^i \right.$$
$$+ \lambda_h^i - \kappa r_h^{i+1} - \frac{1}{\tau} \rho'(\varphi_h^i) v_h^{i+1} \cdot (q_h^{i+2} - q_h^{i+1})$$
$$- (\rho'(\varphi_h^i) v_h^{i+1} - \frac{\rho_2 - \rho_1}{2} m'(\varphi_h^i) \nabla \mu_h^{i+1})(Dq_h^{i+2})^\top v_h^{i+2}$$
$$\left. + 2\eta'(\varphi_h^i) \varepsilon(v_h^{i+1}) : Dq_h^{i+1} + \operatorname{div}(\mu_h^{i+1} q_h^{i+1}) \right](\varphi_\delta^i),$$

$$\eta_{AD\mu,i} := \left[-r_h^i - \operatorname{div}(m(\varphi_h^{i-1}) \nabla \chi_h^i) - \operatorname{div}(\frac{\rho_2 - \rho_1}{2} m(\varphi_h^{i-1})(Dq_h^{i+1})^\top v_h^{i+1}) \right.$$
$$\left. - q_h^i \cdot \nabla \varphi_h^{i-1} \right](\mu_\delta^i),$$

$$\eta_{ADv,i} := \left[-\frac{1}{\tau} \rho(\varphi_h^{i-1})(q_h^{i+1} - q_h^i) - \rho(\varphi_h^{i-1})(Dq_h^{i+1})^\top v_h^{i+1} \right.$$
$$- (Dq_h^i)(\rho(\varphi_h^{i-2}) v_h^{i-1} - \frac{\rho_2 - \rho_1}{2} m(\varphi_h^{i-2}) \nabla \mu_h^{i-1})$$
$$\left. - \operatorname{div}(2\eta(\varphi_h^{i-1}) \varepsilon(q_h^i)) + \chi_h^i \nabla \varphi_h^{i-1} \right](v_\delta^i).$$

By these definitions and Theorem 16, the discretization error with respect to the objective function is then given by

$$
\mathscr{J}(\varphi_h, \mu_h, v_h, u_h) - \mathscr{J}(\varphi, \mu, v, u)
$$

$$
= \sum_{i=0}^{K-1} (\eta_{CM1,i} + \eta_{CM2,i} + \eta_{CM3,i} + \eta_{CM4,i} + \eta_{CH1,i} \tag{13.49}
$$

$$
+ \eta_{CH2,i} + \eta_{NS,i} + \eta_{AD\varphi,i} + \eta_{AD\mu,i} + \eta_{ADv,i}).
$$

We point out that the integral structure of these error terms allows a patchwise evaluation on the underlying mesh. Apart from the weights φ_δ^i, μ_δ^i and v_δ^i and χ_δ^i, q_δ^i, r_δ^i, respectively, the primal-dual-weighted error estimators only contain discrete quantities. In order to obtain a fully a-posteriori error estimator the weights are approximated involving a local higher-order approximation based on the respective discrete variables.

13.4.2 The Numerical Realization

For a numerical realization we discretize problem (P_Ψ) in space using a sequence of meshes $(\mathscr{T}^i)_{i=1}^K$ and introduce fully discrete sequences of functions using linear finite elements for φ, μ, and p and quadratic finite elements for v, yielding fully discretized variables φ_h, μ_h, p_h, and v_h. Note that we introduce the pressure p as a primal variable.

We further introduce the following approximation of Ψ_0

$$
\Psi_0^s(\varphi) := \frac{s}{2} \left(\max(0, \varphi - 1)^2 + \min(\varphi + 1)^2 \right), \ s > 0.
$$

The resulting fully discrete optimization problem is then solved using the steepest descent method for a sequence $s_n \to \infty$, mimicking the approach from Theorem 14. Especially we define the multipliers arising in Theorem 14 using Ψ_0 as given in Theorem 15.

The overall procedure is given in the subsequent Algorithm 1.

Here, the outer loop describes the refinement of the grids $(\mathscr{T}^i)_{i=1}^K$ using the error estimator given in (13.49). When the for-loop breaks, then we have found an approximate optimal control on the current sequence of grids that solves the system (13.40)–(13.46) sufficiently well in the sense that the complementarity conditions (13.45), (13.46) are satisfied up to a given tolerance tol_c. Then, in line 10, we evaluate the error indicators $\eta_T^i := \eta_{CM1,i}|_T + \eta_{CM2,i}|_T + \eta_{CM3,i}|_T + \eta_{CM4,i}|_T + \eta_{CH1,i}|_T + \eta_{CH2,i}|_T + \eta_{NS,i}|_T + \eta_{AD\varphi,i}|_T + \eta_{AD\mu,i}|_T + \eta_{ADv,i}|_T$ for all time steps i and

Data: Initial data: $\varphi_{-1}, \varphi_0, v_0, N_{\max}$

```
 1 repeat
 2 |   for l = 1, … do
 3 |   |   solve (P_ψ) using steepest descent method;
 4 |   |   if complementarity conditions (13.45), (13.46) are satisfied up to a tolerance tol_c
   |   |   then
 5 |   |   |   break;
 6 |   |   else
 7 |   |   |   increase s_n;
 8 |   |   end
 9 |   end
10 |   calculate the error indicators and find the set M_r of cells to refine and the set M_c of
   |   cells to coarsen;
11 |   Adapt (T^i)_{i=1}^K using M_r and M_c;
12 until ∑_{i=1}^K |T^i| < N_max;
```

Algorithm 1: The overall solution procedure

for all cells $T \in \mathscr{T}^i$ and choose $\mathscr{M}_r$ as the set with smallest cardinality, such that

$$\sum_{T \in \mathscr{M}_r} \eta_T \geq \theta^r \sum_{i=1}^{K} \sum_{T \in \mathscr{T}^i} \eta_T$$

with a parameter $0 < \theta^r < 1$ using a greedy marking algorithm. We mark all cells in $\mathscr{M}_r$ for refinement. As in [65] we further choose $\theta^c \in (0, 1)$ and define

$$M_c := \left\{ T \in (\mathscr{T}^i)_{i=1}^K \,|\, \eta_T \geq \frac{\theta^c}{N} \sum_{i=1}^{K} \sum_{T \in \mathscr{T}^i} \eta_T \right\},$$

where $N := \sum_{i=1}^K |\mathscr{T}^i|$. Thus, we use the well-known Dörfler marking procedure, where we refine a given proportion of the estimated error. We stress that we do not perform Dörfler marking on each time instance separately, but, as the representation (13.49) suggests, we perform a marking over all cells in the space-time cylinder. We point out that we have to use a locally refined initial grid in order to get a meaningful initial resolution of the interface. This prevents us from using a very coarse grid initially. As a consequence, we also need to introduce a coarsening strategy, where we mark cells for coarsening, if they contain an error that is smaller than θ^c times the mean error. We repeat this outer adaptation unless a given total amount of cells $N_{\max}$ is reached, summed over all cells, see line 12.

The inner loop, i.e. lines 2–9, solves (P_ψ) using the steepest descent method from the GNU scientific library [52]. Thereafter we check whether the complementarity conditions are sufficiently well approximated by the current Moreau–Yosida relaxed system. For this we evaluate the terms (13.45)–(13.46) for all time instances. If the absolute value of all these terms is smaller then a given tolerance tol_c, we accept the solution and proceed with the adaptation step. If any of these terms has an absolute

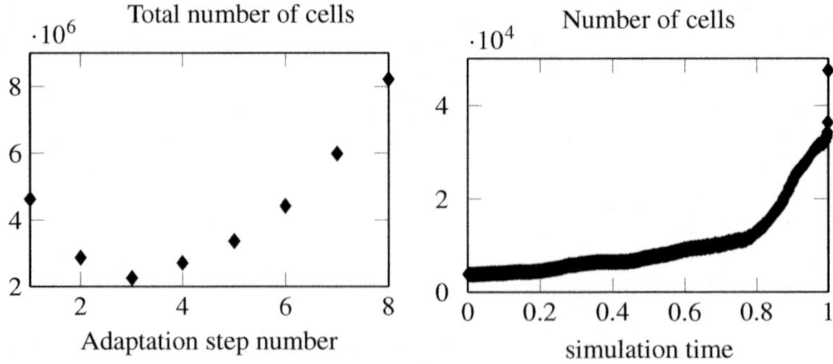

Fig. 13.2 The evolution of the total number of cells, i.e. $\sum_{i=1}^{K} NC(\mathcal{T}^i)$, where $NC(\mathcal{T}^i)$ denotes the number of cells of the triangulation $\mathcal{T}^i$ over the adaptation steps (*left*). We note, that we can not start with an arbitrary coarse mesh, as the interface as least has to be roughly resolved at the initialization of the optimization procedure. On the *right* we depict the distribution of the number of cells over the time horizon. We observe, that the mesh is refined most close to the final time instance, where our optimization aim is located

value larger than tol_c we increase parameter s_n and solve the optimality problem again.

Finally, we shortly illustrate the performance of our algorithm in an example where we aim to prevent a bubble from rising and split it into two bubbles. For more details we refer to [72].

We show the evolution of the total number of cells over the optimization procedure in Fig. 13.2 (left). On the right we show the distribution of the cells over the simulation horizon. These plots clearly show the benefits of using the proposed adaptive concept for the optimization of two-phase fluids.

13.5   Further Aspects and Future Research Directions

In the present section we briefly discuss further aspects and future research directions in simulation and control of variable density multiphase flows.

13.5.1   Optimal Control of Multiphase Flows Based on (13.15)–(13.18)

Based on the results in Sect. 13.2.1 we have a stable time discretization scheme at hand, for which we next state a time discrete optimal control problem. Here we restrict to control by volume forces that act on the fluid. In [44] additionally Dirichlet boundary control and control using the initial value φ_0, which can be seen as an inverse problem, are also investigated. Let us first state additional assumptions on the data.

Assumption 5 *Additionally to Assumption 1 we assume the following for the data:*

A5 W is twice continuous differentiable and there exist $C > 0$ such that $|W''_+(\varphi)| \leq C(1 + |\varphi|^2)$ and $|W''_-(\varphi)| \leq C(1 + |\varphi|^2)$.

A6 For ease of presentation in the following we assume, that ρ and η are affine linear with respect to φ, and that m is constant.

In the following in (13.1) we add an additional volume force Bu on the right hand side of the equation and understand this a control action, that can be chosen to influence the two-phase fluid. Let $U = L^2(0, T; \mathbb{R}^S)$ and $f_i \in L^2(\Omega)^n$, $i = 1, \ldots, S$ be given volume forces. We define $B : U \to L^2(\Omega)^n$ by

$$(Bu)(t, x) = \sum_{i=1}^{S} u_i(t) f_i(x),$$

i.e. our control acts as amplitudes of given volume forces. Especially u is independent of the actual spatial discretization.

To state the time discrete optimization problem, we define $u^k := \tau^{-1} \int_{t_{k-1}}^{t_k} u(t) \, dt$. This can be regarded as an ansatz using piecewise constant ansatz functions and we stress, that this can be obtained by variational discretization [74] of the first order optimality system stated below. Additionally, for the ease of presentation, we assume, that sufficient initial data is available for the two-step scheme, i.e. φ^{-1}, φ^0, μ^0, and v^0 are given functions, see Remark 3.

Now we can formulate the optimal control problem under consideration.

$$\min_{u \in L^2(0,T,\mathbb{R}^S)} \mathcal{J}(\varphi_h, u) := \frac{1}{2} \|\varphi_h - \varphi_d\|^2_{L^2(\Omega)} + \frac{\xi}{2} \|u\|^2_U$$

$$\text{s.t.} \qquad (\mathscr{P}_h)$$

accordingly modified equations (13.15)–(13.18)

Here and in the following we use the abbreviation $\varphi_h := (\varphi_h^k)_{k=1}^K$ and equivalently for the other variables.

Remark 11 Note that we apply a time continuous control $u \in L^2(0, T, \mathbb{R}^S)$ to the fully discrete system (13.15)–(13.18). Note that a simulation of two-phase flow has to obey certain CFL conditions, and thus the time step size τ has to be chosen depending on the actual a-priori unknown velocity field v. By not discretizing the control, we can base our numerical approach on a descent method with respect to the control and adjust the time step size τ during the optimization process without changing the actual control space U.

Based on the energy inequality from Theorem 3 one can show the following result by standard techniques.

Theorem 17 *There exists at least one solution to* $(\mathscr{P}_h)$*, i.e. at least one optimal control.*

By classical KKT-theory one can show the following first order optimality conditions.

Theorem 18 *Let* $u^\star, v_h^\star, p_h^\star, \varphi_h^\star, \mu_h^\star$ *be an optimal solution to* $(\mathscr{P}_h)$*. Then there exist adjoint variables* $p_{v,h}^\star \in H_0^1(\Omega)^n$, $p_{p,h}^\star \in L_{(0)}^2(\Omega)$, $p_{\varphi,h}^\star \in H^1(\Omega)$, $p_{\mu,h}^\star \in H^1(\Omega)$ *such that* (13.15)–(13.18) *and the following system is fulfilled for all* $k = 1, \dots, K$ *and all* $w \in H_0^1(\Omega)^n$, $q \in L_{(0)}^2(\Omega)$, $\Psi \in H^1(\Omega)$, $\Phi \in H^1(\Omega)$, $\tilde{u} \in U$.

$$-\frac{1}{\tau}\left((\frac{\rho^{k-1}+\rho^{k-2}}{2}w, p_{v,h}^k) - (\rho^{k-1}w, p_v^{k+1})\right)$$

$$-a(\rho^k w, v^{k+1}, p_v^{k+1}) - a(\rho^{k-1}v_h^{k-1} + J^{k-1}, w, p_{v,h}^k)$$

$$-(2\eta^{k-1}Dw, Dp_{v,h}^k) - (div(w), p_h^k) - (w\nabla\varphi^{k-1}, p_{\varphi,h}^k) = 0,$$

$$-(div p_v^k, q) = 0,$$

$$-a(J_{\mu_h^k}^k\Psi, v_h^{k+1}, p_{v,h}^{k+1}) + (\Psi\nabla\varphi^{k-1}, p_{v,h}^k) - (m\nabla\Psi, \nabla p_{\varphi,h}^k) + (\Psi, p_{\mu,h}^k) = 0,$$

$$\nabla p_{\varphi,h}^k \cdot \vec{\nu}_\Omega = 0,$$

$$\delta_{kK}(\varphi_h^k - \varphi_d, \Phi) - \frac{1}{\tau}\left(\rho'\frac{v^{k+1}p_v^{k+1} + v^{k+2}p_v^{k+2}}{2}, \Phi\right) + \frac{1}{\tau}\left(\rho'v^{k+1}p_v^{k+2}, \Phi\right)$$

$$-a(\rho'\Phi v_h^k, v^{k+1}, p_v^{k+1}) - (\eta'\Phi Dv^{k+1}, Dp_v^{k+1})$$

$$+(\mu^{k+1}\nabla\Phi, p_v^{k+1}) + (\rho'\Phi g, p_v^{k+1})$$

$$-\frac{1}{\tau}\left((\Phi, p_{\varphi,h}^k) - (P^{k+1}\Phi, p_\varphi^{k+1})\right) - (v^{k+1}\nabla\Phi, p_\varphi^{k+1})$$

$$-\sigma\varepsilon(\nabla\Phi, \nabla p_{\mu,h}^k) - \frac{\sigma}{\varepsilon}(W_+''(\varphi_h^k)\Phi, p_{\mu,h}^k) - \frac{\sigma}{\varepsilon}(W_-''(P^{k+1}\varphi_h^k)P^{k+1}\Phi, p_\mu^{k+1}) = 0,$$

$$\xi\tau u^k + B^\star p_{v,h}^k = 0 \in \mathbb{R}^S.$$

Here by $J_{\mu_h^k}^k\Psi$ *we abbreviate* $-\frac{d\rho}{d\varphi}m\nabla\Psi$*, i.e. the derivative of* J^k *with respect to* μ_h^k *in direction* Ψ*. >From integration by parts we obtain the boundary data*

$$\nabla p_{\mu,h}^k \cdot \vec{\nu}_\Omega = 0.$$

Here $B^\star p_v^k$ *is defined as*

$$B^\star p_{v,h}^k := ((f_l, p_{v,h}^k)_{L^2(\Omega)^n})_{l=1}^S.$$

Here for notational convenience we introduce artificial variables v_h^{K+1}, v_h^{K+2}, $p_{v,h}^{K+1}$, and $p_{v,h}^{K+2}$, and set them as zero.

Remark 12 We note that the prolongation operator P^k enters the adjoint equation acting on the test function Φ.

Concerning a numerical example we refer to [44].

13.5.2 Model Predictive Control

The optimization problem $(\mathscr{P}_h)$ describes so called open-loop control, which relies on the assumption, that the controlled system is not subject to external disturbances. In many practical applications however, such disturbances are present, which require the design of appropriate regulators. From the mathematical point of view this leads to concepts of closed-loop control. Here we propose the use of so called model predictive control (MPC) [51] and especially of the variant called instantaneous control (IC) [33, 73]. For a further discussion and for the application of IC to two-phase flows we refer to [75] closed-loop control of single-density two-phase flows, and to [76] for the case variable density two-phase flows.

13.5.3 Optimal Control with Non-smooth Free Energy Density

In Sect. 13.3, we established a function space version of C-stationarity for the optimal control problem in the case of non-smooth free energy densities. While this is already a beneficial form of stationarity, since most numerical solvers target this type of stationarity, recent results by Jarušek et al. [81] for parabolic variational inequalities suggest that it might be possible to obtain even stronger stationarity conditions such as strong stationarity or B-stationarity conditions. Their method does not require the differentiability of the constraint mapping but rather uses the Lipschitz continuity of the control-to-state operator, which can be established for the Cahn-Hilliard-Navier-Stokes system, in order to bound certain difference quotients. This enables the extraction of weakly convergent subsequences whose limit points prove to be auspicious candidates for the directional derivative of the control-to-state operator. One of the main advantages of strong stationarity is that it also permits the application of novel numerical concepts such as the bundle-free implicit programming technique, cf. [61], which can be used to the design efficient solution algorithms for the optimal control problem.

13.5.4 Model Order Reduction

It turned out and was expected that the numerical effort for optimal control of two-phase fluids due to the involved Navier–Stokes equation is enormous. Here model order reduction techniques are a promising tool to dramatically decrease the overall cost of the optimization process. We aim at so called proper orthogonal decomposition (POD) [77, 89]. It is well known, that POD is well applicable for the solution of the Navier–Stokes equation. Based on a high resolution simulation, a small subspace of the finite element space is constructed during the optimization process. This drastically reduces the overall number of unknowns, thus speeding up the computations. First numerical experiments indicate that POD is also well applicable for the Cahn–Hilliard equation with smooth free energy. In the case of non-smooth free energy densities, adapted schemes as proposed in [10] are required to obtain the desired reduction of unknowns. Combining these building blocks and adding properly adapted dual weighted residual error estimation will provide a highly efficient solver for optimal control problems of two-phase fluids.

Acknowledgements The authors gratefully acknowledge the support of the DFG through the priority programm 1506 "Transport processes at fluidic interfaces" under the grants HI 689_7-1 and HI 1466/2-1. This research was further supported by the Research Center MATHEON through project C-SE5 and D-OT1 funded by the Einstein Center for Mathematics Berlin. In addition, this research was partly supported by the Berlin Mathematical School.

References

1. Abels, H., Breit, D.: Weak solutions for a non-Newtonian diffuse interface model with different densities. arXiv:1509.05663v1 (2015). http://arxiv.org/abs/1509.05663
2. Abels, H., Garcke, H., Grün, G.: Thermodynamically consistent, frame indifferent diffuse interface models for incompressible two-phase flows with different densities. Math. Models Methods Appl. Sci. **22**(3), 1150013(40) (2012)
3. Abels, H., Depner, D., Garcke, H.: Existence of weak solutions for a diffuse interface model for two-phase flows of incompressible fluids with different densities. J. Math. Fluid Mech. **15**(3), 453–480 (2013)
4. Abels, H., Depner, D., Garcke, H.: On an incompressible Navier–Stokes/Cahn–Hilliard system with degenerate mobility. Ann. Inst. H. Poincaré (C) Non Linear Anal. **30**(6), 1175–1190 (2013)
5. Adams, R.A., Fournier, J.H.F.: Sobolev spaces. In: Pure and Applied Mathematics, vol. 140, 2nd edn. Elsevier, Amsterdam (2003)
6. Ainsworth, M., Oden, J.T.: A Posteriori Error Estimation in Finite Element Analysis. Wiley, New York (2000)
7. Aki, G.L., Dreyer, W., Giesselmann, J., Kraus, C.: A quasi-incompressible diffuse interface model with phase transition. Math. Models Methods Appl. Sci. **24**(5), 827–861 (2014)
8. Aland, S.: Time integration for diffuse interface models for two-phase flow. J. Comput. Phys. **262**, 58–71 (2014)
9. Aland, S., Voigt, A.: Benchmark computations of diffuse interface models for two-dimensional bubble dynamics. Int. J. Numer. Methods Fluids **69**, 747–761 (2012)

10. Alla, A., Falcone, E.: An adaptive pod approximation method for the control of advection-diffusion equations (2013). Arxiv: 1302.4072
11. Baňas, L., Nürnberg, R.: Adaptive finite element methods for Cahn–Hilliard equations. J. Comput. Appl. Math. **218**, 2–11 (2008)
12. Baňas, L., Nürnberg, R.: A posteriori estimates for the Cahn–Hilliard equation. Math. Modell. Numer. Anal. **43**(5), 1003–1026 (2009)
13. Bangerth, W., Rannacher, R.: Adaptive Finite Element Methods for Differential Equations. Lectures in Mathematics ETH Zürich. Birkhäuser Verlag, Basel (2003). doi:10.1007/978-3-0348-7605-6. http://dx.doi.org/10.1007/978-3-0348-7605-6
14. Barbu, V.: Optimal control of variational inequalities. In: Research Notes in Mathematics, vol. 100. Pitman (Advanced Publishing Program), Boston (1984)
15. Barbu, V.: Analysis and control of nonlinear infinite-dimensional systems. In: Mathematics in Science and Engineering, vol. 190. Academic, Boston (1993)
16. Becker, R., Kapp, H., Rannacher, R.: Adaptive finite element methods for optimal control of partial differential equations: basic concept. SIAM J. Control Optim. **39**(1), 113–132 (2000) (electronic). doi:10.1137/S0363012999351097, http://dx.doi.org/10.1137/S0363012999351097
17. Becker, R., Kapp, H., Rannacher, R.: Adaptive finite element methods for optimization problems. In: Numerical Analysis 1999 (Dundee). Chapman & Hall/CRC Research Notes in Mathematics, vol. 420, pp. 21–42. Chapman & Hall/CRC, Boca Raton (2000)
18. Benedix, O., Vexler, B.: A posteriori error estimation and adaptivity for elliptic optimal control problems with state constraints. Comput. Optim. Appl. **44**(1), 3–25 (2009). doi:10.1007/s10589-008-9200-y, http://dx.doi.org/10.1007/s10589-008-9200-y
19. Benzi, M., Golub, G., Liesen, J.: Numerical solution of saddle point problems. Acta Numer. **14**, 1–137 (2005)
20. Bergounioux, M.: Optimal control of an obstacle problem. Appl. Math. Optim. **36**(2), 147–172 (1997). doi:10.1007/s002459900058, http://dx.doi.org/10.1007/s002459900058
21. Bergounioux, M., Dietrich, H.: Optimal control of problems governed by obstacle type variational inequalities: a dual regularization-penalization approach. J. Convex Anal. **5**(2), 329–351 (1998)
22. Blank, L., Butz, M., Garcke, H.: Solving the Cahn–Hilliard variational inequality with a semismooth Newton method. ESAIM Control Optim. Calc. Var. **17**(4), 931–954 (2011)
23. Blank, L., Farshbaf-Shaker, M., Garcke, H., Rupprecht, C., Styles, V.: Multi-material phase field approach to structural topology optimization. In: Leugering, G., Benner, P., Engell, S., Griewank, A., Harbrecht, H., Hinze, M., Rannacher, R., Ulbrich, S. (eds.) Trends in PDE Constrained Optimization. International Series of Numerical Mathematics, vol. 165. Birkhäuser Verlag, Basel (2015)
24. Blowey, J.F., Elliott, C.M.: The Cahn–Hilliard gradient theory for phase separation with nonsmooth free energy. Part I: mathematical analysis. Eur. J. Appl. Math. **2**, 233–280 (1991)
25. Boyer, F.: A theoretical and numerical model for the study of incompressible mixture flows. Comput. Fluids **31**(1), 41–68 (2002)
26. Boyer, F., Lapuerta, C., Minjeaud, S., Piar, B., Quintard, M.: Cahn–Hilliard/Navier–Stokes model for the simulation of three-phase flows. Transp. Porous Media **82**(3), 463–483 (2010)
27. Bramble, J., Pasciak, J., Steinbach, O.: On the Stability of the L^2 projection in $H^1(\Omega)$. Math. Comput. **71**(237), 147–156 (2001)
28. Brenner, S.C., Scott, L.R.: The Mathematical Theory of Finite Element Methods. Texts in Applied Mathematics, vol. 15. Springer, Berlin (2008)
29. Brett, C., Elliott, C.M., Hintermüller, M, Löbhard, C.: Mesh adaptivity in optimal control of elliptic variational inequalities with point-tracking of the state. Interfaces Free Bound. **17**(1), 21–53 (2015). doi:10.4171/IFB/332, http://dx.doi.org/10.4171/IFB/332
30. Cahn, J.W., Hilliard, J.E.: Free energy of a nonuniform system. I. Interfacial free energy. J. Chem. Phys. **28**(2), 258–267 (1958)
31. Carstensen, C.: Quasi-interpolation and a-posteriori error analysis in finite element methods. Math. Modell. Numer. Anal. **33**(6), 1187–1202 (1999)

32. Chen, L.: *i*FEM: an innovative finite element method package in Matlab. Available at: ifem.wordpress.com (2008)
33. Choi, H., Temam, R., Moin, P., Kim, J.: Feedback control for unsteady flow and its application to the stochastic Burgers equation. J. Fluid Mech. **253**, 509–543 (1993)
34. Constantin, P., Foias, C.: Navier-Stokes-Equations. The University of Chicago Press, Chicago (1988)
35. Davis, T.A.: Algorithm 832: Umfpack v4.3 - an unsymmetric-pattern multifrontal method. ACM Trans. Math. Softw. **30**(2), 196–199 (2004)
36. Deckelnick, K., Günther, A., Hinze, M.: Finite element approximation of elliptic control problems with constraints on the gradient. Numer. Math. **111**(3), 335–350 (2009). doi:10.1007/s00211-008-0185-3, http://dx.doi.org/10.1007/s00211-008-0185-3
37. Ding, H., Spelt, P.D.M., Shu, C.: Diffuse interface model for incompressible two-phase flows with large density ratios. J. Comput. Phys. **226**(2), 2078–2095 (2007)
38. Ekeland, I., Témam, R.: Convex Analysis and Variational Problems. Classics in Applied Mathematics, vol. 28, English edn. Society for Industrial and Applied Mathematics, Philadelphia (1999). doi:10.1137/1.9781611971088, http://dx.doi.org/10.1137/1.9781611971088. Translated from the French
39. Elliott, C., Stinner, B., Styles, V., Welford, R.: Numerical computation of advection and diffusion on evolving diffuse interfaces. IMA J. Numer. Anal. **31**(3), 786–812 (2011)
40. Feng, X.: Fully discrete finite element approximations of the Navier–Stokes–Cahn–Hilliard diffuse interface model for two-phase fluid flows. SIAM J. Numer. Anal. **44**(3), 1049–1072 (2006)
41. Friedman, A.: Optimal control for variational inequalities. SIAM J. Control Optim. **24**(3), 439–451 (1986). doi:10.1137/0324025, http://dx.doi.org/10.1137/0324025
42. Garcke, H., Lam, K.F., Stinner, B.: Diffuse interface modelling of soluble surfactants in two-phase flow. Commun. Math. Sci. **12**(8), 1475–1522 (2014)
43. Garcke, H., Hecht, C., Hinze, M., Kahle, C.: Numerical approximation of phase field based shape and topology optimization for fluids. SIAM J. Sci. Comput. **37**(4), 1846–1871 (2015)
44. Garcke, H., Hinze, H., Kahle, C.: Optimal Control of time-discrete two-phase flow driven by a diffuse-interface model (2016). arXiv:1612.02283
45. Garcke, H., Hinze, M., Kahle, C.: A stable and linear time discretization for a thermodynamically consistent model for two-phase incompressible flow. Appl. Numer. Math. **99**, 151–171 (2016)
46. Girault, V., Raviart, P.A.: Finite Element Methods for Navier-Stokes Equations. Springer Series in Computational Mathematics. Theory and Algorithms, vol. 5. Springer, Berlin (1986). doi:10.1007/978-3-642-61623-5, http://dx.doi.org/10.1007/978-3-642-61623-5
47. Gross, S., Reusken, A.: Numerical Methods for Two-Phase Incompressible Flows. Springer Series in Computational Mathematics, vol. 40. Springer, Berlin (2011)
48. Grün, G.: On convergent schemes for diffuse interface models for two-phase flow of incompressible fluids with general mass densities. SIAM J. Numer. Anal. **51**(6), 3036–3061 (2013)
49. Grün, G., Klingbeil, F.: Two-phase flow with mass density contrast: stable schemes for a thermodynamic consistent and frame indifferent diffuse interface model. J. Comput. Phys. **257**(A), 708–725 (2014)
50. Grün, G., Guillén-Gonzáles, F., Metzger, S.: On fully decoupled convergent schemes for diffuse interface models for two-phase flow with general mass densities. Commun. Comput. Phys. **19**(5), 1473–1502 (2016)
51. Grüne, L., Pannek, J.: Nonlinear Model Predictive Control. Communications and Control Engineering. Springer, Berlin (2011)
52. GSL - GNU Scientific Library. http://www.gnu.org/software/gsl/
53. Guillén-González, F., Tierra, G.: On linear schemes for a Cahn–Hilliard diffuse interface model. J. Comput. Phys. **234**, 140–171 (2013)
54. Guillén-Gonzáles, F., Tierra, G.: Splitting schemes for a Navier–Stokes –Cahn–Hilliard model for two fluids with different densities. J. Comput. Math. **32**(6), 643–664 (2014)

55. Guillén-González, F., Tierra, G.: Numerical methods for solving the Cahn–Hilliard equation and its applicability to related energy-based models. Arch. Comput. Methods Eng. **22**(2), 269–289 (2015)

56. Günther, A., Hinze, M.: Elliptic control problems with gradient constraints—variational discrete versus piecewise constant controls. Comput. Optim. Appl. **49**(3), 549–566 (2011). doi:10.1007/s10589-009-9308-8, http://dx.doi.org/10.1007/s10589-009-9308-8

57. Guo, Z., Lin, P., Lowengrub, J.S.: A numerical method for the quasi-incompressible Cahn–Hilliard–Navier–Stokes equations for variable density flows with a discrete energy law. J. Comput. Phys. **276**, 486–507 (2014)

58. Hintermüller, M., Hoppe, R.H.W.: Goal-oriented adaptivity in control constrained optimal control of partial differential equations. SIAM J. Control Optim. **47**(4), 1721–1743 (2008). doi:10.1137/070683891, http://dx.doi.org/10.1137/070683891

59. Hintermüller, M., Hoppe, R.H.W.: Goal-oriented adaptivity in pointwise state constrained optimal control of partial differential equations. SIAM J. Control Optim. **48**(8), 5468–5487 (2010). doi:10.1137/090761823, http://dx.doi.org/10.1137/090761823

60. Hintermüller, M., Kopacka, I.: Mathematical programs with complementarity constraints in function space: *C*- and strong stationarity and a path-following algorithm. SIAM J. Optim. **20**(2), 868–902 (2009). doi:10.1137/080720681, http://dx.doi.org/10.1137/080720681

61. Hintermüller, M., Surowiec, T.: A bundle-free implicit programming approach for a class of MPECs in function space. Preprint (2012)

62. Hintermüller, M., Wegner, D.: Optimal control of a semidiscrete Cahn-Hilliard-Navier-Stokes system. SIAM J. Control Optim. **52**(1), 747–772 (2014). doi:10.1137/120865628, http://dx.doi.org/10.1137/120865628

63. Hintermüller, M., Ito, K., Kunisch, K.: The primal-dual active set strategy as a semi-smooth Newton method. SIAM J. Optim. **13**(3), 865–888 (2003)

64. Hintermüller, M., Hoppe, R.H., Iliash, Y., Kieweg, M.: An a posteriori error analysis of adaptive finite element methods for distributed elliptic control problems with control constraints. ESAIM Control Optim. Calc. Var. **14**(3), 540–560 (2008)

65. Hintermüller, M., Hinze, M., Tber, M.H.: An adaptive finite element Moreau–Yosida-based solver for a non-smooth Cahn–Hilliard problem. Optim. Methods Softw. **25**(4–5), 777–811 (2011). doi:10.1080/10556788.2010.549230

66. Hintermüller, M., Hinze, M., Hoppe, R.H.: Weak-duality based adaptive finite element methods for PDE-constrained optimization with pointwise gradient state-constraints. J. Comput. Math **30**(2), 101–123 (2012)

67. Hintermüller, M., Hinze, M., Kahle, C.: An adaptive finite element Moreau–Yosida-based solver for a coupled Cahn–Hilliard/Navier–Stokes system. J. Comput. Phys. **235**, 810–827 (2013)

68. Hintermüller, M., Hoppe, R.H.W., Löbhard, C.: Dual-weighted goal-oriented adaptive finite elements for optimal control of elliptic variational inequalities. ESAIM Control Optim. Calc. Var. **20**(2), 524–546 (2014). doi:10.1051/cocv/2013074, http://dx.doi.org/10.1051/cocv/2013074

69. Hintermüller, M., Mordukhovich, B.S., Surowiec, T.M.: Several approaches for the derivation of stationarity conditions for elliptic MPECs with upper-level control constraints. Math. Program. **146**(1–2, Ser. A), 555–582 (2014). doi:10.1007/s10107-013-0704-6, http://dx.doi.org/10.1007/s10107-013-0704-6

70. Hintermüller, M., Schiela, A., Wollner, W.: The length of the primal-dual path in Moreau-Yosida-based path-following methods for state constrained optimal control. SIAM J. Optim. **24**(1), 108–126 (2014). doi:10.1137/120866762, http://dx.doi.org/10.1137/120866762

71. Hintermüller, M., Keil, T., Wegner, D.: Optimal control of a semidiscrete Cahn-Hilliard-Navier-stokes system with non-matched fluid densities. arXiv preprint arXiv:1506.03591 (2015)

72. Hintermüller, M., Hinze, H., Kahle, C., Keil, T.: A goal-oriented dual-weighted adaptive finite elements approach for the optimal control of a Cahn-Hilliard-Navier-Stokes system. Hamburger Beiträge zur Angewandten Mathematik 2016-29 (2016)

73. Hinze, M.: Instantaneous closed loop control of the Navier–Stokes system. SIAM J. Control Optim. **44**(2), 564–583 (2005)
74. Hinze, M.: A variational discretization concept in control constrained optimization: the linear quadratic case. Comput. Optim. Appl. **30**(1), 45–61 (2005)
75. Hinze, M., Kahle, C.: A nonlinear model predictive concept for the control of two-phase flows governed by the Cahn–Hilliard Navier–Stokes system. In: System Modeling and Optimization, vol. 391. IFIP Advances in Information and Communication Technology (2013)
76. Hinze, M., Kahle, C.: Model predictive control of variable density multiphase flows governed by diffuse interface models. In: Proceedings of the first IFAC Workshop on Control of Systems Modeled by Partial Differential Equations, vol. 1, pp. 127–132 (2013)
77. Hinze, M., Volkwein, S.: Proper orthogonal decomposition surrogate models for nonlinear dynamical systems: error estimates and suboptimal control. In: Dimension Reduction of Large-Scale Systems. Lecture Notes in Computational Science and Engineering, vol. 45, pp. 261–306. Springer, Berlin (2005)
78. Hohenberg, P.C., Halperin, B.I.: Theory of dynamic critical phenomena. Rev. Mod. Phys. **49**(3), 435–479 (1977)
79. Hysing, S., Turek, S., Kuzmin, D., Parolini, N., Burman, E., Ganesan, S., Tobiska, L.: Quantitative benchmark computations of two-dimensional bubble dynamics. Int. J. Numer. Methods Fluids **60**(11), 1259–1288 (2009)
80. Ito, K., Kunisch, K.: Optimal control of elliptic variational inequalities. Appl. Math. Optim. **41**(3), 343–364 (2000). doi:10.1007/s002459911017, http://dx.doi.org/10.1007/s002459911017
81. Jarušek, J., Krbec, M., Rao, M., Sokołowski, J.: Conical differentiability for evolution variational inequalities. J. Differ. Equ. **193**(1), 131–146 (2003). doi:10.1016/S0022-0396(03)00136-0, http://dx.doi.org/10.1016/S0022-0396(03)00136-0
82. Kahle, C.: An L^∞ bound for the Cahn–Hilliard equation with relaxed non-smooth free energy density. arXiv:1511.02618 (2015)
83. Kay, D., Welford, R.: A multigrid finite element solver for the Cahn–Hilliard equation. J. Comput. Phys. **212**, 288–304 (2006)
84. Kay, D., Loghin, D., Wathen, A.: A preconditioner for the steady state Navier–Stokes equations. SIAM J. Sci. Comput. **24**(1), 237–256 (2002)
85. Kay, D., Styles, V., Welford, R.: Finite element approximation of a Cahn–Hilliard–Navier–Stokes system. Interfaces Free Bound. **10**(1), 15–43 (2008). http://www.ems-ph.org/journals/show_issue.php?issn=1463-9963&vol=10&iss=1
86. Kim, J.: A continuous surface tension force formulation for diffuse-interface models. J. Comput. Phys. **204**(2), 784–804 (2005)
87. Kim, J., Kang, K., Lowengrub, J.: Conservative multigrid methods for Cahn-Hilliard fluids. J. Comp. Phys. **193**, 511–543 (2004)
88. Krumbiegel, K., Rösch, A.: A virtual control concept for state constrained optimal control problems. Comput. Optim. Appl. **43**(2), 213–233 (2009). doi:10.1007/s10589-007-9130-0, http://dx.doi.org/10.1007/s10589-007-9130-0
89. Kunisch, K., Volkwein, S.: Galerkin proper orthogonal decomposition methods for parabolic problems. Numer. Math. **90**(1), 117–148 (2001). doi:10.1007/s002110100282, http://dx.doi.org/10.1007/s002110100282
90. Li, R., Liu, W., Ma, H., Tang, T.: Adaptive finite element approximation for distributed elliptic optimal control problems. SIAM J. Control Optim. **41**(5), 1321–1349 (2002). doi:10.1137/S0363012901389342, http://dx.doi.org/10.1137/S0363012901389342
91. Liu, W., Yan, N.: A posteriori error estimates for distributed convex optimal control problems. Adv. Comput. Math. **15**(1–4), 285–309 (2002) (2001). doi:10.1023/A:1014239012739, http://dx.doi.org/10.1023/A:1014239012739. A posteriori error estimation and adaptive computational methods
92. Lowengrub, J., Truskinovsky, L.: Quasi-incompressible Cahn–Hilliard fluids and topological transitions. Proc. R. Soc. A **454**(1978), 2617–2654 (1998)

93. Luo, Z.Q., Pang, J.S., Ralph, D.: Mathematical Programs with Equilibrium Constraints. Cambridge University Press, Cambridge (1996). doi:10.1017/CBO9780511983658, http://dx.doi.org/10.1017/CBO9780511983658
94. Mignot, F.: Contrôle dans les inéquations variationelles elliptiques. J. Funct. Anal. **22**(2), 130–185 (1976)
95. Mignot, F., Puel, J.P.: Optimal control in some variational inequalities. SIAM J. Control Optim. **22**(3), 466–476 (1984). doi:10.1137/0322028, http://dx.doi.org/10.1137/0322028
96. Oono, Y., Puri, S.: Study of phase-separation dynamics by use of cell dynamical systems. I. Modeling. Phys. Rev. A **38**(1), 434–463 (1988)
97. Otto, F., Seis, C., Slepčev, D.: Crossover of the coarsening rates in demixing of binary viscous liquids. Commun. Math. Sci. **11**(2), 441–464 (2013)
98. Outrata, J., Kočvara, M., Zowe, J.: Nonsmooth Approach to Optimization Problems with Equilibrium Constraints. Theory, Applications and Numerical Results. Nonconvex Optimization and Its Applications, vol. 28. Kluwer Academic Publishers, Dordrecht (1998). doi:10.1007/978-1-4757-2825-5, http://dx.doi.org/10.1007/978-1-4757-2825-5
99. Repin, S.: A Posteriori Estimates for Partial Differential Equations. Radon Series on Computational and Applied Mathematics, vol. 4. Walter de Gruyter GmbH & Co. KG, Berlin (2008). doi:10.1515/9783110203042, http://dx.doi.org/10.1515/9783110203042
100. Rösch, A., Wachsmuth, D.: A-posteriori error estimates for optimal control problems with state and control constraints. Numer. Math. **120**(4), 733–762 (2012). doi:10.1007/s00211-011-0422-z, http://dx.doi.org/10.1007/s00211-011-0422-z
101. Scheel, H., Scholtes, S.: Mathematical programs with complementarity constraints: stationarity, optimality, and sensitivity. Math. Oper. Res. **25**(1), 1–22 (2000). doi:10.1287/moor.25.1.1.15213, http://dx.doi.org/10.1287/moor.25.1.1.15213
102. Schmidt, A., Siebert, K.G.: Design of Adaptive Finite Element Software: The Finite Element Toolbox ALBERTA. Lecture Notes in Computational Science and Engineering, vol. 42. Springer, Berlin (2005)
103. Schneider, R., Wachsmuth, G.: A posteriori error estimation for control-constrained, linear-quadratic optimal control problems. SIAM J. Numer. Anal. **54**(2), 1169–1192 (2016)
104. Sethian, J.A.: Theory, algorithms, and applications of level set methods for propagating interfaces. Acta Numer. **5**, 309–395 (1996)
105. Shen, J., Yang, X.: A phase-field model and its numerical approximation for two-phase incompressible flows with different densities and viscosities. SIAM J. Sci. Comput. **32**(3), 1159–1179 (2010)
106. Tiba, D.: Optimal Control of Nonsmooth Distributed Parameter Systems. Lecture Notes in Mathematics, vol. 1459. Springer, Berlin (1990). doi:10.1007/BFb0085564, http://dx.doi.org/10.1007/BFb0085564
107. Verfürth, R.: A posteriori error analysis of space-time finite element discretizations of the time-dependent Stokes equations. Calcolo **47**(3), 149–167 (2010). doi:10.1007/s10092-010-0018-5, http://dx.doi.org/10.1007/s10092-010-0018-5
108. Vexler, B., Wollner, W.: Adaptive finite elements for elliptic optimization problems with control constraints. SIAM J. Control Optim. **47**(1), 509–534 (2008). doi:10.1137/070683416, http://dx.doi.org/10.1137/070683416
109. Wachsmuth, G.: Towards M-stationarity for optimal control of the obstacle problem with control constraints. SIAM J. Control Optim. **54**(2), 964–986 (2016). doi:10.1137/140980582, http://dx.doi.org/10.1137/140980582

Chapter 14
Dynamical Stability of Diffuse Phase Boundaries in Compressible Fluids

Heinrich Freistühler and Matthias Kotschote

Abstract Aiming at an understanding of the nonlinear stability of moving fluidic phase boundaries, this project has provided (a) solution theories for three basic systems of nonlinear partial differential equations that model two-phase fluid flow as well as (b) results on the existence of corresponding traveling waves and the spectrum of the operators that result from linearizing the PDEs about these waves.

The three models are the Navier-Stokes-Korteweg (NSK), the Navier-Stokes-Allen-Cahn (NSAC), and the Navier-Stokes-Cahn-Hilliard (NSCH) equations. For compressible NSAC and NSCH, the theories of strong solutions obtained seem to be the first ones in the literature. For NSK, a new theory of strong solutions has been developed, which in particular provides an alternative to the 'quasi-incompressible' approach that Abels et al. pursue for NSCH in the case of two separately incompressible phases of different density.

While for NSK the existence of traveling waves representing phase boundaries was known before, corresponding results have been newly obtained for NSAC and NSCH. For all three contexts, the project has established the spectral stability of these traveling waves. Viscous shock waves providing useful heuristic inspiration, fluidic interfaces corresponding to phase boundaries turn out to have their own flavour.

14.1 Scope

14.1.1 Introduction

Mathematically, the temporospatial behaviour of compressible fluids is described by one or the other system of conservation form or balance form,

$$\partial_t U + \nabla \cdot \mathcal{F}[U] = 0 \quad \text{or} \quad \partial_t U + \nabla \cdot \mathcal{F}[U] = \mathcal{G}[U], \tag{14.1}$$

H. Freistühler • M. Kotschote (✉)

Department of Mathematics and Statistics, University of Konstanz, 78457 Konstanz, Germany

e-mail: heinrich.freistuehler@unikonstanz.de; matthias.kotschote@uni-konstanz.de

© Springer International Publishing AG 2017

D. Bothe, A. Reusken (eds.), *Transport Processes at Fluidic Interfaces*,
Advances in Mathematical Fluid Mechanics, DOI 10.1007/978-3-319-56602-3_14

of partial differential equations in $(x, t) \in \mathbb{R}^n \times \mathbb{R}$. To start we recapitulate that the dynamics of compressible one-phase fluids is modelled by the Navier-Stokes equations,

$$\partial_t \rho + \nabla\cdot(\rho u) = 0,$$

$$\partial_t(\rho u) + \nabla\cdot(\rho u \otimes u - \mathbf{T}) = 0, \tag{14.2}$$

$$\partial_t \mathcal{E} + \nabla\cdot((\mathcal{E}\mathbf{I} - \mathbf{T})u) - \nabla\cdot(\beta\nabla\theta) = 0,$$

where $\rho > 0$ and $u \in \mathbb{R}^n$ are the fluid's density and velocity, and

$$\mathcal{E} \equiv \rho(E + \tfrac{1}{2}|u|^2) \quad \text{and} \quad \mathbf{T} \equiv -p\mathbf{I} + \mathbf{S} \tag{14.3}$$

are the total energy and the total Cauchy stress; the specific internal energy $E = E(\rho, S)$, the pressure $p = \rho^2\partial_\rho E(\rho, S)$, and the temperature $\theta = \partial_S E(\rho, S)$ are given functions of ρ and the entropy S. Viscous stress

$$\mathbf{S} = 2\eta(Du)^{s0} + \zeta\nabla\cdot u\,\mathbf{I}, \quad (Du)^{s0} \equiv \tfrac{1}{2}\left(Du + (Du)^\top\right) - \tfrac{1}{n}\nabla\cdot u\,\mathbf{I}, \tag{14.4}$$

and heat flux, $-\beta\nabla\theta$, are quantified by means of the coefficients η, ζ, β of shear viscosity, bulk viscosity, and thermal conductivity, all three given functions of ρ and θ. The three equations in (14.2) express the conservation of mass, momentum, and energy.

The project that we here report on has been studying systems of partial differential equations that model *two-phase* flows, with the perspective goal of understanding the stability properties of moving diffuse phase boundaries. The three models that the project studies are the Navier-Stokes-Korteweg system, the Navier-Stokes-Allen-Cahn system, and the Navier-Stokes-Cahn-Hilliard system. In this section we present these models, explain the mathematical description of moving diffuse phase boundaries, review related literature, and sketch the plan of the paper.

14.1.2 *Navier-Stokes-Korteweg and Diffuse Phase Boundaries*

We begin with the classical Navier-Stokes-Korteweg system (NSK). Reading

$$\partial_t \rho + \nabla\cdot(\rho u) = 0,$$

$$\partial_t(\rho u) + \nabla\cdot(\rho u \otimes u - \bar{\mathbf{T}}) = 0, \tag{14.5}$$

$$\partial_t \bar{\mathcal{E}} + \nabla\cdot((\bar{\mathcal{E}}\mathbf{I} - \bar{\mathbf{T}})u) - \nabla\cdot(\beta\nabla\theta) = 0,$$

NSK looks exactly like the one-phase system (14.2), but now the specific internal energy $\bar{E}$ that occurs in

$$\bar{\mathcal{E}} \equiv \rho\left(\bar{E} + \frac{1}{2}|u|^2\right) \tag{14.6}$$

depends also on the gradient of the density,

$$\bar{E} = \bar{E}(\rho, \theta, \nabla\rho) = \check{\bar{E}}(\rho, \theta, |\nabla\rho|^2) \tag{14.7}$$

and the Cauchy stress,

$$\bar{\mathbf{T}} \equiv -p\mathbf{I} + \mathbf{K} + \mathbf{S}, \tag{14.8}$$

comprises the *Korteweg tensor*

$$\mathbf{K} = \theta\left[\rho\nabla\cdot\left(\partial_{\nabla\rho}\left(\frac{\rho}{\theta}\bar{F}\right)\right)\mathbf{I} - \nabla\rho \otimes \partial_{\nabla\rho}\left(\frac{\rho}{\theta}\bar{F}\right)\right]. \tag{14.9}$$

In (14.9), $\bar{F}$ is the fluid's Helmholtz energy $\bar{F}(\rho, \theta, \nabla\rho)$, which is related to $\bar{E}$ by a Legendre transform

$$\bar{F} = \bar{E} - \theta S \quad \text{where} \quad \theta = \partial_S\bar{E}, \ S = -\partial_\theta\bar{F}.$$

All models we consider have their *isothermal versions*, in which temperature (/entropy) is not a variable.[1] In particular, the isothermal Navier-Stokes-Korteweg model[2] is given by the equations

$$\partial_t\rho + \nabla\cdot(\rho u) = 0,$$
$$\partial_t(\rho u) + \nabla\cdot(\rho u \otimes u - \bar{\mathbf{T}}) = 0, \tag{14.10}$$

where now

$$\bar{\mathbf{T}} \equiv -p\mathbf{I} + \mathbf{K} + \mathbf{S},$$

with

$$\mathbf{K} = \rho\nabla\cdot\left(\partial_{\nabla\rho}\left(\rho\bar{F}\right)\right)\mathbf{I} - \nabla\rho \otimes \partial_{\nabla\rho}\left(\rho\bar{F}\right) \quad \text{and} \quad \bar{F} = \bar{F}(\rho, \nabla\rho).$$

---

[1]The isothermal versions can be viewed as constant-temperature limits of the "full" ("non-isothermal") versions, as the heat conductivity β tends to infinity.

[2]We do not separately write down the isothermal version of the Navier-Stokes equations (14.2), since it is well-known and of course corresponds to the special case $\bar{F} = \bar{F}(\rho), \mathbf{K} = 0$ of the isothermal NSK model (14.10).

The diffuse phase boundaries this project aims at are special instances of so-called heteroclinic traveling waves. A *traveling wave* in a system (14.1) of conservation or balance laws is a solution to (14.1) of the special form

$$U^{\varphi}(x,t) = \varphi(x \cdot \mathbf{n} - st) \tag{14.11}$$

where $s \in \mathbb{R}$ and $\mathbf{n} \in S^{d-1}$ are the speed and the direction of propagation of the wave. A traveling wave (14.11) is called heteroclinic if the limits

$$U^{\pm} = \lim_{\hat{x} \to \pm\infty} \varphi(\hat{x})$$

exist and are different from each other. It is called *nonlinearly stable* if for all initial data U_0 that are sufficiently close to φ in an appropriate norm, system (14.1) has a unique solution $U(.,t)$ for all $t > 0$ and there exists a "phase shift" $\delta \in \mathbb{R}$ such that, in an appropriate norm,

$$U(.,t) - \varphi(. + \delta - st) \to 0 \quad \text{as} \quad t \to \infty.$$

It is called *spectrally stable* if the operator

$$-\nabla \cdot D\mathcal{F}(\varphi) \quad \text{or} \quad -\nabla \cdot D\mathcal{F}(\varphi) + D\mathcal{G}(\varphi)$$

that corresponds to the linearization

$$V_t + \nabla \cdot (D\mathcal{F}(\varphi)[V]) = 0 \quad \text{or} \quad V_t + \nabla \cdot (D\mathcal{F}(\varphi)[V]) = D\mathcal{G}(\varphi)[V] \tag{14.12}$$

of (14.1) at φ, has all its spectrum contained in the left half of the complex plane. The conceptual connection between nonlinear stability and spectral stability is given by the fact that violation of spectral stability induces "growing modes" for the linearized system (14.12).

We now use the context of isothermal NSK to explain what we will technically understand in this paper by the term *diffuse phase boundary*. For the purposes of the present paper, we will call a heteroclinic traveling wave for a compressible fluid model (14.1) a diffuse phase boundary if it is subsonic, i.e., its relative speed $\tilde{s} = |u \cdot \mathbf{n} - s|$ is smaller than the speed of sound, here $\sqrt{dp/d\rho(\rho^{\pm})}$, at either end. The idea is that this excludes shock waves (cf. [64]). As the simplest example, Fig. 14.1 shows the pressure $P(\tau) = p(1/\tau)$ as a function of the specific volume τ for an isothermal "van der Waals" gas at two different temperatures.

In the simplest setting for NSK, which is, in 1D and in a Lagrangian space variable, y,

$$\partial_t \tau - \partial_y u = 0,$$

$$\partial_t u + \partial_y(P(\tau)) = \eta \partial_y^2 u - \kappa \partial_y^3 \tau,$$

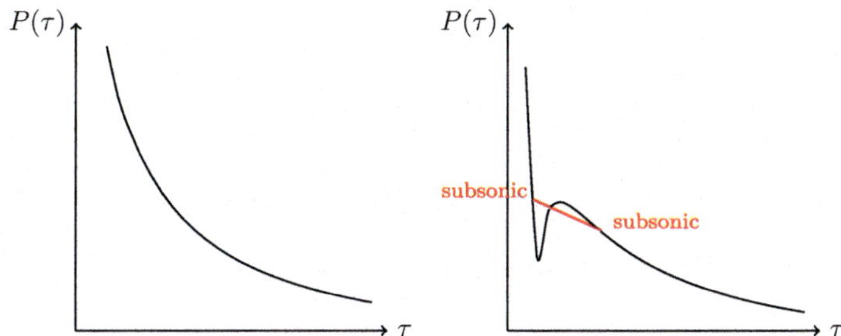

Fig. 14.1 van der Waals law for supercritical (*left*) and subcritical temperature (*right*)

the "p-system" with constant coefficients η and κ of viscosity and capillarity, a relatively simple phase plane analysis shows the subcritical pressure law admits heteroclinic traveling waves that are subsonic at either end—the supercritical pressure law does not [61].

14.1.3 *Navier-Stokes-Allen-Cahn and Navier-Stokes-Cahn-Hilliard*

Phase-Field Models The Navier-Stokes-Korteweg model uses the density of the fluid as the *order parameter*, and it is amazing how far one gets qualitatively in this simple way. However, better modeling is needed for real two-phase fluids, and to express the possibility of phase mixing one uses the concentrations $\chi, 1 - \chi \in (0, 1)$ of two distinguished phases. The resulting *phase-field models* have the form

$$\partial_t \rho + \nabla\cdot(\rho u) = 0,$$
$$\partial_t(\rho u) + \nabla\cdot(\rho u \otimes u - \mathbf{T}) = 0,$$
$$\partial_t \mathcal{E} + \nabla\cdot((\mathcal{E}\mathbf{I} - \mathbf{T})u) - \nabla\cdot(\beta\nabla\theta) = 0,$$
$$\partial_t(\rho\chi) + \nabla\cdot(\rho\chi u) - \mathcal{J} = 0,$$

(14.13)

where again

$$\mathcal{E} \equiv \rho(E + \frac{1}{2}|u|^2) \quad \text{and} \quad \mathbf{T} \equiv -p\mathbf{I} + \mathbf{C} + \mathbf{S} \tag{14.14}$$

are the total energy and the total Cauchy stress. For phase field-models (14.13), one assumes that the internal energy,

$$E(\rho, S, \chi, \nabla\chi) = \check{E}(\rho, S, \chi, |\nabla\chi|^2),$$

and the Helmholtz energy

$$F(\rho, \theta, \chi, \nabla\chi) = \check{F}(\rho, \theta, \chi, |\nabla\chi|^2) = (E - \theta S)(\rho, \theta, \chi, \nabla\chi), \quad \theta = \partial_S E, \ S = -\partial_\theta F \tag{14.15}$$

depend also on χ and $\nabla\chi$, but not on $\nabla\rho$, and $\mathbf{C}$ is the Ericksen tensor [29],

$$\mathbf{C} = -\nabla\chi \otimes \partial_{\nabla\chi}(\rho F). \tag{14.16}$$

Viscous stress $\mathbf{S}$ and heat flux are defined as in NSK. Obviously, the first three equations in (14.13) express the conservation of mass, momentum, and energy. The fourth equation, in view of the first one equivalent to its counterpart

$$\partial_t(\rho(1 - \chi)) + \nabla\cdot(\rho(1 - \chi)u) + \mathcal{J} = 0$$

for the other phase, encodes the exchange between the phases. It can be a balance law or a conservation law, depending on the definition of $\mathcal{J}$. Two options correspond to the possibility of phase transformation and its impossibility, respectively.

Navier-Stokes-Allen-Cahn The Navier-Stokes-Allen-Cahn equations (NSAC) are obtained by closing the system through specifying $\mathcal{J}$ in terms of F as

$$\mathcal{J} = \mathcal{J}_{AC} = \frac{\theta}{\varepsilon}\left(-\partial_\chi\left(\frac{\rho}{\theta}F\right) + \nabla\cdot\left(\partial_{\nabla\chi}\left(\frac{\rho}{\theta}F\right)\right)\right), \tag{14.17}$$

where $\varepsilon > 0$ is a relaxation time. The separate masses $\rho\chi$ and $\rho(1 - \chi)$ are not conserved.

Navier-Stokes-Cahn-Hilliard The Navier-Stokes-Cahn-Hilliard equations result from letting

$$\mathcal{J} = \mathcal{J}_{CH} = \nabla\cdot\mathcal{J}_\mu \quad \text{with} \quad \mathcal{J}_\mu = \gamma\theta\nabla\left(\frac{1}{\theta}\mu\right), \tag{14.18}$$

where

$$\frac{\rho}{\theta}\mu = \partial_\chi\left(\frac{\rho}{\theta}F\right) - \nabla\cdot\left(\partial_{\nabla\chi}\left(\frac{\rho}{\theta}F\right)\right) \tag{14.19}$$

and $\gamma > 0$ is a mobility coefficient. Due to the divergence structure of $\mathcal{J}_{CH}$, the masses of the two phases are conserved.

Isothermal Phase-Field Models Isothermal phase-field models,

$$\partial_t\rho + \nabla\cdot(\rho u) = 0,$$

$$\partial_t(\rho u) + \nabla\cdot(\rho u \otimes u - \mathbf{T}) = 0, \tag{14.20}$$

$$\partial_t(\rho\chi) + \nabla\cdot(\rho\chi u) - \mathcal{J} = 0,$$

differ from the "full" versions (14.13) by the absence of the energy conservation law. In this case the Helmholtz energy,

$$F(\rho, \chi, \nabla \chi) = \check{F}(\rho, \chi, |\nabla \chi|^2), \tag{14.21}$$

does not know a temperature.
Isothermal NSAC is (14.20) with

$$\mathcal{J} = \mathcal{J}_{AC}^{iso} = \frac{1}{\varepsilon} \left(-\partial_\chi (\rho F) + \nabla \cdot \left(\partial_{\nabla \chi} (\rho F) \right) \right). \tag{14.22}$$

Isothermal NSCH is (14.20) with

$$\mathcal{J} = \mathcal{J}_{CH}^{iso} = \nabla \cdot \mathcal{J}_{\mu^{iso}} \quad \text{where} \quad \mathcal{J}_{\mu^{iso}} = \gamma \nabla \mu^{iso}, \tag{14.23}$$

and

$$\rho \mu^{iso} = \partial_\chi (\rho F) - \nabla \cdot \left(\partial_{\nabla \chi} (\rho F) \right). \tag{14.24}$$

14.1.4 Related Literature

Going back to Korteweg [45], the Navier-Stokes-Korteweg equations have later first been studied by Dunn and Serrin [28].[3] The Navier-Stokes-Allen-Cahn equations appear to have first been formulated by Blesgen [17]. The Navier-Stokes-Cahn-Hilliard equations for compressible fluids seem to appear first in Lowengrub and Truskinovsky [56].[4] Solution theories for compressible NSK, EK, NSAC, NSCH were given by Danchin and Desjardins [23], Hattori and Li [41], Benzoni-Gavage et al. [15, 16], Haspot [39, 40], Feireisl et al. [31], Ding et al. [25], and Abels and Feireisl [5]. Solution theories for incompressible two-phase flows have been established by Starovoitov [65], Boyer [20], and Abels and co-authors in [1–4, 8], Prüss et al. in [58, 59], and Gal and Grasselli [37].

The idea to use the existence of a traveling-wave profile as a criterion for moving phase boundaries, including the corresponding groundbreaking consideration about the "p-system with viscosity and capillarity" we have referenced in Sect. 14.1.2, is due to Slemrod [61, 62]. The papers [11–13, 15] by Benzoni-Gavage and collaborators have established a rich picture regarding the existence and spectral stability of diffuse phase boundaries for NSK. A first proof for the nonlinear stability

---

[3]Though with an interesting difference regarding **K**, see [33].
[4]Though with an interesting difference regarding μ, see [33].

of diffuse phase boundaries has recently been given by Chen et al. [22], though in the very simplifying 1D setting. In several space dimensions, it is hard to imagine how to prove dynamical stability without considerations on the spectrum of the linearized operator. The spectral stability approach for diffuse interfaces was used earlier for reaction-diffusion waves [10, 30, 43] and shock waves [36, 38, 69]. Diffuse and sharp interfaces in fluids with two incompressible phases have been studied by Abels and co-authors in [6, 7] and Prüss and co-authors in [19, 60].

The derivation of dynamical equations for multiphase fluids and their interfaces, including notably their thermodynamic justification is a field in active development. We primarily refer to Bothe and Dreyer [18] and references therein. More specifically, Abels et al. [7] and Aki et al. [9] consider fluids with two incompressible phases, while Witterstein [67] and Dreyer et al. [26, 27] study fluids with two compressible phases (cf. also [14]). Our own point of view regarding justification of the Navier-Stokes-Korteweg, the Navier-Stokes-Allen-Cahn, and the Navier-Stokes-Cahn-Hilliard equations for compressible fluids can be found in [33].

14.1.5 Plan of the Paper

Section 14.2 presents solution theories for compressible Navier-Stokes-Korteweg, compressible Navier-Stokes-Allen-Cahn and compressible Navier-Stokes-Cahn-Hilliard, as well as a new theory for compressible Navier-Stokes. Section 14.3 discusses our results on the existence of diffuse phase boundaries for Navier-Stokes-Allen-Cahn and compressible Navier-Stokes-Cahn-Hilliard and their spectral stability.

14.2 Solution Theories

The development of mathematical analysis is typically not rectilinear. Even with the unboundedly extending traveling waves in mind, we developed the solution theories for NSK, NSAC, and NSCH until they reached a somewhat complete state also regarding finite domains, where boundary conditions are important. In this chapter, we present the theories in the richer setting of finite domains, referring the reader to the original research articles for the similar theories we obtained on the full space $\mathbb{R}^n$.

14.2.1 Solution Theory for NSK

The tensor named after Korteweg was proposed by him more than a century ago. Only recently has it become clearer how natural it is to include the Korteweg tensor with the equations of fluid dynamics. Brenner [21] has provided deep arguments for that, from the point of view of employing volume-averaged velocity instead of the often (and here) used mass-averaged velocity. Slemrod has shown that considerations by Gorban and Karlin [44] from the point of view of kinetic theory can be interpreted as a derivation of the Korteweg tensor from microphysics and calls the Korteweg tensor a necessary correction to the Navier-Stokes equations [63, 68].

In this section existence and uniqueness of strong solutions are addressed for the NSK system in its most general form. To be precise, the Helmholtz energy density

$$(\rho, \theta, \nabla\rho) \mapsto \overline{F}(\rho, \theta, \nabla\rho) = \check{F}(\rho, \theta, \phi), \quad \phi = |\nabla\rho|^2,$$

may depend on $|\nabla\rho|$ in any conceivable way. The viscosity coefficients

$$\eta = \eta(\rho, \theta, \phi, |(Du)^{s0}|_2^2), \quad \zeta = \zeta(\rho, \theta, \phi, |\nabla\cdot u\mathbf{I}|_2^2), \quad \phi = |\nabla\rho|^2$$

may also depend on $|\nabla\rho|$ and, in particular, on the norms $|(Du)^{s0}|_2$ and $|\nabla u\mathbf{I}|_2$ of the tracefree symmetric part of the velocity gradient and its trace part, respectively. I.e., the fluid may be non-Newtonian. The heat conductivity coefficient β may depend on ρ, θ, ϕ, and $|\nabla\theta|^2$.

Let $\Omega \subset \mathbb{R}^n$, $n \geq 1$, be a bounded domain, which is filled by the fluid, with compact boundary $\Gamma := \partial\Omega$ of class C^3 decomposing disjointly as

$$\Gamma = \Gamma_0 \dot{\cup} \Gamma_s = \Gamma_d \dot{\cup} \Gamma_n, \tag{14.25}$$

where each set may be empty. The conservation laws (14.5) for momentum, mass, and energy are supplemented boundary conditions and initial condition

$$u(0, x) = u_0(x), \quad \theta(0, x) = \theta_0(x), \quad \rho(0, x) = \rho_0(x), \quad x \in \Omega. \tag{14.26}$$

Two boundary conditions are of interest for the velocity u, namely, the no-slip case

$$u(t, x) = 0, \quad t > 0, \quad x \in \Gamma_0, \tag{14.27}$$

and the pure slip case

$$(u(t, x) \,|\, v(x)) = 0, \quad t > 0, \quad x \in \Gamma_s,$$
$$\mathcal{Q}(v(x))\mathcal{S}(u(t, x)) \cdot v(x) = 0, \quad t > 0, \quad x \in \Gamma_s. \tag{14.28}$$

Here $(a \,|\, b) := \sum_i a_i b_i$ denotes the inner product of $\mathbb{R}^n$, $v(x)$ is the outer normal at the position $x \in \Gamma := \partial\Omega$, and $\mathcal{Q}(v) := \mathbf{I} - v \otimes v$ projects a vector field u on its tangential part. As for boundary conditions for ρ and θ, we consider[5]

$$(\nabla \rho(t,x) \mid v(x)) = 0, \quad t > 0, \quad x \in \Gamma, \tag{14.29}$$

and

$$\theta(t,x) = 0, \quad t > 0, \quad x \in \Gamma_d,$$
$$(\nabla \theta(t,x) \mid v(x)) = 0, \quad t > 0, \quad x \in \Gamma_n. \tag{14.30}$$

In contrast to an opinion previously shared by many, the Navier-Stokes-Korteweg (NSK) equations are of parabolic type and therefore the methods of maximal regularity can be applied! These facts were first established in [47] (in that paper under the assumptions that the fluid be Newtonian and the dependence of its Helmholtz energy on ϕ linear) and [48]. At first glance the NSK system seems to be of mixed, namely, hyperbolic-parabolic type, and, as the equation of conservation of mass is involved, it is no wonder that this point of view was widespread. However, $\nabla \cdot \mathbf{K}$ contributes terms of third order regarding ρ resp. second order terms regarding $v = \nabla \rho$ in the momentum equation,

$$\nabla \cdot \mathbf{K} = \rho \kappa \nabla \Delta \rho + 2\rho \partial_\phi \kappa \nabla^3 \rho : \nabla \rho \otimes \nabla \rho + \rho \partial_\theta (\kappa/\theta) \nabla^2 \theta \cdot \nabla \rho + k^{low}$$

$$= \left(\rho \kappa \mathbf{I} + 2\rho \partial_\phi \kappa v \otimes v \right) : \nabla^2 v + \rho \partial_\theta (\kappa/\theta) \nabla^2 \theta \cdot v + k^{low} \quad \text{with } \kappa = 2\rho \partial_\phi \check{\bar{F}}(\rho, \theta, \phi).$$

The point is now that these third-order terms cause such a coupling between the momentum and mass equations that the system (14.5) turns into a parabolic one. In [47], this fact was revealed by a "direct approach" meaning that so-called full-space and half-space model problems were studied and it thereby turned out that the density satisfies a damped wave equation. This fact illuminates the hyperbolic nature as well as the regularisation property known from parabolic equations. The key issue was to reinterpret the continuity equation,

$$\partial_t \rho + \rho \nabla \cdot u + \nabla \rho \cdot u = 0, \quad x \in \Omega, \quad t \in J, \text{ an interval,}$$

and to find its "leading terms". In fact, if one expects the regularity $\rho \in L_p(J; H_p^3(\Omega))$ in view of the third-order spatial derivatives of ρ in the momentum equation, then it seems only natural to consider the term $\nabla \rho \cdot u$ as a *lower order term*. Note that this point of view is in contrast to the "hyperbolic nature" of this equation. So, it is left to deliberate on $\rho \nabla \cdot u$ in the mass equation. Supposing that this term is of lower order as well, which has been the former point of view, only the time regularity $\rho \in H_p^1(J; L_p(\Omega))$ would follow. But remembering the natural

---

[5]Inhomogeneous boundary data are treatable as well.

regularity for the velocity u,

$$u \in \mathbb{E}_1(J) := H_p^1(J; L_p(\Omega; \mathbb{R}^n)) \cap L_p(J; H_p^2(\Omega; \mathbb{R}^n)),$$

and the embedding

$$H_p^1(J; L_p(\Omega; \mathbb{R}^n)) \cap L_p(J; H_p^2(\Omega; \mathbb{R}^n)) \hookrightarrow H_p^{1/2}(J; H_p^1(\Omega; \mathbb{R}^n)),$$

which follows from the mixed derivative theorem, one sees that the term $\rho \nabla \cdot u$ belongs to $H_p^{1/2}(J; L_p(\Omega)) \cap L_p(J; H_p^1(\Omega))$. If the NSK system behaves like a parabolic problem, it should be possible to transfer this additional time and space regularity to $\partial_t \rho$, meaning that we can also expect to find $\rho \in H_p^{3/2}(J; L_p(\Omega)) \cap H_p^1(J; H_p^1(\Omega))$. In sum we have found for ρ the regularity class

$$H_p^{3/2}(J; L_p(\Omega)) \cap H_p^1(J; H_p^1(\Omega)) \cap L_p(J; H_p^3(\Omega)).$$

Since the regularity $H_p^1(J; H_p^1(\Omega))$ is redundant due to the embedding

$$\mathbb{E}_4(J) := H_p^{3/2}(J; L_p(\Omega)) \cap L_p(J; H_p^3(\Omega)) \hookrightarrow H_p^1(J; H_p^1(\Omega)),$$

the correct regularity class for the density ρ should be just $\mathbb{E}_4(J)$. This was indeed verified in [47] and a groundbreaking change of point of view.

Once parabolicity of the NSK system had been worked out, the next goal was to use [24] to treat the NSK system in its full generality, i.e., allowing for general Helmholtz energy and non-Newtonian fluids. However, NSK is a mixed order system that does not just readily fit in the framework of [24]. Could it be reformulated to suit [24]? It could, and the main idea for that was to augment the vector of unknowns (u, θ, ρ) by the variable $v = \nabla \rho$. Introducing $w = (u, v, \theta, \rho) \in \mathbb{R}^{2n+2}$, a meaningful equation for the additional variable v was needed. We found it by applying ∇ to the continuity equation,[6] which leads to

$$
\begin{aligned}
\partial_t v + \nabla(\rho \nabla \cdot u) &= -\nabla(v \mid u), & (t, x) &\in J \times \Omega, \\
(v \mid \nu) &= 0, & (t, x) &\in J \times \Gamma, \\
v &= \nabla \rho_0, & (t, x) &\in \{0\} \times \Omega.
\end{aligned}
\tag{14.31}
$$

Note that this formulation indicates a structural property of v, namely, v is a potential, and this property is of vital importance. Finally, we still needed a parabolic

---

[6] In the simple case of the one-dimensional problem with *constant* coefficients of viscosity and capillarity, Slemrod found a transformation which turns the equations into a 2×2 semilinear strictly parabolic system [61]. His beautiful "trick" anticipated, by more than 20 years, Brenner's later proposal to use the volume-averaged speed, indeed a sum $u + \alpha \nabla \rho$, as the primary variable [21]. The formulation proposed *here* induces parabolicity in generality and multiple dimensions, at the small price of augmenting the number of unknowns.

equation for ρ. For that purpose we added $-\Delta\rho + \nabla\cdot v$, which is nothing than zero, on the left hand-side of the continuity equation, to obtain

$$\partial_t\rho - \Delta\rho + \nabla\cdot v + \rho\nabla\cdot u = -(v\,|\,u) =: f, \quad (t,x) \in J \times \Omega,$$

$$(\nabla\rho\,|\,v) = 0, \qquad\qquad (t,x) \in J \times \Gamma, \qquad (14.32)$$

$$\rho = \rho_0, \qquad\qquad (t,x) \in \{0\} \times \Omega.$$

The special feature of this parabolic problem is the regularity $\rho \in \mathbb{E}_4(J)$ we had to establish. We proved the following.

Theorem 14.2.1 *Let Ω be an open bounded domain in $\mathbb{R}^n$, $n \geq 1$, with C^3-boundary Γ. Let $J = [0,T]$, $0 < T$, and $1 < p < \infty$, $p \neq 3/2$. Then problem (14.32) with $u, v \in \mathbb{E}_1(J)$ given has exactly one solution ρ in $\mathbb{E}_4(J)$ if and only if the data f, ρ_0 satisfy the following conditions*

1. $f \in H_p^{1/2}(J; L_p(\Omega)) \cap L_p(J; H_p^1(\Omega))$, $\rho_0 \in W_p^{3-2/p}(\Omega)$;
2. $\partial_\nu\rho_{0|\Gamma} = 0$ in $W_p^{2-3/p}(\Gamma)$ in case $p > 3/2$.

The idea of augmenting the original system only works if both formulations are equivalent, but this is indeed the case.

Lemma 14.2.1 *Let J be a time interval and $p \geq 2$. Then (u, θ, ρ) is a strong solution of (14.5),(14.26)–(14.30) on J with initial value (u_0, θ_0, ρ_0) if and only if $w = (u, v, \theta, \rho)$ is a strong solution of*

$$\partial_t(\rho u) + \nabla\cdot(\rho u \otimes u) - \nabla\cdot\overline{\mathbf{T}} = 0, \qquad x \in \Omega, \, t > 0,$$

$$\partial_t v + \nabla(\rho\nabla\cdot u) = -\nabla\,(v\,|\,u), \qquad x \in \Omega, \, t > 0,$$

$$\partial_t\rho - \Delta\rho + \nabla\cdot v + \rho\nabla\cdot u = -(v\,|\,u), \qquad x \in \Omega, \, t > 0, \qquad (14.33)$$

$$\partial_t\overline{\mathcal{E}} + \nabla\cdot\left([\overline{\mathcal{E}}\mathcal{I} - \overline{\mathbf{T}}]\cdot u\right) - \nabla\cdot(\beta\nabla\theta) = 0, \qquad x \in \Omega, \, t > 0,$$

(14.27)–(14.30) on J and with initial condition

$$w(0,x) = w_0(x) = (u_0(x), \nabla\rho_0(x), \theta_0(x), \rho_0(x)), \quad x \in \Omega. \qquad (14.34)$$

Proof Use Eqs. (14.31) and (14.32) to derive a relation for the L_2-norm of $\chi :=$ $v - \nabla\rho$. $\qquad\square$

The point is now that the Korteweg system (14.33), (14.27)–(14.34) with inhomogeneous boundary data can be reformulated as a quasi-linear parabolic problem

$$\mathcal{M}(w)\partial_t w + \mathcal{A}(w,D)w = G(w), \qquad (t,x) \in J \times \Omega,$$

$$\mathcal{B}_k(w,D)w = \sigma_k, \qquad (t,x) \in J \times \Gamma_k, \quad k = 0, s, d, n,$$

$$w = w_0, \qquad (t,x) \in \{0\} \times \Omega$$

with a certain coefficient matrix $\mathcal{M}$, second order differential operator $\mathcal{A}$, boundary operators $\mathcal{B}_k$, and nonlinearity $G(w)$ comprising terms of lower order. Moreover, the linearization of this parabolic problem decouples into two sub-problems, namely a system for (u, v, θ) and an equation for ρ, where solvability of the ρ-equation is ensured by Theorem 14.2.1. For the other sub-problem the results of [24] can be applied, providing a unique solution $(u, v, \theta) \in \mathbb{E}_1(J) \times \mathbb{E}_2(J) \times \mathbb{E}_3(J)$ with

$$\mathbb{E}_1(J) := \mathrm{H}_p^1(J; \mathrm{L}_p(\Omega; \mathbb{R}^n)) \cap \mathrm{L}_p(J; \mathrm{H}_p^2(\Omega; \mathbb{R}^n)),$$

$$\mathbb{E}_2(J) := \{v \in \mathbb{E}_1(J) : v \equiv \nabla\varphi, \ \varphi \in \mathbb{E}_4(J)\},$$

$$\mathbb{E}_3(J) := \mathrm{H}_p^1(J; \mathrm{L}_p(\Omega)) \cap \mathrm{L}_p(J; \mathrm{H}_p^2(\Omega)),$$

$$\mathbb{E}_4(J) := \mathrm{H}_p^{3/2}(J; \mathrm{L}_p(\Omega)) \cap \mathrm{L}_p(J; \mathrm{H}_p^3(\Omega)),$$

where $p \in (1, \infty)$.

One main result on local-in-time solvability, proved in [49], reads as follows.

Theorem 14.2.2 ([49]) *Let Ω be a bounded domain in $\mathbb{R}^n$, $n \geq 1$, with C^3-boundary, $\Gamma := \partial\Omega$ decomposing disjointly as in (14.25) and J denote the compact time interval $[0, T]$, $T > 0$. Let $n + 2 < p < \infty$ and suppose that*

1. $\check{F} \in \mathrm{C}^{3-}(\mathbb{R}_+^3)$ with $-\rho\theta\partial_\theta^2\check{F}(\rho, \theta, \phi) > 0$ for all ρ, $\theta > 0$, $\phi \geq 0$;
2. $\eta, \zeta, \beta \in \mathrm{C}^{2-}(\mathbb{R}_+^4)$, $\kappa \in \mathrm{C}^{3-}(\mathbb{R}_+^3)$ such that

$$a(z, s) > 0, \quad a(z, s) + 2s\partial_s a(z, s) > 0, \quad \forall(z, s) \in \mathbb{R}_+^4, \quad a \in \{\eta, \zeta, \beta\},$$

$$\kappa(y, s) > 0, \quad \kappa(y, s) + 2s\partial_s\kappa(y, s) > 0, \quad \forall(y, s) \in \mathbb{R}_+^3.$$

Then for each initial data (u_0, θ_0, ρ_0) in

$$V := \{(u, \theta, \rho) \in \mathrm{W}_p^{2-2/p}(\Omega; \mathbb{R}^n) \times \mathrm{W}_p^{2-2/p}(\Omega; \mathbb{R}_+) \times \mathrm{W}_p^{3-2/p}(\Omega; \mathbb{R}_+): \ \theta, \rho > 0 \text{ on } \overline{\Omega}\}$$

satisfying the compatibility conditions

$$u_{0|\Gamma_0} = 0, \quad (u_0 \mid v)_{|\Gamma_s} = 0, \quad (\mathcal{Q}(v)\mathbf{S}_{|t=0} \cdot v)_{|\Gamma_s} = 0,$$

$$(\nabla\rho_0 \mid v)_{|\Gamma} = 0, \quad \theta_{0|\Gamma_d} = 0, \quad (\nabla\theta_0 \mid v)_{|\Gamma_n} = 0,$$

(14.35)

there exists a unique solution (u, θ, ρ) of (14.5), (14.6)–(14.9), (14.26)–(14.30) on a maximal interval, which is $J^* = [0, t^*)$, $t^* := t^*(u_0, \theta_0, \rho_0) \in (0, T]$, if the solution is not global. The solution belongs to the maximal regularity class $\mathbb{E}_1(J_0) \times \mathbb{E}_3(J_0) \times \mathbb{E}_4(J_0)$ for each interval $J_0 = [0, T_0]$, $T_0 < t^*$, or to the class $\mathbb{E}_1(J) \times \mathbb{E}_3(J) \times \mathbb{E}_4(J)$ if the solution exists globally. The critical time t^* is characterised by the property

$$\lim_{t \to t^*}(u(t), \theta(t), \rho(t)) \quad \text{does not exist in } V.$$

Moreover, the solution is classical in $(0, t^*) \times \Omega$ and ρ, θ are nonnegative. The map $(u_0, \theta_0, \rho_0) \to (u, \theta, \rho)(t)$ defines a local semiflow on the natural phase space $\Phi := \{\varphi \in V : \varphi$ satisfies $(14.35)\}$.

14.2.2 Solution Theory for Compressible (One-Phase) Navier-Stokes

Sometimes one has to take a step "back". While the original ideas about the project did not contain any plan to think about the standard Navier-Stokes equations for compressible fluids, our thinking on NSAC and NSCH faced difficulties which had nothing to do with diphasicity. The literature offered many well-posedness results for compressible Navier-Stokes, but did not provide a solution theory which would have allowed an elegant transfer to NSAC and NSCH. While we were aiming at a setting which would be suitable as a starting point for a clean maximal-regularity approach, the fact that being able to deal with non-constant coefficients of viscosity and heat conductivity is essential for two-phase fluids[7] was a strong additional, physical motivation to try something new for the compressible Navier-Stokes-Fourier equations for *one*-phase fluids.

Let $\Omega \subset \mathbb{R}^n$ be a bounded domain with compact boundary $\Gamma := \partial\Omega$ of class C^2 decomposing disjointly as in (14.25). The outer unit normal of Γ at position x is denoted again by $\nu(x)$. Furthermore, let $J_0 = [0, T_0]$, $T_0 \in (0, \infty]$, be a time interval. The partial differential equations (14.2)–(14.4) have to be complemented by initial conditions

$$u(0, x) = u_0(x), \quad \theta(0, x) = \theta_0(x), \quad \rho(0, x) = \rho_0(x), \quad x \in \Omega, \tag{14.36}$$

and boundary conditions. Two boundary conditions are of interest for the velocity u, namely, the no-slip case

$$u(t, x) = h_0(t, x), \quad t > 0, \quad x \in \Gamma_0, \tag{14.37}$$

and the pure slip case

$$(u(t, x) \mid \nu(x)) = h_{s1}(t, x), \quad t > 0, \quad x \in \Gamma_s,$$
$$\mathcal{Q}(\nu(x))\mathcal{S}(u(t, x)) \cdot \nu(x) = h_{s2}(t, x), \quad t > 0, \quad x \in \Gamma_s. \tag{14.38}$$

---

[7]As the different phases will rarely have identical values for these coefficients. Indeed it is well-known that the viscosity and heat conductivity coefficients also of real one-phase fluids show strong dependence on temperature and density [42].

As for the temperature θ, we prescribe Dirichlet or Neumann boundary condition

$$\theta(t, x) = g_d(t, x), \quad t > 0, \quad x \in \Gamma_d,$$
$$(\nabla\theta(t, x) \mid \nu(x)) = g_n(t, x), \quad t > 0, \quad x \in \Gamma_n. \tag{14.39}$$

To state the main result, we still need assumptions on the coefficients and the Helmholtz energy. The viscosity coefficients η, ζ and the heat conductivity coefficient β may depend on ρ and θ. Moreover, they have to be subject to positivity and regularity conditions,

$$\eta, \zeta, \beta \in C^2(\mathbb{R}_+^2), \quad \eta(z), \zeta(z), \beta(z) > 0, \quad \forall z \in (0, \infty)^2, \tag{14.40}$$

where $\mathbb{R}_+^2 := \mathbb{R}_+ \times \mathbb{R}_+$. The internal energy density E has to satisfy

$$E \in C^{3-}(\mathbb{R}_+^2), \quad \partial_S^2 E(\rho, S) > 0, \quad \forall \rho, S > 0. \tag{14.41}$$

As we are looking for strong solutions in the L_p-sense, the following regularity class seems to be natural on the basis of the differential equations in (14.2)–(14.4). We are interested in unique solutions

$$(u, \theta, \rho) \in \mathbb{E}(J) = \mathbb{E}_1(J) \times \mathbb{E}_2(J) \times \mathbb{E}_3(J),$$

where

$$\mathbb{E}_1(J) := H_p^1(J; L_p(\Omega; \mathbb{R}^n)) \cap L_p(J; H_p^2(\Omega; \mathbb{R}^n)),$$

$$\mathbb{E}_2(J) := H_p^1(J; L_p(\Omega)) \cap L_p(J; H_p^2(\Omega)),$$

$$\mathbb{E}_3(J) := C^1(J; L_p(\Omega)) \cap C(J; H_p^1(\Omega))$$

and $p \in (1, \infty)$. We shall also need the function spaces

$$Y_{j,k}(J; E) := W_p^{1-j/2-1/2p}(J; L_p(\Gamma_k; E)) \cap L_p(J; W_p^{2-j-1/p}(\Gamma_k; E)),$$
$$j = 0, 1, \quad k = 0, s, d, n.$$

The main result of this section is

Theorem 14.2.3 ([50]) *Let Ω be a bounded domain in $\mathbb{R}^n$, $n \geq 1$, with compact C^2-boundary Γ decomposing disjointly as mentioned above. Furthermore, let $J_0 = [0, T_0]$ with $T_0 \in (0, \infty]$, and $p \in (\hat{p}, \infty)$ with $\hat{p} := \max\{n, 2\}$. Assume (14.40), (14.41). Then for each initial data (u_0, θ_0, ρ_0) in*

$$V := \{(u, \theta, \rho) \in W_p^{2-2/p}(\Omega; \mathbb{R}^n) \times W_p^{2-2/p}(\Omega) \times H_p^1(\Omega) :$$

$$(u(y) \mid \nu(y)) \geq 0, \quad \forall y \in \Gamma, \quad \rho(x) > 0, \quad \theta(x) > 0, \quad \forall x \in \overline{\Omega}\}$$

and boundary data h_0, $h_s = (h_{s1}, h_{s2})$, g_d, g_n satisfying the regularity

$$h_0 \in Y_{0,0}(J_0; \mathbb{R}^n), \quad h_{s1} \in Y_{0,s}(J_0), \quad h_{s2} \equiv 0, \quad g_d \in Y_{0,d}(J_0), \quad g_n \equiv 0,$$

$$\forall t \in J_0: \ (h_0(t,x) \,|\, \nu(x)) \geq 0, \ \forall x \in \Gamma_0, \quad h_{s1}(t,x) \geq 0, \ \forall x \in \Gamma_s, \quad g_d(t,x) > 0, \ \forall x \in \Gamma_d$$

and the compatibility conditions

$$u_{0|\Gamma_0} = h_{0|t=0}, \quad (u_0 \,|\, \nu)_{\Gamma_s} = h_{s1|t=0}, \quad \theta_{0|\Gamma_d} = g_{d|t=0}, \quad (\nabla \theta_0 \,|\, \nu)_{|\Gamma_n} = 0, \tag{14.42}$$

there exists a unique solution (u, θ, ρ) of (14.2)–(14.4), (14.36)–(14.39) on a maximal time interval, which is $J^* = [0, t^*)$, $t^* := t^*(u_0, \theta_0, \rho_0) \in (0, T_0]$ if the solution is not global. The solution belongs to the class $\mathbb{E}(J)$ for each interval $J = [0, T]$, $T < t^*$, or to the class $\mathbb{E}(J_0)$ if the solution exists globally. The maximal time t^* is characterised by the property:

$$\lim_{t \to t^*} (u(t), \theta(t), \rho(t)) \quad \text{does not exist in } V.$$

The map $(u_0, \theta_0, \rho_0) \to (u, \theta, \rho)(t)$ generates a local semiflow on the phase space $\Phi := \{\varphi \in V : \varphi \text{ satisfies } (14.42)\}$.

The proof of this well-posedness result basically splits into establishing self-mapping and contraction of a certain mapping defined on the solution space $\mathbb{E}(J)$. However, in contrast to the case of the Navier-Stokes-Korteweg system discussed in the previous section, the Navier-Stokes-Fourier system (14.2) does constitute a mixed type, namely hyperbolic-parabolic, problem and is thus *more* involved.

The first idea of the proof is to regard the density as placeholder for the solution operator of the continuity equation. More precisely, given $u \in \mathbb{E}_1(J)$ with $(u \,|\, \nu)_{|\partial\Omega} \geq 0$ and $\rho_0 \in H_p^1(\Omega)$, Kato's theory of evolution families makes it possible to solve the continuity equation

$$\partial_t \rho + \nabla \cdot (\rho u) = 0, \quad (t, x) \in J \times \Omega,$$

$$\rho(0) = \rho_0, \quad x \in \Omega$$

uniquely such that $\rho \in \mathbb{E}_3(J)$ and $\|\rho\|_{\mathbb{E}_3(J)} \leq c(\|\rho_0\|_{H_p^1(\Omega)}, \|u\|_{\mathbb{E}_1(J)})$ holds. This gives rise to the solution operator

$$L : H_p^1(\Omega) \times \mathbb{E}_1(J) \to \mathbb{E}_3(J), \quad (\rho_0, u) \mapsto L[u]\rho_0 = \rho$$

depending on u nonlinearly and on ρ_0 linearly. Taking up this point of view, i.e. replacing ρ by $L[u]\rho_0$ in the momentum and energy equation, the starting problem is reduced to a nonlocal, fully nonlinear parabolic problem for (u, θ). This problem is then locally solved by means of a fixed point argument, i.e. one linearizes this parabolic problem around a given reference state and defines the fixed point mapping according to

$$\Xi : \mathbb{E}_1(J) \times \mathbb{E}_2(J) \to \mathbb{E}_1(J) \times \mathbb{E}_2(J), \quad (\hat{u}, \hat{\theta}) \mapsto (u, \theta) =: \Xi(\hat{u}, \hat{\theta}),$$

where (u, θ) denotes the solution of the linear problem. Maximal L_p-regularity is very useful to show this mapping behaviour of Ξ. Note that highest order terms of ρ in the momentum and energy equation involve $\nabla \rho \in C(J; L_p(\Omega; \mathbb{R}^n))$, which perfectly fits together with the basic space $L_p(J; L_p(\Omega; \mathbb{R}^n)) \times L_p(J; L_p(\Omega))$ considered for these equations and the concept of maximal L_p-regularity.

The property of contraction is, however, not so easily obtained, as we encounter the "hyperbolic nature" of the density $\rho = L[u]\rho_0$ in the sense that contraction of the fixed point mapping Ξ cannot be proved in the regularity class $\mathbb{E}_1(J) \times \mathbb{E}_2(J)$. In fact, studying that system of partial differential equations, which differences of solutions satisfy, one easily sees that this problem contains terms not having enough regularity to prove "good" estimates. This relates to the theory of quasilinear symmetric hyperbolic systems where the contraction mapping principle cannot be applied in its usual setting. The idea to resolve this difficulty consists in considering contraction in a rougher topology. According to Majda [57], this idea goes back to Kato and Lax.

To become more precise and to see what the stumbling stones are, let $\overline{w} = (\overline{u}, \overline{\theta}) \in \mathbb{E}_1(J) \times \mathbb{E}_2(J)$ with $\overline{w}(0) = (u_0, \theta_0)$ and $(\overline{u} \,|\, v)_{|\Gamma} = 0$ denote a given reference state and

$$\Sigma := \{w = (u, \theta) \in \mathbb{E}_1(J) \times \mathbb{E}_2(J) : (u \,|\, v) \geq 0 \text{ on } \Gamma, \ w_{|t=0} = (u_0, \theta_0),$$

$$\|w - \overline{w}\|_{\mathbb{E}_1(J) \times \mathbb{E}_2(J)} \leq R\}, \quad R \in (0, 1),$$

be mapped by Ξ to itself.[8] Let $(u_i, \theta_i) \in \Sigma$ and $\rho_i = L[u_i]\rho_0$, then $\varrho := \rho_1 - \rho_2$ and $v := u_1 - u_2$ satisfy

$$\partial_t \varrho + \nabla \cdot (\varrho u_1) = -\nabla \cdot (\rho_2 v), \quad (t, x) \in J \times \Omega,$$

$$\varrho(0) = 0, \quad x \in \Omega.$$

The point is now that $\nabla \varrho$ appears in the momentum and energy equation and the norm $\|\nabla \varrho\|_{L_p(J; L_p(\Omega; \mathbb{R}^n))}$ should be estimated by $\|v\|_{\mathbb{E}_1(J)}$. However, having in mind the regularities $\rho_1, \rho_2 \in \mathbb{E}_3(J)$ and $u_1, u_2 \in \mathbb{E}_1(J)$ as well as choosing $p \in (1, \infty)$ large enough, one easily verifies that the right-hand side $\nabla \cdot (\rho_2 v)$ only belongs to $L_p(J; L_p(\Omega))$, since the term $\nabla \rho_2 \in C(J; L_p(\Omega; \mathbb{R}^n))$ is involved. This regularity is not enough to obtain $\varrho \in \mathbb{E}_3(J)$ by the equation itself and to gain an estimate of $\|\varrho\|_{L_p(J; H_p^1(\Omega))}$ in terms of $\|v\|_{\mathbb{E}_1(J)}$. This lack of regularity always occurs and

---

[8]In order that the idea of Kato and Lax works with this set, one has first to show that Σ is a closed subset regarding the larger function space for which contraction can be proved, see [50, Lemma 3.1].

cannot be resolved by considering higher regularities. Instead, one can prove the (contraction) inequality

$$\|\varrho\|_{C(J;L_2(\Omega))} \leq e^{\|u_1\|_{L_1(J;C^1(\overline{\Omega};\mathbb{R}^n))}} \int_0^T \|\nabla\cdot(\rho_2 v)(s)\|_{L_2(\Omega)}\, ds \leq c_1 \|v\|_{L_1(J;H_2^1(\Omega;\mathbb{R}^n))}$$

$$\leq c_1 T^{1/2} \|v\|_{L_2(J;H_2^1(\Omega;\mathbb{R}^n))}, \quad J = [0, T],$$

where the embedding $H_2^1(\Omega;\mathbb{R}^n) \hookrightarrow L_{\frac{2p}{p-2}}(\Omega;\mathbb{R}^n)$ entered, and the constant $c_1 > 0$ is independent of u_1 and ρ_1, since $\|u_1\|_{E_1(J)} \leq \|u_1 - \overline{u}\|_{E_1(J)} + \|\overline{u}\|_{E_1(J)} < 1 + \|\overline{u}\|_{E_1(J)}$. Now, one has to find a setting for momentum and energy equation which does not require $\nabla\varrho \in L_p(J;L_p(\Omega;\mathbb{R}^n))$ and allows to estimate solutions. This was the actual challenge for proving well-posedness—for the Navier-Stokes-Fourier system *and* for NSAC/NSCH. We refer to [50].

14.2.3   Solution Theory for NSAC[9]

Although the NSAC system (14.13)–(14.16), (14.17) is thermodynamically and mechanically consistent for any Helmholtz energy

$$(\rho, \theta, \chi, |\nabla\chi|^2) \mapsto \check{F}(\rho, \theta, \chi, |\nabla\chi|^2),$$

for the well-posedness result we have to slightly restrict the generality, assuming

$$F(\rho, \theta, \chi, \nabla\chi) = F_0(\rho, \theta, \chi) + F_{mix}(\rho, \theta, \chi, |\nabla\chi|^2),$$

$$F_{mix}(\rho, \theta, \chi, |\nabla\chi|^2) = \theta\left(W(\chi) + \frac{\delta(\rho)}{2}|\nabla\chi|^2\right), \tag{14.43}$$

where F_{mix} denotes the so-called mixing energy and a prototypical form of W is given by

$$W(\chi) = k_1[\chi\ln(\chi) + (1 - \chi)\ln(1 - \chi)] + k_2\chi(1 - \chi) \tag{14.44}$$

with $k_1 \in (0, \infty)$, $k_2 \in \mathbb{R} \setminus \{0\}$, suggesting several scenarios depending on the sign of k_2. In fact, $k_2 < 0$ means that interactions between different phases are more favorable in order to lower the mixing energy, while $k_2 > 0$ produces a separation of the phases. Such a structure of F naturally arises by relating the energy of the fluid mixture to equations of state of the separate phases, cf. again [35]. Finally, the parameter δ, which may depend on the density, measures the interface thickness.

---

[9]The 'quasi-incompressible' case of NSAC is covered by NSK, cf. [33, 34].

The proof of well-posedness for the NSAC system runs as for the Navier-Stokes-Fourier equations, since it just couples them with an additional parabolic equation (the 'convective Allen-Cahn' equation) and from the mathematical point of view this circumstance plays a minor role. Therefore, we omit details and minor specific difficulties of the proof in this survey.

Let $\Omega \subset \mathbb{R}^n$ be a bounded domain with compact boundary $\Gamma := \partial\Omega$ of class C^2 decomposing disjointly as

$$\Gamma = \Gamma_d \,\dot{\cup}\, \Gamma_s,$$

where each set may be empty. The outer unit normal of Γ at position x is denoted again by $\nu(x)$. Furthermore, let $J_0 = [0, T_0]$, $T_0 \in (0, \infty]$, be a time interval. The partial differential equations (14.13) have to be complemented by initial conditions

$$u(0, x) = u_0(x), \quad \theta(0, x) = \theta_0(x), \quad \chi(0, x) = \chi_0(x), \quad \rho(0, x) = \rho_0(x), \quad x \in \Omega. \tag{14.45}$$

As for the boundary conditions, we keep close to the situation as for the Navier-Stokes-Fourier system. We prescribe no-slip or pure slip condition for u,

$$u(t, x) = h_d(t, x), \quad t > 0, \quad x \in \Gamma_d,$$

$$(u(t, x) \,|\, \nu(x)) = h_{s1}(t, x), \quad t > 0, \quad x \in \Gamma_s, \tag{14.46}$$

$$\mathcal{Q}(\nu(x))\mathbf{S}(u(t, x)) \cdot \nu(x) = 0, \quad t > 0, \quad x \in \Gamma_s.$$

The boundary conditions for θ and χ are of Dirichlet or Neumann type,

$$\theta(t, x) = g_d(t, x), \quad \chi(t, x) = l_d(t, x), \quad (t, x) \in J \times \Gamma_d, \tag{14.47}$$

and

$$(\nabla\theta(t, x) \,|\, \nu(x)) = 0, \quad (\nabla\chi(t, x) \,|\, \nu(x)) = 0, \quad (t, x) \in J \times \Gamma_s. \tag{14.48}$$

We note that for technical reasons, only homogenous boundary conditions can be taken into account in the pure slip and Neumann case. We are able to consider non-constant coefficients that have to satisfy the regularity assumptions

$$\eta, \zeta, \beta, \varepsilon \in \mathbf{C}^{2-}(\mathbb{R}_+^2 \times [0, 1]), \quad \delta \in \mathbf{C}^{2-}(\mathbb{R}_+) \tag{14.49}$$

and the positivity assumptions

$$\eta(\xi), \zeta(\xi), \beta(\xi), \varepsilon(\xi) > 0, \quad \forall \xi \in (0, \infty)^2 \times [0, 1], \quad \delta(z) > 0, \quad \forall z \in (0, \infty). \tag{14.50}$$

As for the parts F_0 and W of the Helmholtz energy, we require that

$$F_0 \in C^{3-}(\mathbb{R}_+^2 \times [0,1]), \quad -\theta \partial_\theta^2 F_0(\rho,\theta,\chi) > 0, \quad \forall\,(\rho,\theta,\chi) \in (0,\infty)^2 \times [0,1],$$

$$W \in C^1(0,1), \quad \lim_{\chi \searrow 0} \partial_\chi (F_0 + \theta W) > 0 > \lim_{\chi \nearrow 1} \partial_\chi (F_0 + \theta W)$$

$$(14.51)$$

in order that χ stays in $[0,1]$ and the heat equation does not degenerate. Notice that if there exists a smooth solution χ and the conditions above apply as well as the boundary data l_d lies in $(0,1)$, the parabolic maximum principle implies $\chi \in (0,1)$.

The main result in case of bounded domains concerns existence and uniqueness of strong solutions $(u,\theta,\chi,\rho) \in \mathbb{E}(J)$ with $\mathbb{E}(J) = \mathbb{E}_1(J) \times \mathbb{E}_2(J) \times \mathbb{E}_3(J) \times \mathbb{E}_4(J)$ and

$$\mathbb{E}_1(J) = \mathrm{H}_p^1(J; \mathrm{L}_p(\Omega; \mathbb{R}^n)) \cap \mathrm{L}_p(J; \mathrm{H}_p^2(\Omega; \mathbb{R}^n)),$$

$$\mathbb{E}_2(J) = \mathrm{H}_p^1(J; \mathrm{L}_p(\Omega)) \cap \mathrm{L}_p(J; \mathrm{H}_p^2(\Omega)), \quad \mathbb{E}_3(J) = \mathbb{E}_2(J),$$

$$\mathbb{E}_4(J) = \mathrm{H}_p^2(J; \mathrm{H}_p^{-1}(\Omega)) \cap \mathrm{C}^1(J; \mathrm{L}_p(\Omega)) \cap \mathrm{C}(J; \mathrm{H}_p^1(\Omega)).$$

Theorem 14.2.4 ([51]) *Let Ω be a bounded domain in $\mathbb{R}^n$, $n \geq 1$, with compact C^2-boundary Γ decomposing disjointly as $\Gamma = \Gamma_d \,\dot\cup\, \Gamma_s$, $J_0 = [0,T_0]$ with $T_0 \in (0,\infty]$ and $p \in (n+2,\infty)$. Assume (14.49)–(14.51). Then for each initial data $(u_0,\theta_0,\chi_0,\rho_0)$ in*

$$V := \{(u,\theta,\chi,\rho) \in \mathrm{W}_p^{2-2/p}(\Omega; \mathbb{R}^n) \times \mathrm{W}_p^{2-2/p}(\Omega) \times \mathrm{W}_p^{2-2/p}(\Omega) \times \mathrm{H}_p^1(\Omega) :$$

$$(u(y)\,|\,v(y)) \geq 0, \quad \forall y \in \Gamma, \quad \rho(x) > 0, \quad \theta(x) > 0, \quad \chi(x) \in [0,1], \quad \forall x \in \overline{\Omega}\}$$

and boundary data

$$h_d \in Y_{0,d}(J_0; \mathbb{R}^n), \quad h_{s1} \in Y_{0,s}(J_0; \mathbb{R}_+), \quad (h_d(t,x)\,|\,v(x))_{|\Gamma_d} \geq 0, \quad \forall(t,x) \in J \times \Gamma_d,$$

$$g_d \in Y_{0,d}(J_0; \mathbb{R}), \quad g_d(t,x) > 0, \quad \forall(t,x) \in J \times \Gamma_d,$$

$$l_d \in Y_{0,d}(J_0; \mathbb{R}), \quad l_d(t,x) \in (0,1), \quad \forall(t,x) \in J \times \Gamma_d$$

satisfying the compatibility conditions

$$u_{0|\Gamma_d} = h_{d|t=0}, \quad (u_0\,|\,v)_{|\Gamma_s} = h_{s1|t=0}, \quad QS_{|t=0} \cdot v_{|\Gamma_s} = 0,$$

$$\theta_{0|\Gamma_d} = g_{d|t=0}, \quad \chi_{0|\Gamma_d} = l_{d|t=0}, \quad (\nabla\theta_0\,|\,v)_{|\Gamma_s} = 0, \quad (\nabla\chi_0\,|\,v)_{|\Gamma_s} = 0,$$

$$(14.52)$$

there is a unique solution (u,θ,χ,ρ) of (14.13)–(14.17), (14.45)–(14.48) on a maximal time interval, which is $J^* = [0,t^*)$, $t^* := t^*(u_0,\theta_0,\chi_0,\rho_0) \in (0,T_0]$ if the solution is not global. The solution belongs to the class $\mathbb{E}(J)$ for each interval $J = [0,T]$, $T < t^*$, or to the class $\mathbb{E}(J_0)$ if the solution exists globally. The maximal *time t^* is characterised by the property:*

$$\lim_{t \to t^*} (u, \theta, \chi, \rho)(t) \quad \textit{does not exist in } V.$$

Moreover, χ stays in $(0, 1)$ and, if additionally $\partial_\chi F_0(\rho, 0, \chi) = 0$ and $\mathbf{S} : (Du)^s \geq 0$, the temperature θ is nonnegative. The solution map $(u_0, \theta_0, \chi_0, \rho_0) \to (u, \theta, \chi, \rho)(t)$ generates a local semiflow on the phase space $\Phi := \{\varphi := (u_0, \theta_0, \chi_0, \rho_0) \in V : \varphi$ satisfies (14.52) $\}$.

14.2.4 Solution Theory for NSCH[10]

For the isothermal case (14.20), (14.14)–(14.16), (14.23), (14.24) existence and uniqueness of strong solutions to NSCH have been proved in [55], while the system (14.13)–(14.16) with temperature has been studied in [53]. While many similarities between NSCH and NSAC exist and therefore the approach for proving well-posedness is the same, NSCH is technically different, as the convective Cahn-Hilliard equation is of fourth order. To keep the survey readable, we consider here only isothermal NSCH. Adding the energy equation and its treatment were almost straightforward for NSK, compressible Navier-Stokes, and NSAC, but in [53] it has been seen that NSCH with temperature is distinctly more involved than one might initially think.

As in the previous section, we assume that the Helmholtz energy is of the form

$$F(\rho, \chi, \nabla\chi) = F_0(\rho, \chi) + \tfrac{\delta(\rho,\chi)}{2}|\nabla\chi|^2 \tag{14.53}$$

and has the regularity

$$F_0 \in C^5(\mathbb{R}_+^2). \tag{14.54}$$

The double-well potential W that was present in (14.43) is now subsumed under F_0. In the NSAC case we have also assumed some structural conditions to show $\chi \in [0, 1]$, which is not so evident in the Cahn-Hilliard case, since there is no maximum principle for fourth order problems.

Let $\Omega \subset \mathbb{R}^n$ be a bounded domain with boundary $\Gamma := \partial\Omega$ of class C^4—this is necessary due to the Cahn-Hilliard equation—, decomposing disjointly as

$$\Gamma = \Gamma_d \,\dot\cup\, \Gamma_s,$$

where each set may be empty. Furthermore, let $J = [0, T]$ be a compact time interval. The partial differential equations (14.20), (14.23) have to be complemented

---

[10]The 'quasi-incompressible' case of NSCH is covered by a variant of NSK, cf. [33, 53].

by initial conditions

$$u(0,x) = u_0(x), \quad \chi(0,x) = \chi_0(x), \quad \rho(0,x) = \rho_0(x), \quad x \in \Omega. \qquad (14.55)$$

Two natural boundary conditions are of interest for u, namely the no-slip condition

$$u = 0, \quad (t,x) \in J \times \Gamma_d \qquad (14.56)$$

and the pure slip condition

$$(u(t,x) \mid v(x)) = 0, \quad \mathcal{Q}(v(x))\mathbf{S}(u(t,x)) \cdot v(x) = 0, \quad (t,x) \in J \times \Gamma_s \qquad (14.57)$$

with $\mathcal{Q}(v(x)) := \mathbf{I} - v(x) \otimes v(x)$ projecting a vector field on the boundary to its tangential part. Another physically relevant situation is the presence of tangential friction at the boundary, which can be captured by the so-called Navier boundary condition,

$$(u(t,x) \mid v(x)) = 0, \quad (t,x) \in J \times \Gamma_s,$$
$$\mathcal{Q}(v(x))\mathbf{S}(u(t,x)) \cdot v(x) + \alpha \mathcal{Q}(v(x))u(t,x) = 0, \quad (t,x) \in J \times \Gamma_s, \qquad (14.58)$$

where $\alpha \geq 0$ is an empirical constant. As boundary conditions for χ, we consider

$$(\nabla\mu(t,x) \mid v(x)) = 0, \quad (\nabla\chi(t,x) \mid v(x)) = 0, \quad (t,x) \in J \times \Gamma, \qquad (14.59)$$

meaning that no diffusion through the boundary occurs and the diffuse interface is orthogonal to the boundary of the domain. Observe that the first boundary condition is of nonlinear type, since

$$\rho\mu = \partial_\chi(\rho F) - \nabla\cdot(\partial_{\nabla_\chi}(\rho F)) = \partial_\chi(\rho F_0) + \tfrac{1}{2}\rho|\nabla\chi|^2\partial_\chi\delta - \nabla\cdot(\rho\delta(\rho,\chi)\nabla\chi).$$

The viscosity, mobility, and capillarity coefficients may depend on ρ and χ. Thus we need again assumptions regarding regularity and positivity; we assume

$$\eta, \zeta, \delta \in C^4(\mathbb{R}_+^2), \, \gamma \in C^2(\mathbb{R}_+^2) : \quad \eta(\xi), \zeta(\xi), \gamma(\xi), \delta(\xi) > 0, \quad \forall \xi \in (0,\infty)^2. \qquad (14.60)$$

Again, we are looking for strong solutions in the L_p-setting. More precisely, in the isothermal case we seek solutions $(u,\chi,\rho) \in \mathbb{E}(J) := \mathbb{E}_1(J) \times \mathbb{E}_2(J) \times \mathbb{E}_3(J)$ where

$$\mathbb{E}_1(J) := H_p^{3/2}(J; L_p(\Omega; \mathbb{R}^n)) \cap H_p^1(J; H_p^2(\Omega; \mathbb{R}^n)) \cap L_p(J; H_p^4(\Omega; \mathbb{R}^n)),$$

$$\mathbb{E}_2(J) := H_p^1(J; L_p(\Omega)) \cap L_p(J; H_p^4(\Omega)),$$

$$\mathbb{E}_3(J) := H_p^{2+1/4}(J; L_p(\Omega)) \cap C^1(J; H_p^2(\Omega)) \cap C(J; H_p^3(\Omega)),$$

and

$$p \in (\hat{p}, \infty), \quad \hat{p} := \max\{4, n\}.$$

As these function spaces are not obvious and the situation is indeed involved, let us motivate this choice of solution classes. We first look at the convective Cahn-Hilliard equation which is to be considered in $L_p(J; L_p(\Omega))$. There exist many maximal L_p-regularity results with the regularity class $\mathbb{E}_2(J)$ for the Cahn-Hilliard equation (the non-convective case), thus this choice seems not to be so surprising. The primary difficulty of the Cahn-Hilliard equation in (14.20), (14.23), (14.24) is the presence of a third order term of ρ entailing the requirement $\rho \in L_p(J; H^3_p(\Omega))$. On the other hand, ρ only obtains its regularity from ρ_0 and $\nabla \cdot u$ entering as data in the continuity equation. We therefore have to demand $\rho_0 \in H^3_p(\Omega)$ and $u \in L_p(J; H^4_p(\Omega; \mathbb{R}^n))$, where the latter claim forces us to consider the momentum equation in the base space $L_p(J; H^2_p(\Omega; \mathbb{R}^n))$. This approach causes a strong coupling between the momentum equation and the Cahn-Hilliard equation. To see this, let $\chi \in \mathbb{E}_2(J)$ be given. Due to the mixed derivative theorem we have

$$\chi \in \mathbb{E}_2(J) = H^1_p(J; L_p(\Omega)) \cap L_p(J; H^4_p(\Omega)) \hookrightarrow H^{1/2}_p(J; H^2_p(\Omega)).$$

Thus the natural regularity class for $\nabla \cdot \mathbf{C}$, in view of second order terms of χ occurring therein, is the space

$$\mathcal{X}_1(J) := H^{1/2}_p(J; L_p(\Omega; \mathbb{R}^n)) \cap L_p(J; H^2_p(\Omega; \mathbb{R}^n)).$$

Considering $\nabla \cdot \mathbf{C}$ as "input" for the momentum equation, the right choice for the base space can only be $\mathcal{X}_1(J)$. Then, from the maximal L_p-regularity point of view, one may expect that u belongs to the space $\mathbb{E}_1(J)$. Once we have $u \in \mathbb{E}_1(J)$, the continuity equation should provide $\rho \in \mathbb{E}_3(J)$. (This may not be so obvious, but it is possible to prove this regularity and no more regularity can be transferred to ρ.) Observe that the Navier-Stokes equations and the Cahn-Hilliard equation are now strongly coupled, as $\partial_t u$, $\nabla \cdot \mathbf{S}$, and $\nabla \cdot \mathbf{C}$ are of the same order; unlike in the case of incompressible fluids for which $u \in H^1_p(J; L_p(\Omega; \mathbb{R}^n)) \cap L_p(J; H^2_p(\Omega; \mathbb{R}^n))$ is completely sufficient. Indeed, in the incompressible case, i.e. $\rho = const$ and w.l.o.g. $\rho = 1$, the density is given and thus there is no need to consider high regularities of u. On the other hand, the pressure now becomes an unknown. Moreover, the tensor $\mathbf{C}$ simplifies to

$$\mathbf{C} = \mathbf{C}_{inc} = -\delta(\chi)\nabla\chi \otimes \nabla\chi.$$

The momentum equation can be considered in $L_p(J; L_p(\Omega; \mathbb{R}^n))$ and $\nabla \cdot \mathbf{C}_{inc}$ becomes a lower order term, as

$$\nabla \cdot \mathbf{C}_{inc} \in \mathcal{X}_1(J) \hookrightarrow\hookrightarrow L_p(J; L_p(\Omega; \mathbb{R}^n)).$$

Back to the compressible case! The basic tool to prove the theorem below is again the contraction mapping principle and one proceeds exactly as for the NSAC system. The only difference to NSAC is the nonlinear boundary condition $\partial_\nu \mu = 0$ regarding χ. It seems that for the derivation of the desired contraction property one requires that the identity $\partial_\nu \mu = 0$ is preserved under the fixed point mapping. To cope with this difficulty we add the variable μ, that is we work with triples (u, χ, μ) and view the reduced problem for (u, χ) as a problem for (u, χ, μ). Since $\chi \in \mathbb{E}_2(J)$ and

$$\mu = \partial_\chi F_0 + \tfrac{1}{2}|\nabla\chi|^2 \partial_\chi \delta - \rho^{-1}\nabla\cdot(\delta\rho\nabla\chi) ,$$

the natural regularity class of μ is given by

$$\mu \in \mathbb{E}_\mu(J) := \mathrm{H}_p^{1/2}(J; \mathrm{L}_p(\Omega)) \cap \mathrm{L}_p(J; \mathrm{H}_p^2(\Omega)).$$

Following the approach described above well-posedness for the NSCH system can be proved.

Theorem 14.2.5 ([55]) *Let $\Omega \subset \mathbb{R}^n$ be a bounded domain with compact C^4-boundary Γ decomposing disjointly as $\Gamma = \Gamma_d \dot\cup \Gamma_s$, $J_0 = [0, T_0]$ with $T_0 \in (0, \infty]$, and $p \in (\hat{p}, \infty)$. Let further the assumptions (14.53), (14.54), (14.60) and*

(i) $(u_0, \chi_0, \rho_0) \in V := \mathrm{W}_p^{4-\frac{2}{p}}(\Omega; \mathbb{R}^n) \times \mathrm{W}_p^{4-\frac{4}{p}}(\Omega) \times \{\varphi \in \mathrm{H}_p^3(\Omega) : \varphi(x) > 0, \ \forall x \in \overline{\Omega}\};$

(ii) the subsequent compatibility conditions hold:

$$u_{0|\Gamma_d} = 0, \quad (u_0 \mid \nu)_{|\Gamma_s} = 0, \quad Q\mathbf{S}(u)_{|t=0,\Gamma_s} \cdot \nu_{|\Gamma_s} = 0, \quad \partial_\nu \chi_0 = 0,$$

$$\partial_\nu \mu(\rho_0, \chi_0) = 0,$$

$$-\nabla\cdot\mathbf{S}_{|t=0,\Gamma_d} = (\nabla\cdot\mathbf{C})_{|t=0,\Gamma_d} \in \mathrm{W}_p^{2-\frac{3}{p}}(\Gamma_d; \mathbb{R}^n),$$

$$-\left(\nabla\cdot\mathbf{S}_{|t=0} \mid \nu\right)_{|\Gamma_s} = (\nabla\cdot\mathbf{C} - \rho\nabla u \cdot u \mid \nu)_{|t=0,\Gamma_s} \in \mathrm{W}_p^{2-\frac{3}{p}}(\Gamma_s),$$

$$-Q\mathbf{S}(\nabla\cdot\mathbf{S})_{|t=0} \cdot \nu_{|\Gamma_s} = Q\mathbf{S}(\nabla\cdot\mathbf{C} - \rho\nabla u \cdot u)_{|t=0,\Gamma_s} \cdot \nu_{|\Gamma_s} \in \mathrm{W}_p^{1-\frac{3}{p}}(\Gamma_s; \mathbb{R}^n)$$

be satisfied. Then the system (14.20), (14.14), (14.16), (14.21), (14.23), (14.24), (14.55)–(14.59) possesses a unique strong solution (u, χ, ρ) on a maximal time interval $J^* = [0, t^*)$, $t^* := t^*(u_0, \chi_0, \rho_0) \in (0, T_0]$ if the solution is not global. The solution belongs to the class $\mathbb{E}(J)$ for each interval $J = [0, T]$, $T < t^*$, or to the class $\mathbb{E}(J_0)$ if the solution exists globally. The maximal time t^* is characterised by the property:

$$\lim_{t\to t^*} (u(t), \chi(t), \rho(t)) \quad \text{does not exist in } V.$$

Moreover, the solution map $(u_0, \chi_0, \rho_0) \mapsto (u, \chi, \rho)(\cdot)$ *generates a local semiflow on the phase space* $\Phi = \{(v, c, \varrho) \in V : (v, c, \varrho)$ *satisfies the compatibility condition* $(ii)\}$.

14.3 Existence and Spectral Stability of Diffuse Phase Boundaries

14.3.1 Existence

Just as for NSK (cf. Sect. 14.1.2), diffuse phase boundaries are possible for NSAC and NSCH only under certain assumptions on the constitutive law of the fluid, here its Helmholtz energy F. In this subsection we consider a parametrized family of fluids, with Helmholtz energies[11]

$$F(\tau, \chi, \nabla\chi, \vartheta) = f(\tau, \chi, \vartheta) + \frac{1}{2}\delta(\vartheta)|\nabla\chi|^2, \tag{14.61}$$

where $\vartheta \in \Theta$ is the family parameter, ranging in some open interval $\Theta \subset \mathbb{R}$. We characterize a generic situation in which heteroclinic traveling waves emerge in a bifurcation ('phase transition') at a critical value ϑ_* of ϑ. One possible interpretation of ϑ is that of a temperature, so that (14.61) would correspond to a class of isothermal fluids; in that interpretation, f could, e.g., be of the form

$$f(\tau, \chi, \vartheta) = \hat{f}(\tau, \chi) + \vartheta W(\chi)$$

with a reduced mixing entropy W such as in (14.44), cf. [17]. Generally, we make the natural assumptions

$$f_{\tau\tau}(\tau, \chi, \vartheta) > 0, \quad f_{\chi\chi}(\tau, \chi, \vartheta) > 0, \ (\tau, \chi, \vartheta) \in (0, \infty) \times (0, 1) \times \Theta \tag{14.62}$$

and

$$\lim_{\chi \searrow 0} f_\chi(., \chi, .) = -\infty, \quad \lim_{\chi \nearrow 1} f_\chi(., \chi, .) = +\infty. \tag{14.63}$$

The phase transition is now directly associated with the formation, for ϑ varying across ϑ_*, of a domain of nonconvexity for the otherwise convex function f. We write

$$p(\tau, \chi, \vartheta) = -f_\tau(\tau, \chi, \vartheta), \quad q(\tau, \chi, \vartheta) = -f_\chi(\tau, \chi, \vartheta)$$

---

[11]For this subsection it is convenient to use the specific volume $\tau \equiv 1/\rho$ in place of ρ.

and uniquely define $C : (0, \infty) \times \Theta \to (0, 1)$ through

$$q(\tau, C(\tau, \vartheta)) = 0.$$

The precise assumption we need is that at a certain critical point (τ_*, ϑ_*), the restricted determinant

$$\Delta(\tau, \vartheta) \equiv (f_{\tau\tau}f_{\chi\chi} - f_{\tau\chi}^2)(\tau, C(\tau, \vartheta), \vartheta)$$

satisfies

$$\Delta(\tau_*, \vartheta_*) = \Delta_\tau(\tau_*, \vartheta_*) = 0, \quad \Delta_{\tau\tau}(\tau_*, \vartheta_*) > 0, \quad \Delta_\vartheta(\tau_*, \vartheta_*) > 0. \tag{14.64}$$

For the rest of this subsection, we restrict attention to diffuse phase boundaries with small mass flux

$$m = \rho(u \cdot \mathbf{n} - s).$$

If m is different from 0, fluid continually crosses the interface, undergoing an actual phase transformation. If $m = 0$ however, the "no-flux" boundary travels or rests with the fluid and no phase transformation takes place.

The next two theorems result from a careful analysis of the ODE system that describes the traveling waves of (14.20).

Theorem 14.3.1 ([32] Maxwell States and No-Flux Phase Boundaries for NSAC and NSCH) *Assume (14.61) with (14.62), (14.63) and (14.64). Then, with $\tilde{\vartheta} < \vartheta_*$ sufficiently close to the critical value ϑ_*, the following holds for every $\vartheta \in (\tilde{\vartheta}, \vartheta_*]$. There are locally uniquely determined fluid states $(\underline{\tau}_0, \underline{\chi}_0), (\overline{\tau}_0, \overline{\chi}_0)$, depending continuously on ϑ, such that*

(i)

$$q(\underline{\tau}_0, \underline{\chi}_0) = q(\overline{\tau}_0, \overline{\chi}_0) = 0,$$

$$p(\underline{\tau}_0, \underline{\chi}_0) = p(\overline{\tau}_0, \overline{\chi}_0)$$

with

$$(\underline{\tau}_0, \underline{\chi}_0) = (\overline{\tau}_0, \overline{\chi}_0) \quad if \quad \vartheta = \vartheta_*,$$

and
(ii) if $\vartheta < \vartheta_*$, then

$$\underline{\tau}_0 > \overline{\tau}_0$$

and the system (14.20) with (14.22) (NSAC) or (14.23) (NSCH) admits a no-flux (m = 0) phase boundary

$$(\vec{\tau}(x), 0, \vec{\chi}(x)) \quad \text{with} \quad (\vec{\tau}(-\infty), \vec{\chi}(-\infty)) = (\underline{\tau}_0, \underline{\chi}_0),$$
$$(\vec{\tau}(\infty), \vec{\chi}(\infty)) = (\overline{\tau}_0, \overline{\chi}_0)$$

and (equivalently via $x \mapsto -x$) a no-flux phase boundary

$$(\overleftarrow{\tau}(x), 0, \overleftarrow{\chi}(x)) \quad \text{with} \quad (\overleftarrow{\tau}(-\infty), \overleftarrow{\chi}(-\infty)) = (\overline{\tau}_0, \overline{\chi}_0),$$
$$(\overleftarrow{\tau}(\infty), \overleftarrow{\chi}(\infty)) = (\underline{\tau}_0, \underline{\chi}_0).$$

Theorem 14.3.2 ([32] Phase Transforming Phase Boundaries for NSAC) *For sufficiently small mass flux m,*

(i) there is a continuous family of (left endstate, profile, right endstate) triples

$$(\vec{\tau}_m^-, \vec{u}_m^-, \vec{\chi}_m^-), (\vec{\tau}_m, \vec{u}_m, \vec{\chi}_m), (\vec{\tau}_m^+, \vec{u}_m^+, \vec{\chi}_m^+),$$

with $\vec{u}_0^-, \vec{u}_0, \vec{u}_0^+ = 0$ and $\vec{\tau}_0 = \vec{\tau}, \vec{\chi}_0 = \vec{\chi}$ corresponding to the no-flux boundary case

$$(\underline{\tau}_0, 0, \underline{\chi}_0), (\vec{\tau}, 0, \vec{\chi}), (\overline{\tau}_0, 0, \overline{\chi}_0),$$

such that $(\vec{\tau}_m, \vec{u}_m, \vec{\chi}_m)$ represents a phase transforming phase boundary that is densifying if $m > 0$ and rarefying if $m < 0$;

(ii) there is a continuous family of (left endstate, profile, right endstate) triples

$$(\overleftarrow{\tau}_m^-, \overleftarrow{u}_m^-, \overleftarrow{\chi}_m^-), (\overleftarrow{\tau}_m, \overleftarrow{u}_m, \overleftarrow{\chi}_m), (\overleftarrow{\tau}_m^+, \overleftarrow{u}_m^+, \overleftarrow{\chi}_m^+)$$

with $\overleftarrow{u}_0^-, \overleftarrow{u}_0, \overleftarrow{u}_0^+ = 0$ and $\overleftarrow{\tau}_0 = \overleftarrow{\tau}, \overleftarrow{\chi}_0 = \overleftarrow{\chi}$ corresponding to the no-flux boundary case

$$(\overline{\tau}_0, 0, \overline{\chi}_0), (\overleftarrow{\tau}, 0, \overleftarrow{\chi}), (\underline{\tau}_0, 0, \underline{\chi}_0),$$

such that $(\overleftarrow{\tau}_m, \overleftarrow{u}_m, \overleftarrow{\chi}_m)$ represents a phase transforming phase boundary that is densifying if $m < 0$ and rarefying if $m > 0$.

The reader recalls that for NSCH the phases are conserved, (14.18), so that all phase boundaries are no-flux.

14.3.2 *Spectral Stability*

The project has established spectral stability of no-flux ($m = 0$) diffuse phase boundaries for NSK, NSAC, and NSCH. The object of interest is the operator, L, that is obtained by linearizing the PDE system around its traveling wave,

$$\left(\hat{\rho}(\hat{x}), 0, \hat{\chi}(\hat{x})\right), \quad \hat{x} := x \cdot \mathbf{n}.$$

(We assume that the speed s of the traveling wave is 0. Due to Galilean invariance, this means no loss of generality.) We could show that (i) the point spectrum of L lies in the union of the open left half-plane and $\{0\}$, and (ii) the essential spectrum, lying also in left half plane, can be described near the imaginary axis by three curves, the non-positive real line and two parabolas that are axisymmetric with respect to $\mathbb{R}$ and have vertex zero. Points of the essential spectrum, which are far away from the imaginary axis, are negative real numbers. Finally, zero belongs to the point spectrum only for space dimension one and thus it is embedded into the essential spectrum. Although this circumstance makes investigations on the eigenvalue zero more complicated, we were able to also prove its simplicity.

We report results obtained in [46] on isothermal NSAC. Concerning both isothermal NSCH and isothermal NSK, analogous investigations have been carried out and astonishingly the results are quite similar, i.e. the structure of the spectrum is exactly the same [54]. Also, the situations for the non-isothermal NSAC, NSK, and NSCH are similar.

We assume a Helmholtz energy

$$F(\tau, \chi, \nabla\chi) = f(\tau, \chi) + \frac{1}{2}\delta|\nabla\chi|^2 \text{ with a constant } \delta > 0.$$

In the spectral analysis it has turned out that the energy ρF plays an important role. We therefore use

$$\Psi(\rho, \chi) \equiv \rho f(1/\rho, \chi).$$

We require

$$
\begin{aligned}
&(i) \quad \Psi \in C^3((0, \infty) \times [0, 1]),\\
&(ii) \quad \Psi_{\rho\rho}(\rho, \chi) > 0, \quad \Psi_{\chi\chi}(\rho, \chi) > 0, \quad \forall(\rho, \chi) \in (0, \infty) \times (0, 1).
\end{aligned}
\tag{14.65}
$$

Properties of the diffuse phase boundary, that are needed for the spectral analysis, are

$$\hat{\rho} \in C^1(\mathbb{R}), \ \hat{\chi} \in C^2(\mathbb{R}) : \quad \lim_{\hat{x}\to\pm\infty} \hat{\rho}(\hat{x}) = \hat{\rho}^\pm, \quad \lim_{\hat{x}\to\pm\infty} \hat{\chi}(\hat{x}) = \hat{\chi}^\pm,$$

$$0 < \hat{\rho}^- < \hat{\rho}^+, \quad 0 < \hat{\chi}^- < \hat{\chi}^+ < 1, \quad \hat{\chi}'(\hat{x}) > 0, \quad \forall\hat{x} \in \mathbb{R}.$$

$$(iii) \quad 1 - \frac{\delta}{\hat{\rho}(\hat{x})\Psi_{\rho\rho}(\hat{\rho}(\hat{x}), \hat{\chi}(\hat{x}))}|\hat{\chi}'(\hat{x})|^2 > 0, \quad \forall \hat{x} \in \mathbb{R},$$

$$(iv) \quad \det\left(\Psi''(\hat{\rho}^{\pm}, \hat{\chi}^{\pm})\right) > 0. \tag{14.66}$$

Note that conditions (14.65) and (14.66) are satisfied for the diffuse phase boundaries we established in Theorem 14.3.1 for $F = F(.,.,.,\vartheta)$ as in (14.61) with ϑ close to and smaller than ϑ_*.

We finally need positivity assumptions for the coefficients

$$\theta, \delta, \varepsilon > 0, \quad \eta(\rho, \chi), \zeta(\rho, \chi) > 0, \quad \forall (\rho, \chi) \in (0, \infty) \times [0, 1]. \tag{14.67}$$

Introducing the deviation

$$w(t, x) := (\varrho(t, x), v(t, x), c(t, x)) = (\rho(t, x) - \hat{\rho}(\hat{x}), u(t, x) - 0, \chi(t, x) - \hat{\chi}(\hat{x}))$$

from the standing wave $\hat{w}(\hat{x}) = (\hat{\rho}(\hat{x}), 0, \hat{\chi}(\hat{x}))$ and linearizing the isothermal NSAC system (14.20), (14.22) around it, one obtains the linear problem

$$\partial_t \varrho + \nabla \cdot (\hat{\rho} v) = f_1, \qquad x \in \mathbb{R}^n,$$

$$\varepsilon[\hat{\rho}\partial_t c + (\nabla \hat{\chi} \mid \hat{\rho} v)] + \mathcal{A}(D)(\varrho, c) = f_2, \qquad x \in \mathbb{R}^n, \tag{14.68}$$

$$\hat{\rho}\partial_t v - \nabla \cdot \mathcal{S}(D)v + \hat{\rho}\nabla \mathcal{B}(D)(\varrho, c) - \nabla \hat{\chi} \mathcal{A}(D)(\varrho, c) = f_3, \qquad x \in \mathbb{R}^n,$$

where the differential operators $\mathcal{A}, \mathcal{B}, \mathcal{S}$ are defined as follows

$$\mathcal{A}(D)(\varrho, c) = -\nabla \cdot (\delta\hat{\rho}\nabla c + \delta\nabla\hat{\chi}\varrho) + \hat{\Psi}_{\rho\chi}\varrho + \hat{\Psi}_{\chi\chi}c,$$

$$\mathcal{B}(D)(\varrho, c) = \hat{\Psi}_{\rho\rho}\varrho + \hat{\Psi}_{\rho\chi}c + \delta\nabla\hat{\chi} \cdot \nabla c,$$

$$\mathcal{S}(D)v = 2\hat{\eta}D^{s0}v + \hat{\zeta}\nabla \cdot v\,\mathbf{I}.$$

Here we have set $\hat{\Psi} = \Psi(\hat{\rho}, \hat{\chi})$, $\hat{\eta} = \eta(\hat{\rho}, \hat{\chi})$, and $\hat{\zeta} = \zeta(\hat{\rho}, \hat{\chi})$. Associating the linear problem (14.68) with $\partial_t w + Lw = f$, we are interested in the spectrum/resolvent set of the operator

$$L : D(L) \subset X \to X, \quad X := H^1_2(\mathbb{R}^n) \times L_2(\mathbb{R}^n) \times L_2(\mathbb{R}^n; \mathbb{R}^n)$$

with domain $D(L) = H^1_2(\mathbb{R}^n) \times H^2_2(\mathbb{R}^n) \times H^2_2(\mathbb{R}^n; \mathbb{R}^n)$.

Lemma 14.3.1 ([46]) *Let $n \geq 1$, $m = 0$ and the assumptions (14.65)–(14.67) be satisfied. Then the point spectrum $\sigma_p(-L)$ of $-L$ can be characterized as follows:*

1. If $n = 1$ then $\sigma_p(-L) \subset \mathbb{C}_- \cup \{0\}$, where $\mathbb{C}_- = \{z \in \mathbb{C} : \Re ez < 0\}$. Zero is a simple eigenvalue and its associated eigenvector w^0 is given by $(\hat{\rho}', 0, \hat{\chi}')$. In this *case the null space $N(L) = span\{w^0\}$ is the tangent space of $\mathcal{E} = \{\hat{w}(\cdot + \sigma) = (\hat{\rho}(\cdot + \sigma), 0, \hat{\chi}(\cdot + \sigma)) : \sigma \in \mathbb{R}\}$.*

2. If $n \geq 2$ we even have $\sigma_p(-L) \subset \mathbb{C}_-$.

Moreover, the resolvent set $\rho(-L)$ of $-L$ contains the open sector $\Sigma_{\frac{\pi}{2}+\theta,a}$[12] with some $\theta \in (0, \pi/2)$ and some $a \geq 0$.

The next result concerns the essential spectrum $\sigma_{ess}(-L)$ of $-L$, i.e. the set of those complex numbers that are not Fredholm points. More precisely, let $T : D(T) \subset X \to X$ be a linear operator. We put $\Phi_T = \{\lambda \in \mathbb{C} : \lambda I - T \text{ is Fredholm}\}$, $\Phi_T^0 = \{\lambda \in \Phi_T : \text{ind}(\lambda I - T) = 0\}$,[13] and $\sigma_{ess}(T) = \mathbb{C} \backslash \Phi_T$. To track down the essential spectrum we follow [66], in which equivalence of Fredholm points and invertibility of so-called limiting problems is proved (see [66, Theorem 6.3]). Our situation is the simplest case that can happen, namely we have to study that eigenvalue problem $\kappa w = -\hat{L}w$, where $\hat{L}$ is obtained by freezing the coefficients of L at $\hat{x} = \pm\infty$. In doing so, one can prove the following characterisation of the essential spectrum.

Lemma 14.3.2 ([46]) *Let the assumptions of Lemma 14.3.1 be satisfied. Then the essential spectrum $\sigma_{ess}(-L)$ of $-L$ is contained in $\mathbb{C}_- \cup \{0\}$. More precisely, $\sigma_{ess}(-L)$ consists of the union of three curves, namely the non-negative real line $\mathbb{R}_- := \{z \in \mathbb{C} : \Im m z = 0, \Re e z \leq 0\}$ and curves E^+ and E^-, that can be described near the imaginary line by $E_\epsilon^\pm := E^\pm \cap B_\epsilon(0)$, $\epsilon > 0$ sufficiently small, where*

$$E_\epsilon^\pm = \{z \in \mathbb{C} : z = i\omega_0^\pm \cdot |\xi| - \Omega^\pm(|\xi|) \cdot |\xi|,$$

$$z = -i\omega_0^\pm \cdot |\xi| - \overline{\Omega^\pm(|\xi|)} \cdot |\xi|, \ 0 \leq |\xi| < \epsilon\}$$

and $\omega_0^\pm := \sqrt{\hat{\rho}^\pm \det((\hat{\Psi}^\pm)'')/\hat{\Psi}_{\chi\chi}^\pm}$. The analytic functions $\Omega^\pm : \mathbb{R}_+ \to \mathbb{C}$ satisfy $\Omega^\pm(0) = 0$ and $\Re e\, \Omega^\pm(|\xi|) > 0$ for all $0 < |\xi| < \epsilon$. Sufficiently far away from the imaginary line the essential spectrum only consists of negative real numbers, in particular the curves $E^\pm$ go back to the negative real line.

The picture Fig. 14.2 illustrates the location of the essential spectrum of $-L$ near the imaginary line. With this result at hand we are able to characterize those subset of $\mathbb{C}$ which is the connected set of $\mathbb{C} \backslash \sigma_{ess}(-L)$ containing $\Sigma_{\pi/2+\theta,a}$. Denoting this set by Δ we have $\partial\Delta = E^+ \cup \{z \in \mathbb{C} : \Im m z = 0, \Re e z \leq z_0\}$ with $z_0 = \min\{z : z \in E^+ \cap \mathbb{R}_-\} < 0$. Since $\text{ind}(\cdot)$ is constant on connected sets of $\mathbb{C}$ and $\rho(-L) \cap \Delta \neq \varnothing$, we immediately deduce $\text{ind}(\kappa + L) = n(\kappa + L) - d(\kappa + L) = 0$ for all $\kappa \in \Delta$. Moreover, as the integers $n(\kappa + L)$ and $d(\kappa + L)$ are also constant except for singular

---

[12]The set $\Sigma_{\theta,a} \subset \mathbb{C}$ denotes the open sector with vertex $a \in \mathbb{R}$, opening angle 2θ, which is symmetric with respect to the real axis $\mathbb{R}$, i.e. $\Sigma_{\theta,a} := \{z \in \mathbb{C} \setminus \{a\} : |arg(z - a)| < \theta\}$.

[13]As usual we use the notation: $T : X \to Y$ is called Fredholm if T is closed and the integers $n(T) = \dim N(T)$ and $d(T) = \text{codim}(T) = \dim(Y/R(T))$ are finite. The index $\text{ind}(T)$ of T is defined as the integer $n(T) - d(T)$.

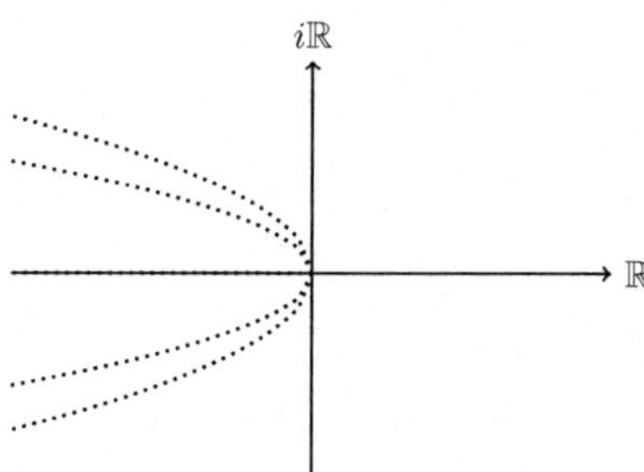

Fig. 14.2 Essential spectrum of $-L$ for no-flux diffuse phase boundaries

points $\kappa_j \in \Delta$, there holds

$$n(\kappa + L) = d(\kappa + L) = 0, \quad \forall \kappa \in \Delta,\ \kappa \neq \kappa_j,$$

$$n(\kappa_j + L) = d(\kappa_j + L) = r_j, \quad 0 < r_j < \infty.$$

Thus Δ is a subset of $\rho(-L)$ except for κ_j, which are in $\sigma_{discr}(-L)$, i.e., isolated points of $\sigma(-L)$; the κ_j are isolated eigenvalues with finite (algebraic) multiplicities and their geometric multiplicities are r_j. Due to Lemma 14.3.1 we know that $\sigma_p(-L) \cap \overline{\mathbb{C}}_+$ is either empty (if $n \geq 2$) or equal $\{0\}$ (if $n = 1$). As in any case $0 \in \sigma_{ess}(-L)$ we finally obtain $\overline{\mathbb{C}}_+ \backslash \{0\} \cup \Sigma_{\pi/2+\theta,a} \subset \rho(-L)$ with some $\theta \in (0, \pi/2)$ and $a > 0$ (see Lemma 14.3.1). We summarize these investigations in

Lemma 14.3.3 ([46]) *The connected set $\Delta \subset \mathbb{C}$, as described above, only consists of regular points, i.e., points that either lie in $\rho(-L)$ or in $\sigma_{discr}(-L)$. The set $\overline{\mathbb{C}}_+ \backslash \{0\} \cup \Sigma_{\pi/2+\theta,a}$ with $\theta \in (0, \pi/2)$ and $a > 0$ as in Lemma 14.3.1, is a subset of $\rho(-L)$.*

14.3.3 Conclusion

The project has led to modern solution theories for the Navier-Stokes-Korteweg, Navier-Stokes-Allen-Cahn, and the Navier-Stokes-Cahn-Hilliard systems, three fundamental models of two-phase fluid dynamics. As for most empirically known two-phase fluids at least one of the phases, and thus the mixed fluid, is compressible, it seems worthwhile emphasizing again that also for Navier-Stokes-Allen-Cahn and Navier-Stokes-Cahn-Hilliard, our theories concern the case of *compressible* fluids.

For the Navier-Stokes-Allen-Cahn and Navier-Stokes-Cahn-Hilliard models, the project has newly shown the existence of traveling waves that represent diffuse phase boundaries, and identified a generic "van der Waals" scenario for the fluid's constitutive law to permit bifurcation of such waves at a critical value of a parameter such as temperature. For no-flux diffuse phase boundaries (recall that for Navier-Stokes-Cahn-Hilliard all phase boundaries are no-flux), we have established spectral stability.

While technical objectives of the project have thus been met, its long-term goal, a proof for the nonlinear stability of moving phase boundaries in the Navier-Stokes-Allen-Cahn and Navier-Stokes-Cahn-Hilliard models, has not yet been reached. However, in view of the results reported here and [52], the stage seems set.

We are grateful to D. Bothe and A. Reusken for coordinating the most inspiring DFG Priority Programme "Transport Processes at Fluidic Interfaces" which has made this work possible. While we were originally focussed on compressible phases, the stronger emphasis of the programme on incompressible fluids has encouraged us to consider also fluids with two incompressible phases and thus find out that for that case, our previously given solution theory for compressible Navier-Stokes-Korteweg is a viable alternative to the 'quasi-incompressible' approach. We thank notably H. Abels for interesting discussions.

References

1. Abels, H.: On a diffuse interface model for two-phase flows of viscous, incompressible fluids with matched densities. Arch. Ration. Mech. Anal. **194**, 463–506 (2009)
2. Abels, H.: Longtime behavior of solutions of a Navier-Stokes/Cahn-Hilliard system. In: Nonlocal and Abstract Parabolic Equations and Their Applications. Banach Center Publ., vol. 86, pp. 9–19. Polish Acad. Sci. Inst. Math., Warsaw (2009)
3. Abels, H.: Existence of weak solutions for a diffuse interface model for viscous, incompressible fluids with general densities. Commun. Math. Phys. **289**, 45–73 (2009)
4. Abels, H.: Strong well-posedness of a diffuse interface model for a viscous, quasi-incompressible two-phase flow. SIAM J. Math. Anal. **44**, 316–340 (2012)
5. Abels, H., Feireisl, E.: On a diffuse interface model for a two-phase flow of compressible viscous fluids. Indiana Univ. Math. J. **57**, 659–698 (2008)
6. Abels, H., Röger, M.: Existence of weak solutions for a non-classical sharp interface model for a two-phase ow of viscous, incompressible fluids. Ann. Inst. H. Poincaré Anal. Non Linéaire **26**, 2403–2424 (2009)
7. Abels, H., Garcke, H., Grün, G.: Thermodynamically consistent, frame indifferent diffuse interface models for incompressible two-phase flows with different densities. Math. Models Methods Appl. Sci. **22**, 1150013 (2012)
8. Abels, H., Depner, D., Garcke, H.: On an incompressible Navier-Stokes/Cahn-Hilliard system with degenerate mobility. Ann. Inst. H. Poincaré Anal. Non Linéaire **30**, 1175–1190 (2013)
9. Aki, G.L., Dreyer, W., Giesselmann, J., Kraus, C.: A quasi-incompressible diffuse interface model with phase transition. Math. Models Methods Appl. Sci. **24**, 827–861 (2014)
10. Alexander, J., Gardner, R., Jones, C.: A topological invariant arising in the stability analysis of travelling waves. J. Reine Angew. Math. **410**, 167–212 (1990)
11. Benzoni-Gavage, S.: Stability of multi-dimensional phase transitions in a van der Waals fluid. Nonlinear Anal. **31**, 243–263 (1998)
12. Benzoni-Gavage, S.: Stability of subsonic planar phase boundaries in a van der Waals fluid. Arch. Ration. Mech. Anal. **150**, 23–55 (1999)
13. Benzoni-Gavage, S.: Linear stability of propagating phase boundaries in capillary fluids. Physica D **155**, 235–273 (2001)
14. Benzoni-Gavage, S., Freistühler, H.: Effects of surface tension on the stability of dynamical liquid-vapor interfaces. Arch. Ration. Mech. Anal. **174**, 111–150 (2004)
15. Benzoni-Gavage, S., Danchin, R., Descombes, S., Jamet, D.: Structure of Korteweg models and stability of diffuse interfaces. Interfaces Free Bound. **7**, 371–414 (2005)

16. Benzoni-Gavage, S., Danchin, R., Descombes, S.: On the well-posedness for the Euler-Korteweg model in several space dimensions. Indiana Univ. Math. J. **56**, 1499–1579 (2007)
17. Blesgen, T.: A generalisation of the Navier-Stokes equations to two-phase-flows. J. Phys. D: Appl. Phys. **32**, 1119–1123 (1999)
18. Bothe, D., Dreyer, W.: Continuum thermodynamics of chemically reacting fluid mixtures. (English summary) Acta Mech. **226**, 1757–1805 (2015)
19. Bothe, D., Prüss, J.: On the two-phase Navier-Stokes equations with Boussinesq-Scriven surface fluid. J. Math. Fluid Mech. **12**, 133–150 (2010)
20. Boyer, F.: Mathematical study of multiphase flow under shear through order parameter formulation. Asymptot. Anal. **20**, 175–212 (1999)
21. Brenner, H.: Kinematics of volume transport. Physica A **349**, 11–59 (2005)
22. Chen, Z., He, L., Zhao, H.: Nonlinear stability of traveling wave solutions for the compressible fluid models of Korteweg type. J. Math. Anal. Appl. **422**, 1213–1234 (2015)
23. Danchin, R., Desjardins, B.: Existence of solutions for compressible fluid models of Korteweg type. Ann. Inst. H. Poincaré Anal. Non Linéaire **18**, 97–133 (2001)
24. Denk, R., Hieber, M., Prüss, J.: Optimal L^p- L^q-estimates for parabolic boundary value problems with inhomogeneous data. Math. Z. **257**, 193–224 (2007)
25. Ding, S., Li, Y., Luo, W.: Global solutions for a coupled compressible Navier-Stokes/Allen-Cahn System in 1D. J. Math. Fluid Mech. **15**, 335–360 (2012)
26. Dreyer, W., Giesselmann, J., Kraus, C., Rohde, C.: Asymptotic analysis for Korteweg models. Interfaces Free Bound. **14**, 105–143 (2012)
27. Dreyer, W., Giesselmann, J., Kraus, C.: A compressible mixture model with phase transition. Physica D **273/274**, 1–13 (2014)
28. Dunn, J.E., Serrin, J.: On the thermomechanics of interstitial working. Arch. Ration. Mech. Anal. **88**, 95–133 (1985)
29. Ericksen, J.L.: Liquid crystals with variable degree of orientation Arch. Ration. Mech. Anal. **113**, 97–120 (1990)
30. Evans, J.W.: Nerve axon equations. I–IV. Indiana Univ. Math. J. **22**, 75–90 (1972)
31. Feireisl, E., Rocca, E., Petzeltova, H., Schimperna, G.: Analysis of a phase-field model for two-phase compressible fluids. Math. Models Methods Appl. Sci. **20**, 1129–1160 (2010)
32. Freistühler, H.: Phase transitions and traveling waves in compressible fluids. Arch. Ration. Mech. Anal. **211**, 189–204 (2014)
33. Freistühler, H., Kotschote, M.: Phase-field and Korteweg-type models for the time-dependent flow of compressible two-phase fluids. Arch. Ration. Mech. Anal. **224**, 1–20 (2017)
34. Freistühler, H., Kotschote, M.: Models of two-phase fluid dynamics à la Allen-Cahn, Cahn-Hilliard, and . . . Korteweg! Confluentes Math. **7**, 57–66 (2015)
35. Freistühler, H., Kotschote, M.: Internal structure of dynamic phase-transition fronts in a fluid with two compressible or incompressible phases. Bull. Inst. Math. Acad. Sin. (N.S.) **10**, 541–552 (2015)
36. Freistühler, H., Szmolyan, P.: Spectral stability of small-amplitude viscous shock waves in several space dimensions. Arch. Ration. Mech. Anal. **195**, 353–373 (2010)
37. Gal, C.G., Grasselli, M.: Asymptotic behavior of a Cahn-Hilliard-Navier-Stokes system in 2D. Ann. Inst. Henri Poincaré Anal. Non Linéaire **27**, 401–436 (2010)
38. Gardner, R., Zumbrun, K.: The gap lemma and geometric criteria for instability of viscous shock profiles. Commun. Pure Appl. Math. **51**, 797–855 (1998)
39. Haspot, B.: Existence of global weak solution for compressible fluid models of Korteweg type. J. Math. Fluid Mech. **13**, 223–249 (2011)
40. Haspot, B.: Existence of global strong solution for Korteweg system with large infinite energy initial data. J. Math. Anal. Appl. **438**, 395–443 (2016)
41. Hattori, H., Li, D.: The existence of global solutions to a fluid dynamic model for materials of Korteweg type. J. Partial Differ. Equ. **9**, 323–342 (1996)
42. Jeans, J.: Introduction to the Kinetic Theory of Gases. Cambridge University Press, Cambridge (1940)

43. Jones, C.K.R.T.: Stability of the travelling wave solution of the FitzHugh-Nagumo system. Trans. Am. Math. Soc. **286**, 431–469 (1984)
44. Karlin, I.V., Gorban, A.N.: Hydrodynamics from Grad's equations: what can we learn from exact solutions? Ann. Phys. (Leipzig) **11**, 783–833 (2002)
45. Korteweg, D.J.: Sur la forme que prennent les équations des mouvements des fluides si l'on tient compte des forces capillaires par des variations de densité. Arch. Néer. Sci. Exactes II **6**, 1–24 (1901)
46. Kotschote, M.: Spectral analysis for travelling waves in compressible two-phase fluids of Navier-Stokes-Allen-Cahn type. J. Evol. Equ. **17**, 359–385 (2017)
47. Kotschote, M.: Strong solutions for a compressible fluid model of Korteweg type. Ann. Inst. Henri Poincaré Anal. Non Linéaire **25**, 679–696 (2008)
48. Kotschote, M.: Strong well-posedness for a Korteweg-type model for the dynamics of a compressible non-isothermal fluid. J. Math. Fluid Mech. **12**, 473–484 (2010)
49. Kotschote, M.: On compressible non-isothermal fluids of non-Newtonian Korteweg-type. SIAM J. Math. Anal. **44**, 74–101 (2012)
50. Kotschote, M.: Strong solutions to the compressible non-isothermal Navier-Stokes equations. Adv. Math. Sci. Appl. **22**, 319–347 (2012)
51. Kotschote, M.: Strong solutions to the Navier-Stokes equations for a compressible fluid of Allen-Cahn type. Arch. Ration. Mech. Anal. **206**, 489–514 (2012)
52. Kotschote, M.: Existence and time-asymptotics of global strong solutions to dynamic Korteweg models. Indiana Univ. Math. J. **63**, 21–51 (2014)
53. Kotschote, M.: Mixing rules and the Navier-Stokes-Cahn-Hilliard equations for compressible heat-conductive fluids. Bull. Braz. Math. Soc. (N.S.) **47**, 457–471 (2016)
54. Kotschote, M.: Spectral analysis for travelling waves in compressible two-phase fluids of Navier-Stokes-Cahn-Hilliard and Navier-Stokes-Korteweg type (in preparation)
55. Kotschote, M., Zacher, R.: Strong solutions in the dynamical theory of compressible fluid mixtures. Math. Models Methods Appl. Sci. **25**, 1217–1256 (2015)
56. Lowengrub, J., Truskinovsky, L.: Quasi-incompressible Cahn-Hilliard fluids and topological transitions. Proc. R. Soc. Lond. A **454**, 2617–2654 (1998)
57. Majda, A.: Compressible Fluid Flow and Systems of Conservation Laws in Several Space Variables. Springer, New York (1984)
58. Prüss, J., Simonett, G.: On the two-phase Navier-Stokes equations with surface tension. Interfaces Free Bound. **12**, 311–345 (2010)
59. Prüss, J., Shibata, Y., Shimizu, S., Simonett, G.: On well-posedness of incompressible two-phase flows with phase transitions: the case of equal densities. Evol. Equ. Control Theory **1**, 171–194 (2012)
60. Prüss, J., Shimizu, S., Wilke, M.: Qualitative behaviour of incompressible two-phase flows with phase transitions: the case of non-equal densities. Commun. Partial Differ. Equ. **39**, 1236–1283 (2014)
61. Slemrod, M.: Admissibility criteria for propagating phase boundaries in a van der Waals fluid. Arch. Ration. Mech. Anal. **81**, 301–315 (1983)
62. Slemrod, M.: Dynamics of first order phase transitions. In: Phase Transformations and Material Instabilities in Solids (Madison, Wis., 1983). Publ. Math. Res. Center Univ. Wisconsin, vol. 52, pp. 163–203. Academic, Orlando, FL (1984)
63. Slemrod, M.: Chapman-Enskog implies viscosity-capillarity. Quart. Appl. Math. **70**, 613–624 (2012)
64. Smoller, J.: Shock Waves and Reaction-Diffusion Equations. Springer, New York (1994)
65. Starovoitov, V.N.: The dynamics of a two-component fluid in the presence of capillary forces. Math. Notes **62**, 244–254 (1997)
66. Volpert, V.: Elliptic Partial Differential Equations. Volume 1: Fredholm Theory of Elliptic Problems in Unbounded Domains. Monographs in Mathematics, vol. 101. Birkhäuser/Springer, Basel (2011)
67. Witterstein, G.: Phase change flows with mass exchange. Adv. Math. Sci. Appl. **21**, 559–611 (2011)

68. Wolchover, N.: Famous fluid equations are incomplete. Quanta Magazine, July 2015. https://www.quantamagazine.org/20150721-famous-fluid-equations-are-incomplete
69. Zumbrun, K.: Multidimensional stability of planar viscous shock waves. In: Advances in the Theory of Shock Waves. Progress in Nonlinear Differential Equations and Their Applications, vol. 47, pp. 307–516. Birkhäuser, Boston, MA (2001)

Part III
Experimental and Numerical Investigation of Interfacial Phenomena

The third part of this book reports on combined experimental and numerical investigations of certain interfacial phenomena. Many such phenomena are related to variable surface tension effects and two chapters are concerned with such influences either from adsorbed surface active agents, so-called surfactants, or the dependence of surface tension on the composition of the involved bulk phases. In both cases, the possibly inhomogeneous interfacial tension leads to Marangoni stresses which strongly influence the flow field close to the interface, sometimes inducing strong convection and even leading to complex pattern formation. Surfactants include certain macro-molecules and even small solid particles may adsorb at fluid interfaces. In such cases, the interfacial flow field can be employed for the separation of enantiomers, a further topic within this part. In case of more complex interfaces composed of lipid layers, different phases may form on the interface and phase change phenomena are possible. A final topic in this part is the structure and the dynamics of thin liquid-liquid films which form if certain liquid polymers are placed on another liquid polymer. This part on experimental and numerical investigations of interfacial phenomena consists of the following contributions, which all emerged from tandem projects, combining experimental and computational work.

Chapter 15.   C. Pesci, H. Marschall, T. Kairaliyeva, V. Ulaganathan, R. Miller, D. Bothe, *Experimental and Computational Analysis of Fluid Interfaces Influenced by Soluble Surfactant*

Chapter 16.   K. Eckert, T. Köllner, K. Schwarzenberger and T. Boeck, *Complex Patterns and Elementary Structures of Solutal Marangoni Convection: Experimental and Numerical Studies*

Chapter 17.   O. Boyarkin, S. Burger, T. Franke, T. Fraunholz, R. Hoppe, S. Kirschler, K. Lindner, M. Peter, F. Strobl, and A. Wixforth, *Transport at Interfaces in Lipid Membranes and Enantiomer Separation*

Chapter 18.   S. Jachalski, D. Peschka, S. Bommer, R. Seemann, B. Wagner, *Structure Formation in Thin Liquid-Liquid Films*

Below, the main topics of these contributions are summarized.

In Chap. 15, the *influence of soluble surfactant* is studied for two fundamental processes: the growing of a droplet from a capillary and the rise of a single bubble in a surfactant solution. Both setups are investigated both *experimentally and* also by means of *direct numerical simulations*. The growing droplet experiment is a standard procedure to measure dynamic interfacial tension by capillary pressure tensiometry, but the droplet shape is influenced by the internal flow and pressure fields, thus requiring detailed modeling and numerical simulations in order to reveal the hydrodynamical impact on the droplet dynamics. Based on an ALE-interface-tracking method, the experimental results can be described and analyzed concerning the droplet's inner flow. This allows detecting limiting growth rates above which the standard evaluation of the experimental data becomes inaccurate and paves the way for an improved experimental analysis. The same method is applied to simulate rising bubbles with ad- and desorption of surfactant from the ambient solution, showing good agreement to the experimentally measured transient rise velocities. As for the droplets, the detailed numerical simulations provide the basis for simplified models which still cover the main surfactant effect but are computationally cheaper.

Chapter 16 is concerned with *Marangoni convection phenomena* which appear during the transfer of a solute between two liquid layers. For such systems, instabilities may occur mainly due to two different effects. On one side, mass transfer leads to local density difference and buoyancy effects can than trigger the Rayleigh instability. On the other side, due to the dependence of surface tension on the local composition of the adjacent liquid bulk phases, the transfer of a chemical component locally changes surface tension which induces tangential flow, the so-called Marangoni convection. The appearance of such instabilities depends in particular on transport properties, namely the solute diffusivity and the liquid viscosities. Employing *specifically designed and controlled experiments in combination with numerical simulations*, this chapter reports on the investigation of two prototype systems. In the first system, characterized by a stable density stratification, a stationary Marangoni instability leads to the formation of hierarchical cellular convection patterns, showing a rich, nonlinear dynamics. Due to the planar two-layer geometry of this system, pseudo-spectral methods can be used to perform efficient numerical simulations which help to understand the complex pattern evolution. The second scenario examines a system with unstable density stratification which is built during the solute transfer. By means of combined experimental and numerical investigations, the *mechanism behind eruptions at the interface has been be clarified* to result from a coupling of Rayleigh and Marangoni convection.

In Chap. 17, the *dynamics and formation* of differently ordered lateral phases of *interfacial lipid layers* is investigated, both by *experiments and by simulation*. Similarly, the dynamics of objects embedded in an air-water interface is studied and the *surface-acoustic-wave-actuated separation of enantiomers* (chiral objects) on the surface of the carrier fluid is demonstrated. It turns out that the dynamics and the separation of the phases do not only depend on parameters such as temperature, mobilities and line tension but also on the mechanics of the lipid layers subjected to exterior forces as, for instance, compression, extensional and shear forces in film-balance experiments. The modeling of these systems is based on the incompressible Navier-Stokes equations with a *viscoelastic stress* term and a capillary term, a convective Jeffrey (Oldroyd) equation of viscoelasticity, and the Cahn-Hilliard equation with a transport term. The numerical simulations are based on discontinuous-Galerkin methods for the Cahn-Hilliard equation. Results from validation of the model and verification of the simulation method against experimental data are provided. Furthermore, the feasibility of enantiomer separation by surface-acoustic-wave-generated vorticity patterns is shown both experimentally and by means of numerical simulations.

In Chap. 18, the problem of a liquid polymer film dewetting from another liquid polymer substrate is studied. The main emphasis is placed on the direct *comparison of results from mathematical modeling, rigorous analysis, numerical simulation and experimental investigations of rupture, dewetting dynamics and equilibrium patterns of a thin liquid-liquid system.* The experimental system employs as a model system a thin polystyrene/polymethylmethacrylate bilayer of a few hundred nanometers. This polymer system allows for in situ observation of the dewetting process by means of atomic force microscopy and for a precise ex situ imaging of the liquid-liquid interface. The molecular chain length of the used polymers has been chosen in such a way that the polymers can be considered as Newtonian liquids, while by increasing the chain length, the rheological properties of the polymers can be also tuned to display a viscoelastic flow behavior. Systematic asymptotic derivation of thin film models are carried out and the predictions based on these models are compared to experimental results. System parameters like contact angle and surface tensions are determined from the experimental data and the results are used for quantitative comparison. An excellent agreement for transient drop shapes on their way towards equilibrium, as well as dewetting rim profiles and dewetting dynamics is obtained.

Chapter 15
Experimental and Computational Analysis of Fluid Interfaces Influenced by Soluble Surfactant

Chiara Pesci, Holger Marschall, Talmira Kairaliyeva,
Vamseekrishna Ulaganathan, Reinhard Miller, and Dieter Bothe

Abstract The present contribution is the result of a collaboration between the Max Planck Institute of Colloids and Interfaces and the Technical University of Darmstadt (MMA group). The main objective is to give a quantitative description of fluid interfaces influenced by surfactants, comparing experimental and computational results. Surfactants are amphiphilic molecules subject to ad- and desorption processes at fluid interfaces. In fact, they accumulate at the interface, modifying the respective interfacial properties. Since these interfaces are moving, continuously deforming and expanding, the local time-dependent interfacial coverage is the most relevant quantity. The description of such processes poses severe challenges both to the experimental and to the simulation sides. Two prototypical problems are considered for comparison between experiments and simulations: the formation of droplets under the influence of surfactants and rising bubbles in aqueous solutions contaminated by surfactants. Direct Numerical Simulations (DNS) provide valuable insights into local quantities such as local surfactant distribution and surface tension, but at high computational costs and restricted to short time frames. On the other hand, experiments can give global quantities necessary for the validation of the numerical procedures and can afford longer time frames. The two methodologies thus yield complementary results which help to understand such complex interfacial phenomena.

C. Pesci • H. Marschall • D. Bothe (✉)
Fachbereich Mathematik, Mathematical Modeling and Analysis, Technische Universität
Darmstadt, Alarich-Weiss-Strasse 10, 64287 Darmstadt, Germany
e-mail: pesci@mma.tu-darmstadt.de; marschall@mma.tu-darmstadt.de;
bothe@mma.tu-darmstadt.de

T. Kairaliyeva • V. Ulaganathan • R. Miller
Max Planck Institute of Colloids and Interfaces, Am Mühlenberg 1, 14424 Potsdam, Germany
e-mail: Talmira.Kairaliyeva@mpikg.mpg.de; Vamseekrishna.Ulaganathan@mpikg.mpg.de;
miller@mpikg.mpg.de

© Springer International Publishing AG 2017
D. Bothe, A. Reusken (eds.), *Transport Processes at Fluidic Interfaces*,
Advances in Mathematical Fluid Mechanics, DOI 10.1007/978-3-319-56602-3_15

15.1   Introduction

Surfactants are essential ingredients in many technological processes such as foaming or emulsification. Their main action is based on the fact that due to their amphiphilic character they assemble at liquid interfaces and modify the respective interfacial properties. In contrast to liquid-gas interfaces, there is an additional complication in the understanding of processes at the interface between two fluids due to the distribution of surfactants between the two bulk phases. Modern technologies are in most cases based on highly dynamic processes. This is true, for example, in the membrane emulsification, where drops are formed at pores of a membrane [82]. The same applies in the process of foam formation, where bubbles are generated via a porous plate which then rise and form a foam layer [23]. The energy needed to form the drops or the bubbles is proportional to the tension σ of the interfacial layer. In the mentioned processes, the drops or bubbles are formed rather quickly so that the interface is strongly expanded and the dynamic interfacial tension at a respective effective adsorption time is the relevant parameter [84]. Also the interfacial rheology has a significant impact on the processes of emulsification and foaming, and on any of the subsequent processes of coalescence between the emulsion droplets or rupture of foam lamella [95].

The thermodynamic foundation of adsorption processes for surfactants at the water/air interface is on a quantitative level with respect to the equations of state. Improvement of the agreement between models and experimental dilatational rheology data was achieved recently by introducing an intrinsic compressibility [46].

Investigations of highly dynamic processes in interfacial layers require a deep knowledge of the hydrodynamics in the adjacent phase, because all experimental tools and the situations in practice refer to dynamic conditions, like growing or oscillating drops and bubbles, compressions and expansions of interfacial layers. The drop formation at pores or capillary tips is a topic dealt with by a number of groups, for example, in [5, 76, 77, 96]. For simple surfactant systems at least a semi-quantitative description of short time adsorption data is possible [40]. The problem, however, not considered in the surface science point of view is the distribution along the interface. Under dynamic conditions, a homogeneous distribution is obviously only a rough approximation and gradients have to be considered. Such gradients lead to Marangoni effects which affect the local stress balance and, hence, the flow field. The impact of the liquid flow on the surfactant transport in the bulk and to/from the interface plus transport along the interface have been new subjects to be dealt with in this project, both experimentally and theoretically. For example, the process of a growing drop of an aqueous solution was simulated by Computational Fluid Dynamics and it was clearly shown that under certain conditions the surfactant is not homogeneously distributed along the drop surface [20]. Consequently, the surface tension is not constant along the drop surface and the classical Gauss-Laplace Equation (GLE) is not applicable to determine the tension from the profile of the dynamic drop [44]. Depending on the rate of deformation, the interfacial layers can be more or less inhomogeneous, which generates a Marangoni flow in the adjacent bulk phase [41]. There are many experiments on rising bubbles, the behavior of which is discussed on the basis of such inhomogeneous adsorption layers [25] and also some numerical simulations have been performed and new simulations are under way [26, 31]. The main target of such studies is the scientific foundation of flotation processes. In this connection, also the formation of so-called rear stagnant caps at the surface of rising bubbles was discussed [54]. Equivalent experiments for settling or rising drops in surfactant solutions are more and more intensively studied. Due to mass transfer, interfacial instabilities are generated based on Marangoni convection, as discussed for example by the groups of Kraume [97] or Eckert [85].

The dynamic interfacial tension can be measured most efficiently by the capillary pressure tensiometry, pioneered by Passerone et al. [67]. We recently established high level experiments, dedicated to growing and oscillating drops. These experiments are suitable to determine quantitatively the dynamic properties of interfacial layers between water and different oil phases [40]. In particular the adaptation of the capillary pressure module ODBA (SINTERFACE Technologies, Berlin) to short times and high frequencies was the subject of recent projects. There is little systematic work on the characterization of properties for interfaces between two liquids [7], and only during the last 10–15 years the adsorption of surfactants at water/oil interfaces was studied carefully, most of all the cationic surfactants CnTAB (alkyl trimethyl ammonium bromide) [57, 60, 74]. Until recently, the impact of the oil phase on the adsorption behavior of surfactants was not yet quantitatively understood. In [29] it was shown quantitatively that the oil molecules are not simply build into the surfactant adsorption layer at the interface and then squeezed out at higher surface pressure. It could be demonstrated that on the contrary the oil molecules co-adsorb with the surfactants and the structure and composition of the interfacial layer is the result of the competition of the two compounds. First experiments have also been performed on the dynamic characteristics taking into account different chain lengths of the alkane being the oil phase [61]. However, the mechanisms of the co-adsorbing oil molecules in the formation or relaxation processes have not been understood yet.

Besides the required experimental analysis, modern numerical methods for the simulation of two-phase flows with full resolution of the interface dynamics provide a basis for a computational analysis of the local interplay of transport and transformation processes. To tap the full potential of this theoretical approach, rigorous methods for the simulation of the influence of (soluble) surfactant on moving and deforming fluidic interfaces have to be improved and further developed. While integral values, e.g. for the surface tension, can be extracted from the experimental data, detailed local information cannot be easily accessed. Therefore, numerical simulations play an important role in understanding the interfacial transport phenomena and sorption processes. Direct Numerical Simulations (DNS) are performed to give valuable insights about local surfactant distribution, both on the interface and in its vicinity together with the corresponding local surface tension profiles.

The continuum physical model of two-phase flows under the influence of surfactant consists of the two-phase Navier-Stokes equations with appropriate interfacial jump conditions and the surfactant transport equations in the bulk and on the interface coupled via sorption processes. Thus, coupled non-linear transport processes (PDEs, mostly advection-diffusion equations, in the bulk and on the interface) have to be solved, while the interface itself is deformable and the whole domain is moving. The mathematical model is described in details in [14, 70]. Therefore, only the main assumptions and equations are summarized in Sect. 15.3.1.

From a numerical point of view the description of multiphase flows under the influence of surfactant is a challenging task. In the literature, a variety of methods for the DNS of two-phase flows under the influence of surfactant exists. A brief overview of the state-of-the-art of the numerical methods will be given in the following paragraph, while the description of the critical points and limitations to simulate multiphase flows will be addressed in Sects. 15.3.2 and 15.5. There are two general categories of methods to simulate two-phase systems with fluidic interfaces: Eulerian interface-capturing methods and Lagrangian interface-tracking methods. While the first describe the interface implicitly by an additional color or marker function, the latter represent the fluid interface in an explicit manner via a computational surface mesh or a set of connected marker particles.

Eulerian interface-capturing methods use a fixed Eulerian grid and introduce a marker field to represent one of the phases or the interface itself. The evolution of the marker field is tracked by solving an advection equation. Then, for some of these methods, it is necessary to reconstruct the interface from the marker field. Well known representatives of interface-capturing methods are Volume-of-Fluid (VOF) methods [36], Level-Set methods, which typically describe the position of the interface with a signed distance function [65], hybrid coupled LS/VOF methods [87], and diffuse-interface methods [6]. All these methods have been enhanced to describe insoluble or soluble surfactant transport. Within the Level-Set methods, first, Adalsteinsson and Sethian [1] and Xu and Zhao [99] described a method for solving partial differential equations on moving interfaces. Xu et al. then coupled the interfacial transport method to a flow solver for incompressible Stokes flow to simulate the influence of an insoluble surfactant on droplet-droplet interactions [100]. This method was extended to three-dimensional problems in [101]. Another example of a level-set method developed by Lahrenfeld and Reusken applied to the solution of interface problems and PDEs on surfaces can be found in Chap. 2. Various authors extended the VOF method to solve for insoluble [38, 78] or soluble [4] surfactants. All these contributions use the so-called Continuum Surface Force model [15] to evaluate the surface tension force. Recently, Fleckenstein and Bothe [31] proposed a simplified model, deriving a momentum jump condition that models the influence of surfactants at fluid interfaces due to Marangoni stresses. Within Hybrid methods, Yang and James [102], basing their work on [87], track the surfactant mass by applying an Arbitrary Lagrangian-Eulerian method. Other hybrid methods dealing with surfactants can be found in [17] and [2]. Diffuse Interface methods have been enhanced to handle surfactants for instance in [3, 66, 89] and in Chaps. 10 and 11.

Lagrangian interface-tracking methods resolve the fluid interface with either a deformable surface mesh (boundary integral methods [73]), a set of connected Lagrangian marker particles [91], or a boundary fitted surface mesh subjected to Lagrangian advection [63, 94]. Also for these methods, extensions to handle the presence of surfactants were introduced. Boundary Integral methods have been used to simulate the effect of surfactants on droplet or bubble deformation by different authors, among them the most recent works can be found in [8, 30, 42, 86]. Ganesan and Tobiska, in [33, 34] and in Chap. 1, considered insoluble and soluble surfactants on the basis of a second-order iso-parametric surface Finite Element method on moving meshes. Based on the Front Tracking method [91], Zhang et al. [103] introduced a 2D-axisymmetric method to account for soluble surfactants. Muradoglu and Tryggvason [62] propose a combined front-tracking/finite-difference method to handle soluble surfactants. The method is 2D-axisymmetric, too, and has been applied to rising bubbles by Tasoglu et al. [88]. Lai et al. [49] account for insoluble surfactants within an Immersed Boundary method [72]. Within the ALE interface tracking methods, Li [50] and Hameed et al. [35] investigate the influence of insoluble surfactants. The method by Tornberg and Engquist [90], capable to handle insoluble surfactants in 2D simulations, was recently extended by Khatri and Tornberg [45] to handle soluble ones. Finally, Tuković and Jasak [93] developed a methodology capable to handle soluble surfactants on arbitrary meshes and being second order accurate in space and time.

The present work is based on the Arbitrary Lagrangian-Eulerian (ALE) interface-tracking method, originally presented by Muzaferija and Perić [63] and further developed by Tuković and Jasak [94]. The Finite Volume based ALE interface-tracking framework is utilized in combination with the Finite Area method [93] for the simulation of interfacial transport processes. In other words, the transport equations in the bulk are discretized via the Finite Volume method, while the transport processes on the interface are solved via the Finite Area method. Regarding the coupling between surfactant transport processes in the bulk and on the interface, a sorption library has been developed, distinguishing between fast (diffusion controlled) and slow (kinetically controlled) sorption processes, since different implementation strategies for the two cases are required. The deformation of the interface is taken into account via an ALE reference frame. Since the motion of the interface implies distortions of the volumetric mesh, an automatic mesh motion [39] is applied, in combination with re-meshing algorithms [59, 75, 83], when necessary, to avoid mesh degeneration. The overall procedure is based on the open source library OpenFOAM, an object-oriented C++ library for numerical simulation of Computational Continuum Mechanics and Computational Fluid Dynamics.

While the thermodynamics of adsorption at the water/air interface can be described quantitatively, there are a number of peculiarities of the water/oil interfaces. Especially the solubility of surfactants in both liquid phases is an extra item that has to be considered in more detail [51]. Not only the adsorption characteristics of one and the same surfactant can be very different at the water/air and water/oil interface [60]. Most of all, the non-equilibrium properties, such as the dilatational rheology [61], change a lot when passing from one to the other interface. In addition, the interactions of the adsorbing molecules with the solvent on both sides of the interface play a crucial role. Hence, it has to be shown how far approaches developed for liquid/gas interfaces can be applied to interfaces between two immiscible liquids. Moreover, it has to be analyzed which additional features must be considered in a quantitative description of all processes going on at water/oil interfaces [29, 74], which, however, will not be discussed in detail here.

This chapter aims at describing the most recent state of the art on the dynamic behavior of interfaces between two fluid phases from the experimental as well as the theoretical point of view. The experiments are mainly done with three experimental techniques: the drop profile analysis tensiometry, capillary pressure tensiometry and velocimetry of rising bubbles in surfactant solutions. All these experimental methods have limitations in their range of applicability and simulations provide the opportunity to define these limits or even allow a correct quantitative analysis of the obtained experimental data.

Regarding the numerical simulations, we limited our investigations to gas/liquid systems contaminated by surfactants. The final goal of this research work is to improve the understanding of the surfactant effects. With this aim in mind, the mathematical and numerical models for two-phase flows were further developed and a rigorous validation of the methodology assured about the range of validity of the results obtained. The numerical results are then compared to the experimental ones.

The main results of this tandem project and which problems remained open for future research are addressed in the discussion in Sect. 15.6 at the end of this chapter.

15.2 Materials and Experimental Set-Up

15.2.1 Surfactant Properties

The main surfactants used in the presented experiments are members of the homologous series of the non-ionic alkyl dimethyl phosphine oxides (C_nDMPO) and the cationic alkyl trimethyl ammonium bromides (C_nTAB). Both surfactants can be purchased from Sigma/Aldrich and the aqueous solutions are prepared with milli-Q water. The adsorption behavior at the solution/air interface can be well described by the Frumkin adsorption model, given by the following two equations [32]

$$\Pi = -\frac{RT}{\omega_0}\left[\log(1-\theta) + a\theta^2\right],$$ (15.1)

$$bc = \frac{\theta}{1-\theta}e^{-2a\theta}.$$ (15.2)

Here c is the surfactant bulk concentration, Π is the surface pressure $\Pi = \sigma_0 - \sigma$ (σ_0 is the surface tension of pure water and σ is the surface tension of the surfactant solution), ω_0 is the molar area of an adsorbed surfactant, θ is the surface coverage given by the product of adsorbed amount c^Σ times ω_0, a is an interaction parameter between adsorbed molecules in the interfacial layer, and b is the so-called adsorption activity constant. In Figs. 15.1 and 15.2 the surface tension isotherms $\sigma(\log c)$ are presented for the homologous series of C_nDMPO and C_nTAB, respectively. The solid curves represent the best fit of the Frumkin model to the experimental data, which were taken from literature on C_nDMPO [10, 53] and C_nTAB [9]. A systematic analysis of experimental data on the basis of various adsorption models was presented in [27]. The data for the cationic surfactants are presented as a

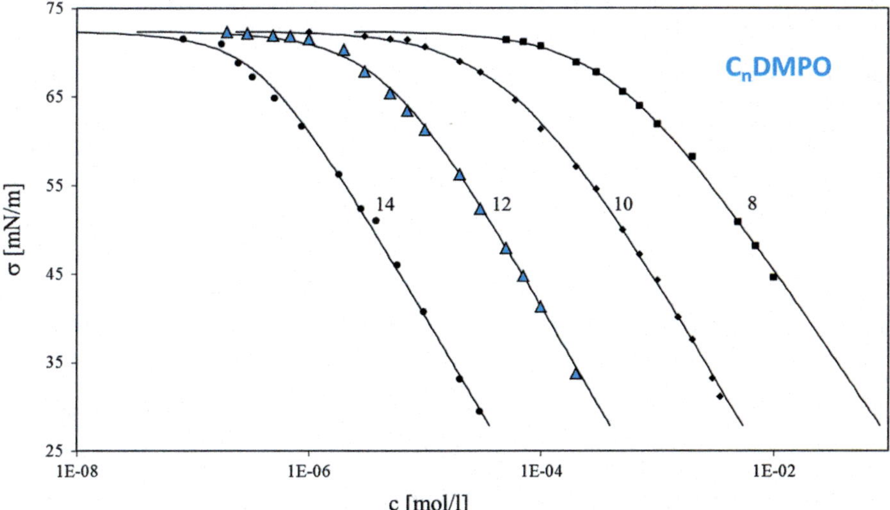

Fig. 15.1 Equilibrium surface tension isotherms for C_nDMPO solutions, *circle*—C_{14}DMPO, *triangle*—C_{12}DMPO, *diamond*—C_{10}DMPO, *square*—C_8DMPO, the theoretical curves are calculated from the Frumkin adsorption model; data taken from [53]

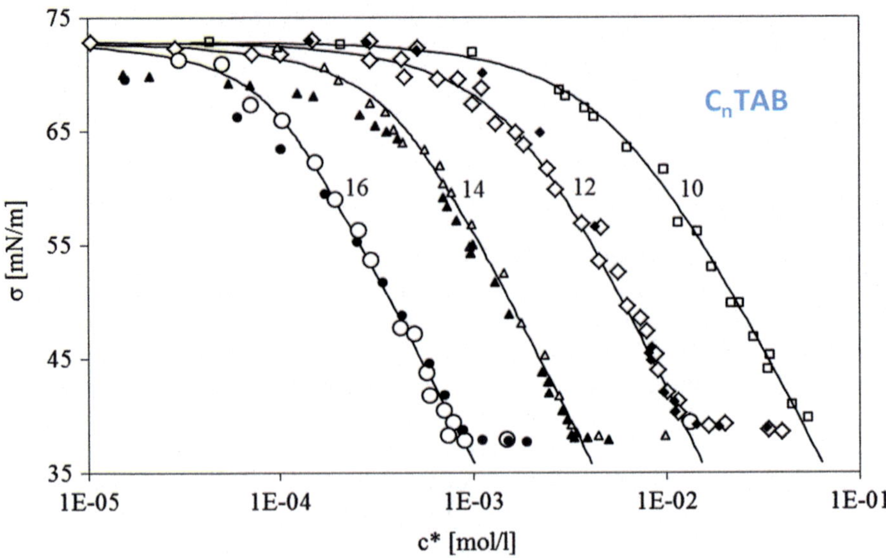

Fig. 15.2 Dependence of surface tension for C$_n$TAB solutions as a function of the mean ionic activity c*; different *symbols* refer to different sources for the data (cf. [27]), *numbers* denote the number of carbon atoms, the theoretical curves were calculated from the Frumkin model using the parameters listed in Table 15.2

Table 15.1 Model parameters for C$_n$DMPO using the Frumkin adsorption model

C$_n$DMPO	ω_0 $(10^5\ \mathrm{m^2/mol})$	a (–)	b (l/mol)
C$_8$DMPO	2.9	0.0	$2.5 \cdot 10^3$
C$_{10}$DMPO	2.5	−0.2	$2.2 \cdot 10^4$
C$_{12}$DMPO	2.4	0.4	$1.5 \cdot 10^5$
C$_{14}$DMPO	2.5	0.8	$7.6 \cdot 10^5$

Table 15.2 Model parameters for C$_n$TAB using the Frumkin adsorption model

C$_n$TAB	ω_0 $(10^5\ \mathrm{m^2/mol})$	a (–)	b (l/mol)
C$_{10}$TAB	1.6	0.6	94.2
C$_{12}$TAB	1.5	0.8	263
C$_{14}$TAB	1.6	1.1	846
C$_{16}$TAB	1.6	1.3	3050

function of the mean activity c^* as defined by

$$c^* = f_{\pm} \left(c_{\mathrm{C_nTAB+NaBr}}\, c_{\mathrm{C_nTAB}} \right)^{1/2} \tag{15.3}$$

which is identical with the molar concentration $c_{\mathrm{C_nTAB}}$ when no electrolyte NaBr is added. The parameters obtained by the fitting procedure are summarized in Tables 15.1 and 15.2; see [27]. As we can conclude from Figs. 15.1 and 15.2 and also from the data summarized in Tables 15.1 and 15.2, the non-ionic surfactants

C_nDMPO are about 100 times more surface active than the corresponding cationic C_nTAB at the same carbon number n.

The C_nTAB surfactants have also been studied at the water/oil interface. One of the first studies were performed by Medrzycka and Zwierzykowski [57], however, the data analysis was still based on theories developed essentially for the W/A interface. This means that no specific facts had been taken into consideration with respect to the interaction between surfactant molecules and the oil phase. Also the systematic study by Pradines et al. [74] and Mucic et al. [60] was still based on the idea that the oil molecules are simply intercalated in the interfacial layer and depending on the chain length of surfactant and oil these oil molecules can be squeezed out at a certain surface coverage.

The model finally proposed in [28] and further refined in [29] assumed for the first time a competitive adsorption of the surfactant and oil molecules. When we assume that component 1 (oil) is adsorbed from the liquid oil phase, and component 2 (surfactant) adsorbs from its aqueous solution, the following equation of state for the mixed interfacial layer has been obtained in [29]:

$$-\frac{\Pi \omega_0^*}{RT} = \ln(1 - \theta_1 - \theta_2) + a_1\theta_1^2 + a_2\theta_2^2 + 2a_{12}\theta_1\theta_2, \tag{15.4}$$

with

$$\omega_0^* = \frac{\omega_{10}\theta_1 + \omega_{20}\theta_2}{\theta_1 + \theta_2}. \tag{15.5}$$

Here $\omega_{i,0}$ are the molar areas required by the components i ($i = 1$ or 2) at very diluted interfacial layers. The coefficient a_2 represents the interaction between the surfactant molecules. For the oil molecules we get the adsorption isotherm

$$b_1 c_1 = \frac{\theta_1}{1 - \theta_1 - \theta_2} exp\left(-2a_1\theta_1 - 2a_{12}\theta_2\right), \tag{15.6}$$

while for the surfactant adsorption from the aqueous phase the isotherm reads

$$b_2 c_2 = \frac{\theta_2}{1 - \theta_1 - \theta_2} exp\left(-2a_2\theta_2 - 2a_{12}\theta_1\right). \tag{15.7}$$

Here b_1 and b_2 are the parameters representing the surface activity of the two adsorbing species.

The results obtained in [28], using (15.4)–(15.7) and the model parameters given in Table 15.3, are shown in Fig. 15.3 by the continuous curves. The model proposed by Eqs. (15.4)–(15.7) allows describing the experimental data for all C_nTAB and hexane, but also for the other alkanes (heptane, octane, nonane, decane) studied in [60]. Note that the dotted lines shown in Fig. 15.3 and obtained from a modified Frumkin adsorption model also fit the experimental data quite well. However, a huge difference is found in the physical picture of the two models. The one given

Table 15.3 The parameters involved in Eqs. (15.4)–(15.7) for various C_nTAB aqueous solutions at the interface with hexane; according to [28]

C_nTAB	c_1 (mol/l)	b_1 (l/mol)	ω_1 $(10^5\,\mathrm{m^2/mol})$	b_2 $(10^5\,\mathrm{l/mol})$	ω_1 $(10^5\,\mathrm{m^2/mol})$	a_2 (–)	a_{12} (–)
C_{10}TAB	7.58	2.6	3.3	0.43	3.45	0.8	1.1
C_{12}TAB	7.58	2.6	3.2	2.0	3.5	1.3	1.4
C_{14}TAB	7.58	2.6	3.3	12	3.4	1.4	1.6
C_{16}TAB	7.58	2.6	3.2	120	3.2	1.6	1.7

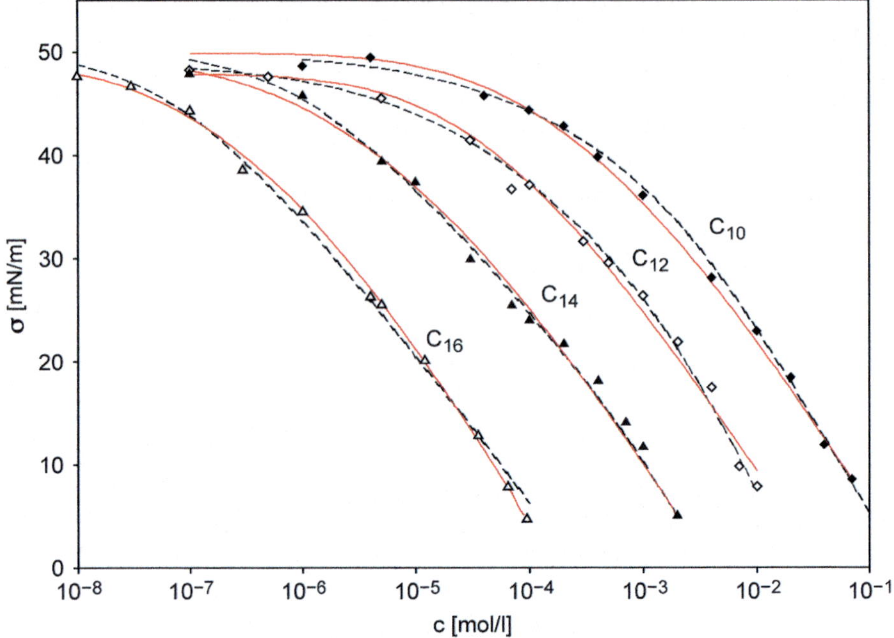

Fig. 15.3 Interfacial tension isotherms of C_{10}TAB (*filled diamond*), C_{12}TAB (*empty diamond*), C_{14}TAB (*filled triangle*) and C_{16}TAB (*empty triangle*) at the interface between an aqueous solution and hexane (in phosphate buffer of 10 mM at pH7); *symbols* are experimental data from [60]; *dashed lines* were calculated with a modified Frumkin model; *solid lines* were calculated from Eqs. (15.4)–(15.7) with the parameters given in Table 15.3, according to [28]

by a competitive adsorption between the surfactant and oil molecules allows to understand quantitatively the effect of the intercalation of oil molecules into the interfacial layer, while the generalized Frumkin model only assumes a change in the interaction between the molecules in the interfacial layer.

15.2.2   Growing Droplet

To measure correct values of surface tension of surfactant solutions, axisymmetric drop shape analysis (ADSA) can be successfully used. The pioneering work was made by Neumann and his team [79]. The principle of this method consists in determining the surface tension of a liquid from the profile of a drop or a bubble [19]. With the development of electronic computers, ADSA technique reached an excellent accuracy due to the systematic work of Neumann and his group [18]. Note, however, that all the work was made for drops and bubbles in rest, i.e. neglecting any change in size during the experiments. Thus, so-called dynamic experiments with growing or oscillating drops/bubbles are actually not correctly described except one disregards any liquid flow inside the drops and outside the drops/bubbles.

Growing drop experiments can be performed by using the Profile Analysis Tensiometer PAT-1 (SINTERFACE Technologies, Berlin, Germany) [52]. The scheme of this instrument is shown in Fig. 15.4. It is composed of a cell where a drop or a bubble is formed at the tip of a vertical or U-shape capillary, a dosing system, a light source and a video camera which is connected to a computer with an integrated digitizer. The coordinates of a drop or bubble taken from a video image by the camera, are fitted with the Gauss-Laplace Equation (GLE), as schematically shown in Fig. 15.5.

The GLE contains several physical quantities, of which the surface tension σ is the only unknown one, while the density difference between the two fluids $\Delta\rho$ and the gravity constant g are known:

$$\sigma\left(\frac{1}{R_1} + \frac{1}{R_2}\right) = \Delta P, \tag{15.8}$$

where R_1 and R_2 are the two principal radii of interface curvature, and ΔP is the pressure difference across the interface. This equation represents also the balance between the surface tension and gravity forces,

$$\Delta P = \Delta P_0 + \Delta\rho g z \tag{15.9}$$

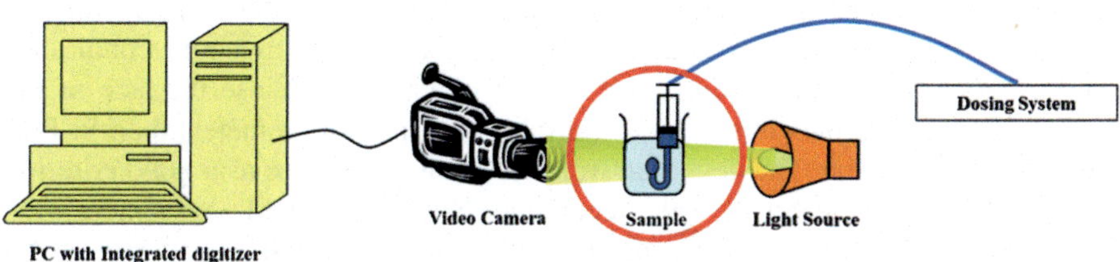

Fig. 15.4 Scheme of a Profile Analysis Tensiometer (according to PAT-1 from SINTERFACE Technologies, Germany)

Fig. 15.5 Coordinates of a
buoyant drop profile (*dots*)
and calculated profiles

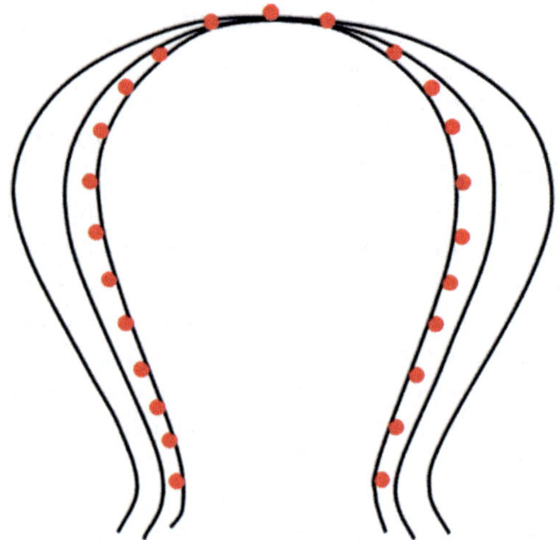

where ΔP_0 is the pressure difference at a reference plane, and z is the vertical height
measured from a reference plane. While the surface tension force tends to make the
drop spherical, the gravity force tends to elongate the drop. The shape of the drop is
therefore the result of these two forces.

There are various strategies for fitting the GLE to the experimental drop/bubble
profiles in order to obtain the interfacial tension. The most frequently used protocol
was proposed by Maze and Burnet [56], which is still in use although new and
better routines are applied now [37]. Although the accuracy of this methodology was
increased by various means, such as sub-pixel resolution of image analysis, faster
and more accurate solution of the GLE, more efficient algorithms for fitting the GLE
to experimental coordinates, systematic errors come into play when it is applied to
drops or bubbles under dynamic conditions. It was extensively discussed in [43]
that the profiles of, for example, growing drops or bubbles deviate systematically
from the Laplacian shape. A correct description would require to take the additional
inertia effects of the inflowing liquid into account and describe the profile of a
growing drop, for instance, by such a modified GLE. The solution of this problem
requires a numerical study of the flow field as performed within this project.

Capillary pressure tensiometry is a second powerful methodology that can
be used to study interfacial surfactant layers at static and dynamic conditions.
The advantage is the less pronounced impact of gravity because the size of the
studied drops or bubbles is much smaller (100–200 μm in diameter) than in PAT
experiments (2–5 mm). The deviation from sphericity can serve as a good criterion
for the limits of applicability of the technique. Note that in PAT experiments, the
mean deviation between the drop profile and the best fit by the GLE is a useful
criterion for the applicability as shown in [43]. As soon as the standard deviation
of the fitting error exceeds a certain limit, typically of the order of the pixel size of
about 1 μm, the method starts to provide only effective interfacial tensions which
can be far from the real physical value.

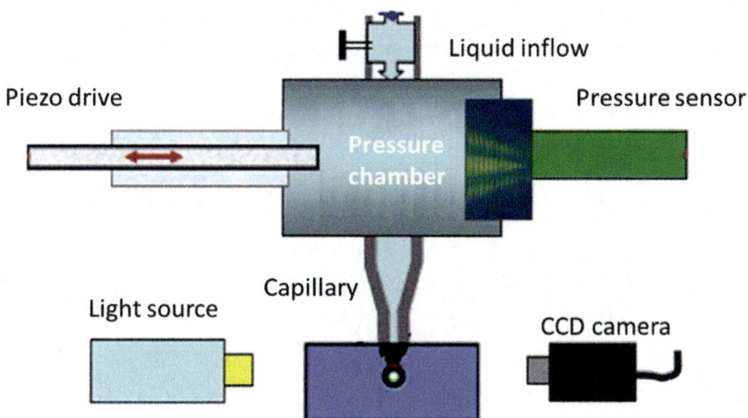

Fig. 15.6 Scheme of a capillary pressure tensiometer, according to [44]

The schematic of a capillary pressure tensiometer, suitable for experiments at the water/air as well as at the water/oil interface, first proposed by Passerone et al. in [67], is, for example, presented in [41]. While the liquid flows into the chamber, the pressure is continuously recorded and corresponds to the capillary pressure inside the growing drop formed at the capillary tip. The capillary tip can be immersed into air or into a liquid, respectively. With the help of the piezo drive, accurate changes of the formed drop, such as small amplitude oscillations, can be generated (Fig. 15.6).

With the same techniques, capillary pressure (ODBA) and drop profile analysis tensiometry (PAT), the liquid/liquid interface can be studied. An example for the dynamic interfacial tension measured for a surfactant adsorbing at the aqueous solution/oil interface by using these two methods is displayed in Fig. 15.7. While the short time data for 10 ms up to about 100 s were obtained with ODBA, the values for adsorption times between 1 and 10,000 s were measured by PAT.

Both sets of data complement each other nicely. The red dotted and black solid curves are fitted with a classical diffusion controlled adsorption kinetics using a Frumkin (solid line) and Langmuir (dashed line) adsorption model, respectively. This type of theoretical data analysis does not require fluid dynamic simulations.

Note that, using the methodology of growing drops (coupled with capillary pressure tensiometry), we could reach dynamic interfacial tensions and adsorption kinetics data for much shorter adsorption times. Such data are, however, presently no yet available. Studies of such type and a direct comparison with simulations of growing drops will be the subject of future work.

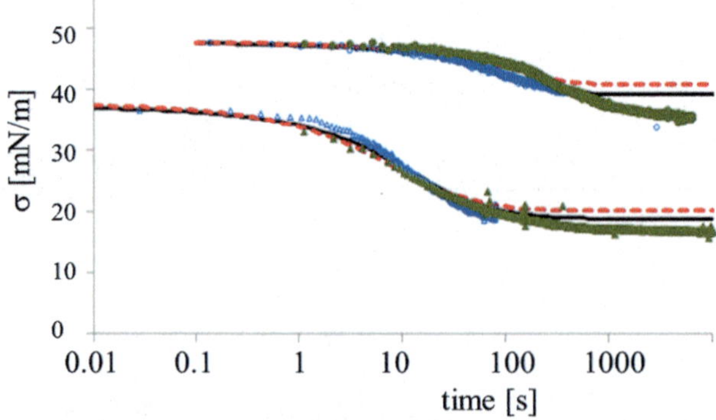

Fig. 15.7 Time dependence of interfacial tension σ for C_{16}TAB on the time t, *blue symbols*—data measured with capillary pressure tensiometry; *green symbols* measured by PAT; *triangles*—$c = 5 \cdot 10^{-6}$ mol/L, *circles*—$c = 2 \cdot 10^{-6}$ mol/L; *lines* were calculated using the fitting parameters summarized in Table 15.3; according to [55]

15.2.3 Rising Bubble

The setup for studies of rising bubbles performed in this project consists of a glass column of square cross section area 40×40 mm and a height of 500 mm as shown in Fig. 15.8. A capillary of inner diameter 0.025 mm is mounted to the bottom of this glass column which is connected to a syringe pump for a controlled air supply. The flow rate is set to level that the time interval between the bubbles is about 12–15 s to be sure that the liquid in the column is quiescent for each bubble. A digital camera is fitted to the calibrated column which could be moved along the axis of the glass column to capture the video of the bubble at any required distance. Finally, a stroboscope is used as light source, producing flashes of light at a desired frequency f. With this set-up to obtain the velocity profiles of rising bubbles, the images extracted from the monitored video are analyzed following a special protocol.

In Fig. 15.9, a video image of a rising bubble is shown. It contains the bubble at different positions, caused by the stroboscopic light. Here the distance L is the distance between the capillary tip and the bubble, and its local velocity U_L at this distance is $\Delta L/t$. The distance between the bottom poles of two subsequent bubble images is ΔL and t is the time between two subsequent flashes from the stroboscope, i.e. $t = 1/f$. The stroboscope light flashes are set up such that only two or three images of a bubble are on a respective snapshot.

The experimental setup was applied to various surfactant solution systems as shown for example in [48, 54, 55]. The general features of such experiments are summarized in Fig. 15.10.

A typical graph with the local velocity of a bubble as a function of the distance from the capillary tip is shown in Fig. 15.10. For a bubble rising in clean water, i.e. when the surface of the bubble is completely mobile, there is an acceleration

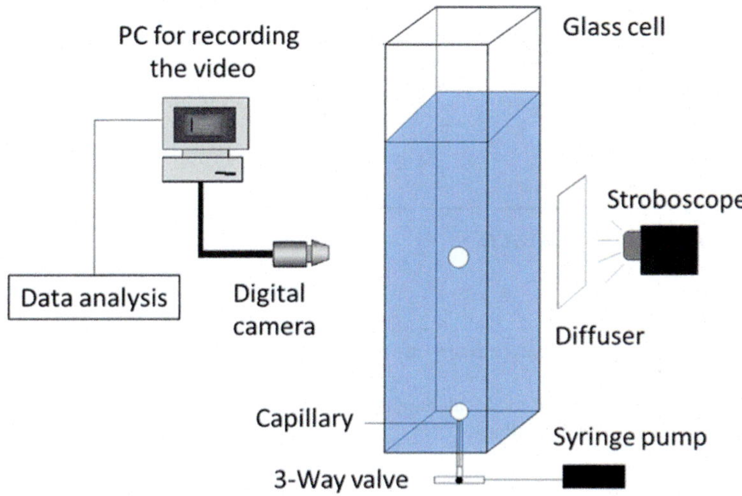

Fig. 15.8 Schematic of the rising bubble instrument with capillary

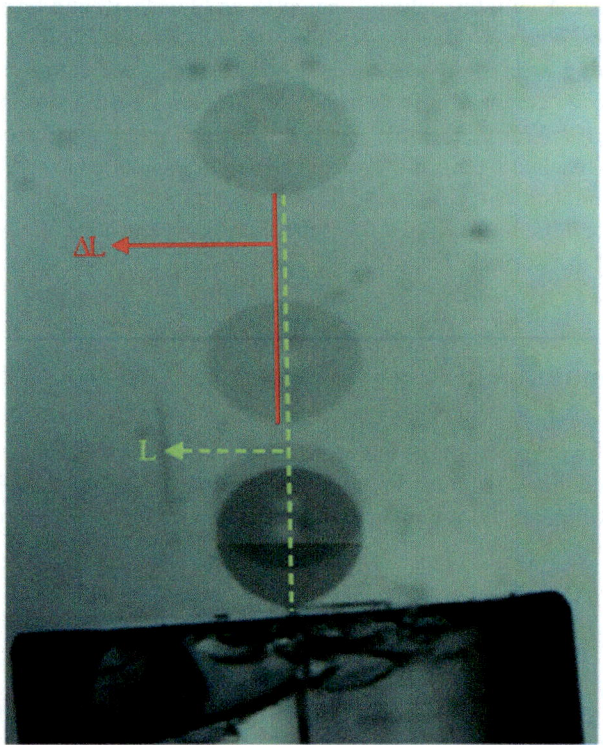

Fig. 15.9 Photo of a rising bubble taken with a stroboscopic light source

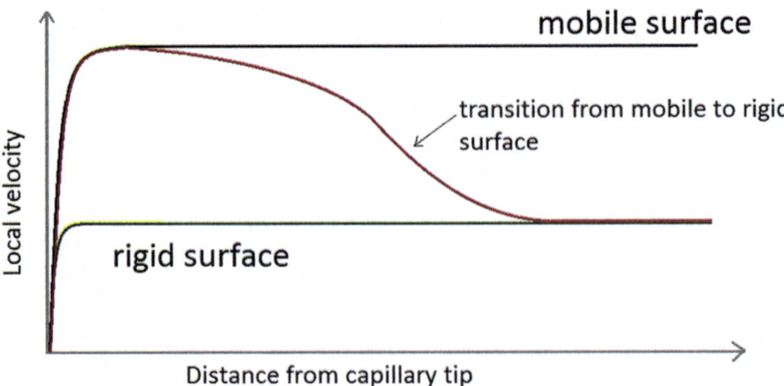

Fig. 15.10 Schematic course of the local velocity profile plotted against the distance travelled by the bubble from the tip of the capillary

followed by a constant velocity which is the terminal velocity. The same is observed for bubbles rising in a surfactant solution of high concentration. Then the bubble surface is quickly covered by surfactant molecules and attains a rigid surface behavior. Again, the velocity increases and levels off at a terminal velocity which, however, is lower than that for a mobile bubble surface. This is due to the drag force acting on a rigid surface. At intermediate concentrations, the bubble surface undergoes a transition from a mobile surface to a rigid surface, as one can see in Fig. 15.10. With increasing concentration, the maximum velocity becomes smaller and is reached later, until the concentration is reached at which the bubble behaves completely as a rigid sphere. This concentration represents the limit of the experimental technique, as above this surfactant concentration the same velocity profile would be observed.

15.3 Computational Analysis Set-Up

In this section an overview of the mathematical and numerical model is presented on which the simulations are based. In particular, the continuum physical and numerical modeling of two-phase flows under the effect of surfactant is described.

15.3.1 Mathematical Model

Consider a fluid domain Ω containing two immiscible fluids, separated by a deformable interface. The mathematical model for two-phase flows (free-surface flows are considered a sub-case of this) employs a sharp interface representation, meaning that the interface is represented by a surface of zero thickness. The deformable interface, $\Sigma(t)$, divides the domain into two sub-domains, $\Omega^+(t)$ and $\Omega^-(t)$, corresponding to the two bulk phases. The presence of surfactant in the bulk phases and on the interface is taken into account. The governing equations are based on the conservation of mass, momentum and surfactant molar mass under the hypothesis of

- incompressible Newtonian fluids
- isothermal conditions
- absence of phase change and chemical reactions
- relative normal motion of the interface due to sorption and mass transfer neglected
- no-slip at the interface
- negligible inertia of the adsorbed surfactant
- dilute surfactant concentration in the bulk, but possibly being non-dilute on the interface.

Special attention is given to the description of the adsorption processes at the interface and the surfactant transport in the bulk phase and on the interface. Two different situations can be considered: diffusion-controlled sorption (fast) and kinetically-controlled sorption (slow) [21, 70]. In the first case the sorption process is much faster than the diffusive transport, while in the latter case the diffusive transport is faster than the sorption process, typically due to the presence of a kinetic barrier. Thus the transfer rate will be determined in two different ways, namely

- diffusion-controlled (fast) sorption:

$$s_i^\Sigma = -[\![\mathbf{j}_i \cdot \mathbf{n}_\Sigma]\!], \tag{15.10}$$

where $\mathbf{j}_i$ are the diffusive fluxes and $\mathbf{n}_\Sigma$ the normal to the interface. Equation (15.10) follows from the local balance of mass exchange with simplifications according to the above mentioned assumptions. In the case of fast (as compared to kinetically-controlled transport) sorption, the ad- and desorption rates are locally equilibrated, i.e.

$$s_i^{\mathrm{ads}}\big(c_{i|\Sigma}, c_i^\Sigma\big) = s_i^{\mathrm{des}}\big(c_i^\Sigma\big), \tag{15.11}$$

where $c_{i|\Sigma}$ is the concentration of surfactant in the bulk phase close to the interface and c_i^Σ is the concentration of surfactant on the interface. This equality leads to an additional local relationship between c_i^Σ and $c_{i|\Sigma}$ (the so-called *adsorption isotherm*) which needs to be accounted for in the numerical solution. In the case of fast sorption, the interfacial surfactant mass balance can be viewed as a dynamical coupling condition between the Dirichlet- and the Neumann-data of the bulk concentration field. The numerical method needs to map this into a discretized form.

- kinetically-controlled (slow) sorption:

$$s_i^{\Sigma} = s_i^{\text{ads}}\left(c_{i|\Sigma}, c_i^{\Sigma}\right) - s_i^{\text{des}}\left(c_i^{\Sigma}\right), \tag{15.12}$$

where $s_i^{\text{ads}}\left(c_{i|\Sigma}, c_i^{\Sigma}\right)$ describes the rate of adsorption and $s_i^{\text{des}}\left(c_i^{\Sigma}\right)$ the desorption rate. Note that the rate of adsorption is a function of the bulk concentration near the interface and the concentration of the adsorbed species, while the desorption rate is usually assumed to be a function of the adsorbed species only.

In both cases, the effect of surfactant on the interfacial surface tension is described by the surface tension equation of state, already presented in one of its specialized forms for the Frumkin model; see (15.1), (15.2). Such equations, derived from the Gibbs adsorption equation, relate the bulk molar fraction ($x = c/c_{tot}$), the concentration of the adsorbed form c^{Σ} and the change of surface tension $d\sigma$ according to

$$d\sigma = -RTc^{\Sigma}d(\ln x). \tag{15.13}$$

Under these assumptions, the full sharp-interface model of a two-phase flow with soluble surfactants reads as follows. Below, $\mathbf{S}^{\text{visc}}$ indicates the viscous stress tensor.

Bulk Equations in $\Omega^{\pm}(t)$

$$\nabla \cdot \mathbf{v} = 0, \tag{15.14}$$

$$\partial_t(\rho\mathbf{v}) + \nabla \cdot (\rho\mathbf{v} \otimes \mathbf{v}) = -\nabla p + \nabla \cdot \mathbf{S}^{\text{visc}} + \rho\,\mathbf{g}, \tag{15.15}$$

$$\partial_t c_i + \nabla \cdot \left(c_i\mathbf{v} + \mathbf{j}_i\right) = 0. \tag{15.16}$$

Interface Equations on $\Sigma(t)$

$$[\![\mathbf{v}]\!] = 0, \tag{15.17}$$

$$\mathbf{v} \cdot \mathbf{n}_{\Sigma} = \mathbf{v}^{\Sigma} \cdot \mathbf{n}_{\Sigma}, \tag{15.18}$$

$$[\![p\mathbf{I} - \mathbf{S}^{\text{visc}}]\!] \cdot \mathbf{n}_{\Sigma} = \sigma\kappa\mathbf{n}_{\Sigma} + \nabla_{\Sigma}\sigma, \tag{15.19}$$

$$\partial_t^{\Sigma} c_i^{\Sigma} + \nabla_{\Sigma} \cdot \left(c_i^{\Sigma}\mathbf{v}^{\Sigma} + \mathbf{j}_i^{\Sigma}\right) = s_i^{\Sigma}, \tag{15.20}$$

$$s_i^{\Sigma} + [\![\mathbf{j}_i \cdot \mathbf{n}_{\Sigma}]\!] = 0. \tag{15.21}$$

Initial Conditions at $t = 0$

$$\mathbf{v}(0, \mathbf{x}) = \mathbf{v}_0(\mathbf{x}), \quad \mathbf{x} \in \Omega_\pm(0), \tag{15.22}$$

$$c_i(0, \mathbf{x}) = c_{i,0}(\mathbf{x}), \quad \mathbf{x} \in \Omega_\pm(0), \tag{15.23}$$

$$c_i^\Sigma(0, \mathbf{x}) = c_{i,0}^\Sigma(\mathbf{x}), \quad \mathbf{x} \in \Sigma(0). \tag{15.24}$$

Bulk Diffusive Fluxes

$$\mathbf{j}_i = -D_i \, \nabla c_i \quad \text{in } \Omega \text{ (dilute solution).} \tag{15.25}$$

Interface Diffusive Fluxes

$$\mathbf{j}_i^\Sigma = -D_i^\Sigma \, \nabla c_i^\Sigma \quad \text{on } \Sigma \text{ (dilute solution),} \tag{15.26a}$$

$$-\sum_{j \neq i}^{N} \frac{c_j^\Sigma \mathbf{j}_i^\Sigma - c_i^\Sigma \mathbf{j}_j^\Sigma}{D_{ij}^\Sigma} = \mathbf{d}_i^\Sigma \quad \text{on } \Sigma \text{ (non-dilute solution)} \tag{15.26b}$$

with $\mathbf{d}_i^\Sigma = \frac{c_i^\Sigma}{RT} \nabla_\Sigma \mu_i^\Sigma + \frac{y_i^\Sigma}{RT} \nabla_\Sigma \sigma$ the interface thermodynamical driving forces. For more details on non-dilute systems we refer to [11, 12, 22, 80].

The full model comprises the balance equations (15.14)–(15.21) for Newtonian fluids, with diffusive fluxes given by (15.25), (15.26), the interfacial equation of state (15.13) and sorption modeled either by (15.10), (15.11) or by (15.12), completed by appropriate boundary conditions at the outer domain boundary,

$$\mathbf{v} = 0 \quad \text{on } \partial\Omega, \tag{15.27}$$

$$\mathbf{j}_i \cdot \mathbf{n} = 0 \quad \text{on } \partial\Omega. \tag{15.28}$$

We refer to [70] for the details on the sorption modeling and the available sorption models implemented.

15.3.2 Numerical Model

Direct numerical simulations (DNS) are performed to solve the above system of equations within an ALE (Arbitrary Lagrangian Eulerian) Interface-Tracking framework. The interface is represented by a boundary-fitted surface mesh holding surfactant molar mass which is subject to interfacial transport. The boundary mesh is advected in a Lagrangian manner under the enforcement of appropriate jump conditions at the interface (15.14), (15.19), whereas the mesh away from the interface is updated through automatic mesh motion with Laplacian smoothing or re-meshing in order to guarantee mesh validity. This methodology allows for very accurate description of interfacial transport processes and calculation of surface tension force. In addition, parasitic currents around the interface are significantly reduced. Moreover, the "force-conservative" approach implies that the condition for zero net surface tension force on closed surfaces is satisfied exactly. The major drawbacks are large computational costs and the limitation to moderately deformed interfaces without changes in mesh topology, if it is not accompanied by an automatic re-meshing technique; cf. [58].

The model equations are discretized by means of a collocated Finite Volume Method (FVM, [68, 69, 98]) for transport processes in the bulk, and a Finite Area Method (FAM, [93]) for transport processes on the interface. Such a practice fulfills the requirements of conservativeness, boundedness, stability and consistency of the numerical method [22]. The description of the transport of surfactant in the bulk and on the interface is included in the model. To account for sorption processes, several sorption models are included. The method is implemented in OpenFOAM, an open source object-oriented C++ library for numerical simulation of Computational Continuum Mechanics and Computational Fluid Dynamics. For a detailed description of the overall numerical methodology we refer to [70].

The central part of the model is the surfactant transport in the bulk and on the interface. In this section, the attention is focused on the coupling of these equations via the different sorption processes. The overall solution algorithm is depicted in Fig. 15.11.

The constitutive modeling of sorption processes has been introduced in Sect. 15.3.1 and it differs considerably for fast and slow sorption. Common to both types is the evaluation of the surface tension as a function of the concentration of the adsorbed surfactant.

- For diffusion controlled (fast) sorption processes, the ad- and desorption rates are locally equilibrated, as stated in Eq. (15.11). This equality leads to an additional local relationship between c^Σ and $c_{|\Sigma}$, say

$$c^\Sigma = f(c_{|\Sigma}). \tag{15.29}$$

Various sorption models $f(c_{|\Sigma})$ are available to describe this relation; see [70] for the full range of models implemented. From the sorption model, the value of $c_{|\Sigma} = f^{-1}(c^\Sigma)$ is taken as a Dirichlet boundary condition for the surfactant bulk equation (15.16). After the solution of the bulk equation with this Dirichlet data, the source term for the surface concentration equation is computed as stated in (15.10), i.e.

$$s^\Sigma = \mathbf{j} \cdot \mathbf{n}_\Sigma = -D(\mathbf{n} \cdot \nabla c)_\Sigma =: s_{\text{fast}}^\Sigma. \tag{15.30}$$

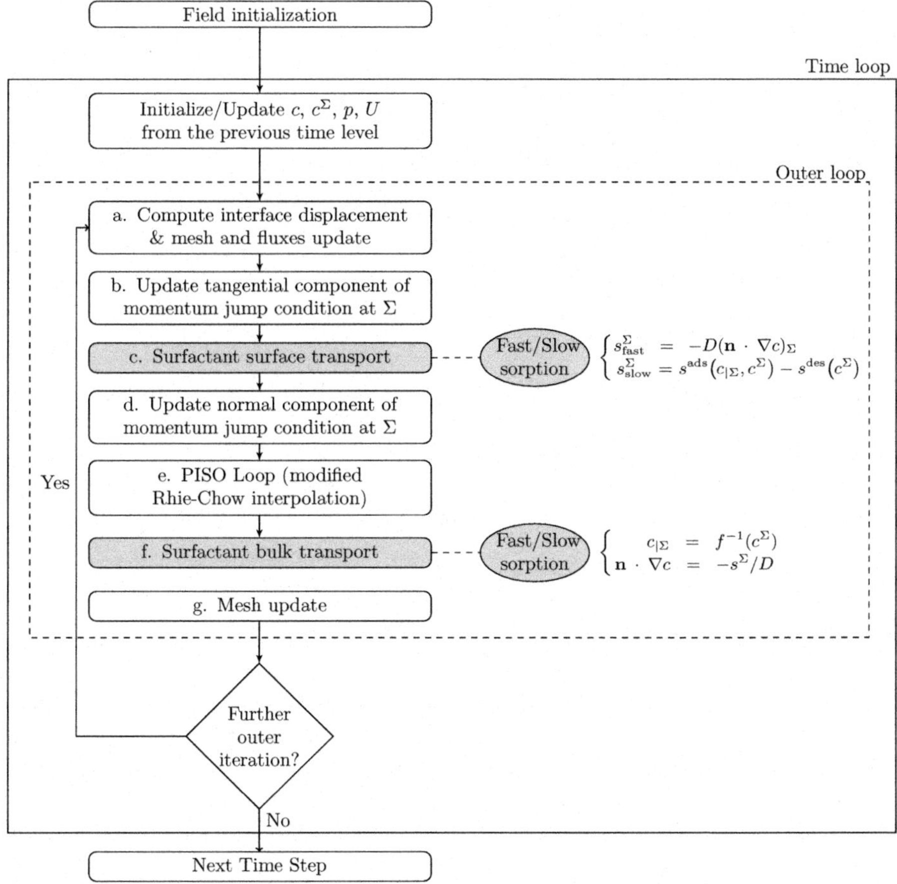

Fig. 15.11 Algorithm overview

In this way, the Neumann data from the new bulk field is transferred to the interfacial molar mass balance. Then the surface transport equation (15.20) is solved to obtain the new surface concentration field of the surfactant species.

- For kinetically controlled (slow) sorption processes, Henry and Langmuir models are available as in [70]. The source term s^Σ is given directly by the adopted sorption model, as a balance between adsorption and desorption rates, cf. (15.12), i.e.

$$s^\Sigma = s^{\text{ads}}\left(c_{|\Sigma}, c^\Sigma\right) - s^{\text{des}}\left(c^\Sigma\right) =: s^\Sigma_{\text{slow}}. \tag{15.31}$$

In this case, the coupling between the interface and the bulk flow is achieved by enforcing the discretized boundary condition (15.21) as a Neumann boundary condition for the bulk equation, i.e.

$$\mathbf{n} \cdot \nabla c = -\frac{s^{\Sigma}}{D} \quad \text{at } \Sigma. \tag{15.32}$$

After solving the interfacial and bulk surfactant transport equations, the surface tension $\sigma = \sigma(c^{\Sigma})$ is updated according to the sorption model chosen.

15.4 Validation

The code used for the simulations is based on the original interface tracking by Tuković and Jasak [94]. The validation and verification of this base code can be found in the PhD thesis by Tuković, [92]. Since the original methodology was improved and extended, a further validation for the modified parts was necessary. In the following, a brief overview of the updated validation is reported.

15.4.1 *Hydrodynamics of Rising Bubbles*

After revising the boundary conditions at the interface in case of two-phase flows and including necessary corrections in the Finite Area methodology,[1] a further validation of the hydrodynamics for rising bubble was necessary. A careful literature survey, looking for the most reliable data in case of super purified water, suggested to compare the simulation results to the experimental ones provided by Duineveld in [24].

The parameters needed to set up the simulations are the fluid properties of the liquid and the gas phases at a given temperature. From the experiment, the working temperature is $T = 293$ K; the liquid phase is pure water with $\rho_A = 998.3 \,\text{kg/m}^3$, $\mu_A = 1 \cdot 10^{-3} \,\text{kg/(ms)}$; the gas phase is air with $\rho_B = 1.205 \,\text{kg/m}^3$, $\mu_B = 1.82 \cdot 10^{-5} \,\text{kg/(ms)}$. The surface tension between pure water and air at the given temperature is $\sigma_0 = 0.0727 \,\text{N/m}$. The domain is divided into two sub-domains, one representing the gas phase and the other one representing the liquid phase. The two sub-domains are coupled at the interface. The meshes used for the simulation consist of polyhedral cells in the gas phase and prismatic cells with polyhedral base in the bulk phase, as can be seen from Fig. 15.12. The interface consists of polyhedral faces with edge length around 50 μm and boundary layer cell thickness around 30 μm. This mesh resolution is a compromise between computational time requirements

---

[1]Correction for the surface gradient computation (*GaussFaGrad*) within the Finite Area method; included the corrected Rhie-Chow interpolation from Tuković et al. [94]; flux correction on the free surface to satisfy the DGCL on the surface, too.

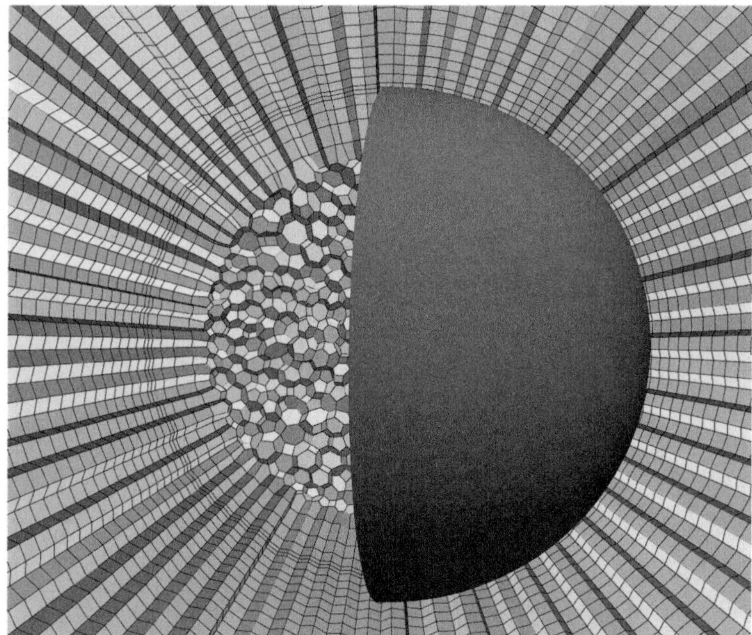

Fig. 15.12 Example of domain discretization for the 3D validation cases

and accuracy of the results; moreover, it has to be underlined that such accuracy is enough to get a correct value of the terminal rise velocity, but not an accurate bubble path, as reported also in [94].

The comparison with respect to the rising velocities, Fig. 15.13 (squares), and the aspect ratios, Fig. 15.13 (circles), for different bubble sizes is reported below. These two graphs show that there is a very good agreement between experimental and simulation results. The small deviation from the experimental results in terms of aspect ratio for the bigger bubbles can be addressed to the relatively coarse mesh resolution.

15.4.2 Surface Transport

An extensive validation of the surface transport is reported in the paper by Dieter-Kissling et al. [22], where a variety of test-cases is set out in detail and compared against the respective analytical solutions, which are not reproduced here for brevity.

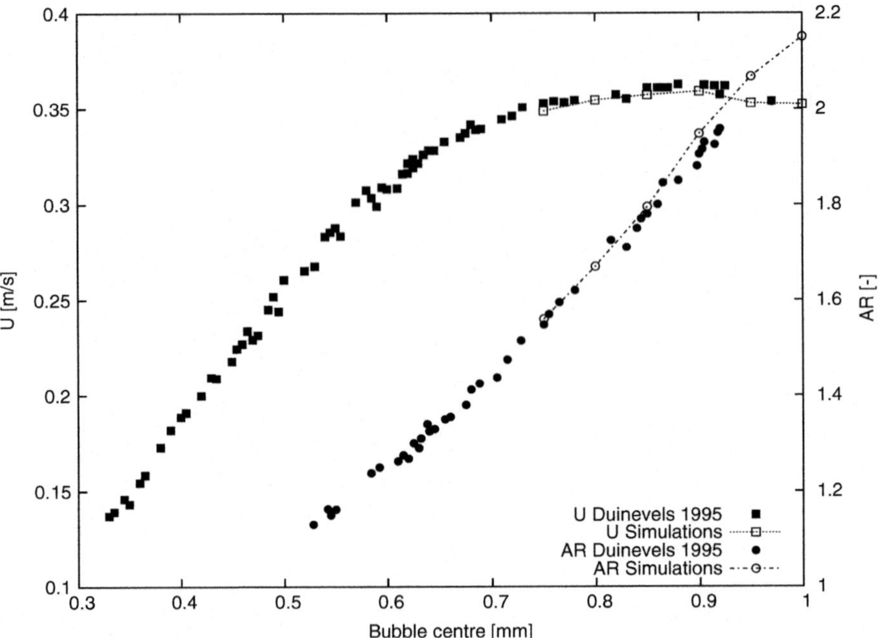

Fig. 15.13 Simulation results compared to experimental ones from Duineveld [24], terminal rise velocity and aspect ratio

15.4.3 Sorption Processes

For validation regarding fast and slow sorption processes, a simplified test case is considered, cf. [62], but here not only the slow sorption case is considered, but also the fast one and the exact analytical solution is derived. The test case, in analogy with a typical transient heat conduction problem [16], consists of a spherical domain of radius r_0 where the surfactant is only transported by diffusion in the bulk. At the initial time, the surface of the sphere is clean. The surfactant can be adsorbed either via a *fast* or a *slow* mechanism. For both cases, an analytical solution for $c(t, r)$ and $c^\Sigma(t)$ can be derived. Due to space limitations the derivation of the analytical solutions, the test case set-up and the full comparison between the analytical and simulation results are not reported here, but can be found in [71]. Only the results of the comparison between the analytical solution and the numerical results for the surfactant bulk concentration are reported below.

Figure 15.14 shows the comparison between the analytical solution (solid lines) and the simulation results (markers) at different times for the slow sorption case. The concentration in the bulk is progressively decreasing with time; the surfactant leaving the bulk phase is then adsorbed on the interface. There is a very good agreement between the analytical solution and the simulation results. The results for the fast sorption validation case are represented in Fig. 15.15. The initial distribution of the surfactant in the bulk phase here is linear, the surfactant is then adsorbed

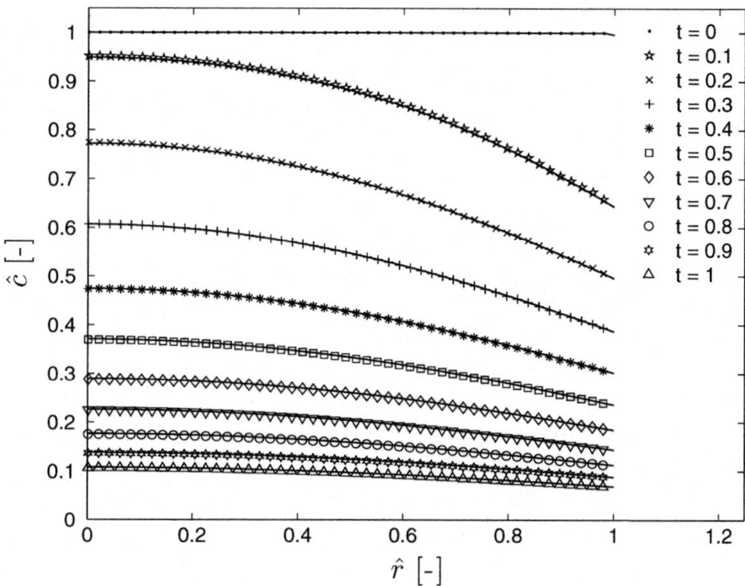

Fig. 15.14 Profiles of normalized bulk concentration $\hat{c}(\hat{r}, \hat{t})$ in case of slow sorption, the *continuous lines* represent the analytical solution

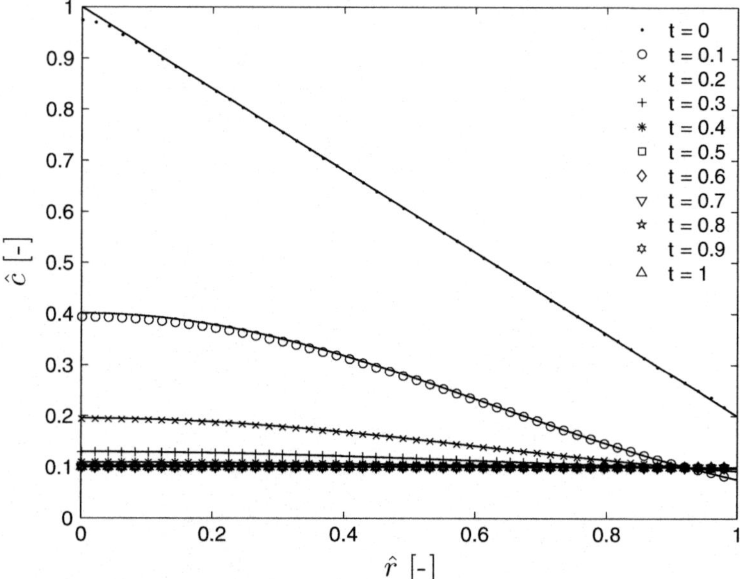

Fig. 15.15 Profiles of normalized bulk concentration $\hat{c}(\hat{r}, \hat{t})$ in case of fast sorption, the *continuous lines* represent the analytical solution

on the interface, the bulk concentration decreases accordingly, until it reaches an equilibrium value. As can be seen from the figure, also for the fast sorption case, there is a very good agreement between the analytical solution and the simulation results.

15.5 Applications/Results

15.5.1 Growing Droplet

In the following, the latest experimental results are reported in terms of integral quantities such as droplet volume and mean surface tension. The simulation results are compared with the experimental ones in terms of droplet shapes and volumes, both for clean and contaminated cases. At the end of the section, the numerical results are presented for growing droplets under the effect of surfactants, showing local surfactant concentrations and surface tension.

15.5.1.1 Set-Up

The experimental set-up has been described in detail in Sect. 15.2.2. Here, the emphasis is put on the simulation set-up. A schematic representation of the case set-up is given in Fig. 15.16a. The orifice is represented by a cylindrical tube. At the inlet an inflow boundary condition with a parabolic velocity profile is applied, where the flow rates from the experiments are known. The free surface is pinned at

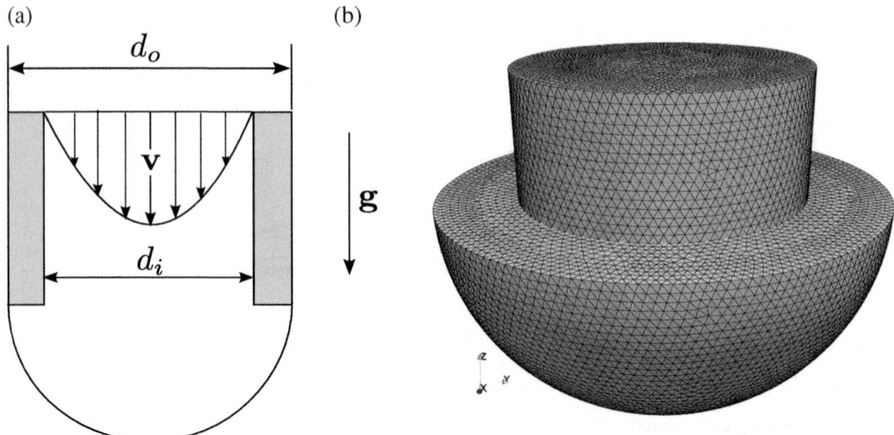

Fig. 15.16 Set-up for the simulation of droplet formation processes. (**a**) Schematic representation of the geometry. (**b**) Tetrahedral mesh

the outer edge of the capillary, as it has been observed in the experiments; see [20]. The hydrodynamics is described by (15.14) and (15.15), completed by appropriate initial and boundary conditions at the free surface boundary; see (15.17)–(15.19) for a free surface flow problem. In the first set of experiments and simulations, a tube with an inner diameter of $d_i = 0.45\,\text{mm}$ and an outer diameter of $d_o = 0.7\,\text{mm}$ has been used. A relatively high inflow rate is applied in order to study the dynamics of the droplet formation process. Here, the liquid phase consists of pure water, the contaminated case is addressed below. The physical properties of the clean liquid phase are $\rho = 1000\,\text{kg/m}^3$, $\mu = 1 \cdot 10^{-3}\,\text{kg/ms}$, $\sigma_0 = 0.072\,\text{N/m}$.

The spatial domain is discretized with tetrahedral cells with average edge length of $4 \cdot 10^{-5}$ m, Fig. 15.16b shows the mesh used for the simulations. The free surface, represented by a surface mesh, is deformable. Automatic mesh motion and dynamic re-meshing allow the mesh to follow the free surface motion, maintaining a sufficient quality of the computational cells. This set-up is addressed in Sect. 15.5.1.2 as the "first set-up".

A second experimental set-up ("second set-up") is considered, where a different capillary has been adopted. The tube has an inner diameter of $d_i = 1.0\,\text{mm}$ and an outer diameter of $d_o = 1.5\,\text{mm}$. In these experiments much smaller inflow rates are applied, such that the drops can be assumed to have Laplacian shapes, although they grow, and to allow focusing on the surfactant effects. Two different liquid phases are considered: pure water for the clean case and water with surfactant ($C_{14}TAB$) for the contaminated one. A new set of experiments with a higher inflow rate (but small enough for the applicability of the GLE), $\dot{V} = 7.85\,\text{mm}^3/\text{s}$ has been performed. To facilitate the comparison with the numerical results, the $C_{12}DMPO$ has been used. In fact the surfactant considered is non-ionic and hence allows to disregard ionic effects that are not considered in the numerical model. The physical properties of the liquid phase are the one at the given temperature of 296 K, i.e. $\rho = 997.3\,\text{kg/m}^3$, $\mu = 9.3 \cdot 10^{-4}\,\text{kg/(ms)}$ and $\sigma_0 = 0.0724\,\text{N/m}$.

15.5.1.2   Insights from the Experiments

First Set-Up: Addressing PAT Applicability

The first set-up has been already tested and compared to experiments in [20]. Here we only recall a few main findings. Most of the attention was given to the study of applicability of Drop Profile Analysis Tensiometry (PAT) under dynamic conditions.

Within the PAT procedure, the surface tension is evaluated by fitting to the Gauss Laplace Equation, which is based on the hypothesis of rotational symmetry of the pressure profile inside the droplet. Such study was fundamental to address the validity of the PAT results for dynamic droplet formations. Simulation results of droplet formation at relatively high inflow rates were compared to the experimental ones; see [20]. The main findings from this comparison are reported here for completeness.

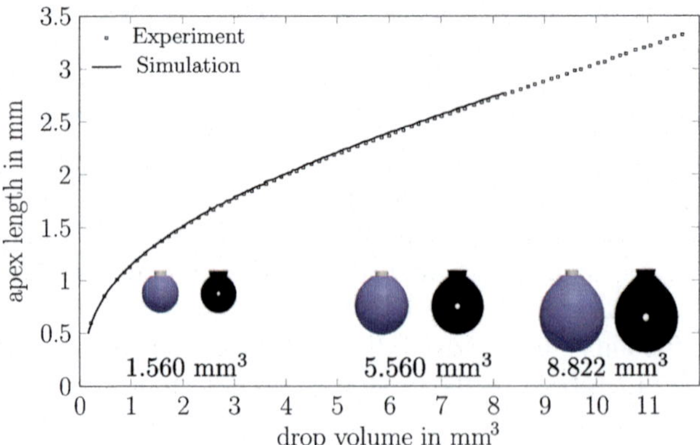

Fig. 15.17 Droplet formation at $\dot{V} = 10\,\mathrm{mm^3/s}$. Comparison of numerical (*solid line*) and experimental profiles (*squares*) at characteristic drop volumes. Republished with permission of Elsevier from [20]; permission conveyed through Copyright Clearance Center, Inc.

Three different flow rates have been considered, $\dot{V} = 10$, 25 and $42\,\mathrm{mm^3/s}$. The droplets have been analyzed regarding the drop shape, in particular comparing the experimentally obtained displacement of the apex of a growing drop at different flow rates to the corresponding one from the simulations. As shown in [20], the results are in very good agreement in a broad range of flow rates. As an example of this comparison, Fig. 15.17 is reported here from [20]. In the case with $\dot{V} = 10\,\mathrm{mm^3/s}$, the numerical results match the experimental ones very accurately. Such a good agreement also serves as a validation for both the numerical and the experimental side.

It is interesting to note that, for higher flow rates, already at an early stage of the formation process the flow cannot be considered rotational symmetric anymore, as visible in Fig. 15.18 which is reproduced from [20]. In fact, for $\dot{V} = 42\,\mathrm{mm^3/s}$ the inflow jet is not fully dissipated before approaching the droplet's surface, resulting in a non-uniform pressure distribution. The assumption of rotational symmetry for the evaluations of the surface tension by fitting to a Gauss Laplace profile becomes inaccurate. Therefore, it can be stated that the fitting to Gauss Laplace profiles is only valid for static droplets, generated at a low flow rate, here $\dot{V} = 10\,\mathrm{mm^3/s}$.

A quantitative evaluation of these results has been made considering the product of the local curvature and the surface tension with respect to the polar angle; see Fig. 15.19 which is reproduced from [20]. The curvature at the bottom of the drop is increased and the shape is no longer spherical. Moreover, the interfacial pressure profiles calculated from the local curvature (capillary pressure) and computed from the simulation results (one sided pressure limit) are similar, but the simulations provide smaller absolute values. This difference is attributed to shear effects at the bottom of the droplet. Summarizing, it can be stated that the standard GLE-based analysis cannot work under dynamic conditions without a correction of the Gauss Laplace Equation accounting for inertia and shear effects at high flow rates.

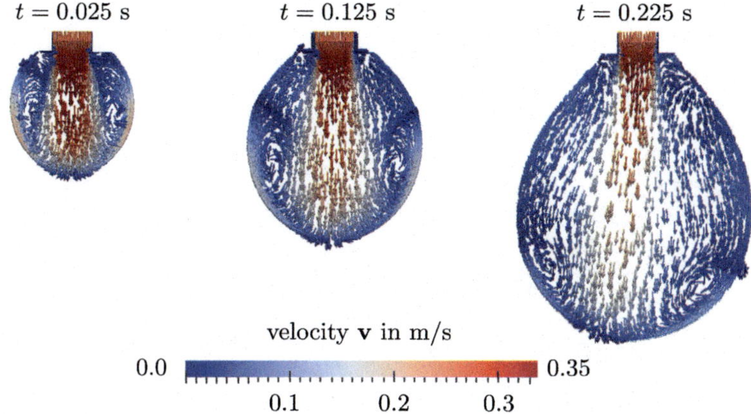

Fig. 15.18 Flow field inside the droplet during the formation process. The vectors are *colored* according to the velocity magnitude. Republished with permission of Elsevier from [20]; permission conveyed through Copyright Clearance Center, Inc.

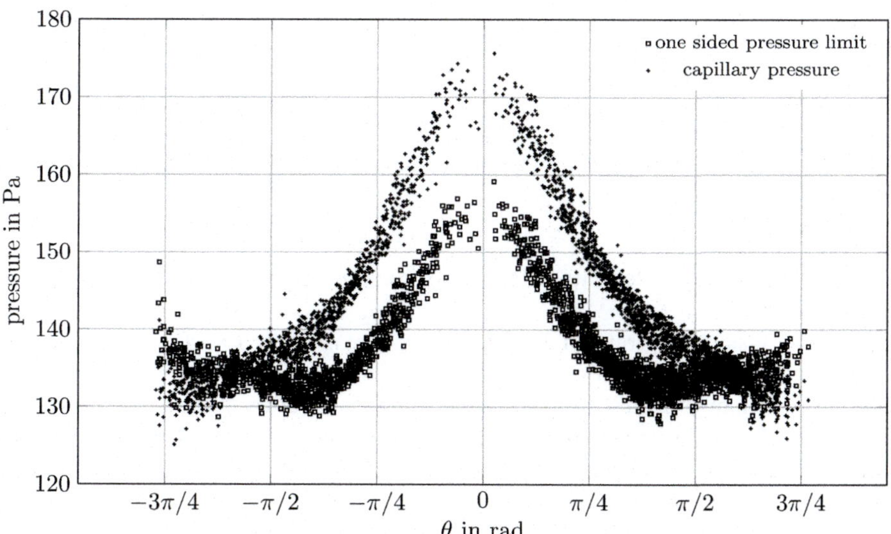

Fig. 15.19 Pressure distribution at the apex of the droplet at a flow rate of $\dot{V} = 42\,\mathrm{mm}^3/\mathrm{s}$. Republished with permission of Elsevier from [20]; permission conveyed through Copyright Clearance Center, Inc.

Second Set-Up: Low Inflow Rate

Experimental data obtained for two aqueous solutions of $C_{14}TAB$ as measured by PAT in the mode of growing drops are shown in Figs. 15.20 and 15.22, while the measured change of the drop volume corresponding to Fig. 15.20 is shown in

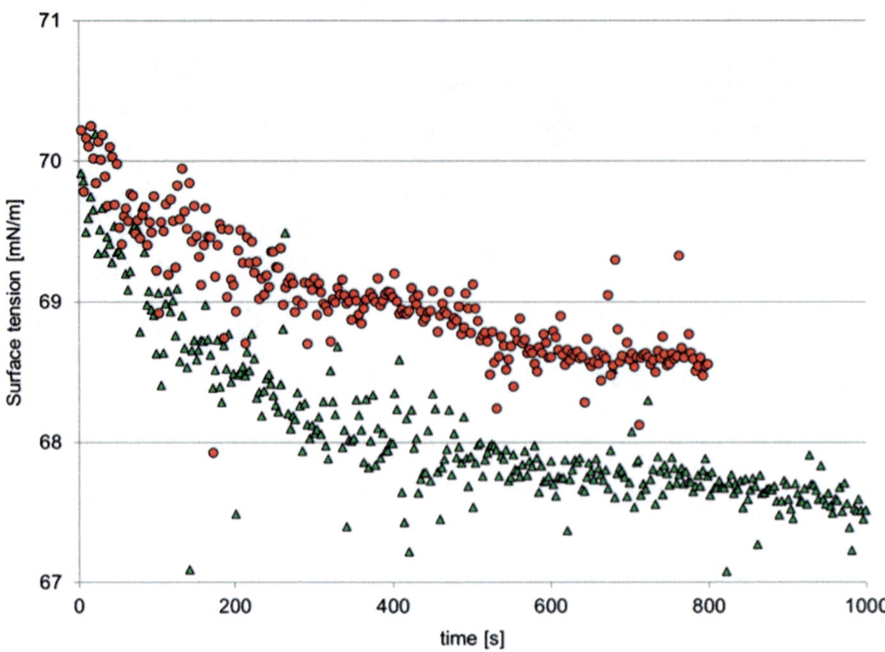

Fig. 15.20 Dynamic interfacial tension for an aqueous $3 \cdot 10^{-4}$ mol/l C_{14}TAB solution measured with PAT using the growing drop mode; flow rates: *circles*—0.025 mm^3/s and *triangles*—0.0125 mm^3/s

Fig. 15.21. The two sets of data in Fig. 15.20 correspond to two different liquid flow rates.

For the faster flow rate of 0.025 mm^3/s, the measured surface tensions are significantly higher than those obtained for the lower flow rate. This is due to the continuous expansion of the drop surface leading to a dilution of the adsorption layer and, hence, higher surface tension. The evolution of the dynamic surface tension is due to a competition between adsorption of the surfactant molecules with time (leads to a decrease in σ) and an interfacial expansion due to the growth of the drop (leads to an increase in σ). For the higher C_{14}TAB solution the measured dynamic surface tensions are shown in Fig. 15.22, where the absolute changes within the same time range are less. This is due to the larger adsorption rate for this surfactant solution with a three times higher bulk concentration.

Note that it is assumed for these experiments that the measured profiles and, consequently, the determined surface tensions are physically correct, i.e. the liquid inflow into the drops during the measurements do not influence the drop profiles. In [43], as also explained above, it was demonstrated that at much higher drop growth rates the resulting drop profiles are of dynamic character and therefore the use of the GLE for fitting these profiles did not lead to physically correct surface tension values.

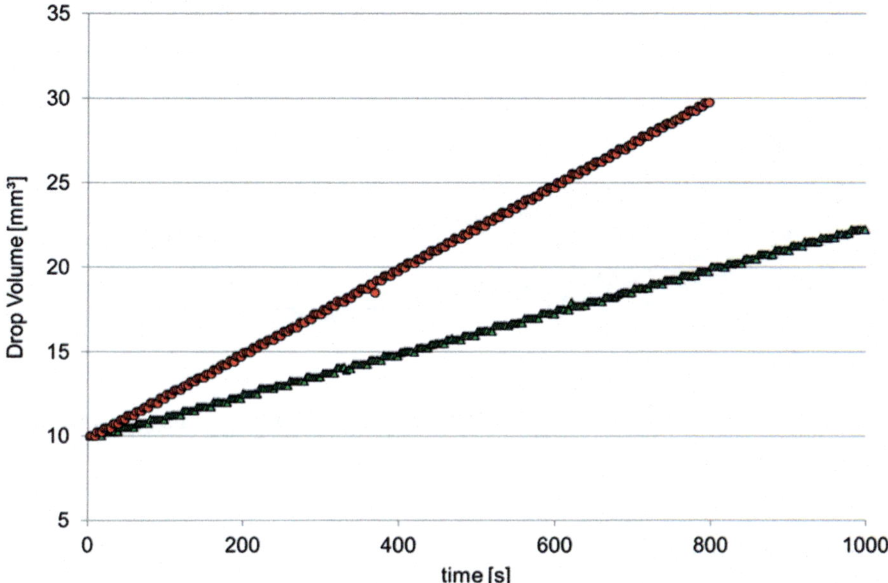

Fig. 15.21 Change of the drop volume with time measured with PAT using the growing drop mode; flow rates: *circles*—0.025 mm^3/s and *triangles*—0.0125 mm^3/s

Second Set-Up: Higher Inflow Rate

As anticipated in the case set-up, to facilitate the comparison with the numerical results another set of experiments was performed with a higher inflow rate and C_{12}DMPO as surfactant. The concentration of surfactant in the solution is $7.5 \cdot 10^{-5}$ mol/l, the initial volume of the droplet is $V = 2\,\text{mm}^3$. The experimental results are presented in Figs. 15.23 and 15.24. Since the inflow rate is constant and the same for the clean and the contaminated case, no difference in the droplet volumes at the same time instances is to be expected. Such behavior is confirmed by the experimental data (triangles) in Fig. 15.23. On the other hand, referring to the mean surface tension values (circles) in Fig. 15.23, the expected reduction in surface tension due to the presence of surfactant can be observed. In Fig. 15.24 the droplet shapes corresponding to four of the first time instances are represented and then used below for comparison with the simulations. As Fig. 15.24 shows, there is not a sensible difference appreciable with the naked eye between the clean and the contaminated cases. At this early stage of droplet formation, only the mean surface tension values can give information about the presence of surfactant in the solution.

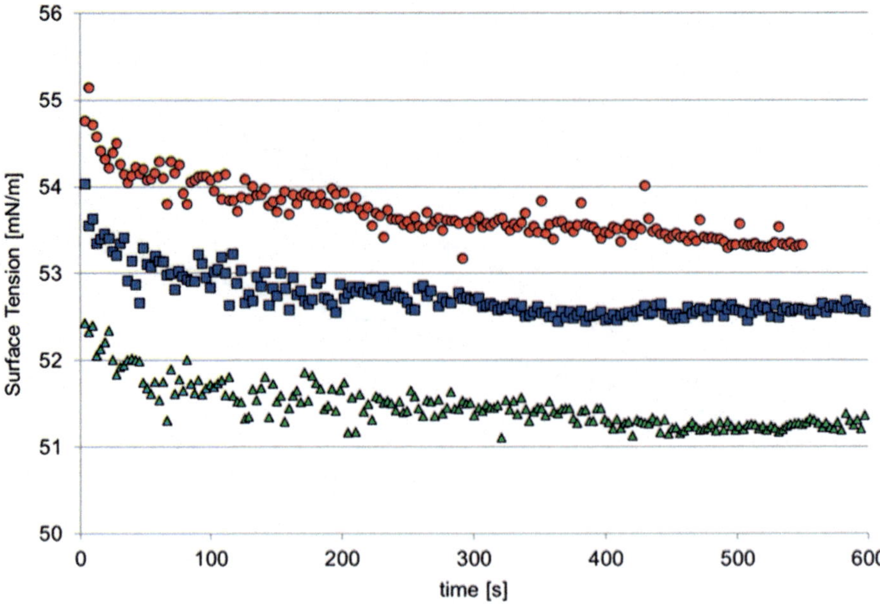

Fig. 15.22 Dynamic interfacial tension for an aqueous $1 \cdot 10^{-3}$ mol/l C$_{14}$TAB solution measured with PAT using the growing drop mode; flow rates: *circles*—0.025 mm^3/s, *squares*—0.01875 mm^3/s and *triangles*—0.0125 mm^3/s

15.5.1.3   Comparison: Experiments and Simulations

In this section the shapes of the clean and contaminated droplets obtained with the simulations are compared to the experimental ones. In Fig. 15.25, the gray points on the left correspond to the clean experimental data, the black points on the right come from the experiment with surfactant. The continuous lines are the simulation results for the clean (gray) and contaminated (black) cases where, for each time step, the droplets have the same volume. First of all, from Fig. 15.25 it can be said that the experimental and numerical results are in very good agreement. However, it is more important to notice that via the simulations it is possible to show the small change in droplet shapes for equivalent volumes resulting from the surfactant contamination. A close look to the simulation results reveals that the contaminated drops have slightly more elongated shapes, while the clean ones show more spherical shapes next to the capillary. These differences in the shapes are not directly visible from the experiments, but, as said above, these small differences actually determine the change in surface tension, thus explaining the difficulties of the method. This first comparison has shown the limitations of the experimental methodology used above when considering short physical times required by the simulations. For future comparisons, it will be necessary to review either the experimental methodology

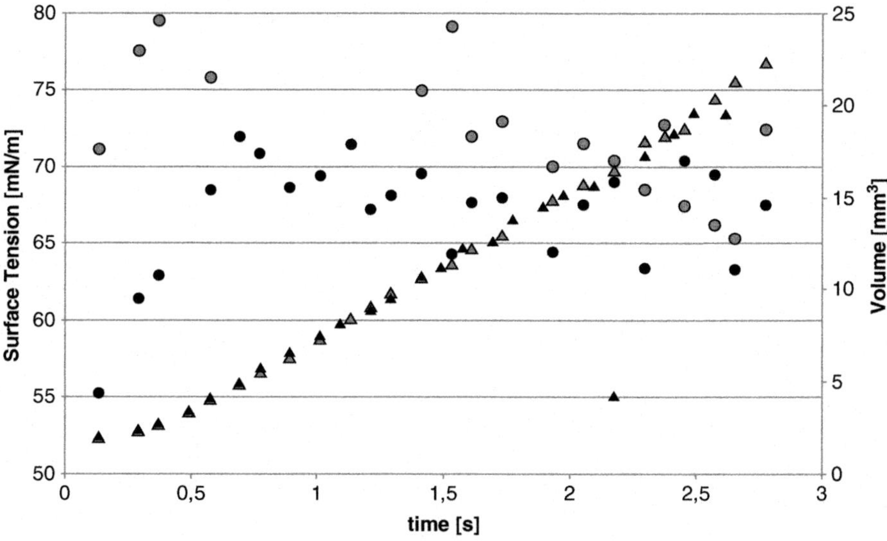

Fig. 15.23 Mean surface tension—*circles*, and droplet volume—*triangles*; clean (*gray*) and contaminated (*black*) cases

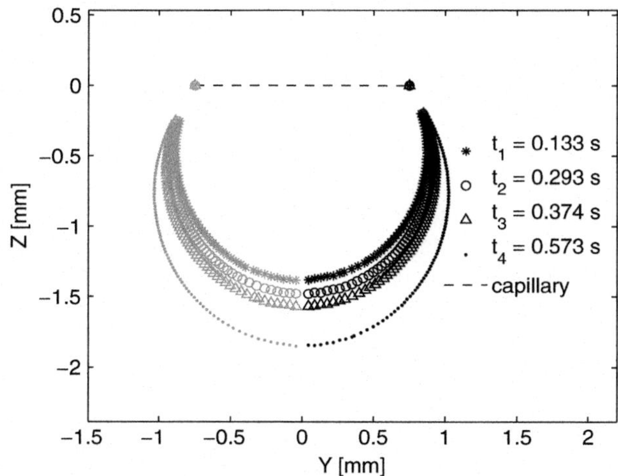

Fig. 15.24 Droplet shapes, clean (*gray*) and contaminated (*black*) cases

or the set-up, to find a compromise between the physical time to simulate and the capability to show interesting surfactant effects.

As a first step to compare numerical and experimental results for the cases contaminated by surfactants, a simplified model for the surfactant has been used: the surface tension has been modified uniformly according to the value given by the Langmuir surface tension equation of state with the given surfactant concentration.

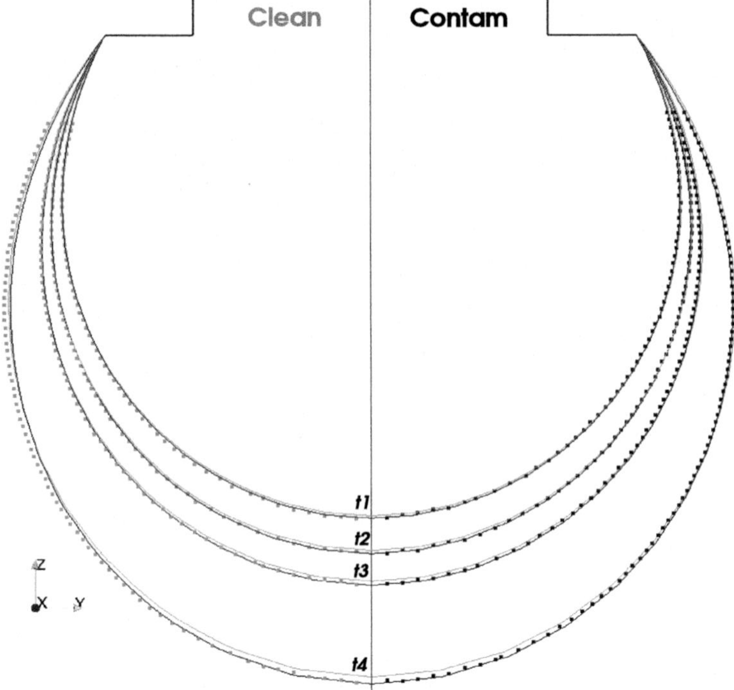

Fig. 15.25 Droplet shapes for the clean and contaminated cases, simulation (*solid line*) and experiment (*dots*)

While the outcome of the comparison with the experimental results is satisfying, the next step is to solve the full surfactant transport both in the bulk and on the interface. The solution of the full set of equations should provide an even better match with the experiments.

15.5.1.4 Insights from the Simulation

The surfactant transport in growing droplets has already been tackled within this project and the results for a first test case can be found in [21]. In this paper, the first set-up described above is used and different sorption processes and their influence on the droplet formation process are compared. In detail, Henry and Langmuir sorption processes are considered for a single surfactant species, while the Langmuir sorption is applied to a surfactant mixture, too. Here, the results for the Langmuir model are reported as an exemplifying case of the capability of the solver. The interested reader is referred to [21] for the case set-up and the full report about the results.

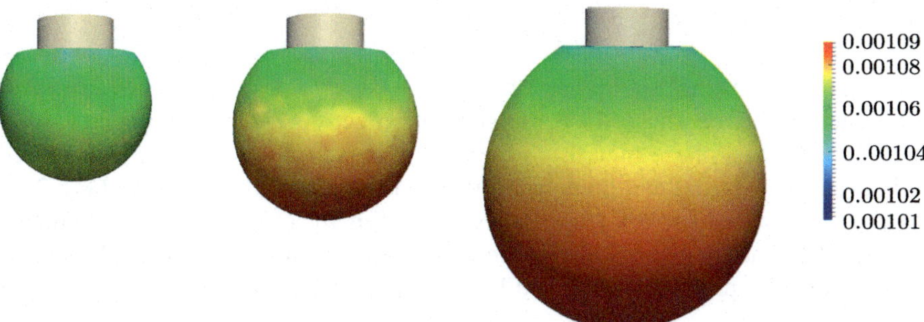

Fig. 15.26 Concentration profile (mol/m^2) during a droplet formation process assuming Langmuir sorption at $t = 0$, 0.03, 0.11 s. Republished with permission of Elsevier, from [21]; permission conveyed through Copyright Clearance Center, Inc.

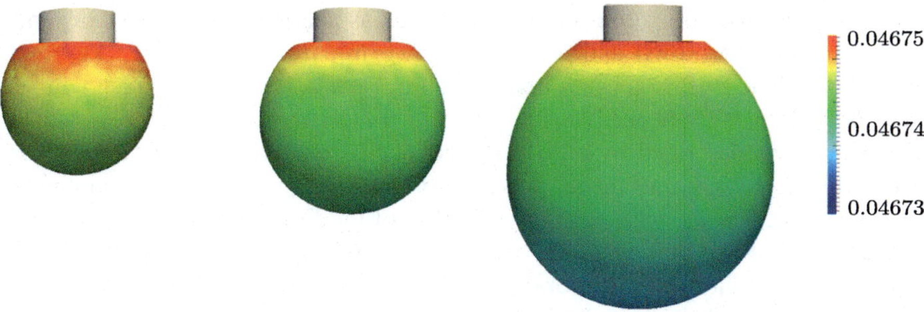

Fig. 15.27 Surface tension profile (N/m) during a droplet formation process assuming Langmuir sorption at $t = 0$, 0.03, 0.11 s. Republished with permission of Elsevier, from [21]; permission conveyed through Copyright Clearance Center, Inc.

As can be seen from Figs. 15.26 and 15.27 from [21], local data as surfactant concentration and surface tension can be easily accessed by the simulations. From the two figures, however, it can also be noticed that the surface fields are not very smooth. One main reason is the numerical treatment of the contact line with variable contact angle which has severe impact onto the local curvature computation.

15.5.2   Rising Bubble

The second prototypical problem considered is a rising bubble in aqueous solution contaminated by surfactants. In this section the simulation results are compared with the experimental ones in terms of rise velocities for different surfactant initial concentrations.

15.5.2.1 Set-Up

While the experimental set-up has already been described in detail in Sect. 15.2.3, the attention will be focused on the simulation set-up and some related issues. In this case a two-phase model is necessary and Eqs. (15.14)–(15.21) are solved with appropriate initial and boundary conditions, i.e. (15.22)–(15.24) and (15.27)–(15.28). The surfactant diffusive fluxes are given by (15.25) and (15.26a), the change in surface tension follows the interfacial equation of state (15.13) and the fast sorption process is modeled by (15.29) and (15.21) or in the discretized form (15.30).

A single air bubble of diameter $d_B = 1.45\,\mathrm{mm}$ rising in water is considered. Experimental data are collected both in case of pure water and contaminated water. As surfactant, the nonionic dodecyl-dimethyl-phosphine-oxide (C_{12}DMPO) is used, considering different initial uniform concentrations. The working temperature is 296 K. The physical properties of the liquid and gas phases are $\rho_A = 997.3\,\mathrm{kg/m^3}$, $\mu_A = 9.3 \cdot 10^{-4}\,\mathrm{kg/(ms)}$ and $\rho_B = 1.1965\,\mathrm{kg/m^3}$, $\mu_B = 1.83 \cdot 10^{-5}\,\mathrm{kg/(ms)}$, respectively. The surface tension between pure water and air at the given temperature is $\sigma_0 = 0.0724\,\mathrm{N/m}$. It is known from the experiments that the surfactant effects are well described by a diffusion-controlled model, that is a fast sorption model. The *fast* Langmuir model is employed for the simulation. It is tempting to view the Langmuir model as a simplified case of the Frumkin model presented in Sect. 15.2.1 with the parameter $a = 0$ and the Langmuir constant $a_L = 1/b$. But note that, even if the Langmuir constant a_L and the Frumkin parameter $1/b$ have the same physical meaning, they always have different values in order for the Langmuir model to compensate for $a = 0$ instead of $a = 0.4$ as mentioned for the C_{12}DMPO in Sect. 15.2.1. The value of a_L for the surfactant considered here is taken from [27], where, for the Langmuir model, $b = 206\,\mathrm{m^3/mol}$.

The adsorption isotherm and the surface tension equation of state for the Langmuir model are given as

$$c^{\Sigma} = c_{\infty}^{\Sigma} \frac{c/a_L}{1 + c/a_L}, \tag{15.33}$$

$$\sigma = \sigma_0 + RTc_{\infty}^{\Sigma}\ln\left(1 - \frac{c^{\Sigma}}{c_{\infty}^{\Sigma}}\right). \tag{15.34}$$

The parameters to model the surfactant are

- the diffusion coefficients in the bulk phase, $D = 5 \cdot 10^{-10}\,\mathrm{m^2/s}$, and on the interface, $D^{\Sigma} = 5 \cdot 10^{-10}\,\mathrm{m^2/s}$. Note that the diffusivity on the interface is not known from experiments, thus only an estimate can be employed;
- the Langmuir equilibrium constant, $a_L = 4.85 \cdot 10^{-3}\,\mathrm{mol/m^3}$;
- the saturated surfactant concentration on the interface, $c_{\infty}^{\Sigma} = 4.17 \cdot 10^{-6}\,\mathrm{mol/m^2}$.

In Table 15.4, an overview of the experiments and the performed simulations is given. As can be seen from Table 15.4, not all the cases investigated by the

Table 15.4 Experiments/Simulation matrix and corresponding terminal rise velocities

c_0 (mol/m^3)	Full transient 3D simulation	U_T^{exp} (m/s)	U_T^{sim} (m/s)
0	✓	0.351	0.352
$5 \cdot 10^{-4}$		0.172	–
$1 \cdot 10^{-3}$		0.155	–
$2 \cdot 10^{-3}$		0.152	0.154
$5 \cdot 10^{-3}$		0.154	0.155
$8 \cdot 10^{-3}$		0.158	0.157
$1 \cdot 10^{-2}$		0.165	0.158
$2 \cdot 10^{-2}$		0.161	0.156
$5 \cdot 10^{-2}$	✓	0.160	0.157

experiments can be fully reproduced by the simulation with the current version of the solver. The biggest issue here is the requirement on the mesh resolution due to the presence of a species boundary layer that has to be resolved. In other words, for diffusion-dominated problems, the diffusion coefficient has a direct impact on the mesh resolution requirements: the smaller the diffusivity is, the thinner the species boundary layer around the bubble is and the more difficult to estimate the gradients at the interface is. Under-resolved meshes will produce wrong estimates of the gradients leading to unphysical results. This requirement is imposing a much more restrictive constraint on the mesh resolution than from the hydrodynamics. To properly resolve the diffusion processes in the boundary layer, at least three cells have to be present there. From the correlation for the Sherwood number depending on the Péclet number, one can estimate the boundary layer thickness, as shown by (15.37) below. After computing the Péclet number as

$$\text{Pe} = \frac{d_{\text{B}} U_{\text{T}}}{\text{D}}, \tag{15.35}$$

where d_{B} is the bubble diameter, U_{T} the terminal rise velocity and D the surfactant diffusivity, and the Sherwood number from the correlation by Oellrich et al. [64],

$$Sh = 2 + \frac{0.232 \text{Pe}^{1.72}}{1 + 0.205 \text{Pe}^{1.22}}, \tag{15.36}$$

the species boundary layer thickness can be estimated as

$$\delta_{tot}^{BL} \approx \frac{d_{\text{B}}}{\text{Sh}}. \tag{15.37}$$

This means that for the diffusivity of C_{12}DMPO, the boundary layer thickness is about 1.2 μm, with the first three cells' thickness around 0.4 μm. This would result in a mesh with a huge number of cells, that the solver is currently not capable to handle, even with adaptive grid refinement. To overcome this problem, one of the options is to increase the diffusivity of the transported species in the bulk, but of course this procedure has some drawbacks. First of all, the distribution of the surfactant in the bulk will not be anymore the real one, since altering the diffusivity implies a modification of the physics. In other words, the species boundary layer will be thicker and the surfactant distribution in the bulk will be smoothed. With this in mind, considerations about the surfactant distribution in the bulk can still be made. Secondly, in case of fast sorption where the explicit source term contains directly the bulk diffusion coefficient, increasing the diffusivity will change the source term speeding up the adsorption of the surfactant on the interface at the beginning of the simulation. A sensitivity study for the diffusion coefficient showed that for high initial concentration of surfactant, the resulting bubble rise velocity is still in agreement with the experiments; see Fig. 15.29. This is due to the fact that the concentration of surfactant in the bulk is so high that the bubble surface is quickly fully covered. The bubble will not reach a peak velocity, but it will behave as a rigid surface as already anticipated by Fig. 15.10 and displayed in Fig. 15.29 (contaminated case). At intermediate surfactant concentration and with an increased diffusivity, since the sorption process is accelerated, the bubble surface will be covered by surfactant quicker and the transition from mobile to rigid surface will not be captured by the simulations. In these cases only the steady state solution could be compared, see Table 15.4. The same sensitivity study for the diffusion coefficient has been done for one of these intermediate surfactant concentrations, the results are reported in Fig. 15.30 and they show that with the use of the physical diffusivity, the method is able to catch the correct velocity transient. These results and their relevance for the success of this project will be further explained in Sect. 15.5.2.3.

15.5.2.2 Insights from the Experiments

In Fig. 15.28, the profiles of rising bubbles in aqueous solutions of dodecyl-dimethyl-phosphine-oxide (C_{12}DMPO) are summarized. As expected from the general picture described in Fig. 15.10, for extremely low (pure water) and sufficiently high surfactant concentrations simple velocity profiles are observed, i.e. after a range of acceleration the bubble velocity levels off at a constant terminal velocity. For intermediate concentration we observe the typical course of the velocity changes with a range of acceleration, passing through a velocity maximum, followed by a decrease and a levelling at a respective terminal velocity. This final velocity corresponds to a bubble with a rigid surface layer behavior. The higher the surfactant concentrations are, the earlier the velocity maximum appears. Also the absolute values in the velocity maximum decrease with increasing surfactant concentration. The data in Fig. 15.28 therefore provides an excellent example for the behavior of gas bubbles of a given size (about 1.45 mm in diameter). Note that all velocity values are averaged values from 10 to 20 different bubbles. Due to experimental shortcomings, the size of the individual bubbles varies in the range of 1–2%. Also the determination of the bubble velocity has experimental errors. Hence, each data point has a respective uncertainty; on average the accuracy is $\pm 5\%$.

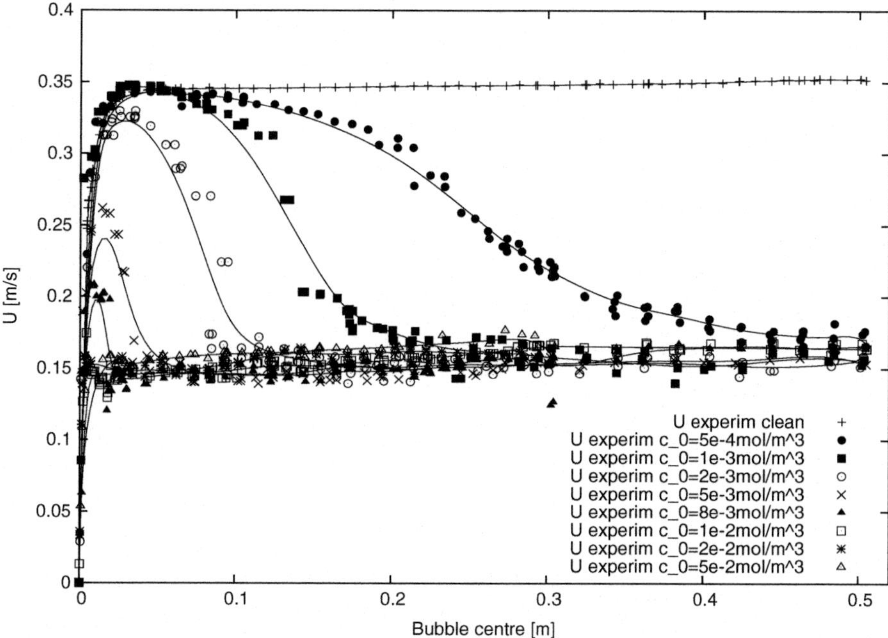

Fig. 15.28 Rising velocity U as a function of the travelled distance (measure for the bubble centre) for different solutions of the nonionic surfactant C_{12}DMPO, *solid line*—fitted curve

15.5.2.3 Comparison: Experiments and Simulations

In Fig. 15.29, the simulation results for the clean and the most contaminated case are compared to the experiments. As can be seen from the graphs, there is a very good agreement between the numerical and experimental results. The agreement for the clean case is confirming that the hydrodynamics of rising bubbles is well captured by the simulations, but it is more interesting to see how the simulations are properly reproducing the transient of the velocity for the contaminated case.

As said before, the increased surfactant bulk diffusivity has an effect on the transient of the rise velocity, thus, in most of the cases, only the terminal rise velocities can be compared. The values of the terminal rise velocities (steady state) resulting from the experiments and from the simulations are compared in Table 15.4. The comparison confirms that the steady state is reached by the simulations and the results obtained by the numerics are in good agreement with the experiments, even if the transient is not properly captured by the simulations within this set-up.

A parameter study on the bulk diffusivity has been conducted and the results for the initial transient of the simulations are shown. For this study, only one surfactant concentration from the experiments has been selected, $c_0 = 5 \cdot 10^{-3}$ mol/m^3 and four different diffusivities have been chosen. For the three smaller diffusivities, a finer mesh has been used.

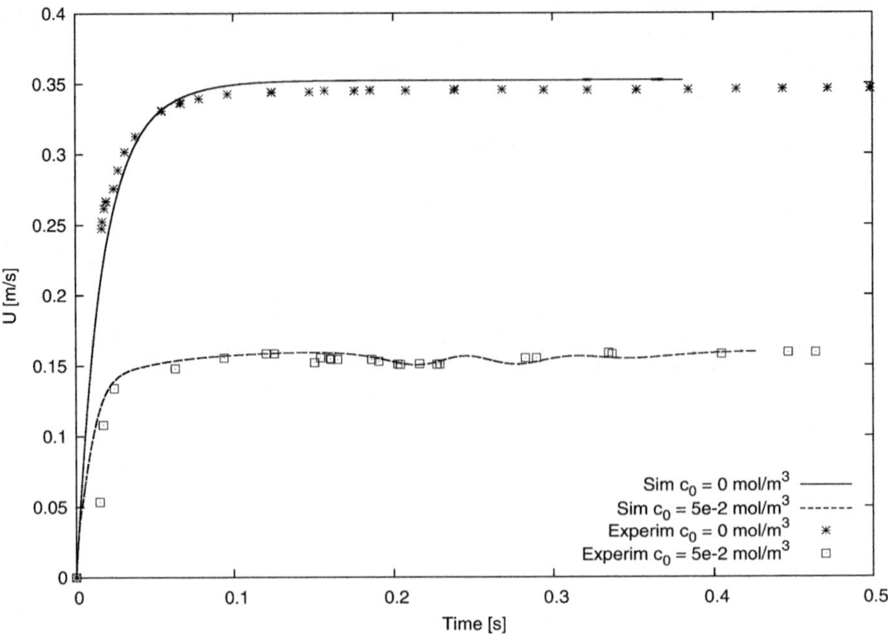

Fig. 15.29 Bubble rise velocities under the effect of surfactant—limiting cases

As can be seen from Fig. 15.30, the smaller the diffusivity is, the better the velocity peak is approached. This trend confirms our assumptions regarding the surfactant transport in the bulk in relation to the mesh resolution and the adsorption process for the intermediate cases. In fact, using the physical diffusivity of the C_{12}DMPO, i.e. $D = 5 \cdot 10^{-10}\,m^2/s$, the correct velocity peak value is reached, followed by a decrease in the rise velocity, as expected from the experiments. These results are very promising and encouraging to solve the mesh resolution issue to get the expected transient behavior of the rising bubble. It is also important to underline that in the cases with $D = 5 \cdot 10^{-9}\,m^2/s$ and $D = 5 \cdot 10^{-10}\,m^2/s$ the mesh was still not fine enough, thus the velocity peak is approached, but the full dynamic cannot be captured because of under-resolution of the mesh and related numerical stability issues. Further improvements of the numerics are in progress, in order to be capable to handle very fine meshes that fully resolve the species (surfactant) boundary layer and to model the species boundary layer via an approximated solution, as done for example by means of sub-grid scale modeling for mass transfer problems, see [13].

15.5.2.4 Insights from the Simulation

From the simulation results local data can be easily accessed. It is well known that the presence of surfactant, changing the interfacial properties, i.e. the surface

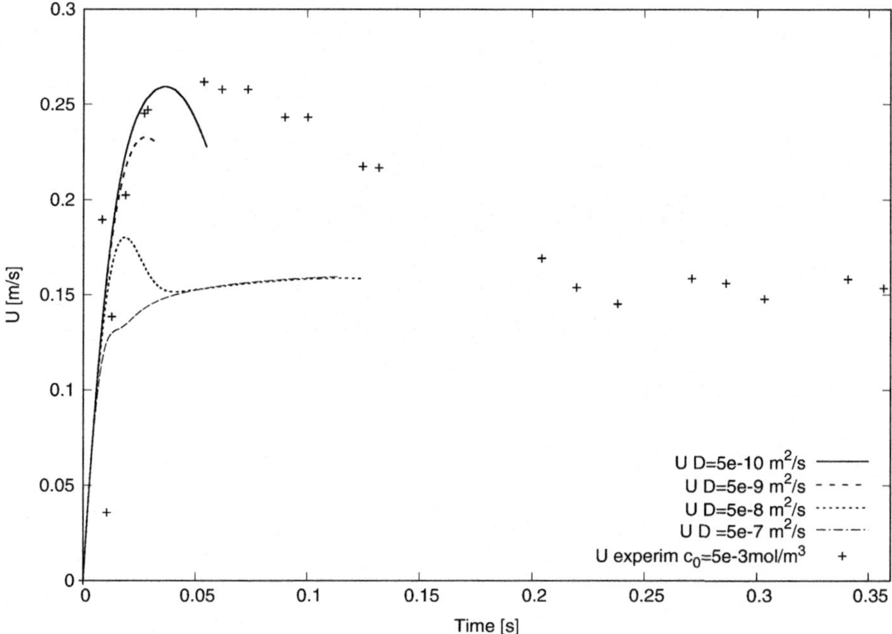

Fig. 15.30 Bubble rise velocities under the effect of surfactant—different diffusivities

tension, has a strong effect on the flow field. Figure 15.32 shows the comparison between the clean and the contaminated case ($c_0 = 5 \cdot 10^{-2}\,\text{mol/m}^3$) in terms of the velocity field. In pure water the bubble is rising straight without recirculation regions in the wake; on the other hand, in contaminated water, the bubble is rising much slower, following a helical path and with vortices in its wake. As expected, the contaminated bubble is also much less deformed; see Figs. 15.31 and 15.32.

Thanks to the Finite Area method and the explicit representation of the deformable interface, it is also possible to visualize the surfactant distribution on the interface and in the bulk as shown in Fig. 15.33. The relative surface tension is accessible, too; see Fig. 15.34. For these cases, the most contaminated bubble is considered ($c_0 = 5 \cdot 10^{-2}\,\text{mol/m}^3$, $D = 5 \cdot 10^{-7}\,/\text{m}^2/\text{s}$). Here it is important to underline that the distribution of surfactant in the liquid phase depends on the diffusivity chosen for the simulation. For these cases the highest diffusivity was chosen in order to have a fully resolved case. But the results shown here are then only valid for the fully contaminated case and only exemplify the data that can be accessed by the simulations. Figure 15.33 shows the surfactant concentration on the interface. As can be seen from the values of c^Σ, the bubble is fully contaminated and the average surface concentration is close to the equilibrium one. In fact, for this case the Langmuir adsorption isotherm gives as equilibrium value for c^Σ the

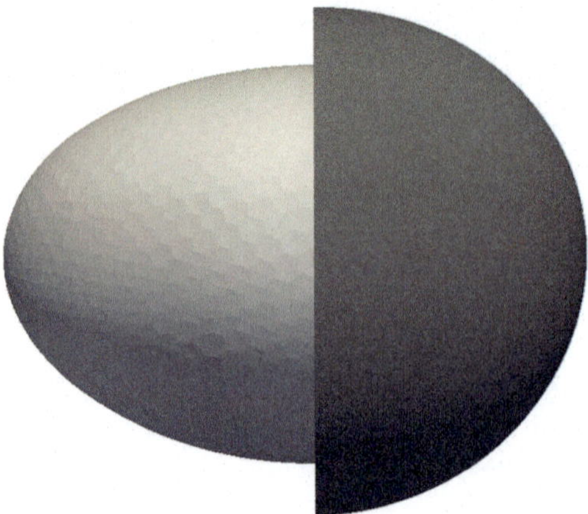

Fig. 15.31 Bubble shapes, *left side* of a clean and *right side* of a surfactant covered bubble, ref. Fig. 15.29, t = 0.338 s

(a)

(b)

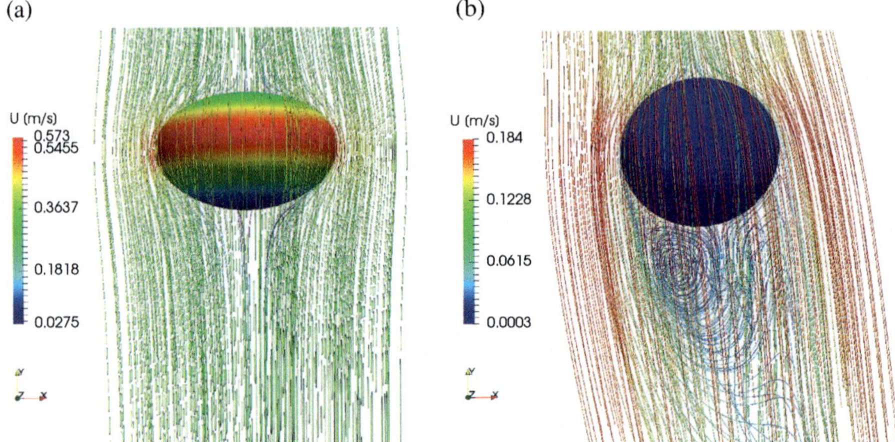

Fig. 15.32 Flow field (relative velocities), ref. Fig. 15.29, t = 0.338 s. (**a**) Clean case. (**b**) Contaminated case

relation

$$c_{eq}^{\Sigma} = c_{\infty}^{\Sigma} \frac{c|_{\Sigma}/a_L}{1 + c|_{\Sigma}/a_L} = 3.8 \cdot 10^{-6} \, /\text{mol}/\text{m}^2 \tag{15.38}$$

with $c|_{\Sigma} = c_0 = 5 \cdot 10^{-2} \, \text{mol}/\text{m}^3$. On the left of Fig. 15.33, the velocity field (magnitude) is reported, where once more the recirculation regions in the bubble wake are visible. On the right of Fig. 15.33 the surfactant distribution in the bulk is shown. In this figure, the regions where ad- and desorption occurs are clearly visible.

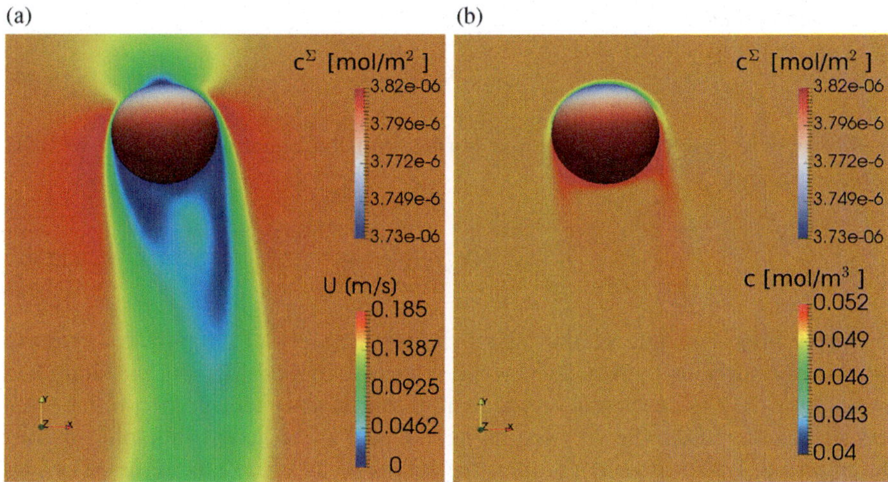

Fig. 15.33 Contaminated case $c_0 = 5 \cdot 10^{-2}$ mol/m^3, t = 0.338 s. (**a**) Surfactant distribution and velocity field. (**b**) Surfactant distribution on Σ and in Ω

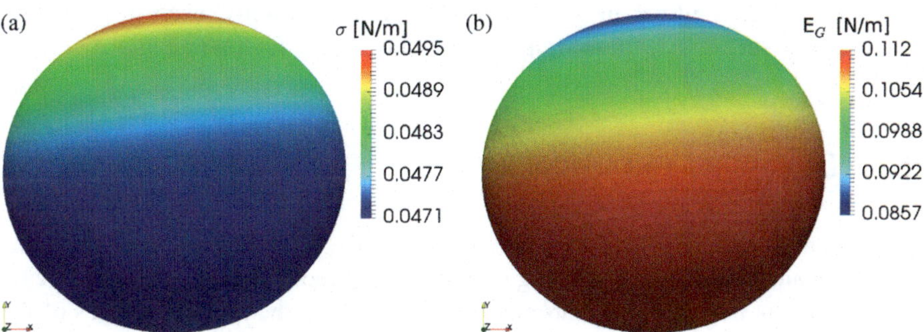

Fig. 15.34 Contaminated case $c_0 = 5 \cdot 10^{-2}$ mol/m^3, t = 0.338 s. (**a**) Surface tension. (**b**) Gibbs' elasticity

In the vicinity of the upper part of the bubble the surfactant concentration is smaller. This means that the surfactant is adsorbed at the top of the bubble, is transported on the interface towards the bottom and then desorbed at the rear part of the bubble, where the bulk concentration hence becomes higher. The surfactant is then transported within the bulk, with its bulk distribution resembling the recirculation regions of the velocity field.

In Fig. 15.34a the local surface tension distribution is shown. The surface tension distribution mirrors the surfactant distribution on the interface: the higher the surfactant concentration, the lower the surface tension. It is interesting to observe that the surface tension is reduced by a maximum of 35% with respect to the clean surface tension ($\sigma_0 = 0.0724$ N/m) on more than half of the bubble surface. Figure 15.34b shows the local distribution of the Gibbs' elasticity [47], where the latter is computed as

$$E^{\Sigma} = \sigma'\left(c^{\Sigma}\right) c^{\Sigma}. \tag{15.39}$$

In the specific case of the Langmuir surface tension equation of state, the local Gibbs' elasticity is

$$E^{\Sigma} = RTc_{\infty}^{\Sigma} \frac{c^{\Sigma}}{c_{\infty}^{\Sigma} - c^{\Sigma}}. \tag{15.40}$$

E^{Σ} gives important information about the applicability of simplified models to estimate the surfactant effect, such as the stagnant cap model [81] or the simplified model from [31]. Even for this very high initial surfactant concentration which leads to a rigid surface from the very beginning, the Gibbs' elasticity is not uniform over the interface. Under those conditions, the stagnant cap model, for instance, would give a good approximation of the results. Nevertheless, there are cases where the dynamics of the contaminated rising bubble (e.g. high Reynolds number, larger bubbles rising into water) is producing non-axisymmetric surfactant distribution on the interface, resulting in a non uniform Gibbs' elasticity, even in azimuthal direction. For such cases, these simplified models would not hold anymore.

15.6   Discussion and Outlook

The tandem project was split into two main periods, the first one focusing on the simulation and experimental investigation of free surface flows. The reference case considered was the process of droplet formation. First, the hydrodynamics of the process was studied and numerical results were compared to the experimental ones, as shown in [20]. For small inflow rates, a good agreement between the numerical and experimental results is shown. For high inflow rates, the problem of applicability of the Gauss-Laplace equation to correctly evaluate the surface tension is pointed out.

The second step involved the introduction of surfactants in the system under investigation. Experimentally, the effect of surfactants on the droplet formation process is addressed in [40, 43, 44]. Numerically, the presence of surfactants in the system implied quite some challenges and required the development of a framework capable to deal with multicomponent surfactant systems. In such systems the surfactant species cannot be considered dilute anymore and the solution of the Maxwell-Stefan equations is required. All the details about the numerical methodology are described in [22], while the results from this methodology applied on the droplet formation case can be found in [21]. The comparison between numerical and experimental results in presence of surfactants remained an open issue from the first research period and was partially addressed in this chapter. The second period of the tandem project focused mainly on two-phase flows under the influence of surfactant, considering, as a test case, rising bubble in a liquid phase. The presence of a second phase in the system introduced some numerical issues that had to be faced. Meanwhile, a *sorption library* was created, where various different sorption models were implemented to cover a wider range of phenomena that can now be simulated. From the experimental side, the techniques to study rising bubbles were further improved and the results can be found in [26, 48, 55]. The first numerical results for rising bubbles can be found in [70] and in [71], where a rigorous validation of the simulation of adsorption mechanism is included. The comparison between numerical and experimental results for two-phase flows is reported in the former sections.

To summarize, within this joined project, the continuum physical model and the numerical methodology to simulate two-phase flows under the influence of soluble surfactant were improved and further developed to achieve new level of knowledge on the adsorption dynamics of surfactants at liquid interfaces. In the former sections the results were presented for two exemplifying cases, the droplet formation process and the rising bubble, comparing the simulation results to the experimental ones. A good agreement between the two was found, reaching a new state of the art both experimentally and from a theoretical point of view. Nevertheless, further improvements to both sides, experiments and numerics, are needed. From an experimental point of view, as pointed out in Sect. 15.5.1 and above, an extension to the Gauss-Laplace equation for shear effects is needed to allow the evaluation procedure of the surface tension to be applied to more dynamic droplet formation conditions. With respect to the numerics, the full system of transport equations of the surfactants has to be solved and compared to the experiments for the droplet formation process. Moreover, as already described in Sect. 15.5.2, to simulate cases such as rising bubbles in contaminated water with realistic surfactant bulk diffusivities, a solution to the mesh restrictions has to be found. Among the possibilities to solve this issue, there are two ideas: the first one is a *dual mesh* approach, i.e. to solve the hydrodynamics on a coarser mesh and the surfactant bulk transport on a refined mesh with proper mapping of the velocity field onto the refined mesh. The second possibility is to verify the applicability of a sub-grid scale model (SGS), a by now standard approach in mass transfer problems [13], to approximate the *surfactant boundary layer* in the vicinity of the bubble.

Once this numerical issue is solved, the research, from the modeling point of view, can focus on liquid/liquid interfaces with surfactant soluble in both phases. In this field, as mentioned previously in this chapter, very little has been done so far. Thus a systematic study with Direct Numerical Simulations side by side to the experiments would give novel contributions and a better understanding of the complex phenomena occurring at (contaminated) liquid-liquid interfaces.

Acknowledgements The authors thank the German Research Foundation (DFG) for financial support within the Priority Program SPP1506 "Transport Processes at Fluidic Interfaces" [BO1879/9-2, Mi418/18-2] and the Technische Universität Darmstadt HHLR (High Performance Computer center) for the computational resources.

References

1. Adalsteinsson, D., Sethian, J.: Transport and diffusion of material quantities on propagating interfaces via level set methods. J. Comput. Phys. **185**(1), 271–288 (2003)
2. Adami, S., Hu, X., Adams, N.: A conservative SPH method for surfactant dynamics. J. Comput. Phys. **229**(5), 1909–1926 (2010)
3. Aland, S., Lowengrub, J., Voigt, A.: A continuum model of colloid-stabilized interfaces. Phys. Fluids **23**(6), 062103 (2011)
4. Alke, A., Bothe, D.: 3D numerical modelling of soluble surfactant at fluid interfaces based on the Volume-of-Fluid method. Fluid Dyn. Mater. Process. **5**(4), 345–372 (2009)
5. Ambravaneswaran, B., Wilkes, E., Basaran, O.: Drop formation from a capillary tube: comparison of one-dimensional and two-dimensional analyses and occurrence of satellite drops. Phys. Fluids **14**, 2606–2621 (2002)
6. Anderson, D.M., McFadden, G., Wheeler, A.A.: Diffuse interface in fluid mechanics. Annu. Rev. Fluid Mech. **30**, 139–165 (1998)
7. Aveyard, R., Haydon, D.: Thermodynamic properties of aliphatic hydrocarbon-water interfaces. J. Colloid Sci. **61**, 2255–2261 (1965)
8. Bazhlekov, I., Anderson, P., Meijer, H.: Numerical investigation of the effect of insoluble surfactants on drop deformation and breakup in simple shear flow. J. Colloid Interface Sci. **298**(1), 369–394 (2006)
9. Bergeron, V.: Disjoining pressures and film stability of alkyltrimethylammonium bromide foam films. Langmuir **13**, 3474–3482 (1997)
10. Blunk, D., Tessendorf, R., Buchavzov, N., Strey, R., Stubenrauch, C.: Purification, surface tensions and miscibility gaps of alkyldimethyl and alkyldiethylphosphine oxides. J. Surfactant Deterg. **10**, 155–165 (2007)
11. Bothe, D.: On the Maxwell-Stefan approach to multicomponent diffusion. In: Escher, J., Guidotti, P., Hieber, M., Mucha, P., Prüss, J.W., Shibata, Y., Simonett, G., Walker, C., Zajaczkowski, W. (eds.) Parabolic Problems. Progress in Nonlinear Differential Equations and Their Applications, vol. 80, pp. 81–93. Springer, Basel (2011)
12. Bothe, D.: On the multi-physics of mass transfer across fluid interfaces. In: Schindler, F.P., Kraume, M. (eds.) 7th International Workshop - IBW7 on Transport Phenomena with Moving Boundaries and More, Fortschr.-Ber. VDI Reihe 3, vol. 947, pp. 1–23. VDI-Verlag, Düsseldorf (2015)
13. Bothe, D., Fleckenstein, S.: A volume-of-fluid-based method for mass transfer processes at fluid particles. Chem. Eng. Sci. **101**, 283–302 (2013)
14. Bothe, D., Prüss, J., Simonett, G.: Well-posedness of a two-phase flow with soluble surfactant. In: Chipot, M., Escher, J. (eds.) Nonlinear Elliptic and Parabolic Problems, pp. 37–61. Birkhäuser, Basel (2005)
15. Brackbill, J., Kothe, D., Zemach, C.: A continuum method for modeling surface tension. J. Comput. Phys. **100**(2), 335–354 (1992)
16. Cengel, Y.A., Boles, M.A.: Thermodynamics: An Engineering Approach. McGraw-Hill, New York (2001)
17. Ceniceros, H.D.: The effects of surfactants on the formation and evolution of capillary waves. Phys. Fluids **15**(1), 245–256 (2003)
18. Chen, P., Kwok, D., Prokop, R., del Rio, O., Susnar, S., Neumann, A.: Axisymmetric drop shape analysis (ADSA) and its applications. In: Möbius, D., Miller, R. (eds.) Drops and Bubbles in Interfacial Research. Studies in Interface Science, vol. 6, pp. 61–138. Elsevier, Amsterdam (1998)
19. del Rio, O., Neumann, A.: Axisymmetric drop shape analysis: computational methods for the measurement of interfacial properties from the shape and dimensions of pendant and sessile drops. J. Colloid Interface Sci. **196**(2), 136–147 (1997)
20. Dieter-Kissling, K., Karbaschi, M., Marschall, H., Javadi, A., Miller, R., Bothe, D.: On the applicability of Drop Profile Analysis Tensiometry at high flow rates using an Interface Tracking method. Colloids Surf. A Physicochem. Eng. Asp. **441**, 837–845 (2014)

21. Dieter-Kissling, K., Marschall, H., Bothe, D.: Direct numerical simulation of droplet formation processes under the influence of soluble surfactant mixtures. Comput. Fluids **113**, 93–105 (2015)
22. Dieter-Kissling, K., Marschall, H., Bothe, D.: Numerical method for coupled interfacial surfactant transport on dynamic surface meshes of general topology. Comput. Fluids **109**, 168–184 (2015)
23. Drenckhan, W., Saint-Jalmes, A.: The science of foaming. Adv. Colloid Interf. Sci. **222**, 228–259 (2015)
24. Duineveld, P.C.: The rise velocity and shape of bubbles in pure water at high Reynolds number. J. Fluid Mech. **292**, 325–332 (1995)
25. Dukhin, S., Miller, R., Loglio, G.: Drops and Bubbles in Interfacial Research, vol. 6, pp. 367–432. Elsevier, Amsterdam (1998)
26. Dukhin, S., Lotfi, M., Kovalchuk, V., Bastani, D., Miller, R.: Dynamics of rear stagnant cap formation at the surface of rising bubbles in surfactant solutions at large Reynolds and Marangoni numbers and for slow sorption kinetics. Colloids Surf. A **492**, 127–137 (2016)
27. Fainerman, V., Miller, R., Aksenenko, E., Makievski, A.: Equilibrium adsorption properties of single and mixed surfactant solutions. In: Fainerman, V., Möbius, D., Miller, R. (eds.) Surfactants Chemistry, Interfacial Properties, Applications. Studies in Interface Science, vol. 13, pp. 189–285. Elsevier, Amsterdam (2001)
28. Fainerman, V., Mucic, N., Pradines, V., Aksenenko, E., Miller, R.: Adsorption of alkyltrimethylammonium bromides at water/alkane interfaces – competitive adsorption of alkanes and surfactants. Langmuir **29**, 13783–13789 (2013)
29. Fainerman, V., Aksenenko, E., Mucic, N., Javadi, A., Miller, R.: Thermodynamics of adsorption of ionic surfactants at water/alkane interfaces. Soft Matter **10**, 6873–6887 (2014)
30. Feigl, K., Megias-Alguacil, D., Fischer, P., Windhab, E.: Simulation and experiments of droplet deformation and orientation in simple shear flow with surfactants. Chem. Eng. Sci. **62**(12), 3242–3258 (2007)
31. Fleckenstein, S., Bothe, D.: Simplified modeling of the influence of surfactants on the rise of bubbles in VOF-simulations. Chem. Eng. Sci. **102**, 514–523 (2013)
32. Frumkin, A.N.: Die Kapillarkurve der höheren Fettsäuren und die Zustandsgleichung der Oberflächenschicht. Z. Phys. Chem. **116**, 466–484 (1925)
33. Ganesan, S., Tobiska, L.: A coupled arbitrary Lagrangian-Eulerian and Lagrangian method for computation of free surface flows with insoluble surfactants. J. Comput. Phys. **228**(8), 2859–2873 (2009)
34. Ganesan, S., Tobiska, L.: Arbitrary Lagrangian-Eulerian finite-element method for computation of two-phase flows with soluble surfactants. J. Comput. Phys. **231**(9), 3685–3702 (2012)
35. Hameed, M., Siegel, M., Young, Y.N., Li, J., Booty, M., Papageorgiou, D.: Influence of insoluble surfactant on the deformation and breakup of a bubble or thread in a viscous fluid. J. Fluid Mech. **594**, 307–340 (2008)
36. Hirt, C.W., Nichols, B.D.: Volume of Fluid (VOF) method for the dynamics of free boundaries. J. Comput. Phys. **39**, 201–225 (1981)
37. Hoorfar, M., Neumann, A.: Axisymmetric drop shape analysis (ADSA). In: Neumann, A., David, R., Zuo, Y., Raton, B. (eds.) Applied Surface Thermodynamics, pp. 107–174. CRC, New York, London (2011)
38. James, A.J., Lowengrub, J.: A surfactant-conserving Volume-of-Fluid method for interfacial flows with insoluble surfactant. J. Comput. Phys. **201**, 685–722 (2004)
39. Jasak, H., Tuković, Z.: Automatic mesh motion for the unstructured Finite Volume method. Trans. FAMENA **30**(2), 1–20 (2006)
40. Javadi, A., Krägel, J., Pandolfini, P., Loglio, G., Kovalchuk, V., Aksenenko, E., Ravera, F., Liggieri, L., Miller, R.: Short time dynamic interfacial tension as studied by the growing drop capillary pressure technique. Colloids Surf. A **365**, 62–69 (2010)
41. Javadi, A., Karbaschi, M., Bastani, D., Ferri, J., Kovalchuk, V., Kovalchuk, N., Javadi, K., Miller, R.: Marangoni instabilities for convective mobile interfaces during drop exchange: experimental study and CFD simulation. Colloids Surf. A **441**, 846–854 (2014)

42. Jin, F., Stebe, K.: The effects of a diffusion controlled surfactant on a viscous drop injected into a viscous medium. Phys. Fluids **19**(11), 112103 (2007)
43. Karbaschi, M., Bastani, D., Javadi, A., Kovalchuk, V., Kovalchuk, N., Makievski, A., Bonaccurso, E., Miller, R.: Drop profile analysis tensiometry under highly dynamic conditions. Colloids Surf. A **413**, 292–297 (2012)
44. Karbaschi, M., Rahni, M.T., Javadi, A., Cronan, C., Schano, K., Faraji, S., Won, J., Ferri, J., Krägel, J., Miller, R.: Dynamics of drops – formation, growth, oscillation, detachment, and coalescence. Adv. Colloid Interface Sci. **222**, 413–424 (2015)
45. Khatri, S., Tornberg, A.K.: An embedded boundary method for soluble surfactants with interface tracking for two-phase flows. J. Comput. Phys. **256**, 768–790 (2014). doi:10.1016/j.jcp.2013.09.019. http://www.sciencedirect.com/science/article/pii/S0021999113006244
46. Kovalchuk, V., Miller, R., Fainerman, V., Loglio, G.: Dilational rheology of adsorbed surfactant layers—role of the intrinsic two-dimensional compressibility. Adv. Colloid Interface Sci. **114–115**, 303–313 (2005)
47. Kralchevsky, P., Danov, K., Broze, G., Mehreteab, A.: Thermodynamics of ionic surfactant adsorption with account for the counterion binding: effect of salts of various valency. Langmuir **15**, 2351–2365 (1999)
48. Krzan, M., Zawala, J., Malysa, K.: Development of steady state adsorption distribution over interface of a bubble rising in solutions of n-alkanols (C5, C8) and n-alkyl trimethyl ammonium bromides (C8, C12, C16). Colloids Surf. A **298**, 42–51 (2007)
49. Lai, M.C., Tseng, Y.H., Huang, H.: An immersed boundary method for interfacial flows with insoluble surfactant. J. Comput. Phys. **227**(15), 7279–7293 (2008)
50. Li, J.: The effect of an insoluble surfactant on the skin friction of a bubble. Eur. J. Mech. B Fluids **25**(1), 59–73 (2006)
51. Liggieri, L., Ravera, F., Ferrari, M., Passerone, A., Miller, R.: Adsorption kinetics of alkylphosphine oxides at water/hexane interface. 2. Theory of the adsorption with transport across the interface in finite systems. J. Colloid Interface Sci. **186**, 46–52 (1997)
52. Loglio, G., Pandolfini, P., Miller, R., Makievski, A., Ravera, F., Liggieri, L.: Drop and bubble shape analysis as a tool for dilational rheological studies of interfacial layers. In: Möbius, D., Miller, R. (eds.) Novel Methods to Study Interfacial Layers. Studies in Interface Science, vol. 11, pp. 439–483. Elsevier, Amsterdam (2001)
53. Lunkenheimer, K., Haage, K., Miller, R.: On the adsorption properties of surface-chemically pure aqueous solutions of n-alkyl-dimethyl and n-alkyl-diethyl phosphine oxides. Colloids Surf. **22**, 215–224 (1987)
54. Malysa, K., Krasowska, M., Krzan, M.: Influence of surface active substances on bubble motion and collision with various interfaces. Adv. Colloid Interface Sci. **114–115**, 205–225 (2005)
55. Małysa, K., Zawala, J., Krzan, M., Krasowska, M.: Bubbles rising in solutions; local and terminal velocities, shape variations and collisions with free surface. In: Miller, R., Liggieri, L. (eds.) Bubble and Drop Interfaces. Progress in Colloid and Interface Science, vol. 2. CRC, Taylor & Francis, Boca Raton, London (2011)
56. Maze, C., Burnet, G.: A non-linear regression method for calculating surface tension and contact angle from the shape of a sessile drop. Surf. Sci. **13**, 451–470 (1969)
57. Medrzycka, K., Zwierzykowski, W.: Adsorption of alkyltrimethylammonium bromides at the various interfaces. J. Colloid Interface Sci. **230**, 67–72 (2000)
58. Menon, S., Schmidt, D.P.: Conservative interpolation on unstructured polyhedral meshes: an extension of the Supermesh approach to cell-centered Finite-Volume variables. Comput. Methods Appl. Mech. Eng. **200**, 2797–2804 (2011)
59. Mooney, K., Menon, S., Schmidt, D.: A computational study of viscoelastic drop collisions. In: ILASS – Americas 22nd Annual Conference on Liquid Atomization and Spray Systems, Cincinnati (2010)

60. Mucic, N., Kovalchuk, N., Aksenenko, E., Fainerman, V., Miller, R.: Adsorption layer properties of alkyl trimethylammonium bromides at interfaces between water and different alkanes. J. Colloid Interface Sci. **410**, 181–187 (2013)
61. Mucic, N., Kovalchuk, N., Pradines, V., Javadi, A., Aksenenko, E., Krägel, J., Miller, R.: Dynamic properties of CnTAB adsorption layers at the water/oil interface. Colloids Surf. A **441**, 825–830 (2014)
62. Muradoglu, M., Tryggvason, G.: A front-tracking method for computation of interfacial flows with soluble surfactants. J. Comput. Phys. **227**(4), 2238–2262 (2008)
63. Muzaferija, S., Perić, M.: Computation of free-surface flows using the Finite-Volume method and moving grids. Numer. Heat Transfer Part B Fundam. **32**(4), 369–384 (1997)
64. Oellrich, L., Schmidt-Traub, H., Brauer, H.: Theoretische Berechnung des Stofftransports in der Umgebung einer Einzelblase. Chem. Eng. Sci. **28**(3), 711–721 (1973)
65. Osher, S., Sethian, J.: Fronts propagating with curvature dependent speed – algorithms based on Hamilton-Jacobi formulations. J. Comput. Phys. **79**, 12–49 (1988)
66. Park, J., Hulsen, M., Anderson, P.: Numerical investigation of the effect of insoluble surfactant on drop formation in microfluidic device. Eur. Phys. J. Appl. Phys. Spec. Top. **222**, 199–210 (2013)
67. Passerone, A., Liggieri, L., Rando, N., Ravera, F., Ricci, E.: A new experimental-method for the measurement of the interfacial-tension between immiscible fluids at zero bond number. J. Colloid Interface Sci. **146**, 152–162 (1991)
68. Patankar, S.: Numerical Heat Transfer and Fluid Flow: Computational Methods in Mechanics and Thermal Science. Hemisphere Publishing, Washington, DC (1980)
69. Perić, M., Kessler, R., Scheuerer, G.: Comparison of Finite-Volume numerical methods with staggered and collocated grids. Comput. Fluids **16**(4), 389–403 (1988)
70. Pesci, C., Dieter-Kissling, K., Marschall, H., Bothe, D.: Finite Volume/Finite Area Interface Tracking method for two-phase flows with fluid interfaces influenced by surfactant. In: Rahni, M.T., Karbaschi, M., Miller, R. (eds.) Progress in Colloid and Interface Science. CRC, Taylor & Francis, Boca Raton, London (2015)
71. Pesci, C., Marschall, H., Bothe, D.: Computational analysis of single rising bubbles influenced by soluble surfactant (2017, in preparation)
72. Peskin, C.: The immersed boundary method. Acta Numer. **11**, 479–517 (2002)
73. Pozrikidis, C.: Boundary Integral and Singularity Methods for Linearized Viscous Flow. Cambridge University Press, Cambridge (1992)
74. Pradines, V., Fainerman, V., Aksenenko, E., Krägel, J., Mucic, N., Miller, R.: Alkyltrimethy-lammonium bromides adsorption at liquid/fluid interfaces in the presence of neutral phosphate buffer. Colloids Surf. A **371**, 22–28 (2010)
75. Quan, S., Schmidt, D.: A moving mesh Interface Tracking method for 3D incompressible two-phase flows. J. Comput. Phys. **221**, 761–780 (2007)
76. Rahni, M.T., Karbaschi, M., Miller, R. (eds.): Computational Methods for Complex Liquid-Fluid Interfaces. Progress in Colloid and Interface Science, vol. 5. CRC/Taylor & Francis, Boca Raton/London (2016)
77. Rayner, M., Trägardh, G., Trägardh, C.: The impact of mass transfer and interfacial expansion rate on droplet size in membrane emulsification processes. Colloids Surf. A **266**, 1–17 (2005)
78. Renardy, Y.Y., Renardy, M., Christini, V.: A new Volume-Of-Fluid formulation for surfactants and simulations of drop deformations under shear at a low viscosity ratio. Eur. J. Fluid Mech. B **21**, 49–59 (2002)
79. Rotenberg, Y., Boruvka, L., Neumann, A.: Determination of surface tension and contact angle from the shapes of axisymmetric fluid interfaces. J. Colloid Interface Sci. **93**(1), 169–183 (1983)
80. Sagis, L.: The Maxwell-Stefan equations for diffusion in multiphase systems with intersecting dividing surfaces. Physica A **254**(3), 365–376 (1998)
81. Savic, P.: Circulation and distortion of liquid drops falling through a viscous medium. Technical Report, National Research Council Canada, Ottawa (1953)

82. Schadler, V., Windhab, E.: Continuous membrane emulsification by using a membrane system with controlled pore distance. Desalination **189**, 130–135 (2006)
83. Schmidt, D., Dai, M., Wang, H., Perot, J.: Direct interface tracking of droplet deformation. Atomization Sprays **12**(5–6), 721–735 (2002)
84. Schröder, V., Behrend, O., Schubert, H.: Effect of dynamic interfacial tension on the emulsification process using microporous, ceramic membranes. J. Colloid Interface Sci. **202**, 334–340 (1998)
85. Sczech, R., Eckert, K., Acker, M.: Convective instability in a liquid-liquid system due to complexation with a crown ether. J. Phys. Chem. A **112**, 7357–7364 (2008)
86. Severino, M., Campana, D., Giavedoni, M.: Effects of a surfactant on the motion of a confined gas-liquid interface. The influence of the Peclet number. Lat. Am. Appl. Res. **35**(3), 225–232 (2005)
87. Sussman, M., Puckett, E.G.: A coupled Level Set and Volume-of-Fluid method for computing 3D incompressible two-phase flows. J. Comput. Phys. **162**, 301–337 (2000)
88. Tasoglu, S., Demirci, U., Muradoglu, M.: The effect of soluble surfactant on the transient motion of a buoyancy-driven bubble. Phys. Fluids **20**(4), 040805 (2008)
89. Teigen, K., Song, P., Lowengrub, J., Voigt, A.: A diffuse-interface method for two-phase flows with soluble surfactants. J. Comput. Phys. **230**(2), 375–393 (2011)
90. Tornberg, A.K., Engquist, B.: The segment projection method for interface tracking. Commun. Pure Appl. Math. **56**(1), 47–79 (2003)
91. Tryggvason, G., Bunner, B., Esmaeli, A., Juric, D., Al-Rawahi, N., Tauber, W., Han, J., Nas, S., Jan, Y.J.: Front tracking method for the computation of multiphase flow. J. Comput. Phys. **169**, 708–759 (2001)
92. Tuković, Z.: Finite Volume method on domains of varying shape (in Croatian). Ph.D. thesis, Faculty of Mechanical Engineering and Naval Architecture, University of Zagreb (2005)
93. Tuković, Z., Jasak, H.: Simulation of free-rising bubble with soluble surfactant using moving mesh Finite Volume/Area method. In: 6th International Conference on CFD in Oil & Gas, Metallurgical and Process Industries SINTEF/NTNU, Trondheim (2008)
94. Tuković, Z., Jasak, H.: A moving mesh Finite Volume Interface Tracking method for surface tension dominated interfacial fluid flow. Comput. Fluids **55**, 70–84 (2012)
95. van der Graaf, S., Nisisako, T., Schroen, C., van der Sman, R., Boom, R.: Lattice Boltzmann simulations of droplet formation in a T-shaped microchannel. Langmuir **22**, 4144–4152 (2006)
96. van der Zwan, E., Schröen, K., van Dijke, K., Boom, R.: Visualization of droplet break-up in pre-mixmembrane emulsification using microfluidic devices. Colloids Surf. A **277**, 223–229 (2006)
97. Wegener, M., Grünig, J., Stüber, J., Paschedag, A., Kraume, M.: Transient rise velocity and mass transfer of a single drop with interfacial instabilities - experimental investigations. Chem. Eng. Sci. **62**, 2967–2978 (2007)
98. Weller, H., Tabor, G., Jasak, H., Fureby, C.: A tensorial approach to computational continuum mechanics using object-oriented techniques. Comput. Phys. **12**, 620–631 (1998)
99. Xu, J.J., Zhao, H.: An Eulerian formulation for solving partial differential equations along a moving interface. J. Sci. Comput. **19**, 573–594 (2003)
100. Xu, J.J., Li, Z., Lowengrub, J., Zhao, H.: Numerical study of surfactant-laden drop-drop interactions. Commun. Comput. Phys. **10**(2), 453–473 (2011)
101. Xu, J.J., Huang, Y., Lai, M.C., Li, Z.: A coupled immersed interface and level set method for three-dimensional interfacial flows with insoluble surfactant. Commun. Comput. Phys. **15**(2), 451–469 (2014)
102. Yang, X., James, A.: An arbitrary Lagrangian-Eulerian (ALE) method for interfacial flows with insoluble surfactants. Fluid Dyn. Mater. Process. **3**(1), 65–96 (2007)
103. Zhang, J., Eckmann, D., Ayyaswamy, P.: A front tracking method for a deformable intravascular bubble in a tube with soluble surfactant transport. J. Comput. Phys. **214**(1), 366–396 (2006)

Chapter 16
Complex Patterns and Elementary Structures of Solutal Marangoni Convection: Experimental and Numerical Studies

Kerstin Eckert, Thomas Köllner, Karin Schwarzenberger, and Thomas Boeck

Abstract The transfer of a solute between two liquid layers is susceptible to convective instabilities of the time-dependent diffusive concentration profile that may be caused by the Marangoni effect or buoyancy. Marangoni instabilities depend on the change of interfacial tension and Rayleigh instabilities on the change of liquid densities with solute concentration. Such flows develop increasingly complex cellular or wavy patterns with very fine structures in the concentration field due to the low solute diffusivity. They are important in several applications such as extraction or coating processes. A detailed understanding of the patterns is lacking although a general phenomenological classification has been developed based on previous experiments. We use both highly resolved numerical simulations and controlled experiments to examine two exemplary systems. In the first case, a stationary Marangoni instability is counteracted by a stable density stratification producing a hierarchical cellular pattern. In the second case, Rayleigh instability is opposed by the Marangoni effect causing solutal plumes and eruptive events with short-lived Marangoni cells on the interface. A good qualitative and acceptable quantitative agreement between the experimental visualizations and measurements and the corresponding numerical results is achieved in simulations with a planar interface, and a simple linear model for the interface properties, i.e. no highly

K. Eckert (✉)
Institute of Process Engineering and Environmental Technology, Technische Universität Dresden, Dresden, Germany

Institute of Fluid Dynamics, Helmholtz-Zentrum Dresden-Rossendorf, Dresden, Germany
e-mail: kerstin.eckert@tu-dresden.de

T. Köllner • T. Boeck
Institute of Thermodynamics and Fluid Mechanics, TU Ilmenau, Helmholtzring 1, 98684 Ilmenau, Germany
e-mail: thomas.koellner@tu-ilmenau.de; thomas.boeck@tu-ilmenau.de

K. Schwarzenberger
Institute of Process Engineering and Environmental Technology, Technische Universität Dresden, Dresden, Germany
e-mail: karin.schwarzenberger@tu-dresden.de

© Springer International Publishing AG 2017
D. Bothe, A. Reusken (eds.), *Transport Processes at Fluidic Interfaces*,
Advances in Mathematical Fluid Mechanics, DOI 10.1007/978-3-319-56602-3_16

specific properties of the interface are required for the complex patterns. Simulation results are also used to characterize the mechanisms involved in the pattern formation.

16.1 Introduction

16.1.1 *Marangoni Effect and Marangoni Flows*

The Marangoni effect refers to a mostly small-scale flow driven by local inhomogeneities of interfacial tension σ either caused by temperature (thermocapillary flow) or by concentration gradients (solutocapillary flow). Historically the discovery of this effect goes back to Thomson [112] and Marangoni [72] while the first systematic study of the resulting patterns in plane liquid layers, exposed to a steady temperature gradient, was done in the seminal work by Bénard [7]. Pearson's [80] linear stability theory applied to Bénard's experiments revealed that gradients in surface tension are the driving force in thin layers and provided the critical conditions for the onset of what is now called surface-tension-driven Bénard convection.

Motivated by early experimental observations on surprising spontaneous interfacial motions during mass transfer [57, 63, 101, 104], Sternling and Scriven [109] tackled the mass-transfer analogue of Pearson's problem in a similarly important paper in 1959 which is detailed later on. Since then, an ongoing research can be noticed in this area which is mainly motivated by two facts.

(1) First, there is an enigmatic variety of patterns emerging from the Marangoni effect making such systems to paradigmatic objects of study for pattern formation. The structures found comprise hexagonal [7, 54], polygonal [5, 22, 78], square [33, 89] or round [65] convection cells, eruptions [21, 64] and oscillations [56] as well as different sorts of waves. Recently found new structures include circular patterns in volatile binary liquids [118] or the emission of small satellite droplets from pulsating larger droplets [83]. Also chemically reactive systems are affected, e.g. by influencing the shape of propagating autocatalytic reaction fronts [87] or the front velocity [102], see the recent focus issue [30]. Furthermore, we refer to reviews of thermocapillary [25, 88] and solutocapillary instabilities [24, 96].

(2) The second source for the continuing interest is the seemingly ubiquitous occurrence of the Marangoni effect both in established techniques, such as extraction [40], distillation [85], boiling [73], crystal growth [90], drying [74, 115] or wetting [20, 62], and newly emerging technologies such as the self-assembly of patterns of functional materials [10]. To illustrate briefly the relevance of the Marangoni effect we take the extraction processes as a first example. Here, the mass transfer enhancement due to the Marangoni effect [5, 18, 114] is an important but critical issue in the correct scale-up of pilot plants designed for the extraction of particular species. A second example is the drying of lacquers or polymer solutions on which many industrial applications such as coating or printing are based. The goal of a smooth surface finish is sometimes not achieved because

the solvent evaporation can trigger Marangoni convection which may in turn cause marked surface corrugations [6, 115]. There are also newly emerging technologies in which polymer/carbon nanotubes composite films appear as a versatile object for numerous applications. In [47] it was shown that the electrical conductivity and the light transmittance of such films can be improved by controlling the solutal Marangoni instability. Another promising technique is the evaporation of sessile droplets which is attractive for the deposition of rather different types of particles (polymers, DNA etc.) in self-assembled patterns. Numerous pattern morphologies have been observed since the original work on the single coffee ring stain [31], see e.g. [10]. All of them show a complex interplay between pinning-depinning events of the contact line and Marangoni flows to and away from the contact line arising from solvent evaporation. Criteria for the onset of Marangoni convection in evaporating sessile droplets were established recently [70]. Computer simulations of evaporating pinned sessile water droplets of submicrometer size [99] have shown that the thermal Marangoni flow loses its importance only for very small droplets with diameters $L \leq 10^{-5}$ m. Coalescence of neighboring sessile droplets of different but miscible liquids is suppressed by a Marangoni flow that sucks liquid out of the neck which connects both droplets, thereby counteracting the capillary-driven flow into the neck [48].

Progress in the—by now—large field of research on Marangoni flows was possible thanks to modern experimental and numerical methods. Besides advanced microscopic methods such as wide-field fluorescence microscopy, the growing drop capillary pressure tensiometry [45] has to be mentioned as a tool to measure interfacial changes even on short timescales. Furthermore, an extension of the validity limits of the drop profile tensiometry, an originally static technique based on fitting the Gauss-Laplace equation, towards fast growing drops can be expected from the recent combination of computational fluid dynamics and experiments [32]. Novel numerical techniques such as Volume-of-Fluid [69, 84], phase-field methods [15], level-set [113] or pseudospectral formulations [50] in combination with high-performance computing now allow one to perform fully three-dimensional simulations of the mass transfer in liquid-liquid systems.

Despite of this progress, and while new fields of research are established, there are still numerous unresolved problems in Marangoni flows. The complex interaction of different patterns and their transition at high Marangoni driving force, or the delayed onset of Marangoni convection at convective mobile interfaces [46] are examples of new fields while the apparently well understood tears-of-wine phenomenon still presents some unresolved issues. As pointed out in [111], the simple explanation in terms of shear force at the air-liquid interface pulling fluid of higher alcohol concentration, hence lower σ, toward the top of the film where σ is higher due to alcohol evaporation might be questionable. However, such debates keep the subject alive and further evolving dynamically.

The present chapter reviews our works on solutal Marangoni convection, performed within the priority program 1506 of the Deutsche Forschungsgemeinschaft to obtain a refined understanding of Marangoni convection patterns [50, 52, 53, 68,

94–96, 98]. To achieve this, we use well-defined model systems which are mainly liquid-liquid systems separated by a plane interface which are supplemented to a lesser extent by works on droplets. Liquid-liquid systems with known material properties are excellent candidates for an in-depth comparison between experiments and numerics which is at the core of the project. Indeed, the lack of geometrical constraints in two or at least one dimension renders those systems accessible to highly accurate pseudospectral methods.

Before explaining the methods used in the projects in Sect. 16.2, we start with an introduction into mechanisms, instability criteria and main features of solutal Marangoni convection to work out the open issues which were addressed in the course of the project. The main results are reviewed in the Sects. 16.3 and 16.4 where Sect. 16.3 describes the key features of Marangoni convection in the case where mass transfer creates a stable density stratification in both phases. By contrast, Sect. 16.4 addresses the interaction between a Rayleigh convection, resulting from an unstable density stratification, and Marangoni convection. An outlook concludes this work in Sect. 16.5.

16.1.2 Fundamentals of Transient Solutal Convection Patterns

16.1.2.1 Mechanisms of Marangoni and Rayleigh Convection

To understand the main mechanisms for a solutal convection we consider in Fig. 16.1a an immiscible two-layer system shortly after initialization of mass transfer from phase (2) → (1). In this figure, dark color refers to a high concentration c of the solute. At the interface, a positive perturbation of solute concentration is visualized. Furthermore, assume that solute decreases interfacial tension, i.e. $d\sigma/dc < 0$. Inevitably, interfacial tension gradients will create a flow that spreads solute tangentially away from this initial variation. The ensuing flow carries fluid from the bulk to the interface. The flow can be considered symmetric to the interface

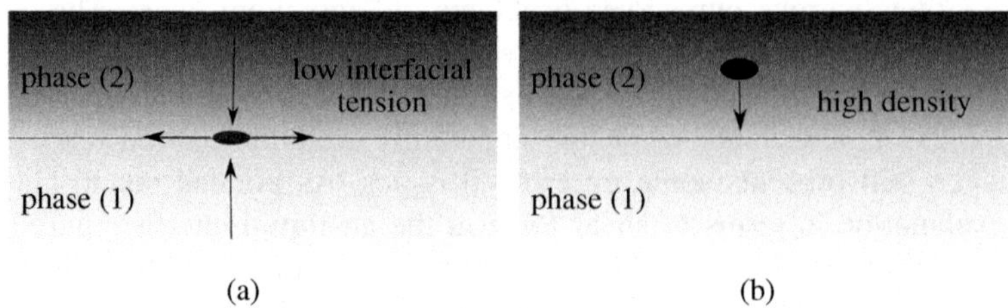

Fig. 16.1 Sketch of Marangoni effect (**a**) and Rayleigh effect (**b**). *Dark color* depicts high concentration and *arrows* depict fluid motion. Solute increases density but lowers interfacial tension

if vertical boundaries are far away, the ensuing flow is not strong (Stokes flow), and density gradients can be neglected.[1]

The subsequent impact on the interfacial concentration depends on the relative size of diffusivities $D^{(1)}$ and $D^{(2)}$ of the solute in the two layers since the assumed perturbation could be amplified or damped by the ensuing motion. If transport is *from* the phase with *lower* diffusivity *into* the phase with *higher* diffusivity, the interfacial concentration will decrease from the point where bulk fluid impinges the interface, consequently, concentration is highest at the inflow and convection is amplified. This behavior is explained by considering the approximately parallel flow of fluid near the interface, here the phase with stronger diffusion will force the interfacial concentration closer to its bulk value than the other phase with lower diffusivity. In the sketched situation of transport $(2) \rightarrow (1)$, Marangoni instability requires $D^{(2)} < D^{(1)}$, because in this way concentration will decrease from the point of inflow. These considerations are formalized in the classical analysis of Sternling and Scriven, cf. Sect. 16.1.3 for a stationary Marangoni instability.

The second source for convection is sketched in Fig. 16.1b. Here, again a transport of solute from the top to the bottom is depicted. A portion of fluid with higher concentration is sketched that—taking the solutal expansion coefficient $\beta_c^{(i)} > 0$—has a higher density than fluid at the same vertical level. Necessarily, a flow sets in that lowers the dense fluid against the lighter fluid. A self-sustained motion, i.e. Rayleigh convection, might be triggered. In the sketched situation the perturbation is amplified since a dense layer (dark color) overlays a light layer. However, if the direction of transport is reversed or the dependence of density on solute changes its sign ($\beta_c^{(i)} < 0$), the diffusive equilibration might be stable. Note that this is a rather simple picture since the coupling between both effects and oscillatory instabilities are neglected, for a more comprehensive account consult [77].

16.1.3 Sternling and Scriven Criteria for Instability

The foundation for a theoretical understanding of the structures arising from solutal Marangoni instability was laid by Sternling and Scriven [109]. Figure 16.2 summarizes the most relevant cases for liquid-liquid systems on the left-hand side and for liquid-gas systems on the right-hand side. The particular instability regimes in dependence on the mass transfer direction of a conventional surfactant ($d\sigma/dc < 0$) and the ratio between the most important material parameters, kinematic viscosity

---

[1]In this work, liquid/liquid layers will be considered exclusively. They usually have high Schmidt numbers $Sc^{(i)} = \nu^{(i)}/D^{(i)} > 1000$. This is not the case for gas/liquid systems or for the thermal problem, which can give rise to oscillatory instabilities due to the additional time scales.

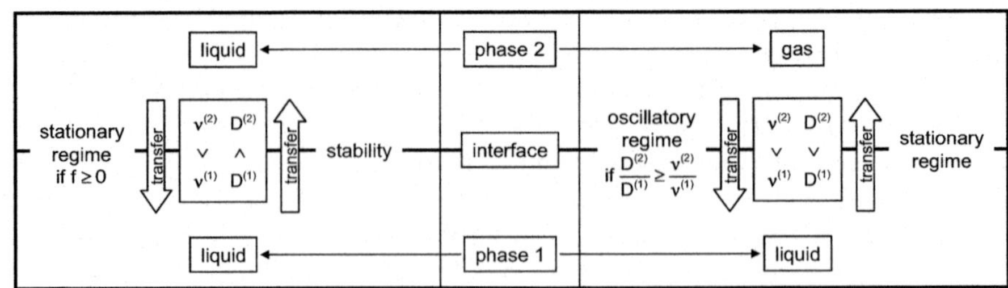

Fig. 16.2 Overview of conditions according to Sternling and Scriven [109]. Reprinted from [96], with permission from Elsevier

Table 16.1 Basic non-dimensional parameters for the two-layer system

Nondimensional quantity	Symbol
Marangoni number	$Ma = \dfrac{c_0 \alpha_c \sigma_{ref} d^{(1)}}{\rho^{(1)} v^{(1)} D^{(1)}}$
Grashof number	$G = \dfrac{c_0 \beta_c^{(1)} g (d^{(1)})^3}{(v^{(1)})^2}$
Schmidt number (1)	$Sc = v^{(1)}/D^{(1)} = Sc^{(1)}$
Schmidt number (2)	$Sc^{(2)} = v^{(2)}/D^{(2)} = Scv/(D^{(1)}D)$
Partition coefficient	$H = K_1/K_2$
Density ratio	$\rho = \rho_{ref}^{(2)}/\rho_{ref}^{(1)}$
Kinematic viscosity ratio	$v = v^{(2)}/v^{(1)}$
Diffusivity ratio	$D = D^{(2)}/D^{(1)}$
Ratio of expansion coefficients	$\beta = \beta_c^{(2)}/\beta_c^{(1)}$
Layer height ratio	$d = d^{(2)}/d^{(1)}$
Dynamic viscosity ratio	$\mu = v\rho$

$v^{(i)}$ and mass diffusivity $D^{(i)}$, are indicated. Furthermore, the value of the parameter

$$f = \frac{D+1}{2} + \frac{1/H+1}{2\left(1/(HD)+1\right)} +$$
$$+ \frac{1/\rho+1}{4Sc^{(2)}\left(1/(v\rho)+1+\alpha_{NS}\mu_s/(2\mu^{(2)})\right)} + \frac{Dv-1}{4Sc^{(2)}\left(D-1\right)} \tag{16.1}$$

plays a certain role. D, v, ρ, H and $Sc^{(2)}$ refer to the ratios of diffusivities, kinematic viscosities and densities, to the partition coefficient and the Schmidt number of the second phase. These quantities are summarized in Table 16.1. $\mu^{(2)}$ is the dynamic viscosity of phase 2 and μ_s refers to the intrinsic surface viscosity which is frequently neglected for estimation of the instability behavior. Finally, α_{NS} stands for the wave number of neutral stability.

The liquid-liquid system is stable when the solute in the donating phase has a higher diffusivity than in the accepting phase provided that viscosity of the latter is higher than that of the donating phase. In the opposite case, stationary instability sets in, if $f \geq 0$ holds. Adapting this situation to the much higher diffusivities in gases, a liquid-gas system shows a stationary instability under a viscosity ratio similar to the above one, when the solute transfer is from the liquid to the gas. If the direction of mass transfer is reversed under these conditions, the oscillatory mode of instability occurs. For other combinations not shown in Fig. 16.2 both oscillatory and stationary modes are possible.

The simple rule of Sternling and Scriven can be used as a first estimate to predict the stability behavior of the most Marangoni-sensitive systems provided no parasitic buoyancy-driven convection is present which can markedly modify the system's behavior (see Sect. 16.4). Of course non-linear effects by established convection or geometrical constraints may lead to other instability effects, which are out of the scope of the linear instability of the diffusive state [109].

The issue of stability in the two-layer setup has been studied thoroughly in subsequent works on different levels of complexity, see [25, 77] for an overview. Stability results for two finite layers, assuming linear basic profiles, were obtained for a deformable interface in [86, 108] and with inclusion of the Rayleigh effect in [43]. Recently, also time-dependent and nonlinear basic states received some interest. For this see the paper of Sun [110] and references therein, which however mostly rely on a "Biot boundary condition" for the solute transport at the interface, since they are tailored to a liquid/gas interface.

A study of a nonlinear basic state in the full two-layer setup has been performed by Gross and Hixon [39]. However, they used different initial and boundary conditions, i.e. initially zero concentrations in both layers with a fixed concentration at the boundaries. Further linear stability analyses include the effects of adsorbed surfactants [18, 106] and heat of solution [27, 105, 107]. The linear marginal threshold for convection growing out of the transient diffusive profile between two infinite layers has been solved in [49].

16.1.4 Solutal Marangoni Convection in Presence of a Stable Density Stratification: Experiments and Pattern Classification

The most frequently studied regime is that of Marangoni convection with a stable density stratification [4, 37, 64, 64, 66, 78, 91, 92]. This might be due to its complex but also highly ordered patterns, see Fig. 16.3 and Sect. 16.3. The underlying process of mass transfer and the corresponding patterns driven by the Marangoni effect were experimentally observed via schlieren or shadowgraph methods [76].

To overcome the vague description as "interfacial turbulence" in early scientific works [109], Orell and Westwater [78] made a first attempt to classify experimentally observed patterns in terms of polygonal cells, stripes and ripples. In a joint initiative of H. Linde, one of the pioneers of experimental Marangoni studies, and two of the authors a complete classification of the highly complex and unsteady patterns was proposed half a century later in the form of a few hypotheses [68], based on a broad range of experiments:

L_1 Interfacial convection is built up of three basic structures: (a) Marangoni roll cells (b) relaxation oscillations and (c) synchronized relaxation oscillation waves.

L_2 Each of these structures may occur in n different hierarchy steps of different size, which we call nth order, referring to the number of substructures which are embedded. Substructure(s) of all three types can occur in any of the three patterns.

L_3 Driving force of all these structures is the Marangoni shear stress, $d\sigma/dc \cdot \partial c/\partial x$, operating on different length scales.

L_4 Interfacial convection can consist of numerous periodic cycles of amplification and decay of the three basic structures. The complexity in large containers, whose size exceeds the largest wavelength, arises from the fact that structures of different types or of a different hierarchy might occur simultaneously in different regions of the container.

The first hypothesis of Linde et al. (L_1) proposes three main structures that should cover all Marangoni convection patterns in mass transfer systems with a stable density stratification.

The first structure is the Marangoni *roll cell* RC, sketched in the left column of Fig. 16.3. These RCs are similar to the polygonal cells studied in Bénard-Marangoni convection [16, 33, 54]. They form a relatively stable polygonal network, which is driven by low interfacial tension in the cell centers and high interfacial tension at the boundaries. This interfacial tension distribution results from the inflow in the cell centers and outflow at the cell borders.

The second structure identified by Linde is the *relaxation oscillation* of Marangoni roll cells-ROs, cf. middle column of Fig. 16.3. This pattern is proposed to have the same origin as the RC but is highly unsteady. Individual cells grow

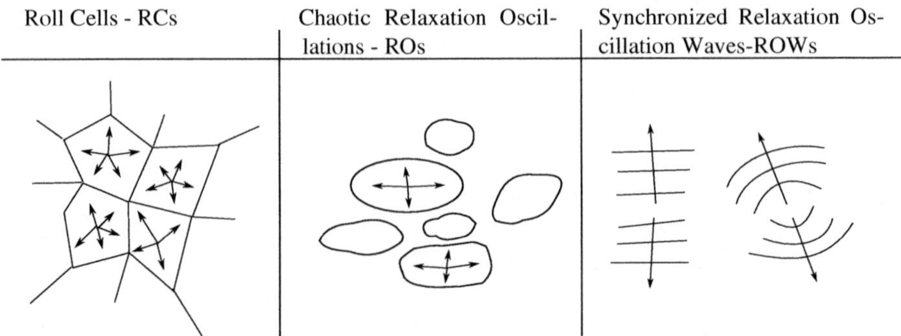

Fig. 16.3 Sketch of basic Marangoni patterns according to the theory of Linde [68]. *Lines* depict the highest contrast in experimental shadowgraph images (high concentration gradients) and *arrows* fluid velocity at the interface

fast in size and thereby compress neighboring structures. After the spreading of an individual cell, the motion will cease (relaxation). Note that this pattern shares features to the observed eruptions in Rayleigh unstable systems (see next Sect. 16.1.5), but has been observed [68] in systems where Rayleigh convection is not likely to be amplified.

The third basic structure according to Linde are the *synchronized relaxation oscillations waves*—ROWs. Mostly, they appear as aligned (straight or circular concentric) convection cells that move in a common direction. The concentric type has been termed *ripples* by Orell and Westwater [78]. Both types are divided into approximately equidistant relaxation zones [94] and can occur either as a single structure or in the form of a substructure (see L_2).

Until recently, only the simplest Marangoni roll cells without substructure have been thoroughly investigated by experiments and two-dimensional simulations, including recent work [17, 23, 35, 38, 71, 100]. Especially, no one-to-one comparison between experimental observations and simulations of the three-dimensional problem has been carried out. In this view, several open questions remained unsolved. They motivated the investigations carried out in our tandem project during the first 4 years of work (cf. Sect. 16.3):

(M_1)   Is the model of Sect. 16.2.2 able to describe experimentally observed patterns [64, 68, 78], triggered by mass transfer? Or, are other physical effects (e.g. adsorption of surfactants, interfacial deformation, heat of solution, wall effects) necessary to understand the pattern formation?

(M_2)   Which physical mechanisms are responsible for the appearance of basic structures (L_1), hierarchy formation (L_2) and "periodic cycles" (L_4)? Thus, can we confirm the hypotheses of Linde?

(M_3)   Can we perform a quantitative one-to-one comparison between experimental observation and numerical simulation, and what is its outcome?

(M_4)   How do physical parameters influence Marangoni convection, and can we make further fundamental predictions for this equilibration process?

16.1.5   *Interplay Between Solutal Rayleigh and Marangoni Convection in Presence of an Unstable Density Stratification*

As already introduced in Sect. 16.1.2.1, Rayleigh convection may appear as a counterpart to Marangoni convection in systems in which the mass transfer is accompanied by the formation of an unstable density stratification. An early observation of such a system was published by Kroepelin and Neumann [58]. They considered the transport of acetic acid from ethyl acetate into water inside a Hele-Shaw cell. They observed Rayleigh convection in both phases and vigorous movements at the interface, which they called eruptions (they used the German word "Eruptionen"). They also mentioned the analogy of these Rayleigh induced eruptions with eruptions triggered by forced convection. The experiments on the plane interface were motivated from their former experiments at pendant drops [57].

Orell and Westwater [78] also studied this combination of materials, but at an extended interface. They characterized the interfacial structures as "interfacial turbulence, chaotic and unorganized activity", but did not provide images of this regime. Due to this "unorganized" appearance this convective regime received less attention and its structures are often just termed as "interfacial turbulence". Further similar observations were published from Berg and Morig's experiments [9] of a benzene-chlorobenzene-water system with the transport of acetic acid between both phases (in a Hele-Shaw cell).

Schwarz [91, 92] paid more attention to this convection regime and published observations in the liquid-liquid system cyclohexanol/water+1-propanol. For this, he reported on erratic motions of circular spreadings (German "Spreitungen"), which he also called eruptions [92]. These structures of interfacial convection showed a close similarity to the dynamics reported on the ROs (Fig. 16.3) of Linde but are obviously triggered by Rayleigh convection. Schwarz reported such eruptions for numerous systems in his doctoral thesis [91].

In a broader context, numerous authors have studied the interplay between buoyant convection and the Marangoni effect. For instance, early studies on thermal convection revealed the damping effect of dissolved surfactant [8, 26, 79, 117]. Recently, studies (see review [56]) on the localized addition of surfactants under a surface showed an oscillatory behavior.

Similar to the eruptions in the layered systems, the coupling of buoyant convection and the Marangoni effect manifests in droplet geometries, e.g., Lappa et al. observed structures like eruptions (they called them shooting) at dissolving drops [59, 60]. Also, a comprehensive study of dissolving droplets has been performed by Agble and coworkers [1–3] in binary systems. They showed that additionally introduced surfactants trigger irregular Marangoni convection in otherwise purely buoyant convection.

Despite this interest, the detailed theoretical reproduction of classically observed eruptions from mass transfer experiments [9, 58, 78] is still lacking. Therefore—as in the former Marangoni case—no one-to-one comparison between experimental observations of complex and time-dependent structures and theoretical predictions is available. Hence, the efforts of the tandem project during the second period were devoted to understanding the interplay between Rayleigh and Marangoni convection and to resolving the following open questions:

(R_1) Is a simple plane layer model able to reproduce the experimentally observed structures?

(R_2) What physical mechanisms are responsible for the observed eruptions?

(R_3) Can we perform a one-to-one comparison between experimental observation and numerical simulation?

(R_4) How do physical parameters fundamentally influence the equilibration process?

16.2 Methods

16.2.1 Experimental Setup

The model system consists of two layers separated by a plane interface. The binary phases are in thermodynamic equilibrium due to mutual saturation to reduce effects from phase changes and multi-component diffusion. The diffusing substance is solved in one of the phases with a molar concentration of c_0.

As the experimental conditions have to agree with the simulations as far as possible, the preparation of the two-phase system demands an appropriate setup. On the one hand, disturbances of the Marangoni flow by the superposition of the phases should be minimized. On the other hand, the procedure of superposition should be completed quickly so that the desired step-like initial concentration profile can be maintained. These requirements are to a good degree realized by the assembly shown in Fig. 16.4. A similar experimental setup has already been used by Linde and coworkers [67, 92].

Each phase is filled in a glass cuvette (A, B) with an inner size of $L \times W \times H = 60\,\mathrm{mm} \times 60\,\mathrm{mm} \times 20\,\mathrm{mm}$. The box height H corresponds to the parameters $d^{(1)}$, $d^{(2)}$ that are used in the mathematical model, discussed in Sect. 16.2.2. As indicated by the bold horizontal arrow in Fig. 16.4, both phases are joined by sliding cuvette B with the lighter phase over cuvette A with the heavier one. After this procedure of superposition, the whole apparatus is carefully introduced into the shadowgraph optics (beam path sketched by the vertical line arrows).

The experimental shadowgraph images result from the deflection of light that is caused by the dependence of the refractive index on solute concentration. Hence, they visualize the structures in the concentration field and serve to follow the evolution of the flow patterns and their length scales. Furthermore, some information on the flow field can be extracted from the shadowgraph images by the optical flow procedure which detects the movement of structures in the concentration field [98]. More precise velocity measurements were performed by means of particle image velocimetry (PIV) in [52, 94]. The cubical cuvettes described above are reduced to a quasi two dimensional Hele-Shaw geometry in Sect. 16.3.2.1.

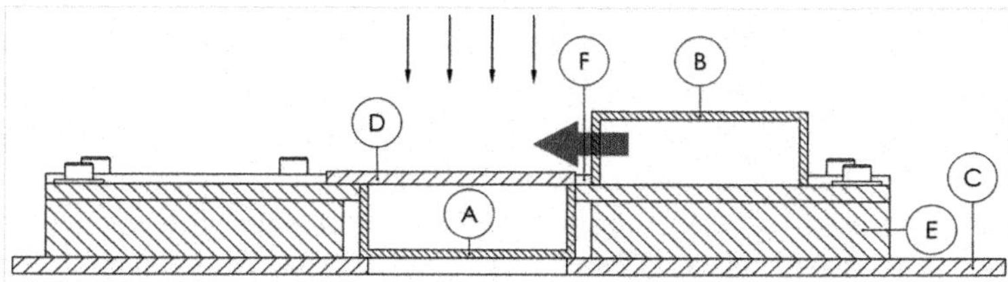

Fig. 16.4 Experimental setup to superimpose the two phases. Reprinted from [50], with permission from AIP Publishing

Fig. 16.5 Sketch of the two-layer-mass-transfer system. Initially, solute is either solely in the top or bottom layer with corresponding direction of transport

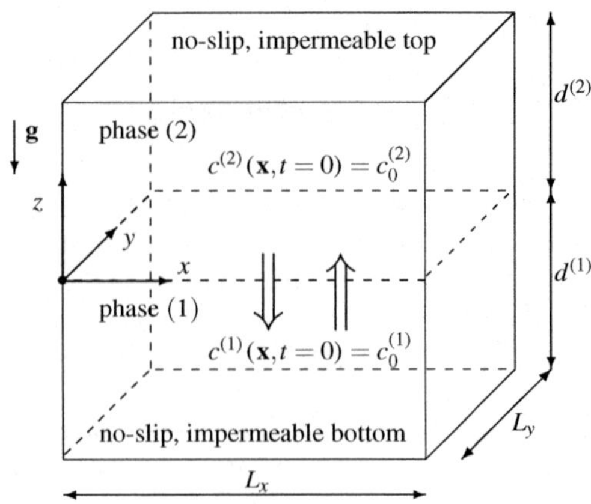

16.2.2 Governing Equations

We consider two immiscible liquid phases separated by an interface and confined between planar, impermeable, and rigid walls at top and bottom. The bottom layer (first phase) is denoted with the superscript (1) and the top layer (second phase) with superscript (2). Their respective thickness is denoted by $d^{(i)}$. Both layers are isothermal, viscous and incompressible with constant material properties. Changes in density due to concentration variation of a solute are treated by the Boussinesq approximation. The geometry is sketched in Fig. 16.5.

The interface is assumed as planar. It is located at $z = 0$ and does not deform. Convection arises from the instability associated with the diffusive transfer of the solute from one liquid to the other, which proceeds from the layer (i) with initial solute concentration $c_0^{(i)} = c_0 > 0$ to the other layer (j) ($j \neq i$) with $c_0^{(j)} = 0$. Interfacial tension (σ) and density ($\rho^{(i)}$) in layer (i) are assumed to depend linearly on dimensional solute concentration ($\tilde{c}^{(i)}$), i.e.

$$\rho^{(i)} = \rho_{ref}^{(i)} + \rho_{ref}^{(i)} \beta_c^{(i)} \tilde{c}^{(i)}, \tag{16.2}$$

$$\sigma = \sigma_{ref} + \sigma_{ref} \alpha_c \tilde{c}^{(1)}. \tag{16.3}$$

Note that a tilde is added for the dimensional concentration to distinguish it from the nondimensional concentration. The solutal expansion coefficients are $\beta_c^{(i)}$ and the coefficient of interfacial tension change is α_c. Kinematic viscosities $\nu^{(i)}$ and diffusivities $D^{(i)}$ of the solute are constant in each phase. The concentration at the interface is supposed to be in thermodynamic equilibrium with the local excess concentration of the interface Γ. In Henry's model, the latter is proportional to the concentrations adjacent to the interface, i.e. $\Gamma = K_1 \tilde{c}^{(1)}$ and $\Gamma = K_2 \tilde{c}^{(2)}$. The partition coefficient or Henry's constant H is defined by

$$\tilde{c}^{(1)} = K_2 / K_1 \tilde{c}^{(2)} = \tilde{c}^{(2)} / H \tag{16.4}$$

and approximated by means of a correlation method [61]. The material properties provide non-dimensional parameters summarized in Table 16.1.

In both layers the momentum transport is modeled by the incompressible Navier-Stokes-Boussinesq equations and the solute transport by an advection-diffusion equation. The nondimensional equations are derived by introducing the following scales. Mass is measured in multiples of $\tilde{M} = \rho_{ref}^{(1)}(d^{(1)})^3$, time in viscous units $\tilde{T} = (d^{(1)})^2/v^{(1)}$, length in multiples of the lower layer height $\tilde{L} = d^{(1)}$ and molar concentration of solute relative to the initial concentration c_0. The computational domain in Fig. 16.5 has a horizontal size of $l_x \times l_y$ and is assumed periodic in the two horizontal directions. The vertical dimensions are $-1 \leq z \leq 0$ for the lower phase and $0 \leq z \leq d$ for the upper phase. The nondimensional equations are [77]

$$\partial_t \mathbf{u}^{(1)} = -\mathbf{u}^{(1)} \cdot \nabla \mathbf{u}^{(1)} - \nabla p_d^{(1)} + \Delta \mathbf{u}^{(1)} - c^{(1)} G \mathbf{e}_z, \tag{16.5}$$

$$\nabla \cdot \mathbf{u}^{(1)} = 0, \tag{16.6}$$

$$\partial_t \mathbf{u}^{(2)} = -\mathbf{u}^{(2)} \cdot \nabla \mathbf{u}^{(2)} - \frac{1}{\rho} \nabla p_d^{(2)} + v \Delta \mathbf{u}^{(2)} - c^{(2)} G \beta \mathbf{e}_z, \tag{16.7}$$

$$\nabla \cdot \mathbf{u}^{(2)} = 0, \tag{16.8}$$

$$\partial_t c^{(1)} = -\mathbf{u}^{(1)} \cdot \nabla c^{(1)} + \frac{1}{Sc^{(1)}} \Delta c^{(1)}, \tag{16.9}$$

$$\partial_t c^{(2)} = -\mathbf{u}^{(2)} \cdot \nabla c^{(2)} + \frac{D}{Sc^{(1)}} \Delta c^{(2)}, \tag{16.10}$$

where the dimensional molar concentration is $\tilde{c}^{(i)} = c^{(i)} \cdot c_0$ and the dimensional velocity in phase i is $\mathbf{u}^{(i)} \cdot \tilde{L}/\tilde{T}$, respectively.

The matching conditions at the plane interface ($z = 0$) are:

$$u_x^{(1)} = u_x^{(2)}, u_y^{(1)} = u_y^{(2)}, u_z^{(1)} = u_z^{(2)} = 0, \partial_z c^{(1)} = D\partial_z c^{(2)}, c^{(1)}H = c^{(2)}, \tag{16.11}$$

$$\frac{Ma}{Sc^{(1)}} \partial_x c^{(1)} = -\mu \partial_z u_x^{(2)} + \partial_z u_x^{(1)}, \quad \frac{Ma}{Sc^{(1)}} \partial_y c^{(1)} = -\mu \partial_z u_y^{(2)} + \partial_z u_y^{(1)}. \tag{16.12}$$

These relations arise from the continuity of velocity, the conservation of solute in the bulk, Henry's model and the shear stress balance at the interface. The boundary conditions are periodic in the x-y dimension, no-slip and impermeable boundaries are supposed for the solid walls at the bottom ($z = -1$) and top ($z = d$). The various non-dimensional parameters are defined in Table 16.1.

The shadowgraph technique used in the experiments is mimicked by averaging the horizontal Laplacian of the concentration distribution over both layers [76],

$$s(x, y) = \int_{-1}^{0} (\partial_x^2 + \partial_y^2) c^{(1)}(\mathbf{x}, t) dz + \int_{0}^{d} (\partial_x^2 + \partial_y^2) c^{(2)}(\mathbf{x}, t) dz. \tag{16.13}$$

A better approach would be to take the different refraction indices in both layers into account. This has not been done since these parameters have not been determined. Moreover, the *experimental* shadowgraph records are affected by slight deflections of the interface and nonlinearities of the refractive index as well as of the optical devices. Nonetheless, the quantity $s(x, y)$ permits a visual comparison of the emerging structures with the experiments.

16.2.3 Numerical Method

The planar two-layer geometry is suitable for the application of pseudospectral methods [19, 82]. A significant advantage of the pseudospectral discretization is that the solution of large linear systems of equations, which appear as a result of most other discretization strategies, is avoided. As a result, a very efficient numerical scheme is obtained. However, this advantage comes at the cost of low flexibility in terms of geometry and boundary conditions. The main application of pseudospectral discretizations is on domains with at least one periodic direction where the other boundaries conform to coordinate lines. This includes hexahedral domains for cartesian coordinates, or cylindrical or spherical shells with the respective cylindrical or spherical coordinates.

The particular pseudospectral scheme that has been used has been comprehensively presented in [12]. A significant variety of different flows have been simulated with this code (slightly adapted and extended for the specific problem), e.g. turbulent magnetohydrodynamic channel flow [13] and chemical-reaction-driven buoyant convection [51]. To study mass transfer in the two-layer system, temperature in [12] is replaced by the concentration field, and the respective boundary conditions for this field are modified [49].

Both planar layers are treated as separate computational domains that are coupled at the interface. In each layer, the fields are expanded in truncated Fourier series in the two periodic horizontal directions x, y. The vertical direction z is expanded in Chebyshev polynomials T_p of order p. The smallest wavenumbers for the x,y directions are $k_{x0} = 2\pi/l_x$, $k_{y0} = 2\pi/l_y$. For example, the expansion of the vertical velocity in the upper layer, where $(x, y, z) \in [0, l_x] \times [0, l_y] \times [0, d]$, reads:

$$u_z^{(2)}(x, y, z, t) = \sum_{m=-N_x/2}^{N_x/2-1} \sum_{n=-N_y/2}^{N_y/2-1} \sum_{p=0}^{N_z} e^{imk_{x0}x + ink_{y0}y} T_p(2z/d - 1)\hat{u}_z^{(2),m,n,p}. \qquad (16.14)$$

The data output in physical space and the calculation of the nonlinear terms is done on the Gauss-Lobatto collocation points. For layer (2) they are:

$$\left(x_i, y_j, z_k\right) = \left(i \cdot l_x/N_x, \; j \cdot l_y/N_y, \; d \cdot 0.5(1 + \cos(k\pi/N_z^{(2)}))\right), \qquad (16.15)$$

with $i \in \{0, 1, \ldots, N_x\}, j \in \{0, 1, \ldots, N_y\}, k \in \{0, 1, \ldots, N_z^{(2)}\}$. The velocity field is represented through the poloidal-toroidal decomposition, whereby incompressibility is automatically satisfied.

The time step is adjusted according to the current grid Courant-Friedrichs-Levy (CFL) number,

$$C_g = \max \left\{ \frac{u_x \delta t}{\Delta x}, \frac{u_y \delta t}{\Delta y}, \frac{u_z \delta t}{\Delta z} \right\}. \tag{16.16}$$

This number is calculated every time step by dividing the local displacement $u_\alpha \delta t$ by the local values Δx, Δy, Δz of the spacing between collocation points. We force C_g to be smaller than a constant C_b and to be larger than $C_b/2$. It turned out that $C_b = 0.3$ is appropriate for a stable time-stepping scheme, i.e. $C_b = 0.3$ was consistently used. The velocity field is initialized (at $t = 0$) with pseudorandom numbers. Specifically, the gridpoint values of the vertical velocity component $u_z(x_i, y_j, z_k)$ and the vertical vorticity $\nabla \times \mathbf{u}(x_i, y_j, z_k) \cdot \mathbf{e}_z$ are uniformly distributed between $[0, 1 \cdot 10^{-3}]$. The solute is initialized with a homogeneous concentration of unity either in layer 1 or layer 2 and zero in the other layer.

The numerical resolution requirements are quite severe due to the high Schmidt number Sc of the system. A very high vertical resolution near the interface is required to capture the vertical structures properly. The horizontal resolutions (N_x, N_y) are set to resolve the similar fine solute structures in the horizontal directions. The main resolution criterion in our studies was the absence of visible oscillations in the solution variables on the scale of the grid step, i.e. the spacing between collocation points. Such oscillations typically appear when a hydrodynamic field is poorly resolved, and may be enhanced due to aliasing errors when products of hydrodynamic variables are computed in physical space. Synthetic shadowgraph images $s(x, y)$ are fairly sensitive to such a lack of resolution since they involve second-order derivatives. The particular resolution will be given in the result sections.

The resolution requirements are difficult to predict beforehand: For the interfacial tension driven flows we observed two mechanisms that require special care. The first is the amplification of Marangoni roll cells, which is described in Sect. 16.3.1. The primary instability mechanism can amplify rather small flow structures with sizes that are difficult to predict. E.g., in [49], we made an attempt to predict the smallest convection cells that one has to expect. The second mechanism are velocity gradients generating so-called eruptions in the systems where buoyancy and Marangoni convection lead to an oscillatory behavior. The eruptions can create a high concentration gradient during the propagation of a concentration front. In both processes, the precise treatment of solute diffusion adjacent to the interface is essential, because it governs the source for convection, i.e. the gradients in interfacial solute concentration.

16.2.4 Parallelization Strategy

Parallelization of the numerical scheme with message passing interface (MPI) is based on a domain decomposition in one horizontal direction, which requires transposition of the array of expansion coefficients across the MPI tasks in order to compute the Fast Fourier Transforms (FFT). All other crucial operations are local in the $x - y$ or $k_x - k_y$ plane. The number of grid points in each direction is a power of two because only base two FFTs are used. The three-dimensional FFT consists of a sequence of one-dimensional FFTs in each coordinate direction. To perform them on the complete three-dimensional array representing one hydrodynamic variable, the array is decomposed into equal slices assigned to individual MPI processes for computation. Cuts are made perpendicular to one of the horizontal directions. The one-dimensional FFT with respect to the intact directions can be computed by each process irrespective of the others. The one-dimensional FFT in the cut direction clearly requires data exchange between processes. A computation of the one-dimensional FFT on a vector, i.e. a one-dimensional array of numbers, that is distributed across the processes, by exchanging data as needed is possible [93]. However, it requires fairly involved logic since the Fourier transform is non-local. To avoid this, it is customary to transpose the complete array, which puts the "cut vectors" back together so that they reside in the local memory of the individual process [44]. This naturally requires the array to be cut in another direction (unless we allow for extra storage). One can visualize this in two dimensions. Figure 16.6 sketches the algorithm for the case of four processes assuming an equal number of data points in the x- and y-directions.

The initial distribution of the data in Fig. 16.6a is such that each process can perform the transforms on the vectors (indicated by the arrows) with respect to x. In order to be able to transform with respect to y, the data have to be transposed to obtain the mirror image of the array with respect to the main diagonal as reflection plane as shown in Fig. 16.6d. Two steps are needed to perform this transposition. In the first step indicated in Fig. 16.6b, we divide the data into square sub-arrays of equal size (obtained by slicing in y direction) and transpose each sub-array locally (i.e. by each process without communication) as required for the mirror image. After that, the sub-arrays are exchanged between the processes (Fig. 16.6c), which gives the desired result.

Because data are swapped among the MPI processes it is natural to pair them for data exchange. The data exchange is complete when all possible process pairs have been exhausted. Process pairs are grouped into disjoint sets, which can then communicate at the same time without interference. This leads naturally to an exchange in stages. For the four process cases, three stages are required. One possible pattern could consist of the process pairs $(0, 1)$ and $(2, 3)$ in the 1st stage, the pairs $(0, 2)$ and $(1, 3)$ in the 2nd stage and finally the pairs $(0, 3)$ and $(1, 2)$ in the 3rd stage. It is obvious that this can be generalized to any even number M of processes. The data exchange will then require $M - 1$ stages. Details on the implementation are given in [11]. We also note that the scalability of the code might

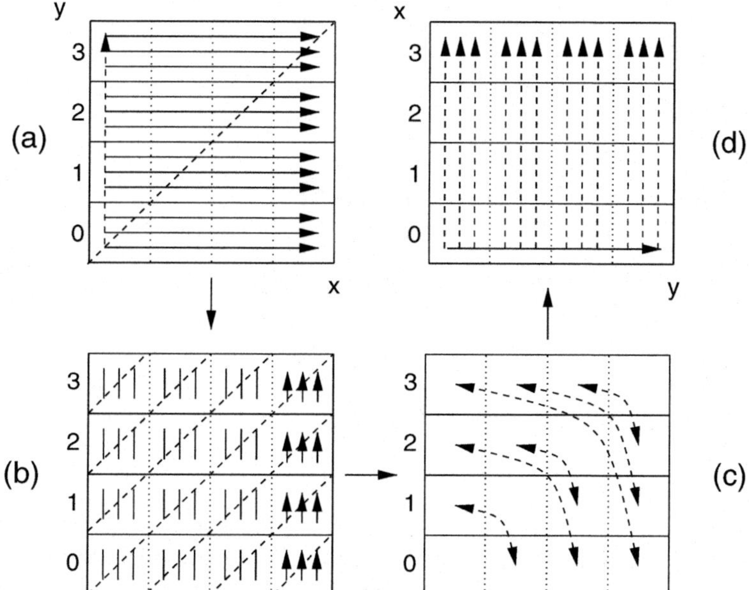

Fig. 16.6 Data array transposition across processes for parallelized FFT. Initial distribution of data is indicated by the *full arrows* in (**a**), where the *frame boxes* correspond to the individual processes numbered from 0 to 3. The *dashed arrow* indicates that one-dimensional transforms in y direction cannot be computed this way. Transposition about the diagonal (shown as *dashed line*) proceeds by local sub-array transposition (**b**) and swapping sub-arrays between processes (**c**) in order to arrive at the desired distribution (**d**)

be further improved by implementing a decomposition approach also for the second horizontal direction [81].

16.2.5 Parallel Efficiency

The speed-up due to the parallel execution is demonstrated in this section. As a test case, we performed one hundred time steps with a resolution of $N_x = N_y = 2048$, $N_z^{(2)} = N_z^{(1)} = 128$. This amounts to 1.07×10^9 collocation points. We simulated this test problem with different numbers n of MPI processes (one per CPU core) and measured the respective wall clock time t_w. The parallel efficiency $\eta(n)$ can be defined as

$$\eta = \frac{8 t_w(8)}{n \times t_w(n)} \tag{16.17}$$

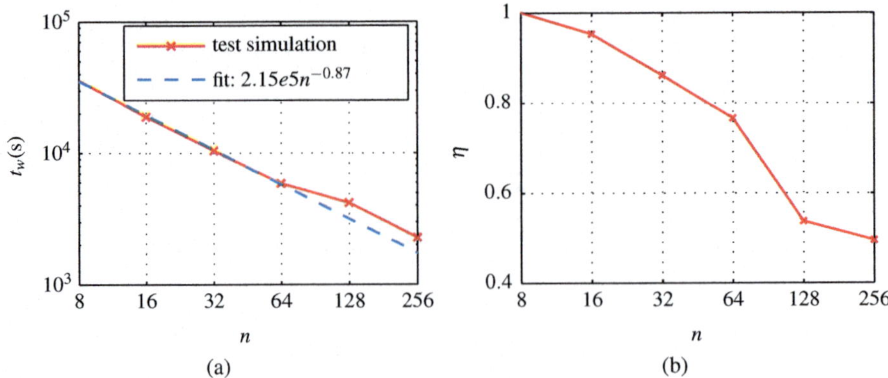

(a)   (b)

Fig. 16.7 Speed-up test of the pseudospectral code performed on the computing cluster of TU Ilmenau

where the reference value $n = 8$ is used as the smallest number of cores. Theoretically, the parallel efficiency should be close to one (ideal speed-up) since the code has almost no sequential parts. However, this consideration ignores the time needed for inter-process communication. The parallel efficiency should, therefore, be smaller than one and, moreover, decrease with n. The speed-up test is carried out on the local cluster (mapacc3) of TU Ilmenau. Each node, i.e. computer with shared memory, has a main memory of 64 GB with four *AMD OPTERON 6134 2.3* GHz CPUs. Each CPU has 8 cores, i.e. there are 32 cores per node. We fixed the number of nodes to four for all cases except for $n = 256$ where eight nodes with all 32 cores per node are used. We always allocated the full node to the job in order to prevent uncontrolled side effects from other processes. Furthermore, results are averaged over three independent runs to account for the variance in wall clock time. The averaged wall clock time was $t_w(n = 8) = 35,573$ s, $t_w(n = 16) = 18,681$ s, $t_w(n = 32) = 10,336$ s, $t_w(n = 64) = 5815$ s, $t_w(n = 128) = 4146$ s, $t_w(n = 256) = 2249$ s. Figure 16.7a displays those points together with a power law fit of $t_w = 2.15 \times 10^5 n^{-0.87}$. Figure 16.7b shows the parallel efficiency, which decreases with the number of utilized processes or cores. Note that for higher resolution a performance gain could be achieved by implementing an additional shared memory parallelization with OpenMP.

16.3 Marangoni Convection with Stable Density Gradient

The case with stable density gradient requires a system where the transfer of the solute reduces the density near the interface in the lower layer (1) and increases the density in the upper layer (2). At the same time, the conditions for stationary Marangoni instability identified by Sternling and Scriven have to be met. A suitable system is composed of a lower aqueous phase (1) covered by layer of cyclohexanol (2) where butanol is transferred into the aqueous phase. The material properties of this system are given in [50]. The nondimensional groups independent of the geometry and concentration are $Sc = 2400$, $\rho = 0.96$, $\nu = 16.7$, $D = 0.14$, $\beta = 0.75$, $H = 31$.

16.3.1 Three-Dimensional Flows in Wide Domains

Experimental and numerical investigations are carried out with layer heights $d^{(1)} = d^{(2)} = 20\,\text{mm}$, i.e. $d = 1$. Acceleration due to gravity is $g = 9.81\,\text{m/s}^2$. For a reference case the initial molar concentration of butanol is $\tilde{c}_0 = 0.82$ mol/l, which corresponds to a volume concentration of 7.5%. The resulting Marangoni and Grashof numbers are $Ma = -2.4 \times 10^8$ and $Gr = -7.67 \times 10^5$. The spatial resolution is $N_z^{(1)} = 256$, $N_z^{(2)} = 512$ for the vertical direction and $N_x = N_y = 2048$ for the horizontal directions.

16.3.1.1 Evolution of Hierarchical Patterns

We first describe the pattern evolution for the reference case with a volume concentration of butanol equal to 7.5%. The evolution is captured by shadowgraph images in the experiment. In the simulations we record the quantity s defined in Eq. (16.13) to enable a direct comparison with experimental images. Figures 16.8 and 16.9 display the quantity s from a highly resolved simulation of the reference configuration.

For the present physical configuration, $s(x, y)$ basically displays the horizontal solute distribution near the interface (due to highest horizontal solute gradients there) preferentially for the top layer. Bright regions ($s(x, y) > 0$) in the synthetic shadowgraph images can be considered as locations with a gain in solute by horizontal diffusion, which is mostly equal to a low concentration and hence high surface tension. This interpretation is simply deduced by comparing the definition (16.13) with the diffusion equation for the solute. Furthermore, the following correlations were observed: (bright region) $\leftrightarrow$ (flow from the interface to the bulk, high interfacial tension) and (dark regions) $\leftrightarrow$ (flow from the bulk to the interface, low interfacial tension). Note that the grayscale is not generic. It depends on the specific experimental configuration. At present, the grayscale is simply adapted to the experimental images.

The evolution illustrated by Figs. 16.8 and 16.9 is divided into two main phases. The rapid phase *I* is called *initial spreading*, which encompasses two short sub-phases. The first subphase *Ia* is that of *exponential growth*. The second subphase *Ib* named *saturation* begins after a maximum in interfacial velocity. It is characterized by a significant reduction in mean interfacial velocity and ends approximately at time t=0.07. Thereafter, the long phase *II*, named *hierarchy formation and*

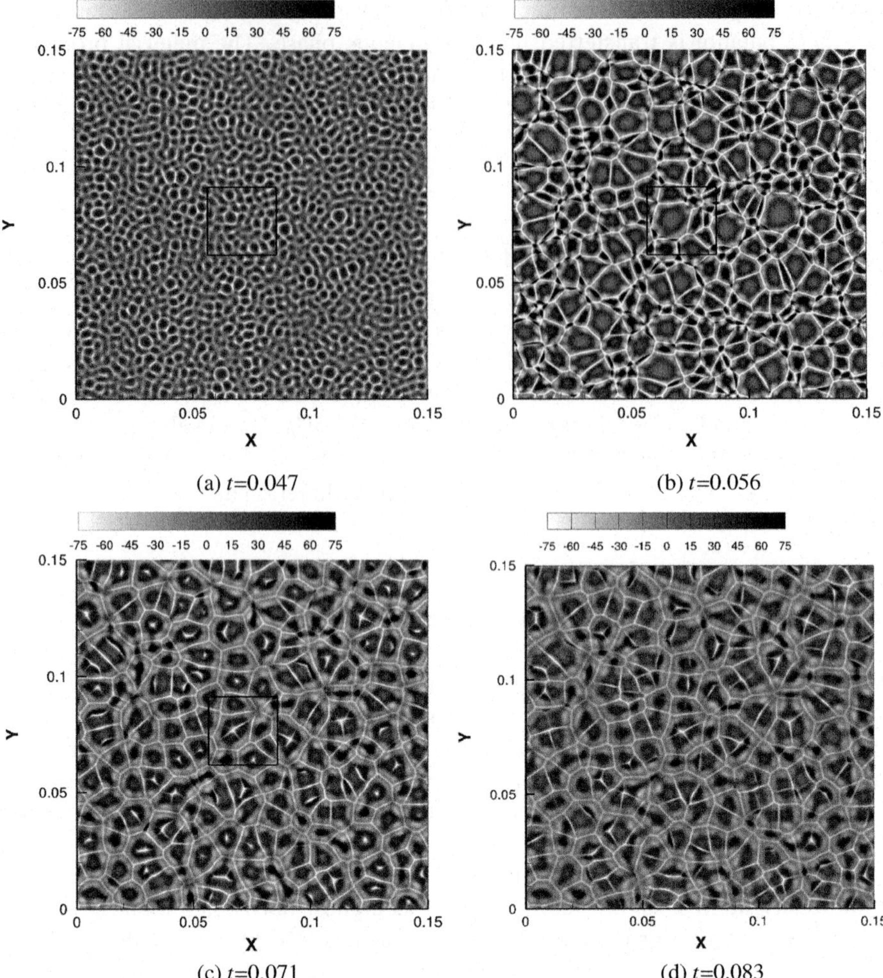

(a) t=0.047

(b) t=0.056

(c) t=0.071

(d) t=0.083

Fig. 16.8 Synthetic shadowgraph images $-s(x, y)$ from simulation of reference case show the pattern formation at start-up of interfacial convection: (**a**) the end of phase *Ia*—exponential growth, (**b**) phase *Ib* with initial RO mark the saturation of perturbation growth, (**c,d**) transition to phase *II* by breakdown of initial RO. (**a–c**) Reprinted from [50], with permission from AIP Publishing

coarsening, starts by substructuring of the cells in Fig. 16.8c and continues with only minor changes in the mean velocity.

Figure 16.8a shows phase *Ia* with a spotted pattern of small cells, being nuclei for a vivid spreading of cells. According to the classification by Linde et al. [68] (see Sect. 16.1.4), these growing structures are called (initial) relaxation oscillations—ROs.

The next subphase *Ib* saturation (Fig. 16.8b) shows strongly growing cells and cells that are either squeezed or incorporated by the stronger advective motion

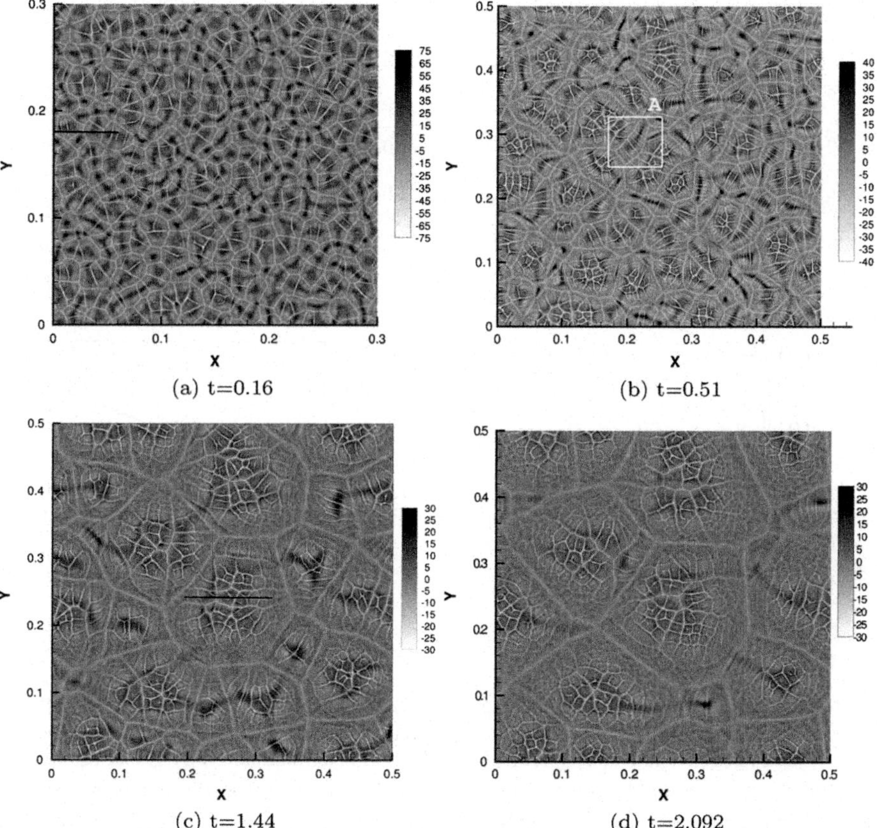

Fig. 16.9 Synthetic shadowgraph images $(-s(x, y))$ in phase *II* from simulation of reference case. Note the change in the domain size from (**a**) to (**b**) and also the adaption of the grayscale. (**a–c**) Reprinted from [50], with permission from [96], with permission from Elsevier

of their neighbors. Due to the balance of equally strong neighbors, the extensive spreading motion comes to a halt with a maximum in cell sizes. The cell borders orient in mostly straight lines and thus build a polygonal pattern.

Phase II, termed *hierarchy formation and coarsening*, is initiated by the breakdown of the initial ROs, where the biggest cells (e.g. in the region marked with a square in Fig. 16.8) split into a network of smaller polygonal, more persistent cells.

In Fig. 16.9a–d, the more vigorous cells grow and develop an internal substructure of smaller Marangoni cells. In line with [68], we term these polygonal patterns with substructure *Marangoni roll cells of second order* RC-IIs. The enclosed cellular substructures and the individual cells without substructure are called *Marangoni roll cells of first order*, RC-Is [68]. They constitute the lowest level of the hierarchy. In the early stage, presented in Fig. 16.9a, an unambiguous assignment of the individual cells to a hierarchy order is hardly possible due to the continuous evolution of length scales. However, from Fig. 16.9b–d the beginning spatial hierarchy formation

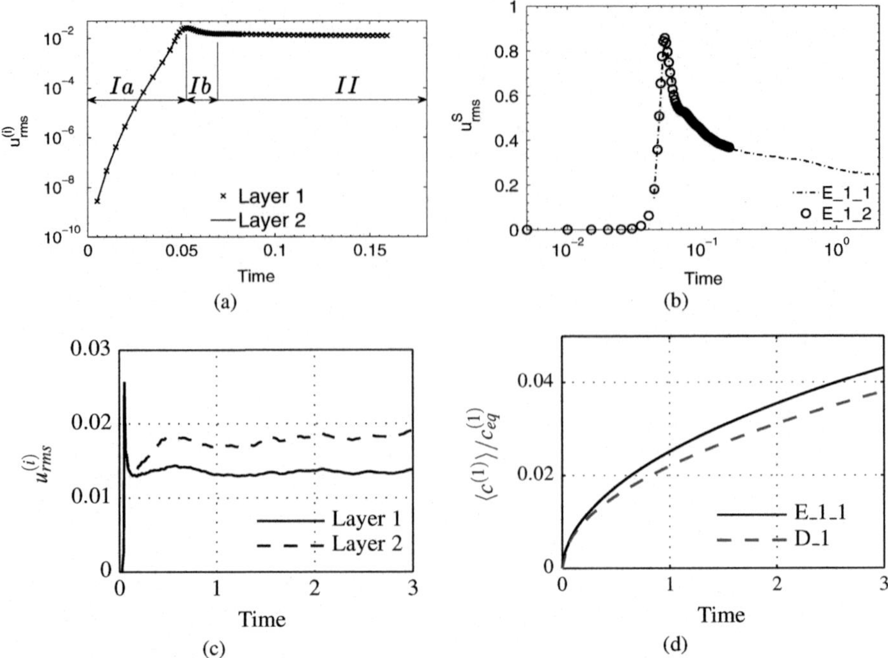

Fig. 16.10 Simulation of reference configuration: (**a**) Root-mean-square velocity in the bulk, (**b**) rms velocity at the interface for two different numerical resolutions (no visible difference), (**c**) rms velocity, (**d**) averaged solute concentration in the bottom layer normalized by global equilibrium value for the reference simulation (E_1_1) and pure diffusion (D_1). The global maximum of $u_{rms}^{(i)}$ and $u_{rms}^{(s)}$ is around t=0.053. (**a–c**) Reprinted from [50], with permission from AIP Publishing

is clearly discernible because the RC-IIs steadily increase in size and the separation of scales between RC-Is and RC-IIs grows.

Another type of pattern, in contrast to the closed polygonal RC-Is, is observed in Fig. 16.9b–d. Particularly in those RC-IIs that are about to shrink and disappear, e.g. white **A** mark in Fig. 16.9b, arrays of aligned, straight solute fronts are visible. According to its wavelike appearance, this pattern is referred to as *relaxation oscillation waves*—ROWs [68], see also Fig. 16.3. This pattern is not that persistent and appears favored in host RC-II that have a distinctly elongated geometry.

In the late stage in Fig. 16.9d the periodicity length of the domain becomes of the same order as the RC-IIs and consequently influences their behavior. At this time, only approximately 4% of the maximum transferable amount of solute was transferred.

Figure 16.10a–c depicts the root-mean-square velocities $u_{rms}^{(i)}$, $u_{rms}^{(s)}$ in the bulk and at the interface, which are defined as

$$u_{rms}^{(i)}(t) = \sqrt{\langle \mathbf{u}^{(i)} \cdot \mathbf{u}^{(i)} \rangle_{xyz}}, \quad u_{rms}^{(s)}(t) = \sqrt{\langle \mathbf{u} \cdot \mathbf{u} \rangle_s}. \tag{16.18}$$

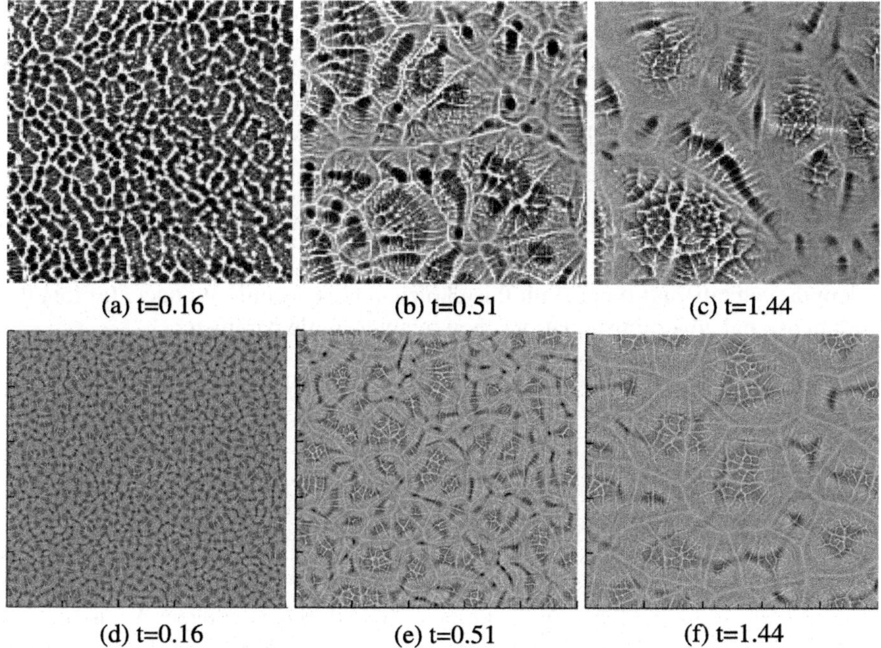

<table>
<tr><td align="center">(a) t=0.16</td><td align="center">(b) t=0.51</td><td align="center">(c) t=1.44</td></tr>
<tr><td align="center">(d) t=0.16</td><td align="center">(e) t=0.51</td><td align="center">(f) t=1.44</td></tr>
</table>

Fig. 16.11 Growing size of RC-IIs in reference case: (**a,b,c**) experimental shadowgraph images from [50] (window of approx. 0.5×0.5) and (**d,e,f**) simulated pattern ($-s(x, y)$) in a domain of 0.5×0.5 (cf. Fig. 16.9). Reprinted from [50], with permission from [96], with permission from Elsevier

They are obtained by averaging over the entire layer $\langle\rangle_{xyz}$ and over the interface $\langle\rangle_s$. Figure 16.10d shows the amount of solute transferred to the lower layer normalized with the global equilibrium value in the bottom layer $c_{eq}^{(1)} = 1/(H+1)$ in comparison to the purely diffusive case. The different sub-phases Ia,Ib in the early pattern evolution can be clearly identified in Fig. 16.10a,b. The second phase II shows only minor variations in the velocity and a gradual increase in the mean solute concentration in the accepting layer.

Figure 16.11 compares experimental shadowgraph images with the simulations. There is a good qualitative agreement regarding cell shapes and hierarchy formation although the pattern evolution appears to be somewhat faster in the experiment. The initial stage of the evolution cannot be accessed in the experiment due to the lack of optical access during the overlaying procedure.

16.3.1.2 Effect of Initial Concentration

In addition to the reference case, different initial solute concentrations were explored in the experiments and in the simulations. The early evolution proceeds in a similar manner, i.e. with the same stages and qualitatively similar cell pattern.

This similarity is not accidental but based on an approximate scaling invariance of the basic equations. It can be identified when one uses the so-called interface units based on the characteristic interfacial tension difference $\Delta\sigma = |\sigma_{ref}\alpha_c c_0|$ computed with the initial butanol concentration c_0. These interface units are independent of the layer height. We define them by

$$L_{int} = \frac{\mu^{(1)}\nu^{(1)}}{\Delta\sigma}, \qquad U_{int} = \frac{\Delta\sigma}{\mu^{(1)}}, \qquad T_{int} = L_{int}/U_{int} \qquad (16.19)$$

for length, velocity and time. With this choice, and c_0 as unit of concentration, the nondimensional governing equations then retain a single parameter

$$Mo = \frac{GSc^3}{|Ma|^3} = \frac{c_0\beta_c^{(1)}(\mu^{(1)})^3\nu^{(1)}g}{(\Delta\sigma)^3} \qquad (16.20)$$

multiplying the buoyancy term, which contains c_0. The Morton number Mo is a natural parameter, e.g. for the terminal velocity of bubbles rising in a liquid due to buoyancy. Based on interface units, the interfacial boundary conditions complementing the equations are also free of parameters containing c_0. In contrast to the viscous units, the location of the outer boundaries now depends on L_{int}, i.e. c_0. However, they can be assumed to be at infinity for times that are short compared to the time required for equilibration (vertical diffusion time $(d^{(i)})^2/D^{(i)}$).

In the limit $Mo = 0$, a modification of the initial concentration would then clearly result in an identical evolution if all quantities are measured in interface units. This is verified in the simulations for finite Mo for integral quantities such as characteristic velocities and length scales of the flows, provided that the time is sufficiently small. For larger times, the buoyancy term becomes appreciable and the scaling invariance of the problem with respect to c_0 no longer holds. We also remark that the detailed evolution, i.e. exact location, shape and size of cells, depends on the initial perturbation fields, which, in general, did not scale between different c_0 according to the interface units.

16.3.1.3 Flow Structure and Mechanisms

The transfer of solute naturally leads to a coarsening since an increasingly larger volume is depleted in solute (in the donating layer) or enriched in solute (in the accepting layer). In phase II of pattern evolution, the largest characteristic length scale in the flow therefore increases with time. In contrast to a purely diffusive evolution, the Marangoni effect continuously generates small-scale features due to a local instability. It occurs in the so-called inflow regions of the largest cells.

This characteristic feature is illustrated in Fig. 16.12 showing a vertical section of a second-order Marangoni roll cell (RC-II) at time $t = 1.436$, i.e. well into phase II. The large-scale outer flow consists of two counter-rotating eddies, which show a fairly uniform concentration and extend up to $z \approx 0.05$ in the top and $z \approx -0.03$

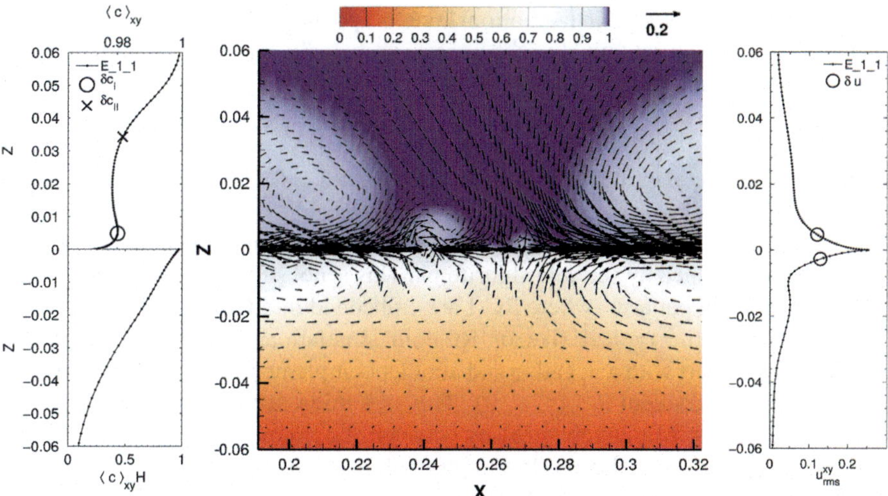

Fig. 16.12 Marangoni roll cells in simulation of reference case (E_1_1): Horizontally averaged concentration (*left*, **a**) and rms velocity profiles (*right*, **c**) together with a vertical cross section of a *RC-II with a substructure of smaller RC-Is* at t=1.436 for y=0.24 (*center*, **b**), i.e. along the *black line* in Fig. 16.9c. The *color* depicts the concentration distribution in the upper layer $c^{(2)}$ and the scaled concentration in the lower layer $c^{(1)}H$. Vectors for velocity are plotted on a much coarser grid than used in the simulation rendering the large-scale flow. The individual markers in the profiles (**a**) and (**c**) represent the collocation points. δc_I, δc_{II}, δu represent length scales based on the mean profiles. Reprinted from [50], with permission from [96], with permission from Elsevier

in the bottom layer. In the central part of Fig. 16.12 the flow is directed towards the interface, and the concentration is close to the uniform initial values. In this region the smaller RC-I are generated by the instability mechanism and subsequently advected by the larger RC-II eddies. The structure of the horizontally averaged concentration and rms velocity profiles that are also shown in Fig. 16.12 reflect this particular hierarchical flow structure. In the upper phase ($z > 0$) the concentration profiles have two characteristic gradients separated by turning points. The first one near the interface reflects the small-scale RC-I. The second gradient connects the *mixed fluid* near the interface with almost unchanged bulk fluid, and originates from the larger RC-II. The measures δc_I and δc_{II} defined in [50] with the slope of mean concentration profile and the amount of transferred solute characterize these gradients adequately. In the lower layer there is no such distinct concentration profile due to the significantly higher solute diffusivity. The profiles of horizontally averaged velocity shown in Fig. 16.12 are defined by

$$u_{rms}^{xy}(z) = \sqrt{\langle \mathbf{u} \cdot \mathbf{u} \rangle_{xy}}. \tag{16.21}$$

The highest velocities are reached at the interface. In contrast to the concentration, the presence of RC-I in addition to RC-II cannot be directly identified. The smaller vertical extent of the RC-II in the bottom layer is apparent. The shape of the velocity profile also agrees qualitatively with measured velocity profiles of substructured Marangoni cells in a Hele-Shaw geometry [94].

16.3.1.4 Characteristic Horizontal Length Scales

For a quantitative analysis the different horizontal length scales apparent in Fig. 16.9 need to be determined by suitable algorithms. This is not straightforward on account of the hierarchical patterns, where RC-I are embedded into RC-II.

To determine the typical size of RC-I in simulations the interfacial area is separated into a part with a negative and another complementary part with a positive vertical velocity gradient $\partial_z u_z(z = 0)$. This definition accounts for the source of convection, i.e. the Marangoni stresses between the inflow region ($\partial_z u_z < 0$, low interfacial tension) and the outflow region ($\partial_z u_z > 0$, high interfacial tension). Specifically, we produce a binary distribution $I_0(\partial_z u_z)(x, y)$ of white $I_0(\partial_z u_z)(x, y) = 1$ for $\partial_z u_z > 0$ (outflow) and black $I_0(\partial_z u_z)(x, y) = 0$ for $\partial_z u_z \leq 0$ (inflow). An example of such a binarized image is shown in Fig. 16.13. A characteristic feature in Fig. 16.13 is the general bounding of the black inflow regions by the white outflow regions. These connected black subareas are identified as individual cells. The characteristic horizontal length scale, λ_{RC-I}, of the RC-Is is calculated from the

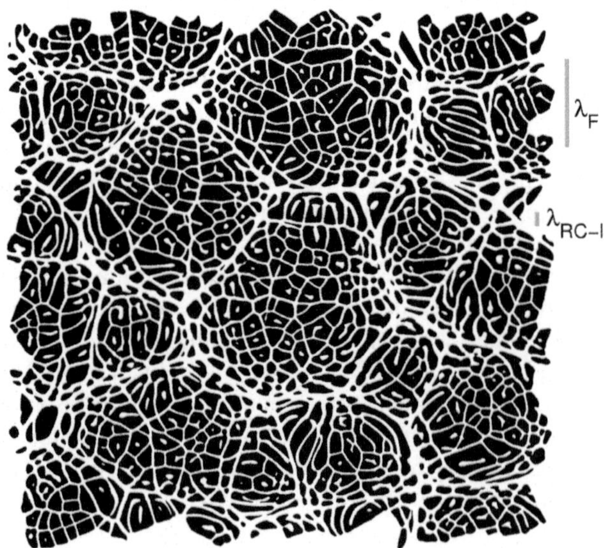

Fig. 16.13 Vertical velocity gradient in binary representation $I_0(\partial_z u_z(z = 0, t = 1.44))$ and calculated length scales (λ_F, λ_{RC-I}) for reference case. Reprinted from [50], with permission from AIP Publishing

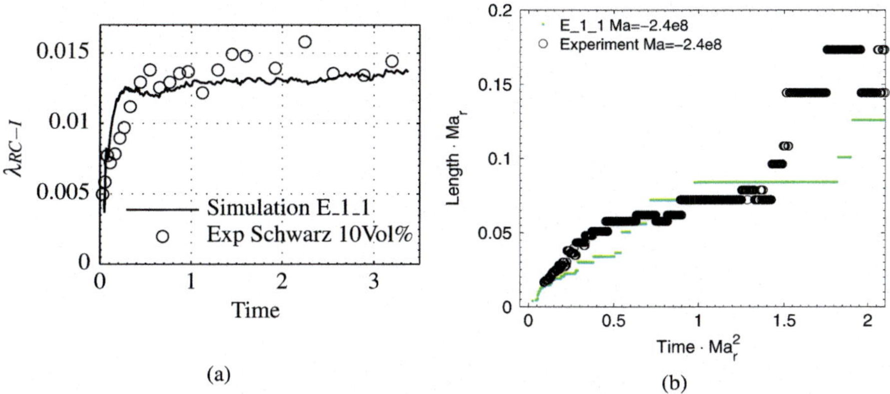

Fig. 16.14 Evolution of length scales for reference case: (**a**) λ_{RC-I} of the RC-Is from simulation and measurements of the mean RC-Is size by Schwarz (see Fig. 56 in [91]). (**b**) λ_F for size of RC-IIs from simulation and experiment [50]. The jumps result from averaging the spectrum around wavelength l_x/i for $i \in (1, 2, 3, \ldots, N_x/2)$. Reprinted from [50], with permission from AIP Publishing

individual cell areas A_j as the mean of the circular diameters. Details of the image processing are described in [50].

The mean RC-IIs size was quantified by the dominant Fourier modes in the shadowgraph image and denoted as λ_F (with the subscript F for Fourier). The calculation is also explained in [50]. Note that we defined λ_F in a particular way to extract a length scale from experimental as well as simulated images that indeed represent RC-II sizes in agreement with the visual impression. This is because the power spectrum of $s(x, y)$ peaks for wavenumbers that are related to the width of cell boundaries, which is not the desired length and also not accurately resolved in the experimental images.

The mean diameter of the RC-Is for the reference simulation is presented in Fig. 16.14a. This diameter grows until it reaches a fairly constant value $\lambda_{RC-I} \approx 0.013$, which agrees well with the corresponding measurement of Schwarz [91] (circles), see his Fig. 56. The end of the RC-Is growth phase coincides with the occurrence of the RC-IIs. According to our analysis in Sect. 16.3.1.2, we tried to account for the concentration change by the rescaling of time and length. It turns out that there is an effectively identical behavior of the RC-I sizes for different concentrations, especially in the early stage.

The quantity λ_F reflects the RC-IIs size after t=0.3 (simulation labelled E_1_1 for reference case) when the RC-II regime starts to dominate. The same evaluation procedure was carried out for the experimental shadowgraph records, cf. Fig. 16.11. Experiments with a lower concentration of solute could not be used due to poor contrast of the shadowgraph images. The growth of λ_F is shown in Fig. 16.14b. It agrees between the simulations (dots) and the observations from the reference experiment (black circles). However, the experimental length scales exceed the ones from the simulations in the early course of the evolution. This is in line with the visual impression from Fig. 16.11. The length ratio λ_F/λ_{RC-I} is related to the number of subcells in a RC-II and increases with time. More details on the transformation of cells in the course of the pattern evolution are given in [50].

16.3.2 Flows in Hele-Shaw Geometry

16.3.2.1 Experiment

Since the basic mechanism of Marangoni instability does not depend on an intrinsically three-dimensional mechanism, the flow evolution in two dimensions should not be fundamentally different from that in three dimensions. To realize an approximately two-dimensional situation in experiments, a Hele-Shaw (HS) cell is commonly used, where the liquid-liquid system is placed in a narrow gap between two glass plates. Such a cell provides simplified access to the vertical structure of measurement quantities recorded by optical methods. By varying the orientation of the HS cell, it is possible to reveal information about the influence of gravity [34, 52, 66, 94, 102].

Figure 16.15 sketches the experimental HS setup (a) and the computational domain (b), which is drawn as a gray inset in (a) for comparison. A spacer made of polytetrafluoroethylene (PTFE) foil, whose inner contour is shown as a dashed line in Fig. 16.15a, acts as a container for the liquids. The shape of the PTFE foil was optimized to provide a robust filling procedure for different two-phase systems [103]. The gap width 2ε, see Fig. 16.15b, is set by the thickness of the foil.

16.3.2.2 Hele-Shaw Model

According to the experimental setup, the nondimensional plate distance is $2\varepsilon/d^{(1)} = 2\gamma^{-1/2}$ where $\gamma = \frac{(d^{(1)})^2}{\varepsilon^2}$ is the ratio of squared layer height to half plate distance, cf. Fig. 16.15b. In x and z-direction, the geometry is defined according to the three-dimensional domain in Sect. 16.2.2.

The flow in a HS gap is typically simulated by gap-averaged hydrodynamic models [14, 17, 38]. In our work [52] we adopt the formulation of [17] based on the following two assumptions. First, the nondimensional velocity field $\mathbf{u}^{(i)}(x, y, z, t)$ in layer$^{(i)}$ has only two non-zero components ($u_y^{(i)} = 0$) with a parabolic dependence on y, i.e.

$$\mathbf{u}^{(i)}(x, y, z) = \frac{3}{2}(1 - y^2\gamma)(v_x^{(i)}(x, z)\mathbf{e}_x + v_z^{(i)}(x, z)\mathbf{e}_z). \tag{16.22}$$

In what follows, we are concerned with velocity fields that are gap-averaged over y and denoted by $\mathbf{v}^{(i)}(x, z, t) = \langle \mathbf{u}^{(i)} \rangle_y$. The second assumption is that solute

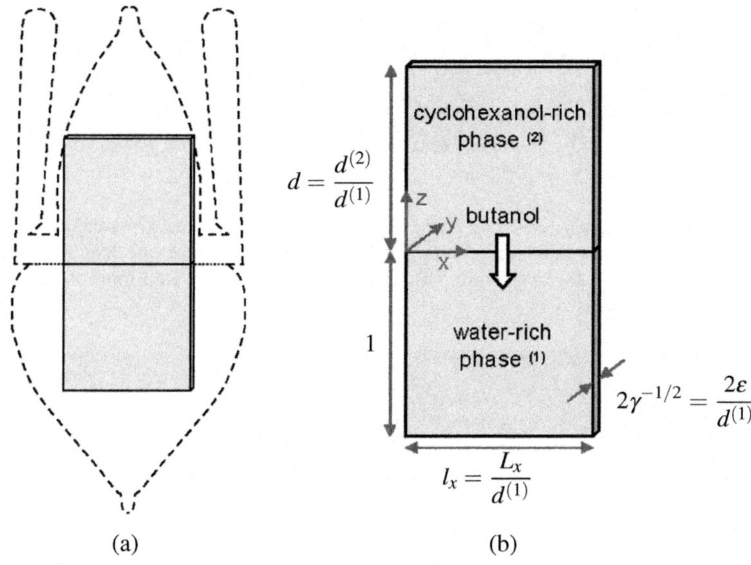

Fig. 16.15 Hele-Shaw (HS) geometry as experimental setup (**a**) according to [103] with *gray inset* of numerical domain as defined in [52]; detailed numerical domain (**b**) in dimensionless lengths according to [52] (scaled by lower layer height $d^{(1)}$). Reprinted from [98], with permission from CRC Press

concentration is constant across the gap, i.e. it does not depend on the y-coordinate: $c^{(i)} = c^{(i)}(x, z, t)$.

With this, the three-dimensional Navier-Stokes-Boussinesq and advection-diffusion equations for the solute yield the two-dimensional HS model:

$$\partial_t \mathbf{v}^{(1)} = -\frac{6}{5}\mathbf{v}^{(1)} \cdot \nabla \mathbf{v}^{(1)} - \nabla p_d^{(1)} + \Delta \mathbf{v}^{(1)} - c^{(1)} G \mathbf{e}_z - 3\gamma \mathbf{v}^{(1)}, \tag{16.23}$$

$$\partial_t \mathbf{v}^{(2)} = -\frac{6}{5}\mathbf{v}^{(2)} \cdot \nabla \mathbf{v}^{(2)} - \frac{1}{\rho}\nabla p_d^{(2)} + \nu \Delta \mathbf{v}^{(2)} - c^{(2)} G \beta \mathbf{e}_z - 3\gamma \nu \mathbf{v}^{(2)}, \tag{16.24}$$

$$\nabla \cdot \mathbf{v}^{(1)} = 0, \quad \nabla \cdot \mathbf{v}^{(2)} = 0, \tag{16.25}$$

$$\partial_t c^{(1)} = -\mathbf{v}^{(1)} \cdot \nabla c^{(1)} + \frac{1}{Sc^{(1)}} \Delta c^{(1)}, \quad \partial_t c^{(2)} = -\mathbf{v}^{(2)} \cdot \nabla c^{(2)} + \frac{D}{Sc^{(1)}} \Delta c^{(2)}. \tag{16.26}$$

The additional damping with prefactor 3γ in Eq. (16.23) is well known from Darcy's law.

The matching conditions at the interface as well as the boundary conditions are adapted from the three-dimensional case. The set of equations underlying the HS model again is solved by the pseudo-spectral algorithm described in Sect. 16.2.2.

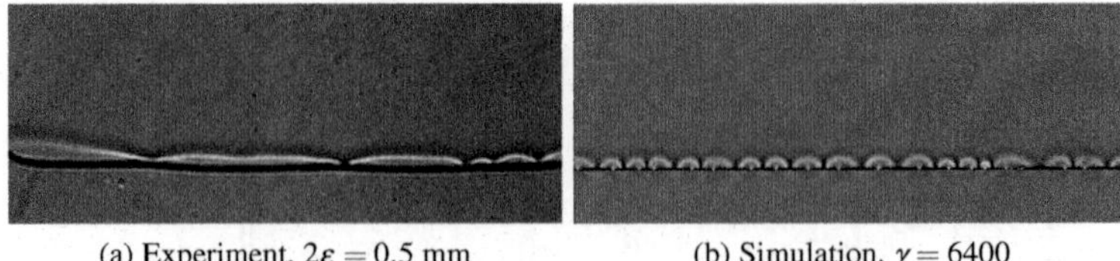

(a) Experiment, $2\varepsilon = 0.5$ mm   (b) Simulation, $\gamma = 6400$

Fig. 16.16 Pattern of mainly RC-I in a *thin* HS cell at $t = 1.0$. The experimental and numerical shadowgraph images show the same domain with horizontal extent of one length unit $l_x = 1$ (20 mm). (**a**) Experiment, $2\varepsilon = 0.5$ mm. (**b**) Simulation, $\gamma = 6400$. Reprinted from [98], with permission from CRC Press

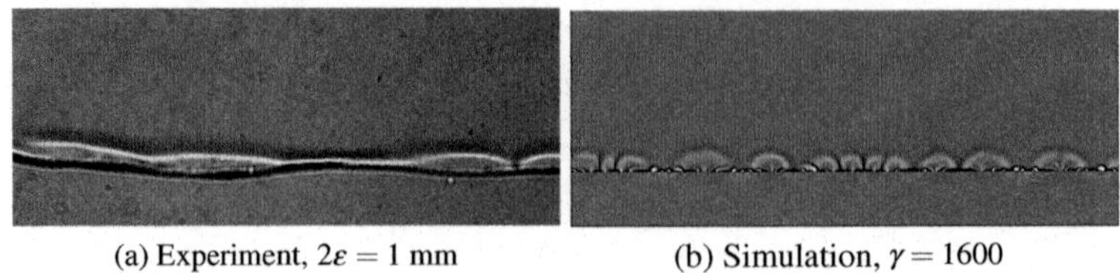

(a) Experiment, $2\varepsilon = 1$ mm   (b) Simulation, $\gamma = 1600$

Fig. 16.17 Pattern of RC-II in a *thick* HS cell at $t = 1.0$. The experimental and numerical shadowgraph images show the same domain with horizontal extent of one length unit $l_x = 1$ (20 mm). (**a**) Experiment, $2\varepsilon = 1$ mm. (**b**) Simulation, $\gamma = 1600$. Reprinted from [98], with permission from CRC Press

16.3.2.3 Effect of the Gap Width

In our study [52] we focus on the question how well the experimental observations can be captured by the HS model. For this purpose, two configurations differing in plate distance (= gap size) 2ε and correspondingly different friction coefficients γ are examined, namely

- *Thin* HS cell: $2\varepsilon = 0.5$ mm, $\gamma = \dfrac{(d^{(1)})^2}{\varepsilon^2} = 6400$,
- *Thick* HS cell: $2\varepsilon = 1$ mm, $\gamma = \dfrac{(d^{(1)})^2}{\varepsilon^2} = 1600$.

The shadowgraph distributions from experiment and simulation are used for a qualitative comparison here, while a quantitative evaluation can be found in [52]. Figure 16.16 shows the pattern in a 0.5 mm gap while the pattern for the doubled plate distance of 1 mm is depicted in Fig. 16.17. With increasing plate distance, the influence of wall friction in the HS model decreases due to the lower friction factor γ. Besides, three-dimensional effects (e.g. edge convection [9]) are expected to be more pronounced in the experiments with large gap.

The general structure of the Marangoni roll cells in the HS configuration can be seen nicely in Fig. 16.16. In the upper organic phase, the region of mixed fluid (poor in solute, i.e. butanol) is bordered by a dark-grey rim to fluid rich in solute at the top. A second boundary at the bottom is given by the interface which manifests in the experimental image (a) as a solid black rim due to the meniscus formed between both liquid phases (detailed below). The dark-grey rim of the mixed fluid is bent toward the interface at the inflow regions, where butanol-rich fluid from the bulk flows to the interface. Apparently, the horizontal size of the roll cells (distance from inflow to inflow) is larger in the experiments (a) than in the simulation (b). Qualitatively however, the experimental flow structures are well represented by the simulation [52]. In both cases, Marangoni roll cells without internal substructure develop (RC-I).

For the doubled plate distance in Fig. 16.17, the hierarchical pattern of small and large roll cells (RC-II) appears well visible in the numerical shadowgraph image (b). This can be explained by the decreasing influence of wall friction which gives rise to an intensified flow [52]. In Fig. 16.17, the patterns reach further into the bulk and the inflow region has a broader horizontal extent compared to Fig. 16.16. This enhanced advection of solute again leads to high gradients at the interface so that small scales can be produced according to the theory [109]. In the experimental image (a), the small substructure of RC-I is only faintly visible by a dim horizontal modulation near the interface: Contrary to the supposed plane interface in the numerical model, a concave meniscus (viewed from the aqueous layer) across the plates is formed in the experiments. As a result, the small substructures are shadowed in the experimental images due to the deflection of light at this meniscus.

Furthermore, it seems likely that the curved interface is a main cause for the observed differences in pattern size between simulation and experiment [52]. Apart from the concave meniscus, the position of the interface also changes in lateral direction as a result of the experimental filling procedure, cf. Fig. 16.17a. This imposes a higher concentration on those parts of the interface which reach deeper into the delivering phase. The resulting Marangoni convection accelerates the experimental evolution compared to the simulations. Three-dimensional flow effects, i.e. a variation of concentration across the gap and deviations from the parabolic velocity profile, are another source of discrepancies [52]. In addition to the flow originating from the meniscus convection, Marangoni cells may emerge which are smaller than the gap size. Therefore, it is expected that Marangoni cells are also amplified across the gap ($\partial_y c \neq 0$). This effect is particularly important at a large plate distance leading to the pronounced deviations between experiment and simulation in Fig. 16.17. On this basis, we can conclude that the HS model is expected to be more precise for systems where large Marangoni cells without substructure develop, e.g. in [35]. The qualitative agreement of structures encourages further studies based on the HS configuration. An extension of the model to include interfacial deformations has the potential to approach the experimental situation more closely and to estimate the contribution of interface deformations.

476   K. Eckert et al.

16.4 Marangoni Convection with Unstable Density Gradient

16.4.1 Eruptive Regime of Rayleigh-Marangoni Convection

As already introduced in Sect. 16.1.5, mass transfer of surface active substances
is frequently accompanied by the formation of an unstable density stratification.
The resulting Rayleigh convection gives rise to a further complexity of the
pattern which was responsible for an early labeling as *interfacial turbulence*. To
provide a better understanding of the underlying mechanisms we have analyzed an
exemplary mass transfer system via simulation and experiment for which water+2-
propanol/cyclohexanol was used similar to [92], see Fig. 16.18. For a compilation
of the material data we refer to [49, 53]. The non-dimensional groups characterizing
the particular case described here are $Sc = 1348$, $\rho = 0.96$, $v = 20.74$, $D = 0.082$,
$\beta = 0.92$, $H = 1.6$, $d = 1$, $Ma = -6.8 \times 10^6$ and $Gr = -1.81 \times 10^5$. In contrast
to the foregoing section, this system is stable with respect to stationary Marangoni
instability studied before, but susceptible to Rayleigh convection at both sides of
the interface. Based on this configuration, we show that the dominant *eruptive
regime* results from a two-step process. Rayleigh convection continuously produces
localized regions of elevated concentration at the interface that in turn trigger a
vigorous Marangoni flow. The spatial resolution is $N_z^{(1)} = 256$, $N_z^{(2)}=512$ for the
vertical direction and $N_x = N_y = 1024$ for the horizontal directions.

Let us first consider the base state when the solute is only transported by
diffusion, i.e. when Rayleigh and Marangoni effects are disregarded, i.e. $Ma = G =
0$. The corresponding solute profiles for successive times are shown in Fig. 16.18b.
This base state is known to be stable with respect to the stationary Marangoni
instability [109] because 2-propanol lowers the interfacial tension and diffuses faster

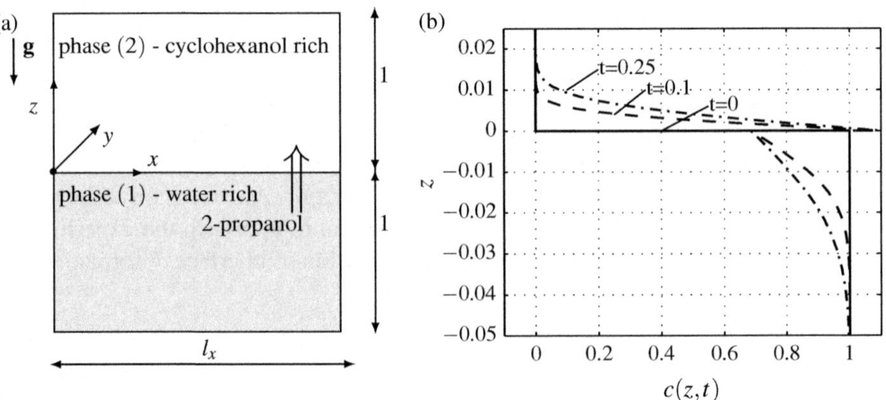

Fig. 16.18 Eruptive convection: (**a**) sketch of the numerical domain, (**b**) simulated concentration
profiles for pure diffusion, i.e. $G = Ma = 0$. Reprinted from [53], with permission from
Cambridge University Press

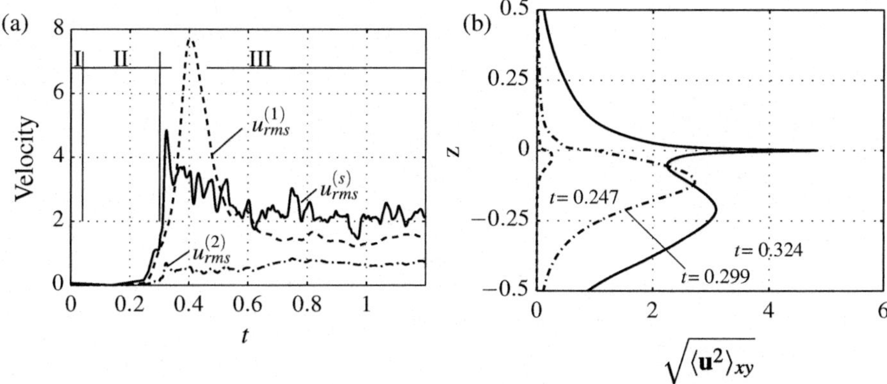

Fig. 16.19 Temporal evolution of convection in the simulation: (**a**) rms velocities, (**b**) horizontally averaged velocity at three time instances. Reprinted from [53] with permission from Cambridge University Press

in the phase that delivers solute. However, since 2-propanol decreases the density in both phases, a less dense concentration boundary layer develops above the interface and a denser one below the interface. This density configuration is susceptible to a Rayleigh instability that leads to buoyant convection.

The subsequent evolution is monitored in Fig. 16.19a by means of the root-mean-square velocity $u_{rms}^{(i)}$ in each layer and at the interface $u_{rms}^{(s)}$ [Eq. (16.18)]. A clear division into three regimes can be recognized. Regime I ($0 < t \lesssim 0.04$) refers to the diffusive base state, which is free of convection. The diffusive regime ends with the onset of the primary Rayleigh instability. We defined the time of onset as the earliest time for which the average of rms velocities $(u_{rms}^{(1)} + u_{rms}^{(2)})/2$ is growing.

The growth of the Rayleigh instability governs the second regime II ($0.04 \leq t < 0.3$). It lasts until the growth of $u_{rms}^{(s)}(t)$ stops for a moment. At this point, the eruptive regime III ($0.3 \leq t$) is initiated. The interfacial velocity grows rapidly to reach a maximum at $t = 0.32$. The intense interfacial flow can also be noted as a high peak at $z = 0$ in the corresponding velocity profile ($t = 0.324$) in Fig. 16.19b. After this vivid initial phase, regime III reaches a chaotic state that changes on a slower time scale, i.e. the changes in $u_{rms}^{(i)}$ and $u_{rms}^{(s)}$ are less pronounced in Fig. 16.19a.

The primary Rayleigh instability forms a polygonal pattern in top view as shown in the upper row of Fig. 16.20 ($t = 0.299$). The experimental pattern in the right column of Fig. 16.20 appears similar but does not possess this pronounced polygonal structure as in the simulations. By comparison with the simulation data, we see that bright schlieren in the experiment result from solute transferred to the upper layer that remains still localized near the interface (A-mark). This less dense fluid rises only slowly all along the sides of the polygons in the upper layer. Before fluid rises preferably in the vertices, eruptions are initialized and break up this ordered polygonal pattern. Due to the faster Rayleigh convection in the lower layer, the downwelling plumes of denser fluid are mostly located at the x, y position

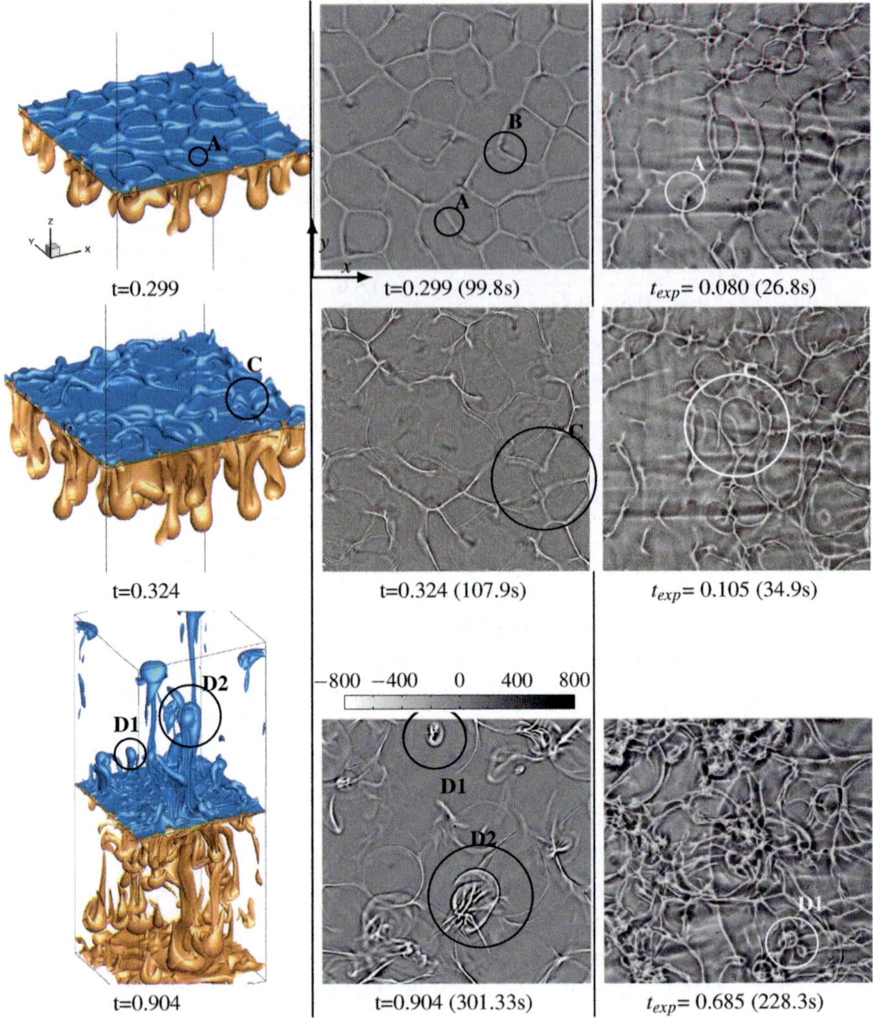

Fig. 16.20 Simulated and experimental patterns: (*left*) simulated isosurfaces $c^{(1)} = 0.97$, $c^{(2)} = 0.2$; (*center*) simulated synthetic shadowgraph image $s(x, y)$ in a domain of 0.75 (15 mm) $\times$ 0.75 (15 mm); (*right*) experimental shadowgraph image of identical size. *Circles* with labels refer to the explanations in the text. Reprinted from [53], with permission from Cambridge University Press

of a cell vertex, as observed in the synthetic shadowgram of the simulations (B-mark). In the experimental shadowgram, these falling plumes are not clearly visible, probably because of a lower refractive index change. Finally, eruptions manifest as expanding circular schlieren in the shadowgraph images of Fig. 16.20, second row (see C-marks).

An eruption cycle proceeds in three characteristic stages. First, eruptions are *initialized* at a specific point. The location of this point results from the complex interaction between the growing portions of solute-rich fluid in the upper layer and the downwelling plumes in the lower layer. In this process, the solute-rich fluid (in the upper phase) that gathered near the interface is transported *towards* the interface now, manifesting as a diverging radial motion in both experimental and numerical shadowgraph images. Second, the spreading motion accelerates by the continuing inflow of solute-rich fluid and adopts a state of *maximum speed*. Third, the eruptions *decay*, i.e. the spreading of solute ceases by interfering with the neighboring structures.

After the eruptive regime is established, slowly rising plumes of enriched fluid in the top layer (marked D in Fig. 16.20) have a strong signature in the shadowgrams. During their rise, they do not change position or size significantly. They cause distinct expanding circular shapes when they hit the top wall. This process depends on the layer height, i.e. on the values of Ma and G. The influence of these parameters is the subject of ongoing work.

16.4.2 Relaxation Oscillations of Marangoni Convection at Drops

In contrast to the eruptive Rayleigh-Marangoni convection, described previously, which is triggered by an *unstable density gradient*, we next present a second system where Marangoni convection *locally reverses an initially stable density gradient*. As a result, an intriguing coupling between Marangoni and Rayleigh convection in the form of relaxation oscillations occurs [97].

The experimental system, paraffin oil with 5 vol% 2-propanol over water, the material data of which are documented in [97], is shown in Fig. 16.21a. The mass transfer of the surface-active 2-propanol in the two-layer system is prone to Marangoni convection. The resulting large-scale convection cell is visible as an arc of light and dark gray above the interface. Additionally, small-scale relaxation oscillations take place at paraffin oil drops which were purposefully placed in the aqueous phase. These drops of a diameter 0.1–0.5 mm sit on the glass wall of a Hele-Shaw cell with a gap width of 1 mm. An exemplary drop is marked by a white rectangle in Fig. 16.21a.

A characteristic feature at such droplets is the occurrence of periodic re-amplifications of Marangoni convection as documented in Fig. 16.21b (described in detail later). Within one cycle of re-amplification, distinct stages of convection are passed, cf. Fig. 16.22. The cycle begins with the start-up of Marangoni convection in Fig. 16.22a. Two small vortices form at either side of the drop and are amplified to the state depicted in Fig. 16.22b, where the peak of Marangoni convection is reached. The flow is obviously caused by the conditions in the surrounding of the drop. Due to the mass transfer of 2-propanol across the overlying interface of the two-layer system, the drop is placed in a vertical concentration gradient. Hence it is subjected to high solute concentration at the upper side, implying both low interfacial tension σ and density $\rho^{(1)}$, and low concentration at the bottom, i.e. both higher σ and $\rho^{(1)}$. Therefore, Marangoni convection at the drop surface is directed

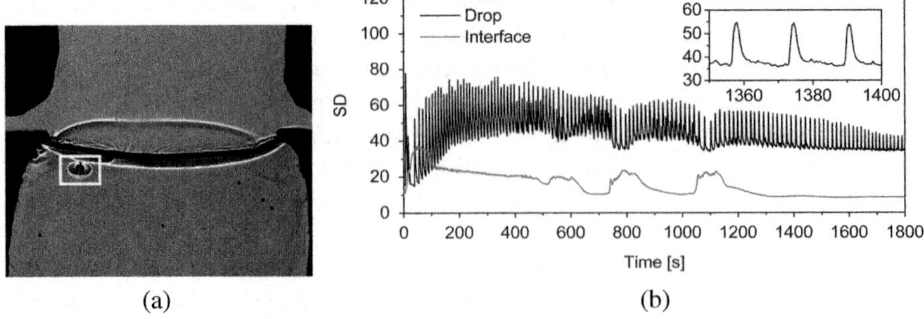

Fig. 16.21 Marangoni convection in the paraffin oil/water system with mass transfer of 2-propanol: Shadowgraph image (**a**) of the experimental system in the Hele-Shaw cell. The width of the shown window is 20.7 mm. The *white rectangle* in (**a**) marks a drop with periodic Marangoni convection, see (**b**) and Fig. 16.22. Temporal evolution (**b**) of standard deviation (SD) of *gray value* distribution for the drop (*black curve*) and the overlying interface of the two-layer system (*gray curve*). The *inset* details three periods of relaxation oscillations at the drop. Reprinted from [97], with permission from Elsevier

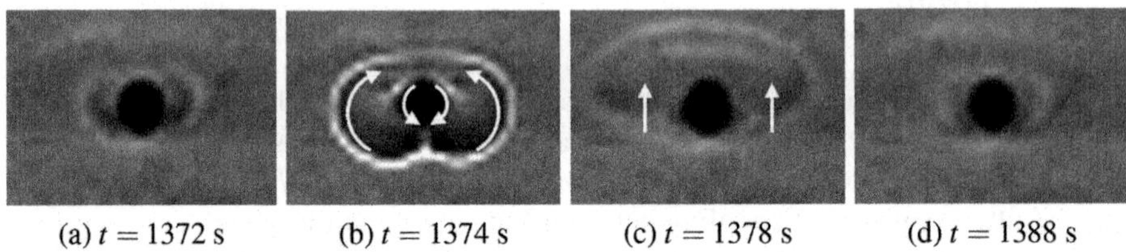

(a) $t = 1372$ s   (b) $t = 1374$ s   (c) $t = 1378$ s   (d) $t = 1388$ s

Fig. 16.22 Cycle of relaxation oscillation at a drop of paraffin oil immersed in the aqueous phase, see Fig. 16.21a. *White arrows* indicate the direction of flow in the active (**a,b**) and relaxation phase (**c,d**). The width of the shown window is 2.4 mm. Reprinted from [97], with permission from Elsevier

downwards, entraining lighter fluid rich in solute to lower regions of higher density. The initially stably stratified aqueous phase is stirred, and the additional buoyancy accelerates the rise of the lighter fluid at some distance from the drop surface. As a result, an intense mixing of the fluid, visible in Fig. 16.22b, takes place which exhausts the vertical concentration gradient in the vicinity of the drop and leads to the breakdown of Marangoni convection. In the subsequent relaxation phase (Fig. 16.22c), the initial concentration gradient is restored by both the buoyancy-driven convection, which carries the mixed fluid upwards, and diffusion. The end of the relaxation phase is shown in Fig. 16.22d, followed by another cycle of Marangoni convection. A similar mechanism was already observed in [55] for other fluid combinations.

The intensity of Marangoni convection which disturbs the concentration gradient in the surrounding of the drop can be characterized by the standard deviation of the gray value distribution of the shadowgraph images, see Fig. 16.21b. The inset zooms in on three periods of relaxation oscillations, indicating their sawtooth-like progression. The steep rise in standard deviation corresponds to the short, impulsive stage of Marangoni convection. The relaxation phase lasts longer due to the slower processes of diffusion and buoyant convection that are responsible for the restoration of the Marangoni driving force. The main diagram shows about one hundred periods of regular relaxation oscillations over half an hour. As the drop is located close to the interface of the two-layer system, its relaxation oscillations are influenced to a certain extent by the large-scale Marangoni convection. To illustrate this, the intensity of convection at the interface is plotted for comparison in Fig. 16.21b (gray curve). It can be seen that during active Marangoni convection of the two-layer system, i.e. a rise of the gray curve, the frequency of the relaxation oscillations at the drop decreases. Under this additional mixing, it will take longer until the concentration gradient at the drop for the onset of the next oscillation is available again. Furthermore, simulations based on a phase-field model support the mechanism explained above [97].

16.5   Conclusion and Outlook

The first part of the project was devoted to the case of stationary Marangoni instability under a stabilizing density stratification. A comprehensive numerical and experimental study was carried out to provide a deeper understanding of former experimental observations by Schwarz, and to verify the classification of patterns by Linde et al. (Sect. 16.1.4).

Simulations successfully reproduced a two-level hierarchy of convection cells that consist of large, slowly growing cells (called RC-IIs according to the Linde et al. classification), which host smaller rapidly changing cells (named RC-Is). Hence, the main hypotheses of Linde et al. could be effectively applied. While the main hypotheses of Linde et al. could be verified, the last hypothesis L_4 that the integral activity of convection may fluctuate on a large time scale in "cycles" was not encountered in the present material systems. The experimental results have agreed in most aspects with the simulated results. Deviations were within a range that could be expected from the uncertainty of material properties and experimental conditions. In this way, it has been demonstrated that multiscale flow patterns can indeed be described by the simple model from Sect. 16.2.2.

The formation of hierarchical patterns has been explained by the action of coarsening and local instability. It has been reasoned that the RC-IIs grow as a result of advancing equilibration of the butanol distribution (coarsening). The RC-IIs induce a distribution of solute near the interface that is susceptible to a continuous creation of smaller Marangoni cells (local instability).

If buoyant effects and a geometrical confinement can be excluded, a change in the initial solute concentration leaves nondimensional equations and boundary conditions unchanged when the so-called interface units are used. In this case there is a complete similarity in the patterns and their evolution between different initial solute concentrations. It could be shown that this similarity applies in the simulations at early times, i.e. one obtains identical patterns if length and time are appropriately scaled. This behavior is perturbed and eventually eliminated by buoyancy effects, which become important at later stages of the pattern formation.

Additionally, we performed simulations for a two-dimensional Hele-Shaw model and compared these to our experiments in four different configurations. For the thin (0.5 mm gap) and vertical HS cell, the numerical results qualitatively represented the experiments, but the time evolution in the simulations appeared retarded as in the three-dimensional simulation. Furthermore, it seems that three-dimensional flow effects have to be expected for a horizontal orientation and a large gap width (1 mm).

In the second part of the project the focus was laid on systems in which the density stratification, evolving during mass transfer, played an important role, see Sect. 16.4. For Marangoni stable systems we could clarify the mechanism of the eruptive regime in the form of a coupling of Rayleigh convection, which produces localized regions of elevated concentration, and Marangoni convection, triggered by these concentration gradients. A comparison with the experiments showed a good qualitative agreement, but experimentally measured velocities were roughly two times as high as the simulated ones. For a second generic situation, i.e. droplets placed in a bulk concentration gradient, we were able to provide a sound explanation of the observed relaxation oscillations of Marangoni convection.

Based on the understanding gained in the project, we finally wish to comment on a number of issues worth to be followed in future works.

A *first* suggestion is to reduce the deviation between simulations and experiments for a particular chemical system by (1) more accurate values of material properties and (2) extension of the physical model in several respects, e.g. physical properties depending on composition, interfacial deformation, phase changes, wall effects. However, the inclusion of some physical mechanisms may distinctly decrease numerical efficiency. In this respect, the Hele-Shaw system seems to be a promising configuration to simulate the effect of a curved interface and solid walls.

A *second* suggestion is to consider mass transport in a binary system of partially miscible liquids. Fundamentally, the physical modeling appears easier since only two substances are involved. However, due to the phase rule, there is no degree of freedom for the composition at the interface (if pressure and temperature are constant). Thus, Marangoni convection is triggered by the heat of solution. Classical experiments [28, 41, 75, 116] and theoretical works [27, 29, 42, 105] are available in literature.

As a *third* direction, there are still a number of open questions in the present theoretical and numerical framework:

- The parameter analysis in the Rayleigh-Marangoni case with eruptions showed a power-law behavior whose range of applicability is yet unclear [49]. Its examination could be performed by further simulations and theoretical work.
- The classification of patterns by Linde et al. (Sect. 16.1.4) still addresses features that we have not seen in the present material systems. These features are the cyclic behavior L_4 and relaxations oscillations (ROs) beside those at the onset of convection. Consequently, these features might be candidates to assess further the hypotheses of Linde et al.

- The study of more complex systems should go in hand with improvements in the numerical methods in terms of efficiency. On the one hand, the present code could be better parallelized by including a domain decomposition in both horizontal directions. On the other hand, the undesired clustering of collocation points at the top and bottom could be reduced by using another discretization for the z-direction or a mapping of collocation points [19, 82].
- The last point concerns the modeling of surfactants for which the authors already made an attempt to include them into the present numerical code. The main new physical feature that has to be modeled is the adsorbed amount of solute $\Gamma(x, y, t)$ on the interface [36]. For a plane interface this amounts to revised matching conditions for the bulk concentration field in the form of

$$\partial_t \Gamma + \partial_x(u_x \Gamma) + \partial_y(u_y \Gamma) - D_g \Delta(\Gamma) = D^{(2)} \partial_z c^{(2)}(z = 0) - D^{(1)} \partial_z c^{(1)}(z = 0), \tag{16.27}$$

which poses some numerical difficulties due to its nonlinearity (e.g. by $\partial_y(u_y \Gamma)$). However, the numerous applications of surfactants suggest that this extension is worth doing.

Acknowledgements Financial support by the Deutsche Forschungsgemeinschaft in the framework of the Priority Program 1506 (grants Ec201/2 and Bo1668/6) is gratefully acknowledged. Furthermore, we thank the computing center (UniRZ) of TU Ilmenau and FZ Jülich (NIC) for access to its parallel computing resources. We are also grateful to Prof. H. Linde for numerous fruitful discussions.

References

1. Agble, D.K.: Interfacial mass transfer in binary-liquid systems. Ph.D. thesis, Imperial College, University of London, London (1998)
2. Agble, D., Mendes-Tatsis, M.: The effect of surfactants on interfacial mass transfer in binary liquid-liquid systems. Int. J. Heat Mass Transf. **43**(6), 1025–1034 (2000)
3. Agble, D., Mendes-Tatsis, M.: The prediction of Marangoni convection in binary liquid-liquid systems with added surfactants. Int. J. Heat Mass Transf. **44**(7), 1439–1449 (2001)
4. Bakker, C.A.P., van Buytenen, P.M., Beek, W.J.: Interfacial phenomena and mass transfer. Chem. Eng. Sci. **21**, 1039–1046 (1966)
5. Bakker, C., van Vlissingen, F.F., Beek, W.: The influence of the driving force in liquid-liquid extraction - a study of mass transfer with and without interfacial turbulence under well-defined conditions. Chem. Eng. Sci. **22**(10), 1349–1355 (1967)
6. Bassou, N., Rharbi, Y.: Role of Bénard- Marangoni instabilities during solvent evaporation in polymer surface corrugations. Langmuir **25**(1), 624–632 (2008)
7. Bénard, H.: Les tourbillons cellulaires dans une nappe liquide.-Méthodes optiques d'observation et d'enregistrement. J. Phys. Theor. Appl. **10**(1), 254–266 (1901)
8. Berg, J., Acrivos, A.: The effect of surface active agents on convection cells induced by surface tension. Chem. Eng. Sci. **20**(8), 737–745 (1965)
9. Berg, J., Morig, C.: Density effects in interfacial convection. Chem. Eng. Sci. **24**, 937–946 (1969)

10. Bi, W., Wu, X., Yeow, E.: Unconventional multiple ring structure formation from evaporation-induced self-assembly of polymers. Langmuir **28**(30), 11056–11063 (2012)
11. Boeck, T.: Bénard-Marangoni convection at low Prandtl number: results of direct numerical simulations. Ph.D. thesis, TU-Ilmenau (2000)
12. Boeck, T., Nepomnyashchy, A., Simanovskii, I., Golovin, A., Braverman, L., Thess, A.: Three-dimensional convection in a two-layer system with anomalous thermocapillary effect. Phys. Fluids **14**, 3899 (2002)
13. Boeck, T., Krasnov, D., Zienicke, E.: Numerical study of turbulent magnetohydrodynamic channel flow. J. Fluid Mech. **572**, 179–188 (2007)
14. Boos, W., Thess, A.: Thermocapillary flow in a Hele-Shaw cell. J. Fluid Mech. **352**, 305–330 (1997)
15. Borcia, R., Bestehorn, M.: Phase-field model for Marangoni convection in liquid-gas systems with a deformable interface. Phys. Rev. E **67**(6), 066307 (2003)
16. Bragard, J., Velarde, M.: Bénard–Marangoni convection: planforms and related theoretical predictions. J. Fluid Mech. **368**(1), 165–194 (1998)
17. Bratsun, D., De Wit, A.: On Marangoni convective patterns driven by an exothermic chemical reaction in two-layer systems. Phys. Fluids **16**, 1082–1096 (2004)
18. Brian, P.L.T., Vivian, J.E., Mayr, S.T.: Cellular convection in desorbing surface tension-lowering solutes from water. Ind. Eng. Chem. Fundam. **10**(1), 75–83 (1971)
19. Canuto, C., Hussaini, M., Quarteroni, A., Zang, T.: Spectral Methods in Fluid Dynamics. Springer, Berlin (1988)
20. Cazabat, A., Heslot, F., Troian, S., Carles, P.: Fingering instability of thin spreading films driven by temperature gradients. Nature **346**(6287), 824–826 (1990)
21. Changxu, L., Aiwu, Z., Xigang, Y., et al.: Experimental study on mass transfer near gas–liquid interface through quantitative schlieren method. Chem. Eng. Res. Des. **86**(2), 201–207 (2008)
22. Chauvet, F., Dehaeck, S., Colinet, P.: Threshold of Bénard-Marangoni instability in drying liquid films. Europhys. Lett. **99**, 34001 (2012)
23. Chen, S., Fu, B., Yuan, X., Zhang, H., Chen, W., Yu, K.: Lattice Boltzmann method for simulation of solutal interfacial convection in gas-liquid system. Ind. Eng. Chem. Res. **51**, 10955–10967 (2012)
24. Chen, J., Yang, C., Mao, Z.S.: The interphase mass transfer in liquid–liquid systems with Marangoni effect. Eur. Phys. J. Spec. Top. **224**(2), 389–399 (2015)
25. Colinet, P., Legros, J., Velarde, M., Prigogine, I.: Nonlinear Dynamics of Surface-Tension-Driven Instabilities. Wiley Online Library (2001)
26. Davenport, I., King, C.: Marangoni stabilization of density-driven convection. Chem. Eng. Sci. **28**(2), 645–647 (1973)
27. De Ortiz, E.P., Sawistowski, H.: Interfacial stability of binary liquid-liquid systems - I. Stability analysis. Chem. Eng. Sci. **28**(11), 2051–2061 (1973)
28. De Ortiz, E.P., Sawistowski, H.: Interfacial stability of binary liquid-liquid systems - II. Stability behaviour of selected systems. Chem. Eng. Sci. **28**(11), 2063–2069 (1973)
29. De Ortiz, E., Sawistowski, H.: Stability analysis of liquid-liquid systems under conditions of simultaneous heat and mass transfer. Chem. Eng. Sci. **30**(12), 1527–1528 (1975)
30. De Wit, A., Eckert, K., Kalliadasis, S.: Introduction to the focus issue: chemo-hydrodynamic patterns and instabilities. Chaos-Woodbury **22**(3), 037101 (2012)
31. Deegan, R., Bakajin, O., Dupont, T., Huber, G., Nagel, S., Witten, T.: Capillary flow as the cause of ring stains from dried liquid drops. Nature **389**(6653), 827–829 (1997)
32. Dieter-Kissling, K., Karbaschi, M., Marschall, H., Javadi, A., Miller, R., Bothe, D.: On the applicability of drop profile analysis tensiometry at high flow rates using an interface tracking method. Colloids Surf. A Physicochem. Eng. Asp. **441**, 837–845 (2012)
33. Eckert, K., Bestehorn, M., Thess, A.: Square cells in surface-tension-driven Bénard convection: experiment and theory. J. Fluid Mech. **356**, 155–197 (1998)
34. Eckert, K., Acker, M., Shi, Y.: Chemical pattern formation driven by a neutralization reaction. I. Mechanism and basic features. Phys. Fluids **16**(2), 385–399 (2004)

35. Eckert, K., Acker, M., Tadmouri, R., Pimienta, V.: Chemo-Marangoni convection driven by an interfacial reaction: pattern formation and kinetics. Chaos **22**(3), 037112–037112 (2012)
36. Edwards, D., Brenner, H., Wasan, D.: Interfacial Transport Processes and Rheology. Butterworth-Heinemann, Oxford (1991)
37. Golovin, A.: Mass transfer under interfacial turbulence: kinetic regulaties. Chem. Eng. Sci. **47**(8), 2069–2080 (1992)
38. Grahn, A.: Two-dimensional numerical simulations of Marangoni-Bénard instabilities during liquid-liquid mass transfer in a vertical gap. Chem. Eng. Sci. **61**, 3586–3592 (2006)
39. Gross, B., Hixson, A.: Marangoni instability with unsteady diffusion in the undisturbed state. Ind. Eng. Chem. Fundam. **8**(2), 288–296 (1969)
40. Hanson, C.: Recent Advances in Liquid-Liquid Extraction. Elsevier, Amsterdam (2013)
41. Heines, H., Westwater, J.: The effect of heats of solution on Marangoni convection. Int. J. Heat Mass Transf. **15**(11), 2109–2117 (1972)
42. Ho, K.L., Chang, H.C.: On nonlinear doubly-diffusive Marangoni instability. AIChE J. **34**(5), 705–722 (1988)
43. Imaishi, N., Fujikawa, K.: Theoretical study of the stability of two-fluid layers. J. Chem. Eng. Jpn. **7**(2), 81–87 (1974)
44. Jackson, E., She, Z.S., Orszag, S.A.: A case study in parallel computing: I. Homogeneous turbulence on a hypercube. Journal of Scientific Computing **6**(1), 27–45 (1991)
45. Javadi, A., Krägel, J., Pandolfini, P., Loglio, G., Kovalchuk, V., Aksenenko, E., Ravera, F., Liggieri, L., Miller, R.: Short time dynamic interfacial tension as studied by the growing drop capillary pressure technique. Colloids Surf. A Physicochem. Eng. Asp. **365**(1), 62–69 (2010)
46. Javadi, A., Karbaschi, M., Bastani, D., Ferri, J., Kovalchuk, V., Kovalchuk, N., Javadi, K., Miller, R.: Marangoni instabilities for convective mobile interfaces during drop exchange: experimental study and CFD simulation. Colloids Surf. A Physicochem. Eng. Asp. **441**, 846–854 (2014)
47. Jun, S., Lee, H.: The effect of Bénard-Marangoni convection on percolation threshold in amorphous polymer-multiwall carbon nanotube composites. Curr. Appl. Phys. **12**(2), 467–472 (2012)
48. Karpitschka, S., Riegler, H.: Noncoalescence of sessile drops from different but miscible liquids: hydrodynamic analysis of the twin drop contour as a self-stabilizing traveling wave. Phys. Rev. Lett. **109**(6), 066103 (2012)
49. Köllner, T.: Simulations of solutal Marangoni convection in two liquid layers: complex and transient patterns. Ph.D. thesis, Technische Universität Ilmenau (2016)
50. Köllner, T., Schwarzenberger, K., Eckert, K., Boeck, T.: Multiscale structures in solutal Marangoni convection: three-dimensional simulations and supporting experiments. Phys. Fluids **25**, 092109 (2013)
51. Köllner, T., Rossi, M., Broer, F., Boeck, T.: Chemical convection in the methylene-blue–glucose system: optimal perturbations and three-dimensional simulations. Phys. Rev. E **90**(5), 053004 (2014)
52. Köllner, T., Schwarzenberger, K., Eckert, K., Boeck, T.: Solutal Marangoni convection in a Hele-Shaw geometry: impact of orientation and gap width. Eur. Phys. J. Spec. Top. **224**(2), 261–276 (2015)
53. Köllner, T., Schwarzenberger, K., Eckert, K., Boeck, T.: The eruptive regime of mass-transfer-driven Rayleigh-Marangoni convection. J. Fluid Mech. **719**, R4, 1–12 (2016)
54. Koschmieder, E., Biggerstaff, M.: Onset of surface-tension-driven Bénard convection. J. Fluid Mech. **167**, 49–64 (1986)
55. Kostarev, K., Shmyrov, A., Zuev, A., Viviani, A.: Convective and diffusive surfactant transfer in multiphase liquid systems. Exp. Fluids **51**(2), 457–470 (2011)
56. Kovalchuk, N.: Spontaneous oscillations due to solutal Marangoni instability: air/water interface. Cent. Eur. J. Chem. **10**(5), 1423–1441 (2012)
57. Kroepelin, H., Neumann, H.: Über die Entstehung von Grenzflächeneruptionen bei dem Stoffaustausch an Tropfen. Naturwissenschaften **43**(15), 347–348 (1956)

58. Kroepelin, H., Neumann, H.: Eruptiver Stoffaustausch an ebenen Grenzflächen. Naturwissenschaften **44**(10), 304–304 (1957)
59. Lappa, M., Piccolo, C.: Higher modes of the mixed buoyant-Marangoni unstable convection originated from a droplet dissolving in a liquid/liquid system with miscibility gap. Phys. Fluids **16**(12), 4262–4272 (2004)
60. Lappa, M., Piccolo, C., Carotenuto, L.: Mixed buoyant-Marangoni convection due to dissolution of a droplet in a liquid–liquid system with miscibility gap. Eur. J. Mech. B. Fluids **23**(5), 781–794 (2004)
61. Leo, A., Hansch, C.: Linear free energy relations between partitioning solvent systems. J. Org. Chem. **36**(11), 1539–1544 (1971)
62. Levich, V.G.: Physicochemical Hydrodynamics. Prentice Hall, Upper Saddle River, NJ (1962)
63. Linde, H.: Untersuchungen über den Stoffübergang an ebenen und waagerechten flüssigflüssig-Phasengrenzen unter der Bedingung starker erzwungener Konvektion bei verschiedenartigem Einsatz grenzflächenaktiver Stoffe. Fette, Seifen, Anstrichmittel **60**(9), 826–829 (1958)
64. Linde, H., Schwarz, E.: Untersuchungen zur Charakteristik der freien Grenzflächenkonvektion beim Stoffübergang an fluiden Grenzen. Zeitschrift für physikalische Chemie **224**, 331–352 (1963)
65. Linde, H., Schwarz, E.: Über grossräumige Rollzellen der freien Grenzflächenkonvektion. Monatsberichte Deutsche Akademie der Wissenschaften zu Berlin **7**, 330–338 (1964)
66. Linde, H., Pfaff, S., Zirkel, C.: Strömungsuntersuchungen zur hydrodynamischen Instabilität flüssig-gasförmiger Phasengrenzen mit Hilfe der Kapillarspaltmethode. Zeitschrift für physikalische Chemie **225**, 72–100 (1964)
67. Linde, H., Schwarz, E., Gröger, K.: Zum Auftreten des Oszillatorischen Regimes der Marangoniinstabilität beim Stoffübergang. Chem. Eng. Sci. **22**(6), 823–836 (1967)
68. Linde, H., Schwarzenberger, K., Eckert, K.: Pattern formation emerging from stationary solutal Marangoni instability: a roadmap through the underlying hierarchic structures. In: Rubio, R., Ryazantsev, Y., Starov, V., Huang, G.X., Chetverikov, A., Arena, P., Nepomnyashchy, A., Ferrus, A., Morozov, E. (eds.) Without Bounds: A Scientific Canvas of Nonlinearity and Complex Dynamics. Understanding Complex Systems, vol. 5. Springer, Berlin (2013)
69. Ma, C., Bothe, D.: Direct numerical simulation of thermocapillary flow based on the volume of fluid method. Int. J. Multiphase Flow **37**(9), 1045–1058 (2011)
70. MacDonald, B., Ward, C.: Onset of Marangoni convection for evaporating liquids with spherical interfaces and finite boundaries. Phys. Rev. E **84**(4), 046319 (2011)
71. Mao, Z., Lu, P., Zhang, G., Yang, C.: Numerical simulation of the Marangoni effect with interphase mass transfer between two planar liquid layers. Chin. J. Chem. Eng. **16**(2), 161–170 (2008)
72. Marangoni, C.: Über die Ausbreitung der Tropfen einer Flüssigkeit auf der Oberfläche einer anderen. Ann. Phys. **219**, 337–354 (1871)
73. Marek, R., Straub, J.: The origin of thermocapillary convection in subcooled nucleate pool boiling. Int. J. Heat Mass Transf. **44**(3), 619–632 (2001)
74. Marra, J., Huethorst, J.: Physical principles of Marangoni drying. Langmuir **7**(11), 2748–2755 (1991)
75. Mendes-Tatsis, M., De Ortiz, E.P.: Spontaneous interfacial convection in liquid-liquid binary systems under microgravity. Proc. R. Soc. Lond. A Math. Phys. Sci. **438**(1903), 389–396 (1992)
76. Merzkirch, W.: Flow Visualization. Academic, New York (1987)
77. Nepomnyashchy, A., Legros, J., Simanovskii, I.: Interfacial Convection in Multilayer Systems. Springer, Berlin (2006)
78. Orell, A., Westwater, J.W.: Spontaneous interfacial cellular convection accompanying mass transfer: ethylene glycol-acetic acid-ethyl acetate. AIChE J. **8**, 350–356 (1962)
79. Palmer, H., Berg, J.: Hydrodynamic stability of surfactant solutions heated from below. J. Fluid Mech. **51**(2), 385–402 (1972)

80. Pearson, J.: On convection cells induced by surface tension. J. Fluid Mech. **4**(05), 489–500 (1958)
81. Pekurovsky, D.: P3DFFT: a framework for parallel computations of Fourier transforms in three dimensions. SIAM J. Sci. Comput. **34**(4), C192–C209 (2012)
82. Peyret, R.: Spectral Methods for Incompressible Viscous Flow. Springer, Berlin (2002)
83. Pimienta, V., Brost, M., Kovalchuk, N., Bresch, S., Steinbock, O.: Complex shapes and dynamics of dissolving drops of dichloromethane. Angewandte Chemie International Edition **50**, 10728–10731 (2011)
84. Popinet, S.: Gerris: a tree-based adaptive solver for the incompressible Euler equations in complex geometries. J. Comput. Phys. **190**(2), 572–600 (2003)
85. Proctor, S., Biddulph, M., Krishnamurthy, K.: Effects of Marangoni surface tension forces on modern distillation packings. AIChE J. **44**(4), 831–835 (1998)
86. Reichenbach, J., Linde, H.: Linear perturbation analysis of surface-tension-driven convection at a plane interface (Marangoni instability). J. Colloid Interface Sci. **84**(2), 433–443 (1981)
87. Rongy, L., Assemat P., De Wit, A.: Marangoni-driven convection around exothermic autocatalytic chemical fronts in free-surface solution layers. Chaos **22**(3), 037106–037106 (2012)
88. Schatz, M.F., Neitzel, G.P.: Experiments on thermocapillary instabilities. Annu. Rev. Fluid Mech. **33**(1), 93–127 (2001)
89. Schatz, M.F., VanHook, S.J., McCormick, W., Swift, J., Swinney, H.L.: Time-independent square patterns in surface-tension-driven Bénard convection. Phys. Fluids **11**, 2577 (1999)
90. Schwabe, D., Scharmann, A.: Some evidence for the existence and magnitude of a critical Marangoni number for the onset of oscillatory flow in crystal growth melts. J. Cryst. Growth **46**(1), 125–131 (1979)
91. Schwarz, E.: Hydrodynamische Regime der Marangoni-Instabilität beim Stoffübergang über eine fluide Phasengrenze. Ph.D. thesis, HU Berlin (1968)
92. Schwarz, E.: Zum Auftreten von Marangoni-Instabilität. Wärme- und Stoffübertragung **3**, 131–133 (1970)
93. Schwartztrauber, P.N.: Multiprocessor FFTs. Parallel Comput. **5**, 197–210 (1987)
94. Schwarzenberger, K., Eckert, K., Odenbach, S.: Relaxation oscillations between Marangoni cells and double diffusive fingers in a reactive liquid-liquid system. Chem. Eng. Sci. **68**, 530–540 (2012)
95. Schwarzenberger, K., Köllner, T., Linde, H., Odenbach, S., Boeck, T., Eckert, K.: On the transition from cellular to wave-like patterns during solutal Marangoni convection. Eur. Phys. J. Spec. Top. **219**, 121–130 (2013)
96. Schwarzenberger, K., Köllner, T., Linde, H., Boeck, T., Odenbach, S., Eckert, K.: Pattern formation and mass transfer under stationary solutal Marangoni instability. Adv. Colloid Interf. Sci. **206**, 344–371 (2014)
97. Schwarzenberger, K., Aland, S., Domnick, H., Odenbach, S., Eckert, K.: Relaxation oscillations of solutal Marangoni convection at curved interfaces. Colloids Surf. A Physicochem. Eng. Asp. **481**, 633–643 (2015)
98. Schwarzenberger, K., Köllner, T., Boeck, T., Odenbach, S., Eckert, K.: Hierarchical Marangoni roll cells: experiments and direct numerical simulations in three and two dimensions. In: Rahni, M.T., Karbaschi, M., Miller, R. (eds.) Computational Methods for Complex Liquid-Fluid Interfaces. Progress in Colloid and Interface Science, vol. 5. CRC, Boca Raton (2015)
99. Semenov, S., Starov, V., Rubio, R., Velarde, M.: Computer simulations of evaporation of pinned sessile droplets: influence of kinetic effects. Langmuir **28**(43), 15203–15211 (2012)
100. Sha, Y., Chen, H., Yin, Y., Tu, S., Ye, L., Zheng, Y.: Characteristics of the Marangoni convection induced in initial quiescent water. Ind. Eng. Chem. Res. **49**, 8770–8777 (2010)
101. Sherwood, T., Wei, J.: Interfacial phenomena in liquid extraction. Ind. Eng. Chem. **49**, 1030–1034 (1957)
102. Shi, Y., Eckert, K.: Acceleration of reaction fronts by hydrodynamic instabilities in immiscible systems. Chem. Eng. Sci. **61**, 5523–5533 (2006)

103. Shi, Y., Eckert, K.: A novel Hele-Shaw cell design for the analysis of hydrodynamic instabilities in liquid-liquid systems. Chem. Eng. Sci. **63**, 3560–3563 (2008)
104. Sigwart, K., Nassenstein, H.: Vorgänge an der Grenzfläche zweier flüssiger Phasen. Naturwissenschaften **42**, 458–459 (1955)
105. Slavtchev, S., Mendes, M.: Marangoni instability in binary liquid-liquid systems. Int. J. Heat Mass Transf. **47**(14–16), 3269–3278 (2004)
106. Slavtchev, S., Hennenberg, M., Legros, J.C., Lebon, G.: Stationary solutal Marangoni instability in a two-layer system. J. Colloid Interface Sci. **203**(2), 354–368 (1998)
107. Slavtchev, S., Kalitzova-Kurteva, P., Mendes, M.: Marangoni instability of liquid-liquid systems with a surface-active solute. Colloids Surf. A Physicochem. Eng. Asp. **282–283**, 37–49 (2006)
108. Smith, K.: On convective instability induced by surface-tension gradients. J. Fluid Mech. **24**(02), 401–414 (1966)
109. Sternling, C., Scriven, L.: Interfacial turbulence: hydrodynamic instability and the Marangoni effect. AIChE J. **5**(4), 514–523 (1959)
110. Sun, Z.: Onset of Rayleigh–Bénard–Marangoni convection with time-dependent nonlinear concentration profiles. Chem. Eng. Sci. **68**(1), 579–594 (2012)
111. Tadmor, R.: Marangoni flow revisited. J. Colloid Interface Sci. **332**(2), 451–454 (2009)
112. Thomson, J.: On certain curious motions observable at the surfaces of wine and other alcoholic liquors. Lond. Edinb. Dublin Philos. Mag. J. Sci. **10**(67), 330–333 (1855)
113. Wang, Z., Lu, P., Wang, Y., Yang, C., Mao, Z.S.: Experimental investigation and numerical simulation of Marangoni effect induced by mass transfer during drop formation. AIChE J. pp. n/a–n/a (2013)
114. Wegener, M., Paschedag, A., Kraume, M.: Mass transfer enhancement through Marangoni instabilities during single drop formation. Int. J. Heat Mass Transf. **52**(11–12), 2673–2677 (2009)
115. Yiantsios, S., Serpetsi, S., Doumenc, F., Guerrier, B.: Surface deformation and film corrugation during drying of polymer solutions induced by Marangoni phenomena. Int. J. Heat Mass Transf. **89**, 1083–1094 (2015)
116. Ying, W., Sawistowski, H.: Interfacial and mass transfer characteristics of binary liquid-liquid systems. In: Proceedings of the International Solvent Extraction Conference, vol. 2, pp. 840–851 (1971)
117. Zeren, R., Reynolds, W.: Thermal instabilities in two-fluid horizontal layers. J. Fluid Mech. **53**(02), 305–327 (1972)
118. Zhang, J., Oron, A., Behringer, R.: Novel pattern forming states for Marangoni convection in volatile binary liquids. Phys. Fluids **23**, 072102 (2011)

Chapter 17
Transport at Interfaces in Lipid Membranes and Enantiomer Separation

Oleg Boyarkin, Stefan Burger, Thomas Franke, Thomas Fraunholz, Ronald H.W. Hoppe, Simon Kirschler, Kevin Lindner, Malte A. Peter, Florian Strobl, and Achim Wixforth

Abstract We study the dynamics and formation of differently ordered lateral phases of interfacial lipid layers for two types of lipid systems, a vesicle-supported bilayer and a Langmuir–Blodgett monolayer, both in experiment and by simulation.

O. Boyarkin • T. Fraunholz
Institute of Mathematics, University of Augsburg, 86159 Augsburg, Germany
e-mail: oleg.boyarkin@math.uni-augsburg.de; Thomas.fraunholz@math.uni-augsburg.de

S. Burger • S. Kirschler • K. Lindner
Institute of Physics, University of Augsburg, 86159 Augsburg, Germany

T. Franke (✉)
Institute of Physics, University of Augsburg, Universitätsstraße 1, 86159 Augsburg, Germany

Nanosystems Initiative Munich, Schellingstrasse 4, 80799 München, Germany

Biomedical Engineering, School of Engineering, College of Science & Engineering, University of Glasgow, Oakfield Avenue, Glasgow G12 8LT, United Kingdom
e-mail: thomas.franke@physik.uni-augsburg.de

R.H.W. Hoppe
Institute of Mathematics, University of Augsburg, 86159 Augsburg, Germany

Department of Mathematics, University of Houston, Houston, TX 77204-3008, USA
e-mail: hoppe@math.uni-augsburg.de

M.A. Peter (✉)
Institute of Mathematics, University of Augsburg, 86159 Augsburg, Germany

Augsburg Centre for Innovative Technologies, University of Augsburg, 86159 Augsburg, Germany
e-mail: malte.peter@math.uni-augsburg.de

F. Strobl
Institute of Physics, University of Augsburg, 86159 Augsburg, Germany

Nanosystems Initiative Munich, Schellingstrasse 4, 80799 München, Germany

A. Wixforth
Institute of Physics, University of Augsburg, 86159 Augsburg, Germany

Augsburg Centre for Innovative Technologies, University of Augsburg, 86159 Augsburg, Germany

Nanosystems Initiative Munich, Schellingstrasse 4, 80799 München, Germany
e-mail: achim.wixforth@physik.uni-augsburg.de

© Springer International Publishing AG 2017
D. Bothe, A. Reusken (eds.), *Transport Processes at Fluidic Interfaces*,
Advances in Mathematical Fluid Mechanics, DOI 10.1007/978-3-319-56602-3_17

Similarly, we investigate the dynamics of objects embedded in a simpler interface given by an air–water surface and demonstrate the surface-acoustic-wave-actuated separation of enantiomers (chiral objects) on the surface of the carrier fluid. It turns out that the dynamics and the separation of the phases do not only depend on parameters such as temperature, mobilities and line tension but also on the mechanics of the lipid layers subjected to exterior forces as, for instance, compression, extensional and shear forces in film-balance experiments. Since the mechanical behavior of lipid layers is viscoelastic, we use a modeling approach based on the incompressible Navier–Stokes equations with a viscoelastic stress term and a capillary term, a convective Jeffrey (Oldroyd) equation of viscoelasticity, and the Cahn–Hilliard equation with a transport term. The numerical simulations are based on C^0-interior-penalty discontinuous-Galerkin methods for the Cahn–Hilliard equation. Model-validation results and the verification of the simulation results by experimental data are presented. The feasibility of enantiomer separation by surface-acoustic-wave-generated vorticity patterns is shown both experimentally and through numerical simulations. This technique is cost-effective and provides an extremely high time resolution of the dynamics of the separation process compared to more traditional approaches. The experimental setup is an enhanced Langmuir–Blodgett film balance with a surface-acoustic-wave-generated vorticity pattern of the fluid, where model enantiomers (custom-made photoresist particles) float on the surface of the carrier fluid. For the simulations, we propose a finite element immersed boundary method (FEIBM) for deformable enantiomers and a fictitious-domain approach based on a distributed Lagrangian multiplier finite element immersed boundary method (DLM-FEIBM) for rigid chiral objects, both of which lead to simulation results consistent with experiments.

17.1 Introduction

Biological cells are complex viscoelastic and soft objects which are composed of various components and compartments such as the plasma membrane, the nucleus, the endoplasmatic reticulum and a number of other cell organelles. All these compartments are surrounded by a bilayer membrane to separate the interior from its fluid environment and to regulate cell communication and maintain homeostasis. The biological concept of organelle-enclosing bilayers is a generic feature of all eukaryotic cells. However, on the molecular level these bilayers are far more heterogeneous and made of different amphiphilic molecules which form the "fluid scaffold" of the membrane. Therein, various types of macromolecules such as proteins and larger aggregates such as ion channels are embedded. This model was first described by Singer and Nicolson in 1972 [62] (cf. Fig. 17.1) and has since then been refined and extended in several ways. One of those refinements is based on the observation that the bilayer membrane can undergo different phase-separation processes from a homogeneous membrane to a laterally structured heterogeneous membrane [42, 51]. This decomposition is clearly a result of the multicomponent composition of the membrane from many types of lipids and a small amphiphilic molecule called cholesterol.

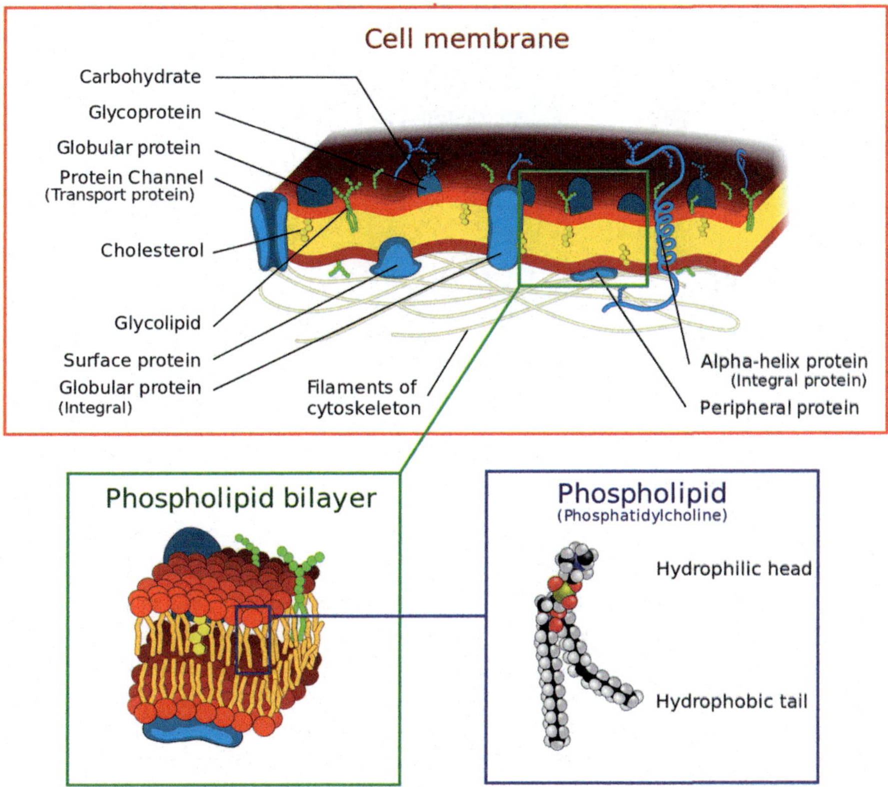

Fig. 17.1 Schematic sketch of a part of a cell membrane (adapted from: https://en.wikipedia.org/wiki/Cell_membrane)

Lipids are one of the key building blocks of the membrane. They allow the cell to maintain morphological consistency and structural order in a mechanical as well a thermodynamic sense. Lipid membranes consist of two monolayers of lipid molecules, each molecule is made up of a polar hydrophilic end and a nonpolar hydrophobic tail. When exposed to water, the lipids tend to organize to form bilayers owing to the hydrophobic effect. Under appropriate conditions, the preferred shape of membranes is a closed shell, a so-called vesicle (cf. Fig. 17.2). Vesicles can be created artificially in a relatively easy way, since lipids spontaneously self-assemble to form a closed bilayer structure and they have been analyzed thoroughly both theoretically and experimentally as model cell membranes. Such artificially created containers made of closed bilayers have also been used to deliver drugs, e.g. in liposomes. It is thus of vital importance to gain a profound understanding of the functioning of lipid membranes and how to manipulate them in order to comprehend their behavior and to ensure the designated effect of drugs. An important step towards this goal is to introduce mathematical models of the processes and to verify

Fig. 17.2 Structure of a lipid (*left*), many of which form a closed shell, a so-called vesicle, under appropriate conditions (*right*) [28]

them by experimental observations. A key feature of the system is that the layers are essentially complex two-dimensional objects in a three-dimensional environment.

Biological lipid bilayers are mainly made up of phospholipids and behave like liquid crystals. The arrangement of the lipids can be changed depending on the situation (e.g., temperature) so that the membrane can exhibit several phases. The two liquid phases are distinctly ordered and are referred to as the liquid-ordered phase and the liquid-disordered phases. Because of the coexistence of phases, such a membrane can be thought of as a two-dimensional binary fluid. Below the characteristic temperature T_m, which is a function of the concentrations of the liquid-ordered phase and the liquid-disordered phase, we can observe the phase-separation process. The kinetics of this separation process depend on whether the system is cooled to a metastable or an unstable region. In the metastable region, the decomposition of the phases is characterized by nucleation and growth. Cooling to the unstable region results in spinodal decomposition. Both types of separation have been observed experimentally.

Phase separation processes can be modeled by the Cahn–Hilliard equation [13]. In Sect. 17.2, as a first step, we compare qualitatively and quantitatively the results of experiments and simulations for temperature-induced lipid phase separation in artificially created vesicles neglecting mechanical effects. It is shown that the Cahn–Hilliard equation is an appropriate model in this context.

Since lipid layers often show a viscoelastic behavior, the Cahn–Hilliard equation has to be coupled with a viscoelastic fluid-flow model such as a convective Jeffreys (Oldroyd) model (see [39, 44] for a detailed description). We are interested in experiments and numerical simulations where a lipid monolayer is spread onto the surface of a Langmuir–Blodgett trough filled with water. Phase separation is studied for a surface-acoustic-wave-actuated fluid flow in the trough featuring four counter-rotating vortices at the upper surface of the trough (cf. Fig. 17.3). In this case, the numerical simulation is based on a system of equations consisting of

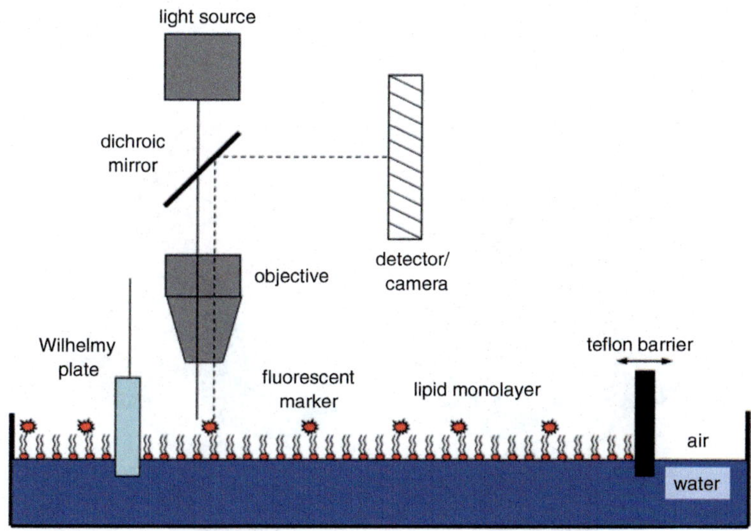

Fig. 17.3 Film balance experimental setup equipped with a microscopic observation unit. The total area of the lipid monolayer can be controlled by the motion of a teflon barrier

the incompressible Navier–Stokes equations with a viscoelastic stress term and a capillary term, a convective Jeffreys (Oldroyd) equation of viscoelasticity, and the Cahn–Hilliard equation with a transport term [8, 9, 17]. This system will be referred to as the NS/CJ/CH system. Both experimental measurements and numerical simulations for this system will be addressed in Sects. 17.3 and 17.4, respectively.

Finally, we explore the experimental acoustic control and understanding of microflows that we gained in the previous sections to a practical problem of immense importance—the separation of enantiomers. Enantiomers are chiral geometric objects, where an object is said to be chiral if it is not identical to its mirror image. Since the word "chiral" stems from the Greek "$\chi \varepsilon \iota \rho$", which means "hand", one distinguishes enantiomers by their handedness (right- resp. left-handedness; cf. Fig. 17.4).

In chemistry, chirality refers to a molecule, which is not superposable on its mirror image. Compounds consisting of molecules of the same handedness are called enantiopure or unichiral, whereas compounds consisting of the same amount of right- and left-handed enantiomers are referred to as racemic. Since the chemical synthesis of enantiomers usually gives rise to racemic compounds, enantiomer separation plays a significant role in agrochemical, electronic, and pharmaceutical as well as food, flavor and fragrance industries. Traditional separation technologies are based on gas or high-pressure liquid chromatography, capillary electrophoresis, or nuclear magnetic resonance, but most of them are slow and require costly chiral media.

A different approach uses the fact that enantiomers drift in microflows with a direction depending on their chirality [46, 48]. Enantiomer separation using a

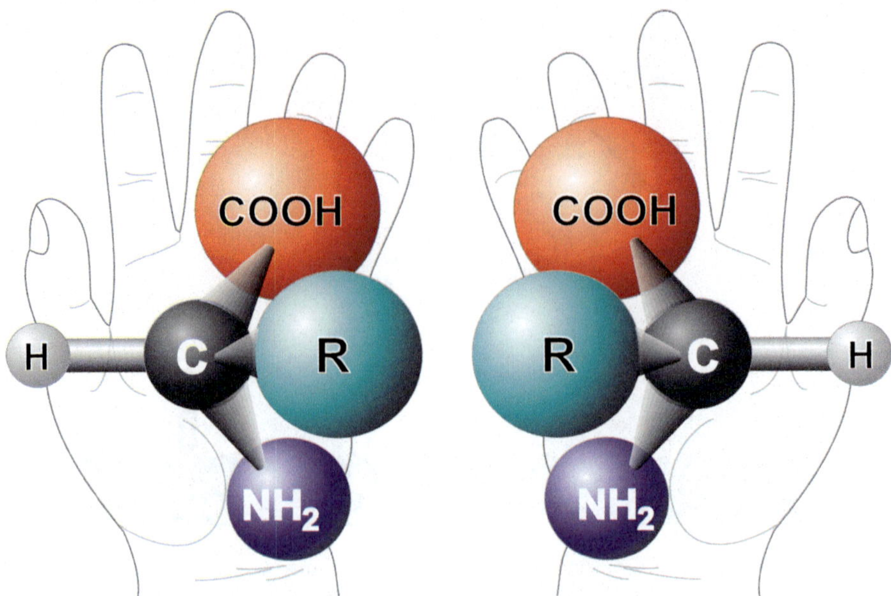

Fig. 17.4 Left- and right-handed enantiomers (Source: https://en.wikipedia.org/wiki/Chirality_(chemistry))

quadrupolar force field generated by surface acoustic waves has been theoretically predicted in [43] for simple idealized chiral objects.

In Sect. 17.5, we report both on an experimental setup for surface-acoustic-wave-actuated enantiomer separation and the mathematical modeling and numerical simulation of this setup by the finite element immersed boundary method and the distributed Lagrangian multiplier finite element immersed boundary method. The results confirm that the setup is capable of separating enantiomers by their handedness.

17.2   Phase Separation and Decomposition in a Lipid Bilayer Membrane

This section deals with the phase-separation process of a bilayer system, which is caused by lateral demixing of its multiple molecular components into different domains and describes the demixing in the framework of a Cahn–Hilliard model.

Domain structures can occur on length scales, which are amenable to optical light microscopy (several micrometers), though it is assumed that they also develop at scales of 10–100 nm and have been associated within the so-called "lipid raft hypothesis" [36, 61] in biological cells. This hypothesis links the formation of a phase-separated membrane and its structure to biological functions, e.g. to control adhesion and cell signaling. It is hypothesized that the decomposition of the phase is directly related to molecular organization of proteins and other macromolecules, e.g. for adhesion, and is essential for physiological and cellular processes.

Here, to reduce the cell-membrane complexity, we use Giant Unilamellar Vesicles (GUVs) as a simple model system, which is also accessible to light microscopy. GUVs are composed of a well-controlled number of different lipid types and cholesterol but still display the basic properties of cell membranes. Over the past years, there has been extensive experimental effort undertaken to study ternary lipid systems [2, 4, 68, 70] consisting of two types of phosphatidylcholines (PC) such as 1,2-Dipalmitoyl-sn-glycero-3-phosphocholine (DPPC) and 1,2-Dioleoyl-sn-glycero-3-phosphocholine (DOPC) and cholesterol as the third component. The phase separation of this system has been found to occur in two different ways, either by a nucleation process with subsequent ripening of the discrete domains into larger domains or by a spinodal decomposition process [69].

From a theoretical perspective, there have been a number of studies to reproduce the experimental findings qualitatively. However, quantitative comparisons of theoretical simulations with experimental data are rare [42]. Here, we address this shortcoming by directly comparing the temporal development of liquid-ordered domains in a GUV bilayer system. We quantitatively relate the optically observed domains in our experiments with a numerical simulation based on the Cahn–Hilliard equation.

17.2.1 Vesicle Experiments

We produce GUVs with diameters of up to $200\,\mu m$ using the electroformation method as described by Angelova et al. [1]. In short terms, a small volume ($5\,\mu l$) of a $1\,mg/ml$ total lipid in chloroform solution is deposited onto an ITO coated microscope slide and the solvent is evaporated. A second ITO slide is positioned opposite to the slide with the dried lipid spot and the gap between two slides is filled with distilled water or an aqueous solution containing sucrose. Then, an electric field of $10\,Hz$ and $10\,V/cm$ is applied and kept constant overnight (at least $6\,h$) and the GUVs form during this period. In all experiments, we use a mixture of $60\,mol\%$ lipid and $40\,mol\%$ cholesterol. Among the lipid part, we use the ratio of DOPC:DPPC:DPPG (5:4:1). Note that in addition to the commonly used DOPC/DPPC system, we added small traces of negatively charged 1,2-Dipalmitoyl-sn-glycero-3-phospho-rac-(1-glycerol) (DPPG). In previous experiments, we found that using DPPG enables supercooling of the vesicles and prevents undesired and uncontrolled spontaneous prematuring of the vesicles. The lipids and cholesterol ($>99\%$) were purchased from Sigma Aldrich (Munich, Germany) and were used without further purification. The visualization of domains was enhanced by the fluorescence marker Texas Red (TR) which was also obtained from Sigma Aldrich. For observations, we used a Zeiss Axiovert 200 inverted fluorescence microscope (Zeiss, Oberkochen, Germany) and a CCD camera (Hamamatsu, Herrsching, Germany) for video recording. All experiments were performed in a custom-made and temperature-controlled fluidic chamber and observations were directly done after

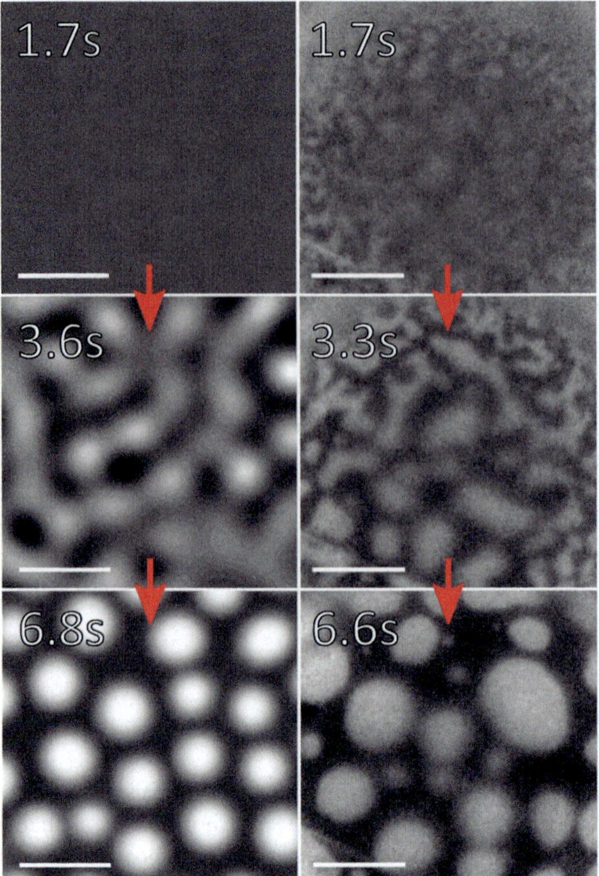

Fig. 17.5 Comparison of the temporal evolution of domain patterns obtained from a numerical simulation (*left*) and in the corresponding experiment (*right* micrographs). Each set of images shows the same bilayer changing its domain morphology during the phase decomposition. The numbers in the images indicate the time that has passed since the onset of phase decomposition. The time difference between the images is chosen such that each image shows a representative pattern that is observed in the distinct phases of decomposition illustrated in Fig. 17.6. The scalebar is 10 μm. Reprinted with permission from [11]. Copyright 2013 American Chemical Society

GUV formation. Therefore, the GUVs were supercooled below the critical phase-transition temperature T_m and then illuminated with the microscope lamp to observe the spontaneous phase decomposition of the bilayer membrane. A typical temporal sequence of fluorescence micrograph frames is depicted in the right column of Fig. 17.5.

In the first micrograph, a continuous and still diffuse pattern of domains can be seen, which closely resembles what is expected in the regime of spinodal decomposition. In the second frame, discrete yet connected domains have been already formed, which are, however, still far away from their circular equilibrium shape. In the further course of the experiment (micrograph 3), the domains quickly relax into a circular shape and start to grow mainly by coalescing with neighboring domains [11].

17.2.2 Cahn–Hilliard Simulations

The challenge for the numerical simulation is to reproduce both the qualitative topological features as well as the quantitative temporal development and the domain size correctly. To model bilayers, a number of different methods have been applied including Monte–Carlo methods [38] or molecular-dynamics simulations [3]. In the past, phase-field models have been successfully used to reproduce complex microstructures [14]. Here, we use a Cahn–Hilliard model, which has been applied to a number of diverse problems before [6, 66] and which is discussed elsewhere in more detail [23, 52].

We capture the temporal decomposition into two different phases by introducing the mole fraction c as an order parameter of the system. The values $c = 1$ and $c = 0$ indicate the pure phases and are represented in white and black, respectively, in the simulation pattern in Fig. 17.5(left column). We determine the order parameter c from an equation of Cahn–Hilliard type

$$\frac{\partial c}{\partial t} - \nabla \cdot \left[M \nabla \big(f'(c) - \epsilon^2 \Delta c \big) \right] = 0. \tag{17.1}$$

Here, $f = f(c)$ is the local free energy per unit of area. In the first approach, we use a double-well potential $f(c) := \varphi c^2 (c - 1)^2$ with a scaling parameter φ. The parameter M is the mobility which depends on the diffusion constants D_0 and D_1 of both phases according to $kTM = D_0(1-c)+D_1 c$, where k is the Boltzmann constant and T is temperature. For simplicity, we assume the same diffusion coefficients for both phases described by an effective constant D [54], i.e. $M = D$ in numerical value. The parameter ϵ describes the energy cost of the boundaries of the two phases by the energy term $\frac{\epsilon^2}{2}|\nabla c|^2$. In this model, the physical properties of the system are captured by the three control parameters D $[\frac{\text{m}^2}{\text{s}}]$, φ $[\frac{\text{J}}{\text{m}^2}]$, and ϵ^2 [J]. We explore the solution of the Cahn–Hilliard equation and the temporal development of the phases and their domain size systematically on these control parameters and compare them with the experiments. The simulations were performed using a C^0-interior-penalty discontinuous-Galerkin finite-element method [29, 71], which has proved to be stable and efficient for such kind of problems. This method was particularly chosen to account for the fourth-order spatial derivatives of the Cahn–Hilliard equation. In this way, it was not necessary to introduce a coupled system of partial differential equations and instead we could use the standard second-order Lagrangian finite elements. We solve the nonlinear discrete equation with periodic boundary conditions by a Newton method. The triangulation was carefully chosen to avoid artificial symmetry effects, which could have been introduced by the discretization otherwise.

17.2.3 Comparison of Cahn–Hilliard Simulations and Experiments

We match the simulation to the microscopic observation and find optimal values for the control parameters D $[\frac{m^2}{s}]$, φ $[\frac{J}{m^2}]$ and ϵ^2 [J] by comparing the temporal development of mean domain sizes in both experiment and simulation. Since the domain size is difficult to determine directly from a distribution of sizes we analyze the micrographs and simulation plots with a 2D Fast Fourier Transform (FFT) and use the structure factor $\mathcal{S}(q)$ of a given mode q to determine the mean domain size $1/\langle q \rangle$ from

$$\langle q \rangle := \frac{\mathcal{S}(q)q}{\sum_q \mathcal{S}(q)} \tag{17.2}$$

The structure factor is obtained from the image grey value $a(p_j)$ at p_j depending on position p_j and the scaling factor N given by the number of q_j. For comparison, we plot the mean domain size against time as shown in Fig. 17.6(upper-left image) for parameter values of $D = 1.0\,\mu m^2/s$, $\varphi = 4.0\,J/m^2$ and $\epsilon^2 = 1.0 \times 10^{-12}\,J$ together with the experimental domain size. At first, it is apparent that the domain growth occurs in three distinct temporal periods in the simulations as well as in the experiments. It can be characterized by a slow initial domain growth, followed by accelerated growth and then, finally, a reduced growth approaching the final domain size. In the first period, which ends at $t \approx 2\,s$ for the simulations and at $t \approx 1.5\,s$ for the experiments, we observe the spinodal decomposition of the membrane, which is also depicted in the upper panel of Fig. 17.5. The second period lasts about 3 s in the experiment and about 2.5 s in the simulation and is characterized by an enhanced domain growth rate of a factor of 8. Domain growth in this period is dominated by coalescence of adjacent small domains in the experiments while, in the simulations, domain formation still occurs as depicted in the middle panel of Fig. 17.5. In the last period, which sets in at $t \approx 4.5\,s$ (the lower panel of Fig. 17.5), matured and circular equilibrium domain shapes have been formed with a significant decrease in growth rate. The mean diameter at the end of the considered interval is approximately $5.0\,\mu m$ for both simulations and experiments. These qualitative and quantitative agreements of our simulations with the experimental data confirm that we capture the basic physics of the decomposition.

However, there are also quantitative differences, which we address in the following systematic parameter study as shown in the plots of Fig. 17.6. Therefore, we keep two of the control parameters constant and vary the remaining parameter. The sensitivity of the domain growth on the diffusion parameter D is plotted in the top right image of Fig. 17.6. As qualitatively expected, the domain growth rate increases with increasing values of D because lipids are more mobile and can achieve the energetically favorable states faster. The simulation parameter D is related to the physical lateral diffusion coefficient in lipid bilayers and has been measured to be of the order of 1–$20\,\mu m^2/s$ using various types of methods and lipid

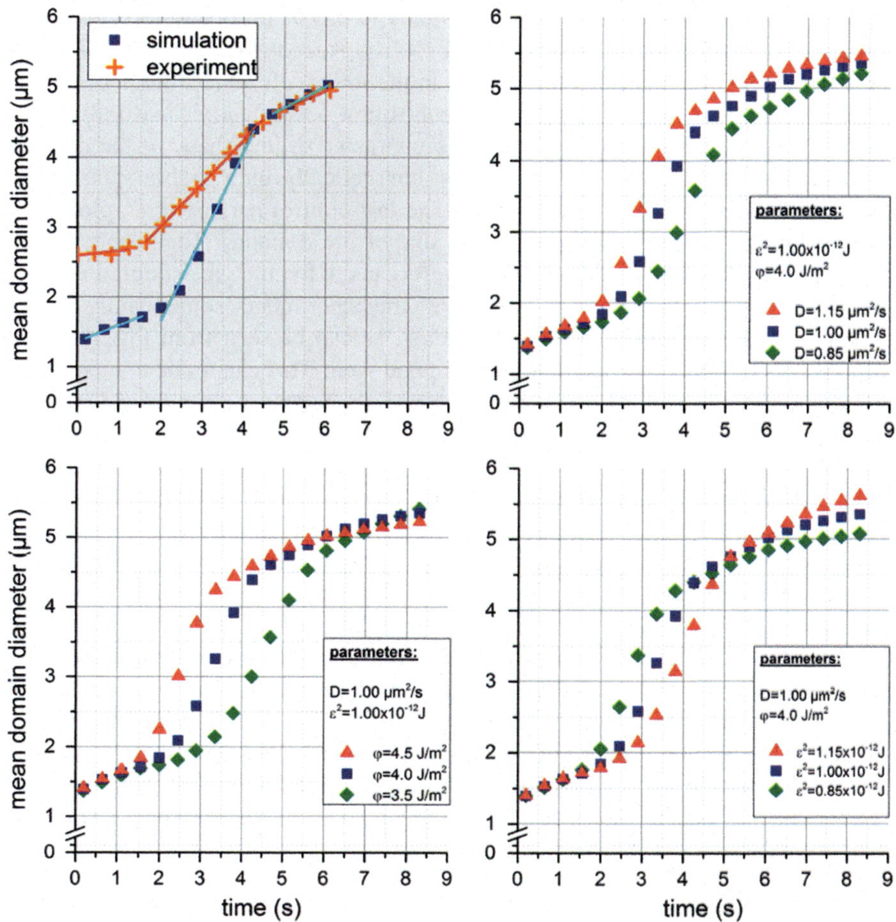

Fig. 17.6 *Top left:* Evolution of the mean domain diameter during the phase transition. *Orange crosses* mark experimental data while *blue squares* represent results of numerical simulations ($D = 1\,\frac{\mu m^2}{s}$, $\epsilon^2 = 1 \times 10^{-12}$ J, $\varphi = 4\,\frac{J}{m^2}$). *Top right:* Results of numerical simulations with varying mobility parameter D. *Bottom left:* Results of numerical simulations with varying local free energy scaling parameter φ. *Bottom right:* Results of numerical simulations with varying effective line tension parameter ϵ^2. Note that the curve represented by *blue squares* is identical in all panels. Reprinted with permission from [11]. Copyright 2013 American Chemical Society

systems [20, 26, 40, 54]. The values for the ternary system containing the lipids DOPC, DPPC and cholesterol have been given by Orädd [54] and were used to guide this parameter study. The other two control parameters φ and ϵ^2 are addressed in the lower-right and -left image of Fig. 17.6, respectively. It is important to point out that unlike D, these two parameters cannot directly be correlated to physically measured values from the literature. Although, for example, the parameter ϵ^2 controls the energy associated with the boundary of the two domains, it cannot be identified with the line tension [57]. To relate φ and ϵ^2 directly to physically-measured quantities, another model would be necessary, which clearly exceeds the frame of this study but is interesting to consider in the future. In the lower-left image, the dependence on φ indicates that for larger φ the decomposition occurs earlier and the rate of growth is enhanced. Since the parameter φ characterizes the shape of the double-well potential and increases its steepness, energetically unfavorable states relax more rapidly into the equilibrium states. The last control parameter ϵ^2 plotted in the right image strongly affects the final size of the domains. Together with the experimentally measured value, it was therefore taken to guide the simulations.

However, despite the good agreement with the experiments, some features of the domain growth still remain challenging. First, as can be seen from Fig. 17.6, the initial values of domain sizes do not correspond well. This cannot be compensated by variation of the control parameters because, as becomes clear from Fig. 17.6, the initial domain size only weakly depends on the chosen values of D, φ and ϵ^2. We hypothesize that the larger value for the size in the experiment is attributed to incomplete mixing in the beginning and may also be affected by the limited resolution of the optical system as well as nonlinear partitioning of the Texas Red marker between the two phases. Another characteristic difference to the experiments is that, in the second domain growth period, the experiment is mainly driven by coalescence while, in the simulations, coalescence occurs as well but seems to be less dominant. We can speculate that one reason for this behavior is the role of the traces of negatively charged DPPG used in the experiments. Moreover, our model does not account for the mechanical effects in the system and the mechanical stress that certainly impact the values of the control parameters. We address the dynamics of a lipid system in the subsequent section by compressing a lipid monolayer on a Langmuir film balance and address the dynamic effects that include viscoelastic shear properties.

17.3 Probing Mechanics of Lipid Monolayers by Acoustic Actuation

In the previous section, we focused on the decomposition of a lipid bilayer system. One of the limitations was that mechanical effects were not taken into account. Here, we use a slightly different system, a lipid monolayer, spread on a Langmuir trough film balance, and investigate the viscoelastic response to an external mechanical excitation.

There has already been a vast amount of work dedicated to investigate lipid monolayers and bilayers [55] (see also the book [51] for a better understanding of the physical properties of these systems). For monolayers, mechanical properties such as the area elasticity and dilatational viscosity have been studied intensively depending on the thermodynamic state of lipid monolayers. Most natural lipids typically show a first-order transition between a liquid-expanded (LE) and a liquid-condensed (LC) phase. In monolayers, this transition becomes apparent in

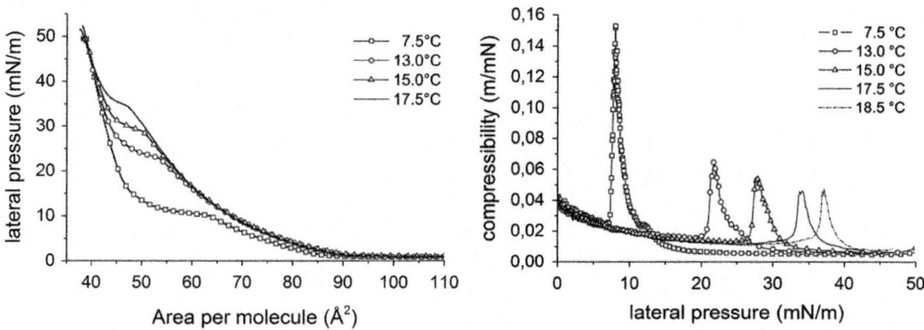

Fig. 17.7 (**a**) Typical isotherms for DPPC monolayers at different temperatures; (**b**) The area elasticity of the monolayer approaches a maximum at the phase transition (adapted from [63])

a pronounced plateau in the surface pressure vs. molecular area isotherm, which is commonly recorded on a Langmuir film balance, see Fig. 17.7.

In a Langmuir film balance, the lateral pressure is measured as a function of the molecular area and is given as the difference in surface pressure in the presence and the absence of the lipid monolayer. Thermodynamically, the lateral pressure is the derivative of the surface free energy with respect to the area per molecule. The isothermal compressibility and the isobaric thermal expansivity as material parameters are then given as the partial derivatives of the molecular area with respect to the pressure and temperature, respectively. The horizontal section in Fig. 17.7a suggests that the transition is discontinuous and of first order. In the vicinity of a critical state (near the phase transition) the thermodynamic fluctuations of a system approach a maximum [63].

Since these fluctuations are closely related to the susceptibilities and hence linked to the material properties, parameters like area elasticity, heat capacity or electrical conductivity change dramatically and often take extreme values in the transition region. Whereas these phenomena have been investigated intensively for the area properties, much less work has been done to investigate the shear properties of these systems. Here, we focus on the shear properties of lipid monolayers in the vicinity of their LE–LC transition. The LC-regime has been investigated extensively due to its outstanding relevance for the physics of breathing and due to its good accessibility for established rheological methods [15, 24, 41]. In contrast, lipid layers in the liquid-expanded regime do not show any shear elasticity and have very low shear viscosities. The ratio of drag forces arising from the surface to those arising from the subphase underneath is usually described by the Boussinesq number

$$\text{Bo} = \frac{\text{surface drag}}{\text{subphase drag}} = \frac{\eta_s}{\eta_0} \cdot \frac{R}{A}, \tag{17.3}$$

where η_s and η_0 are the viscosities of inter- and subphase, respectively, and the ratio R/A describes the geometry of the probe used, i.e. the ratio of contact perimeter

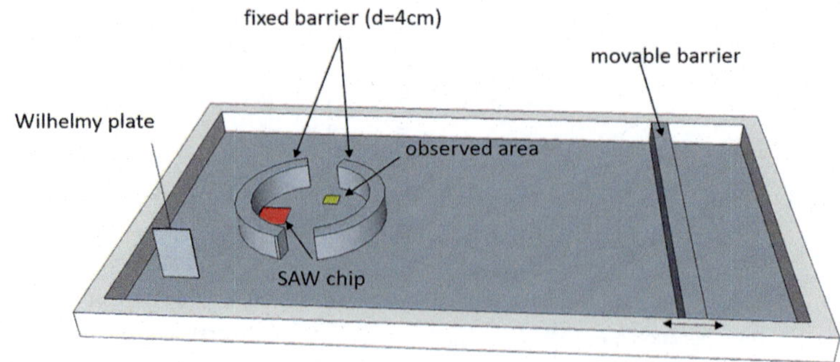

Fig. 17.8 Top view of the trough geometry. Two additional semicircular barriers decouple the observed area from large-scale flow excited by external disturbance. The SAW chip (*red*) is mounted at the bottom such that the fluid streamline with maximal velocity crosses the observed monolayer area

with the interphase and contact area with the subphase [10]. For direct shear measurements, where R/A is limited by technical means, low surface viscosities are difficult to measure. While some work has been done for LE-phase of lipid monolayers at constant surface pressure [60], very little data is available for the main transition regime [21, 56, 58, 59]. Especially the influence of the monolayer microstructure on the shear properties has not been investigated systematically, even though it is well known that the structure of LC-domains in the phase-transition regime is sensitive to their growth velocity and impurities as for example tracer particles or fluorophore molecules [33, 49, 50].

To probe the monolayers mechanically, we introduce here a new method for investigating the shear viscosity and elasticity of lipid monolayers as functions of surface pressure. We show that both viscous and elastic properties depend on the total area of LC-domains as well as on their microstructure.

For all experiments, a customized Langmuir trough from NIMA was used. The surface area can be adjusted by a movable teflon barrier. The surface pressure is measured using a Wilhelmy plate and a NIMA force sensor. A focused interdigital acoustic transducer was installed at the bottom of the trough for the excitation of an acoustic flow in the observed region as shown in Figs. 17.3 and 17.24. The chip is powered by a high frequency generator and an amplifier at its resonance frequency with an input amplitude of 9 dBm. The trough is covered by a plastic lid to protect the interface from potential air motion. Additionally, several teflon barriers spanning the full height of the trough were introduced to suppress large scale streaming (see Fig. 17.8). The temperature of the teflon trough and the subphase was stabilized by a heating bath at 23 °C. The monolayer was observed through a glass window with an upright fluorescent microscope equipped with a Photron FastCam camera.

Dipalmitoyl-sn-glycero-phosphatidylcholine (DPPC) from Avanti Polar Lipids (Alabaster, USA) and Texas Red-DHPE from Invitrogen were used without further purification. All other chemicals were purchased from Sigma Aldrich (Germany). Ultrapure water ($>18\,\mathrm{M\Omega}$ cm, PureAqua, Tuttlingen, Germany) was used for all

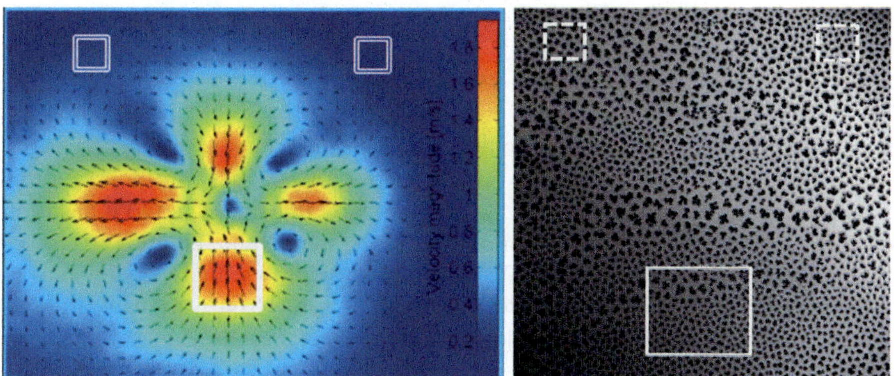

Fig. 17.9 Typical flow profile obtained from the PIV analysis. The *larger solid frame* indicates the region of interest, the *double lined squares* indicate the reference fields

aqueous solutions. A solution of 10 M HEPES buffer and 90 mM NaCl was adjusted to a pH of 7.0 and was used as subphase for all experiments and shields electrostatic interactions. DPPC and Texas-Red-DHPE with a molar ratio of 0.5% were dissolved in chloroform with an overall concentration of 1 mg/ml. Lipid monolayers were formed by slowly pipetting a fixed amount of lipid solution onto the fluid surface. Before starting the experiments, the monolayer was equilibrated for 10 min to allow for the evaporation of chloroform traces.

After equilibration, the monolayer was compressed slowly until first LC domains appear. From here, the pressure was increased in steps of 0.5 mN/m until the domains were strongly compressed. Afterwards the barrier motion was reversed and the area was increased again in steps of 0.5 mN/m. At each step, the monolayer was equilibrated at constant area for 5 min before the SAW flow was repeatedly turned on and off for time intervals of 5 s. To track the flow, videos of the domain motion were captured with a frame rate of 125 fps (see Fig. 17.9 for typical flow profiles).

Captured videos were first analyzed by a script based on the open source PIVlab toolkit22 to obtain the velocity field for the complete field of view at different time steps. Twenty frames were used for the particle imaging velocimetry (PIV) analysis of one time step to reduce motion and data processing artefacts. Hence, the velocity information for further processing has a temporal resolution of 160 ms. A region of interest of 50×50 pixels in one center stream line of the quadrupolar flow field was chosen and the y-component of the velocity field was averaged spatially. We refer to the velocity obtained in this way as v_y in the following. In the same way, a reference velocity $v_{\text{ref},y}$ was determined by spatially averaging the flow fields in the two opposing corners of the field of view.

The structure of the (dark) LC-domains within the region of interest was analyzed by using ImageJ-scripts. The area ratio $\alpha = \frac{A_{\text{LC}}}{A_{\text{LE}} + A_{\text{LC}}}$ was determined by "thresholding" and the connectivity ζ by the BoneJ plugin from ImageJ. The latter value is derived from the Euler characteristic of the indexed picture and measures the "porosity" of the LC-domains (e.g. small, isolated domains will result in very negative ζ values).

17.3.1 Experimental Results and Discussion

17.3.1.1 Velocity Amplitude and Viscosity

In a first set of experiments, we periodically excite acoustic streaming by repeatedly switching the SAW on and off for 5 s at a period of 10 s. As a response to this excitation, the differential velocity $\Delta v_y = v_y - v_{\text{ref,y}}$ in the selected ROI has been analyzed for different thermodynamic states of the membrane. Figure 17.10 gives an overview of the measured data. We introduced the differential velocity Δv_y to compensate vibration artefacts in the experimental setup from various sources as small vibration of the damping table or air flow. Since usually $\Delta v_y \gg v_{\text{ref,y}}$, the error introduced here is very small.

One can clearly see from Fig. 17.10 that the maximum amplitude of Δv_y decreases with higher surface pressure as can be expected as consequence of increasing surface viscosity at higher surface pressures. For quantification, we derive $v_{y,\text{max}}$ by averaging over ten data points from each period just before the SAW is switched off. When we increase the surface area after compression into the LE-regime, we observe very strong hysteresis effects and there is no clear correlation to the measured surface pressure (see Fig. 17.11a). Hysteresis behavior is well known and has been reported in the literature [55]. The reason for this is the domain growth dynamics of these systems, which was intensively discussed [51], particularly the fact that labeled lipid monolayers may have to be regarded as a two-component system. If we choose the relative area of LC domains α instead of the surface pressure π as the independent variable for further analysis, the correlation becomes much more significant, as can be seen in Fig. 17.11b.

In Fig. 17.12, we summarize the data obtained from a large number of experiments. A clear trend can be observed, even though there is quite a high scattering of the data points. The reasons for this become apparent when we consider the domain structure of the monolayer at different data points. As pointed out by Möhwald and others, the diffusion of dye molecules or other impurities controls the typical fractal domain growth [33, 49, 50]. Additional factors like history

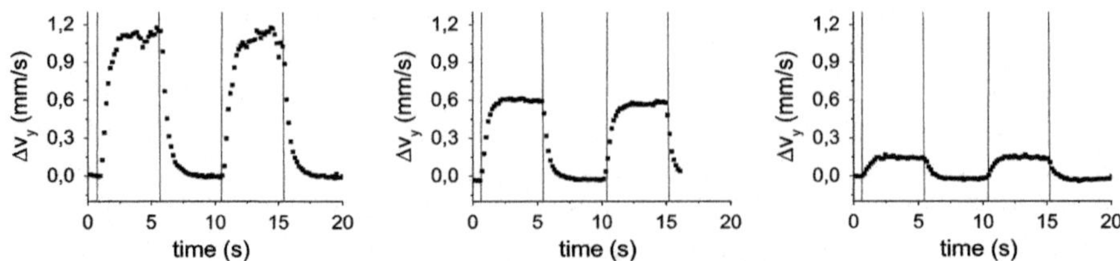

Fig. 17.10 Δv_y as a function of time for different membrane states (ratio of the LC phase), i.e. different area ratios. (a) $\alpha = 50\%$ (b) $\alpha = 60\%$ (c) $\alpha = 80\%$. With rising LC-ratio, the velocity amplitude decreases significantly as a result of increasing viscosity. The SAW was switched on and off for a duration of 5 s at the power of 9 dBm. The vertical lines tag the time points when the SAW power is switched on/off

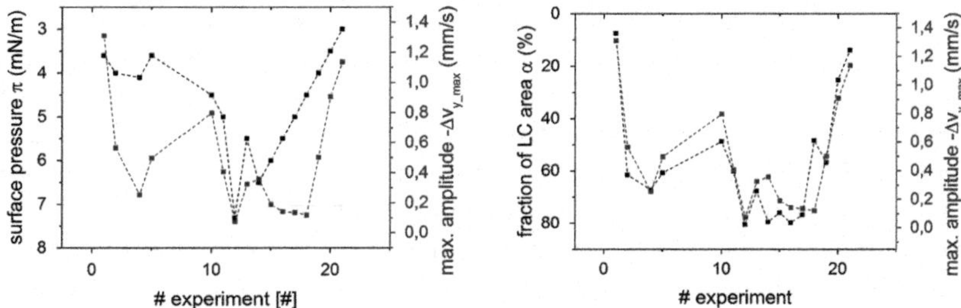

Fig. 17.11 (**a**) A comparison of surface pressure and Δv_y shows only moderate correlation. However, the maximum velocity amplitude seems to lag behind the surface pressure. (**b**) In contrast the graphs for the relative area of LC-domains and $v_{y,\mathrm{max}}$ match very well. As a guide to the eye we connected the subsequent data points by lines to simplify the comparison of the two graphs

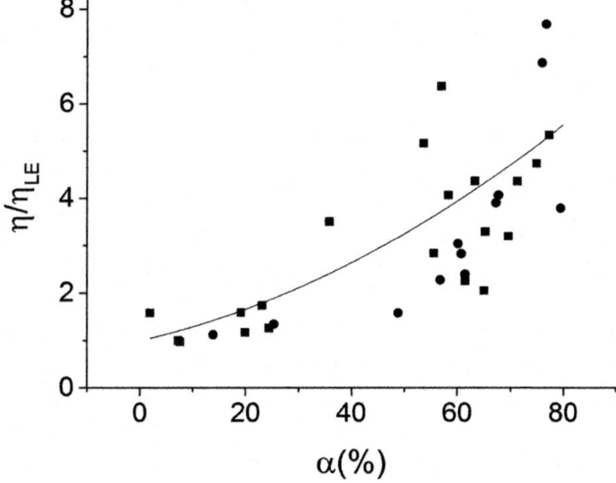

Fig. 17.12 Relative viscosity η/η_{LE} as a function of the LC-area proportion α. The strong scattering of the measured values is most likely a consequence of a variation in domain microstructure. The continuous line is a fit using Einstein's equation (17.4) with $C = 4$ and indicates the increase of the viscosity ratio with α

dependence and mechanical disturbances lead to the fact that the microstructure of lipid monolayers is not fully controlled in the transition regime. Therefore, it is not surprising that different domain structures lead to different shear properties. If we regard the monolayer in the transition regime as a system, which consists of more or less solid-like LC-rafts floating in a fluid LE-matrix, it is obvious that its fine structure affects the flow. However, as a first attempt, we treat the LC-domains as weakly interacting circular discs and consider the system as a 2-dimensional quasi-continuous (emulsion-like) medium. The viscosity of such systems can usually be described empirically by a modified Einstein's equation [22, 32]

$$\frac{\eta}{\eta_0} = 1 + 2.5\alpha + C\alpha^2. \tag{17.4}$$

Here, η_0 is the viscosity of the fluid LE-phase. We further assume that $\triangle v_y \propto F_y/\eta \propto 1/\eta$, where F_y denotes the y-component of the SAW-streaming force field, which engages on the interface, we can fit our data with (17.4) after normalization with the maximum measured value for $\triangle v_y$, which represents $1/\eta$. For the solid line in Fig. 17.12, we choose $C = 4$. An empirical description of the system as a 2D colloid is thus feasible. For a more quantitative description, however, a precise analysis of the flow field as well as a refined model for the relation between surface viscosity and domain structure is necessary.

17.3.1.2 Domain Structure and Viscoelasticity

So far, the lipid monolayer was treated as purely viscous without considering elastic effects. However, a closer look at the data in Fig. 17.10 reveals that there is an elastic contribution to the waveform of $\triangle v_y$. This becomes much more pronounced for smaller SAW power, i.e. reduced excitation force (see Fig. 17.13).

This elastic behavior can be observed in many experiments. However, the quantitative strength of it varies significantly between experiments and different ROIs in one experiment. Figure 17.14 shows a typical picture captured during an experiment. There are two clearly distinguished regions of different domain morphology. Region one consists of isolated LC islands (disconnected), whereas region two exhibits a mesh-like domain structure (connected). Both morphologies were stable without mechanical disturbance and have similar LC proportion. However, the formation of morphology 2 was a result of a preceding SAW excitation and demonstrates that the domain structure can be influenced by slight disturbances. If one observes the dynamics of regions like the ones shown under flow excitation, one realizes that regions with unconnected domains behave more viscous, whereas those with connected domains show strong viscoelastic characteristics. One possibility for a quantification of the domain structure is the connectivity ζ, which can be determined

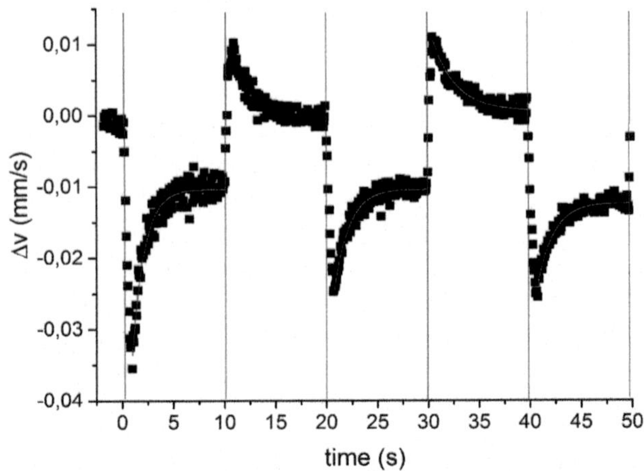

Fig. 17.13 $\triangle v_y$ as a function of time for small SAW power. The reversal of the flow direction indicates a pronounced elastic contribution to the shear properties of the lipid monolayer

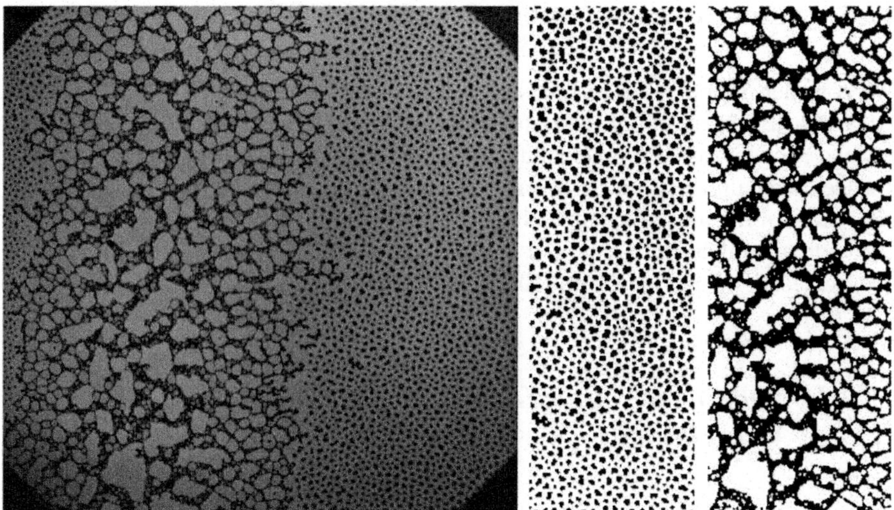

Fig. 17.14 Coexistence of two regions of very distinct morphology: (*left*) original picture; (*middle*) "thresholded" detailed picture region 1 (disconnected): $\zeta = -458$; (*right*) "thresholded" detailed picture region 2 (connected): $\zeta = +208$. The field of view in the orignial image (*left*) is 318 μm

by the BoneJ plugin for ImageJ as shown in Fig. 17.14. This clearly demonstrates that the shear viscosity of lipid monolayers in the LC–LE coexistence region depends strongly on the area proportion of LC domains as well as on the structure of the domains. A more thorough quantitative analysis of the dependence of the material properties on the connectivity is a very promising direction for further experimental and analytical studies of this highly interesting and complex system and remains the subject of ongoing work.

The development of a numerical model that considers the mutual influence of mechanical stress and domain structure is a first step towards a better understanding of the viscoelastic shear properties of lipid membranes in the vicinity of their phase transition. From a physical point of view, the simplest mechanical model that can cover the observed dynamics is an extended Maxwell model as proposed by Letherish [45] for viscoelastic fluids. The combination of such a simple material model with a dynamic simulation of the quadrupolar force field for the SAW excitation is the subject of the following Section and can be a basis for later data analysis. A second step should be the identification of relations between the mechanical material parameters and the phase-separation behavior of lipid membranes and a first attempt to develop such a model is presented in the next Section.

17.4 Numerical Simulation of the NS/CJ/CH System

We consider a monolayer of lipids spread onto the upper surface of a water-filled Langmuir–Blodgett trough and study phase separation under the influence of a surface-acoustic-wave-actuated fluid flow, which generates four counter-rotating vortices at the upper surface. The process can be described by the NS/CJ/CH system consisting of the incompressible Navier–Stokes equations with a viscoelastic stress term and a capillary term coupled with the convective Jeffreys (Oldroyd) equation of viscoelasticity and the Cahn–Hilliard equation with a transport term.

17.4.1 The Convective Jeffreys (Oldroyd) Model of Viscoelasticity

Classical models of viscoelastic fluids are given by one-dimensional mechanical systems composed by springs and dashpots (cf., e.g., [67]). In particular, a Maxwell element consists of a spring and a dashpot in series. We denote by σ the stress and by ε the strain, and we refer to G as the elastic modulus of the spring, to η as the viscosity of the dashpot, and to $\tau_{\mathrm{rel}} = \eta/G$ as the relaxation time. The constitutive equation for the Maxwell element reads

$$\tau_{\mathrm{rel}} \, \frac{\partial \sigma}{\partial t} + \sigma = \eta \, \frac{\partial \varepsilon}{\partial t}. \tag{17.5}$$

On the other hand, a Voigt element consists of a spring and a dashpot in parallel. Its constitutive equation is given by

$$\sigma = G\varepsilon + \eta \, \frac{\partial \varepsilon}{\partial t}. \tag{17.6}$$

The Zener model [72] is a Voigt element in series with a spring and also referred to as the standard three-parameter Voigt model [67] with the constitutive equation

$$a \, \frac{\partial \sigma}{\partial t} + \sigma = G\varepsilon + \eta \, \frac{\partial \varepsilon}{\partial t}, \tag{17.7}$$

where the additional parameter a is related to the elastic modulus of the additional spring.

The anti-Zener model consists of a Voigt element in series with a dashpot of viscosity η_{f} (cf. Fig. 17.15). Denoting by $\tau_{\mathrm{rel}} = (\eta + \eta_{\mathrm{f}})/G$ the relaxation time and by $\tau_{\mathrm{ret}} = \eta/G$ the retardation time, its constitutive equation is given by

$$\tau_{\mathrm{rel}} \, \frac{\partial \sigma}{\partial t} + \sigma = \eta_{\mathrm{f}} \left(\frac{\partial \varepsilon}{\partial t} + \tau_{\mathrm{ret}} \, \frac{\partial^2 \varepsilon}{\partial t^2} \right). \tag{17.8}$$

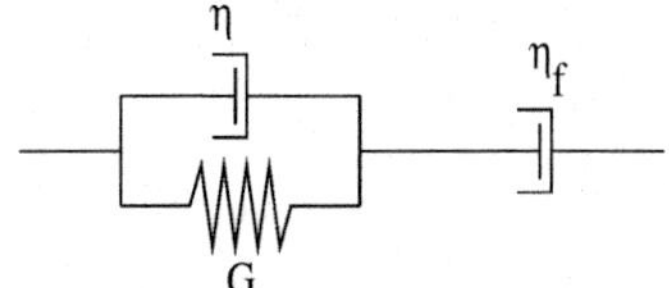

Fig. 17.15 The anti-Zener model also known as the non-standard three-parameter Voigt model or the Jeffreys model

The anti-Zener model is also known as the non-standard three-parameter Voigt model [67] or the Jeffreys model [37].

We introduce the solvent (Newtonian) viscosity η_s and the elastic viscosity η_e according to

$$\eta_s = \eta_f \frac{\tau_{\text{ret}}}{\tau_{\text{rel}}}, \quad \eta_e = \eta_f \left(1 - \frac{\tau_{\text{ret}}}{\tau_{\text{rel}}}\right). \tag{17.9}$$

Then, the stress σ can be split into two parts by means of

$$\sigma = \eta_s \frac{\partial \varepsilon}{\partial t} + T. \tag{17.10}$$

The first part $\eta_s \frac{\partial \varepsilon}{\partial t}$ is a Newtonian stress. The second part T is the so-called extra stress, which satisfies

$$\tau_{\text{rel}} \frac{\partial T}{\partial t} + T = \eta_e \frac{\partial \varepsilon}{\partial t}. \tag{17.11}$$

The explicit solution of (17.11) is given by

$$T(t) = \exp(-t/\tau_{\text{rel}})T(0) + \frac{\eta_e}{\tau_{\text{rel}}} \int_0^t \exp(-(t-s)/\tau_{\text{rel}}) \frac{\partial \varepsilon}{\partial s}(s) \, ds.$$

Setting $T(0) = 0$, we obtain

$$\sigma(0) = \eta_s \frac{\partial \varepsilon}{\partial t}(0),$$

i.e., at time $t = 0$, the material experiences an immediate viscous response with the effective viscosity coefficient η_s. For the stress σ, we thus get

$$\sigma(t) = \eta_s \frac{\partial \varepsilon}{\partial t}(t) + \frac{\eta_e}{\tau_{\text{rel}}} \int_0^t \exp(-(t-s)/\tau_{rel}) \frac{\partial \varepsilon}{\partial s}(s) \, ds. \tag{17.12}$$

By inverting (17.12) we obtain the dependence of the rate of strain on the history of the stress

$$\frac{\partial \varepsilon}{\partial t}(t) = \frac{1}{\eta_s} \sigma(t) - \frac{1}{\eta \tau_{\text{ret}}} \int_0^t \exp(-(t-s)/\tau_{\text{ret}})\, \sigma(s)\, \mathrm{d}s. \tag{17.13}$$

In terms of the displacement u, we have $\varepsilon = \partial u/\partial x$ for the strain and $v = \partial u/\partial t$ for the velocity. It follows that $\partial \varepsilon/\partial t = \partial v/\partial x = \partial^2 u/(\partial x \partial t)$. Hence, the anti-Zener (Jeffreys) model (17.8) can be written as

$$\tau_{\text{rel}} \frac{\partial \sigma}{\partial t} + \sigma = \eta_f \left(\frac{\partial v}{\partial x} + \tau_{\text{ret}} \frac{\partial}{\partial t}(\frac{\partial v}{\partial x}) \right). \tag{17.14}$$

This equation can be easily generalized to higher dimensions. We refer to $\mathbf{u} = (u_1, \cdots, u_d)$, $d \in \{2, 3\}$, as the displacement vector, to $\mathbf{v} = \partial \mathbf{u}/\partial t$ as the velocity, and to $\boldsymbol{\sigma} = (\sigma_{ij})_{i,j=1}^d$ as the stress tensor. Further, let $\mathbf{D}(\mathbf{v}) = (\nabla \mathbf{v} + \nabla \mathbf{v}^T)/2$ be the stretch tensor and D/Dt be any convective tensorial time derivative. Then, the d-dimensional anti-Zener (Jeffreys) model reads

$$\tau_{\text{rel}} \frac{D\boldsymbol{\sigma}}{Dt} + \boldsymbol{\sigma} = 2\eta_f \left(\mathbf{D}(\mathbf{v}) + \tau_{\text{ret}} \frac{D}{Dt}(\mathbf{D}(\mathbf{v})) \right). \tag{17.15}$$

As a particular convective tensorial time derivative, we consider the one-parameter family of Gordon–Showalter derivatives [31]. For a tensor-valued function $\mathbf{G}$, this family is given by

$$\frac{D\mathbf{G}}{Dt} = \frac{\partial \mathbf{G}}{\partial t} + \mathbf{v} \cdot \nabla \mathbf{G} - (a\mathbf{D}(\mathbf{v}) + \mathbf{W}(\mathbf{v}))\mathbf{G} - \mathbf{G}(a\mathbf{D}(\mathbf{v}) - \mathbf{W}(\mathbf{v})), \tag{17.16}$$

where $\mathbf{W}(\mathbf{v})$ stands for the spin tensor $\mathbf{W}(\mathbf{v}) = (\nabla \mathbf{v} - \nabla \mathbf{v}^T)/2$. The parameter a, which defines the family, is supposed to lie in the interval $[-1, +1]$.

As in the one-dimensional case, the stress tensor $\boldsymbol{\sigma}$ can be split into a Newtonian stress and an extra-stress according to

$$\boldsymbol{\sigma} = 2\eta_s \mathbf{D}(\mathbf{v}) + \mathbf{T}.$$

The extra-stress $\mathbf{T}$ satisfies the so-called Oldroyd viscoelastic equation [53]

$$\tau_{\text{rel}} \frac{D\mathbf{T}}{Dt} + \mathbf{T} = 2\eta_e \mathbf{D}(\mathbf{v}).$$

Equation (17.15) with the Gordon–Showalter derivatives (17.16) is known as the convective Jeffreys or Oldroyd model. In particular, the choice $a = 1$ results in the so-called Oldroyd-B fluid model.

For two-dimensional problems, the local existence of a weak solution and the global existence under the assumption of small data have been shown in [25]. For the parameter $a = 0$ the global existence of a weak solution without the assumption of small data has been established in [47]. More recent global existence results can be found in [18] which, however, do not include the case $a = 1$ (Oldroyd-B model).

17.4.2 The NS/CJ/CH System

We assume that the fluid volume in the Langmuir–Blodgett trough occupies a domain $(-\ell,\ \ell)^2 \times (0,\ h)$, where $(-\ell,\ \ell)^2$ is the cross-section of the trough and h is its height. We consider the viscoelastic fluid flow on the upper surface $(-\ell,\ \ell)^2 \times \{h\}$ of the trough over the time interval $[0, T]$. As a model, we choose the NS/CJ/CH system which is a system of parabolic partial differential equations consisting of the incompressible Navier–Stokes equations for the velocity $\mathbf{v}$ and pressure p coupled with a regularized convective Jeffreys (Oldroyd) model for the extra-stress $\boldsymbol{\sigma}$, and the Cahn–Hilliard equation for the order parameter c (concentration of one of the components of the binary mixture). The system will be given in the dimensionless form. To this end, we introduce L_{ref}, v_{ref}, ρ_{ref}, and η_{ref} as a reference length, velocity, density, and dynamic viscosity, respectively. The computational domain is $\Omega = (-\ell/L_{\mathrm{ref}},\ \ell/L_{\mathrm{ref}})^2$, which is $\Omega = (-1,\ 1)^2$, if we choose $L_{\mathrm{ref}} = \ell$. We denote by $\mathbf{n}_\Gamma$ the exterior unit normal vector on the boundary $\Gamma = \partial\Omega$, and we set $Q = \Omega \times (0,\ T/T_{\mathrm{ref}})$ and $\Sigma = \Gamma \times (0, T/T_{\mathrm{ref}})$. We further denote by $T_{\mathrm{ref}} = L_{\mathrm{ref}}/v_{\mathrm{ref}}$ the reference time. In dimensionless form, the NS/CJ/CH system represents an initial–boundary value problem given by the system of partial differential equations

$$\rho(\frac{\partial \mathbf{v}}{\partial t} + \mathbf{v}\cdot\nabla\mathbf{v}) - \frac{1}{\mathrm{Re}}\eta_s\Delta\mathbf{v} + \nabla p - \nabla\cdot\boldsymbol{\sigma} - \frac{1}{2}\mathrm{Ca}\,\mu(c)\nabla c = \mathbf{f}\quad \text{in } Q, \qquad (17.17a)$$

$$\nabla\cdot\mathbf{v} = 0\quad \text{in } Q, \qquad (17.17b)$$

$$\frac{\partial\boldsymbol{\sigma}}{\partial t} + \mathbf{v}\cdot\nabla\boldsymbol{\sigma} + (\frac{1}{\mathrm{Wi}}\mathbf{I} - \mathbf{M}(a))\boldsymbol{\sigma} - \frac{2}{\mathrm{ReWi}}\eta_e\mathbf{D}(\mathbf{v}) - \kappa\Delta\boldsymbol{\sigma} = 0\quad \text{in } Q, \qquad (17.17c)$$

$$\mathbf{M}(a)\boldsymbol{\sigma} = (a\mathbf{D}(\mathbf{v}) + \mathbf{W}(\mathbf{v}))\boldsymbol{\sigma} + \boldsymbol{\sigma}(a\mathbf{D}(\mathbf{v}) - \mathbf{W}(\mathbf{v})), \qquad (17.17d)$$

$$\frac{\partial c}{\partial t} + \mathbf{v}\cdot\nabla c - \frac{1}{\mathrm{Pe}}\nabla\cdot(M\nabla\mu(c)) = 0\quad \text{in } Q, \qquad (17.17e)$$

$$\mu(c) = F'(c) - \varepsilon^2\Delta c, \qquad (17.17f)$$

with the boundary conditions

$$\mathbf{v} = \mathbf{0},\quad \boldsymbol{\sigma}\mathbf{n}_\Gamma = \mathbf{0},\quad \mathbf{n}_\Gamma\cdot\nabla c = \mathbf{n}_\Gamma\cdot\nabla\mu(c) = 0\quad \text{on } \Sigma, \qquad (17.17g)$$

and the initial conditions

$$\mathbf{v}(0) = \mathbf{0},\quad \boldsymbol{\sigma}(0) = \mathbf{0},\quad c(0) = c_0. \qquad (17.17h)$$

Here, Ca, Pe, Re, and Wi refer to the capillary number, the Péclet number, the Reynolds number, and the Weissenberg number as given by

$$
\mathrm{Pe} = \frac{v_{\mathrm{ref}}}{v_{\mathrm{rel}}}, \qquad \mathrm{Re} = \frac{\rho_{\mathrm{ref}} v_{\mathrm{ref}} L_{\mathrm{ref}}}{\eta_{\mathrm{ref}}},
$$
$$
\mathrm{Ca} = \frac{\eta_{\mathrm{ref}} v_{\mathrm{ref}}}{\gamma}, \quad \mathrm{Wi} = \frac{\tau_{\mathrm{rel}}}{T_{\mathrm{ref}}},
\tag{17.18}
$$

where γ is the surface tension between the two phases, v_{rel} is the characteristic relative velocity between the two phases, and τ_{rel} stands for the relaxation time of the convective Jeffreys model. The surface-acoustic-wave-actuated fluid flow can be modeled by the source term $\mathbf{f}$ in the incompressible Navier–Stokes equations

$$
\mathbf{f} = \begin{cases} \mathbf{f}_{\mathrm{q}}, & \text{if IDT is turned on}, \\[2mm] \mathbf{0}, & \text{if IDT is turned off}, \end{cases}
\tag{17.19}
$$

Here, $\mathbf{f}_{\mathrm{q}}$ represents a quadrupolar force field which can be motivated as follows: The velocity field of an incompressible two-dimensional fluid can be described by means of a scalar function $\Psi_{\mathrm{q}}(x_1, x_2)$ such that the velocity field follows as

$$
\mathbf{v}_{\mathrm{q}}(x_1, x_2) = \mathrm{curl}_{2\mathrm{D}} \Psi_{\mathrm{q}}(x_1, x_2) := \left(\frac{\partial \Psi_q}{\partial x_2}, -\frac{\partial \Psi_q}{\partial x_1} \right)^T.
\tag{17.20}
$$

It is the divergence-free solution of the Stokes equation for a fluid driven by the force density

$$
\mathbf{f}_{\mathrm{q}} = -\eta_{\mathrm{w}} \Delta \mathbf{v}_{\mathrm{q}},
$$

where η_{w} is the viscosity of water. For the surface-acoustic-wave actuated flow field in Ω, we use the quadrupolar stream function

$$
\Psi_{\mathrm{q}}(x_1, x_2) = v_{\mathrm{m}} \frac{\sqrt{3}}{\pi} \frac{\sin(\pi x_1)\sin(\pi x_2)}{(2 - \cos(\pi x_1))(2 - \cos(\pi x_2))},
\tag{17.21}
$$

where v_{m} denotes the maximum value of the velocity magnitude depending on the frequency of the interdigital transducer that generates the surface acoustic waves. With this choice, the fluid flow exhibits a vorticity pattern consisting of four mutually counter-rotating vortices, which is what is experimentally observed (cf. Fig. 17.26 in Sect. 17.5.4).

We further refer to ρ as the mass density of the lipid, to $\mathbf{I}$ as the fourth-order identity tensor, to $\kappa > 0$ as a regularization parameter in Oldroyd's equation, to M as the mobility coefficient in the Cahn–Hilliard equation, to $F(c) = c^4/4 - c^2/2$ as the coarse-grain free energy density function, to c_0 as a given initial concentration, and to $\varepsilon > 0$ as a parameter related to the thickness of the diffuse interface between the two phases (also cf. Sect. 17.2.2 for a thorough discussion of the parameters of the Cahn–Hilliard model). Note that the coarse-grain energy-density function was chosen differently from the one in Sect. 17.2.2 (c.f. [8]). Here, we assume that the pure phases correspond to the value of the order parameter $c = \pm 1$ (rather than $c = 0$ and $c = 1$).

17.4.3 Numerical Solution of the NS/CJ/CH System

We solve the initial–boundary value problem (17.17a)–(17.17h) numerically by a splitting algorithm with respect to a partition of the time interval $[0, \bar{T}], \bar{T} = T/T_{\mathrm{ref}}$, into subintervals $[t_n, t_{n+1}]$ of length $\Delta t = \bar{T}/N, N \in \mathbb{N}, 0 \le n \le N - 1$. We denote by $\mathbf{v}^{(n)}, p^{(n)}, \boldsymbol{\sigma}^{(n)}$, and $c^{(n)}$ approximations of $\mathbf{v}, p, \boldsymbol{\sigma}$, and c at time $t = t_n$ and set the initial values $\mathbf{v}^{(0)} = \mathbf{0}, \boldsymbol{\sigma}^{(0)} = \mathbf{0}, c^{(0)} = c_0$.

Step 1: Solution of the incompressible Navier–Stokes equations

Given $\mathbf{v}^{(n)}, \boldsymbol{\sigma}^{(n)}$, and $c^{(n)}$, we compute $\mathbf{v}^{(n+1)}$ by the Chorin–Temam projection method [16, 64, 65]. This involves the following three fractional steps:

Step 1.1: Computation of a tentative velocity $\tilde{\mathbf{v}}^{(n+1)}$

Compute $\tilde{\mathbf{v}}^{(n+1)}$ as the solution of the boundary value problem:

$$\rho \, \frac{\tilde{\mathbf{v}}^{(n+1)} - \mathbf{v}^{(n)}}{\Delta t} + \rho \mathbf{v}^{(n)} \cdot \nabla \mathbf{v}^{(n)} - \frac{1}{\mathrm{Re}} \, \eta_s \Delta \tilde{\mathbf{v}}^{(n+1)} - \tag{17.22a}$$

$$\nabla \cdot \boldsymbol{\sigma}^{(n)} - \frac{1}{2} \mathrm{Ca} \, \mu(c^{(n)}) \nabla c^{(n)} = \mathbf{f} \quad \text{in } \Omega,$$

$$\tilde{\mathbf{v}}^{(n+1)} = \mathbf{0} \quad \text{on } \Gamma. \tag{17.22b}$$

Step 1.2: Computation of the pressure $p^{(n+1)}$

Compute $p^{(n+1)}$ as the solution of the Poisson problem:

$$-\Delta p^{(n+1)} = -\frac{\rho}{\Delta t} \, \nabla \cdot \tilde{\mathbf{v}}^{(n+1)} \quad \text{in } \Omega, \tag{17.23a}$$

$$\mathbf{n}_\Gamma \cdot \nabla p^{(n+1)} = 0 \quad \text{on } \Gamma. \tag{17.23b}$$

Step 1.3: Computation of the velocity $\mathbf{v}^{(n+1)}$

Compute $\mathbf{v}^{(n+1)}$ as the projection of the tentative velocity $\tilde{\mathbf{v}}^{(n+1)}$ onto the space of divergence-free functions:

$$\rho \, \frac{\mathbf{v}^{(n+1)} - \tilde{\mathbf{v}}^{(n+1)}}{\Delta t} + \nabla p^{(n+1)} = \mathbf{0} \quad \text{in } \Omega, \tag{17.24a}$$

$$\mathbf{v}^{(n+1)} = \mathbf{0} \quad \text{on } \Gamma. \tag{17.24b}$$

Step 2: Solution of the regularized convective Jeffreys model

Compute $\sigma^{(n+1)}$ by the following two fractional steps:

Step 2.1: Computation of a tentative stress $\tilde{\sigma}^{(n+1)}$

Compute $\tilde{\sigma}^{(n+1)}$ as the solution of the regularized transport equation:

$$\frac{\tilde{\sigma}^{(n+1)} - \sigma^{(n)}}{\Delta t} + \mathbf{v}^{(n+1)} \cdot \nabla \tilde{\sigma}^{(n+1)} - \kappa \Delta \tilde{\sigma}^{(n+1)} = \mathbf{0} \quad \text{in } \Omega, \tag{17.25a}$$

$$\mathbf{n}_\Gamma \cdot \kappa \nabla \tilde{\sigma}^{(n+1)} = \mathbf{0} \quad \text{on } \Gamma. \tag{17.25b}$$

Step 2.2: Computation of $\sigma^{(n+1)}$

Compute $\sigma^{(n+1)}$ as the solution of the implicitly-in-time discretized equation (17.17c) without the transport term:

$$\frac{\sigma^{(n+1)} - \tilde{\sigma}^{(n+1)}}{\Delta t} + (\frac{1}{\mathrm{Wi}}\mathbf{I} - \mathbf{M}(a))\sigma^{(n+1)} \tag{17.26a}$$

$$- \kappa \Delta \sigma^{(n+1)} - \frac{2}{\mathrm{ReWi}}\eta_e \mathbf{D}(\mathbf{v}^{(n+1)}) = \mathbf{0} \quad \text{in } \Omega,$$

$$\mathbf{n}_\Gamma \cdot \kappa \nabla \sigma^{(n+1)} = \mathbf{0} \quad \text{on } \Gamma. \tag{17.26b}$$

Step 3: Solution of the Cahn–Hilliard equation

Compute $c^{(n+1)}$ as the solution of

$$\frac{c^{(n+1)} - c^{(n)}}{\Delta t} + \mathbf{v}^{(n+1)} \cdot \nabla c^{(n+1)} - \frac{1}{\mathrm{Pe}}\nabla \cdot \left(M\nabla\mu(c^{(n+1)})\right) = 0 \quad \text{in } \Omega, \tag{17.27a}$$

$$\mathbf{n}_\Gamma \cdot \nabla c^{(n+1)} = \mathbf{n}_\Gamma \cdot \nabla\mu(c^{(n+1)}) = 0 \quad \text{on } \Gamma. \tag{17.27b}$$

The discretization in space is done by finite element approximations with respect to a uniform grid consisting of right isosceles triangles of mesh size $h = 1/64$. In particular, the components of the velocities $\tilde{\mathbf{v}}^{(n)}$ and $\mathbf{v}^{(n)}$ as well as the components of the stress tensors $\tilde{\sigma}^{(n)}$ and $\sigma^{(n)}$ are approximated by C^0-conforming Lagrangian finite elements of polynomial degree 2, whereas the pressure $p^{(n)}$ is approximated by C^0-conforming Lagrangian finite elements of polynomial degree 1 (cf., e.g., [19]). Finally, for the numerical solution of the implicit-in-time discretized Cahn–Hilliard equation we use a C^0-Interior-Penalty Discontinuous-Galerkin method based on C^0-conforming Lagrangian finite elements of polynomial degree 2 (cf. Sect. 17.2.2).

17.4.4 Numerical Results

We have implemented the NS/CJ/CH system (17.17a)–(17.17h) for the parameters and the associated capillary, Péclet, Reynolds, and Weissenberg numbers given in Tables 17.1 and 17.2. The algorithm described in the previous subsection has been applied with a time step size $\Delta t = 1.0 \times 10^{-3}$ and a uniform finite element mesh of mesh size $h = 1/64$.

In a first series of numerical simulations, we investigate the impact of viscoelasticity on the surface-acoustic-wave-generated velocity field at the upper surface of the Langmuir–Blodgett trough. We turn the interdigital transducer (IDT) on and off periodically with period $T_s = 10.0$. In particular, the IDT is turned on at $t = 0, 2T_s, 4T_s, 6T_s$, and turned off at $t = T_s, 3T_s, 5T_s$. Figure 17.16 displays the vorticity pattern consisting of four counter-rotating vortices corresponding to the quadrupolar velocity field $\mathbf{v}_q$ with unit maximal magnitude $v_m = 1.0$ (left) and the computed velocity $\mathbf{v}$ at time $t = 10.00$ (right) at the moment when the interdigital transducer is switched off. Figure 17.17 shows the velocity field at time $t = 11.25$ (left) with the formation of counterflows in a vicinity of the origin and at time $t = 11.50$ (right) with a complete change of the direction of the velocity and accelerated velocity magnitudes owing to the release of stored elastic energy.

The computed concentration profiles at different time moments are shown in Figs. 17.18 and 17.19. The left part of Fig. 17.18 shows c at $t = 9.0$, while the right part corresponds to $t = 10$, the exact time when the IDT is switched off. We can clearly see the movement of the pure phase domains in the direction of the quadrupolar force field. Figure 17.19 shows c at $t = 11.25$ (left) and $t = 12.0$ (right). It can be seen that the direction of the flow is reversed for the short time.

We also choose a test point $\mathbf{x}_0 = (2/32, 14/32)$, which lies in the north-east quadrant near the line of separation of the counter-rotating vortices, and track the change in time of the vertical component v_y of the velocity $\mathbf{v}$ at this point. The results are shown in Fig. 17.20. We observe the typical viscoelastic flow pattern, namely,

- Initially the fluid is at rest (time $t = 0$);
- After switching the IDT on at $t = 0$, v_y attains its maximal value (by modulus). In our case, this level is $|v_y| \approx 0.08$;

Table 17.1 Parameters in the NS/CJ/CH system

ℓ	L_{ref}	T	T_{ref}	T_s	ρ	ρ_{w}
1.0×10^{-2}	1.0×10^{-2}	65	1.0	10	1.0×10^3	1.0×10^3

ρ_{ref}	v_{m}	v_{ref}	v_{rel}	η_{w}	$\eta_{s,\text{phase}1}$	$\eta_{e,\text{phase}1}$
1.0×10^3	1.0×10^{-2}	1.0×10^{-2}	1.0×10^{-4}	1.0×10^{-3}	7.5×10^{-3}	5.0×10^{-3}

$\eta_{s,\text{phase}2}$	$\eta_{e,\text{phase}2}$	η_{ref}	τ_{rel}	γ	κ	a
7.5×10^{-3}	1.0×10^{-3}	1.0×10^{-3}	1.0	1.0×10^{-6}	1.0×10^{-4}	1.0

Table 17.2 Capillary number, Péclet number, Reynolds number, and Weissenberg number

Ca	Pe	Re	Wi
1.0×10^{-2}	1.0×10^2	1.0×10^2	1.0

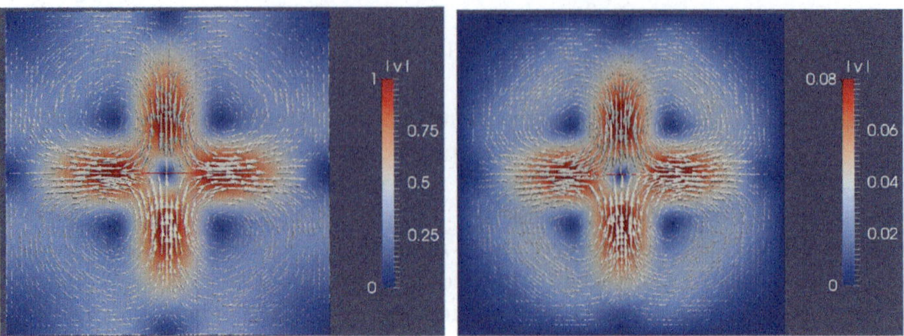

Fig. 17.16 Quadrupolar velocity field $\mathbf{v}_q$ with $v_m = 1.0$ (*left*) and velocity $\mathbf{v}$ at time $t = 10.00$ when the interdigital transducer is switched off (*right*)

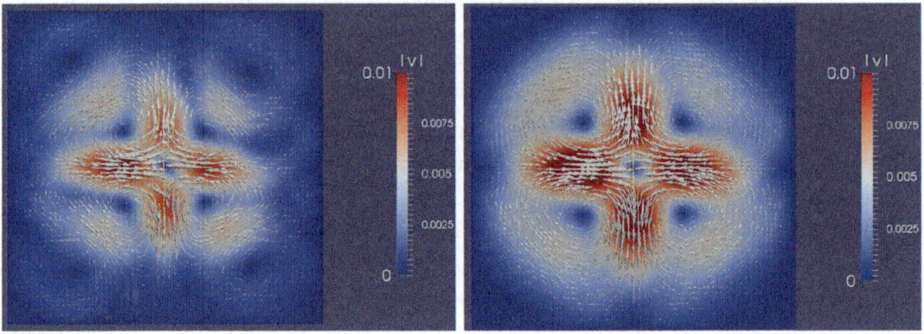

Fig. 17.17 Velocity field at time $t = 11.25$ (*left*) and at time $t = 11.50$ (*right*)

- Then, it stabilizes at $|v_y| \approx 0.07$. The wiggles indicate the effect of the heterogeneity of the domain as well as the elastic contribution;
- At time $t = 10.0$ (corresponds to 1000 in the Figure), we turn the IDT off and observe the flow by inertia, i.e. v_y remains negative and $|v_y|$ decreases;
- After becoming zero, the flow does not stop as it would in the viscous case. Rather, v_y reaches its maximum level $v_y \approx 0.01$ (at this time we observe the maximum magnitude counter-flow).
- Then, the flow decays to zero. Due to elastic effects, it may cross zero back and forth several times with amplitude too small to be visible in the Figure.

Thus, qualitatively we observe the same effect as in the experiments, cf. Fig. 17.13.

In the second series of experiments, we consider only a one-phase fluid. For numerical simulations, we use the system (17.17a)–(17.17c) with the value of the order parameter $c \equiv 1$ (therefore, the term $(1/2)\mathrm{Ca}\mu(c)\nabla c$ in the Eq. (17.17a)

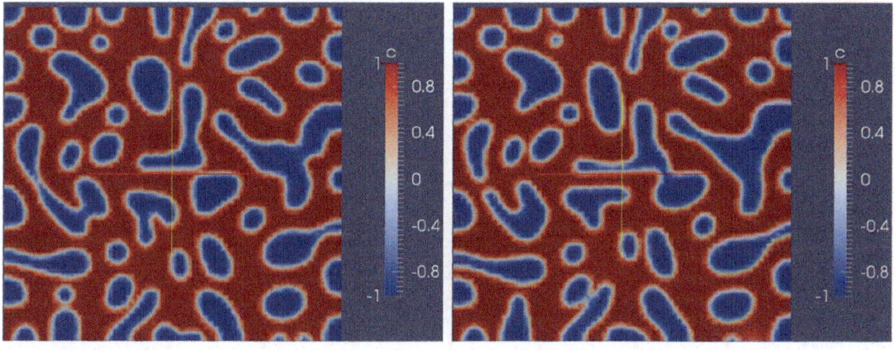

Fig. 17.18 Concentration at time $t = 9.0$ (*left*), and at time $t = 10.0$ (*right*)

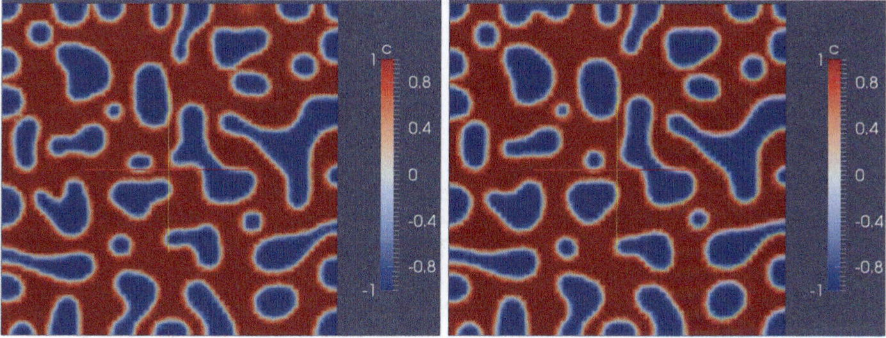

Fig. 17.19 Concentration at time $t = 11.25$ (*left*), and at time $t = 12.0$ (*right*)

vanishes). Again, as in the two-phase case, the interdigital transducer is switched on at times $t = 0, 2T_s, 4T_s, 6T_s$ and switched off at times $t = T_s, 3T_s, 5T_s$ with $T_s = 10.0$.

First, we simulate the flow of the pure viscous liquid ($\eta_e = 0$) and the flow of the liquid-disordered phase of the lipid (phase 2 in the notations of Table 17.1). The history of the y-component of the velocity v_y at the test point $(2/32, 14/32)$ is shown in Figs. 17.21 and 17.22. We can see that v_y does not cross the zero level in the pure viscous case as expected whereas little elastic contribution to the model leads to small positive values which indicate very slow reverse motion.

Second, we simulate the flow of the condensed phase of the lipid (phase 1 in the notations of Table 17.1) and compare the simulation results with corresponding experimental measurements as shown in Fig. 17.23. Qualitatively, the pictures look alike. Again, we see the crossing of zero of the vertical component v_y more expressed than for the liquid-disordered phase. In the experiments, the relaxation is more moderately expressed than in the case of simulations. A possible reason for this is a relatively low Weissenberg number (Wi $= 1.0$) used. For larger values of Wi, a more sophisticated numerical scheme to solve the convective Jeffreys equation would have to be implemented.

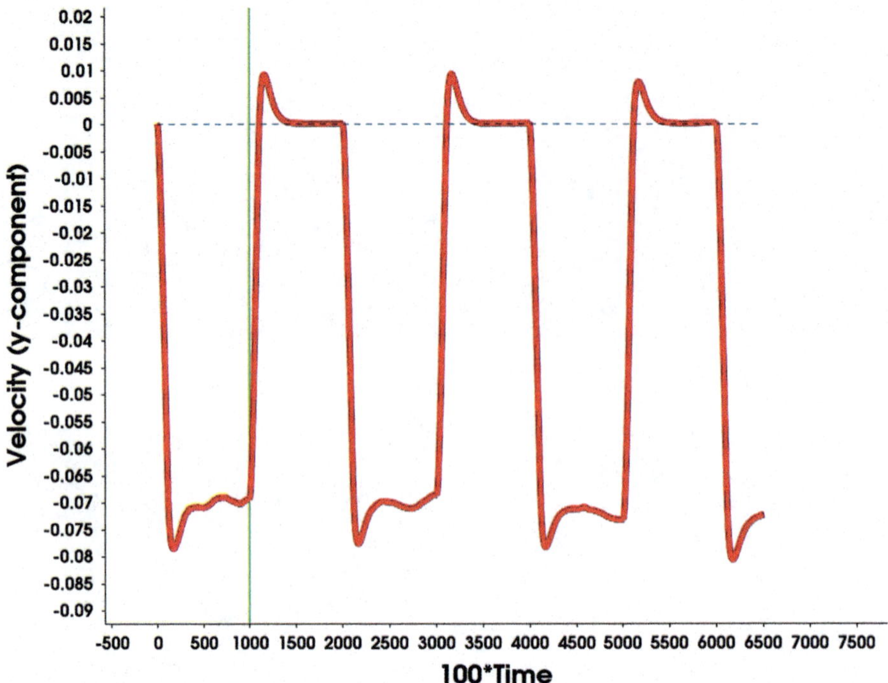

Fig. 17.20 Vertical component v_y of the velocity field v at the test point $x_0 = (2/32, 14/32)$

17.5 Enantiomer Separation

In this section, we report on the experimental setup for surface-acoustic-wave-actuated enantiomer separation (Sect. 17.5.1), the mathematical modeling and numerical simulation based on the finite element immersed boundary method (Sect. 17.5.2) and the distributed Lagrangian multiplier finite element immersed boundary method (Sect. 17.5.3), a comparison of experimental measurements and numerical simulation results (Sect. 17.5.4), and the separation mechanism due to the specific flow pattern created by the surface acoustic waves (Sect. 17.5.5).

17.5.1 Experimental Setup

To drive the microfluidic flow, we apply surface acoustic waves (SAWs) that are generated by an interdigital transducer (IDT). An IDT consists of two interlocking electrodes deposited on a piezoelectric substrate, which are connected to an alternating current generator such that interfering periodic substrate deformations cause the propagation of SAWs. When SAWs couple into a liquid, two fluid jets are

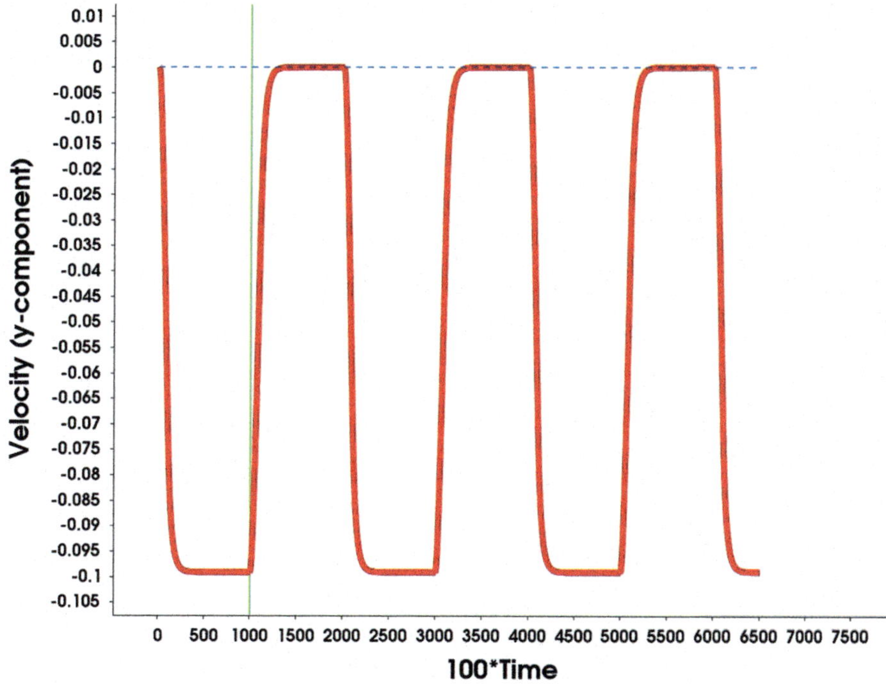

Fig. 17.21 Vertical component of the velocity at the test point. Simulation of the pure viscous liquid

formed, which transport material. These jets drive a bulk streaming, which generates a specific flow pattern consisting of four counter-rotating vortices.

The experimental setup to apply SAW flow to the surface is shown in the photograph in Fig. 17.24. An IDT is connected using a circuit board and a frame of plexiglas (PMMA) is attached to the board to form a miniaturized trough. The top right panel of the same figure shows a schematic sideview of the setup with the IDT on the bottom of the trough with a reservoir of water on top and a particle floating on the surface. Applying an alternating frequency drives the streaming and creates the specific flow pattern.

For the production of the photoresist L-shaped enantiomers, we followed the protocol described by Hernandez and Mason [34] with some modifications. The whole process of manufacturing is shown in Fig. 17.25 and the numbers in parenthesis in the following text also refer to that Figure.

First a sacrificial layer of omnicoat is spincoated on a silicon wafer. (2) In a second spin coating process, SU8-2 photoresist laden with 0.5 mg/ml Nile Red is spun onto the omnicoat layer at 3000 rounds per second achieving a film thickness of about 1.5 μm. (3) After soft baking, the photoresist is exposed using a mask aligner and then baked a second time. (4) After the substrate has cooled to room temperature, the unexposed photoresist is developed using MR-DEV300 leaving

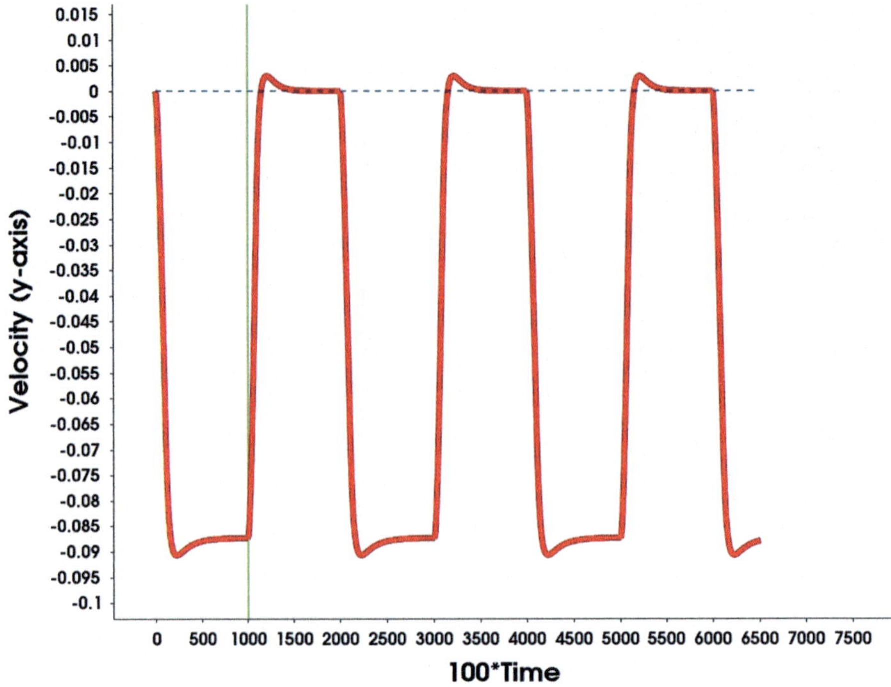

Fig. 17.22 Vertical component of the velocity at the test point. Simulation of the disordered phase of the lipid

the desired particles attached to the sacrificial layer of omnicoat. (5) In order to render the particles hydrophobic on one side, a layer of Trichloro(octadecyl)silane (OTS) is applied by spin coating. (6) To this end, 10 µl OTS are dissolved in 3ml n-hexane and spun onto the particles at 1000 rpm for 10 s. The OTS solution has to be applied after the spin coater has reached its maximum rate of revolution since n-hexane evaporates very quickly thus leaving the OTS scattered on the surface inhomogeneous if the spinning rate is too low. The layer of OTS renders the particles highly hydrophobic on the top side. After these steps, the particles are still firmly attached to the substrate and can be stored in a dark environment until use to prevent bleaching of the fluorescent dye. To remove the particles from the wafer, a lift-off procedure is performed using omnicoat developer. The wafer is immersed in the solution until the omnicoat layer has been sufficiently dissolved, which takes approximately 30 s depending on the geometry of the sample. The wafer is then transferred to the experimental setup and the detached particles can be washed off using pure water. Due to the top side of the particles being considerably more hydrophobic than their bottom side, the orientation of the particles is conserved during the lift off process in most cases and the particles float stably on the surface of the fluid.

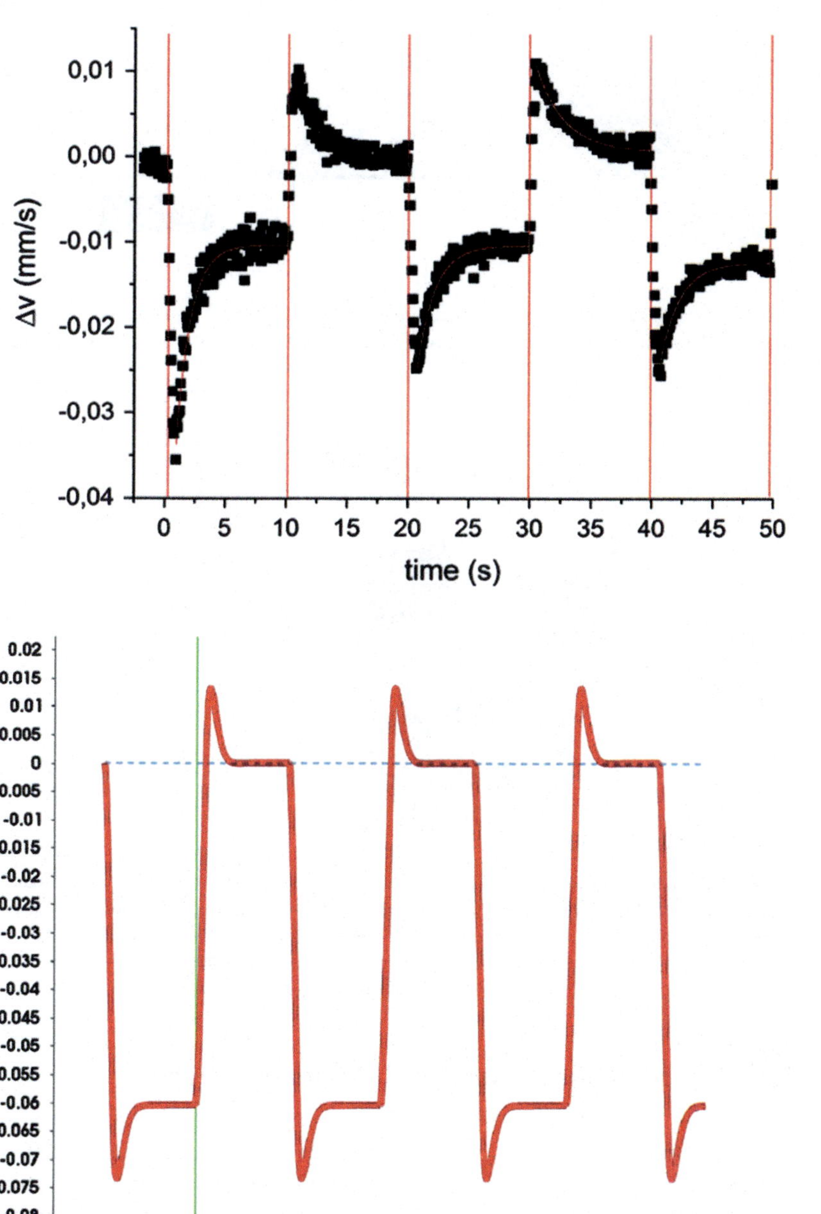

Fig. 17.23 Vertical component of the velocity at the test point. *Top*: experimental measurements. *Bottom*: simulation of the ordered (condensed) phase of the lipid

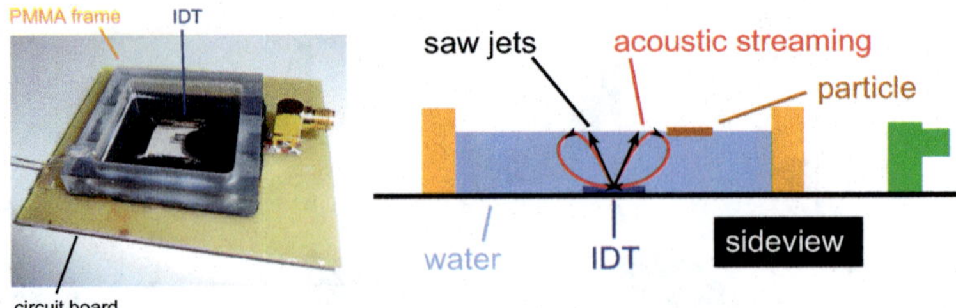

Fig. 17.24 *Left*: Photograph of the experimental setup. *Right*: Schematic sideview, where a particle is placed on the water surface and the IDT is driving the acoustic streaming

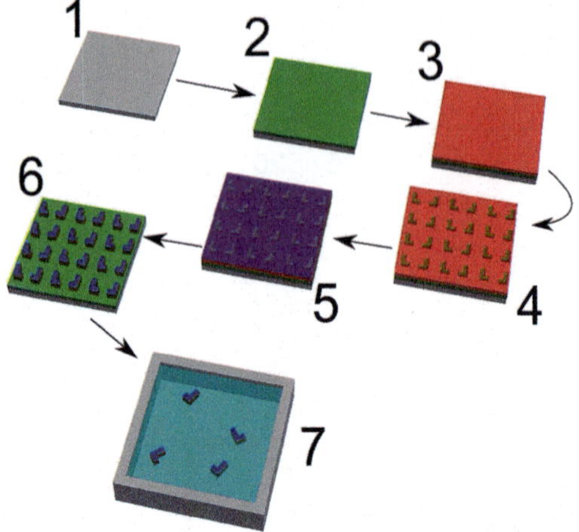

Fig. 17.25 Schematic representation of fabrication of the photoresist particles. (1) Clean silicon wafer. (2) A sacrificial layer of omnicoat is applied (3) SU8-2 is applied. (4) Photoresist is partially exposed. (5) Unexposed SU8-2 is developed. (6) Surface is treated with silane. (7) Lift-off. (8) Particles are brought onto water surface and float stably in well-defined orientation. Reprinted with permission from [12]. Copyright 2015 Walter de Gruyter

17.5.2 Finite Element Immersed Boundary Method

Following the experimental setup as described above, the simulation of the separation of L-shaped enantiomers has been considered by the Finite Element Immersed Boundary Method (FEIBM; cf. [7]). It is based on a coupled system consisting of the incompressible Navier–Stokes equations on the surface Ω of the fluid-filled container and the equations of motion of the enantiomers described with respect to an Eulerian and a Lagrangian coordinate system, respectively. Denoting by $\mathbf{u}, p$ the velocity and the pressure of the carrier fluid, by ρ_f, η_f its density and viscosity, and by $\mathbf{X}$ the position vector on the boundary of length ℓ of an immersed enantiomer, the incompressible Navier–Stokes equations read

$$\rho_{\rm f}\left(\frac{\partial \mathbf{u}}{\partial t} + (\mathbf{u}\cdot\nabla)\mathbf{u}\right) - \eta_{\rm f}\Delta\mathbf{u} + \nabla p = \mathbf{f}_{\rm q} + \mathbf{f}_{\rm g} \quad \text{in } \Omega \times (0,T), \tag{17.28a}$$

$$\nabla\cdot\mathbf{u} = 0 \quad \text{in } \Omega \times (0,T), \tag{17.28b}$$

$$\mathbf{u} = \hat{\mathbf{u}} \quad \text{on } \partial\Omega \times (0,T), \tag{17.28c}$$

$$\mathbf{u}(\cdot,0) = \hat{\mathbf{u}} \quad \text{in } \Omega, \tag{17.28d}$$

whereas the equation of motion is given by

$$\frac{d\mathbf{X}}{dt}(q,t) = \mathbf{u}(\mathbf{X}(q,t),t), \quad q \in [0,\ell],\ t \in [0,T], \tag{17.29a}$$

$$\mathbf{X}(q,0) = \mathbf{X}^0(q), \quad q \in [0,\ell], \tag{17.29b}$$

In (17.28a), the source term $\mathbf{f}_{\rm q}$ stands for the quadrupolar force density, cf. (17.19)–(17.21). On the other hand, the force density $\mathbf{f}_{\rm g}$ reflects the impact of the enantiomers on the carrier fluid according to

$$\langle \mathbf{f}_{\rm g}(t), \mathbf{w}\rangle_{\mathbf{H}^{-1},\mathbf{H}_0^1} = \int_0^L \mathbf{f}_{\rm l}(q,t)\cdot\mathbf{w}(\mathbf{X}(q,t))\,dq, \quad \mathbf{w}\in\mathbf{H}_0^1(\Gamma_s),$$

where $\mathbf{H}_0^1$ is the vector-valued Sobolev space of square-integrable functions with square integrable weak derivative and vanishing boundary trace and $\mathbf{f}_{\rm l}(q,t) = -E'(\mathbf{X}(q,t))$ with E' being the Gâteaux derivative of the total energy E of the immersed body defined by

$$E(t) := E^{\rm e}(t) + E^{\rm b}(t), \ t \in (0,T), \tag{17.30}$$

$$E^{\rm e}(t) := \int_0^L \mathcal{E}^{\rm e}(\mathbf{X}(q,t))\,dq, \quad E^{\rm b}(t) := \int_0^L \mathcal{E}^{\rm b}(\mathbf{X}(q,t))\,dq. \tag{17.31}$$

Here, $\mathcal{E}^{\rm e}(t)$ and $\mathcal{E}^{\rm b}(t)$ are the local energy densities according to

$$\mathcal{E}^{\rm e}(\mathbf{X}(q,t)) = \frac{\kappa_e}{2}\left(\left|\frac{\partial\mathbf{X}}{\partial q}(q,t)\right|^2 - 1\right), \quad \mathcal{E}^{\rm b}(\mathbf{X}(q,t)) = \frac{\kappa_b}{2}\left|\frac{\partial^2\mathbf{X}}{\partial q^2}(q,t)\right|^2, \tag{17.32}$$

where $\kappa_{\rm e} > 0$ and $\kappa_{\rm b} > 0$ denote the elasticity coefficients with respect to elongation–compression and bending.

Based on the variational formulation of the coupled system (17.28), (17.29) in an appropriate function-space setting, for the spatial discretization we have used Taylor–Hood P2/P1-elements for (17.28) with respect to a geometrically conforming simplicial triangulation $\mathcal{T}_h(\Omega)$ of the computational domain Ω and periodic cubic splines for (17.29) with respect to a partitioning of the interval $[0, L]$. Given an equidistant partition of the time interval $[0, T]$, for discretization in time we have used the implicit Euler scheme for (17.28) and the explicit Euler method for (17.29) resulting in a semi-implicit Backward Euler/Forward Euler FE-IB which has to satisfy a CFL-type stability condition (see also [27] for a related application of the FEIBM and [35] for an alternative unconditionally stable fully implicit scheme). For further details we refer to [5].

17.5.3　Distributed Lagrangian Multiplier Finite Element Immersed Boundary Method

In the Distributed Lagrangian Multiplier Finite Element Immersed Boundary Method (DLM-FEIBM) the rigid enantiomer is supposed to occupy a subdomain $B(t) \subset \Omega, t \in [0, T]$. The motion of the rigid enantiomer with density ρ_s and the first Piola–Kirchhoff stress tensor $\mathcal{P}$ are coupled by a distributed Lagrangian multiplier $\boldsymbol{\lambda} \in (H^1(B)^2)^*$. Hence, the DLM-FEIBM represents a fictitious domain approach in the spirit of [30]. In particular, setting $\mathbf{V} := \{\mathbf{v} \in H^1(\Omega)^2 \mid \mathbf{v}|_\Gamma = \hat{\mathbf{u}}\}$, the DLM-FEIBM requires the computation of $\mathbf{u} \in \mathbf{V}, p \in L^2_0(\Omega), \mathbf{X} \in H^1(B)^2$, and $\boldsymbol{\lambda} \in (H^1(B)^2)^*$ such that for all $\mathbf{v} \in H^1_0(\Omega)^2, q \in L^2_0(\Omega), \mathbf{Y} \in H^1(B)^2$, and $\boldsymbol{\mu} \in (H^1(B)^2)^*$ it holds

$$\rho_f \int_\Omega (\frac{\partial \mathbf{u}}{\partial t} + (\mathbf{u} \cdot \nabla)\mathbf{u}) \cdot \mathbf{v}\, dx + \eta_f \int_\Omega \nabla \mathbf{u} : \nabla \mathbf{v}dx \tag{17.33a}$$

$$- \int_\Omega p\nabla \cdot \mathbf{v}dx + \langle \boldsymbol{\lambda}, \mathbf{v}(\mathbf{X}(\cdot, t)) \rangle = \int_\Omega \mathbf{f}_q \cdot \mathbf{v}\, dx + \langle \mathbf{f}_g, v \rangle_{\mathbf{H}^{-1}, \mathbf{H}^1_0},$$

$$\int_\Omega \nabla \cdot uq\, dx = 0, \tag{17.33b}$$

$$(\rho_s - \rho_f) \int_B \frac{\partial^2 \mathbf{X}}{\partial t^2} \cdot \mathbf{Y}\, dx + \int_B \mathcal{P} : \nabla Y\, dx - \langle \boldsymbol{\lambda}, \mathbf{Y} \rangle = 0, \tag{17.33c}$$

$$\langle \boldsymbol{\mu}, \mathbf{u}(\mathbf{X}(\cdot, t)) - \frac{\partial X}{\partial t} \rangle = 0, \tag{17.33d}$$

$$\mathbf{u}(\cdot, 0) = \hat{\mathbf{u}} \text{ in } \Omega, \quad \mathbf{X}(\cdot, 0) = \mathbf{X}^0, \tag{17.33e}$$

where $\langle \cdot, \cdot \rangle$ stands for the duality pairing between $(H^1(B)^2)^*$ and $H^1(B)^2$.

For the numerical solution of the DLM-FEIBM we use inf–sup stable Taylor–Hood P_2/P_1-elements for the approximation of $(\mathbf{u}, p)$, conforming P_1-elements for the approximation of $\mathbf{X}$ and $\boldsymbol{\lambda}$, and the Yanenko–Marchuk fractional step method combined with a Chorin–Marsden splitting for discretization in time. For details we refer to [12].

17.5.4   Comparison of Experimental and Simulation Results

The SAW-induced flow field at the surface and the motion of particles within that flow field can be easily visualized using either bright field or fluorescence microscopy. Figure 17.26(left) displays the experimentally observed flow field in the upper-right quadrant of the surface generated by the SAWs, whereas Fig. 17.26(right) shows the simulated flow pattern generated by the quadrupolar force field.

Moreover, Fig. 17.27(left) shows a micrograph of the experiment and tracked trajectories of two particles (solid colored lines) as they pass through the center of the flow field. The streamlines are visualized by calculating an overlay of many frames. Figure 17.27(right) displays the simulated trajectories based on the DLM-FEIBM.

Fig. 17.26 *Left*: Experimental SAW generated surface streaming profile. *Right*: Simulated vorticity pattern generated by the quadrupolar force field. Reprinted with permission from [12]. Copyright 2015 Walter de Gruyter

Fig. 17.27 *Left*: Experimentally measured trajectories of a square-shaped particle (*red line*) and an L-shaped particle (*yellow line*). *Right*: Simulated trajectories. Reprinted with permission from [12]. Copyright 2015 Walter de Gruyter

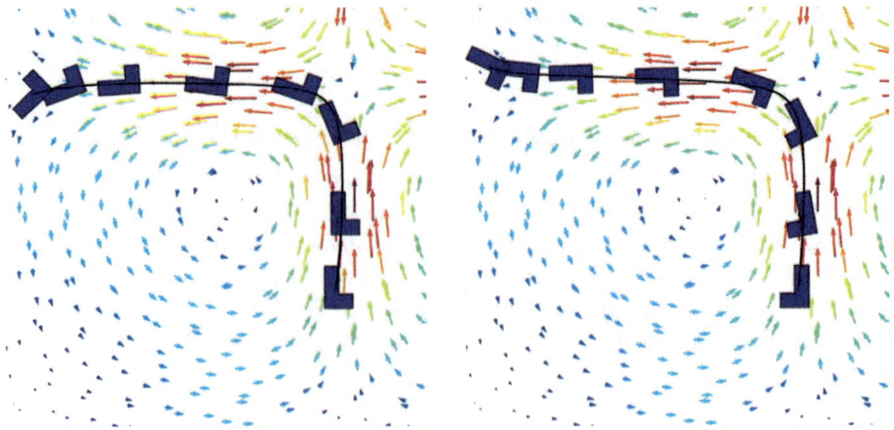

Fig. 17.28 *Left*: Attraction of a right-handed L-shaped enantiomer by the counter-clockwise rotating vortex in the lower-left quadrant. *Right*: Attraction of a left-handed L-shaped enantiomer by the clockwise rotating vortex in the upper-left quadrant. Reprinted with permission from [12]. Copyright 2015 Walter de Gruyter

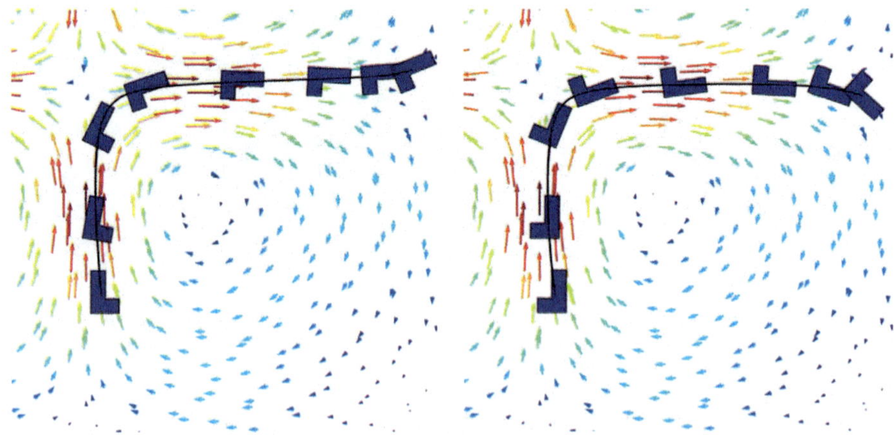

Fig. 17.29 *Left*: Attraction of a right-handed L-shaped enantiomer by the counter-clockwise rotating vortex in the upper-right quadrant. *Right*: Attraction of a left-handed L-shaped enantiomer by the clockwise rotating vortex in the lower-right quadrant

17.5.5   Numerical Simulation of Enantiomer Separation

As far as enantiomer separation is concerned, Fig. 17.28 displays the motion of a right-handed and a left-handed L-shaped enantiomer initially placed between the two counter-rotating vortices in the lower quadrants of the surface of the fluid as simulated using the DLM-FEIBM. As can be seen in Fig. 17.28(left), the right-handed enantiomer gets attracted by the counter-clockwise rotating vortex in the lower-left quadrant. On the other hand, as shown in Fig. 17.28(right), the right-handed enantiomer follows a trajectory that leads to a path around the center of the clockwise rotating vortex in the upper-right quadrant. Likewise, Fig. 17.29 shows that a right-handed (left-handed) enantiomer initially placed a little bit to the right of the middle between the two counter-rotating vortices gets attracted by the counter-clockwise (clockwise) rotating vortex in the upper-right (lower-right) quadrant. This behavior is in accordance with experimental measurements and similar to numerical simulation results that have been obtained for deformable L-shaped enantiomers by an application of the FEIBM [5]. We note that [5] also contains numerical results for multiple enantiomers of different handedness which confirm the separation mechanism described above.

Acknowledgements This work was supported by the DFG Priority Program SPP 1506, by the German Cluster of Excellence "Nanosystems Initiative Munich (NIM)", and by the NSF under grant DMS-1520886.

References

1. Angelova, M., Soléau, S., Méléard, Ph., Faucon, F., Bothorel, P.: Preparation of giant vesicles by external AC electric fields. Kinetics and applications. In: Helm, C., Lösche, M., Möhwald, H. (eds.) Trends in Colloid and Interface Science, vol. VI. Progress in Colloid and Polymer Science, vol. 89, pp. 127–131. Springer, Berlin (1992)
2. Bagatolli, L.A., Gratton, E.: Two photon fluorescence microscopy of coexisting lipid domains in giant unilamellar vesicles of binary phospholipid mixtures. Biophys. J. **78**, 290–305 (2000)
3. Baoukina, S., Mendez-Villuendas, E., Bennett, W., Tieleman, D.: Computer simulations of the phase separation in model membranes. Faraday Discuss. **161**, 63–75 (2013)
4. Baumgart, T., Hess, S., Webb, W.: Imaging coexisting fluid domains in biomembrane models coupling curvature and line tension. Nature **425**, 821–824 (2003)
5. Beleke-Maxwell, K., Franke, T., Hoppe, R.H.W., Linsenmann, C.: Numerical simulation of surface acoustic wave actuated enantiomer separation by the finite element immersed boundary method. Comput. Fluids **112**, 50–60 (2015)
6. Bertozzi, A., Esedoglu, S., Gillette, A.: Inpainting of binary images using the Cahn-Hilliard equation. IEEE Trans. Image Process. **16**, 285–291 (2007)
7. Boffi, D., Gastaldi, L.: A finite element approach for the immersed boundary method. Comput. Struct. **81**, 491–501 (2003)
8. Boyer, F.: A theoretical and numerical model for the study of incompressible mixture flows. Comput. Fluids **31**, 41–68 (2002)
9. Boyer, F., Chupin, L., Fabrie, P.: Numerical study of viscoelastic mixtures through a Cahn-Hilliard flow model. Eur. J. Mech. B Fluids **23**, 759–780 (2004)
10. Brooks, C.F., Fuller, G.G., Frank, C.W., Robertson, C.R.: Transitions in monolayers at the air-water interface. Langmuir **5**, 2450–2459 (1999)
11. Burger, S., Fraunholz, T., Hoppe, R.H.W., Leirer, C., Wixforth, A., Peter, M.A., Franke, T.: Comparative study of the dynamics of lipid membrane phase decomposition in experiment and simulation. Langmuir **29**(25), 7565–7570 (2013)
12. Burger, S., Franke, T., Fraunholz, T., Hoppe, R.H.W., Peter, M.A., Wixforth, A.: Numerical simulation of surface acoustic wave actuated separation of rigid enantiomers by the fictitious domain Lagrange multiplier method. Comput. Methods Appl. Math. **15**(3), 247–258 (2015)

13. Cahn, J.W., Hilliard, J.E.: Free energy of a non-uniform system. I. Interfacial free energy. J. Chem. Phys. **28**, 258–267 (1958)
14. Chen, L.: Phase-field models for microstructural evolution. Annu. Rev. Mater. Res. **32**, 113–140 (2002)
15. Choi, S.Q., Steltenkamp, S., Zasadzinski, J.A., Squires, T.M.: Active microrheology and simultaneous visualization of sheared phospholipid monolayers. Nat. Commun. **2**, 312–316 (2011)
16. Chorin, A.J.: The numerical solution of the Navier-Stokes equations for an incompressible fluid. Bull. Am. Soc. **73**, 928–931 (1967)
17. Chupin, L.: Existence result for a mixture of non-Newtonian flows with stress diffusion using the Cahn-Hilliard formulation. Discrete Continuous Dyn. Syst. B **3**, 45–68 (2003)
18. Chupin, L.: Existence results for the flow of viscoelastic fluids with an integral constitutive law. J. Math. Fluid Mech. **15**, 783–806 (2013)
19. Ciarlet, P.G.: The Finite Element Method for Elliptic Problems. SIAM, Philadelphia (2002)
20. Dietrich, C., Bagatolli, L., Volovyk, Z., Thompson, N., Levi, M., Jacobson, K., Gratton, E.: Lipid rafts reconstituted in model membranes. Biophys. J. **80**, 1417–1428 (2001)
21. Ding, J., Warriner, H.E., Zasadzinski, J.A., Schwartz, D.K.: Magnetic needle viscometer for Langmuir monolayers. Langmuir **18**, 2800–2806 (2002)
22. Einstein, A.: Eine neue Bestimmung der Moleküldimensionen. Ann. Phys. **324**, 289–306 (1906)
23. Elliott, C.: The Cahn-Hilliard model for the kinetics of phase separation. In: Rodrigues, J. (ed.) Mathematical Models for Phase Change Problems, International Series of Numerical Mathematics, vol. 88, pp. 35–74. Birkhauser, Basel (1989)
24. Espinosa, G., López-Montero, I., Monroy, F., Langevin, D.: Shear rheology of lipid monolayers and insights on membrane fluidity. Proc. Natl. Acad. Sci. **108**, 6008–6013 (2011)
25. Fernandéz-Cara, E., Guillén, F., Ortega, R.R.: Some theoretical results concerning non-Newtonian fluids of the Oldroyd kind. Ann. Scuola Norm. Sup. Pisa Cl. Sci. **26**, 1–26 (1998)
26. Filippov, A., Orädd, G., Lindblom, G.: Lipid lateral diffusion in ordered and disordered phases in raft mixtures. Biophys. J. **86**, 891–896 (2004)
27. Franke, T., Hoppe, R.H.W., Linsenmann, C., Schmidt, L., Willbold, C.: Numerical simulation of the motion of red blood cells and vesicles in microfluidic flows. Comput. Vis. Sci. **14**, 167–180, 2011.
28. Fraunholz, T.: Transport at interfaces in lipid membranes and enantiomer separation. PhD dissertation, University of Augsburg, Germany, 2014
29. Fraunholz, T., Hoppe, R.H.W., Peter, M.A.: Convergence analysis of an adaptive interior penalty discontinuous Galerkin method for the biharmonic problem. J. Numer. Math. **23**, 317–330 (2015)
30. Glowinski, R., Pan, T.-W., Hesla, T.I., Joseph, D.D., Periaux, J.: A fictitious domain approach to the direct numerical simulation of incompressible viscous flow past moving rigid bodies: application to particulate flow. J. Comput. Phys. **169**, 363–427 (2001)
31. Gordon, R.J., Showalter, W.R.: Anisotropic fluid theory: a different approach to the dumbbell theory of dilute polymer solutions. Trans. Soc. Rheol. **16**, 79–97 (1972)
32. Guth, E., Simha, R.: Untersuchungen über die Viskosität von Suspensionen und Lösungen. 3. Über die Viskosität von Kugelsuspensionen. Colloid Polym. Sci. **74**, 266–275 (1936)
33. Helm, C., Möhwald, H., Kjaer, K., Als-Nielsen, J.: Phospholipid monolayers between fluid and solid states. Biophys. J. **52**, 381–390 (1987)
34. Hernandez, C.J., Mason, T.G.: Colloidal alphabet soup: monodisperse dispersions of shape-designed lithoparticles. J. Phys. Chem. C **111**, 4477–4480 (2007)
35. Hoppe, R.H.W., Linsenmann, C.: An adaptive Newton continuation strategy for the fully implicit finite element immersed boundary method. J. Comput. Phys. **231**, 4676–4693 (2012)
36. Jacobson, K., Mouritsen, O., Anderson, R.: Lipid rafts: at a crossroad between cell biology and physics. Nat. Cell Biol. **9**, 7–14 (2007)
37. Jeffreys, H.: The Earth, Its Origin, History and Physical Constitution. Cambridge University Press, Cambridge (1924)

38. Jørgensen, K., Mouritsen, O.: Phase separation dynamics and lateral organization of two-component lipid membranes. Biophys. J. **95**, 942–954 (1995)
39. Joseph, D.D.: Fluid Dynamics of Viscoelastic Liquids. Springer, Berlin (1990)
40. Kahya, N., Scherfeld, D., Bacia, K., Poolman, B., Schwille, P.: Probing lipid mobility of raft-exhibiting model membranes by fluorescence correlation spectroscopy. J. Biol. Chem. **278**, 28109–28115 (2003)
41. Krägel, J., Kretzschmar, G., Li, J.B., Loglio, G., Miller, R., Möhwald, H.: Surface rheology of monolayers. Thin Solid Films **284–285**, 361–364 (1996)
42. Krüger, P., Lösche, M.: Molecular chirality and domain shapes in lipid monolayers on aqueous surfaces. Phys. Rev. E **62**, 7031–7043 (2000)
43. Kostur, M., Schindler, M., Talkner, P., Hänggi, P.: Chiral separation in microflows. Phys. Rev. Lett. **96**, 014502-1–014502-4 (2006)
44. Larson, R.G.: The Structure and Rheology of Complex Fluids. Oxford University Press, Oxford (1999)
45. Letherish, W.: The mechanical behavior of Bitumen. J. Soc. Chem. Ind. **59**, 1–26 (1940)
46. Li, P.C.H.: Microfluidic Lab-on-a-Chip for Chemical and Biological Analysis and Discovery. CRC Press, Boca Raton (2006)
47. Lions, P.L., Masmoudi, N.: Global solutions for some Oldroyd models of non-Newtonian fluids. Chinese Ann. Math. Ser. B **21**, 131–146 (2000)
48. Marcos, Fu, H.C., Powers, T.R., Stocker, R.: Separation of microscale chiral objects by shear flow. Phys. Rev. Lett. **102**, 158103-1–158103-4 (2009)
49. Miller, A., Möhwald, H.: Diffusion limited growth of crystalline domains in phospholipid monolayers. J. Chem. Phys. **86**, 4258–4265 (1987)
50. Miller, A., Knol, W., Möhwald, H.: Fractal growth of crystalline phospholipid domains in monomolecular layers. Phys. Rev. Lett. **56**, 2633–2638 (1986)
51. Möhwald, H.: Phospholipid monolayers. In: Lipowsky, R., Sackmann, E. (eds.) Handbook of Biological Physics, vol. 1, pp. 161–211. Elsevier Science, Amsterdam (1995)
52. Novick-Cohen, A.: The Cahn-Hilliard equation: mathematical and modeling perspectives. Adv. Math. Sci. Appl. **8**, 965–985 (1998)
53. Oldroyd, J.G.: On the formulation of rheological equations of state. Proc. R. Soc. A **200**, 523–541 (1950)
54. Orädd, G., Westerman, P., Lindblom, G.: Lateral diffusion coefficients of separate lipid species in a ternary raft-forming bilayer: a Pfg-NMR multinuclear study. Biophys. J. **89**, 315–320 (2005)
55. Pichot, R., Watson, R.L., Norton, I.T.: Phospholipids at the interface: current trends and challenges. Int. J. Mol. Sci. **14**, 11767–11794 (2013)
56. Relini, A., Ciuchi, F., Rolandi, R.: Surface shear viscosity and phase transitions of monolayers at the air-water interface. J. Phys. II **5**, 1209–1221 (1995)
57. Rowlinson, J., Widom, B.: Molecular Theory of Capillarity. Clarendon Press, Oxford (1982)
58. Sacchetti, M., Yu, H., Zografi, G.: In-plane steady shear viscosity of monolayers at the air/water interface and its dependence on free area. Langmuir **9**, 2168–2171 (1993)
59. Sadoughi, A.H., Lopez, J.M., Hirsa, A.H.: Transition from Newtonian to non-Newtonian surface shear viscosity of phospholipid monolayers. Phys. Fluids **25**, 032107 (2013)
60. Sickert, M., Rondelez, F.: Shear viscosity of Langmuir monolayers in the low-density limit. Phys. Rev. Lett. **90**, 126104 (2003)
61. Simons, K., Ikonen, E.: Functional rafts in cell membranes. Nature **387**, 569–572 (1997)
62. Singer, S.J., Nicolson, J.L.: The fluid mosaic model of the structure of cell membranes. Science **175**, 720–735 (1972)
63. Steppich, D., Griesbauer, J., Frommelt, T., Appelt, W., Wixforth, A., Schneider, M.F.: Thermomechanic-electrical coupling in phospholipid monolayers near the critical point. Phys. Rev. E **81**, 061123 (2010)
64. Temam, R.: Remark on the pressure boundary condition for the projection method. Theor. Comput. Fluid Mech. **3**, 181–184 (1991)

65. Temam, R.: Navier-Stokes Equations: Theory and Numerical Analysis. AMS Chelsea Publications, Providence, RI (2000)
66. Tremaine, S.: On the origin of irregular structure in Saturn's rings. Astron. J. **125**, 894–901 (2003)
67. Tschoegl, N.W.: The Phenomenological Theory of Linear Viscoelastic Behavior. Springer, Berlin (1989)
68. Veatch, S., Keller, S.: Organization in lipid membranes containing cholesterol. Phys. Rev. Lett. **89**, 268101 (2002)
69. Veatch, S., Keller, S.: Separation of liquid phases in giant vesicles of ternary mixtures of phospholipids and cholesterol. Biophys. J. **85**, 3074–3083 (2003)
70. Veatch, S., Keller, S.: Seeing spots: complex phase behavior in simple membranes. Biochim. Biophys. Acta **1746**, 172–185 (2005)
71. Wells, G.N., Kuhl, E., Garikipati, K.: A discontinuous Galerkin method for the Cahn-Hilliard equation. J. Comput. Phys. **218**, 860–877 (2006)
72. Zener, C.: Elasticity and Anelasticity of Metals. University of Chicago Press, Chicago (1948)

Chapter 18
Structure Formation in Thin Liquid-Liquid Films

Sebastian Jachalski, Dirk Peschka, Stefan Bommer, Ralf Seemann,
and Barbara Wagner

Abstract We revisit the problem of a liquid polymer that dewets from another liquid polymer substrate with the focus on the direct comparison of results from mathematical modeling, rigorous analysis, numerical simulation and experimental investigations of rupture, dewetting dynamics and equilibrium patterns of a thin liquid-liquid system. The experimental system uses as a model system a thin polystyrene (PS)/polymethylmethacrylate (PMMA) bilayer of a few hundred nm. The polymer systems allow for in situ observation of the dewetting process by atomic force microscopy (AFM) and for a precise ex situ imaging of the liquid-liquid interface. In the present study, the molecular chain length of the used polymers is chosen such that the polymers can be considered as Newtonian liquids. However, by increasing the chain length, the rheological properties of the polymers can be also tuned to a viscoelastic flow behavior. The experimental results are compared with the predictions based on the thin film models. The system parameters like contact angle and surface tensions are determined from the experiments and used for a quantitative comparison. We obtain excellent agreement for transient drop shapes on their way towards equilibrium, as well as dewetting rim profiles and dewetting dynamics.

18.1 Introduction

Even though liquid-liquid dewetting has been investigated to a certain extend in the past, there is still a lack of the underpinning understanding of the precise morphology and dynamics of the interfaces involved in such systems. A fundamental understanding is however crucial for many important nanofluidic problems in

S. Jachalski • D. Peschka • B. Wagner (✉)
Weierstraß Institute, Mohrenstr. 39, 10117 Berlin, Germany
e-mail: sebastian.jachalski@wias-berlin.de; dirk.peschka@wias-berlin.de;
barbara.wagner@wias-berlin.de

S. Bommer • R. Seemann
Experimental Physics, Saarland University, 66123 Saarbrücken, Germany
e-mail: stefan.bommer@physik.uni-saarland.de; r.seemann@physik.uni-saarland.de

© Springer International Publishing AG 2017

531

D. Bothe, A. Reusken (eds.), *Transport Processes at Fluidic Interfaces*,
Advances in Mathematical Fluid Mechanics, DOI 10.1007/978-3-319-56602-3_18

nature and technology ranging from rupture of the human tear film to the interface dynamics of donor/acceptor polymer solutions used in organic solar cells.

Indeed, in contrast to the large body of literature in the field of liquid-solid dewetting, theoretical investigations, after the early fundamental works of Brochard-Wyart et al. [1], are rather limited. Notable exceptions are in particular the works by Pototsky et al. [2, 3], the work by Fisher and Golovin [4, 5] and by Bandyopadhyay et al. [6] and Bandyopadhyay and Sharma [7]. Linear stability analysis and numerical simulations of the short- and long-time evolution have been performed by Pototsky et al. [3], Fisher and Golovin [4], and by Bandyopadhyay et al. [6], even in the presence of surfactants [5]. However, the mathematical theory of the fully non-linear evolution towards rupture of the liquid-liquid system is poorly developed as compared to the liquid-solid dewetting. Similarly, stationary droplet solutions for liquid-liquid systems and their stability have been studied numerically by Pototsky et al. [3]. Generalizations to higher dimensions, rigorous proofs are missing and convergence results are still not completely understood. Some of the first results will be given in this work. Moreover, theoretical and experimental investigations suggest that interfacial slip plays a role between liquid layers [8–12]. A systematic derivation of appropriate thin-film models will be given here.

On the experimental side there are some studies on dewetting and film instabilities of liquid-liquid systems [13]. The instability of the liquid-liquid interface are probed either directly in the reciprocal space by neutron reflectometry but without considering the liquid-air interface, or indirectly by the resulting deformation of the liquid-air interface probed by scanning force microscopy [14]. Other groups studied the breakup and the hole growth of a liquid-liquid system, where the viscosity of one of the liquids is much larger than the viscosity of the other liquid [15] and in a very special case, where the resulting dewetting morphologies are all coated with a thin layer of the underlying liquid [16], whereas the characteristic shape of the liquid-liquid interface was not explored in detail. The shape of an underlying liquid polymethylmethacrylate (PMMA) substrate and the liquid polystyrene (PS) rim profile dewetting from this substrate has been studied first in the pioneering work of the group of G. Krausch [17, 18]. As a result, they found a characteristic rim shape and dewetting dynamics, depending on the relative viscosity of the two liquids. The experimentally observed behavior was claimed to be in agreement with Brochard et al. [1] which is surprising since the dewetting velocity strongly depends on film thickness as we will show and which is not considered in [1]. However, the used polymers in [17, 18] are above the entanglement of the respective chain length and viscoelastic properties cannot be ruled out. Here, we explore the dewetting dynamics and undertake a systematic variation of the physical parameters to make quantitative comparisons with our theoretical models. In addition we develop new thin-film models that include nonlinear viscoelastic rheologies.

18.2   Mathematical Model for the Polymer Liquid-Liquid System

We begin this section by introducing the basic setup and notations. Firstly, notice that we consider a two-dimensional situation with the x-axis pointing in horizontal and the z-axis pointing in vertical direction. Later, we give a remark on the generalisation of the models to three dimensions.

We investigate a system of two layered, immiscible fluids on a flat solid substrate which are surrounded by a gas phase (see Fig. 18.1). The lower liquid, which occupies the domain

$$\Omega_1(t) := \{(x, z) \in \mathbb{R}^2;\ 0 \le z < h_1(x, t)\}, \tag{18.1}$$

we call liquid 1 or layer 1. Mass density ρ_1, viscosity μ_1, pressure p_1 as well as horizontal and vertical velocity components, u_1 and w_1, are associated with this layer. Similarly, the upper liquid, occupying

$$\Omega_2(t) := \{(x, z) \in \mathbb{R}^2;\ h_1(x, t) \le z < h_2(x, t)\}, \tag{18.2}$$

is denoted by liquid 2 or layer 2, with corresponding quantities ρ_2, μ_2, p_2, u_2 and w_2. For Newtonian liquids the stresses in the nth layer, $n = 1, 2$, are given by

$$\tau_n = -p_n I + \mu_n \begin{pmatrix} 2\partial_x u_n & \partial_z u_n + \partial_x w_n \\ \partial_z u_n + \partial_x w_n & 2\partial_z w_n \end{pmatrix}. \tag{18.3}$$

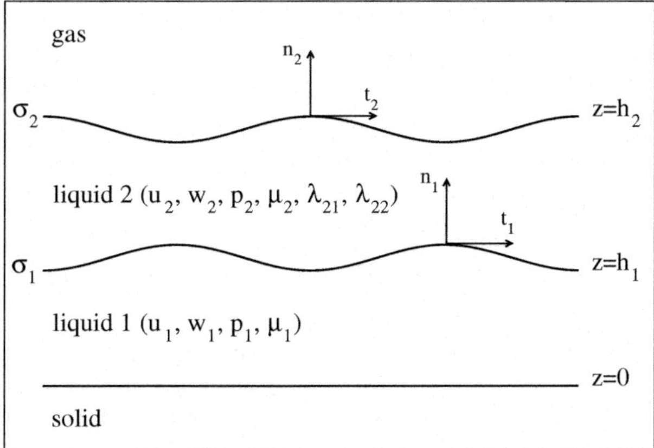

Fig. 18.1   Sketch of a two-layer system

In the case of a viscoelastic upper layer we assume that the symmetric stress tensor τ_2 obeys the corotational Jeffreys model with the constitutive equation

$$\tau_2 + \lambda_{21}\frac{D}{Dt}\tau_2 = \mu_2\left(\dot{\gamma}_2 + \lambda_{22}\frac{D}{Dt}\dot{\gamma}_2\right),\tag{18.4}$$

and where the Jaumann derivative D/Dt is defined by

$$\frac{D\Lambda}{Dt} = \frac{d\Lambda}{dt} + \frac{1}{2}\left(\omega_2\Lambda - \Lambda\omega_2\right),\tag{18.5}$$

for an arbitrary tensor field Λ. The strain rate $\dot{\gamma}_2$ is given by

$$\dot{\gamma}_2 = \begin{pmatrix} 2\partial_x u_2 & \partial_z u_2 + \partial_x w_2 \\ \partial_z u_2 + \partial_x w_2 & 2\partial_z w_2 \end{pmatrix},\tag{18.6}$$

and the vorticity tensor is

$$\omega_2 = \begin{pmatrix} 0 & \partial_x w_2 - \partial_z u_2 \\ \partial_z u_2 - \partial_x w_2 & 0 \end{pmatrix}.\tag{18.7}$$

In this work we assume λ_{21} and λ_{22} to be constant material parameters. The relaxation parameter λ_{21} typically denotes a measure of the time required for the stress to relax to some limiting value, whereas λ_{22} is a measure of the retardation to return to the equilibrium state, see for example [19].

We assume that the system contains three interfaces. The first one between the solid and liquid 1 is located at $z = 0$ and does not change in time t. We call it solid-liquid interface. The tangential and normal vectors of this interface are simply given by $t_s = (1,0)$ and $n_s = (0,1)$. The other two interfaces evolve in time. The one between the two liquids (liquid-liquid interface) is at $z = h_1(x,t)$ while the free surface between liquid 2 and the gas phase (liquid-gas interface) is at $z = h_2(x,t)$. The unit tangential and normal vectors and the curvatures of the liquid-liquid interface (subindex 1) and the liquid-gas interface (subindex 2) are given by

$$n_n = \frac{(-\partial_x h_n, 1)}{\sqrt{1 + (\partial_x h_n)^2}}, \quad t_n = \frac{(1, \partial_x h_n)}{\sqrt{1 + (\partial_x h_n)^2}}, \quad \kappa_n = \frac{\partial_{xx} h_n}{\left(1 + (\partial_x h_n)^2\right)^{3/2}}.\tag{18.8}$$

Moreover, we denote the surface tensions for the liquid-liquid and the liquid-gas interface by σ_1 and σ_2, respectively. For convenience, we also introduce

$$h(x,t) = h_2(x,t) - h_1(x,t),\tag{18.9}$$

the thickness of the top layer as a variable.

Next, we discuss the hydrodynamic equations which describe the evolution of such a system. Then, we introduce a suitable scaling for the variables in this system and obtain a set of nondimensional equations. Finally, using formal asymptotic analysis, we reduce the latter equations to thin film equations for the layer thicknesses, h_1 and h.

18.2.1   Hydrodynamic Equations

In each layer we suppose the Cauchy momentum equations

$$0 = \partial_x u_n + \partial_z w_n, \tag{18.10}$$

$$\rho_n \frac{d}{dt} u_n = -\partial_x p_n + \partial_x \tau_{n,11} + \partial_z \tau_{n,12}, \tag{18.11}$$

$$\rho_n \frac{d}{dt} w_n = -\partial_z p_n + \partial_x \tau_{n,12} + \partial_z \tau_{n,22}, \tag{18.12}$$

where

$$\tau_2 + \lambda_{21} \frac{D}{Dt} \tau_2 = \mu_2 \left(\dot{\gamma}_2 + \lambda_{22} \frac{D}{Dt} \dot{\gamma}_2 \right), \tag{18.13}$$

in the upper layer and

$$\tau_1 = \mu_1 \begin{pmatrix} 2\partial_x u_1 & \partial_z u_1 + \partial_x w_1 \\ \partial_z u_1 + \partial_x w_1 & 2\partial_z w_1 \end{pmatrix}, \tag{18.14}$$

in the lower layer. If $\lambda_{21} = \lambda_{22} = 0$ we are a pure Newtonian case else the upper layer is viscoelastic.

The equations are coupled to each other and to the surrounding solid and gas phase by boundary conditions at the interfaces. At the solid-liquid interface (i.e. $z = 0$), we impose the Navier-slip condition. It says that the tangential component of the velocity is proportional to the shear stress at the interface, or

$$(u_1, w_1) \cdot \boldsymbol{t}_s = \frac{b}{\mu_1} \boldsymbol{n}_s \cdot \tau_1 \cdot \boldsymbol{t}_s. \tag{18.15}$$

The constant b denotes the slip-length. We plug the concrete expressions for $\boldsymbol{t}_s$ and $\boldsymbol{n}_s$ into condition (18.15) and obtain

$$u_1 = b\, \partial_z u_1. \tag{18.16}$$

Besides this we also assume the impermeability condition,

$$w_1 = 0. \tag{18.17}$$

At the liquid-liquid interface, $z = h_1(x,t)$, we have a kinematic condition. It balances the normal component of the velocity of liquid 1 at the interface with the velocity of the interface itself, i.e.

$$(0, \partial_t h_1) \cdot \boldsymbol{n_1} = (u_1, w_1) \cdot \boldsymbol{n_1}. \tag{18.18}$$

Next, we consider capillary forces. These act to reduce the area of the interface and are compensated by the jump of the stress tensors times unit normal vector and also by intermolecular forces, which we explain later,

$$\left(\tau_1 - \tau_2 - \phi'(h)I\right) \cdot \boldsymbol{n_1} = \sigma_1 \kappa_1 \boldsymbol{n_1}.$$

Since the last relation is vector valued we obtain two boundary conditions from it,

$$\boldsymbol{t_1} \cdot \left(\tau_1 - \tau_2 - \phi'(h)I\right) \cdot \boldsymbol{n_1} = 0, \tag{18.19}$$

$$\boldsymbol{n_1} \cdot \left(\tau_1 - \tau_2 - \phi'(h)I\right) \cdot \boldsymbol{n_1} = \sigma_1 \kappa_1. \tag{18.20}$$

These are called tangential and normal stress condition, respectively. At this interface we also suppose a slip condition. In contrast to (18.15) the left hand side of the equation depends on the jump of the velocities,

$$(u_2 - u_1, w_2 - w_1) \cdot \boldsymbol{t_1} = b_1 \left(\frac{1}{\mu_1} + \frac{1}{\mu_2}\right) \boldsymbol{n_1} \cdot \tau_2 \cdot \boldsymbol{t_1}. \tag{18.21}$$

Notice, the factor $(1/\mu_1 + 1/\mu_2)$ could be chosen in a different way. The advantage of our choice is that slip length b_1 has the unit of a length. Furthermore, in the limit $\mu_1 \to \infty$, i.e. liquid 1 becomes solid, condition (18.15) is restored. The impermeability condition at this surface reads

$$(u_2 - u_1, w_2 - w_1) \cdot \boldsymbol{n_1} = 0. \tag{18.22}$$

The free surface $z = h_2(x,t)$ also evolves according to a kinematic condition,

$$(0, \partial_t h_2) \cdot \boldsymbol{n_2} = (u_2, w_2) \cdot \boldsymbol{n_2}. \tag{18.23}$$

Here, the tangential and normal stress conditions are

$$\boldsymbol{t_2} \cdot \left(\tau_2 + \phi'(h)I\right) \cdot \boldsymbol{n_2} = 0, \tag{18.24}$$

$$\boldsymbol{n_2} \cdot \left(\tau_2 + \phi'(h)I\right) \cdot \boldsymbol{n_2} = \sigma_2 \kappa_2. \tag{18.25}$$

Now let us discuss the intermolecular forces we introduced before (18.19). We investigate a situation in which intermolecular interactions in the layered system give contributions to the surface forces. These additional forces can drive dewetting of the upper liquid. On the other hand, we neglect interactions between liquids 1 and 2 with the solid substrate, which might lead to the breakup of layer 1.

The intermolecular potential for the interactions is given by

$$\phi(h) = \frac{8}{3}\phi_* \left(\frac{1}{8} \left(\frac{h_*}{h} \right)^8 - \frac{1}{2} \left(\frac{h_*}{h} \right)^2 \right). \tag{18.26}$$

This potential consists of two competing terms, which represent long-range, h^{-2}, and short-range, h^{-8}, forces. The long-range term is the disjoining pressure contribution from the van-der-Waals potential. This force drives the dewetting. Only when the thickness h becomes very small the short-range term has an impact. In fact, it stabilises and prevents layer 2 from complete rupture, i.e. no interface between liquid 1 and the gas phase appears. There remains a layer of liquid 2 of very small height. This height is associated with the value h_* for which potential (18.26) has a minimum of $\phi_* < 0$. Notice that while the long-range part in (18.26) can be derived from a Lennard-Jones potential, where also other choices for the form of the stabilising part are possible. A discussion referring to this subject can be found e.g. in Oron et al. [20]. In (18.26) the short-range part of the potential is chosen in order to produce a minimum for a particular thickness of the film. The potential (18.26) gives a contribution to the energy of the system. Variations of h_1 and h_2 change this contribution by $-\phi'(h)\delta h_1$ and $\phi'(h)\delta h_2$, respectively, which produces the extra terms $\phi'(h)$ in (18.20) and (18.25).

The main purpose of the intermolecular potential is to account for the interactions responsible for spinodal dewetting as observed in experiments. This feature will be discussed in the linear stability. With the short-range repulsion term such a potential ensures positivity of solutions, which is a major advantage for the analysis. From a modelling point of view, it also allows to set the equilibrium contact angle and to pass to the Γ-limit of zero precursor thickness, as we will discuss for stationary solutions. However, as this limit is still open for time-dependent solutions, we also discuss algorithms for both situations, i.e. global solutions with precursor and free-boundary problems for the expected sharp-interface model.

18.2.2 Nondimensional Problem

Let H and W be typical scales for the height and the vertical velocity components, respectively. Then, we write

$$z = H\tilde{z}, \quad h_n = H\tilde{h}_n, \quad b = H\tilde{b}, \quad b_1 = H\tilde{b}_1 \quad \text{and} \quad w_n = W\tilde{w}_n. \tag{18.27}$$

Analogously, we denote the characteristic scales for the lateral length and the horizontal velocity by L and U,

$$x = L\tilde{x}, \qquad u_n = U\tilde{u}_n. \tag{18.28}$$

For the characteristic time scale T we suppose $T = L/U$ and set

$$t = \frac{L}{U}\tilde{t}, \qquad \lambda_{21} = \frac{L}{U}\tilde{\lambda}_{21}, \qquad \lambda_{22} = \frac{L}{U}\tilde{\lambda}_{22}. \tag{18.29}$$

For the stress tensors we set

$$\begin{pmatrix} \tau_{n,11} & \tau_{n,12} \\ \tau_{n,21} & \tau_{n,22} \end{pmatrix} = \frac{\mu_n}{T} \begin{pmatrix} \tau^*_{n,11} & \frac{L}{H}\tau^*_{n,12} \\ \frac{L}{H}\tau^*_{n,21} & \tau^*_{n,22} \end{pmatrix}. \tag{18.30}$$

Furthermore, we assume that the typical scale for the pressure is equal to the one for the intermolecular forces and denote it by P,

$$p_n = P\tilde{p}_n, \qquad \phi' = P\tilde{\phi}'. \tag{18.31}$$

Be aware that in the following we drop '$\sim$'. At this point we can choose some of the introduced scales freely. In view of the structure of potential (18.26) we set

$$P = \frac{8}{3}\frac{\phi_*}{H}, \tag{18.32}$$

which results in a rather simple form for ϕ',

$$\phi'(h) = \frac{1}{\varepsilon}\left(-\left(\frac{\varepsilon}{h}\right)^9 + \left(\frac{\varepsilon}{h}\right)^3\right), \qquad \text{where} \quad \varepsilon = \frac{h_*}{H}. \tag{18.33}$$

Usually, the minimum point of (18.26), h_*, is much smaller than the characteristic height H. Hence, we suppose $\varepsilon \ll 1$. In the following we use the notations

$$\varepsilon_\ell = \frac{H}{L}, \quad \mathrm{Re} = \frac{\rho_2 UH}{\mu_2}, \quad \rho = \frac{\rho_1}{\rho_2}, \quad \mu = \frac{\mu_1}{\mu_2}, \quad \sigma = \frac{\sigma_1}{\sigma_2}, \quad \alpha = \frac{PH}{\mu_2 U}, \tag{18.34}$$

and we obtain

$$\varepsilon_\ell \rho \mathrm{Re}\left(\partial_t u_1 + u_1\partial_x u_1 + w_1\partial_z u_1\right) = -\alpha\,\varepsilon_\ell\partial_x p_1 + \mu\left(\varepsilon_\ell^2\partial_{xx}u_1 + \partial_{zz}u_1\right), \tag{18.35a}$$

$$\varepsilon_\ell^2 \rho \mathrm{Re}\left(\partial_t w_1 + u_1\partial_x w_1 + w_1\partial_z w_1\right) = -\alpha\,\partial_z p_1 + \mu\left(\varepsilon_\ell^3\partial_{xx}w_1 + \varepsilon_\ell\partial_{zz}w_1\right), \tag{18.35b}$$

$$0 = \partial_x u_1 + \partial_z w_1, \tag{18.35c}$$

and analogously

$$\varepsilon_\ell \operatorname{Re}\left(\partial_t u_2 + u_2\partial_x u_2 + w_2\partial_z u_2\right) = -\alpha\,\varepsilon_\ell\partial_x p_2 + \varepsilon_\ell^2\partial_x \tau_{2,11} + \partial_z\tau_{2,12}, \quad (18.35\mathrm{d})$$

$$\varepsilon_\ell^2 \operatorname{Re}\left(\partial_t w_2 + u_2\partial_x w_2 + w_2\partial_z w_2\right) = -\alpha\,\partial_z p_2 + \varepsilon_\ell\partial_x \tau_{2,21} + \varepsilon_\ell\partial_z\tau_{2,22}, \quad (18.35\mathrm{e})$$

$$0 = \partial_x u_2 + \partial_z w_2, \quad (18.35\mathrm{f})$$

where the stress tensor of the upper liquids fulfil

$$\left(1 + \lambda_{21}\frac{d}{dt}\right)\tau_{2,11} - \lambda_{21}\left(\frac{1}{\varepsilon^2}\partial_z u_2 - \partial_x w_2\right)\tau_{2,12}$$
$$= 2\left(1 + \lambda_{22}\frac{d}{dt}\right)\partial_x u_2 - \lambda_{22}\left(\left(\frac{1}{\varepsilon}\partial_z u_2\right)^2 - (\varepsilon\partial_x w_2)^2\right), \quad (18.35\mathrm{g})$$

$$\left(1 + \lambda_{21}\frac{d}{dt}\right)\tau_{2,22} + \lambda_{21}\left(\frac{1}{\varepsilon^2}\partial_z u_2 - \partial_x w_2\right)\tau_{2,12}$$
$$= 2\left(1 + \lambda_{22}\frac{d}{dt}\right)\partial_z w_2 + \lambda_{22}\left(\left(\frac{1}{\varepsilon}\partial_z u_2\right)^2 - (\varepsilon\partial_x w_2)^2\right), \quad (18.35\mathrm{h})$$

$$\left(1 + \lambda_{21}\frac{d}{dt}\right)\tau_{2,12} + \frac{\lambda_{21}}{2}\left(\partial_z u_2 - \varepsilon^2\partial_x w_2\right)\left(\tau_{2,11} - \tau_{2,22}\right)$$
$$= \left(1 + \lambda_{22}\frac{d}{dt}\right)\left(\partial_z u_2 + \varepsilon^2\partial_x w_2\right) + 2\lambda_{22}\left(\partial_z u_2 - \varepsilon^2\partial_x w_2\right)\partial_x u_2. \quad (18.35\mathrm{i})$$

The boundary conditions at the substrate, i.e. the impermeability and the Navier-slip condition, now read

$$u_1 = b\,\partial_z u_1, \qquad w_1 = 0. \quad (18.36\mathrm{a})$$

At the liquid-liquid interface, $z = h_1$, we get the following nondimensional equations. For the normal, tangential stresses and the kinematic condition,

$$0 = p_1 - p_2 + \phi'(h) + 2\frac{\varepsilon_\ell}{\alpha}\frac{\left(\mu\partial_z u_1 + \varepsilon_\ell^2\mu\partial_x w_1 - \tau_{2,12}\right)\partial_x h_1}{1 + \varepsilon_\ell^2(\partial_x h_1)^2} + \ldots$$
$$- \frac{\varepsilon_\ell}{\alpha}\frac{\left(1 - \varepsilon_\ell^2(\partial_x h_1)^2\right)\left(2\mu\partial_z w_1 - \tau_{2,22}\right)}{1 + \varepsilon_\ell^2(\partial_x h_1)^2} + \sigma\frac{\partial_{xx} h_1}{\left(1 + \varepsilon_\ell^2(\partial_x h_1)^2\right)^{3/2}}, \quad (18.36\mathrm{b})$$

$$0 = \left(\partial_z u_1 + \varepsilon_\ell^2\mu\partial_x w_1 - \tau_{2,12}\right)\left(1 - \varepsilon_\ell^2(\partial_x h_1)^2\right) + \ldots$$
$$- \varepsilon_\ell^2\left(4\mu\partial_x u_1 - 2\tau_{2,11}\right)\partial_x h_1, \quad (18.36\mathrm{c})$$

$$\partial_t h_1 = w_1 - u_1\partial_x h_1. \quad (18.36\mathrm{d})$$

The slip condition becomes

$$(u_2 - u_1) + \varepsilon_\ell^2 (w_2 - w_1) \partial_x h_1 = b_1 \frac{\mu+1}{\mu} \frac{\tau_{2,12}\left(1-\varepsilon_\ell^2(\partial_x h_1)^2\right)-2\varepsilon_\ell^2\tau_{2,11}\partial_x h_1}{\sqrt{1+\varepsilon_\ell^2(\partial_x h_1)^2}}, \quad (18.36e)$$

and the impermeability condition is given by

$$(w_2 - w_1) - (u_2 - u_1) \partial_x h_1 = 0. \quad (18.36f)$$

Finally, at the liquid-gas interface, $z = h_2$, normal and tangential stresses and the kinematic conditions are

$$0 = p_2 - \phi'(h) - \frac{\varepsilon_\ell}{\alpha} \frac{\left(1-\varepsilon_\ell^2(\partial_x h_2)^2\right)\tau_{2,22}-2\tau_{2,12}\partial_x h_2}{1+\varepsilon_\ell^2(\partial_x h_2)^2} + \frac{\partial_{xx} h_2}{\left(1+\varepsilon_\ell^2(\partial_x h_2)^2\right)^{3/2}}, \quad (18.36g)$$

$$0 = \tau_{2,12}\left(1 - \varepsilon_\ell^2 (\partial_x h_2)^2\right) - 2\varepsilon_\ell^2\tau_{2,11}\partial_x h_2, \quad (18.36h)$$

$$\partial_t h_2 = w_2 - u_2\partial_x h_2. \quad (18.36i)$$

Note, to write (18.36g) in this form, without loss of generality, we used the balance

$$\frac{\sigma_2 H}{PL^2} = 1 \quad (18.37)$$

This, together with (18.32), determines parameter ε_ℓ,

$$\varepsilon_\ell = \frac{H}{L} = \sqrt{\frac{8}{3}\frac{\phi_*}{\sigma_2}}. \quad (18.38)$$

To derive thin film equations for the layer thicknesses h_1 and h we assume that $\varepsilon_\ell \ll 1$. In other words, we suppose that the characteristic scale for the height is much smaller than the typical length scale. Equation (18.35a) still depend on several parameters. In the pure Newtonian case we suppose $\lambda_{21} = 0$ and $\lambda_{22} = 0$. While we assume ρ, Re, μ and σ to be of order one w.r.t. ε_ℓ, we consider various magnitudes for the slip lengths b and b_1. For these different magnitudes, we have to choose alternate α's and obtain different models. In the viscoelastic case we assume λ_{21}, λ_{22}, ρ, Re and σ to be of order one but μ of order $O(\varepsilon_\ell^2)$. We note that this order of magnitude of μ is the only choice to incorporate the full nonlinear viscoelastic model into an asymptotically consistent thin-film model.

18.3 Thin-Film Model

For no-slip or small interfacial slip we expect that the profile of the lateral velocity component in layer 1 is parabolic. Therefore, we balance the pressure gradient $\partial_x p_1$ with the dominant viscous term $\partial_{zz} u_1$ in (18.35a),

$$\alpha = \frac{1}{\varepsilon_\ell}. \tag{18.39}$$

This fixes the velocity scale and hence, the capillary number,

$$\mathrm{Ca} = \frac{\mu_2 U}{\sigma_2} = \varepsilon_\ell^3. \tag{18.40}$$

After having all the scales fixed we can now derive a thin-film model from (18.35) and (18.36) assuming that $\varepsilon_\ell \ll 1$ and assuming the solutions can be written in asymptotic expansions in ε_ℓ. Using only the leading order terms in the expansions the derivation of thin-film equations from the underlying hydrodynamic model is straight forward, see e.g. [21]. The coupled scaled system of nonlinear fourth order partial differential equations for the profiles of the free surfaces h_1 and h_2 takes the form

$$\partial_t \mathbf{h} = \nabla \cdot (Q \cdot \nabla \mathbf{p}), \tag{18.41}$$

where $\mathbf{h} = (h_1, h_2)^\mathsf{T}$ is the vector of liquid-liquid interface profile and liquid-air surface profile. The components of the vector $\mathbf{p} = (p_1, p_2)^\mathsf{T}$ are the interfacial pressures given by

$$p_1 = -\sigma \Delta h_1 - \phi_\epsilon'(h_2 - h_1), \qquad p_2 = -\Delta h_2 + \phi_\epsilon'(h_2 - h_1), \tag{18.42}$$

The gradient of the pressure vector is multiplied by the mobility matrix Q which is given by

$$Q = \frac{1}{\mu} \begin{pmatrix} \dfrac{h_1^3}{3} & \dfrac{h_1^3}{3} + \dfrac{h_1^2(h_2 - h_1)}{2} \\[3ex] \dfrac{h_1^3}{3} + \dfrac{h_1^2(h_2 - h_1)}{2} & \dfrac{\mu}{3}(h_2 - h_1)^3 + h_1 h_2(h_2 - h_1) + \dfrac{h_1^3}{3} \end{pmatrix}. \tag{18.43}$$

where $\sigma = \sigma_1/\sigma_2$ and $\mu = \mu_1/\mu_2$ denote surface tension and viscosity ratios for the lower and upper layer and $\epsilon = h_*/h_{max}$ is a small parameter with h_{max} being the maximal distance between the polymer-air and polymer-polymer interface.

The energy functional associated to the gradient flow of the lubrication equation can then be given by

$$E_\varepsilon(h_1, h_2) = \int_0^L \left[\frac{\sigma}{2} |\partial_x h_1|^2 + \frac{1}{2} |\partial_x h_2|^2 + \phi_\epsilon(h_2 - h_1) \right] dx \qquad (18.44)$$

where the potential function ϕ_ϵ denotes the scaled potential with $(n, \ell) = (2, 8)$. The relation to the thin-film equations is $p_i = \delta E_\varepsilon / \delta h_i$.

18.3.1 Mathematical Theory

18.3.1.1 Stationary States

Even though in an actual experiment stationary states represent the late stage of the dewetting process, we begin our analysis with this state, since it allows to identify important quantities, such as the equilibrium Neumann triangle conditions, surface tension and interfacial tensions, by careful comparisons of our mathematical and numerical results with specifically designed experiments. The results of this analysis can then be used in the dynamic models, where other quantities, such as dewetting rates, evolution of interfacial morphologies can be investigated.

We investigated stationary solutions of a thin-film model for liquid two-layer flows with the aim to achieve a rigorous understanding of the contact-angle conditions for such two-layer systems. For this we considered an appropriate energetic formulation that is motivated by its gradient flow structure. We pursued this by investigating a corresponding energy that favors the upper liquid to dewet from the lower liquid substrate, leaving behind a layer of thickness h_*, given by the intermolecular potential

$$\phi(h_2 - h_1) = \frac{\phi_*}{\ell - n} \left[\ell \left(\frac{h_*}{h_2 - h_1} \right)^n - n \left(\frac{h_*}{h_2 - h_1} \right)^\ell \right], \qquad (18.45)$$

where h_1 is the height of the liquid-liquid interface, h_2 the height of the free surface and its minimal value $\phi_* < 0$ is attained at h_*. We note that other energies are possible but this one corresponds more closely to the experimental set-up.

One can then obtain that any positive stationary solution of (18.42)–(18.43) satisfies $\partial_x p_1 = \partial_x p_2 = 0$ in Ω. This in turn is equivalent to

$$\sigma \partial_{xx} h_1 = -\phi_\epsilon'(h_2 - h_1) - \lambda_2 + \lambda_1, \qquad (18.46\text{a})$$

$$\partial_{xx} h_2 = \phi_\epsilon'(h_2 - h_1) - \lambda_1, \qquad (18.46\text{b})$$

where constants λ_2 and λ_1 are Lagrange multipliers associated with conservation of mass. We then first established existence of a global minimizer to the energy functional (18.44) and showed that it satisfies (18.46) with

$$\partial_x h_1 = \partial_x h_2 = \partial_{xxx} h_1 = \partial_{xxx} h_2 = 0 \ \text{ for } \ x \in \partial\Omega. \tag{18.47}$$

Theorem 1 *Let Ω be a bounded domain of class $C^{0,1}$ in $\mathbb{R}^d$, $d \geq 1$ and let $\mathbf{m} = (m_1, m_2)$ with $m_1, m_2 > 0$. Then a global minimizer of $E_\epsilon(\cdot, \cdot)$ defined in (18.44) exists in the class*

$$X_{\mathbf{m}} := \left\{ (h_1, h_2) \in H^1(\Omega)^2 : m_1 = \int_\Omega h_1, \ m_2 = \int_\Omega (h_2 - h_1), \ h_2 \geq h_1 \right\}, \tag{18.48}$$

For $d = 1$ and $\Omega = (0, L)$ the function $h_2 - h_1$ is strictly positive and (h_1, h_2) are smooth solutions to the ODE system (18.46) with (18.47) and

$$\lambda_1 = \frac{1}{L} \int_\Omega \phi'_\epsilon(h_2 - h_1)\, dx, \quad \lambda_2 = 0. \tag{18.49}$$

After proving existence of stationary solutions which is a generalisation of the proof for single-layer thin films [22], we focussed on the limit $h_* \to 0$ via matched asymptotic analysis in order to recapture the Neumann triangle construction together with the corresponding sharp-interface model. Our analysis shows, that the complete matching of the asymptotic solution requires the inclusion of so-called logarithmic switch-back terms. This is also interesting, in view of the fact that as a limiting case our analysis also includes the case of equilibrium droplets on solid substrates.

We then showed existence and uniqueness of the limit $h_* \to 0$ within the framework of Γ-convergence and show

Theorem 2 *For the family of energies E_ε the Γ-limit is*

$$E_0(h_1, h) = \int_\Omega \frac{\sigma}{2}|\nabla h_1|^2 + \frac{1}{2}|\nabla(h_1 + h)|^2 + |\Phi(1)| \chi\{h > 0\}$$

Theorem 3 (Minimizer of Sharp Interface Energy) *Let $\Omega = \mathscr{B}_R(0)$ and $X = \{(h_1, h) \in X_m(\Omega) : h|_{\partial\Omega} = 0\}$ and energy*

$$E(h_1, h) := \int_\Omega \frac{\sigma}{2}|\nabla h_1|^2 + \frac{1}{2}|\nabla(h_1 + h)|^2 + |\Phi(1)|\chi_{\{h>0\}} dx.$$

Then using $\zeta(x) := \alpha(s^2 - |x|^2)^+$ minimizers of E with mass (m_1, m_2) are

$$h_1 = -\frac{1}{\sigma + 1}\zeta(x - x_0) + h_\infty, \qquad h(x) = \zeta(x - x_0),$$

with constant $x_0 \in \Omega$ and $s, \alpha, h_\infty \in \mathbb{R}$. Prescribing the mass (m_1, m_2) fixes s and h_∞, whereas α is fixed by the contact angle (Neumann triangle)

$$\sigma(\nabla h_1)^2 + (\nabla(h_1 + h))^2 = 2|\Phi(1)|, \qquad at \ |x| = s.$$

Comparison with the sharp-interface model obtained from the Γ-limit agrees with the one obtained via matched asymptotics. Our results on the stationary solutions are published in [23].

18.3.1.2 Existence Theory for the Dynamic Problem

While the existence theory for single layer thin film equations is well established, beginning with the seminal paper by Bernis and Friedman [24], for two-layer systems this seems not to be the case. Only recently, Barrett and El Alaoui [25] introduced a finite element scheme for a similar system including surfactants and investigated existence of weak solutions. However their proof relied on the presence of intermolecular forces in the equations.

In [26] we showed existence of weak solution of the dynamic problem of liquid-liquid thin films and in addition prove non-negativity for the system of degenerate parabolic equations:

$$h_{1,t} + (M_{11}p_{1,x} + M_{12}p_{2,x})_x = 0 \quad \text{in } Q_{T_0} = \Omega \times (0, T_0),$$
$$h_t + (M_{21}p_{1,x} + M_{22}p_{2,x})_x = 0 \quad \text{in } Q_{T_0} = \Omega \times (0, T_0),$$

$$(18.50)$$

where

$$p_1 = (\sigma + 1)h_{1,xx} + h_{xx}, \quad p_2 = h_{1,xx} + h_{xx}, \quad M = \frac{1}{\mu} \begin{pmatrix} \frac{1}{3}h_1^3 & \frac{1}{2}h_1^2 h \\ \frac{1}{2}h_1^2 h & \frac{\mu}{3}h^3 + h_1 h^2 \end{pmatrix}.$$

$$(18.51)$$

For the existence proof for (18.50) we introduce a suitable regularised system.

$$h_{1,t} + ((M_{11} + \delta)p_{1,x} + M_{12}p_{2,x})_x = 0 \quad \text{in } Q_{T_0}, \tag{18.52}$$
$$h_t + (M_{21}p_{1,x} + (M_{22} + \varepsilon)p_{2,x})_x = 0 \quad \text{in } Q_{T_0}, . \tag{18.53}$$

where $\delta > 0$ and

$$M = \frac{1}{\mu} \begin{pmatrix} \frac{1}{3}|h_1|^3 & \frac{1}{2}|h_1|^2|h| \\ \frac{1}{2}|h_1|^2|h| & \frac{\mu}{3}|h|^3 + |h_1||h|^2 \end{pmatrix} \tag{18.54}$$

This is parabolic in sense of Petrovskiy. Also, initial conditions $h_{1,0}$ and h_0 are approximated in the $H^1(\Omega)$-norm by $C^{4+\alpha}$ functions $h_{1,0,\delta}$ and $h_{0,\delta}$,

$$h_1(x, 0) = h_{1,0,\delta}(x), \ h(x, 0) = h_{0\delta}(x). \tag{18.55}$$

We assume boundary conditions

$$h_{1,x} = h_{1,xxx} = h_x = h_{xxx} = 0 \quad \text{on } x \in \partial\Omega, . \tag{18.56}$$

For these conditions using a result by Eidelman [27] shows that (18.52),(18.53), (18.55),(18.56) has a unique solution in a small time interval, say in Q_τ for some $\tau > 0$.

The derivation of uniform upper bounds on the $C_{x,t}^{\frac{1}{2};\frac{1}{8}}$-norm of these solutions in Q_τ establishes a priori bounds that allow the conclusion that the solutions can be extended step-by-step to a solution of (18.56), (18.52), (18.53), (18.55) in all of Q_{T_0}. Finally, taking the limit $\varepsilon \to 0$ existence of weak solutions to (18.50) is established.

Moreover, by exploiting the entropy functional G_δ, which is defined by

$$G_\delta(s) = -\int_s^A g_\delta(r)dr, \quad \text{where} \quad g_\delta(s) = -\int_s^A \frac{dr}{(|r|^n + \delta)^{1/2}}, \tag{18.57}$$

non-negativity of the weak solutions is shown in [26].

18.3.2 Numerical Methods for Liquid-Liquid Dewetting

The goal of this subsection is to discuss different ways of solving the aforementioned free boundary problems numerically. Some care will be taken in emphasizing on how the contact line is dealt with in these approaches. To start with, assume that the dynamics of the two liquids is parameterized by a flow map Ψ_t with $\Omega_i(t) = \Psi(t, \Omega_i(0))$ for $i = 1, 2$. Incompressibility implies that the velocity $\mathbf{u} = \partial_t \Psi$ obeys $\nabla \cdot \mathbf{u} = 0$ in the Eulerian reference frame. For fixed time assume that the domains can be parameterized by functions h_1 and h using

$$\Omega_1(t) := \{(x, z) \in \mathbb{R} \times \mathbb{R}^+ : 0 < z < h_1(t, x)\},$$

$$\Omega_2(t) := \{(x, z) \in \mathbb{R} \times \mathbb{R}^+ : h_1(t, x) < z < h_1(t, x) + h(t, x)\}.$$

Based on this representation of a state of the domains we are going to discuss different strategies to solve the transient problem numerically. Geometrical features can be now discussed in terms of h, h_1.

18.3.2.1 Stokes Flow with Free Boundaries

For Newtonian viscous liquids $i = 1, 2$ with viscosities μ_i occupying the domains Ω_i it is straightforward to see that the Stokes system admits a variational formulation. Here we restrict to the situation without slip. There one needs to find a (continuous) velocity $\mathbf{u} \in V = H^1(\Omega_1 \cup \Omega_2; \mathbb{R}^2) \cap \{\mathbf{u} : \nabla \cdot \mathbf{u} = 0\}$, such

$$a(\mathbf{u}, \mathbf{v}) = \sum_{i=1}^{2} \int_{\Omega_i} \frac{\mu_i}{2} \mathbb{D}(\mathbf{u}) : \mathbb{D}(\mathbf{v}) dx = \sum_\alpha \sigma_\alpha \int_{\Gamma_\alpha} \bar{\nabla} \mathrm{id} : \bar{\nabla}\mathbf{v} \, ds = f(\mathbf{v}) \tag{18.58}$$

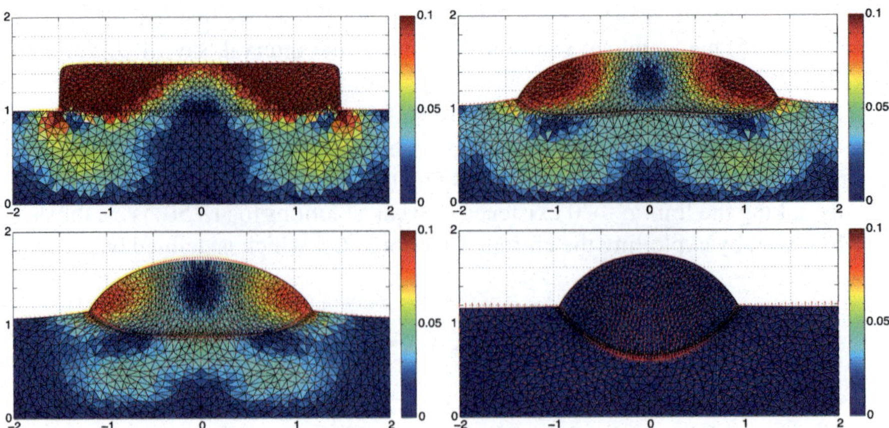

Fig. 18.2 Evolution of a liquid droplet on a liquid substrate into equilibrium $\mu_1 = \mu_2 = \sigma_1 = \sigma_2 = \sigma_3 = 1$. *Colors* indicate $|\mathbf{u}(x, z)|$, whereas *arrows* the direction $\mathbf{u}(x, z)/\|\mathbf{u}\|_\infty$

for all $\mathbf{v} \in V$, where σ_α denotes the surface tension of the interface Γ_α. We used the well-known representation of surface tension by the Laplace-Beltrami-operator, which we can write in two dimensions using tangential gradients $\bar{\nabla}$, cf. [28, 29]. Once the velocity is known the domain is moved using the flow map generated by $\mathbf{u}$. The advantage of (18.58) for the bilayers is that the contact angles (Neumann triangle) can be encoded in energetic structure of the formulation in $f(\mathbf{v})$.

For the numerical discretisation of (18.58) one often employs a finite element (FE) method. Here, one usually transforms the minimization problem into a saddle point problem which, by introducing the pressure as a Lagrange multiplier, enforces the incompressibility $\nabla \cdot \mathbf{u} = 0$. The saddle point problem requires inf-sup stable elements, e.g., Taylor-Hood elements. For this application it makes sense to enrich the pressure space by elements, which allow for a pressure-jump at the liquid-liquid interface. In order to ensure stability of the resulting scheme, one replaces id $\rightsquigarrow$ id $+ \tau\mathbf{u}$ to create a semi-implicit time-discretisation [28]. After the computation of $\mathbf{u}$ one can move the domain (or all vertices $\mathbf{x}_n$ of the underlying FE mesh) with the Lagrangian velocity field using $\mathbf{x}_n(t + \tau) = \mathbf{x}_n(t) + \tau\mathbf{u}_i$. A snapshot of the solution of the upper layer as it evolves into a stationary liquid lens, is shown in Fig. 18.2. However, the great disadvantage of this approach in our context is the potential inefficiency in situations, where we have a separation of length scales as indicated in the thin-film approximation before.[1] In such a situation the horizontal velocity u_x is basically quadratic in z and the vertical component u_z can be neglected. Then, thin-film models such as (18.41) admit an effective description of the Stokes

---

[1] Typical lengths scales in experiments $[x] = 2 \times 10^4$ nm and $[z] = 10^2$ nm.

flow (18.58). The velocity field can be reconstructed from the function h, h_1 and their derivatives with respect to x.

18.3.2.2 Numerical Methods for Thin Film Models

Global Solutions

The corresponding model we need to solve is

$$\partial_t \begin{pmatrix} h \\ h_1 \end{pmatrix} = \nabla \cdot M(h, h_1) \nabla \begin{pmatrix} \pi \\ \pi_1 \end{pmatrix}, \tag{18.59}$$

for $\pi = \delta E / \delta h$ and $\pi_1 = \delta E / \delta h_1$ for some given driving energy $E(h, h_1)$, cf., [30]. Typical energies are of the form

$$E(h, h_1) = \int \frac{\hat{\sigma}}{2} |\nabla (h_1 + h)|^2 + \frac{\hat{\sigma}_1}{2} |\nabla h_1|^2 \mathrm{d}x + V(h, h_1). \tag{18.60}$$

The fact how one is going to treat the support of h_1, h and the question, if (18.59) is still a free boundary problem is reflected in the choice of V. For simplicity we are only going to discuss the dependence on h, the discussion for h_1 or composite terms is entirely analogous. When $V \equiv 0$ the liquid spreads over the liquid substrate with zero contact angle. Similarly as in [31–33] one might expect that one can construct algorithms which preserve non-negativity of solutions. These solution might be even smooth enough to be globally (in space) well-defined. Another case, for which we derived a model earlier, is when

$$V(h) = \int \phi_\varepsilon(h) \, \mathrm{d}x \tag{18.61}$$

with ϕ_ε as in (18.26). These models admit a standard variational formulation, for which the semi-implicit time-discretisation is given by

$$\int h^{k+1} v + h_1^{k+1} v_1 + \sum_{i,j} \tau M_{ij} \nabla \pi_i \nabla v_j \, \mathrm{d}x = \int h^k v + h_1^k v_1 \mathrm{d}x \tag{18.62a}$$

$$\int \pi w + \pi_1 w_1 \mathrm{d}x = \int \hat{\sigma} \nabla (h_1 + h) \cdot \nabla (w + w_1) + \hat{\sigma}_2 \nabla h_1 \cdot \nabla w_1 + \phi'_\varepsilon w \, \mathrm{d}x \tag{18.62b}$$

with $h^k(x) = h(\tau k, x)$, $h_1^k(x) = h_1(\tau k, x)$, and M, V are evaluated at $t = \tau k$. The specific choice of ϕ_ε should generally ensure strictly positive solutions which are defined globally in space. However, the kink in the stationary solution for $\varepsilon \to 0$ already suggests that (local) refinement might be required where the precursor

$h \approx \varepsilon$ meets the support $h \gg \varepsilon$. We solve this problem with standard P1 FEM in one and two space dimensions with natural boundary conditions $\mathbf{n} \cdot \nabla h = 0$ and $\sum_j M_{ij} \mathbf{n} \cdot \nabla \pi_j = 0$.

The Thin-Film Free Boundary Problem

Kriegsmann and Miksis [34] introduced a thin-film model with a sharp triple junction, in which the support of h, i.e., $\omega(t) = \{x : h(t,x) > 0\}$, is part of the unknowns. For quasi-stationary traveling-wave solutions they constructed a scheme to numerically compute h, h_1. Later, Karapetsas et al. [35] developed a scheme to solve the transient scheme with sharp triple-junctions numerically. However, they needed to use mass conservation as a global property to resolve a numerical singularity near the contact line. Now we are going to explain how this problem can be overcome by systematically using local properties of the variational formulation. Similar to the Stokes equation (18.58) we want to find a variational formulation, which enforces contact angles in a natural way and where the contact line motion is contained in a robust way. For single thin layers such an algorithm has been shown to work in higher dimensions [36] and even for zero contact angle [37].

In contrast to globally defined solutions we have $h : \omega \mapsto \mathbb{R}^+$ and $h_1 : \mathbb{R} \mapsto \mathbb{R}^+$, where we expect kinks in h_1 at triple-junctions $\partial\omega$. First note that the driving functional for this model is

$$V(h) = |\omega| = \int \chi(h)\, dx, \qquad \chi(h) = \begin{cases} 1 & h > 0 \\ 0 & \text{else} \end{cases}. \tag{18.63}$$

Since the exact statement of the discrete variational formulation is quite involved, we only state the main differences compared to the standard formulation in (18.62). When we have an evolution of two functions h_1, h encoding domains Ω_1, Ω_2 as explained before, then it is necessary that at the triple junction $(x_c(t), z_c(t))$ we have

$$\lim_{x \searrow x_c} h_1(t,x) = \lim_{x \nearrow x_c} h_1(t,x) = z_c(t), \qquad \lim_{x \searrow x_c} h(t,x) = \lim_{x \nearrow x_c} h(t,x) = 0, \tag{18.64}$$

at all times t. This leads to a condition for time-derivatives $\dot{h}, \dot{h}_1$ of h, h_1

$$\lim_{x \searrow x_c} \dot{h}_1(t,x) + \dot{x}_c \cdot \nabla h_1 = \lim_{x \nearrow x_c} \dot{h}_1(t,x) + \dot{x}_c \cdot \nabla h_1, \tag{18.65}$$

and another condition $\dot{h} + \dot{x}_c \cdot \nabla h = 0$. These are two conditions defining $\dot{x}_c$ and at the same time defining a jump for the time-derivative. This shows that P1 FEM are not suited if we use the Eulerian time-derivatives $\dot{h}, \dot{h}_1$ as unknowns. We solve this dilemma by allowing the time-derivatives to jump at x_c and enforce the constraints on the jump (18.65) using Lagrange multipliers. As a result this

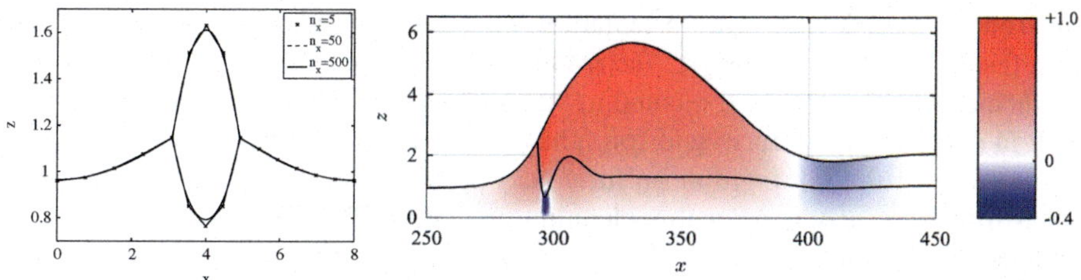

Fig. 18.3 (*left*) Transient solution of sharp triple-junction problem at various spatial resolutions but same moment in time near a droplet solution and (*right*) dewetting rim from sharp triple-junction problem with reconstructed velocity fields

condition also delivers the contact line velocity $\dot{x}_c$ without the need to reconstruct it using conservation of mass.

Another non-standard twist is the proper computation of π. Since we define the pressures π as the derivative with respect to h, h_1, we need to consistently take motion of the domain into account. This is done properly by using Reynolds' transport theorem

$$\frac{d}{dt}\int_\omega f(t,x)\mathrm{d}x = \int_\omega \partial_t f \mathrm{d}x + \int_{\partial\omega} f(\dot{x}_c \cdot \mathbf{n})\mathrm{d}s \tag{18.66}$$

$$= \int_\omega \partial_t f \mathrm{d}x + \int_{\partial\omega} f \frac{\dot{h}}{|\nabla h|}\mathrm{d}s, \tag{18.67}$$

where we used $\mathbf{n} = -\nabla h/|\nabla h|$ and $\dot{h}+\dot{x}_c\cdot\nabla h = 0$. By replacing f with the integrand of the energy $E(h, h_1)$ and setting $V(h) = (-\Sigma)|\omega|$ with spreading coefficient Σ this allows the equilibrium contact angle to be included in the variational formulation. Note that the jump condition (18.65) only affects the FE space for h, h_1, whereas the FE space for the pressures is continuous. In Fig. 18.3 we show exemplary numerical solutions of such a free boundary problem near a stationary state (left) and during dewetting with reconstructed velocity fields (right).

18.4   Experimental Methods and Comparisons to Theoretical Predictions

For the liquid-liquid dewetting experiments thin polystyrene (PS) films are prepared in their glassy state on top of also glassy thin polymethyl methacrylate (PMMA) films which are supported by silicon wafers. In our various experiments presented here, thicknesses of the underlying PMMA substrate are varied from $h_1 \approx 50\,\mathrm{nm}$ to 700 nm, and the thickness of the dewetting PS film h_2 range from about 5 nm to 250 nm. To prepare those samples, first small rectangular pieces of about $2\,\mathrm{cm}^2$ are cut from 5″-wafers with ⟨100⟩ orientation. These silicon rectangles are pre-cleaned by a fast CO_2-stream (snow-jet, Tectra) to remove particles. Subsequently, the pre-cleaned silicon wafers are sonicated in ethanol, acetone and toluene, followed by a bath in peroxymonosulfuric acid (piranha etch) to remove organic contaminations. Remains from the peroxymonosulfuric acid are removed by a careful rinse with hot Millipore™ water. After this cleaning procedure PMMA films are spun from toluene solution on top of the silicon support having a homogeneous thicknesses. To achieve the desired film thickness range, toluene solutions with different polymer concentrations (10–100) mg/ml were used. The resulting film thickness is fine tuned by adjusting the rotation speed between 2000 and 6000 rpm using a spin coater from Laurell Technologies (USA). The acceleration of the spin coater was always set at maximum and the spin coating time was about 120 s to make sure that the solvent evaporated during that time. The top PS films can not be spun directly onto the PMMA and are, in a first step, spun from toluene solution onto freshly cleaved mica sheets, following the same protocol as described for the PMMA film. In a second step, the glassy PS films are transferred from mica onto a Millipore™ water surface and picked up from above with the PMMA coated silicon substrates. During the transfer process, the initially closed PS film ruptures into patches which are transferred onto the PMMA substrate.

For our different experiments presented here, PS and PMMA of different molecular chain weights are used, purchased from Polymer Standard Service Mainz (PSS-Mainz,Germany). PS is used with molecular weights of $M_w = 9.6$ kg/ mol (PS(9.6k)) and $M_w = 64$ kg/mol (PS(64k)), with polydispersities of $M_w/M_n = 1.03$, and $M_w/M_n = 1.05$, respectively. The used PMMA had a molecular weight of $M_w = 9.9$ kg/ mol (PMMA(9.9k)) and a corresponding polydispersities of $M_w/M_n = 1.03$.

The glass transition temperatures of the polymers are $T_{g,PS(9.6k)} = 90 \pm 5\ °C$, $T_{g,PS(64k)} = 100 \pm 5\ °C$, and $T_{g,PMMA(9.9k)} = 115 \pm 5\ °C$. The dewetting experiments are typically conducted at a temperature of $T = 140\ °C$ resulting in PS viscosities of $\mu_{PS(9.6k)} \approx 2.5$ kPa s and $\mu_{PS(64k)} \approx 700$ kPa s and a PMMA viscosity of $\mu_{PMMA(9.9k)} \approx 675$ kPa s. The viscosity values are measured using the self-similarity in stepped polymer films as presented in [38, 39]. The measured values are in good agreement with viscosities extracted from [40, 41] of $\mu_{PS(9.6k)} \approx 2$ kPa s and $\mu_{PMMA(9.9k)} = 675$ kPa s, respectively. For the numerical calculations of transient droplet profiles we used the viscosities measured by us. The liquid/liquid dewetting process is started by heating the sample above the glass transition temperature and monitored in situ by atomic force microscopy (AFM) at 140 °C in Fastscan Mode™ (Bruker, Germany). To additionally determine the shape of the liquid PS/PMMA interface, the dewetting process is stopped at a desired dewetting stage by quenching the sample from the dewetting temperature $T = 140\ °C$ down to room temperature. At room temperature both polymers are glassy and the sample can be easily stored and handled. Subsequently the glassy PS structures are removed

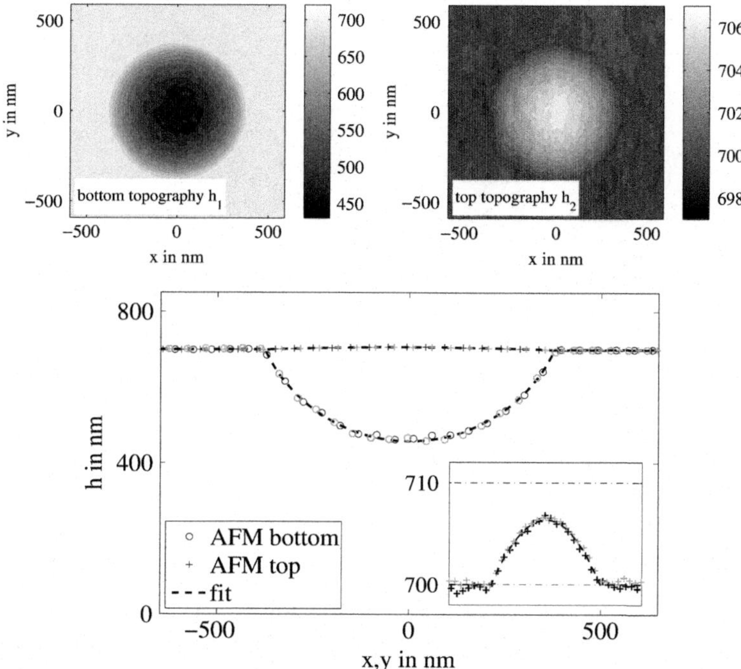

Fig. 18.4 Profile of an equilibrated PS(9.6k) drop swimming on a 700 nm PMMA(9.9k) substrate as determined by AFM. *Top:* The height scales in nanometres are shown to the right of each panel. (*left*) bottom profile h_1 scanned at room temperature and (*right*) top profile h_2 scanned at dewetting temperature. *Bottom:* Two cross sections cut perpendicular through the droplet in x-direction (*dark symbols*) and y-direction (*light symbols*) shown together with the fit using Eq. (18.70) (*dashed line*) in a 1:1 scaling. The inset shows a close up of the top AFM topography with spherical fit. The initially prepared PS and PMMA film thicknesses are 20 and 700 nm, respectively

by a selective solvent (cyclohexane, Sigma Aldrich, Germany) and the formerly PS/PMMA interface is imaged by AFM. The full three dimensional shape of the dewetting PS structures are obtained by combining the subsequently imaged PS/air and PS/PMMA surfaces. The protocol was carefully tested and evaluated to yield accurate results, as described in detail in [42]. To obtain series of such $3D$ snapshots at different times multiple samples with identical film heights are prepared, each stopped at a different dewetting state and imaged by the above described protocol.

An example for an equilibrium $3d$ drop shape obtained by the above described protocol is shown in Fig. 18.4. Using the three dimensional PS drop profiles we derived the contact angles and the surface tensions from the equilibrium shapes. These values will serve as input parameters for the simulation of transient droplet morphologies. The corresponding values found in literature [43, 44] are not precise enough and do not provide conclusive predictions on the sign of the spreading coefficient σ. The analysis of the experimental equilibrium drop shapes is given in [42] and leads to the expression for the Neumann-triangle [45], a local condition

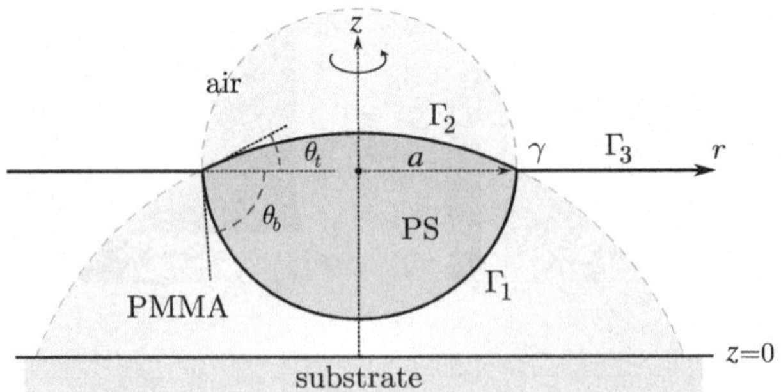

Fig. 18.5 Sketch of an axisymmetric equilibrium droplet

stating that contact angles fulfil the condition

$$\sum_{\alpha=1}^{3} \sigma_\alpha \mathbf{n}_{\Gamma_\alpha} = 0, \tag{18.68}$$

and each interface Γ_α has constant mean curvature H_α. The normalised vector $\mathbf{n}_\alpha$ is tangential to the corresponding interface Γ_α and normal to the contact line γ as indicated in Fig. 18.5. For the planar axisymmetric droplets which we observe in the experiments we have $\mathbf{n}_{\Gamma_1} = -\mathbf{e}_r \cos\theta_b - \mathbf{e}_z \sin\theta_b$, $\mathbf{n}_{\Gamma_2} = -\mathbf{e}_r \cos\theta_t + \mathbf{e}_z \sin\theta_t$, $\mathbf{n}_{\Gamma_3} = \mathbf{e}_r$.

One can easily verify that equilibrium droplets as in Eq. (18.68) only exist if the spreading coefficient $\sigma < 0$ and $\sigma_1, \sigma_2 > -\sigma/2$. From a measured equilibrium configuration we can thus extract the values for the surface tensions from Eq. (18.68) using $\mathbf{n}_{\Gamma_i}$ as follows: If for instance σ_3 is given, then one can determine the other two surface tensions by plugging into solving the linear equation

$$\begin{pmatrix} \cos\theta_t & \cos\theta_b \\ -\sin\theta_t & \sin\theta_b \end{pmatrix} \begin{pmatrix} \sigma_2 \\ \sigma_1 \end{pmatrix} = \begin{pmatrix} \sigma_3 \\ 0 \end{pmatrix}, \tag{18.69}$$

where $\theta_t, \theta_b > 0$ are determined from experimental drop profiles. For contact angles $\theta_t, \theta_b \leq 90°$ a liquid lens has the following axisymmetric equilibrium shape

$$h_2(x, y) = h_\infty + \left(\sqrt{H_2^{-2} - r^2} - \sqrt{H_2^{-2} - a^2} \right), \tag{18.70a}$$

$$h_1(x, y) = h_\infty - \left(\sqrt{H_1^{-2} - r^2} - \sqrt{H_1^{-2} - a^2} \right), \tag{18.70b}$$

for $r \leq a$ and $h_1(x, y) = h_2(x, y) = h_\infty$ for $r > a$. We use the cylindrical coordinates with $r^2 = (x - x_0)^2 + (y - y_0)^2$ and call a the in plane radius of the droplet. A least-squares fit of (18.70) to the measured AFM profiles shown in Fig. 18.4 return the six parameters h_∞, H_1, H_2, a, x_0 and y_0.

Since both interfaces, i.e. h_1 and h_2 are measured independently and the AFM can only measure height differences, h_∞, x_0, y_0 have no absolute value, so one might define $x_0 = y_0 = 0$ and h_∞ as the values set by the preparation of the PMMA layer and as determined independently. Thus even though a is defined absolutely, due to experimental scatter, one finds slightly varying droplet radii a depending on the analysed AFM profiles h_1 or h_2 but which agree within the experimental resolution of ± 10 nm. Using the values for the constant curvatures H_α and the in-plane radius a the contact angles can be directly computed as r-derivatives of the interfaces h_1, h_2 in Eq. (18.70) at $r = a$:

$$\theta_b = \arctan\left(a / \sqrt{H_1^{-2} - a^2}\right), \tag{18.71a}$$

$$\theta_t = \arctan\left(a / \sqrt{H_2^{-2} - a^2}\right). \tag{18.71b}$$

Fitting spherical caps (18.70) to the top and bottom profiles of several droplets, see Fig. 18.4, we obtain a relationship between a and H_1, H_2, respectively which is shown in Fig. 18.6. For constant contact angles Eq. (18.71) suggest that the relationship between curvature and radius must be linear which is true within the accuracy of the experimental data. From the linear relationship between radius and curvature shown in Fig. 18.6 and using Eq. (18.71) we obtain for the top angle $\theta_t \approx (1.98 \pm 0.07)^\circ$ and for the bottom angle $\theta_b \approx (64 \pm 2)^\circ$ where the error is composed from the statistical error in the fit and the systematic error in

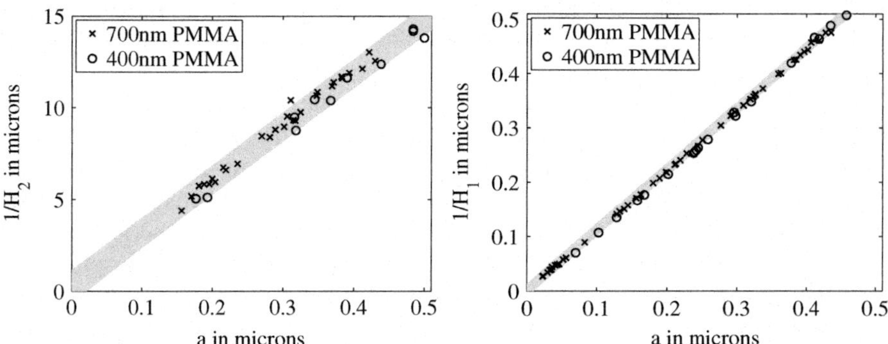

Fig. 18.6 Profiles of PS(9.6k) drops on PMMA(9.9k) substrates. *Top:* Curvature of the top spherical caps and *bottom:* curvature of the bottom spherical caps as a function of droplet radius a measured from equilibrium droplets on 400 nm thick PMMA substrates (*circles*) and 700 nm thick PMMA substrates (*crosses*) with linear fit (*shaded area* 95% confidence interval) gives $H_1^{-1} = (1.11 \pm 0.02)a$ and $H_2^{-1} = (29 \pm 1)a$ and corresponding contact angles $\theta_b = (64 \pm 2)^\circ$ and $\theta_t = (1.98 \pm 0.07)^\circ$

the determination of the droplet radius. Top spherical caps with $a < 150$ nm are not considered in Fig. 18.6 as the height of these drops $\lesssim 2$ nm is comparable to the roughness of the polymer layer. Using Eq. (18.69) we obtain the surface tensions of the PS(9.6k)/PMMA(9.9k) interfaces to

$$\sigma_1 = (0.038 \pm 0.002) \cdot \sigma_3 = (1.22 \pm 0.07) \text{ mN/ m},$$

and of the PS/air interface to

$$\sigma_2 = (0.984 \pm 0.001) \cdot \sigma_3 = (31.49 \pm 0.03) \text{ mN/ m},$$

based on the PMMA(9.9k)/air surface tension $\sigma_3 = 32$ mN/ m at $T = 140\,^{\circ}$C taken from [44]. The corresponding spreading coefficient is

$$\sigma = (-0.022 \pm 0.003)\sigma_3 = (-0.7 \pm 0.1) \text{ mN/ m}.$$

The surface tension for the PS(64k)/PMMA(9.9k) combination was determined similarly to $\sigma_\ell = 32.3$ mN/m [46], where the surface tension $\sigma_{\ell,s} = 1.22 \pm 0.07$ mN/m is unchanged. Note that these values and the modification of σ_s are compatible with the literature, e.g. [44].

18.4.1 Nonequilibrium Droplet Shapes

A purely experimental evaluation of the transient droplet shapes using AFM is limited as one can not continuously image the $3d$ top and bottom shape of a droplet on its journey into equilibrium. In particular the dependence on randomly shaped initial PS patches makes it difficult to describe and understand the morphological evolution on droplets systematically.

To address the question on the dependence of the evolving droplet shapes on the particular choice of the initial configuration theoretically, we choose as initial conditions different cylindrical PS patches of identical volume and fixed thickness of the underlying PMMA film. These patches are then followed numerically towards their respective equilibrium states. A typical example of a time series showing the evolution of the PS droplets with different initial conditions is displayed in Fig. 18.8 for different liquid PS volumes. The chosen initial data correspond to typical droplet volumes observed in our experiments. It is evident from Fig. 18.8 that the characteristic time scale for equilibration strongly depend on the PS volume. For the same dewetting time, a smaller droplet is closer to its equilibrium than a larger one. The results show that for the larger PS volumes (Fig. 18.8) the thicker PS patch quickly develops a characteristic droplet-like shape, with the PS/air interface having almost constant curvature. In contrast, the PS/PMMA interface shows characteristic deformations which are localised around the triple junction and which are evidently different from the equilibrium shape. As discussed before in dewetting experiments and numerical simulations in [17, 47], surface forces make it energetically favourable to pull the triple junction slightly upward. Note however that this upward-deformation has not been observed for any of the equilibrium states studied in the previous section and is characteristic for the transient nature of the droplet shape. During the further equilibration progress the footprint of the droplet is slightly reduced and the corrugations of the PS/PMMA interface grow in amplitude. The pronounced dents of the PS/PMMA interface finally meet each other forming a dome-like shape of the PS/PMMA interface curved towards the air phase, see $t = 3\,$h in the left column and $t = 5\,$min in the right column. Remarkably, the curvature of this dome is opposite to the equilibrium drop shape due to the flow squeezing out the PMMA under the droplet. Provided the thickness of the PMMA layer is sufficiently large, the dome-like shape is finally transferred into its equilibrium shape, i.e. a spherical cap curved towards the solid substrate. In case the PMMA film thickness is below this equilibrium penetration depth, the dome-like interface will flatten and touch the solid substrate $h_1 \to 0$ as $t \to \infty$, whereas the further equilibration is infinitely slow and self-similar in theory. This self-similar rupture $h_1 \to 0$ in infinite time has been discussed previously e.g. by Craster and Matar in [47].

When following the transient droplet shapes for the thinner PS patch (dashed lines in the left column of Fig. 18.8) it is evident that the evolution of the drop morphology rather starts from an axisymmetric rim growing inwards the centre of the patch at $r = 0$. The shape of the PS/PMMA interface that forms close to the triple junction is very similar to that for the thicker patch, whereas the PS/air interface develops differently. The initially corners of the PS/air interface are rounded and develop a characteristic profile which are similar to dewetting rim profiles. The initially prepared film thickness remains constant in the centre of the patch until the rim profiles merge and form a drop like profile similar to that of the thick PS patch.

Surprisingly, the transient drop morphologies for a fixed volume and different start configuration synchronise after a certain time and cannot be distinguished any more on their further way into equilibrium. In the examples presented in Fig. 18.8, the synchronisation occurs after about 45 min for the larger PS volume whereas the synchronisation already occurs after about 1 min for the smaller PS volume. For times larger than the synchronisation time the transient droplet morphologies are independent from the specific initial configuration. Moreover, smaller droplets rather have the chance to develop the typical stationary lens shape not touching the underlying substrate. The general behaviour that drops synchronise onto their way towards equilibrium however is not affected by the PS volume.

Finally, the experimentally obtained drop shapes shall be compared to the theoretical predictions. Similar as in the simulations, early stages of droplet-like configurations observed in experiments will depend on the history of the dewetting process and the initial shape of the PS patches and give rise to complex intermediate states, e.g. see Fig. 18.7. Later on, the experimentally observed droplets become axisymmetric with their specific shape independent from that history— at this point the shape is mostly determined by the droplet volume and the total

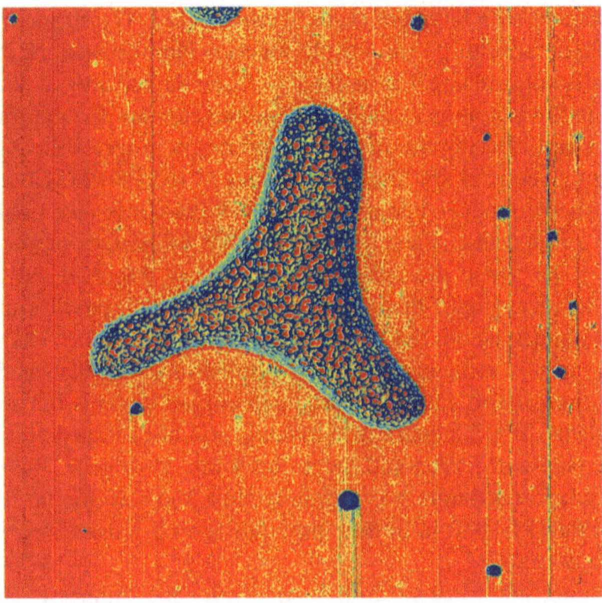

Fig. 18.7 AFM measurement (phase signal) of flower-shaped droplet at early times, where the shape is not yet axisymmetric and depends on the history of the dewetting process

dewetting time. At this point a comparison with simulations makes sense. In the right column of Fig. 18.8 experimentally obtained drop shapes are displayed on top of the theoretical drop shapes for identical volumes after 45 min of dewetting. A visual inspection reveals good agreement of the characteristic morphologies of the transient drop shapes and the time-scales which also emphasises the quality of the (Newtonian) viscosity and the surface tension data. The good agreement between the experimentally determined transient drop shapes and dewetting times indicate moreover that the exact details of the contact angle are not crucial for the drop shape and can be captured precisely by the used thin film model.

18.4.2 Dewetting Rim Profiles

Being able to accurately describe the comparably slow transient drop shapes, we will extend our comparison in the following to the much faster transient rim shapes and their dewetting dynamics. For these experiments the combination of PMMA(9.9k) as liquid substrate and PS(64k) as dewetting liquid is used having about equal viscosities of $\mu \approx 700\,\text{kPa}\,\text{s}$. The largest Weissenberg number $\text{Wi} = \tau\dot{\gamma}$ for the system is calculated, where $\tau = \mu/G$ is the relaxation time with $G = 0.2\,\text{MPa}\,\text{s}$ [48] being the shear modulus of PS. The maximal shear stress for the initial stage of the dewetting is $\dot{\gamma} = 0.05\,\text{s}^{-1}$ as extracted from the numerical simulations. Taking

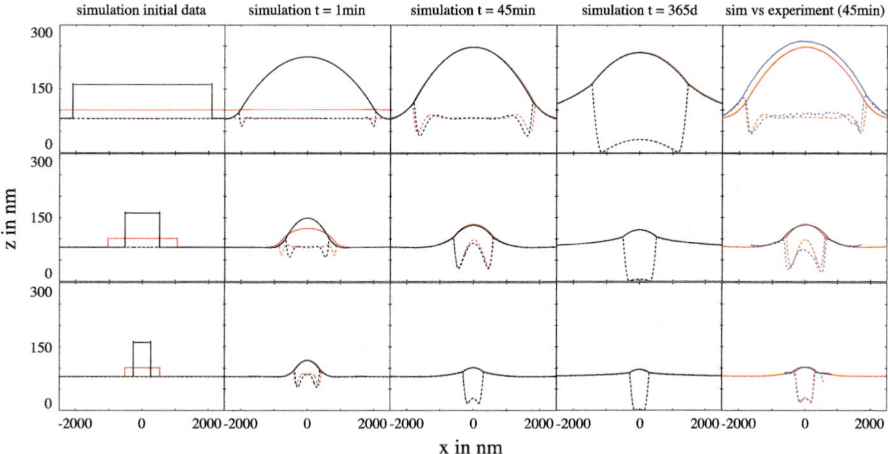

Fig. 18.8 Approach to equilibrium for three different droplet volumes (*top to bottom row*) and two different initial conditions (*red and black*). Initial data in simulations are $h_1(r,0) = \bar{h}_1$ and $h(r,0) = \bar{h}$ for $r < r_0$ and $h(r,0) \approx 0$ (precursor) for $r \geq r_0$ such that the volume $\pi r_0^2 \bar{h}$ matches the experiment. The *right column* compares simulation (*red*) to AFM measurements (*blue*) after 45 min dewetting

all this into account we obtain a Weissenberg number $\mathrm{Wi} = 6.25 \times 10^{-3} \ll 1$ and we can safely assume the polymer as purely Newtonian [49].

While most of the experimental parameters are known with an uncertainty fewer than 4%, the viscosity of the used polymers is the main source of uncertainty with the main effect on the timescale of experiment and simulation. Matching experimental and numerical timescales quantitatively by fitting the experimental contact line dynamics, i.e. x_c as a function of time, cf. Fig. 18.9, we obtain a numerical viscosity $\mu_\ell = 1100\,\mathrm{kPa\,s}$ for both PMMA(9.9k) and PS(64k), which is within experimental accuracy. This viscosity value can be used to quantitatively match experimental and theoretical results for all film thickness ratios and absolute film thicknesses, which are obtained for the same system and at the same temperature. The theoretical prediction that for a fixed film thickness ratio the influence of the absolute height scales linearly is experimentally confirmed by two samples with aspect ratio 1:1 but film thicknesses $\bar{h} \approx 100\,\mathrm{nm}$ and $\bar{h} \approx 240\,\mathrm{nm}$. The dewetting rate appears linear $x_c \sim t$, however, there is no theoretical indication that for aspect ratios and viscosity ratios of order one there should be a power-law dewetting rate. Indeed, further analysis in [46] proves that the velocity slowly decreases over time with transient rates depending on the aspect ratio at hand. This finding confirms previous speculations by Lambooy et al. [50] about the transient nature of the observed dewetting dynamics.

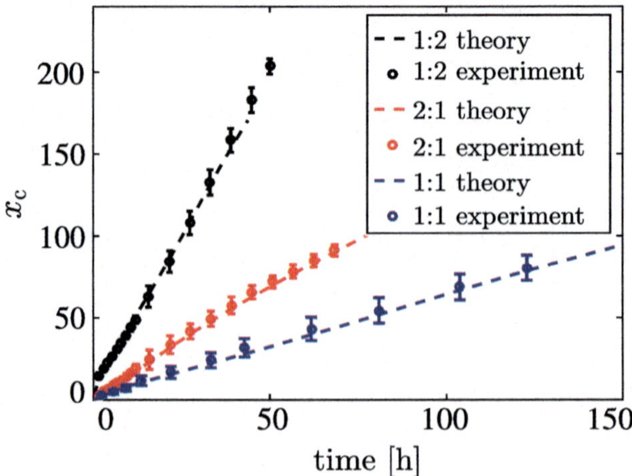

Fig. 18.9 Dewetted distance x_c for aspect ratios 1:1 (240 nm:240 nm), 2:1 (90 nm:45 nm), 1:2 (45 nm:90 nm) from experiment (*circles* with error) and simulation (*dashed lines*)

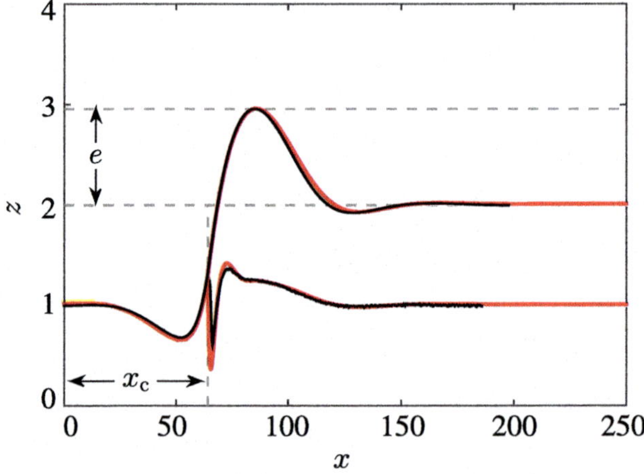

Fig. 18.10 Interfaces from theory (*red*) and experiment (*black*) for an aspect ratio of 1:1 and absolute film thicknesses of 240 μm. The experimental cross section is averaged over 30 scan lines of a straight front

In Fig. 18.10 we show the almost perfect alignment of the experimentally measured and theoretically computed interface profiles at identical dewetting times, for equal PMMA and PS film height. The contact line of the dewetting profile is elevated by the flow, a dynamic feature not observed in stationary droplets for sufficiently thick substrates, cf. Figs. 18.4 and 18.6. The material of the dewetting liquid (PS) accumulates in a rim which, by conservation of mass, grows in time when the liquid retracts from the substrate, cf. Fig. 18.10. Also some material of the substrate (PMMA) is dragged along generating a depletion on the "dewetted side" near the three phase contact line $x < x_c$, and an accumulation of substrate material at the "film side" near the three phase contact line $x > x_c$. Right next to the contact line, some part of the dewetting liquid extends deeply into the substrate and generates a trench and thereby produces additional resistance against the dewetting motion. Note that the size of the trench does not or only weakly depends on the size of the dewetting rim. An equally good agreement between experimental and theoretical rim profiles are obtained also for other film thickness ratios, which can be found in cf. [46]. The only influence of the film thickness ratio on the rim profiles is that for thicker substrates the described features grow, while they shrink for thinner substrates. Away from the rim the interfaces decay in an oscillatory fashion into their prepared constant states $h_1(t, x), h(t, x) \rightarrow \bar{h}_1, \bar{h}$.

We could thus show via quantitative comparisons with experimental results that the thin-film model accurately predicts not only dewetting speeds but also rime shapes of the liquid-liquid dewetting in case of Newtonian liquids and which obey a no-slip boundary condition.

18.5 Role of Interfacial Slip

It has been shown that a for polymer films such as PS that dewets from a substrate coated with a hydrophobic molecular monolayer of grafted polymer chains, the dewetting dynamics may exhibit large "apparent" slip [51, 52]. This has been associated with a distinct motion of the polymer chains within a thin region near the boundary of the substrate, as has been argued in Brochard and De Gennes [53], where they showed that for entangled polymer melts an "apparent" slip length b can be related to a microscopic coil-stretch transition into a disentangled state within a thin boundary layer, where the viscosity is much lower. The effect of such an "apparent" slip was investigated in [54, 55], where they showed that slip can control the morphology, dynamics and stability of the system.

For liquid-liquid systems Lin [56] suggested the possibility of interfacial slip. Experimental evidence of slip at polymer-polymer interfaces was given in Zhao and Macosko [10], who investigated PS on PMMA interfaces. A microscopic theory for immiscible blends was developed in Brochard-Wyart and De Gennes [57] and Ajdari [58], and extended by Goveas and Fredrickson [8] and Adhikari and Goveas [9], investigating entangled, unentangled polymer melts as well as polymer emulsions. In particular, expressions for the interfacial viscosity based on the appropriate chain dynamics in this region were derived and the ratio of the bulk and interfacial viscosity was then related to the size of an "apparent" slip length. In summary one can conclude that for two-layer immiscible polymer films the higher shear rate within a thin interfacial region and the associated interfacial viscosity introduces an apparent velocity discontinuity leading to the concept of "apparent" slip.

In the article [59] we have derived thin-film models for the polymer-polymer-solid substrate system and taking account of slip at the solid-polymer as well as the polymer-polymer interfaces. There are a number of cases to consider, such as weak-slip at the polymer-solid interface and weak-slip at the polymer-polymer interface or the case where we assume strong-slip at both interfaces. Then there are also mixed cases and limiting intermediate-slip cases.

18.5.1 The Strong-Slip Case

As we learn from the derivations of single-layer thin film models (see [55]), another distinguished limit for the slip lengths is the order $O(\varepsilon_\ell^{-2})$. Therefore, we consider slip parameters at the solid-liquid and liquid-liquid interface of the form,

$$b = \frac{\beta}{\varepsilon_\ell^2}, \qquad b_1 = \frac{\beta_1}{\varepsilon_\ell^2}, \tag{18.72}$$

where β and β_1 are of order $O(1)$. Again, we are motivated by the derivations in [55] and assume plug-flow for the vertical velocity component in layer 1. This leads to

$$\alpha = \varepsilon_\ell. \tag{18.73}$$

Hence, the capillary number is $\mathrm{Ca} = \varepsilon_\ell$. We also introduce the reduced Reynolds number Re^* by

$$\mathrm{Re} = \frac{\rho_2 U H}{\mu_2} = \varepsilon_\ell \frac{\rho_2 \sigma_2 H}{\mu_2^2} = \varepsilon_\ell \mathrm{Re}^*. \tag{18.74}$$

We note that the derivation of the thin-film model for the strong slip case involves also the next-to-leading order in the expansions of the variables in order to obtain a closed model.

18.5.1.1 Leading Order Problem

The leading order bulk equations for layer 1 are given by

$$0 = \partial_{zz} u_1^{(0)}, \quad 0 = -\partial_z p_1^{(0)} + \mu \partial_{zz} w_1^{(0)}, \quad 0 = \partial_x u_1^{(0)} + \partial_z w_1^{(0)}, \tag{18.75}$$

and for layer 2 they read

$$0 = \partial_{zz} u_2^{(0)}, \quad 0 = -\partial_z p_2^{(0)} + \partial_{zz} w_2^{(0)}, \quad 0 = \partial_x u_2^{(0)} + \partial_z w_2^{(0)}. \tag{18.76}$$

For the boundary conditions at the substrate, $z = 0$, we obtain

$$\partial_z u_1^{(0)} = 0, \qquad w_1^{(0)} = 0. \tag{18.77}$$

At the liquid-liquid interface $z = h_1^{(0)}$, normal stress, tangential stress and kinematic condition become

$$p_1^{(0)} - p_2^{(0)} + \phi'(h^{(0)}) + \sigma\,\partial_{xx}h_1^{(0)}$$
$$- 2\left(\left(\mu\partial_z w_1^{(0)} - \partial_z w_2^{(0)}\right) - \left(\mu\partial_z u_1^{(0)} - \partial_z u_2^{(0)}\right)\partial_x h_1^{(0)}\right) = 0, \tag{18.78}$$

$$\partial_z\left(\mu u_1^{(0)} - u_2^{(0)}\right) = 0, \tag{18.79}$$

$$\partial_t h_1^{(0)} = w_1^{(0)} - u_1^{(0)}\partial_x h_1^{(0)}. \tag{18.80}$$

The slip condition and the impermeability condition at this interface are given by

$$\partial_z u_2^{(0)} = 0, \qquad \left(w_2^{(0)} - w_1^{(0)}\right) - \left(u_2^{(0)} - u_1^{(0)}\right)\partial_x h_1^{(0)} = 0. \tag{18.81}$$

At the free surface $z = h_2^{(0)}$ we get for the normal, tangential and kinematic condition,

$$p_2^{(0)} - \phi'(h^{(0)}) + \partial_{xx}h_2^{(0)} - 2\left(\partial_z w_2^{(0)} - \partial_z u_2^{(0)}\,\partial_x h_2^{(0)}\right) = 0, \tag{18.82}$$

$$\partial_z u_2^{(0)} = 0, \tag{18.83}$$

$$\partial_t h_2^{(0)} = w_2^{(0)} - u_2^{(0)}\partial_x h_2^{(0)}. \tag{18.84}$$

We observe that the statement

$$u_1^{(0)} = u_1^{(0)}(x, t), \qquad u_2^{(0)} = u_2^{(0)}(x, t), \tag{18.85}$$

results from the first equations in (18.75), (18.76) and boundary conditions (18.77), (18.83). That means that the horizontal velocity components are independent of z. Using this, the continuity equations in (18.75), (18.76) and the impermeability conditions in (18.77), (18.81), we find

$$w_1^{(0)} = -z\partial_x u_1^{(0)}, \tag{18.86}$$

$$w_2^{(0)} = -\left(z - h_1^{(0)}\right)\partial_x u_2^{(0)} - \partial_x u_1^{(0)}h_1^{(0)} + (u_2^{(0)} - u_1^{(0)})\partial_x h_1^{(0)}. \tag{18.87}$$

Combining the second equations in (18.75) and (18.76) with (18.86) we see that the leading order pressures are also independent of z, i.e.

$$p_2^{(0)} = p_2^{(0)}(x, t), \qquad p_1^{(0)} = p_1^{(0)}(x, t). \tag{18.88}$$

Thus we can rewrite the normal stress conditions as

$$p_1^{(0)} = -(\sigma + 1)\partial_{xx}h_1^{(0)} - \partial_{xx}h^{(0)} - 2\mu\partial_x u_1^{(0)}, \tag{18.89}$$

$$p_2^{(0)} = -\partial_{xx}h_1^{(0)} - \partial_{xx}h^{(0)} + \phi'\left(h^{(0)}\right) - 2\partial_x u_2^{(0)}. \tag{18.90}$$

To obtain the latter (18.86) is used, too. Next, we derive equations for the thicknesses $h_1^{(0)}$ and $h^{(0)}$ from the leading order kinematic boundary conditions (18.80), (18.84) and formulas (18.86),

$$\partial_t h_1^{(0)} = -\partial_x \left(u_1^{(0)} h_1^{(0)}\right), \qquad \partial_t h^{(0)} = -\partial_x \left(u_2^{(0)} h^{(0)}\right). \tag{18.91}$$

In contrast to the weak-slip case, we cannot deduce closed forms for $u_1^{(0)}$ and $u_2^{(0)}$ from the leading order system. Therefore we have to look at the next order.

18.5.1.2 Next Order Problem

Here, we only state the equations which are necessary in order to fix $u_1^{(0)}$ and $u_2^{(0)}$, and neglect the complete next order problem.

We start with the next order equations in the bulk,

$$\rho \mathrm{Re}^* \left(\partial_t u_1^{(0)} + u_1^{(0)}\partial_x u_1^{(0)}\right) = -\partial_x p_1^{(0)} + \mu\partial_{xx}u_1^{(0)} + \mu\partial_{zz}u_1^{(1)}, \tag{18.92}$$

$$\rho \mathrm{Re}^* \left(\partial_t w_1^{(0)} + u_1^{(0)}\partial_x w_1^{(0)} + w_1^{(0)}\partial_z w_1^{(0)}\right) = -\partial_z p_1^{(1)} + \mu\partial_{xx}w_1^{(0)} + \mu\partial_{zz}w_1^{(1)}, \tag{18.93}$$

$$0 = \partial_x u_1^{(1)} + \partial_z w_1^{(1)}, \tag{18.94}$$

$$\mathrm{Re}^* \left(\partial_t u_2^{(0)} + u_2^{(0)}\partial_x u_2^{(0)}\right) = -\partial_x p_2^{(0)} + \partial_{xx}u_2^{(0)} + \partial_{zz}u_2^{(1)}, \tag{18.95}$$

$$\mathrm{Re}^* \left(\partial_t w_2^{(0)} + u_2^{(0)}\partial_x w_2^{(0)} + w_2^{(0)}\partial_z w_2^{(0)}\right) = -\partial_z p_2^{(1)} + \partial_{xx}w_2^{(0)} + \partial_{zz}w_2^{(1)}, \tag{18.96}$$

$$0 = \partial_x u_2^{(1)} + \partial_z w_2^{(1)}. \tag{18.97}$$

Moreover, we consider the next order of the slip conditions, both at the solid-liquid interface, $z = 0$,

$$u_1^{(0)} = \beta\, \partial_z u_1^{(1)}, \tag{18.98}$$

and at the liquid-liquid interface, $z = h_1^{(0)}$,

$$u_2^{(0)} - u_1^{(0)} = \beta_1 \frac{\mu + 1}{\mu} \left(\partial_z u_2^{(1)} + \partial_x w_2^{(0)} - 4\partial_x u_2^{(0)} \partial_x h_1^{(0)} \right). \tag{18.99}$$

We also make use of the next order tangential stress boundary conditions at liquid-liquid and the liquid-gas interface,

$$\partial_z \left(\mu u_1^{(1)} - u_2^{(1)} \right) + \partial_x \left(\mu w_1^{(0)} - w_2^{(0)} \right) - 4\partial_x \left(\mu u_1^{(0)} - u_2^{(0)} \right) \partial_x h_1^{(0)} = 0, \tag{18.100}$$

$$\partial_z u_2^{(1)} + \partial_x w_2^{(0)} - 4\partial_x u_2^{(0)} \partial_x h_2^{(0)} = 0. \tag{18.101}$$

Notice, in the equations above we have already used that the z-derivatives of $u_1^{(0)}$ and $u_2^{(0)}$ vanish. When we integrate (18.92) and (18.95) w.r.t. z, we obtain

$$\rho \mathrm{Re}^* h_1^{(0)} \left(\partial_t u_1^{(0)} + u_1^{(0)} \partial_x u_1^{(0)} \right) = h_1^{(0)} \left(-\partial_x p_1^{(0)} + \mu \partial_{xx} u_1^{(0)} \right) + \left(\mu \partial_z u_1^{(1)} \right) \Big|_0^{h_1^{(0)}} \tag{18.102}$$

and, similarly,

$$\mathrm{Re}^* h^{(0)} \left(\partial_t u_2^{(0)} + u_2^{(0)} \partial_x u_2^{(0)} \right) = h^{(0)} \left(-\partial_x p_2^{(0)} + \partial_{xx} u_2^{(0)} \right) + \left(\partial_z u_2^{(1)} \right) \Big|_{h_1^{(0)}}^{h_2^{(0)}}. \tag{18.103}$$

Combining (18.89) and (18.98)–(18.103) leads to

$$\rho \mathrm{Re}^* \left(\partial_t u_1^{(0)} + u_1^{(0)} \partial_x u_1^{(0)} \right) = -\partial_x(-(\sigma + 1)\partial_{xx} h_1^{(0)} - \partial_{xx} h^{(0)})$$

$$+ \frac{4\mu}{h_1^{(0)}} \partial_x(\partial_x u_1^{(0)} h_1^{(0)}) + \frac{\mu(u_2^{(0)} - u_1^{(0)})}{(\mu + 1)\beta_1 h_1^{(0)}} - \frac{\mu u_1^{(0)}}{\beta h_1^{(0)}}, \tag{18.104}$$

$$\mathrm{Re}^* \left(\partial_t u_2^{(0)} + u_2^{(0)} \partial_x u_2^{(0)} \right) = -\partial_x(-\partial_{xx} h_1^{(0)} - \partial_{xx} h^{(0)} + \phi'(h^{(0)}))$$

$$+ \frac{4}{h^{(0)}} \partial_x(\partial_x u_2^{(0)} h^{(0)}) - \frac{\mu(u_2^{(0)} - u_1^{(0)})}{(\mu + 1)\beta_1 h^{(0)}}. \tag{18.105}$$

Recalling (18.91), the full model for the leading order velocity fields the leading order layer thicknesses is given by,

$$\rho \mathrm{Re}^* (\partial_t u_1 + u_1 \partial_x u_1) = -\partial_x(-(\sigma + 1)\partial_{xx} h_1 - \partial_{xx} h)$$

$$+ \frac{4\mu}{h_1} \partial_x(\partial_x u_1 h_1) + \frac{\mu(u_2 - u_1)}{(\mu + 1)\beta_1 h_1} - \frac{\mu u_1}{\beta h_1}, \tag{18.106}$$

$$\partial_t h_1 = -\partial_x(h_1 u_1), \tag{18.107}$$

$$\mathrm{Re}^* \left(\partial_t u_2 + u_2 \partial_x u_2\right) = -\partial_x(-\partial_{xx} h_1 - \partial_{xx} h + \phi'(h))$$
$$+\frac{4}{h}\partial_x(\partial_x u_2 h) - \frac{\mu(u_2 - u_1)}{(\mu + 1)\beta_1 h}, \tag{18.108}$$

$$\partial_t h = -\partial_x(h u_2), \tag{18.109}$$

where we drop the '(0)'.

In many applications, e.g. dewetting of micro- and nanoscopic polymer films, inertia are negligibly small. Therefore, we assume $\mathrm{Re}^* = 0$ in the following. Then, (18.106) reads

$$0 = -\partial_x(-(\sigma + 1)\partial_{xx} h_1 - \partial_{xx} h)$$
$$+\frac{4\mu}{h_1}\partial_x(\partial_x u_1 h_1) + \frac{\mu(u_2 - u_1)}{(\mu + 1)\beta_1 h_1} - \frac{\mu u_1}{\beta h_1}, \tag{18.110}$$

$$\partial_t h_1 = -\partial_x(h_1 u_1), \tag{18.111}$$

$$0 = -\partial_x(-\partial_{xx} h_1 - \partial_{xx} h + \phi'(h)) + \frac{4}{h}\partial_x(\partial_x u_2 h) - \frac{\mu(u_2 - u_1)}{(\mu + 1)\beta_1 h}, \tag{18.112}$$

$$\partial_t h = -\partial_x(h u_2). \tag{18.113}$$

We call (18.110) strong-slip model.

18.5.2 The Intermediate-Slip Case

We consider the limits $\beta, \beta_1 \to 0$ in (18.110) by introducing the scaling

$$u_1 = \beta\, \tilde{u}_1, \quad u_2 = \beta\, \tilde{u}_2, \quad t = \frac{\tilde{t}}{\beta}. \tag{18.114}$$

We obtain

$$0 = -\partial_x(-(\sigma + 1)\partial_{xx} h_1 - \partial_{xx} h)$$
$$+\frac{4\mu\beta}{h_1}\partial_x(\partial_x \tilde{u}_1 h_1) + \frac{\mu\beta(\tilde{u}_2 - \tilde{u}_1)}{(\mu + 1)\beta_1 h_1} - \frac{\mu\tilde{u}_1}{h_1}, \tag{18.115}$$

$$\partial_{\tilde{t}} h_1 = -\partial_x(h_1 \tilde{u}_1), \tag{18.116}$$

$$0 = -\partial_x(-\partial_{xx} h_1 - \partial_{xx} h + \phi'(h)) + \frac{4\beta}{h}\partial_x(\partial_x \tilde{u}_2 h) - \frac{\mu\beta(\tilde{u}_2 - \tilde{u}_1)}{(\mu + 1)\beta_1 h}, \tag{18.117}$$

$$\partial_{\tilde{t}} h = -\partial_x(h \tilde{u}_2). \tag{18.118}$$

Now, let β and β_1 be of order $O(\varepsilon_\ell)$, i.e.

$$\beta = \varepsilon_\ell \tilde{\beta}, \qquad \beta_1 = \varepsilon_\ell \tilde{\beta}_1, \tag{18.119}$$

with $\tilde{\beta}$ and $\tilde{\beta}_1$ order one. Than, the leading order in (18.115) is

$$0 = -\partial_x(-(\sigma + 1)\partial_{xx}h_1 - \partial_{xx}h) + \frac{\mu\tilde{\beta}(\tilde{u}_2 - \tilde{u}_1)}{(\mu + 1)\tilde{\beta}_1 h_1} - \frac{\mu\tilde{u}_1}{h_1}, \tag{18.120}$$

$$\partial_{\tilde{t}}h_1 = -\partial_x(h_1\tilde{u}_1), \tag{18.121}$$

$$0 = -\partial_x(-\partial_{xx}h_1 - \partial_{xx}h + \phi'(h)) - \frac{\mu\tilde{\beta}(\tilde{u}_2 - \tilde{u}_1)}{(\mu + 1)\tilde{\beta}_1 h}, \tag{18.122}$$

$$\partial_{\tilde{t}}h = -\partial_x(h\tilde{u}_2). \tag{18.123}$$

Solving (18.120) and (18.122) for $\tilde{u}_1$ and $\tilde{u}_2$ and plugging the result into (18.121) and (18.123), we obtain (dropping '$\sim$'):

$$\partial_t h_1 = \partial_x\left(M_{11}\partial_x p_1 + M_{22}\partial_x p_2\right), \tag{18.124}$$

$$\partial_t h = \partial_x\left(M_{21}\partial_x p_1 + M_{22}\partial_x p_2\right), \tag{18.125}$$

with the mobility matrix,

$$M = \frac{1}{\mu}\begin{pmatrix} \beta\, h_1^2 & \beta\, h_1 h \\ \beta\, h_1 h & (\beta + (\mu + 1)\,\beta_1)\, h^2 \end{pmatrix}, \tag{18.126}$$

and the pressures

$$p_1 = -(\sigma + 1)\partial_{xx}h_1 + \partial_{xx}h, \tag{18.127}$$

$$p_2 = -\partial_{xx}h_1 - \partial_{xx}h + \phi'(h). \tag{18.128}$$

We refer to (18.124) as intermediate-slip model. Notice, rescaling the time by

$$t = \frac{\mu}{\beta}\tilde{t}, \tag{18.129}$$

we can write the mobility matrix M as

$$M = \begin{pmatrix} h_1^2 & h_1 h \\ h_1 h & (1 + \beta_3)\, h^2 \end{pmatrix}, \quad \text{where} \quad \beta_3 := \frac{(\mu + 1)\,\beta_1}{\beta}. \tag{18.130}$$

18.5.3   Linear Stability: Influence of Slip

We considered linear stability about the flat states to investigate the spinodal wavelength of the unstable modes.

A rather complex scenario arises, where dispersion curves show transitions from dominant long-wave zig-zag modes to shorter wave varicose modes, depending on the relative thicknesses, viscosities, surface tensions of the layers. What is most interesting is that the presence of interfacial slip can completely change the transitions and the wavelengths of the unstable modes, and hence needs to be accounted for when interpreting experimental results. As an example we show two dispersion relations that demonstrate the impact of slip, from weak to strong, on the spinodal wavelength.

Consider the cases of strong slip at solid/polymer interface, with slip length β_1 and strong slip at polymer-polymer interface, with slip length β_2. For the new model

$$\partial_t h = -\partial_x[h u_2], \qquad \partial_t h_1 = -\partial_x(h_1 u_1),$$

$$0 = -\partial_x(\phi'(h) - \partial_{xx}h_1 - \partial_{xx}h) + \frac{4}{h}\partial_x[\partial_x u_2(h)] - \frac{\mu(u_2 - u_1)}{(\mu + 1)\beta_2 h},$$

$$0 = -\partial_x(-(\sigma + 1)\partial_{xx}h_1 - \partial_{xx}h) + \frac{4\mu}{h_1}\partial_x(\partial_x u_1 h_1) + \frac{\mu(u_2 - u_1)}{(\mu + 1)\beta_2 h_1} - \frac{\mu\,u_1}{\beta_1 h_1},$$

we derived the dispersion relation:

$$\omega_{1,2} = -\frac{k^2}{2}Tr(\bar{Q}\cdot E) \pm k^2\sqrt{\frac{Tr(\bar{Q}\cdot E)^2}{4} - Det(\bar{Q}\cdot E)}, \qquad \bar{Q} = -Q_1 T^{-1} Q_2$$

$$Q_1 = \begin{bmatrix} h_1 & 0 \\ 0 & h \end{bmatrix}, \qquad Q_2 = \begin{bmatrix} \beta_1\beta h_1 & 0 \\ 0 & \beta h \end{bmatrix}, \qquad E = \begin{bmatrix} (\sigma + 1)k^2 & k^2 \\ k^2 & k^2 + \phi_h \end{bmatrix}$$

where $\beta = (1 + \mu)\beta_2$. This we compare to the case for weak slip at both interfaces in Fig. 18.11 below, with dispersion relation

$$\omega_{1,2} = -\frac{k^2}{2}Tr(\bar{Q}\cdot E) \pm k^2\sqrt{\frac{Tr(\bar{Q}E)^2}{4} - Det(\bar{Q}E)}$$

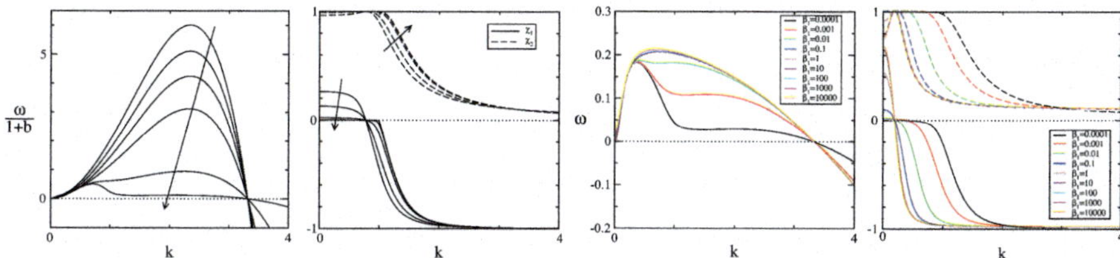

Fig. 18.11 For $h_1 = 10$: Dispersion relations and components of the perturbation vector at both interfaces. For weak-slip with slip length $b = 0, \ldots, 10^4$ (*left*, rescaled by $1 + b$). Arrows point to increasing b. For the strong-slip model for $\beta = 10$, for $\beta_1 = 10^{-4}, \ldots, 10^4$ (*right*)

18.6 The Viscoelastic Case

In this case we consider neither slip nor inertia. Hence, we set $b_1 = b = 0$ as well as $\mathrm{Re} = 0$. Balancing the terms in the stress force conditions yields

$$\frac{\mu_2 U}{\sigma_2} = \varepsilon_\ell, \tag{18.131}$$

and therefore $\alpha = \varepsilon_\ell$. Furthermore we assume

$$\mu := \frac{\mu_1}{\mu_2}\varepsilon^{-2} = O(1), \tag{18.132}$$

keeping the ratio of the surface tensions $\sigma := \sigma_1/\sigma_2 = O(1)$.

We will now show that to leading order in ε the free boundary problem can be integrated and reduced to a system of coupled partial differential equations for the height h, h_1, u_2 and S. To leading order the equations in the bulk of the lower liquid and in the upper liquid are

$$\partial_x u_1 + \partial_z w_1 = 0, \qquad 0 = -\partial_x p_1 + \mu \partial_{zz} u_1, \qquad 0 = -\partial_z p_1 \tag{18.133}$$

and

$$\partial_x u_2 + \partial_z w_2 = 0, \qquad 0 = \partial_z \tau_{2,12}, \qquad 0 = -\partial_z p_2 + \partial_x \tau_{2,12} + \partial_z \tau_{2,22}, \tag{18.134}$$

where

$$\lambda_{21}\partial_z u_2 \tau_{2,12} = \lambda_{22} \left(\partial_z u_2\right)^2 \tag{18.135}$$

and

$$\left(1 + \lambda_{21}\frac{d}{dt}\right)\tau_{2,12} + \frac{\lambda_{21}}{2}\partial_z u_2(\tau_{2,11} - \tau_{2,22}) = \left(1 + \lambda_{22}\frac{d}{dt}\right)\partial_z u_2 + 2\lambda_{22}\partial_z u_2 \partial_x u_2. \tag{18.136}$$

The boundary conditions are

$$u_1 = 0, \qquad w_1 = 0 \tag{18.137}$$

at $z = 0$,

$$\partial_t h_1 = w_1 - \partial_x h_1 u_1, \tag{18.138}$$

$$\tau_{2,12} = 0, \qquad -p_1 + p_2 - \phi'(h) - \tau_{2,22} = \sigma \partial_{xx} h_1, \tag{18.139}$$

$$u_2 - u_1 = 0, \qquad -(u_2 - u_1)\partial_x h_1 + (w_2 - w_1) = 0, \tag{18.140}$$

at $z = h_1$ and

$$\partial_t h_2 = w_2 - \partial_x h_2 u_2, \tag{18.141}$$

$$\tau_{2,12} = 0, \qquad -p_2 + \phi'(h) + \tau_{2,22} = \partial_{xx} h_2 \tag{18.142}$$

at $z = h_2$.

In order to obtain a closed set of equation we also need to account for the relations for the stress tensor from the next order problem, where we have expanded the variables u_i, w_i, p_i, $\tau_{i,jk}$ with $i, j, k \in 1, 2$ as $u_i = u_i^{(0)} + \varepsilon^2 u_i^{(1)} + O(\varepsilon^4)$ and likewise with the other variables. For ease of notation we then dropped the index $^{(0)}$ from the leading order variables. The relations we need are,

$$0 = -\partial_x p_2 + \partial_x \tau_{2,11} + \partial_z \tau_{2,12}^{(1)}, \tag{18.143}$$

$$\left(1 + \lambda_{21}\frac{d}{dt}\right)\tau_{2,11} = 2\left(1 + \lambda_{22}\frac{d}{dt}\right)\partial_x u_2, \tag{18.144}$$

$$\left(1 + \lambda_{21}\frac{d}{dt}\right)\tau_{2,22} = 2\left(1 + \lambda_{22}\frac{d}{dt}\right)\partial_z w_2, \tag{18.145}$$

which hold for $(x, z, t) \in \Omega_2$ as well as

$$(\mu \partial_z u_1 - \tau_{2,12}^{(1)}) - (\tau_{2,22} - \tau_{2,11})\partial_x h_1 = 0, \tag{18.146}$$

at $z = h_1$ and

$$\tau_{2,12}^{(1)} + (\tau_{2,22} - \tau_{2,11})\, \partial_x h_2 = 0, \tag{18.147}$$

at $z = h_2$.

Our first observation is that integrating the leading order momentum balance equations for the upper layer (18.134) w.r.t. z and using the boundary condition (18.142) gives

$$p_2 = \tau_{2,22} - \partial_{xx} h_1 - \partial_{xx} h + \phi'(h). \tag{18.148}$$

Combining this expression with the next order momentum balance (18.143) we obtain

$$0 = \partial_x(\partial_{xx} h_1 + \partial_{xx} h - \phi'(h)) + \partial_x(\tau_{2,11} - \tau_{2,22}) + \partial_z \tau_{2,12}^{(1)}. \tag{18.149}$$

We set

$$\bar{\tau}_2 := \tau_{2,11} - \tau_{2,22}. \tag{18.150}$$

Then integration of Eq. (18.149) gives

$$0 = h\partial_x(\partial_{xx} h_1 + \partial_{xx} h - \phi'(h)) + \int_{h_1}^{h_2} \partial_x \bar{\tau}_2 dz + \tau_{2,12}^{(1)}|_{z=h_2} - \tau_{2,12}^{(1)}|_{z=h_1}. \tag{18.151}$$

If we use this expression together with the next order boundary conditions (18.146)–(18.147) and set

$$S := \frac{1}{4h} \int_{h_1}^{h_2} \bar{\tau}_2 dz, \tag{18.152}$$

we obtain

$$0 = h\partial_x(\partial_{xx} h_1 + \partial_{xx} h - \phi'(h)) + 4\partial_x(hS) - \mu\partial_z u_1|_{z=h_1}. \tag{18.153}$$

In the next step we derive an equation for S. We first note that because of (18.134)–(18.135) and (18.139) $u_2 = u_2(x,t)$ does not depend on z. We combine (18.144) and (18.145) to

$$\left(1 + \lambda_{21}\frac{d}{dt}\right)\bar{\tau}_2 = 4\left(1 + \lambda_{22}\partial_t + \lambda_{22} u_2 \partial_x\right)\partial_x u_2. \tag{18.154}$$

Integration of the left hand side of this equation yields

$$\int_{h_1}^{h_2} \left(1 + \lambda_{21}\tfrac{d}{dt}\right) \bar{\tau}_2 dz \tag{18.155}$$

$$= \int_{h_1}^{h_2} \left(1 + \lambda_{21}\partial_t + \lambda_{21}u_2\partial_x + \lambda_{21}w_2\partial_z\right) \bar{\tau}_2 dz \tag{18.156}$$

$$= \int_{h_1}^{h_2} \left(1 + \lambda_{21}\partial_t + \lambda_{21}u_2\partial_x + \lambda_{21}(-z\partial_x u_2 + \partial_t h_1 + \partial_x(u_2 h_1))\partial_z\right) \bar{\tau}_2 dz \tag{18.157}$$

$$= 4h(1 + \lambda_{21}\partial_t + \lambda_{21}u_2\partial_x)S \tag{18.158}$$

Hence, we obtain the equation for S

$$(1 + \lambda_{21}\partial_t + \lambda_{21}u_2\partial_x)S = (1 + \lambda_{22}\partial_t + \lambda_{22}u_2\partial_x)\,\partial_x u_2. \tag{18.159}$$

The kinematic and impermeability conditions imply the equation for h

$$\partial_t h = -\partial_x(h\,u_2). \tag{18.160}$$

In the last step we consider the evolution of the lower fluid and the interface h_1. From (18.133) and (18.137) we first obtain

$$u_1 = \frac{1}{2\mu}\partial_x p_1 z^2 + c\,z, \tag{18.161}$$

which we use for the evolution equation for h_1

$$\partial_t h_1 = -\partial_x \int_0^{h_1} u_1 dz = -\partial_x \left(\frac{1}{6\mu}h_1^3\partial_x p_1 + \frac{1}{2}h_1^2 c\right). \tag{18.162}$$

and determine the constant c from equation

$$\frac{1}{2\mu}\partial_x p_1 h_1^2 + c\,h_1 = u_2 \tag{18.163}$$

and by using (18.140). We finally obtain the closed system of equations for h_1, h, S and u_2.

$$\partial_t h = -\partial_x(h\,u_2), \tag{18.164}$$

$$\partial_t h_1 = \frac{1}{12\mu}\partial_x \left(h_1^3\partial_x p_1\right) - \frac{1}{2}\partial_x \left(h_1 u_2\right), \tag{18.165}$$

$$0 = -\frac{1}{2}h_1\partial_x p_1 - h\partial_x p_2 + 4\partial_x(hS) - \frac{\mu}{h_1}u_2, \tag{18.166}$$

$$0 = (1 + \lambda_{21}\partial_t + \lambda_{21}u_2\partial_x)S - (1 + \lambda_{22}\partial_t + \lambda_{22}u_2\partial_x)\,\partial_x u_2. \tag{18.167}$$

where

$$p_1 = -(\sigma + 1)\partial_{xx}h_1 - \partial_{xx}h, \quad \text{and} \quad p_2 = -\partial_{xx}h_1 - \partial_{xx}h + \phi'(h). \tag{18.168}$$

The above model is the first model that incorporates the full nonlinear corotational Jeffrey's model into a thin-film theory. Using this model it is now possible to analyse and numerically investigate the nonlinear behaviour and long-time morphological evolution of dewetting liquid-liquid films. This will be subject of future work.

18.7 Conclusion

In this chapter, we considered the liquid-liquid dewetting, where a thin liquid film retracts from an also thin liquid substrate. Mathematical models based on the thin film equation are derived including models that take account of interfacial apparent slip and nonlinear viscoelastic effects. For the Newtonian case, existence results for the stationary and dynamic problems as well as numerical methods for the thin-film as well as the underlying free boundary problem for the Stokes equations were presented.

For the case of Newtonian liquids the theoretical predictions were quantitatively compared to experimental results obtained by polystyrene (PS) dewetting from polymethyl-methacrylate (PMMA). Both polymers were used with sufficiently short molecular chain length and could be considered as Newtonian liquids. The relevant system parameters like viscosity, contact angel and surface tension were determined for the experimental system and used as input parameters for the mathematical model. The quantitative comparison proved that thin-film models are adequate to describe transient dewetting rim and droplet shapes as well as dewetting dynamics which result from a complex interaction of substrate and liquid flow.

Another important problem that is currently being investigated concerns the spinodal dewetting process, investigating the self-similar evolution towards rupture. An interesting question is if this occurs in finite time. In the future we also will carry out comparisons of solutions of our viscoelastic thin-film model with experimental results by repeating our experiments for different chain length of both polymers to vary the rheological properties.

References

1. Brochard Wyart, F., Martin, P., Redon, C.: Liquid/liquid dewetting. Langmuir **9**(12), 3682–3690 (1993)
2. Pototsky, A., Bestehorn, M., Merkt, D., Thiele, U.: Alternative pathways of dewetting for a thin liquid two-layer film. Phys. Rev. E **70**(2), 25201 (2004)

3. Pototsky, A., Bestehorn, M., Merkt, D., Thiele, U.: Morphology changes in the evolution of liquid two-layer films. J. Chem. Phys. **122**, 224711 (2005)
4. Fisher, L.S., Golovin, A.A.: Nonlinear stability analysis of a two-layer thin liquid film: Dewetting and autophobic behavior. J. Colloid Interface Sci. **291**(2), 515–528 (2005)
5. Fisher, L.S., Golovin, A.A.: Instability of a two-layer thin liquid film with surfactants: dewetting waves. J. Colloid Interface Sci. **307**(1), 203–214 (2007)
6. Bandyopadhyay, D., Gulabani, R., Sharma, A.: Instability and dynamics of thin liquid bilayers. Ind. Eng. Chem. Res. **44**(5), 1259–1272 (2005)
7. Bandyopadhyay, D., Sharma, A.: Nonlinear instabilities and pathways of rupture in thin liquid bilayers. J. Chem. Phys. **125**, 054711 (2006)
8. Goveas, J.L., Fredrickson, G.H.: Apparent slip at a polymer-polymer interface. Eur. Phys. J. B **2**(1), 79–92 (1998)
9. Adhikari, N.P., Goveas, J.L.: Effects of slip on the viscosity of polymer melts. J. Polym. Sci. B Polym. Phys. **42**, 1888–1904 (2004)
10. Zhao, R., Macosko, C.W.: Slip at polymer–polymer interfaces: Rheological measurements on coextruded multilayers. J. Rheol. **46**, 145–167 (2002)
11. Lin, Z., Kerle, T., Russell, T.P., Schaffer, E., Steiner, U.: Electric field induced dewetting at polymer/polymer interfaces. Macromolecules **35**(16), 6255–6262 (2002)
12. Zeng, H., Tian, Y., Zhao, B., Tirrell, M., Israelachvili, J.: Friction at the liquid/liquid interface of two immiscible polymer films. Langmuir **25**, 124–132 (2009)
13. Higginsa, A.M., Sferrazza, M., Jones, R.A.L., Jukes, P.C., Sharp, J.S., Dryden, L.E., Webster, J.: The timescale of spinodal dewetting at a polymer/polymer interface. Eur. Phys. J. E **8**, 137–143 (2002)
14. de Silva, J.P., Geoghegan, M., Higgins, A.M., Krausch, G., David, M.O., Reiter, G.: Switching layer stability in a polymer bilayer by thickness variation. Phys. Rev. Lett. **98**(26), 267802 (2007)
15. Segalman, R.A., Green, P.F.: Dynamics of rims and the onset of spinodal dewetting at liquid/liquid interfaces. Macromolecules **32**(3), 801–807 (1999)
16. Slep, D., Asselta, J., Rafailovich, M.H., Sokolov, J., Winesett, D.A., Smith, A.P., Ade, H., Anders, S.: Effect of an interactive surface on the equilibrium contact angles in bilayer polymer films. Langmuir **16**, 2369–2375 (2000)
17. Lambooy, P., Phelan, K.C., Haugg, O., Krausch, G.: Dewetting at the liquid-liquid interface. Phys. Rev. Lett. **76**(7), 1110–1113 (1996)
18. Wang, C., Krausch, G., Geoghegan, M.: Dewetting at a polymer-polymer interface: film thickness dependence. Langmuir **17**(20), 6269–6274 (2001)
19. Hassager, O., Bird, R.B., Armstrong, R.C.: Dynamics of Polymeric Fluids, vol. 1. Wiley, New York (1977)
20. Oron, A., Davis, S.H., Bankoff, S.G.: Long-scale evolution of thin liquid films. Rev. Mod. Phys. **69**(3), 931 (1997)
21. Kostourou, K., Peschka, D., Münch, A., Wagner, B., Herminghaus, S., Seemann, R.: Interface morphologies in liquid/liquid dewetting. Chem. Eng. Process. **50**, 531–536 (2011)
22. Bertozzi, A.L., Grün, G., Witelski, T.P.: Dewetting films: bifurcations and concentrations. Nonlinearity **14**, 1569 (2001)
23. Jachalski, S., Huth, R., Kitavtsev, G., Peschka, D., Wagner, B.: Stationary solutions for two-layer lubrication equations. SIAM J. Appl. Math. **73**(3), 1183–1202 (2013)
24. Bernis, F., Friedman, A.: Higher order nonlinear degenerate parabolic equations. J. Differ. Equ. **83**(1), 179–206 (1990)
25. Barrett, J.W., El Alaoui, L.: Finite element approximation of a two-layered liquid film in the presence of insoluble surfactants. ESAIM: Math. Model. Numer. Anal. **42**(05), 749–775 (2008)
26. Jachalski, S., Kitavtsev, G., Taranets, R.: Weak solutions to lubrication system describing the evolution of bilayer thin films. Commun. Math. Sci. **30**(3), 527–544 (2014)
27. Eidel'man, S.D.: Parabolic Systems. North Holland, Amsterdam (1969)
28. Bänsch, E.: Finite element discretization of the Navier–Stokes equations with a free capillary surface. Numer. Math. **88**(2), 203–235 (2001)

29. Dziuk, G.:   Finite elements for the Beltrami operator on arbitrary surfaces.   In: Partial Differential Equations and Calculus of Variations, pp. 142–155. Springer, Berlin (1988)
30. Huth, R., Jachalski, S., Kitavtsev, G., Peschka, D.: Gradient flow perspective on thin-film bilayer flows. J. Eng. Math. **94**(1), 43–61 (2015)
31. Zhornitskaya, L., Bertozzi, A.L.: Positivity-preserving numerical schemes for lubrication-type equations. SIAM J. Numer. Anal. **37**(2), 523–555 (1999)
32. Grün, G., Rumpf, M.:. Nonnegativity preserving convergent schemes for the thin film equation. Numer. Math. **87**(1), 113–152 (2000)
33. Diez, J.A., Kondic, L.: Computing three-dimensional thin film flows including contact lines. J. Comput. Phys. **183**(1), 274–306 (2002)
34. Kriegsmann, J.J., Miksis, M.J.: Steady motion of a drop along a liquid interface. SIAM J. Appl. Math. **64**(1), 18–40 (2003)
35. Karapetsas, G., Craster, R.V., Matar, O.K.: Surfactant-driven dynamics of liquid lenses. Phys. Fluids **23**(12), 122106–122106 (2011)
36. Peschka, D.: Thin-film free boundary problems for partial wetting. J. Comput. Phys. **295**, 770–778 (2015)
37. Peschka, D.: Numerics of contact line motion for thin films. IFAC-PapersOnLine **48**(1), 390–393 (2015)
38. McGraw, J.D., Salez, T., Bäumchen, O., Raphaël, E., Dalnoki-Veress, K.: Self-similarity and energy dissipation in stepped polymer films. Phys. Rev. Lett. **109**, 128303 (2012)
39. Salez, T., McGraw, J.D., Cormier, S.L., Bäumchen, O., Dalnoki-Veress, K., Raphaël, E.: Numerical solutions of thin-film equations for polymer flows. Eur. Phys. J. E **35**(11), 1–9 (2012)
40. Herminghaus, S., Jacobs, K., Seemann, R.: The glass transition of thin polymer films: some questions, and a possible answer. Eur. Phys. J. E **5**(5), 531–538 (2001)
41. Bäumchen, O., Fetzer, R., Klos, M., Lessel, M., Marquant, L., Hähl, H., Jacobs, K.: Slippage and nanorheology of thin liquid polymer films. J. Phys. Condens. Matter **24**(32), 325102 (2012)
42. Bommer, S., Cartellier, F., Jachalski, S., Peschka, D., Seemann, R., Wagner, B.: Droplets on liquids and their journey into equilibrium. Eur. Phys. J. E **36**(8), 1–10 (2013)
43. Anastasiadis, S.H., Gancarz, I., Koberstein, J.T.: Interfacial tension of immiscible polymer blends: temperature and molecular weight dependence. Macromolecules **21**(10), 2980–2987 (1988)
44. Wu, S.: Surface and interfacial tensions of polymer melts. II. Poly (methyl methacrylate), poly (n-butyl methacrylate), and polystyrene. J. Phys. Chem. **74**(3), 632–638 (1970)
45. Neumann, F.E.: Vorlesung über die Theorie der Capillarität. BG Teubner, Leipzig (1894)
46. Bommer, S., Jachalski, S., Peschka, D., Seemann, R., Wagner, B.: Rates and morphology in liquid-liquid dewetting. WIAS Preprint 2346 (2016)
47. Craster, R.V., Matar, O.K.: On the dynamics of liquid lenses. J. Colloid Interface Sci. **303**(2), 503–516 (2006)
48. Rubenstein, M., Colby, R.H.: Polymer Physics. Oxford University Press, Oxford (2003)
49. Morozov, A.N., van Saarloos, W.: An introductory essay on subcritical instabilities and the transition to turbulence in visco-elastic parallel shear flows. Phys. Rep. **447**(3), 112–143 (2007)
50. Lambooy, P., Phelan, K.C., Haugg, O., Krausch, G.: Dewetting at the liquid-liquid interface. Phys. Rev. Lett. **76**(7), 1110 (1996)
51. Fetzer, R., Jacobs, K., Münch, A., Wagner, B., Witelski, T.P.: New slip regimes and the shape of dewetting thin liquid films. Phys. Rev. Lett. **95**, 127801 (2005)
52. Redon, C., Brzoska, J.B., Brochard-Wyart, F.: Dewetting and slippage of microscopic polymer films. Macromolecules **27**(2), 468–471 (1994)
53. Brochard-Wyart, F., de Gennes, P.G.: Shear-dependent slippage at a polymer/solid interface. Langmuir **8**, 3033–3037 (1992)
54. Kargupta, K., Sharma, A., Khanna, R.: Instability, dynamics and morphology of thin slipping films. Langmuir **20**, 244–253 (2004)
55. Münch, A., Wagner, B., Witelski, T.P.: Lubrication models with small to large slip lengths. J. Eng. Math. **53**, 359–383 (2006)

56. Lin, C.C.: A mathematical model for viscosity in capillary extrusion of two-component polyblends. Polym. J. (Tokyo) **11**, 185–192 (1979)
57. Brochard-Wyart, F., de Gennes, P.-G.: Sliding molecules at a polymer/polymer interface. C.R. Acad. Sci., Ser. II **317**, 13–17 (1993)
58. Ajdari, A.: Slippage at a polymer/polymer interface: entanglements and associated friction. C.R. Acad. Sci., Ser. II **317**, 1159–1163 (1993)
59. Jachalski, S., Münch, A., Peschka, D., Wagner, B.: Impact of interfacial slip on the stability of liquid two-layer films. J. Eng. Math. **86**, 9–29 (2014)

Part IV
Taylor Bubbles: Experiments, Simulation and Validation

A *guiding measure* within this Priority Programme was the realization of a series of systematic Taylor-bubble experiments and the common usage of the obtained data for validation of mathematical models and numerical methods. In this part there are the following contributions:

Chapter 19.   M. Wörner, *Taylor Bubbles in Small Channels: A Proper Guiding Measure for Validation of Numerical Methods for Interface Resolving Simulations*

Chapter 20.   S. Boden, M. Haghnegahdar, U. Hampel, *X-ray Microtomography of Taylor Bubbles with Mass Transfer and Surfactants in Capillary Two-Phase Flow*

Chapter 21.   S. Kastens, C. Meyer, M. Hoffmann, M. Schlüter, *Experimental Investigation and Modelling of Local Mass Transfer Rates in Pure and Contaminated Taylor Flows*

Chapter 22.   S. Aland, A. Hahn, C. Kahle, R. Nürnberg, *Comparative Simulations of Taylor Flow with Surfactants Based on Sharp- and Diffuse-Interface Methods*

Chapter 23.   H. Marschall, C. Falconi, C. Lehrenfeld, R. Abiev, M. Wörner, A. Reusken, D. Bothe, *Direct Numerical Simulations of Taylor Bubbles in a Square Mini-Channel: Detailed Shape and Flow Analysis with Experimental Validation*

This part starts with an *overview* contribution in Chap. 19, in which important characteristics of Taylor bubble flows in small channels are addressed. Furthermore, the specific advantages of Taylor flow as a guiding measure for the Priority Programme SPP 1506 are highlighted. The following four chapters present a summary of important results obtained in the "Taylor Flow" project group which was formed within the SPP 1506 and consisted of researchers from different disciplines. In a joint *interdisciplinary* effort, results are obtained which improve and deepen the understanding of mechanisms and phenomena occurring at fluid interfaces. The researchers that contributed to this part are from eight different research groups, two of which are in experimental fluid dynamics and the other six from the field of modeling and numerical simulation. In Chaps. 20 and 21, different experimental techniques for measuring local Taylor bubble properties and results obtained with these methods are treated. In Chaps. 22 and 23 different modeling and numerical simulation approaches are applied to a Taylor bubble flow problem, and the simulation results are validated by comparing with measurements obtained in the experimental projects. Furthermore, in Chap. 23 a *Taylor flow benchmark problem* is presented which has been developed within the SPP 1506.

Chapter 19
Taylor Bubbles in Small Channels: A Proper Guiding Measure for Validation of Numerical Methods for Interface Resolving Simulations

Martin Wörner

Abstract Taylor bubbles moving in a vertical pipe are elongated, bullet-shaped bubbles that almost fill the channel cross-section and are separated from the wall by a thin liquid film. Taylor bubbles and Taylor flow, which consists of a sequence of Taylor bubbles separated by liquid slugs, are of interest for various technical applications. This article introduces some characteristic features of Taylor bubbles and laminar Taylor flow in small channels to facilitate the understanding of the subsequent chapters in this book. Furthermore, the specific advantages of Taylor flow as guiding measure for the DFG Priority Programme SPP 1506 "Transport Processes at Fluidic Interfaces" are highlighted.

19.1   Introduction

Sir Geoffrey Ingram Taylor (1886–1975) [4] was—together with Ludwig Prandtl (1875–1953)—the probably most influential individual in the field of fluid dynamics in the central period of the last century. Among the various phenomena that today bear his name are Taylor-Couette flow, Rayleigh-Taylor instability, Taylor dispersion, and Taylor bubbles. Concerning the last topic, Taylor studied in two seminal papers the motion of large bubbles rising through tubes [9] and the displacement of liquid in a tube by a bubble [23]. An important earlier contribution on the topic originates from Dumitrescu [10], a student of Prandtl. The earliest photos of Taylor bubbles are probably due to Gibson [13], Fig. 19.1, who noted more than a century ago

> ... when the diameter is about 0.75 that of the tube the bubble begins to adopt a more or less cylindrical form with an ogival head and a flat stern, and the motion becomes steady. Any further increase in the volume is mainly effective in increasing the length of the cylindrical

M. Wörner (✉)

Institute for Catalysis Research and Technology (IKFT), Karlsruhe Institute of Technology (KIT), Engesserstr. 20, 76131 Karlsruhe, Germany

e-mail: martin.woerner@kit.edu

© Springer International Publishing AG 2017

D. Bothe, A. Reusken (eds.), *Transport Processes at Fluidic Interfaces*,
Advances in Mathematical Fluid Mechanics, DOI 10.1007/978-3-319-56602-3_19

Fig. 19.1 Photos of Taylor bubbles rising in a vertical tube (diameter d). *Left*: $d = 1.68$ cm, reprinted from [13], with permission from Taylor & Francis Ltd. *Right*: $d = 7.9$ cm, reprinted from [9], with permission from The Royal Society

portion of the body, the form of the head remaining sensibly unchanged, and the mean diameter, although increasing with length, not altering greatly.

Taylor bubbles are encountered in various industrial applications such as aerated chemical or bio-chemical reactors and boiling of water in nuclear rod bundles, and in natural phenomena such as in volcanic eruptions [22], where they are usually called gas slugs.

With the significant advancement of microfabrication techniques during the last decades, gas-liquid two-phase flows in small channels came into focus in various fields like micro process engineering, lab-on-a-chip systems and material synthesis. In these applications, rather than a single bubble, a sequence of Taylor bubbles is of interest where the neighboring bubbles are separated by liquid slugs, see Fig. 19.2. This flow pattern is known as Taylor flow but is sometimes also referred to as bubble-train flow, segmented flow or capillary slug flow. The hydraulic diameter is typically below a few mm so that gravitational effects are often negligible. The flow is, therefore, usually pressure-driven and, due to the small dimensions, laminar. Taylor flow has distinct advantages especially for chemical process engineering:

- Large interfacial area per unit volume $\rightarrow$ efficient heat and mass transfer.
- Axial segmentation of the liquid phase $\rightarrow$ reduced axial dispersion and narrow residence time distribution.
- Recirculation in liquid slug $\rightarrow$ good mixing and wall-normal convective transport in laminar flow.
- Thin liquid film $\rightarrow$ short diffusion path of educts from the bubble to the catalytic wall.

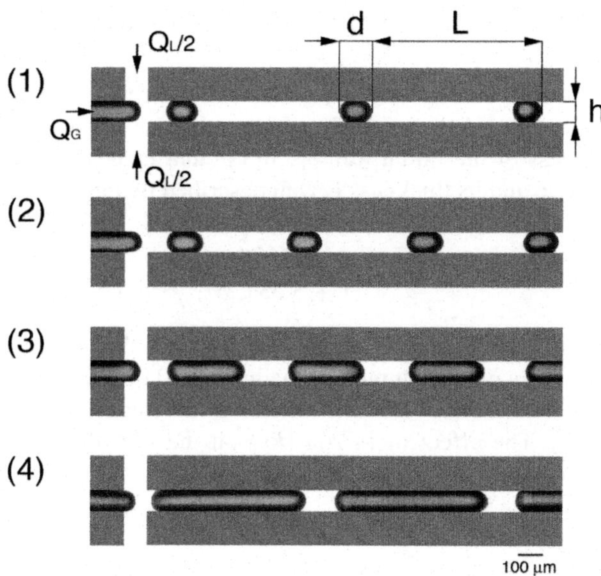

Fig. 19.2 Bubble formation for different volumetric flow rates of water (Q_L) and air (Q_G) in a 100 μm square channel leading to Taylor flow. In subfigures (1)–(4), Q_L is decreased while Q_G is increased. Reprinted figure with permission from [8] © 2005 by the American Physical Society, doi: 10.1103/PhysRevE.72.037302

Several aspects of this list will be detailed in the sequel. For a more complete discussion than is possible in this short overview, and for topics not covered here, such as pressure drop, heat transfer, etc., the interested reader is referred to recent reviews on gas-liquid Taylor flow [2, 14, 15].

19.2   Hydrodynamics

The hydrodynamics of Taylor bubbles in small channels is dominated by viscous forces and surface tension forces. The capillary number $Ca = \mu_L U_B/\sigma$ represents the ratio of both forces and is the relevant non-dimensional parameter. Here, U_B is the bubble velocity, μ_L the dynamic viscosity of the liquid and σ is the coefficient of surface tension. At higher velocities, inertial effects become important which can be characterized by the Reynolds number $Re = \rho_L d_h U_B/\mu_L$. Here, ρ_L is the liquid density and d_h the hydraulic diameter. The influence of gravitational forces can be estimated by the Eötvös number $Eo = (\rho_L - \rho_G)g d_h^2/\sigma$. Due to the proportionality $Eo \propto d_h^2$, the importance of gravity diminishes quickly as the channel size is reduced.

19.2.1 Bubble Shape and Liquid Film Thickness

Figure 19.3a, b shows a sketch of Taylor flow in a circular channel and related geometrical dimensions. A key parameter for technical applications with Taylor flow is the thickness of the liquid film δ_F. In circular channels, the liquid film is azimuthally uniform and its thickness is well described by the relation

$$\frac{\delta_F}{d} = \frac{0.66\,Ca^{2/3}}{1 + 3.33\,Ca^{2/3}} \tag{19.1}$$

This correlation is valid for capillary numbers below about 1, supposed inertia is negligible [3]. For capillary numbers smaller than 0.001, Eq. (19.1) approaches the result $\delta_F/d = 0.66\,Ca^{2/3}$ from Bretherton's lubrication analysis for a semi-infinite bubble [6]. The effect of inertia (Re) on the film thickness is small but non-monotonic, see e.g. [2] for a detailed discussion.

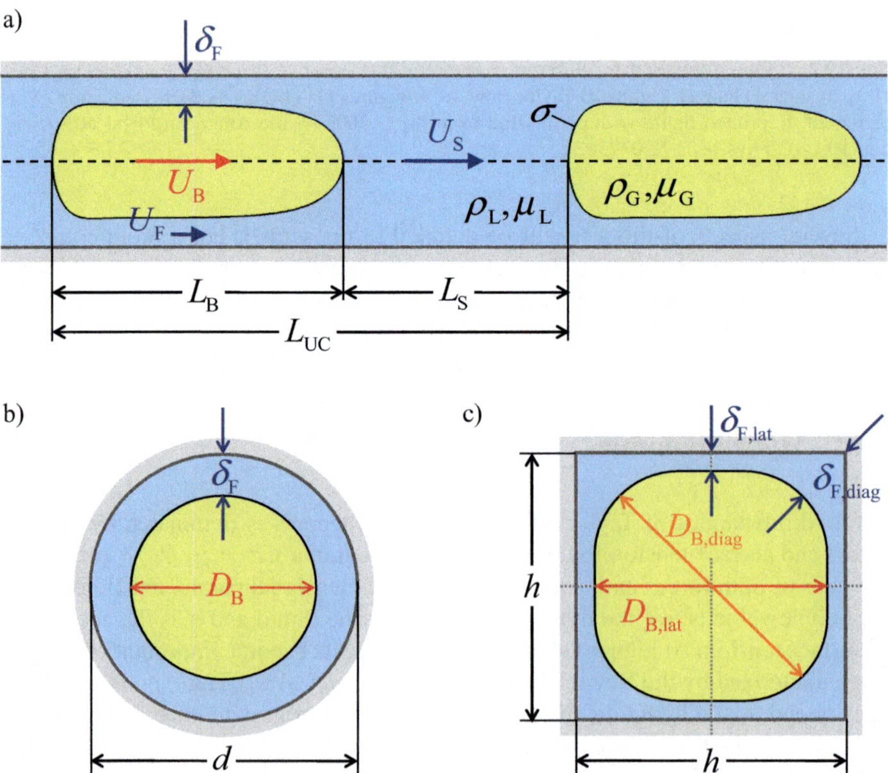

Fig. 19.3 Sketch of Taylor flow. (**a**) Lateral view for Taylor flow in a circular channel. (**b**) Cross-sectional view in the middle of a Taylor bubble in a circular channel ($d_h = d$). (**c**) Cross-sectional view in the middle of a Taylor bubble in a square channel ($d_h = h$)

The capillary number also has a large influence on the shape of the front and rear meniscus of the Taylor bubble. At very small capillary numbers, the bubble front and rear form hemispherical caps. As Ca increases, the curvature of the front meniscus increases so that the bubble nose gets more pointed while the curvature of the rear meniscus decreases (i.e., the bubble rear flattens) and may become even negative (concave shape). While inertia has a small effect on the shape of the front meniscus, the influence of Re on the rear meniscus can be quite substantial [12].

The bubble length L_B and the liquid slug length L_S depend both on the gas and liquid volumetric flow rates, cf. Fig. 19.2, and on the type of device used for bubble generation, see e.g. [15]. Often, T-junctions, Y-junctions or cross-junctions such as in Fig. 19.2 are used. One bubble and one liquid slug form a unit cell of the Taylor flow with length L_{UC}.

In square channels, the liquid film thickness is azimuthally non-uniform and the situation is more complex than in circular tubes. Concerning the bubble shape, two regimes can be distinguished. When the capillary number is larger than about 0.04, the bubble is axisymmetric and its cross-sectional shape is circular. For smaller capillary numbers, the bubble is not axisymmetric. In this case there exist liquid regions in the four channel corners which are connected by thin flat films at the channel sides, see Fig. 19.3c. Correlations for the lateral and diagonal film thickness and bubble diameter are given by Kreutzer et al. [18].

The velocity U_S in Fig. 19.3a denotes the mean liquid axial velocity in a cross-section within the slug, while U_F denotes the mean liquid axial velocity in a cross-section within the film. By a liquid mass balance in a frame of reference moving with the bubble it follows that

$$(U_S - U_B)A = (U_F - U_B)A_F \qquad (19.2)$$

Here, A is the area of the channel cross-section and A_F is the cross-sectional area of the liquid film. Equation (19.2) is valid for circular and square channels and any other cross-sectional channel shapes as well. It reveals that the area of the liquid film and the mean velocity in the liquid film are closely related to the bubble velocity and to U_S, which is in Taylor flow equal to the total superficial velocity $(Q_L + Q_G)/A$ of the two-phase flow.

19.2.2 Recirculation in the Liquid Slug

In his 1961 paper, Taylor [23] proposed qualitative sketches of the flow streamlines in the liquid slug ahead of the bubble. At high Ca, the bubble velocity is larger than the maximum velocity in the liquid slug on the axis of the tube ($U_{L,max}$). In a reference frame co-moving with the bubble, then complete bypass flow occurs, see Fig. 19.4a. At small Ca, it is $U_B < U_{L,max}$ and a recirculation pattern occurs in the tube center, see Fig. 19.4b. Both patterns have later been confirmed experimentally [7] and numerically [20, 25].

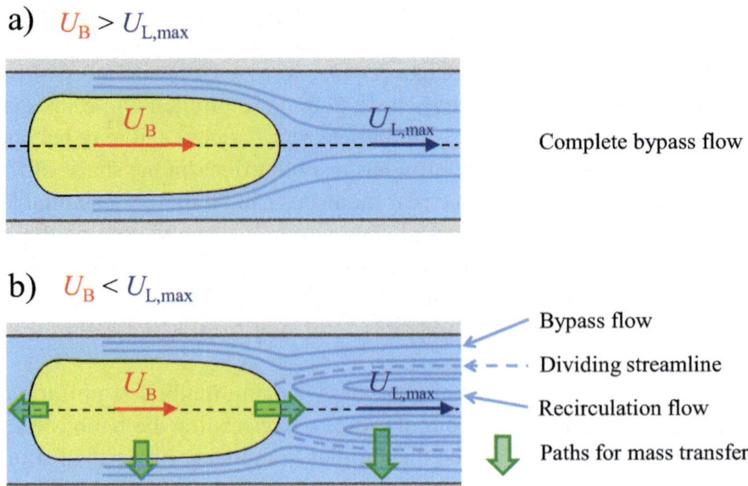

Fig. 19.4 Sketch of possible liquid streamlines in a reference frame co-moving with the bubble. (**a**) Complete bypass flow, (**b**) recirculation flow with paths for mass transfer. Figure adapted from [18, 23]

For a liquid slug with a fully developed laminar velocity profile it is $U_\mathrm{L,max} = C \times U_\mathrm{S}$. The value of the constant C depends on the shape of the channel cross-section. For a circular channel it is $C = 2$ while for a square channel it is $C = 2.096$. The condition for recirculation flow thus becomes $U_\mathrm{L,max} < C \times U_\mathrm{S}$ and occurs at $Ca \approx 0.7$ in horizontal circular tubes.

The cross-sectional regions with bypass flow close to the wall and with recirculation flow in the channel center are separated by the "dividing streamline" [24], see Fig. 19.4b. The position of the dividing streamline is obtained from the condition that the total flow rate within the recirculation area is zero in the moving frame of reference. In a cross-section of a liquid slug with fully developed velocity profile, the size of the recirculation region depends on the velocity ratio $U_\mathrm{B}/U_\mathrm{S}$ and C only, and increases as the velocity ratio decreases [17].

19.3 Mass Transfer and Marangoni Effects

The large interfacial area per unit volume, the thin liquid film, and the recirculation vortex in the liquid slug make Taylor flow attractive for mass transfer applications as well as for heterogeneous chemical reactions, where the channel walls are coated with a catalytic washcoat layer. A detailed review on the latter topic is given by Haase et al. [15]. The mass transfer of chemical species (educts) from the gas bubble to the solid wall takes place by two different paths, see Fig. 19.4b. The first path is given by the mass transfer from the bubble body into the liquid film and through

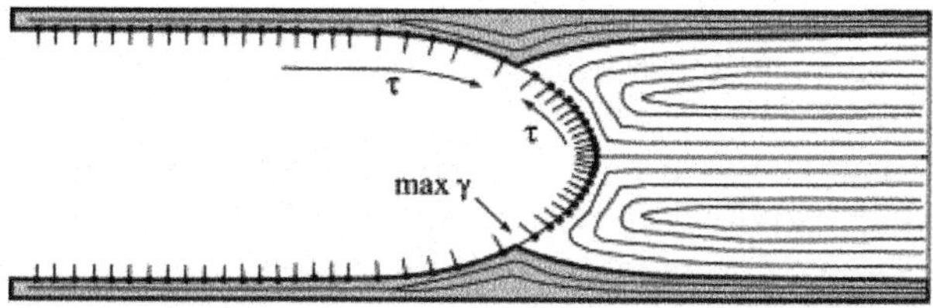

Fig. 19.5 Sketch of the surface concentrations and the interaction with the fluid flow. Reprinted from [18], © (2005), with permission from Elsevier

the liquid film toward the wall. The second path is given by the mass transfer from the front and rear caps of the bubble into the liquid slug and from the liquid slug towards the wall.

In the second path, the educts need to cross the dividing streamline which is possible only by diffusion. Important in this context is the speed at which the liquid in the vortex inside the recirculation zone moves, as this will affect mixing. The intensity of this recirculation can be quantified by the dimensionless recirculation time [24]. This quantity is defined as the ratio of the time needed by a liquid fluid element to move from one end of the liquid slug to the other end, and the time needed by the liquid slug to travel a distance of its own length.

As Taylor flow is dominated by surface tension effects, even small amounts of surface-active agents (surfactants) or contaminants can have a large impact. The presence of surfactants or contaminants on the gas-liquid interface changes surface tension and, therefore, the capillary number. Gradients of the concentration of surfactants cause gradients in surface tension which induces so-called Marangoni stresses. The largest concentration gradients are found near the stagnation rings on the bubble nose where the dividing streamline reaches the interface, see Fig. 19.5. Due to these effects, surfactants and contaminants can locally modify the "boundary conditions" at the interface, which may vary between the limits free-slip and no-slip.

19.4   Guiding Measure Taylor Flow in the SPP 1506

For interface resolving simulations of two-phase flows, various numerical methods are available, see e.g. [26]. Among the most often used methods are the volume-of-fluid method, the level-set method and the front-tracking method. For testing the accuracy of interface resolving simulation methods, artificial test problems such as the rotation of Zalesak's slotted disk are often used. Furthermore, two benchmark configurations which model two-dimensional bubbles rising in liquid columns have been proposed for quantitative comparison of interfacial flow codes [16]. While these test cases are certainly useful, they are essentially pure numerical exercises. For the advancement of numerical methods for interfacial flows towards valuable tools for engineering applications, however, a validation by experimental data for real physical flow problems is desirable. Clearly, there is a lack in literature as suitable local experimental data which allow for a detailed validation on practical flow problems are missing.

One goal of the guiding measure in SPP 1506 was, therefore, to undertake a step to fill this gap by providing such data for the flow of Taylor bubbles in small channels. This specific flow problem was chosen for the following reasons:

- Taylor flow is of practical technical relevance.
- Taylor flow is of fundamental physical interest as it constitutes a prototypical problem for the non-linear interaction between viscous, inertial and surface tension forces under geometric constraints.
- Taylor flow allows the study of hydrodynamics and mass transfer in a relatively simple experimental set-up.
- Taylor flow allows for an increase in the complexity of the flow and bubble shape by variation of the channel cross-section (circular and square).
- Taylor flow hydrodynamics is controlled by one main parameter, i.e., the capillary number. For a certain liquid phase, Ca can be varied by about 1–2 orders of magnitude by variation of the bubble velocity. An even larger variation is possible by using liquids of different viscosity.

For serving as a suitable measure for validation of numerical methods and computer codes, the experiments and measurements—which will be presented in the next two chapters of this book—fulfill the following requirements:

- Experiments are performed in circular [5] and square channels [5, 21] under well-defined and well-documented conditions which allow a detailed recalculation. This encompasses information about:

  - Thermo-physical properties of both phases.
  - Liquid and gas volumetric flow rates of both phases.
  - Geometrical flow parameters such as the volume of a single Taylor bubble and L_B and L_S for Taylor flow. The variation of both lengths in the experiment is sufficiently small to resemble "ideal" Taylor flow.
  - Conditions to be applied at all boundaries of the computational domain.

- Measurements provide detailed *local* experimental data which allow for a quantitative validation of numerical methods and computer codes:

  - Local profiles of bubble shape (in square channel: lateral and diagonal) [5].
  - Local profiles of velocity field in liquid film and liquid slug (in square channel: lateral and diagonal) [21].

For numerical computations, there exist the following advantages (first two bullets) as well as challenges (last bullet) of Taylor flow:

- The numerical simulations can be either 2D axisymmetric (circular channel, which limits the computational costs) or 3D (square channel with optional symmetry assumptions).

- Both, a single Taylor bubble and Taylor flow can well be represented in numerical computations by using either inlet/outlet conditions in combination with a co-moving reference frame (single Taylor bubble) or by considering a unit cell in combination with periodic boundary conditions (ideal Taylor flow).
- Challenging for numerical simulations are the adequate resolution of the thin liquid film and the large local interface curvature at the rear part of the liquid film as well as very thin concentration boundary layers if mass transfer is considered.

An overview of the numerical simulations [1, 11, 19] performed within the SPP 1506 concerning the guiding measure Taylor flow are presented in two subsequent chapters of this book. One chapter covers 2D planar simulations as well as axisymmetric simulations for circular pipes, while the other chapter is on 3D simulations for square channels.

19.5 Conclusions

The laminar flow of Taylor bubbles in small channels is of interest for various technical applications. Taylor flow in circular and square channels is also well suited to study complex interfacial hydrodynamics in confined geometry resulting from the interplay between surface tension, viscous forces and inertia in a relatively simple set-up. Carefully designed experiments on single Taylor bubbles and Taylor flow are well suited for providing local experimental data on the bubble shape and liquid velocity field which are needed for a detailed quantitative validation of numerical methods and computer codes for interface resolving simulations.

In the following chapters, the progress achieved in this context within the SPP 1506 is highlighted. The experimental data and selected numerical data gained in the course of the SPP 1506 guiding measure Taylor flow are provided online on the website of the SPP 1506, see www.dfg-spp1506.de. It is hoped that they will be useful for the entire computational multiphase fluid dynamic community.

The subject is further developed in the DFG Priority Programme SPP 1740 "Reactive Bubbly Flows" where Taylor flow serves as a guiding measure as well. Taylor flow is also attractive to investigate further interfacial phenomena not covered in both SPPs. These include the controlled coalescence of two Taylor bubbles with different volumes. The trailing smaller Taylor bubble is moving in regions with larger velocity as compared to the leading larger bubble so that the distance between both bubbles decreases and finally leads to contact and coalescence. Furthermore, thermal Marangoni effects in Taylor flow could be studied by heating one or more walls of a square channel.

References

1. Aland, S., Boden, S., Hahn, A., Klingbeil, F., Weismann, M., Weller, S.: Quantitative comparison of Taylor flow simulations based on sharp-interface and diffuse-interface models. Int. J. Numer. Methods Fluids **73**, 344–361 (2013)
2. Angeli, P., Gavriilidis, A.: Hydrodynamics of Taylor flow in small channels: a review. Proc. IMechE Part C: J. Mech. Eng. Sci. **222**, 737–751 (2008)
3. Aussillous, P., Quere, D.: Quick deposition of a fluid on the wall of a tube. Phys. Fluids **12**, 2367–2371 (2000)
4. Batchelor, G.: The life and legacy of G.I. Taylor. Cambridge University Press, Cambridge (1994)
5. Boden, S., dos Santos, R.T., Baumbach, T., Hampel, U.: Synchrotron radiation microtomography of Taylor bubbles in capillary two-phase flow. Exp. Fluids **55**, 1–14 (2014)
6. Bretherton, F.P.: The motion of long bubbles in tubes. J. Fluid Mech. **10**, 166–188 (1961)
7. Cox, B.G.: An experimental investigation of the streamlines in viscous fluid expelled from a tube. J. Fluid Mech. **20**, 193–200 (1961)
8. Cubaud, T., Tatineni, M., Zhong, X., Ho, C.-M.: Bubble dispenser in microfluidic devices. Phys. Rev. E **72**, 037302 (2005)
9. Davis, R.M., Taylor, G.I.: The mechanics of large bubbles rising through extended liquids and through liquids in tubes. Proc. R. Soc. Ser. A **200**, 375–390 (1950)
10. Dumitrescu, D.T.: Strömung an einer Luftblase im senkrechten Rohr. Z. Angew. Math. Mech. **23**, 139–149 (1943)
11. Falconi, C.J., Lehrenfeld, C., Marschall, H., Meyer, C., Abiev, R., Bothe, D., Reusken, A., Schlüter, M., Wörner, M.: Numerical and experimental analysis of local flow phenomena in laminar Taylor flow in a square mini-channel. Phys. Fluids **28**, 012109 (2016)
12. Giavedoni, M.D., Saita, F.A.: The rear meniscus of a long bubble steadily displacing a Newtonian liquid in a capillary tube. Phys. Fluids **11**, 786–794 (1999)
13. Gibson, A.H.: On the motion of long air-bubbles in a vertical tube. Philos. Mag. **26**, 952–965 (1913)
14. Gupta, R., Fletcher, D.F., Haynes, B.S.: Taylor flow in microchannels: a review of experimental and computational work. J. Comput. Multiphase Flows **2**, 1–31 (2010)
15. Haase, S., Murzin, D.Y., Salmi, T.: Review on hydrodynamics and mass transfer in mini-channel wall reactors with gas-liquid Taylor flow. Chem. Eng. Res. Des. **113**, 304–329 (1916)
16. Hysing, S., Turek, S., Kuzmin, D., Parolini, N., Burman, E., Ganesan, S., Tobiska, L.: Quantitative benchmark computations of two-dimensional bubble dynamics. Int. J. Numer. Methods Fluids **60**, 1259–1288 (2009)
17. Kececi, S., Wörner, M., Onea, A., Soyhan, H.S.: Recirculation time and liquid slug mass transfer in co-current upward and downward Taylor flow. Catal. Today **147S**, S125–S131 (2009)
18. Kreutzer, M.T., Kapteijn, F., Moulijn, J.A., Heiszwolf, J.J.: Multiphase monolith reactors: chemical reaction engineering of segmented flow in microchannels. Chem. Eng. Sci. **60**, 5895–5916 (2005)
19. Marschall, H., Boden, S., Lehrenfeld, C., Falconi, D.C.J., Hampel, U., Reusken, A., Wörner, M., Bothe, D.: Validation of interface capturing and tracking techniques with different surface tension treatments against a Taylor bubble benchmark problem. Comput. Fluids **102**, 336–352 (2014)
20. Martinez, M.J., Udell, K.S.: Boundary integral analysis of the creeping flow of long bubbles in capillaries. J. Appl. Mech. – Trans. ASME **56**, 211–217 (1989)
21. Meyer, C., Hoffmann, M., Schlüter, M.: Micro-PIV analysis of gas-liquid Taylor flow in a vertical oriented square shaped fluidic channel. Int. J. Multiphase Flow **67**, 140–148 (2014)
22. Seyfried, R., Freundt, A.: Experiments on conduit flow and eruption behavior of basaltic volcanic eruptions. J. Geophys. Res. **105**, 23727–23740 (2000)

23. Taylor, G.I.: Deposition of a viscous fluid on the wall of a tube. J. Fluid Mech. **10**, 161–165 (1961)
24. Thulasidas, T.C., Abraham, M.A., Cerro, R.L.: Flow patterns in liquid slugs during bubble-train flow inside capillaries. Chem. Eng. Sci. **52**, 2947–2962 (1997)
25. Westborg, H., Hassager, O.: Creeping motion of long bubbles and drops in capillary tubes. J. Colloid. Interface Sci. **133**, 135–147 (1989)
26. Wörner, M.: Numerical modeling of multiphase flows in microfluidics and micro process engineering: a review of methods and applications. Microfluid. Nanofluid. **12**, 841–886 (2012)

Chapter 20
X-Ray Microtomography of Taylor Bubbles with Mass Transfer and Surfactants in Capillary Two-Phase Flow

Stephan Boden, Mohammadreza Haghnegahdar, and Uwe Hampel

Abstract Development and application of techniques to measure local properties of dynamic flows is in the focus of the work of the Institute of Fluid Dynamics at HZDR and of the AREVA Endowed Chair of Imaging Techniques in Energy and Process Engineering at TU Dresden. In this paper we report on the application of enhanced X-ray microradiography and microtomography techniques to measure Taylor bubble shapes in micro- and milli-channels. Further, experiments to investigate the mass transport and the influence of surfactants were conducted. The resulting flow structural data will foster meso- and microscalic numerical flow model development for small channel multiphase flow. Data and material of the presented study can be freely downloaded from the website of SPP 1506 (http://www.dfg-spp1506.de/taylor-bubble).

20.1 Introduction

Taylor bubble flow, also known as capillary slug flow, is a desired operation state in microreactor applications due to the frequent change of efficient gas-liquid contacting in the film around the bubbles and the enhanced turbulent mixing in the liquid slugs behind the bubbles [14, 15]. A profound understanding of momentum, heat and mass transfer in capillary two-phase flow is of primary importance form

S. Boden (✉) • U. Hampel
Helmholtz-Zentrum Dresden - Rossendorf, Institute of Fluid Dynamics, Experimental Thermal Fluid Dynamics Division, Bautzner Landstraße 400, 01328 Dresden, Germany

Institute of Power Engineering, AREVA Endowed Chair of Imaging Techniques in Energy and Process Engineering, Institute of Technische Universität Dresden, 01062 Dresden, Germany
e-mail: s.boden@hzdr.de; u.hampel@hzdr.de

M. Haghnegahdar
Helmholtz-Zentrum Dresden - Rossendorf, Institute of Fluid Dynamics, Experimental Thermal Fluid Dynamics Division, Bautzner Landstraße 400, 01328 Dresden, Germany

© Springer International Publishing AG 2017
D. Bothe, A. Reusken (eds.), *Transport Processes at Fluidic Interfaces*,
Advances in Mathematical Fluid Mechanics, DOI 10.1007/978-3-319-56602-3_20

the viewpoint of fundamental science as well as for practical design and operation of new chemical reaction devices, such as monolithic microreactors, miniature heat exchangers, fuel cells and others. Knowledge of flow topology and precise data of the liquid film thickness, bubble shape and liquid velocity profiles around bubbles on the microscopic scale and their dynamics is a decisive input for the development of heat and mass transfer models as well as interface-resolving CFD codes. However, there is only coarse knowledge on Taylor bubble shape in micro- and millichannels available. Recent experimental investigations explored slug length and slug velocity using high-speed imaging, liquid velocity fields using PIV/LIF and film thickness using confocal laser scanning microscopy and laser focus displacement scanning; for a brief review see Boden et al. [6]. But the widely used optical measurement techniques have strong limitations in two-phase flow conditions and particularly are not able to resolve the full 3D shape of the bubbles. Optical techniques are further constrained to transparent channels and fluids and diffraction limits the achievable spatial resolution to about 1 μm.

Within the presented work X-ray radiography and X-ray tomography techniques were developed further to disclose the three-dimensional shape of Taylor bubbles in small channel two-phase flow. The imaging techniques were applied to record the bubble shape for varying flow conditions, i.e. different capillary numbers, without and with mass transfer and under the influence of surface active agents in small and millimeter sized channels with circular and square cross section. In this way numerous data was generated and presented for the further development and validation of hydrodynamic and mass transfer models for this type of flow. The studies were conducted using either a microfocus X-ray imaging device or an X-ray synchrotron radiation source.

20.2 Discussion of Relevant Research

Taylor bubble flow is a desired operation state, in particular, in chemical microre-actor applications due to the frequent change of efficient gas-liquid contacting in the film around and the enhanced micromixing in the liquid slugs behind the bubbles. Mass transfer and hence subsequent reaction macrokinetics is essentially coupled to the local flow conditions. Parameters of practical interest are pressure drop and heat and mass transfer per unit channel length. Mass transfer may be due to chemical reaction, dissolution and absorption of one component into the other or within the channel walls. Capillary slug flow is a target of intensive theoretical and experimental investigation. The interested reader may get more information from reviews given by Kreutzer et al. [15], Angeli and Gavriilidis [3] and Saisorn and Wongwises [17].

The theoretical approaches to describe Taylor bubble flow and related mass transfer using dimensionless numbers and to describe the effects of surface active agents were briefly reviewed in the accompanying work of the authors [6, 11–13].

The fluid dynamics of small channels is mainly governed by the influence of surface tension and to a lesser degree by gravity. Adiabatic two phase-flow is effectively governed by the ratio of viscous forces to surface tension expressed by the capillary number

$$Ca = \frac{\mu U_b}{\sigma},\tag{20.1}$$

where μ is liquid dynamic viscosity, U_b is bubble terminal velocity and σ is surface tension, and furthermore by the Reynolds number

$$Re = \frac{\rho U_b D_h}{\mu}\tag{20.2}$$

where ρ is the liquid density and D_h is the hydraulic diameter of the channel. The probably most prominent feature of a Taylor bubble is the liquid film between the bubble and the channel wall. Such liquid film occurs since the bubble travels at a positive relative speed compared to the superficial velocity of the liquid. Due to the high shear stress in the liquid film [1] the mass transfer in flow direction is enhanced. A first correlation for film thickness δ was given by Bretherton [8] with

$$\delta/D_h = 0.66 Ca^{2/3},\tag{20.3}$$

which was slightly extended by Aussilous and Quéré [5] for round capillaries to

$$\delta/D_h = \frac{0.66 Ca^{2/3}}{\left(1 + 3.33 Ca^{2/3}\right)},\tag{20.4}$$

giving a better fit at higher Ca. Eventually, for rectangular channels, the relative diagonal bubble film thickness is related to the capillary diameter by

$$\sqrt{2} - \frac{2\delta_{diag}}{D_h} = 0.7 + 0.5 \exp\left(-2.25 Ca^{0.445}\right)\tag{20.5}$$

as derived by Kreutzer et al. [15] from experimental observations.

However, liquid film thickness around a Taylor bubble is not a constant, but a function of axial distance behind the bubble front tip, and for channels with non-symmetric cross-section also a function of angular position. With known and constant channel geometry, liquid film thickness and Taylor bubble shape are equivalent. So far, some groups were able to compute plausible Taylor bubbles shapes using analytical models and CFD-based numerical codes, such as the accompanying work in this issue of Advances in Mathematical Fluid Mechanics. However, the exact 3D shape of a Taylor bubble was so far not been measured yet; thus this work attempts to provide data for analytical examination and quantitative comparison.

In the case of mass transfer so far several experimental and theoretical studies are known in which the effect of various parameters on mass transfer between gas bubbles and liquid slugs were investigated, as reviewed by the authors [11, 12]. Mass transfer is characterized by the ratio of the effectively transferred mass due to convection to the mass transfer by pure diffusion,

$$Sh = \frac{k_l d}{D_c} \tag{20.6}$$

where Sh is the Sherwood number, k_l is the liquid side mass transfer coefficient, D_c the gas molecular diffusion coefficient, and d the characteristic length, at which $d = d_{eq}$ is used for Taylor bubble flow with d_{eq} being the sphere-volume equivalent bubble diameter. Available correlations relate Sherwood numbers to Peclet or Reynolds and Schmidt numbers, Pe, Re and Sc, respectively,

$$Sh = f(Pe) \text{ or } Sh = f(Re, Sc) \tag{20.7}$$

with Pe describing the ratio of advective to diffusive transport rates,

$$Pe = \frac{U_b D_h}{D_c}, \tag{20.8}$$

and with Sc being the ratio of momentum diffusivity to mass diffusivity,

$$Sc = \frac{\mu}{\rho D_c}. \tag{20.9}$$

In comparison with the number of studies on the mass transfer from bubbles in infinite liquid, the number of investigations on the effect of the wall on the mass transfer rate is limited. Almost all of these studies are for channels larger than 12 mm inner diameter and so far little attention has been paid to smaller diameters. For small diameters, the importance and role of interfacial forces on the hydrodynamics and mass transfer of bubbles are not negligible and should be considered. Therefore, mass transfer in milli-channels was studied and correlations in form of a modified mass transfer coefficient were presented by the authors.

One of the main factors, which are known to have significant influence both on the hydrodynamics and mass transfer rate of phases, is the presence of surface active agents (surfactants). Surfactants are adsorbed at gas-liquid interfaces and decrease the surface tension. The presence of surfactants in multiphase systems, either in the form of unavoidable impurities or as additives, has great effect on the shape and dynamics of the interfaces. The most relevant studies on the effect of surfactants on the gas-liquid system were reviewed by the authors [13]. It is known, that the presence of surfactants may increase the terminal velocity of the bubbles due to a reduction of surface tension near the bubble nose, and that at certain surfactant concentration the liquid film thickness around a bubble is changed. Further, great effect on mass transfer as well as on drag coefficients of a rising bubble is reported. However, most attention has been paid so far to small spherical bubbles in infinite liquid regarding the effect on the mass transfer rate. Non-spherical large bubbles in small channels, which mainly exist in the form of Taylor bubbles, were subject to only a few studies [4, 18]. Regarding the influence of surfactant on the liquid film thickness for elongated bubbles in capillaries almost all of the investigations are theoretical and according to our knowledge very few experimental evidences could be found [9]. Therefore, the effect of surfactant on the shape, dissolution rate and liquid film thickness of an individual elongated Taylor bubble, whose motion is governed by the channel walls, was investigated by the authors.

20.3 Summary of Main Results

20.3.1 Taylor Bubble Shape in Small Channels

For the generation of high-resolution validation data of Taylor bubble shape, experiments were conducted in small channels with $D_h = 2\,\text{mm}$. Besides the required high spatial resolution of a few micrometers, the fast motion of the moving Taylor bubbles necessitated the application of a high-speed image measurement technique. These demands were satisfied by employing X-ray synchrotron radiography and X-ray synchrotron tomography. The experiment is described in detail by Boden et al. [6], new results with respect to bubble shape dependency on Ca are shown below.

The most suitable experimental setup to conduct such Taylor bubble experiments was identified as follows. A liquid is pumped through a vertically aligned test section (the capillary channel) by means of a pump which produced only little oscillation induced pressure fluctuations in the liquid. Via a T-junction at the bottom of the capillary channel a gas was discretely injected into the flowing liquid using a high-speed injection valve. Using an optical visualization approach it was revealed that it was possible to generate Taylor bubbles at sufficient reproducibility. Different Taylor bubble regimes might be realized by changing the duty cycle parameters of the gas injection valve. To minimize the influence of liquid acceleration during the periodic gas injection, gas bubbles were injected only after the preceding one left the capillary channel. Thus, a single bubble Taylor flow is considered here only. The hydraulic diameter D_h of the capillaries was chosen to be $2\,\text{mm}$, therefore the effect of buoyancy was negligible and only surface tension and viscosity forces acted on the bubbles in such capillary regime. The ratio $\lambda = d_{eq}/D_h$ was chosen to be $\lambda > 1.5$ to ensure fully developed Taylor bubbles, resulting in bubble lengths roughly more than $3D_h$. Experiments using only clean water with a dynamic viscosity of $\mu_{water} = 1\,\text{mPa\,s}$ would not cover Ca and Re numbers suitable for numerical simulations. Therefore, an aqueous 76.9% Glycerol solution with a viscosity of $\mu_{glycerol} = 33.5$ mPa s at 25°C was used. The liquid temperature was continuously monitored at the lower and upper end of the capillary channel; and the dependency of the liquid viscosity on temperature was considered during the analysis afterwards. The bubble velocity was varied between $U_b = 20-320$ mm/s resulting in Capillary numbers $Ca = 0.01-0.13$ with Reynolds numbers well below $Re < 50$. Experiments were carried out using capillaries both with circular and square cross section.

The experiments were conducted at the TOPO/TOMO beamline of the ANKA synchrotron facility in Karlsruhe, Germany. The TOPO/TOMO beamline provided at the time of the studies surpassing quality of high-speed X-ray imaging. The beamline was located on a bending magnet (1.5 T, bending radius 5.6 m) and was operated in white-beam mode. A 1.78 mm by 2.31 mm wide portion of the capillary was projected onto a scintillation screen based X-ray image detector whose camera was read out at rates up to 36,000 radiographic images per second. The effective pixel spacing was 5.6 μm. For the vertical circular capillary channel the projection image is independent on the angular orientation. For the channel with square cross section radiographic projection images were acquired with the channel walls aligned in parallel or orthogonal respectively to the detector plane. To obtain a cross-sectional view of such channel, a tomographic imaging approach was chosen.

Image processing algorithms were developed to extract the bubble shape information from the radiographic images. Inherent noise free imaging by applying longer exposure times was not applicable in order to prevent motion blur. Motion blur would render shape reconstruction almost impossible. Due to the short exposure times, the single images, also called projections, from the synchrotron experiment were corrupted by a fairly amount of noise. Further, due to the small beam diameter, the projections show only parts of the bubble. Image processing therefore comprised of bubble presence detection, velocity measurement, image superposition, three-dimensional image reconstruction if applicable, and interface extraction. Detection of bubble presence in a projection was possible by comparing the mean image brightness to a threshold value. For further processing, the average image showing the capillary only was subtracted from all projections, thus extracting the bubble only information from the data. The bubble's front and rear tip position then was determined by analyzing the brightness distribution along the projected capillary axis. Instantaneous bubble tip velocity was calculated from the successive changes in tip position. Knowing the bubble's tip position and velocity enabled the generation of an almost noise free image of the whole single bubble, as it passed the X-ray beam, by super-positioning of the consecutive projections with respect to their actual position. A careful analysis of the image brightness distribution at the edge of the projected bubble was used to extract the phase boundary between gas and liquid from the projection images. A detailed description of the algorithm can be found in the above mentioned reference [6]. The extracted projected two-dimensional (2D) Taylor bubble shape profiles were plotted with the length scale normalized to D_h. It is known that the curvature of the Taylor bubble's front and rear tip on the channel axis is a function of Ca. Further, the variation of interfacial curvature along the bubble length is made responsible for the capillary pressure in the liquid film around the bubble. Thus, curvature is one of the geometrical target quantities for quantitative comparison with theoretical and numerical models. Therefore, the signed local curvature of the measured 2D Taylor bubble shape profiles in the detector plane, κ_{2D}, were computed according to

$$\kappa_{2D} = \frac{x'z'' - z'x''}{\left(x'^2 + z'^2\right)^{3/2}} \tag{20.10}$$

with $x = x(t)$ and $z = z(t)$ being parametric representation in Cartesian coordinates of the projected bubble shape profile and $'$ and $''$ referring to the first and second derivative with respect to parameter t. For each point $(x(t), z(t))$ on the shape profile the center of curvature is located at a distance $R = 1/\kappa_{2D}$ (the inverse curvature) in the direction normal to the profile, as plotted in Figs. 20.1 and 20.2.

Analyzing the two-dimensional shape profiles from projections is sufficient for channels with circular cross-section assuming rotational symmetry. For square channels, however, a tomographic measurement approach is useful to obtain the full three-dimensional (3D) bubble shape. Tomography requires acquisition of projection images from different angular positions around the channel axis. This might be achieved by rotation of the setup within the X-ray beam. Fast rotation of the setup to observe a single moving Taylor bubble however was not feasible due to mechanical and fluiddynamical restrictions. Therefore, the channel was rotated only slowly, and Taylor bubbles were injected subsequently. The liquid velocity was held constant during the experiment, and from visual inspection it was validated that the Taylor bubbles appeared at the same shape. The tomographic data set thus represented an ensemble average image of the Taylor bubbles. To reduce the amount of generated image data, and to extend the overall measurement time, X-ray images of the liquid filled capillary channel between the passage of two Taylor bubbles were omitted. Therefore, the presence of each Taylor bubble in the channel was detected by two laser light barriers installed below and above the image region of interest. As the bubbles pass through the light beam a trigger signal was generated to produce start or stop trigger impulses. For tomographic image reconstruction, the projection data was preprocessed in such a way, that all acquired bubble images were aligned at a fixed bubble tip position. Image reconstruction was achieved by high-speed graphics processing unit (GPU) accelerated implementation of the Feldkamp reconstruction algorithm mimicking the parallel-beam geometry by application of a long source-object distance. The 3D reconstructed data was analysed with respect to bubble shape using a segmentation technique.

Figure 20.1 shows measured projected shape profiles of the Taylor bubbles' front tips in capillary channels, $D_h = 2\,\text{mm}$, with circular and square cross section. Different Capillary numbers Ca were realized by changing the flow rate of the co-currently flowing Glycerol solution. As Ca is increased, also the liquid film

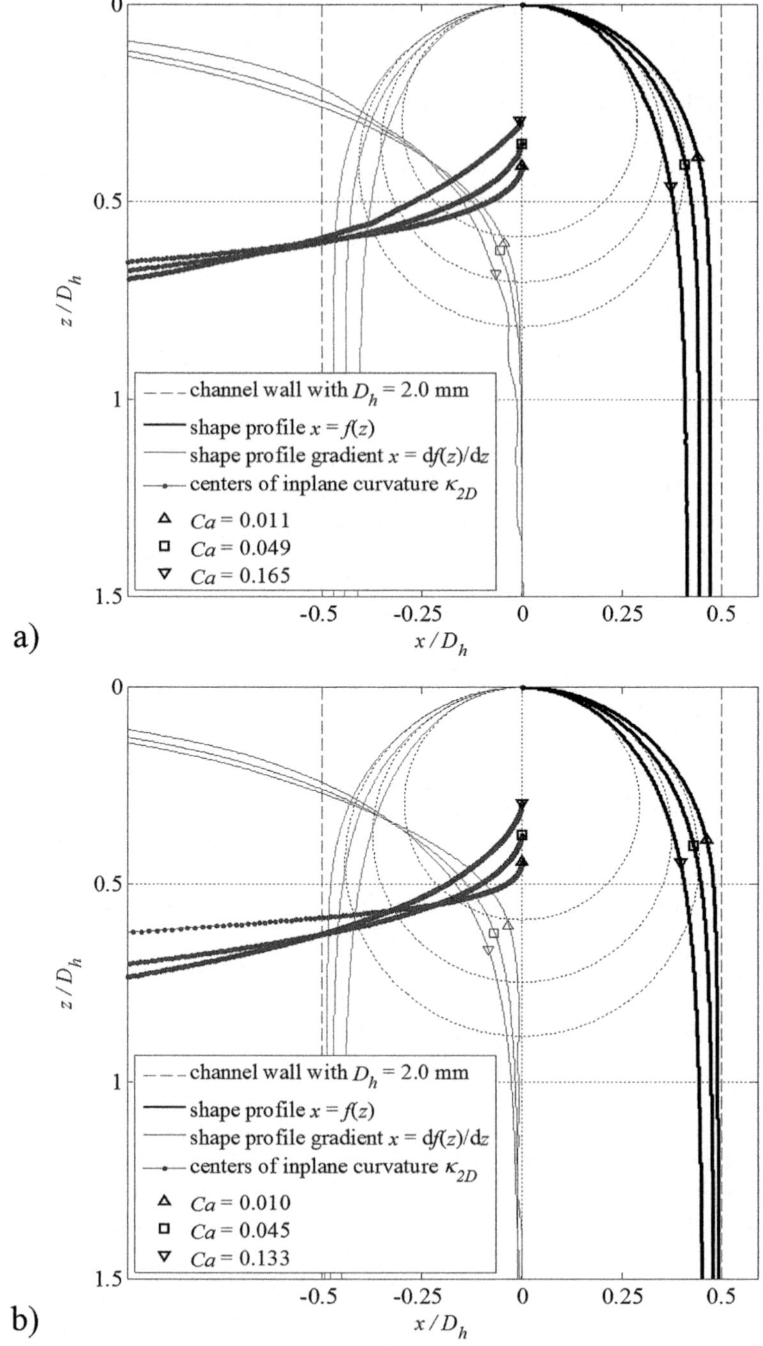

a)

b)

Fig. 20.1 Measured Taylor bubble front tip shapes in channel with (**a**) circular and (**b**) square cross section (parallel projection) at hydraulic diameter of $D_h = 2.0$ mm; and extracted shape profile gradients, centers of 2D curvature (see text) and osculating circle at bubble tip

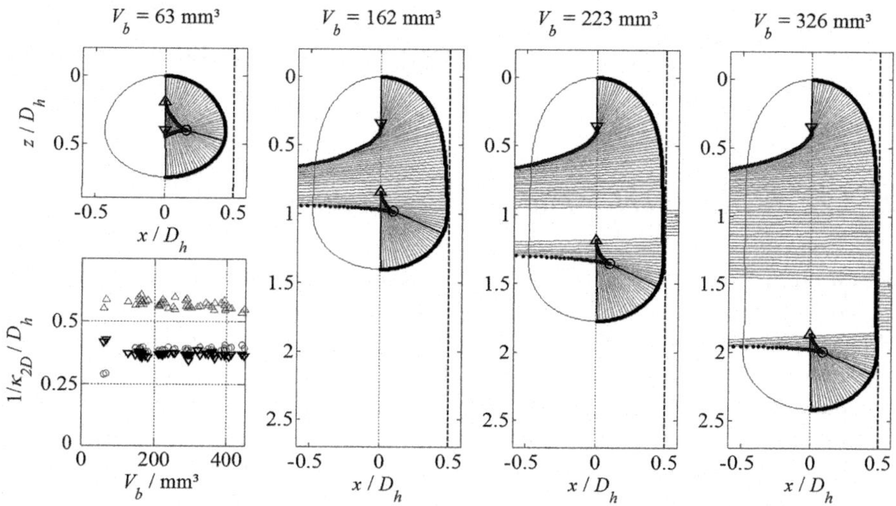

Fig. 20.2 Measured projected 2D shape profiles of gas Taylor bubbles in circular channel, $D_h =$ 6 mm, in countercurrent flow of clean water; and extracted centers of 2D curvature of in-plane shape profiles and corresponding radii of osculating circles with *open inverted triangle, open triangle* and *open circle* being centers of bubble's front tip, rear tip and rear maximum curvatures, respectively; the *dotted straight line* indicates the projected channel boundary

thickness increases accordingly. Such behavior was reported previously, see [15] for example. Also depicted are the corresponding osculating circles at the Taylor bubbles' front tips and the centers of curvature for each point on the shape profile. It becomes obvious that the osculating circles' radii become smaller for increasing Ca, thus curvature increases accordingly. If one follows the projected shape profile from front to rear, the corresponding centers of curvature rapidly depart from the Taylor bubble's vertical axis since the Taylor bubble's shape is confined by the channel walls and since only 2D inplane curvature is considered here. Then, in the liquid film region, 2D curvature becomes so small that centers of curvature are too far away from the channel axis and thus are not shown in Fig. 20.1. Local curvature may be exploited further when analyzing the local flow field and pressure gradients. The slope of the shape profiles with respect to the vertical axis is also shown in Fig. 20.1. The data was smoothed before the derivate was calculated; however, the results are still degraded by image noise. Nonetheless, it is shown that in the case of circular capillary channel cross-section, the liquid film thickness becomes constant $(dx(z)/dz \approx 0)$ already at locations $z \geq 1D_h$ behind the bubble tip. On the other hand, in the case of capillary channels with square cross-section, a constant film thickness cannot be observed for $z \leq 3D_h$, instead there the film thickness gradually decreases notably with $dx(z)/dz < 0$.

The obtained high-resolution Taylor bubble shape data was used for the validation of numerical simulation of Taylor bubble type flow using different numerical techniques [2, 16].

20.3.2 Taylor Bubble Shape in Milli-Channels

Synchrotron X-ray imaging provided high spatial resolution of a few micrometers paired with unsurpassed high-speed imaging of a few thousand images per second. However, access to synchrotron light sources is restricted and thus the number of experiments is limited. Therefore, also the microfocus X-ray imaging technique was qualified for the investigation of Taylor bubble shape in milli-channels. X-ray microfocus radiography enabled the measurement of dynamic processes, such as mass transfer, while X-ray microfocus tomography was utilized to quantitatively calibrate the measurements by exact evaluation of the three-dimensional (3D) Taylor bubble shapes. These techniques are described in detail by Haghnegahdar et al. [11] and Boden et al. [7].

The experimental setup was designed to meet the validation criteria as follows. A microfocus X-ray device was utilized for the visualisation. This enabled to conduct the experiments at laboratory scale at any time while high spatial image resolution still was provided. The X-ray beam of such sources divergently spreads out at large cone angles and, therefore, results in a magnifying projection geometry enabling X-ray visualization for both small and large scale objects. The spatial resolution depends only on the size of the X-ray source's focal spot; however, when low magnification is used the spatial resolution reduces gradually to the pixel size of the employed X-ray image detector. A microfocus X-ray source's beam intensity, however, is by magnitudes lower than that of a synchrotron source; thus, image exposure times were limited to not less than a tenth of a second. This rendered high-speed imaging unfeasible. Taylor bubbles moving at high bubble velocities will become tremendously blurred in the X-ray images. Further, mass transfer processes were considered here (gas bubble dissolution into the liquid). Such processes proceeded at time scales of more than several 10 s, a fast moving Taylor bubble would have tremendously moved during that process most likely leaving the image region of interest. Therefore, a single Taylor bubble was held at fixed position in a vertically aligned channel in a countercurrent liquid flow by a technique described by Schulze and Schlünder [19]. This enabled motion artefact free imaging even at image exposure times of as high as $T_{exp} = 100$ ms. The liquid was stored in a reservoir above the channel, and liquid flow was driven by gravity only to prevent pump induced flow oscillations. A motorized needle valve at the bottom of the channel was used to precisely set the required liquid flow rates. Motorization was necessary since manual access to the experimental setup during X-ray beam time was not possible. The gas was discretely injected into the liquid using a high-speed injection valve as described before, or using a calibrated syringe.

A X-RAY WorX XWT-190HC transmission target X-ray source was used providing a focal spot size of better than $5\,\mu$m for the X-ray parameters used here, which was the spatial resolution limit here. Deionized water with a dynamic viscosity of $\mu_{water} = 1\,$mPa s at a temperature of 25°C was used as liquid. To fixate a Taylor bubble's vertical position in countercurrent liquid flow, the superficial flow velocity u_s has to be at the same magnitude as the terminal velocity U_b of a corresponding rising Taylor bubble in stagnant liquid. In the case of small channels with $D_h \leq 2\,$mm the terminal velocity U_b drops almost to zero. This in turn will result in liquid film thickness of less than $\delta/D_h \leq 0.005$, which is up to now too demanding for numerical simulations. Therefore, hydraulic diameters in the rage of $D_h = 6 \dots 8\,$mm were used. At such D_h a Taylor bubble is freely rising in stagnant liquid at moderate U_b, while the Taylor bubble is still free of capillary waves and shape oscillations. That was verified by visual observations; such disturbances only notably occurred at D_h larger than 8 mm. The precisely manufactured glass channels had circular and square cross section.

Non-dissolving Taylor bubbles were used for measurement of Taylor bubble shape and for quantitative calibration of the measurement technique. Therefore, air was used as gas. Since the liquid (clean and contaminated water) was pre-saturated with ambient air, too, no observable change of bubble shape occurred during the course of each experiment. To extract Taylor bubble shape profiles from the radiographic projections, the same techniques as described above were employed. Multiple single radiographic projections were averaged to reduce the overall noise level and to increase the accuracy of the results. However, two-dimensional (2D) radiographic projections are not sufficient to obtain the true three-dimensional (3D) shape of Taylor bubbles in channels with square cross section. There, rotational symmetry of the bubbles is not given, rather, the bubbles may expand into the corners of the channel. Therefore, a tomographic approach was used and single Talyor bubbles were scanned while the experimental setup rotated within the X-ray beam at low speed. The projection images from various angular positions around the channel were reconstructed using a GPU-accellerated implementation of the reconstruction algorithm proposed by Feldkamp et al. [10]. Extraction of the gas-liquid interface from the reconstructed tomographic dataset revealed the measured true 3D volume and surface area of the gas Taylor bubble.

Figure 20.2 shows measured projected 2D shape profiles of air gas Taylor bubbles in a circular channel, $D_h = 6\,$mm, in countercurrent flow of clean water. The extracted centers of the 2D inplane curvature, κ_{2D}, as in Eq. (20.10), are plotted along with the corresponding radii of osculating circles for each point of the bubble shape profile. Four exemplarily chosen bubbles at different sizes are presented. The centers of curvature of each bubble's front tip, rear tip and at the maximum curvature at the bubble rear are marked in particular using triangle and circle symbols. Their location and magnitude are valuable criteria for quantitative evaluation of numerical models. Corresponding radii of osculating circles, $R = 1/\kappa_{2D}$, are presented for

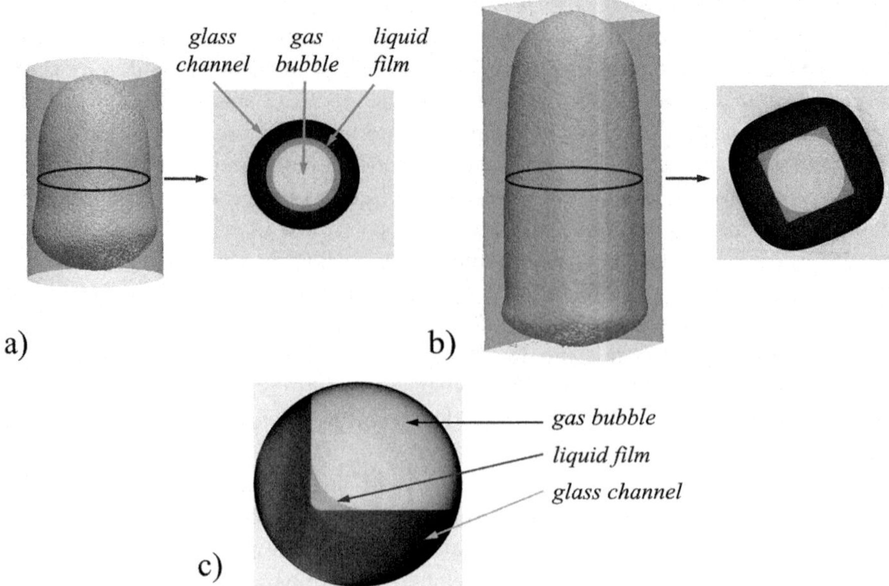

Fig. 20.3 Exemplary microfocus X-ray tomography measurements of Taylor bubble shape in channel, $D_h = 6\,\text{mm}$, with (**a**) circular and (**b**) square cross section, and selected corresponding horizontal slice images; (**c**) higher resolution may be obtained with a local tomography approach measuring only a small part of the channel's cross section

a large range of Taylor bubbles sizes in Fig. 20.2, lower left. It is shown that for Taylor bubbles with $d_{eq} > 1.1D_h$ these particular curvatures are almost independent of bubble size; for small bubbles, which however are no Taylor bubbles anymore, front and maximum rear curvature deviate from the Taylor bubble's values: only the rear curvatures of the two small bubbles presented here are still at the level of long Taylor bubbles. It should be noted, that the velocities of the countercurrently flowing liquid varied for the bubbles presented in Fig. 20.2 (compare with Fig. 20.6c).

The achievements using the microfocus X-ray tomographic measurement technique are demonstrated in Fig. 20.3. Displayed are exemplarily selected micro-tomographic measurements of the shape of non-dissolving Taylor bubbles in milli-channels, $D_h = 6\,\text{mm}$, with circular and square cross section, and selected corresponding horizontal slice images. From the segmented measured tomographic images true 3D bubble volume and size and shape of the interface between the gas and liquid phase immediately can be extracted. The knowledge of the later is crucial in understanding transport processes between the two phases; further, knowledge on the thickness of the liquid film between the Taylor bubble and the channel wall is required which easily can be obtained from 3D measurements. The geometric magnification of the setup was chosen in such way, that a whole Taylor bubble with bubble length $L_b = 3D_h$ still fits into a single projection; thus, effective pixel size was $27.5\,\mu m$. However, in the case of the milli-channel, $D_h = 6\,mm$, with square cross-section and fixed Taylor bubble in countercurrent flow of clean water, the liquid film thickness at the plane channel walls, which we call the thin film region, was well below that resolution and could not be resolved. Therefore, local tomography scans were performed measuring only a small part of the channel's cross section (Fig. 20.3c). From the analysis of the corresponding projection images the liquid film thickness in the thin film region was measured to be approximatively smaller than $5\,\mu m$.

20.3.3   Mass Transport from Taylor Bubbles in Milli-Channels

To foster the knowledge on mass transport processes between a gas Taylor bubble and the surrounding liquid (film), experimental work was conducted. Carbon dioxide (CO_2) gas was used for the mass transfer studies. These studies were conducted in milli-channels, $D_h = 6\ldots8\,mm$, with circular [11] and square [12] cross section in countercurrent flow and in stagnant liquid. Details on the experimental setup, data analysis and results can be found in the given literature references, and are summarized here only briefly.

During the experiments, Taylor bubbles at various sizes were injected into the liquid; and without liquid flow the bubble moved upward by buoyancy into the X-ray beam within less than $5\,s$. Then, the bubble's vertical position was stabilized, and radiographic projection images were captured during the dissolution of the gas Taylor bubble into the liquid. Since a back-diffusion of solved gasses from the liquid to the gas bubble occurred until equilibrium is reached, there is a gas bubble volume left at the end of the dissolution process. The magnitude of this final bubble volume $V_{b,\infty}$ depends on the initial bubble size $V_{b,0}$ since initial bubble size determines the duration of the dissolution process. The longer the dissolution process, the more gas can back-diffuse into the bubble causing larger final bubble volumes.

As the bubble size shrunk during the mass transfer process, changes in buoyancy and drag force caused a change in the bubble's relative velocity resulting in bubble motion, which had to be canceled out by continuous adjustment of the flow rate during the course of the experiments. Since the corresponding control loop was not perfect, slight bubble motion occurred causing vertical blurring in the X-ray projection images. Therefore, the integral radiographic extinction from the background-corrected Taylor bubble-only images was calculated only and served as a measure proportional to the gas bubble's volume; this measure was not susceptible to bubble motion. The measured bubble's integral extinction signal from single projection images was then calibrated to absolute bubble volume using aforementioned tomographic Taylor bubble visualization. Thereby it was assumed, that Taylor bubble shape solely depends on the hydrodynamic configuration, and is independent of the used gas.

In Fig. 20.4 the results of the dynamic observation of dissolving single gas Taylor bubbles with different initial bubble volumes in countercurrent flow of clean water is presented. The bubble volume reduction during the dissolution of CO_2 gas Taylor bubble in channels, $D_h = 6$ mm, with circular and square cross section for different initial gas bubble sizes is shown. These graphs represent the whole dissolution

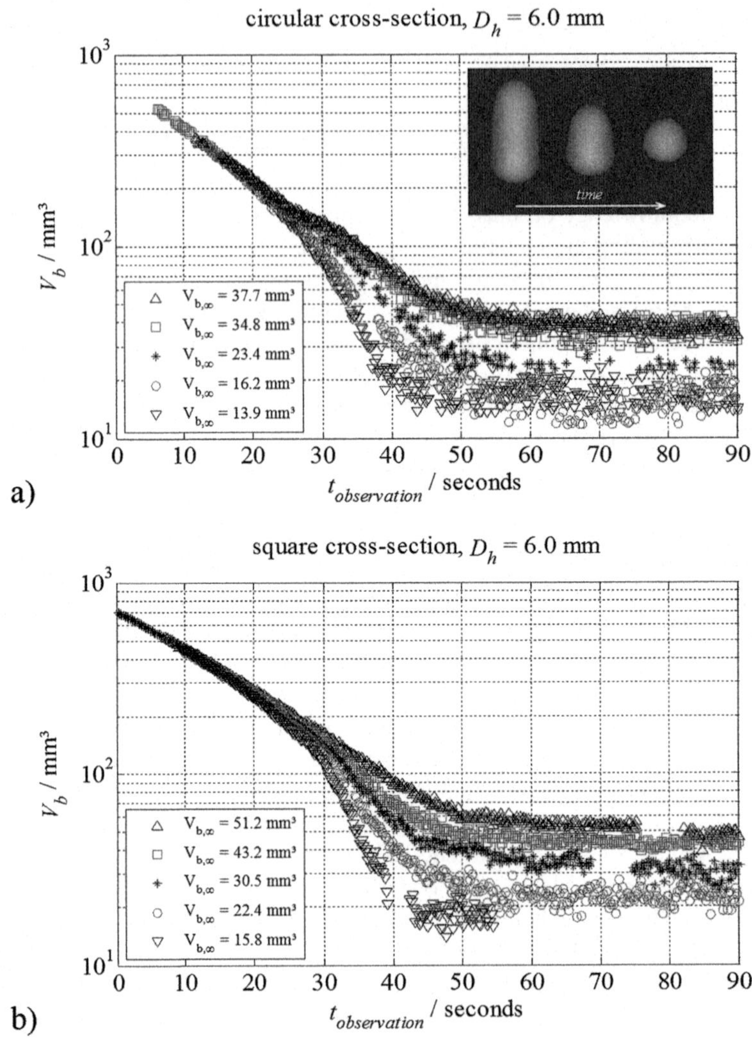

Fig. 20.4 Bubble volume reduction during dissolution of CO_2 gas Taylor bubbles in countercurrent clean water liquid flow in channels with (**a**) circular and (**b**) square cross-section with hydraulic diameters of $D_h = 6.0$ mm for different initial gas bubble sizes $V_{b,0}$. The graphs are artificially plotted in such way, that the observed volume for all Taylor bubbles at $t_{observation} = 0$ s is constant, $V_b(0\,\text{s}) = \text{const}$

process until equilibrium state is reached, which here was roughly after 50 s. A transition in dissolution behavior becomes obvious at about $t_{observation} = 30$ s, which might be related to a bubble shape transition from long Taylor bubble to shorter bubbles with different hydrodynamic behavior. Actually, a transition in instantaneous bubble rise velocity in stagnant liquid of bubbles with corresponding bubble volume can be observed in Fig. 20.6c in clean water, see below.

The bubble dissolution behavior was measured and mass transport coefficients and related dimensionless numbers at the first seconds of the dissolution process were calculated and should serve for validation and development of theoretical and numerical mass transport simulation tools.

20.3.4 Effect of Surfactant on Taylor Bubble Shape and Mass Transfer

Experiments to investigate the effect of surface active agents (surfactants) in the liquid on Taylor bubble shape and on the mass transfer were conducted. Therefore, a surfactant was added to the liquid and studies with non-dissolving and dissolving bubbles were performed. The results are presented by Haghnegahdar et al. [13], and are briefly reviewed here. In particular, additional results regarding the liquid film thickness are presented here first.

The nonionic compound Triton X-100, which is a widely used detergent in laboratories or cleaning materials, was used as surfactant to contaminate the clean water at different concentrations. Studies were conducted in milli-channels, $D_h = 6$ mm, with circular cross section. Carbon dioxide was used as gas during the mass transfer studies. The experiments were conducted as before, and the same tools developed before were used to analyse the data with respect to Taylor bubble shape, liquid film thickness and Taylor bubble rise velocity. Mass transfer coefficients were computed and presented. The microfocus X-ray radiography was a valuable technique, since it provided Taylor bubble shape profiles for direct comparison. A change in Taylor bubble shape indirectly indicates a change of the surrounding hydrodynamic conditions.

Figure 20.5 gives a qualitative impression of the changed Taylor bubble shape due to the contamination of the liquid in countercurrent flow conditions. To enhance the perception of the bubble shape, the images are horizontally stretched by factor 4/3, and brightness level contour lines are added to the image. Easily observable is the necking at the Taylor bubble's rear part in the contaminated case. A more quantitative description of the surfactant effect is given in Fig. 20.6a–c. The measured quantities of instantaneous liquid film thicknesses at the bubble's center (which here means at a vertical position in the midst between front and rear bubble tip), the instantaneous minimum liquid film thicknesses at the bubble's rear and the instantaneous bubble rise velocities of dissolving CO_2 gas Taylor

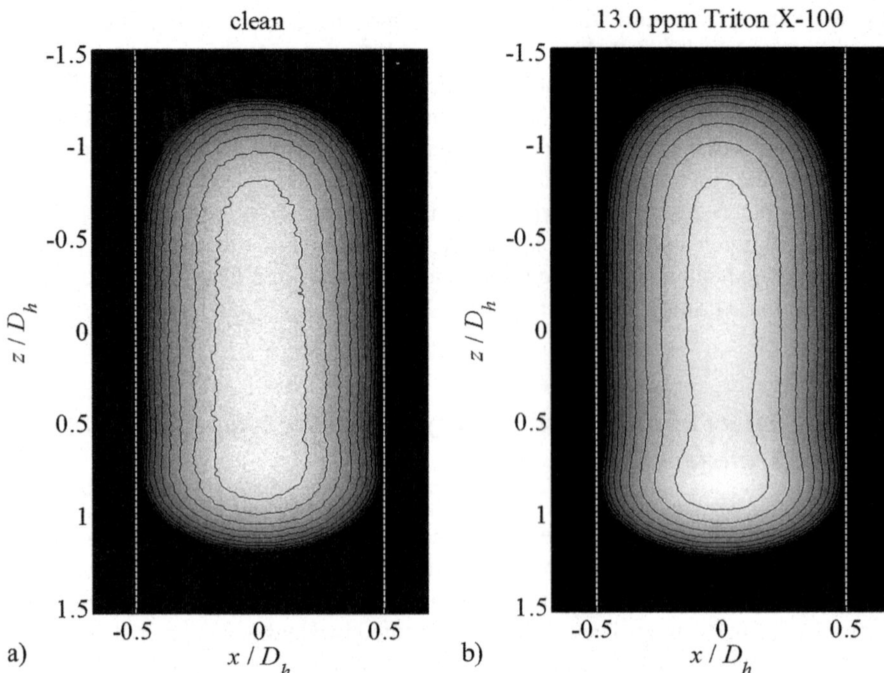

Fig. 20.5 Processed radiographic extinction image of air gas Taylor bubbles, $V_b = 320\,\text{mm}^3$, in a circular channel, $D_h = 6\,\text{mm}$, in countercurrent flow of (**a**) clean and (**b**) 13.0 ppm Triton X-100 contaminated water; *dotted white lines* indicate projected channel wall; brightness level contour lines are added to enhance perception of bubble shape, and the horizontal axis is stretched by factor 4/3

bubbles at different sizes in a vertical circular channel, $D_h = 6\,\text{mm}$, are plotted for comparison. The added surfactant has a thickening effect on the liquid film of the Taylor bubbles. Only for small bubbles, not any longer being Taylor bubbles and thus rising at increased velocities, the surfactant at the given concentration does not change the liquid film thickness. The increased liquid film thickness in turn causes the bubble to rise at higher velocities. However, for large (long) Taylor bubbles with $V_b > 350\,\text{mm}^3$, the minimum liquid film thickness at the bubble's rear seems to be unaffected from the surfactant at the given concentration. And also the rise velocity of such bubbles is similar to Taylor bubbles in the clean liquid. It is therefore concluded, that the Taylor bubbles minimum film thickness effectively controls the bubble's rise velocity for large (long) bubbles.

The obtained Taylor bubble shape profiles and quantification of related parameters such as liquid film thickness and bubble rise velocity are highly valuable to be used for validation of numerical CFD simulation tools.

Fig. 20.6 Measured instantaneous (**a**) liquid film thickness at the bubble center in the midst between front and rear tip, (**b**) minimum liquid film thickness at bubble rear, and (**c**) bubble rise velocity of dissolving CO_2 gas Taylor bubbles at different sizes rising in stagnant clean and 13.0 ppm Triton X-100 contaminated water in vertical circular channel, $D_h = 6$ mm, as function of bubble size

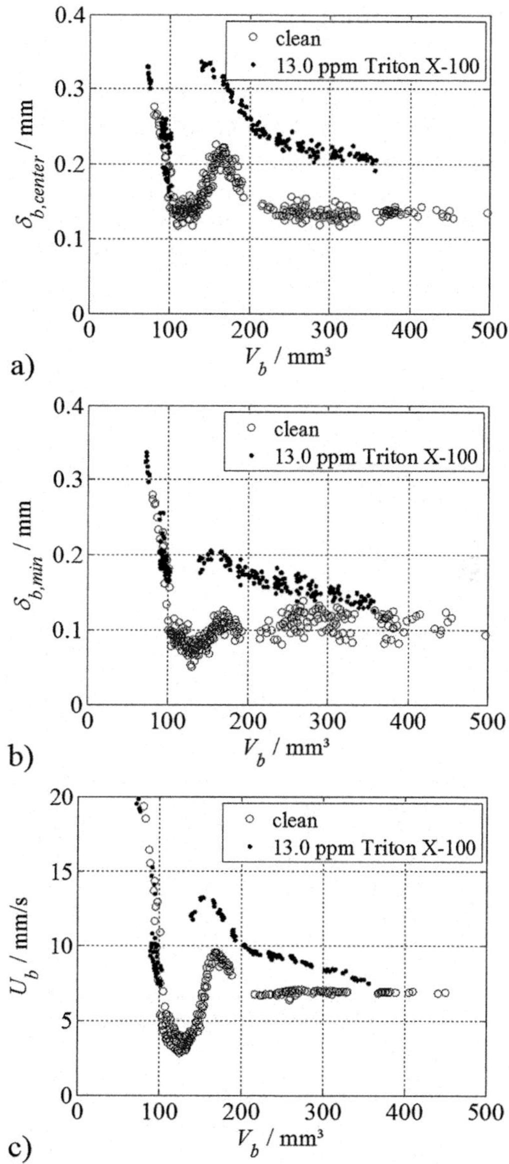

20.4 Outlook

The presented work demonstrated, that using X-ray radiographic and tomographic techniques high-resolution structural data of Taylor bubble shape for different hydrodynamic configurations and with mass transfer and the effect of surfactants was obtained and analyzed. A first validation of numerical techniques for simulation of Taylor bubble shape with sharp interface, diffusive interface, interface capturing and tracking techniques has been conducted. The available data is now ready to be used for studying the performance of the numerical tools with respect to mass transport and surfactant effects.

Acknowledgements The authors acknowledge the German Research Foundation (DFG) for funding the presented work within the projects HA3088/7-1 and HA3088/7-2, which were embedded within the priority program "Transport processes at fluidic interfaces", SPP 1506. The authors further acknowledge the provision of beam time by the synchrotron radiation source ANKA, which was partially funded by the German Federal Ministry of Education and Research by grant 05K10CKB.

References

1. Agonstini, B., et al.: Elongated bubbles in microchannels. Part I: experimental study and modeling of elongated bubble velocity. Int. J. Multiph. Flow **34**, 590–601 (2008)
2. Aland, S., et al.: Quantitative comparison of Taylor flow simulations based on sharp-interface and diffuse-interface models. Int. J. Numer. Methods Fluids **73**, 344–361 (2013). doi:10.102/fld.3802
3. Angeli, P., Gavriilidis, A.: Hydrodynamics of Taylor flow in small channels: a review. Proc. Inst. Mech. Eng. C J. Mech. Eng. Sci. **222**, 737–751 (2008)
4. Aoki, J., et al.: Effects of surfactants on mass transfer from single carbon dioxide bubbles in vertical pipes. Chem. Eng. Technol. **38**, 1955–1964 (2015)
5. Aussilous, P., Quéré, D.: Quick deposition of fluid on the wall of a tube. Phys. Fluids **12**, 2367–2371 (2000)
6. Boden, S., et al.: Synchrotron radiation microtomography of Taylor bubbles in capillary two-phase flow. Exp. Fluids **55**, 1768 (2014). doi:10.1007/s00348-014-1768-7
7. Boden, S., et al.: Measurement of Taylor bubble shape in square channel by microfocus X-ray computed tomography for investigation of mass transfer. Flow Meas. Instrum. **53**, 49–55 (2017). doi:10.1016/j.flowmeasinst.2016.06.004
8. Bretherton, F.: The motion of long bubbles in tubes. J. Fluid Mech. **10**, 166–188 (1961)
9. Daripa, P., Pasa, G.: The effect of surfactant on the motion of long bubbles in horizontal capillary tubes. J. Stat. Mech. Theory Exp. **2011**, L02003 (2011)
10. Feldkamp, L.A., et al.: Practical cone-beam algorithm. J. Opt. Soc. Am. A **1**, 612–619 (1984)
11. Haghnegahdar, M., et al.: Investigation of mass transfer in milli-channels using high-resolution microfocus X-ray imaging. Int. J. Heat Mass Transfer **93**, 653–664 (2016). doi:10.1016/j.ijheatmasstransfer.2015.10.033
12. Haghnegahdar, M., et al.: Mass transfer measurement in a square milli-channel and comparison with results from a circular channel. Int. J. Heat Mass Transfer **101**, 251–260 (2016). doi:10.1016/j.ijheatmasstranfer.2016.05.014

13. Haghnegahdar, M., et al.: Investigation of surfactant effect on the bubble shape and mass transfer in a milli-channel using high-resolution micofocus X-ray imaging. Int. J. Multiphase Flow **87**, 184–196 (2016). doi:10.1016/j.ijheatmasstransfer.2016.09.010
14. Hessel, V., et al.: Gas and liquid and gas and liquid and solid microstructured reactors: contacting principles and applications. Ind. Eng. Chem. Res. **44**, 9750–9769 (2005)
15. Kreutzer, M.T., et al.: Multiphase monolith reactors: chemical reaction engineering of segmented flow in microchannels. Chem. Eng. Sci. **60**, 5895–5916 (2005)
16. Marschall, H., et al.: Validation of interface capturing and tracking techniques with different surface tension treatments against a Taylor bubble benchmark problem. Comput. Fluids **102**, 336–352 (2014). doi:10.1016/j.compfluid.2014.06.030
17. Saisorn, S., Wongwises S.: A review of two-phase gas-liquid adiabatic flow characteristics in micro-channels. Renew. Sustain. Energy Rev. **12**, 824–838 (2008)
18. Sardeing, R., et al.: Effect of surfactants on liquid-side mass transfer coefficients in gas-liquid systems: a first step to modeling. Chem. Eng. Sci. **61**, 6249–6260 (2006)
19. Schulze, G., Schlünder, E.U.: Physical absorption of single gas bubbles in degassed and preloaded water. Chem. Eng. Process. **19**, 27–37 (1985)

Chapter 21
Experimental Investigation and Modelling of Local Mass Transfer Rates in Pure and Contaminated Taylor Flows

Sven Kastens, Christoph Meyer, Marko Hoffmann, and Michael Schlüter

Abstract In many industrial applications of chemical and bio-chemical engineering, new insights into mass transfer processes across fluidic interfaces are of high interest. Mass transfer processes across gas-liquid interfaces have been investigated for decades to understand the coupling of hydrodynamics and mass transport processes and to describe and correlate them for various gas-liquid flow apparatus and process parameters. The investigation of the linked transport processes and the understanding of their interaction is fundamental for the optimization of multiphase reactors and for the validation of numerical simulations, which are pointing at problems of higher complexity during the last years. One challenge for the investigation of gas-liquid flows is the highly stochastic behaviour of gas bubbles rising in liquids under turbulent flow conditions. For the investigation of local mass transfer processes at fluidic interfaces and the validation of numerical simulations, more well-defined and reproducible conditions are necessary. A suitable setup to study mass transfer at fluidic interfaces under well-defined and reproducible conditions is the gas-liquid flow through a small, straight capillary, called "Taylor bubble" for single bubbles and "Taylor flow" for bubbles in a chain. Taylor flows and Taylor bubbles have ideal properties for detailed investigation on the influence of hydrodynamics and mass transfer at clean and contaminated interfaces, where the shape oscillations are suppressed and the Taylor bubbles are self-centering within vertical channels. Therefore, in this work the local hydrodynamics and mass transfer processes in Taylor flows and at Taylor bubbles have been investigated with laser measurement techniques, to obtain a deeper insight into mass transfer processes at fluidic interfaces. Furthermore, experimental data for the guiding measure "Taylor flow" has been provided. The guiding measure has been established within the SPP 1506 to generate a reliable data basis for the validation of numerical simulations.

S. Kastens • C. Meyer • M. Hoffmann • M. Schlüter (✉)
Institute of Multiphase Flows, Hamburg University of Technology, Eißendorfer Str. 38, 21073 Hamburg, Germany
e-mail: sven.kastens@tuhh.de; christoph.meyer@tuhh.de; marko.hoffmann@tuhh.de; michael.schlueter@tuhh.de

© Springer International Publishing AG 2017

D. Bothe, A. Reusken (eds.), *Transport Processes at Fluidic Interfaces*, Advances in Mathematical Fluid Mechanics, DOI 10.1007/978-3-319-56602-3_21

21.1 Introduction

The appearance of Taylor flow is found in a variety of practical applications, such as multiphase monolith micro-reactors [26], heat-exchanger reactors [34] and fuel cells [9]. The superior heat and mass transfer properties are characteristic for this type of flow. Primarily short diffusion paths, large interfacial area, narrow residence time distribution and recirculation within the liquid slugs enhance transport processes in Taylor flows.

The leading edge LE, the trailing edge TE and the liquid film between channel wall and bubble are indicated in Fig. 21.1. An obvious characteristic length scale of the Taylor bubble is its length λ_B. The region between the LE and the TE of two neighbouring Taylor bubbles is the liquid slug that is defined by the slug length λ_S. One of the main quality of Taylor bubbles is the approximately similar dimensions of all of them (i.e. length, width and curvature). In addition the distance between successive Taylor bubbles (i.e. slug length) is nearly constant. In micro- and minichannels Taylor bubbles typically have diameters corresponding to the channel width. However, there exists a thin liquid film between bubble and channel wall (see Fig. 21.1).

To study hydrodynamics and mass transfer at fluidic interfaces, Taylor flows and Taylor bubbles are very useful because of their well-defined flow conditions and high reproducibility. In vertical channels, it depends on the ratio of buoyancy and surface tension whether the Taylor bubble rises driven by buoyancy and induces a flow field by itself around the bubble or whether it needs to be transported through the channel by the liquid flow driven by pressure difference. For Taylor bubbles with an Eötvös number higher than Eo > 4, the buoyancy force dominates the interfacial tension and therefore, the bubble will rise. The bubble is rising by buoyancy as long as the pipe diameter is larger than the critical inner diameter $D_{crit.}$ calculated by the critical Eötvös number $Eo_{Dcrit.} \approx 4$ [16, 43].

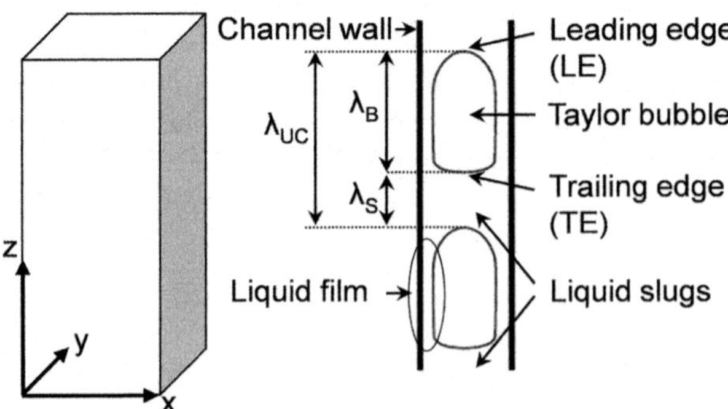

Fig. 21.1 Schematic draft of a Taylor flow with two Taylor bubbles [31]

The Eötvös number is defined as the ratio of the buoyancy force to the interfacial tension force according to

$$Eo_D = \frac{(\rho_L - \rho_G)g D_h^2}{\sigma} \tag{21.1}$$

where σ is the interfacial tension, ρ_L and ρ_G are the densities of the liquid and gas phases, g is the magnitude of the gravitational acceleration, D_h is the hydraulic diameter of a channel. For a water-air system, the critical pipe diameter that leads to a rising bubble through buoyancy is about $D_{crit} = 5.4$ mm.

21.1.1 Hydrodynamics in Taylor Flows

The hydrodynamics of Taylor flows is still discussed controversially, especially concerning the flow within the liquid film and the influence of surfactants.

Taylor flows were originally studied experimentally by Fairbrother and Stubbs [11], Taylor [37] and theoretically by Bretherton [7]. Taylor postulated the presence of a recirculation flow pattern in the slug. The existence of these recirculating flow patterns was experimentally confirmed by Thulasidas et al. [39]. The latter author used particle image velocimetry (PIV) to study Taylor bubbles. Nowadays, PIV is an incontestably measurement technique, but in the period wherein [39] published their paper PIV was still a highly controversial measurement technique. The PIV observations of these scientists enabled the determination of the velocity profile at various (streamwise) locations in the liquid slugs. The study of Thulasidas et al. [39] was based on Taylor flow in both, circular-shaped capillaries (diameter of 2 mm) and square-shaped channels for a wide range of capillary numbers (10^{-4} < Ca < 0.8). The capillary number Ca = $\mu U_B/\sigma$ is considered as the main dimensionless parameter in Taylor flows and it represents the ratio between viscous and surface tension forces.

After developing the first micro particle image velocimetry (μPIV) setup by Santiago et al. [35], numerous researchers carried out experiments with Taylor bubbles using μPIV [14, 15, 18, 23, 30, 40]. The study by Fouilland et al. [14] described an experimental investigation of time-resolved μPIV in order to study the flow characteristics of a gas-liquid system for flow regimes spanning Taylor flow to annular flow (i.e. not exclusively in the liquid film between gas bubbles and channel wall). In the annular flow, when the gas superficial velocity is high compared to the liquid superficial velocity, the inertia of the gas can become important. The velocities of the liquid (water) and gas (nitrogen flow in the Fouillands study are in the range of 0.35–8.65 m s^{-1} (40 < Re_G < 1000) and 0.071–0.18 m s^{-1} (120 < Re_L < 300), respectively. Higher bubble (or slug) Reynolds numbers (Re = $\rho D_h U_B/\mu$) have an influence on the Taylor bubbles shape.

Fouilland et al. used a circular-section tube to study horizontal co-current gas-liquid two-phase flow. Significant bubble asymmetry was observed for the Taylor flow in three circular channels of different internal diameters (d $=$ 1.12, 1.69, 2.12 mm). Leung et al. [28] described the effects of gravity on the bubble behavior for Taylor flows in millimeter-sized channels. In square channels, the bubble shape is axisymmetric for Ca $>$ 0.04 [25] and is complex for Ca $<$ 0.04. For the non-axisymmetric case, when the gas bubble displaced liquid in the channel corners, the thickness of the liquid film is non uniform. The determination of the thickness of the liquid is practically challenging. The liquid film thickness is essential in many applications, i.e. as catalyst coating in monolith reactors [24] or mass transfer from the channel wall to the liquid [5, 26].

Two different methods exist measuring the liquid film thickness experimentally [17]. The film thickness is directly measured by using high quality images [4]. For the other—indirect—method, the bubble velocity is determined experimentally and then the film thickness is calculated using the continuity condition. For the calculation of the film thickness the knowledge of the velocity profile in the liquid film is required, which is commonly unknown a priori. The thickness of the liquid film and the flow field in the liquid film are key parameters in order to get insights in the hydrodynamics of Taylor flow.

As the liquid film in micro- or minichannels is in the range of tens of microns, the velocity profile in the liquid film is a tremendous challenge. Some researchers have treated the film as being stagnant [36, 42], others assumed a fully developed annular flow velocity profile or a fully-developed velocity profile in the liquid film with a no shear boundary condition at the interface [38]. Agostini et al. [1] observed that for an elongated bubble the average velocity in the liquid layer has to be counter current to the bubble velocity because the bubbles travel faster than the liquid slug ahead, thus some liquid has to be evacuated by the liquid layer around the bubble.

One reason for the contradictory results for the hydrodynamics in Taylor flows might be the challenge of a stationary Taylor bubbles formation (i.e. without velocity fluctuations). The study of [40] shows that velocity fluctuations of Taylor bubbles in a microchannel are caused by pressure fluctuations during the formation of the bubbles.

The hydrodynamic of Taylor flows, particular the flow field in the liquid film but also the influence of surfactants, is still uncertain.

Although the surface tension forces dominate over buoyancy in small channels, it exists a limited number of publications studying the influence of surfactants on the Taylor flow. Olgac and Muradoglu [33] investigated numerically the effect of surfactants on the liquid film thickness. The authors found out, that surfactants generally have a thickening effect on the film thickness and that surfactants reduce the mobility of the interface due to increasing Marangoni stresses.

Hayashi and Tomiyama [20] numerically studied the surfactants effects on the terminal velocity of a Taylor bubble. They used the surfactant properties of Triton X-100 to investigate the influence of the adsorption process. They showed, that Triton X-100 accumulates mostly in the trailing edge while the leading edge and also the side regions are almost free from surfactants where the shear flow at a rising taylor bubble is directed downwards and the resulting maragoni force is pointing against the shear. In a Taylor flow, the internal circulations in the liquid slugs, show more complex local shear flows at the interface, which will influence the interfacial surfactant distribution along the bubble.

21.1.2 Mass Transfer in Taylor Flows

Another key parameter that strongly depends on the hydrodynamics in Taylor flows is the mass transfer performance. Mass transfer processes at Taylor flows in small channels ($D_h < D_{crit.}$) have been investigated by van Baten and Krishna [5] using CFD and compared with experimental data obtained by Vandu et al. [41]. For volumetric mass transfer coefficients $k_L a$ for O_2-water Taylor flows in vertical capillaries ($D_h = 1-3$ mm) a good agreement between experiments and simulations has been obtained. Their $k_L a$ model considers the pipe diameter as a very important parameter, since the CFD result showed that the major contribution to mass transfer is the contribution of liquid film, whose thickness strongly depends on the pipe diameter [29]. Despite the fact that the liquid film geometry is strongly affected by the geometry of the channel cross section, Vandu et al. [41] did not confirm a significant influence on the volumetric mass transfer coefficient for a O_2-water system.

When $D_h > D_{crit.}$, it is known that the rise velocities v_B of Taylor bubbles are independent of the bubble volumes. The independence of v_B from the bubble volume facilitates the fixation of Taylor bubbles at a certain position within the vertical channel by a downward flow. Filla [13] has shown the applicability of bubble fixation by a downward flow and has measured the liquid-side mass transfer of single CO_2 Taylor bubbles with a photographic technique in a vertically tapered channel. Esteves and de Carvalho [10] conducted experiments on the mass transfer of very long single Taylor bubbles, of which the ratio of bubble length to D_h is larger than 10, in large channels of $D_h \geq 32$ mm. Hosoda et al. [21] carried out experiments on the mass transfer from single CO_2 bubbles in vertical pipes of $12.5 \leq D_h \leq 25$ mm. They have developed the Sherwood number correlation for Taylor bubbles rising in a vertical pipe

$$Sh = \frac{k_L d_{eq}}{D_L} = (0.49\lambda^2 - 0.69\lambda + 2.1)Pe^{1/2} \qquad (21.2)$$

where k_L is the mass transfer coefficient, D_L the diffusion coefficient and λ the diameter ratio, which is the ratio of sphere-volume equivalent bubble diameter d_{eq} to D_h. Here the Peclet number Pe is defined by

$$Pe = \frac{v_B d_{eq}}{D_L} \qquad (21.3)$$

Equation (21.2) is valid for $\lambda \geq 0.6$ and applicable to the prediction of the long-term dissolution process of single CO_2 Taylor bubbles in water. Aoki et al. investigated the influence of various concentrations of Triton X-100 and 1-octanol on the rise velocity and mass transfer coefficient of bubbles in larger channels and compared their experiments with numerical simulations [2, 3]. The effect on rise velocity was much smaller than on the mass transfer and for small bubbles stronger than for Taylor bubbles. In their case, it was notable, that the ratio of the "clean" interfacial area to the total interfacial area is increasing, for increasing bubble volumes, where the mass transfer coefficients nearly converge for Taylor bubbles in clean and contaminated systems. But for intermediate and small bubbles, the barrier effect of the accumulated surfactant layers, where strongly reducing the diffusive mass transfer across the interface.

Though several studies on the mass transfer of Taylor bubbles in channels were carried out as discussed above, our knowledge on the mass transfer from single buoyancy-driven Taylor bubbles in smaller channel diameters ($D_{crit.} \leq D_h \leq 12.5\,mm$) and on the effects of channel cross sections are still rudimentary. Furthermore, the influence of contaminants (e.g. surfactants) on the mass transfer performance of buoyancy-driven Taylor bubbles in small channels, where the influence of interfacial tension becomes higher, is still unknown.

21.2 Experimental Setup and Methods

For the detailed investigation of Taylor flows, two experimental setups have been designed which enable the investigation of well-defined and reproducible Taylor bubbles in vertical glass capillaries [22, 31]. Whereas for the experimental investigation of the hydrodynamics in Taylor flows a pressure-driven system with square cross section ($D_h = 2.1\,mm$) is appropriate (setup "Taylor flow"), for the detailed investigation of the local mass transfer performance a buoyancy driven Taylor bubble in a capillary of circular cross section ($D_h = 6\,mm$) is more useful (setup "Taylor bubble"). Both experimental setups will be introduced and compared in the following chapters.

21.2.1 Experimental Setup for the Investigation of Taylor Flows

The experimental setup used for the generation of well-defined Taylor flows is shown in Fig. 21.2. The experimental observations are carried out in a vertical oriented setup to avoid gravitational influences on the velocity fields, both in the liquid slug and in the liquid film. The liquid is fed from a pressure container (UCON AG Containersysteme KG) into the bubble generation unit to prevent fluctuations

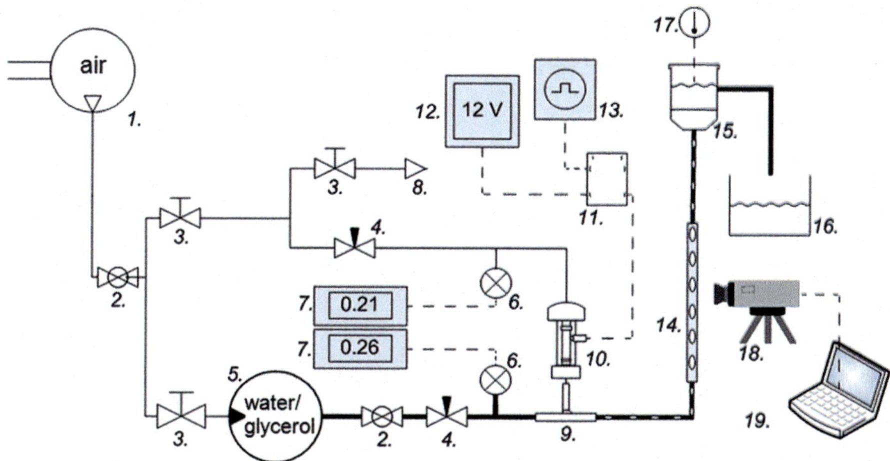

Fig. 21.2 Scheme of a Taylor flow experimental setup: 1. Compressed air supply, 2. Ball valve, 3. Needle valve, 4. Needle valve with fine control valve, 5. Pressure container, 6. Pressure sensor, 7. Digital pressure gauge (bar), 8. Vent, 9. Bubble generation unit, 10. Solenoid injection valve (EV 14, Bosch Engineering GmbH), 11. Input signal unit, 12. Power supply, 13. Frequency generator, 14. Glass capillary, 15./16. Outlet, 17. Thermometer, 18. High-speed video camera, 19. PC

in the bubble velocity by fluctuating pressures. The bubbles are introduced in accordance with a method of [32] by a solenoid injection valve (EV 14, Bosch Engineering GmbH) with a certain, well-defined frequency into a compensation pipe, to create stable and reproducible Taylor bubbles [31]. To avoid buoyancy-driven bubble rise, the square channel is chosen with a hydraulic diameter of approx. $D_h = 2\,mm\,(Eo < 4)$. The accurate diameter has been confirmed by a scanning electron microscope measurement (ZEISS LEO 1530 Gemini) to be $D_h = 2.1\,mm$. For accurate observations of the velocity field within the liquid film, refractive index matching (RIM) has been used. Therefore the channel is surrounded by a duct made of borosilicate glass. The gap is filled with a DMSO (Dimethyl sulfoxide, Sigma Aldrich)-water solution for refractive index matching. For an optical access into the channel with minimized optical disturbance, the concentration and temperature of the DMSO solution was kept at 95 wt.% and $T_{DMSO} = 20 \pm 1.0°C$, respectively, to obtain the same refractive index as borosilicate glass (n = 1.473).

The bubble velocity U_B, the bubble length λ_B and the slug length λ_S have been measured with a high-speed camera (Optronis GmbH, 1000 fps). Based on these optical measurements the bubble velocity is determined over numerous frames.

The observations of the fluid velocities are made with the help of a micro Particle Image Velocimetry (μPIV) system (ILA GmbH). The inverted microscope (Olympus IX71) has been tilted by 90° to enable μPIV measurements at a vertical capillary. Fluorescent particles (3.1 and 1.5 μm polystyrene particles, microparticles GmbH) are excited by two consecutive laser pulses (wave-length 532 nm, Nd:YAG-Laser, Quantel). The measurements are made with a 4× and a 10× microscope

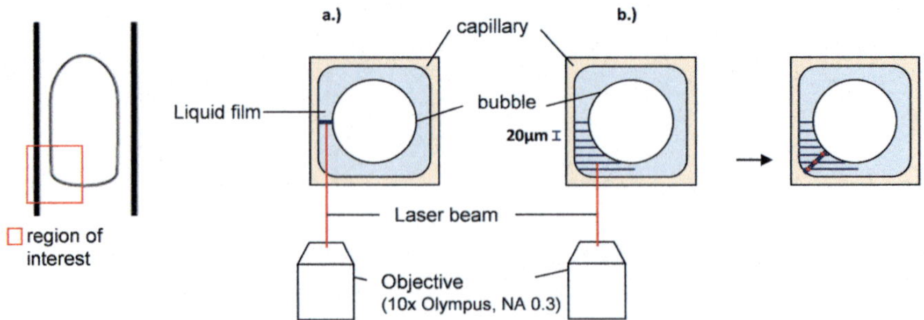

Fig. 21.3 *Left*: Schematic lateral view on the bubble, indicating the region of interest in the liquid film; *Right*: Cross-sectional view of the test section, i.e. μPIV measurements for the velocity fields inside the liquid film: (**a**) lateral, (**b**) construction of the diagonal film

objective with numerical apertures of 0.1 and 0.3. After capturing the images by the CCD camera PCO Sensicam QE, the flow field is analysed by a cross correlation algorithm (software VidPIV 4.7, ILA GmbH). The interrogation size is chosen at 32 pixels × 32 pixels with an overlap of 50%, resulting in a vector-to-vector spacing of about 25 μm [31].

All μPIV measurements have been performed in the measurement frame according to Fig. 21.3. The setup is positioned on a compact precision xystage (OWIS, PKTM 100). By means of the X-Y stage it is possible to traverse the setup in such a way that the lateral film can be measured in several steps (i.e. 20 μm steps, see Fig. 21.3b). To obtain the diagonal film, several points along the plane diagonal are extracted (marked red in Fig. 21.3b).

As liquid phase a glycerol-water mixture (78 wt.% glycerol) has been used and air as gaseous phase. The quality of the purified water is quantified by conductivity measurements less than $\kappa = 1.2\,\mu S\,cm^{-1}$. All measurements have been controlled in temperature by $T_L = 20 \pm 0.5\,°C$.

21.2.2 Experimental Setup for the Investigation of Taylor Bubbles

The experimental apparatus for the investigation of Taylor bubbles consists of a vertical channel, an upper tank, a valve, one/two high-speed video cameras or an LED light source as shown in Fig. 21.4.

The test section consists out of a vertical glass channel of $L = 300\,mm$ length. Four circular pipes, $D_h = 5.5, 6, 7$ and 8 mm, and two square ducts, $D_h = 6$ and 8 mm, were used to investigate the effects of channel geometry on hydrodynamics and mass transfer. All channels are surrounded by a duct made of borosilicate glass, filled with a 97% wt. DMSO water solution, to enable refractive index matching in accordance to the experimental setup "Taylor flows".

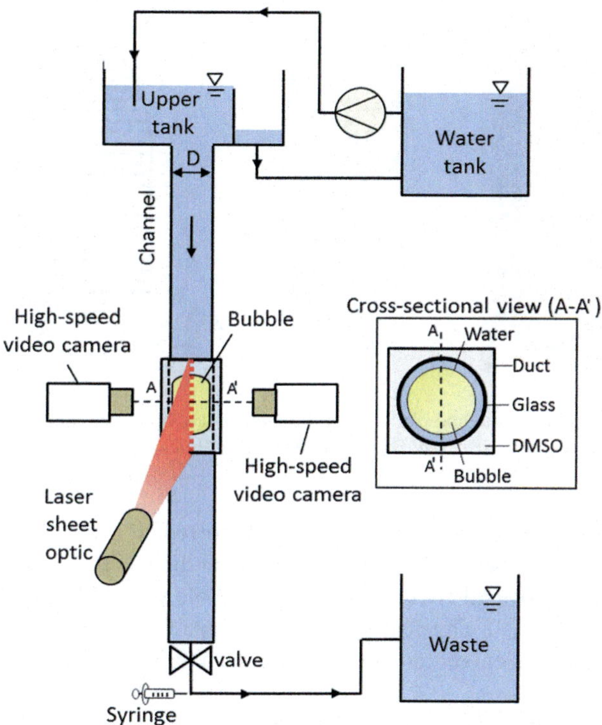

Fig. 21.4 Setup for experiments with single Taylor bubbles in stagnant liquid or counter current flow

A predetermined amount of CO_2 (99.995 vol.% purity, Westfalen AG) is stored in a precise gas-tight syringe (Hamilton 1001) and injected into the setup, filled with deionized water. The conductivity of the water is less than $\kappa = 1.2\ \mu S\ cm^{-1}$. The liquid temperature is set by a thermostat at $T_L = 25 \pm 0.5\ °C$.

The mass transfer coefficient of the Taylor bubbles is measured in a downward flow. Therefore, the liquid flow rate is regulated by a valve so as to keep the bubble suspended at 360 mm below the water surface in the upper tank. This results in a constant hydrostatic pressure condition during the observed dissolution process.

The images of each bubble are recorded at a fixed position in the centre of the observation field of a high-speed video camera (Optronis GmbH, frame rate: 250 frame s^{-1}, exposure time: 1 ms, spatial resolution: 0.08 mm pixel^{-1}), which is placed right-angled to the pipe. Figure 21.5 schematically shows the image processing method for instantaneous bubble volume and diameter measurements. An original bubble image (Fig. 21.5a) is transformed into a binary image (Fig. 21.5b). Since the refractive indices of the water ($n = 1.333$) flowing inside the channel and the glass channel ($n = 1.473$) were different, the angle of reflection differs from the angle of incidence. This is why the camera observed a virtual projection of the bubble with smaller dimensions, where the real bubble would have been

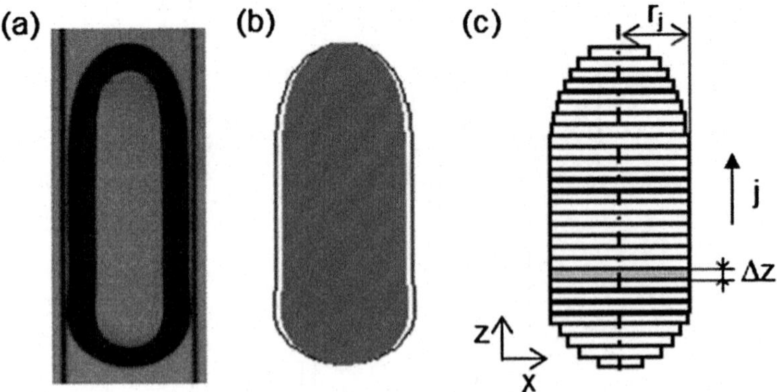

Fig. 21.5 Reconstruction method for Taylor bubble acquisition (reproduced from [22]). (**a**) Original image. (**b**) Binary (*gray region*) and corrected (*out line*) images. (**c**) Reconstruction

slightly underestimated. The binary image was corrected using Snell's law, where the real bubble position is shown with the outer line (Fig. 21.5b). The method assumes that all horizontal cross-sections of a bubble were circular. The height Δz of each circular disk is a physical length of one pixel. The resultant circular disks are piled up in the vertical direction to reconstruct a three-dimensional bubble shape (Fig. 21.5c).

The sphere-volume equivalent bubble diameter is evaluated using

$$d_{eq} = \left(\sum_{j=1}^{N} (6r_j^2 \Delta z) \right)^{1/3} \tag{21.4}$$

where N is the total number of pixels in the vertical direction, j the index denoting the pixel number in the vertical direction and r the disk radius. This post-processing method was validated by analysing images of a rigid sphere of $d = 5.0\,\text{mm}$ in diameter falling through a $D_h = 6.0\,\text{mm}$ capillary. The uncertainty of the diameter measurement has been estimated for a confidence of 95% by less than 1%. Accurate d_{eq} for bubbles in square ducts, however, cannot be obtained by using this method, since their horizontal cross-sections were not axisymmetric [6, 19]. For this reason, a linear regression curve $d_{eq} = a\,d_{eq}^* + b$ has been used, that takes into account the visually measured d_{eq}^* and the real d_{eq}, known by the injected volume. The constants a and b are 1.10 and -0.528 for $D_h = 6.0\,\text{mm}$, and 1.09 and -0.593 for $D_h = 8.0\,\text{mm}$, respectively.

The rise velocities v_B of bubbles in stagnant liquid in the channels have been measured using the vertical position $z(t)$ of a bubble nose where t is the time. The z data were fitted with a linear function using the least square method to obtain v_B. Uncertainty estimated at 95% confidence in v_B was 0.05% [22].

21.2.2.1 Evaluation of Mass Transfer Coefficients

The mass transfer coefficient k_L is evaluated from the decreasing rate of d_{eq} of a shrinking CO_2 bubble, like in [22]. The rate of change in the mole M of CO_2 inside a bubble is given by

$$\frac{dM}{dt} = -k_L A (C_S - C) \tag{21.5}$$

where $A(= \pi d_{ed}{}^2)$ is the bubble surface area calculated with the equivalent bubble diameter d_{eq}, C_S the CO_2 concentration at the bubble interface ($34\,\text{mol m}^{-3}$) and C the CO_2 concentration in water. The bubble surface area is estimated by the sphere volume equivalent area due to the established investigations of bubbles in vertical channels and to ensure the comparability of the results from small bubbles to Taylor bubbles. The C_S can be calculated according to Henry's law, which is given by

$$pX = \frac{C_S}{C_S + C_V} H \tag{21.6}$$

where p is the pressure inside the bubble, X the mole fraction of CO_2, C_V the molar concentration of water ($55.4\,\text{kmol m}^{-3}$) and H the Henry constant ($166\,\text{MPa}$ for CO_2 at $T_L = 298\,\text{K}$). Since the dissolution rates were very small in all the experimental runs, X can be postulated to be unity. C_V is much larger than C_S, i.e. $C_V/C_S \approx 1600$, so that C_S is given by

$$C_S = \frac{C_V p}{H} \tag{21.7}$$

The pressure p is given by

$$p = p_a + \rho_{LG} z \tag{21.8}$$

where p_a is the atmospheric pressure and z is the distance from the free surface to the bubble nose, i.e. $z = 360\,\text{mm}$. Equation (21.8) presumes that the contribution of surface tension is negligible. Substituting Eq. (21.7) into Eq. (21.5) and neglecting C, yields

$$k_L = -\frac{1}{\pi d_{eq}{}^2} \frac{H}{C_V p} \frac{dM}{dt} \tag{21.9}$$

By assuming that CO_2 is an ideal gas, we can express dM/dt in terms of p and d_{eq} as follows:

$$\frac{dM}{dt} = \frac{\pi}{6RT_L} \frac{d(p d_{eq}{}^3)}{dt} \tag{21.10}$$

where R is the universal gas constant. Substituting Eq. (21.10) into Eq. (21.9) and evaluating $d(pd_{eq}^3)/dt$ by using a centred difference scheme, yields

$$k_L = \frac{H(p_2 d_{eq,2}{}^3 - p_1 d_{eq,1}{}^3)}{6RT_L C_V p_{12} d_{eq,12}{}^2 (t_2 - t_1)} \qquad (21.11)$$

where the subscripts, 1, 2 and 12, represent the time t_1, t_2 and $t_{12} = (t_1 + t_2)/2$, respectively. Since p is constant in the experiments, Eq. (21.11) becomes

$$k_L = \frac{H(d_{eq,2}{}^3 - d_{eq,1}{}^3)}{6RT_L C_V d_{eq,12}{}^2 (t_2 - t_1)} \qquad (21.12)$$

The bubble diameters, $d_{eq,1}$ and $d_{eq,2}$, at t_1 and t_2 are obtained from a linear regression for measured time evolution of d_{eq} [22].

21.2.2.2 Measurement of Local Velocity and Concentration Fields

For the investigation of the local velocity and concentration fields around the Taylor bubbles, an optical entrance for the laser sheet is requested for the extinction of the fluorescence dye and the fluorescence particles, as well for up to two high-speed video cameras for the data acquisition. The PIV equipment has been arranged by the Intelligent Laser Application GmbH (ILA, Germany). A high-speed double pulsed Nd:YLF Laser (Quantronix Darwin Duo 527–100M, f = 600 Hz) is used as the exciting light source and a laser sheet optic established the required quasi-2D plane illumination of the flow around the Taylor bubble. Three micrometre linear tables LT60 from OWIS enabled the accurate adjustment of the laser sheet and the channel, before the high-speed cameras (PCO dimax HS2/4) are focused on the light sheet. A Nikon Nikkor 105 mm lens is used for the global view of the inflow, the whole bubble and the wake. A long range distance microscope (Infinity K2 DistaMax, CF-3) enabled the local investigation of the transport processes and the effects of surfactants directly at the interface with high spatial resolutions up to 4.2 μm/pixel. For the post-processing of the PIV-images the commercial software PIVview2C v3.6 (PIVTEC GmbH, Germany) is used with a FFT correlation method, where a multigrid refinement is used with a final 32×32 pixel grid size at 8×8 pixel step size enabled a velocity field estimation with 160 μm/vector for the global view. For the local investigation a resolution down to 4.2 μm/pixel has been realized that leads to spatial resolution of approx. 33 μm/vector. The usage of a high-speed pulsed laser enables to capture equidistant images, where 100 images are the base of the velocity field estimation for high accuracy. To capture the flow field fluorescent particles have been added to the liquid like mentioned before.

The concentration field in the vicinity of the Taylor bubbles has been visualized by means of Laser-induced fluorescence, due to the change of the local pH, while dissolved CO_2 is dissociating in the liquid phase. The used fluorescence dye

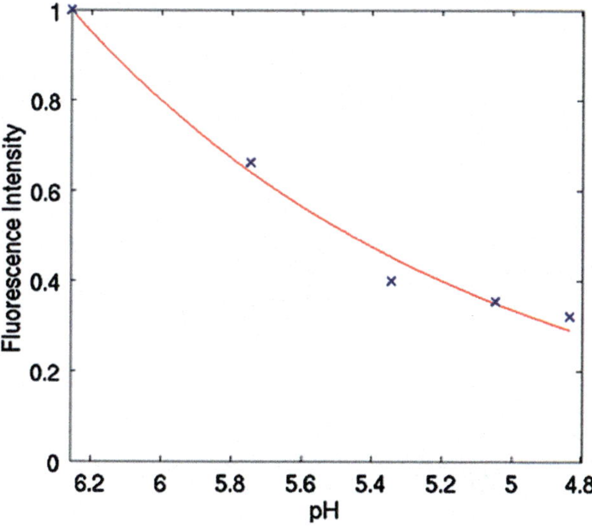

Fig. 21.6 Typical calibration curve for the dependency between pH-value and fluorescence intensity

fluorescein has a fluorescence of lower intensity at those areas where the pH value is reduced. The highest sensitivity of the dye is shown between $5 < pH < 7$. With the help of accurate calibration images which capture the fluorescence intensity in the original setup in its dependency on the liquid pH value (Fig. 21.6), 2D-pH maps around Taylor bubbles can be reconstructed.

21.3 Experimental Results

21.3.1 Hydrodynamics at Taylor Bubbles in Taylor Flows

First investigations have been performed to understand the local hydrodynamics in Taylor flows without contaminations and especially the controversially discussed local hydrodynamics within the liquid film. For this reason, the experimental setup described in Fig. 21.2 has been used with the vertical glass capillary ($D_h = 2$ mm) and square cross section. The square cross section has been chosen to enable high-resolved numerical simulation by the mathematical groups within the DFG priority programme 1506. All flow fields within the 2 mm capillary have been measured by means of Micro-Particle Image Velocimetry according to the arrangement given in Fig. 21.3a. The experimental conditions and characteristic properties of the system are given in Table 21.1.

Figure 21.7 shows a typical flow-field at the trailing edge of a Taylor bubble in a Taylor flow. A local Couette-shaped velocity profile is experimentally confirmed

Table 21.1 Operational conditions and characteristic properties for clean Taylor flows

Property	Value
Liquid phase	Water/glycerol
Viscosity	0.05 Pa s
Density	1.21 kg/l
Surface tension	59.3 mN/m
Gaseous phase	Air
Reynolds number	10.3
Capillary number	0.18
Eötvös number	0.88
Bubble length	4.3 mm
Slug length	0.7 mm

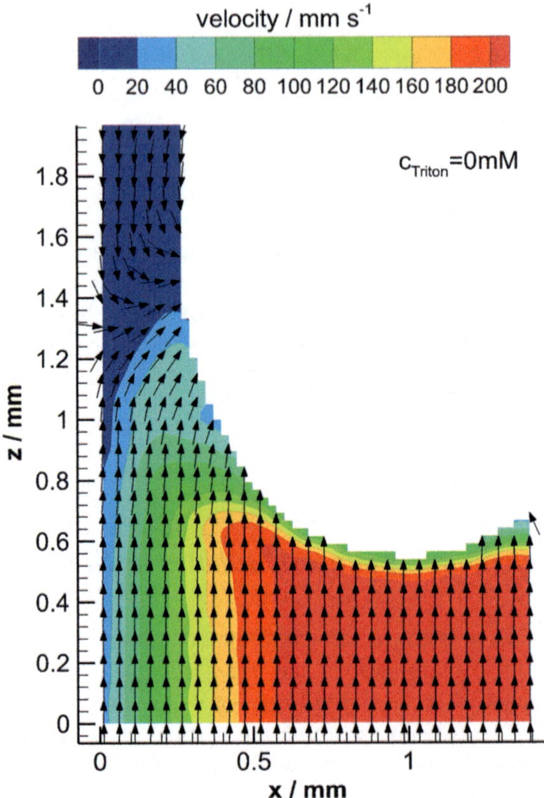

Fig. 21.7 Lateral velocity field at the trailing edge of a Taylor bubble in a Taylor flow $c_{\text{Triton}} = 0\,\text{mM}$

in the liquid film at mid-bubble length (upper part of Fig. 21.7). The maximum absolute film velocity at the interface is 30 times lower than the bubble velocity. This confirms a no-slip condition at the moving interface and toroidal inner gas circulation for pressure-driven Taylor flows in small capillaries (D_h < $D_{crit.}$). Despite the fact, that a co-current Taylor flow is investigated, a local backflow can be observed within the liquid film at the end of each bubble. To fulfil the continuity equation, this is only possible by assuming a complex three-dimensional flow field within the liquid film, even though the thickness of the film is less than $\delta_F = 200\,\mu m$.

These experimental results have been used for the validation of numerical simulations by three mathematical groups within the DFG priority programme 1506 [12]. All groups have confirmed the experimental results of a local backflow (Fig. 21.8).

With the detailed numerical simulations, it is possible to explain the local backflow by a temporal reversal of the wall shear stress during the passage of the Taylor bubbles. Furthermore, it becomes clear by numerical simulation, that the local backflow is only possible in case of a clean, non-contaminated interface that moves downward, driven by the shear forces at the interface. On the other hand, in contaminated systems with less interface mobility, the region of local backflow should be less dominant. Further investigations by means of μ-PIV have been performed and are discussed in the next chapter.

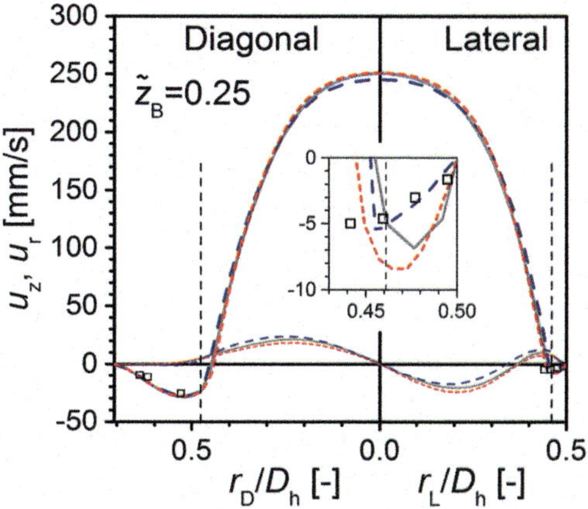

Fig. 21.8 Comparison between numerical and experimental results of the velocity profile within the liquid film (reproduced from [12])

21.3.2 *Effects of Contaminations on Hydrodynamics in Taylor Flows*

In order to study the influence of surfactants on the hydrodynamics, Triton X-100 is added to the glycerol-water mixture (78 wt.% glycerol). The concentration of the surfactant to the glycerol-water mixture is $c_{Triton} = 0.1$ mM and $c_{Triton} = 0.3$ mM, respectively. The clean system, without additional Triton X-100, is referred to as $c_{Triton} = 0$ mM and serves as reference liquid. The other operational conditions and characteristic properties of the system are given in Table 21.2.

Figure 21.9 shows the influence of surfactants on the flow field at the trailing edge of a Taylor bubble in Taylor flow, measured by means of μ-PIV. It becomes clear that in comparison to the flow field with clean interface, described above, the local backward flow does not appear with the influence of surfactants (i.e. $c = 0.1$ mM and $c = 0.3$ mM respectively). Apparently due to the influence of the surfactants, the mobility of the interface is hindered and therefore, because of the no-slip condition at the immobile bubble interface, the liquid is pulled upwards. This can be confirmed by experimental results of the flow field within the diagonal film in the capillary, according to Fig. 21.3b. In Fig. 21.10, left, a strong downward flow can be observed due to the mobile interface, whereas there is almost no backflow in the contaminated system (Fig. 21.10, right).

According to the influence of the surfactants, the interface mobility seems to be hindered leading to a different bubble shape and higher shear forces at the interface. This causes lower velocities of the Taylor bubbles in contaminated Taylor flows. The influence of different concentrations of surfactants on Taylor flows is summarized in Table 21.3. Even if the surfactants are changing multiple parameters at once, the influence on the rise velocity is evident.

It is strongly expected that the influence of surfactants on the hydrodynamics at Taylor bubbles in Taylor flows will change the mass transfer performance at the Taylor bubbles as well. Accordingly, further investigations have been performed

Table 21.2 Operational conditions and characteristic properties of clean and contaminated Taylor flows

$c_{tritonX-100}$	0 mM	0.1 mM	0.3 mM
Bubble velocity U_B (mm/s)	212	172	188
Bubble length λ_B (mm)	4.3	4.5	4
Liquid slug length λ_S (mm)	0.7	1	1.7
Unit cell length λ_{UC} (mm)	5	5.5	5.7
Viscosity ν (Pa s)	0.05	0.05	0.05
Density ρ (kg/l)	1.2	1.2	1.2
Surface tension σ (mN/m)	59	44	38
Reynolds number Re	10.3	8.8	9.3
Capillary number Ca	0.18	0.18	0.24
Temperature ($^\circ$C)	20	20	20

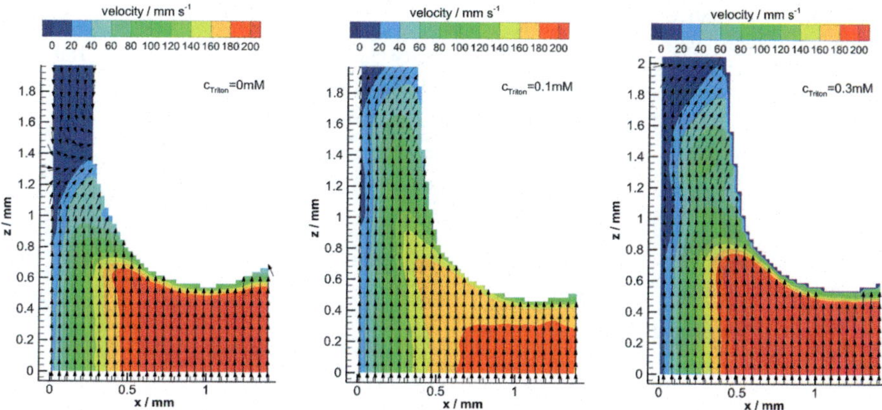

Fig. 21.9 Comparison of the flow field at the trailing edge of a Taylor bubble with different surfactant concentration. $\sigma(c_{\text{tritonX}-100} = 0\,\text{mM}) = 58.3\,\text{mN/m}$, $\sigma(c_{\text{tritonX}-100}=0.1\,\text{mM}) = 44.1\,\text{mN/m}$, $\sigma(c_{\text{tritonX}-100}=0.3\,\text{mM}) = 38\,\text{mN/m}$

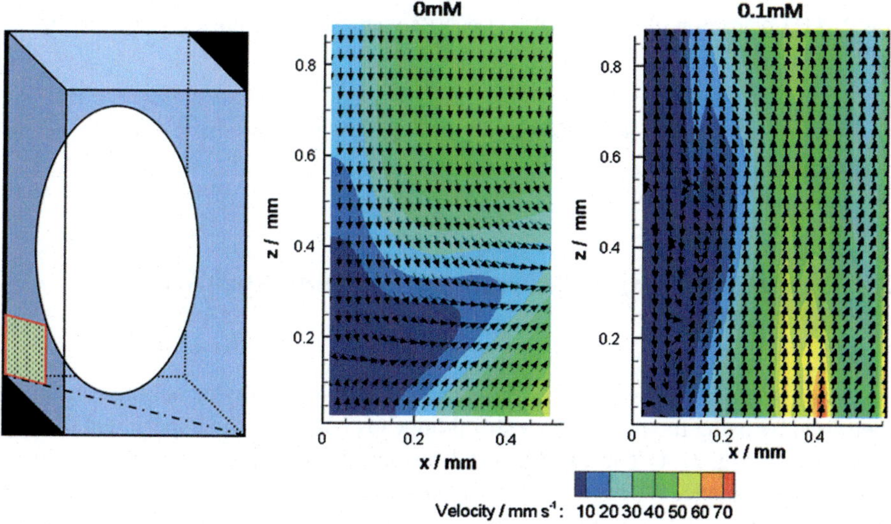

Fig. 21.10 Comparison of the diagonal film at the trailing edge of a Taylor bubble in a pure system ($\sigma(c_{\text{tritonX}-100} = 0\,\text{mM}) = 58.3\,\text{mN/m}$) related to a contaminated system $\sigma(c_{\text{tritonX}-100} = 0.1\,\text{mM}) = 44.1\,\text{mN/m}$)

Table 21.3 Quantitative comparison between pure and contaminated Taylor flows ($T = 20°C$)

$c_{tritonX-100}$	0 mM	0.1 mM	0.3 mM
Bubble velocity U_B (mm/s)	212	172	188
Bubble length λ_B (mm)	4.3	4.5	4
Slug length λ_S (mm)	0.7	1	1.7
Reynolds number	10.3	8.8	9.3
Capillary number	0.18	0.18	0.24

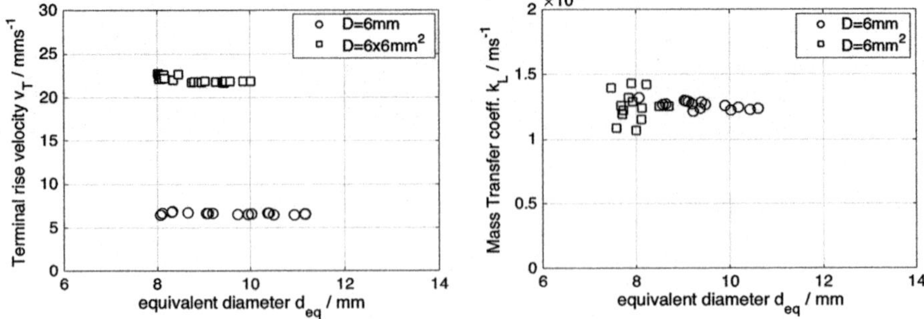

Fig. 21.11 Comparison of rise velocities and mass transfer coefficients for Talyor-Bubbles in vertical channels ($D_h = 6$ mm) with square and circular cross section [22]

at single rising and fixed Taylor bubbles for detailed studies of the mass transfer performance.

21.3.3 Mass Transfer at Clean and Contaminated Taylor Bubbles

The mass transfer performance at clean and contaminated Taylor bubbles has been measured by using the method of a shrinking CO_2 bubble in the experimental setup described in Fig. 21.4. The experimental results of the rise velocity and the mass transfer performance for capillaries with a diameter of $D = 6$ mm and different cross section shape are shown in Fig. 21.11.

The results for the mass transfer coefficients show the interesting phenomenon, that the values for Taylor bubbles in square and circular capillaries are in the same range, even if the rise velocity in the capillary with the square cross section is more than three times higher than in the one with the circular cross section. This stands in contradiction to the conventional assumption that the mass transfer coefficient (Sherwood number) is directly related to the bubble Reynolds number. Obviously, the mass transfer performance is dominated by the liquid film only, that is in direct contact with the bubble interface. The liquid phase and velocity profile in larger distances to the interface, e.g. in the edges of the capillary with square cross section, does not affect the performance. This can be confirmed by comparing the

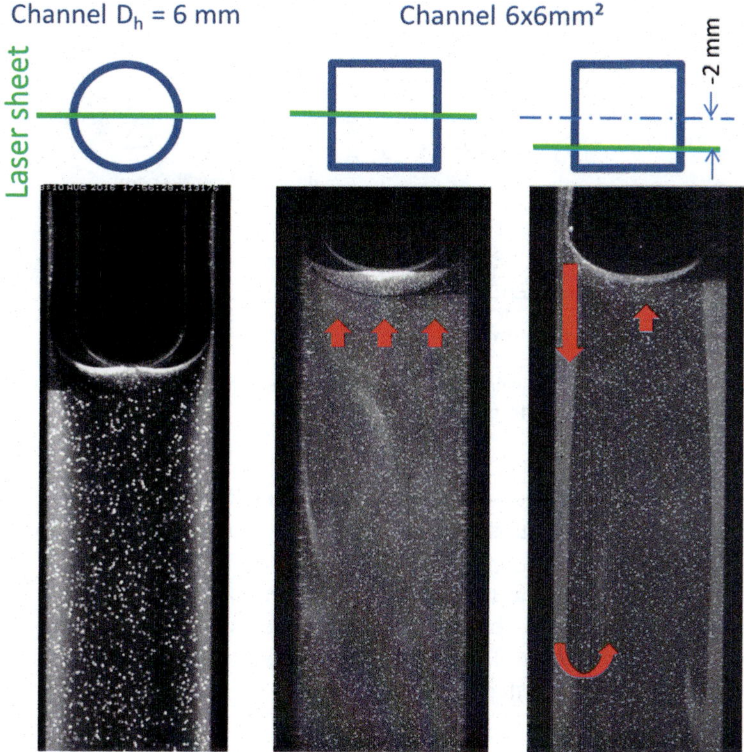

Fig. 21.12 Wake structures visualized by LIF and PIV techniques to show the influence of the channel geometry on the flow situation around the bubble

concentration and flow fields in the wake of a Taylor bubble in channels with circular and square cross section (Fig. 21.12).

In the circular cross section, a smooth rotation-symmetric wake appears, where CO_2-enriched liquid (dark) is transported in a layer flow next to the "fresh" liquid (bright) without strong mixing. Clearly only a very thin layer of the liquid film is enriched with CO_2, whereas a large amount is passing the Taylor bubble unaffected (bright). The flow structure in the channel with the square cross section shows a much more mixed situation (Fig. 21.12, middle). Obviously, a large amount of liquid passes the bubble in the edges of the cross section unaffected, visible as thick bright film passing the bubble (Fig. 21.12, right). The strong velocity gradients lead to strong 3D vortices and a high mixing rate in the wake of the bubble. The strongly asymmetric bubble shape and liquid film thickness in the square channel, could only be estimated by Boden et al. with X-ray tomography [6] to be less than $\delta = 30\,\mu$m at the channel face, because it is too thin for reliable measurement.

This result can be reproduced for several other diameters of circular and square cross section capillaries. Figure 21.13 shows the result for six different capillaries. Whereas a strong influence of the capillary diameter on the mass transfer

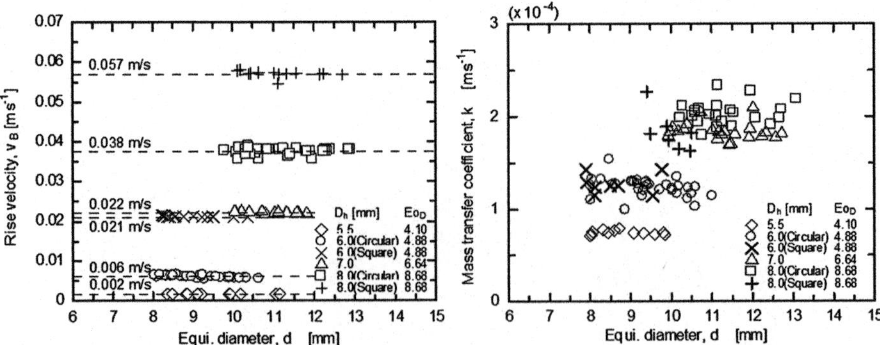

Fig. 21.13 Comparison of rise velocities and mass transfer coefficients for Taylor bubbles in different capillaries (reproduced from [22])

performance can be observed—the larger the channel diameter, the faster the Taylor bubble and the higher the mass transfer coefficient—the shape of the cross section has a negligible influence.

These results indicate that the effect of cross-sectional geometries on the mass transfer from single Taylor bubbles in a mini channel is small and the mass transfer might be expressed in terms of the Sherwood number using the hydraulic diameter D_h as the characteristic length as follows:

$$Sh_D = \frac{k_L D_h}{D_L} \tag{21.13}$$

Judging from the analogy between the momentum and mass transfers, Sh_D should be correlated in terms of Eo_D [27, 43]. The redefined Eötvös number takes into account the fluid property, interfacial tension, and the channel diameter. These parameters predefine the Taylor bubble shape by the curvature of the bubble nose, the thickness of the fluidic falling film and therefore the interfacial area [8]. Hence the measured Sh_D are plotted against Eo_D as shown in Fig. 21.14. Sh_D for single Taylor bubbles in larger pipes of $D_h = 12.5$, 18.2 and 25 mm quoted from [21] are also shown in the figure. As expected, all the data collapse onto a single curve, and Sh_D are well correlated with

$$Sh_D = 290 Eo_D^{0.524} - \left[\frac{1.23}{(Eo_D + 1)^{0.0517}}\right]^{50.1} \tag{21.14}$$

This empirical correlation is valid for mass transfer from single CO_2 Taylor bubbles rising in deionized water through vertical small and intermediate channels of $5.5\,\text{mm} \leq D_h \leq 25.0\,\text{mm}$. The applicable ranges of dimensionless groups are $4 < Eo_D < 85$, $13 < \text{Re}(= \rho_L v_B d_{eq} \nu^{-1}) < 4.6 \times 10^3$, $6.2 \times 10^3 < \text{Pe}(= v_B d_{eq} D_L^{-1}) < 2.2 \times 10^6$, where Re is the bubble Reynolds number and ν the liquid viscosity (reproduced from [22]).

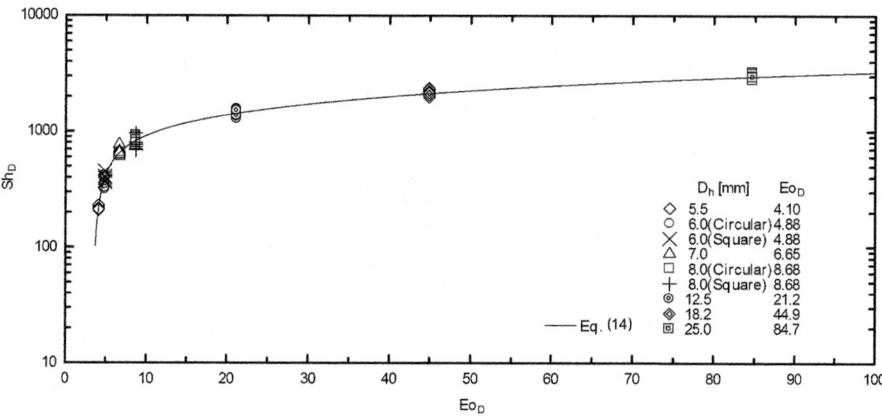

Fig. 21.14 Sherwood number correlation for Taylor bubbles (Reproduced from [22])

The influence of surfactants on Taylor bubbles becomes first obvious by analysing the bubble shapes in capillaries with circular cross section. The distribution of surfactants at the bubble interface depends on the adsorption and desorption kinetics, the local shear forces and the Marangoni forces at the interface. Mainly, the surfactants are adsorbed at the bubble front and sheared to the lower part of the bubble, where the surfactants accumulate. Depending on their interfacial saturation concentration and the shear forces induced by the liquid phase, the surfactants will desorb to the bulk phase again. The local surfactant distribution difference leads to Marangoni forces, which are directed against the shear flow by the liquid film [20]. The local surfactant concentration is affecting the interfacial tension, and lower local interfacial tension leads to bubbles that are more easily deformed by the resulting forces. In the case of single Taylor bubbles at low Eötvös numbers, this leads to a larger bubble length and larger film thicknesses (see Fig. 21.15).

This larger film thickness in case of contaminated Taylor bubbles leads to higher rise velocities for smaller Taylor bubbles ($d_{eq} < 11$ mm), because the drag coefficient of a smaller projected area is lower for a contaminated bubble and the liquid phase can pass the bubble more easily (Fig. 21.16). This effect decreases with increasing bubble volume, because the drag coefficient by friction become more dominant for longer bubbles. At bubble equivalent diameters of approx. $d_{eq} > 11$ mm, the influence of surfactants on the rise velocity can be neglected and the rise velocity is similar to the non-contaminated case mentioned above. The influence of the Triton X-100 contamination ($c_{Triton} = 10$ mmol/m^3) on the mass transfer is therefore significant, where the mass transfer coefficient is reduced nearly by factor $1/2$, even though the rise velocity is faster. This means, that the reduction of the mass transfer must be explained by local phenomena at the interface, which shows the importance of local information by PIV, LIF, numerical and molecular dynamic simulations to understand and describe these phenomena (see Acknowledgement).

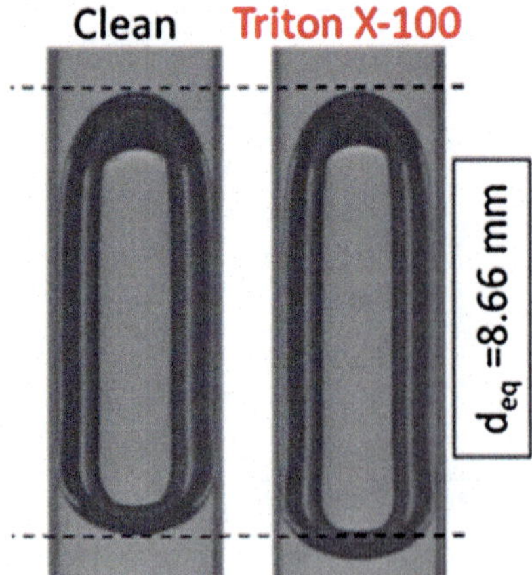

Fig. 21.15 Shape of Taylor bubbles in clean and contaminated water in D = 6 mm channel

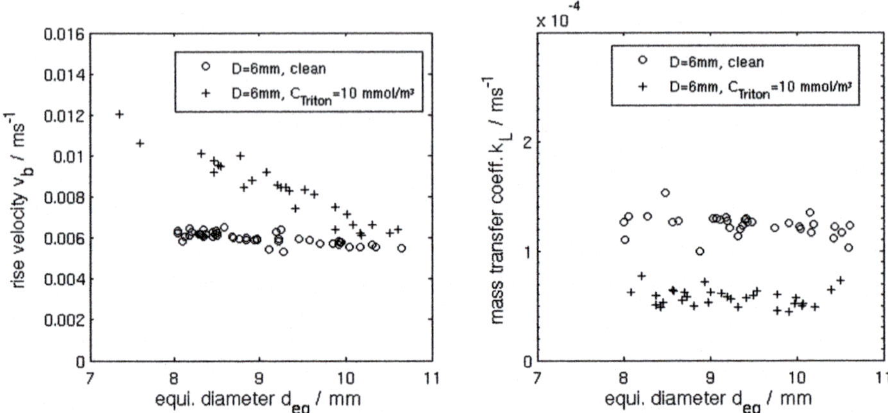

Fig. 21.16 Rise velocities and mass transfer coefficients of Taylor bubbles in clean and contaminated water in $D_h = 6$ mm (circular cross section)

The higher rise velocity of contaminated bubbles requires a higher counter-current flow to fix the bubble in the observation window, which leads to different hydrodynamic conditions in the vicinity of the Taylor bubble. This can be seen by means of the PIV flow-field measurements in Fig. 21.17. The clean Taylor bubble shows a smaller film thickness and therefore a lower rise velocity with slower counter-current flow velocity (Fig. 21.17, upper image), whereas the contaminated system shows the higher parabolic velocity profile (Fig. 21.17, lower image). The images have been rotated by 90° for a better visibility of the vector field.

To study the influence of the local flow field on the mass transfer performance in more detail, the local concentration field enables some deep insights. The concentration field in the vicinity of a Taylor bubble rising in a vertical channel can be

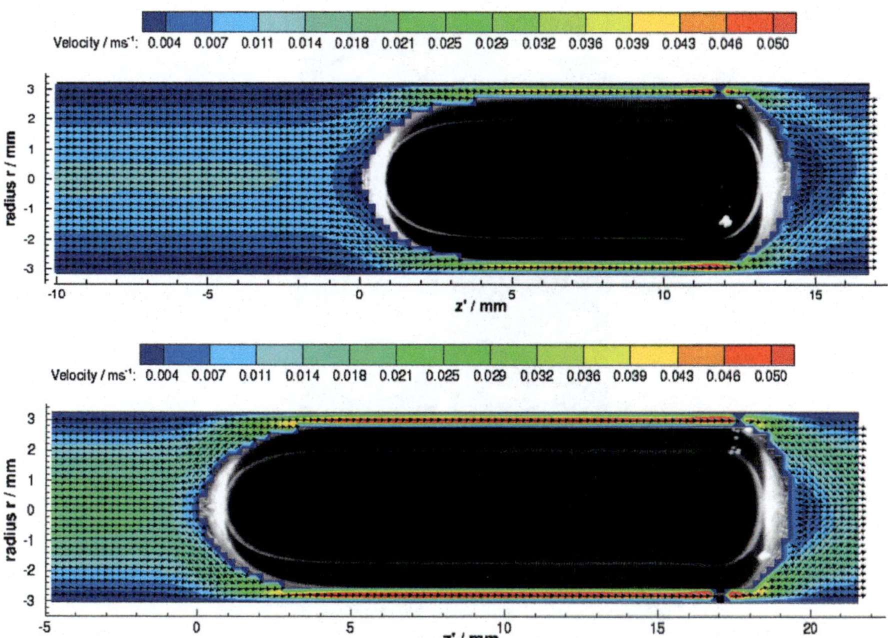

Fig. 21.17 Velocity fields around Taylor bubbles in clean (*upper image*) and contaminated (10 mmol/m^3 Triton X-100) water (*lower image*)

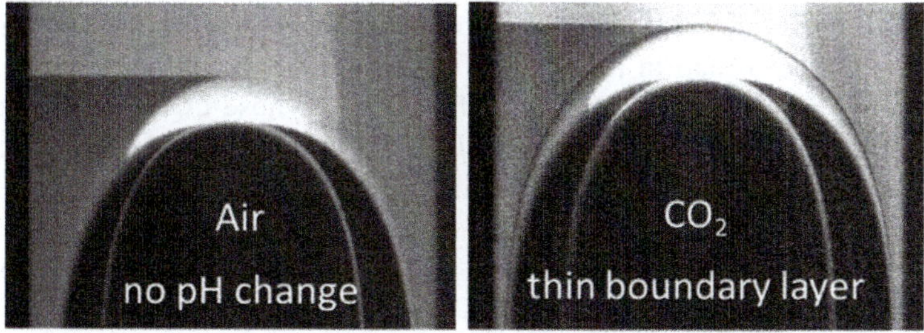

Fig. 21.18 Visualization of the CO_2 concentrations boundary layer at the top of a CO_2 Taylor bubble in comparison with an Air-bubble with constant pH value at the interface

visualized by means of Laser-induced fluorescence, according to the experimental setup described before. This technique enables the detection of the concentration boundary layer directly at the bubble interphase due to the reduction in fluorescence intensity. In Fig. 21.18, the front of an air-bubble and of a CO_2 bubble is shown, where the dark line at the CO_2 interface indicates the concentration boundary layer. In accordance with Prandtl's boundary layer theory, the concentration boundary layer thickness decreases with increasing velocity if the annular liquid film between bubble and wall is accelerated.

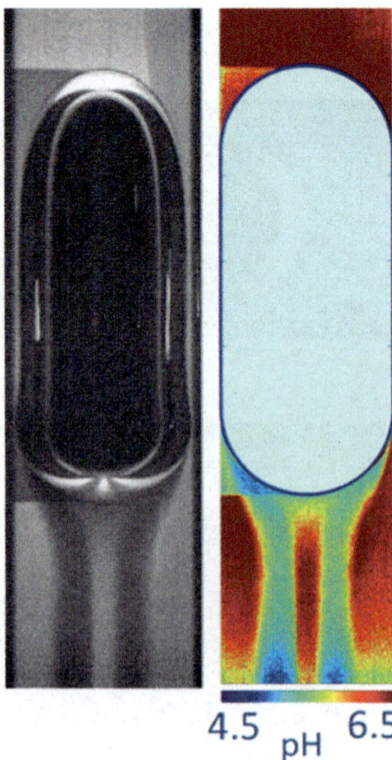

Fig. 21.19 pH-Map of a rising Taylor bubble measured by means of Laser Induced Fluorescence with no additional Triton X-100 ($D_h = 6\,\mathrm{mm}$)

This image of gray scales can be converted by using the calibration curve (Fig. 21.6) into a pH-map in the vicinity of rising Taylor bubbles (Fig. 21.19). It can be seen nicely how the CO_2 is transferred at the nose of the bubble into the hydrodynamic boundary layer that flushes the CO_2 within the liquid film downwards into the wake. Interestingly, on the central axis of the bubbles, an area with lower CO_2 concentration exists, that refers to a region of very poor mixing (dead zone). This could be caused by small amounts of surface active agents, which could have been accumulated at the bubble rear, because the streamlines don't merge in a stagnation point at the bubble rear. A perfectly clean watery system is nearly impossible to investigate experimentally under realistic conditions. But the strong influence of contaminations on the local transport processes opens a wide research field for further collaborations between numerical and experimental groups of this priority programme.

21.4   Conclusion and Outlook

Within this project very accurate and reliable experimental data has been provided for the modeling of transport processes at fluidic interfaces and the validation of numerical simulations. It has been shown experimentally, that even within the very thin film of less than $200\,\mu m$ around a Taylor bubble, local backflow appears. The comparison with numerical simulations of three mathematical groups within SPP 1506 has confirmed this backflow quantitatively and pointed out, that contaminations within the experiments are not affecting the local velocity field at the interface significantly.

Due to the new insights into local mass transfer processes at single Taylor bubbles within the DFG priority programme 1506, it has been shown that Taylor bubbles can be fixed within a capillary in counter-current flow for a long time without any disturbance. Depending on the Eötvös number (capillary diameter), well-defined different flow structures in the wake of the Taylor bubble can be applied in a highly reproducible way. Therefore, Taylor bubbles are an ideal system to investigate chemical gas-liquid reactions in a very detailed manner by using different analytical systems like UV-Vis, Raman, IR, NMR, LIF. Because of this outstanding performance, the Taylor bubble topic has been transferred to the new DFG priority programme 1740 "Reactive Bubbly Flows" where the guiding measure "Taylor Flow" from SPP 1506 will be modified for the investigation of parallel and consecutive chemical reactions.

Acknowledgements The authors gratefully acknowledge the financial support provided by the German Research Foundation within the priority programme "Fluidic Interfaces", DFG SPP 1506.

We would like to thank all other research groups of our project for the fruitful collaboration and discussion about the validation of these complex transport processes at fluidic interfaces.

We would like to thank Prof. Akio Tomiyama, Kobe University, Japan for the intensive collaboration and the PhD exchange of Mr. Shogo Hosoda in 2013, Mr. Jiro Aoki in 2015 and Mr. Sven Kastens in 2016. Our fruitful discussions and experimentally and numerical investigations of the effects of surfactants on transport processes lead to several joint talks on international conferences and publications.

We would like to thank Prof. Dr.-Ing. Irina Smirnova and Dr.-Ing. Sven Jakobtorweihen for setting up molecular dynamic simulations to investigate the mass transfer of CO_2 molecules across a contaminated gas-liquid interface with surfactants.

Last but not least, we would like to thank the students who have worked on this project during their Bachelor or Master thesis or research projects: Krischan Sandmann, Caroline Otto and Maximilian Garbe.

Nomenclature

Symbol	Name	Unit
A	Area	m^2
c_{Triton}	Triton X-100 concentration	$mol\,m^{-3}$
Ca	Capillary number	–
C_s	Molar interfacial concentration	$mol\,m^{-3}$
C	Molar concentration in the bulk	$mol\,m^{-3}$
C_v	Molar concentration of water	$mol\,m^{-3}$
d	Diameter	m
d_{eq}	Sphere volume equivalent diameter	m
d_p	Particle diameter	m
D	Channel diameter	m
D_h	Hydraulic channel diameter	m
$D_{crit.}$	Critical channel diameter	m
D_L	Diffusion coefficient	$m^2\,s^{-1}$
Eo	Eötvös number	–
f	Frequency	s^{-1}
g	Gravitational constant	$m\,s^{-2}$
H	Henry constant	Pa
k_l	Mass transfer coefficient	$m\,s^{-1}$
L	Length	m
M	Moles	mol
n	Refractive index	–
NA	Numerical aperture	–
P	Pressure	Pa
Pe	Peclet number	–
r	Radius	m
R	Gas constant	$J\,mol^{-1}\,K^{-1}$
Re	Reynolds number	–
Sh	Sherwood number	–
t	Time	s

(continued)

Symbol	Name	Unit
T	Temperature	°C
U_B	Bubble travel velocity	m s^{-1}
u_r	Liquid velocity in r direction	m s^{-1}
u_z	Liquid velocity in z direction	m s^{-1}
v_B	Bubble rise velocity	m s^{-1}
X	Mole fraction	–
z'	Relative height	m
δ	Film thickness	m
κ	Conductivity	$\mu\text{S cm}^{-1}$
λ_S	Slug length	m
λ_S	Bubble length	m
λ_{UC}	Unit cell length	m
λ	Diameter ratio	-
ρ_l	Liquid density	kg m^{-3}
ρ_g	Gas density	kg m^{-3}
σ	Interfacial tension	N m^{-1}
ν	Viscosity	Pa

Abbreviations

Symbol	Name
CFD	Computational fluid dynamics
CO_2	Carbon dioxide
DMSO	Dimethyl sulfoxide
fps	Frames per second
LE	Leading edge
LED	Laser emitting diode
LIF	Laser induced fluorescence
O_2	Oxygen
PIV	Particle image velocimetry
TE	Trailing edge
WD	Working distance

References

1. Agostini, B., Revellin, R., Thome, J.R.: Elongated bubbles in microchannels. Part I: experimental study and modeling of elongated bubble velocity. Int. J. Multiphase Flow **34**, 590–601 (2008)
2. Aoki, J., Hayashi, K., Tomiyama, A.: Mass transfer from single carbon dioxide bubbles in contaminated water in a vertical pipe. Int. J. Heat Mass Transfer **83**, 652–658 (2015)
3. Aoki, J., Hayashi, K., Hosokawa S., Tomiyama, A.: Effect of surfactants on mass transfer from single carbon dioxide bubbles in vertical pipes. Chem. Eng. Technol. **38**, 1955–1964 (2015)
4. Aussillous, P., Quéré, D.: Quick deposition of a fluid on the wall of a tube. Phys. Fluids **12**, 2367–2371 (2000)
5. van Baten, J.M., Krishna, R.: CFD simulations of mass transfer from Taylor bubbles rising in circular capillaries. Chem. Eng. Sci. **59**, 2535–2545 (2004)
6. Boden, S., dos Santos Rolo, T., Baumbach, T., Hampel, U.: Synchrotron radiation microtomography of Taylor bubbles in capillary two-phase flow. Exp. Fluids **55**, 1–14 (2014)
7. Bretherton, F.P.: The motion of long bubbles in tubes. J. Fluid Mech. **10**, 166–188 (1961)
8. Brown, R.A.S.: The mechanics of large gas bubbles in tubes. Can. J. Chem. Eng. **43**, 217–223 (1965)
9. Buie, C.R., Santiago, J.G.: Two-phase hydrodynamics in a miniature direct methanol fuel cell. Int. J. Heat Mass Transf. **52**, 5158–5166 (2009)
10. Esteves, M.T.S., de Carvalho, J.R.G.: Liquid-side mass transfer coefficient for gas slugs rising in liquids. Chem. Eng. Sci. **48**, 3497–3506 (1993)
11. Fairbrother, F., Stubbs, A.E.: Studies in electro-endosmosis. Part VI. The "bubble tube" method of measurement. J. Chem. Soc. **119**, 527–529 (1935)
12. Falconi, C.J., Lehrenfeld, C., Marschall, H., Meyer, C., Abiev, R., Bothe, D., Reusken, A., Schlüter, M., Wörner, M.: Numerical and experimental analysis of local flow phenomena in laminar Taylor flow in a square mini-channel. Phys. Fluids **28**, 1–23 (2016)
13. Filla, M.: Gas absorption from a slug held stationary in downflowing liquid. Chem. Eng. J. **22**, 213–220 (1981)
14. Fouilland, T.S., Fletcher, D.F., Haynes, B.S.: Film and slug behaviour in intermittent slug-annular microchannel flows. Chem. Eng. Sci. **65**, 5344–5355 (2010)
15. Fries, D.M., Waelchli, S., von Rohr, P.R.: Gas–liquid two-phase flow in meandering microchannels. Chem. Eng. J. **135**, 37–45 (2008)
16. Gibson, A.H.: On the motion of long air bubbles in a vertical tube. Philos. Mag. Ser. **26**, 952–966 (1913)
17. Gupta, R., Fletcher, D.F., Haynes, B.S.: Taylor flow in microchannels: a review of experimental and computational work. J. Comput. Multiphase Flows **2**, 1–32 (2010)
18. Günther, A., Khan, S.A., Thalmann, M., Trachsel, F., Jensen, K.F.: Transport and reaction in microscale segmented gas–liquid flow. Lab Chip **4**, 278–286 (2004)
19. Haghnegahdar, M., Boden, S., Hampel, U.: Mass transfer measurement in a square milli-channel and comparison with results from a circular channel. Int. J. Heat Mass Transfer. **101**, 251–260 (2016)
20. Hayashi, K., Tomiyama, A.: Effects of surfactant on terminal velocity of a Taylor bubble in a vertical pipe. Int. J. Multiphase Flow **39**, 78–87 (2012)
21. Hosoda, S., Abe S., Hosokawa S., Tomiyama A.: Mass transfer from a bubble in a vertical pipe. Int. J. Heat Mass Transf. **69**, 215–222 (2014)
22. Kastens, S., Hosoda, S., Schlüter, M., Tomiyama, A.: Mass transfer from single Taylor bubbles in mini channels. Chem. Eng. Technol. **38**, 1925–1932 (2015)
23. King, C., Walsh, E., Grimes, R.: PIV measurements of flow within plugs in a microchannel. Microfluid. Nanofluid. **3**, 463–472 (2007)
24. Kolb, W.B., Cerro, R.L.: Coating the inside of a capillary of square cross section. Chem. Eng. Sci. **46**, 2181–2195 (1991)

25. Kreutzer, M.T.: Hydrodynamics of Taylor flow in capillaries and monolith reactors. Dissertation, TU Delft (2003)
26. Kreutzer, M.T., Kapteijn, F., Moulijn, J.A., Heiszwolf, J.J.: Multiphase monolith reactors: chemical reaction engineering of segmented flow in microchannels. Chem. Eng. Sci. **60**, 5895–5916 (2005)
27. Kurimoto, R., Hayashi, K., Tomiyama, A.: Terminal velocities of clean and fully-contaminated drops in vertical pipes. Int. J. Multiphase Flow **49**, 8–23 (2013)
28. Leung, S.S.Y., Gupta, R., Fletcher, D.F., Haynes, B.S.: Gravitational effect on Taylor flow in horizontal microchannels. Chem. Eng. Sci. **69**, 553–564 (2012)
29. Llewellin, E.W., Del Bello, E., Taddeucci, J., Scarlato, P., Lane, S.J.: The thickness of the falling film of liquid around a Taylor bubble. Proc. R. Soc. A **468**, 1041–1064 (2012)
30. Malsch, D., Kielpinski, M., Merthan, R., Albert, J., Mayer, G., Köhler J.M., Süße H., Stahl M., Henkel T.: μPIV-analysis of Taylor flow in micro channels. Chem. Eng. J. **135**, 166–172 (2008)
31. Meyer, C., Hoffmann M., Schlüter M.: Micro-PIV analysis of gas–liquid Taylor flow in a vertical oriented square shaped fluidic channel. Int. J. Multiphase Flow **67**, 140–148 (2014)
32. Ohl, C.D.: Generator for single bubbles of controllable size. Rev. Sci. Instrum. **72**, 252–254 (2001)
33. Olgac, U., Muradoglu M.: Effects of surfactant on liquid film thickness in the Bretherton problem. Int. J. Multiphase Flow **48**, 58–70 (2013)
34. Roudet, M., Loubiere, K., Gourdon, C., Cabassud, M.: Hydrodynamic and mass transfer in inertial gas-liquid flow regimes through straight and meandering millimetric square channels. Chem. Eng. Sci. **66**, 2974–2990 (2011)
35. Santiago, J.G., Wereley, S.T., Meinhart, C.D., Beebe, D.J., Adrian, R.J.: A particle image velocimetry system for microfluidics. Exp. Fluids **25**, 316–319 (1998)
36. Suo, M., Griffith, P.: Two phase flow in capillary tubes. J. Basic Eng. **86**, 576–582 (1964)
37. Taylor, G.I.: Deposition of a viscous fluid on the wall of a tube. J. Fluid Mech. **10**, 161–165 (1961)
38. Thulasidas, T.C., Abraham, M.A., Cerro, R.L.: Bubble-train flow in capillaries of circular and square cross section. Chem. Eng. Sci. **50**, 183–199 (1995)
39. Thulasidas, T.C., Abraham, M.A., Cerro, R.L.: Flow patterns in liquid slugs during bubble-train flow inside capillaries. Chem. Eng. Sci. **52**, 2947–2962 (1997)
40. van Steijn, V., Kreutzer, M.T., Kleijn, C.R.: Velocity fluctuations of segmented flow in microchannels. Chem. Eng. J. **135**, 159–165 (2008)
41. Vandu, C.O., Liu, H., Krishna, R.: Mass transfer from Taylor bubbles rising in single capillaries. Chem. Eng. Sci. **60**, 6430–6437 (2005)
42. Warnier, M.J.F., Rebrov, E.V., De Croon, M., Hessel, V., Schouten, J.C.: Gas hold-up and liquid film thickness in Taylor flow in rectangular microchannels. Chem. Eng. J. **135**, 153–158 (2008)
43. White, E.T., Beardmore, R.H.: The velocity of rise of single cylindrical air bubbles through liquid contained in vertical tubes. Chem. Eng. Sci. **17**, 351–361 (1962)

Chapter 22
Comparative Simulations of Taylor Flow with Surfactants Based on Sharp- and Diffuse-Interface Methods

Sebastian Aland, Andreas Hahn, Christian Kahle, and Robert Nürnberg

Abstract We present a quantitative comparison of simulations based on diffuse- and sharp-interface models for two-phase flows with soluble surfactants. The test scenario involves a single Taylor bubble in a counter-current flow. The bubble assumes a stationary position as liquid inflow and gravity effects cancel each other out, which makes the scenario amenable to high resolution experimental imaging. We compare the accuracy and efficiency of the different numerical models and four different implementations in total.

22.1 Introduction

Taylor bubbles are long bubbles of gas in capillary tubes filled with a liquid. Nowadays, Taylor bubbles play a role in many technical applications, for example, in monolith structures that can be found in catalytic converters [52], multiphase monolith reactors [42, 50], or microfluidic channels [31, 47]. As these applications

S. Aland (✉)
Institute of Scientific Computing, TU Dresden, 01062 Dresden, Germany
e-mail: sebastian.aland@tu-dresden.de

A. Hahn
Fakultät für Mathematik, Otto-von-Guericke University Magdeburg, 39106 Magdeburg, Germany
e-mail: Andreas.Hahn@ovgu.de

C. Kahle
Fakultät für Mathematik, Technical University Munich, Boltzmannstraße 3, 85748 Garching, Germany
e-mail: christian.kahle@ma.tum.de

R. Nürnberg
Department of Mathematics, Imperial College London, London SW7 2AZ, UK
e-mail: robert.nurnberg@imperial.ac.uk

© Springer International Publishing AG 2017

D. Bothe, A. Reusken (eds.), *Transport Processes at Fluidic Interfaces*,
Advances in Mathematical Fluid Mechanics, DOI 10.1007/978-3-319-56602-3_22

typically require bubbles of identical size, shape and distance to each other, the correct prediction of bubble hydrodynamics is of great value. Numerical simulations aim in this direction and provide a means to quantitatively compute the bubble dynamics and hence to optimize the desired flow properties on a computer.

Analysis of bubble speed has shown particular dependence on the thickness of the liquid film that occurs between the bubble and the capillary wall. Using lubrication approximation theory, Bretherton [22] was the first to give an analytical relation between film thickness and bubble speed. Contrary to the expectation, his result underestimated the film thickness for low capillary numbers, where it should be precise [11]. Explaining the discrepancy by means of viscosity, inertia, intermolecular forces or surface roughness failed. However, impurities of surface active agents (surfactants) are known to affect the film thickness, as surfactants soften the interface and induce additional Marangoni forces. In [49] Bretherton's theory was extended to a case with trace amounts of surfactants present. They showed that the film thickness could increase by a maximum factor of $4^{\frac{2}{3}}$, which could explain the discrepancy of Bretherton's results.

These observations show that the inclusion of surfactant dynamics is crucial for the correct prediction of Taylor flow. However, numerical simulations of Taylor flow with surfactants appeared only recently [33, 48], due to the complicated interplay of hydrodynamics, surface evolution and surfactant dynamics. In [33] the effect of surfactant on the terminal velocity of a rising Taylor bubble in a vertical pipe was studied using an interface tracking method. In [48] the effect of surfactant on the motion of a Taylor bubble moving through a capillary tube was investigated numerically using a finite-difference front tracking method.

The rigorous assessment of simulation accuracy, requires benchmarking, i.e., a quantitative code-to-code comparison with physically relevant physical parameters and practically relevant output parameters. In [9] it is stressed that Taylor flow poses a perfect two-phase flow problem for benchmarking, due to its sensitivity to a wide range of surface tensions, and since it admits non-trivial stationary solutions that are amenable to high quality experimental imaging. Aland et al. [8] and later Marschall et al. [45] have already presented such benchmark studies for Taylor flow without surfactants, and both provided highly accurate reference data in two and three dimensions, as well as comparison to experiments.

The goal of this chapter is to extend the work of [8] to include soluble surfactants, and hence to provide the first benchmark study of two-phase flow with surfactants. Therefore we simulate a single Taylor bubble in a counter-current flow scenario, where the bubble assumes a stationary state as buoyancy and liquid inflow balance each other. We examine the effect of surfactant amount, surfactant elasticity and Peclet number on bubble shape, speed and film thickness. Moreover, we analyze surfactant profiles along the surface and in the liquid bulk. Four independent numerical codes are compared, two based on a sharp-interface model, and two based on a diffuse-interface model.

The structure of the chapter is as follows. In Sect. 22.2 we present the employed sharp- and diffuse-interface models. The hydrodynamic part of the benchmark problem is summarized in Sect. 22.3, including an overview of the results from [8]. Finally, in Sect. 22.4, we present the new results including surfactants and assess the accuracy and efficiency of the numerical codes.

22.2 Mathematical Models

In this section, the mathematical models for the diffuse- and the sharp-interface models are given.

22.2.1 Sharp Interface Model

The sharp-interface model consists of the Navier–Stokes equations, a convection-diffusion equation in the bulk and a convection-diffusion equation on the interface.

The domain Ω consists of the two phases, where $\Omega_1(t)$ is the inner phase and $\Omega_2(t)$ the outer phase. The phases are immiscible, i.e. $\Omega_1(t) \cap \Omega_2(t) = \emptyset$, and they are separated by the interface $\Gamma(t) = \overline{\Omega}_1(t) \cap \overline{\Omega}_2(t)$. The Navier–Stokes equations in the time interval $(0, T]$ are

$$
\nu \left(\frac{\partial \mathbf{u}}{\partial t} + (\mathbf{u} \cdot \nabla) \mathbf{u} \right) - \nabla \cdot \mathbb{S}(\mathbf{u}, p) = \mathrm{Fr}^{-1} \nu \mathbf{g}, \quad \nabla \cdot \mathbf{u} = 0 \quad \text{in } \Omega_i(t), \ i = 1, 2,
$$

(22.1)

where $\mathbf{u}$ is the velocity and p is the pressure field, Fr is the Froude number, $\mathbf{g} = (0, -1)^T$ is the gravity vector and ν the dimensionless density. The dimensionless stress tensor $\mathbb{S}$ is given by

$$
\mathbb{S}(\mathbf{u}, p) = \mathrm{Re}^{-1} \left(\nabla \mathbf{u} + (\nabla \mathbf{u})^T \right) - p\mathbb{I}.
$$

(22.2)

The dimensionless density ν and the Reynolds number Re are piecewise constant with respect to the phases,

$$
\mathrm{Re} = \begin{cases} Re_2 \eta_2/\eta_1 , & \text{in } \Omega_1 \\ Re_2 , & \text{in } \Omega_2 \end{cases}, \nu = \begin{cases} \rho_1/\rho_2 , & \text{in } \Omega_1 \\ 1 , & \text{in } \Omega_2 \end{cases}, Re_2 = \frac{\rho_2 V d}{\eta_2}, Fr = \frac{V^2}{gd}, \quad (22.3)
$$

where V is the characteristic velocity, d is the characteristic length, η_i is the dynamic viscosity of phase i, ρ_i is the density of phase i, and g is the gravitational acceleration.

On the interface the following matching conditions are specified. The surface tension force balances the jump in the stress tensor,

$$[|\mathbb{S}(\mathbf{u}, p)|] \cdot \mathbf{n}_\Gamma = \nabla_\Gamma \cdot (\sigma \mathbb{P}_\Gamma) \quad \text{on } \Gamma(t), \tag{22.4}$$

where $[|f|] = f|_{\Omega_1} - f|_{\Omega_2}$ denotes the jump over Γ, $\mathbf{n}_\Gamma$ is the unit normal on Γ, pointing into Ω_2, $\nabla_\Gamma \cdot$ is the surface divergence on Γ, $\sigma = \sigma(c_\Gamma)$ is the nondimensional surface tension which is dependent on the surface surfactant concentration c_Γ, and $\mathbb{P}_\Gamma = \mathbb{I} - \mathbf{n}_\Gamma \otimes \mathbf{n}_\Gamma$ denotes the projection onto the tangential space of Γ. The continuity condition for the velocity $\mathbf{u}$ and the interface velocity $\mathbf{w}$ are

$$[|\mathbf{u}|] = 0, \quad \mathbf{w} \cdot \mathbf{n}_\Gamma = \mathbf{u} \cdot \mathbf{n}_\Gamma \quad \text{on } \Gamma(t). \tag{22.5}$$

The Langmuir law for the surface tension coefficient reads

$$\sigma(c_\Gamma) = \text{We}^{-1}(1 + \text{E}\ln(1 - c_\Gamma)), \tag{22.6}$$

where $\text{We} = \rho_2 V^2 d\sigma_0^{-1}$ is the Weber number and $\text{E} = RT_\infty c_\Gamma^\infty \sigma_0^{-1}$ is the surface elasticity, with the universal gas constant R, the equilibrium temperature T_∞, the maximum packing surface surfactant concentration c_Γ^∞, and the surface tension coefficient σ_0.

The surfactant transport in the bulk and on the surface is modelled by convection-diffusion-reaction equations. The surfactant is considered to be soluble into the liquid phase Ω_2 only,

$$\frac{\partial c}{\partial t} + \mathbf{u} \cdot \nabla c = \text{Pe}^{-1} \Delta c \qquad\qquad \text{in } \Omega_2(t), \tag{22.7}$$

$$\text{Pe}^{-1} \mathbf{n}_\Gamma \cdot \nabla c = -S(c, c_\Gamma) \qquad\qquad \text{on } \Gamma(t), \tag{22.8}$$

$$\dot{c}_\Gamma + c_\Gamma \nabla_\Gamma \cdot \mathbf{u} = \text{Pe}_\Gamma^{-1} \Delta_\Gamma c_\Gamma + S(c, c_\Gamma) \qquad\qquad \text{on } \Gamma(t), \tag{22.9}$$

where c is the bulk surfactant concentration, $\dot{c}_\Gamma$ denotes the material derivative of c_Γ and $\Delta_\Gamma = \nabla_\Gamma \cdot \nabla_\Gamma$ denotes the Laplace–Beltrami operator on Γ. In addition, $\text{Pe} = V d D^{-1}$ and $\text{Pe}_\Gamma = V d D_\Gamma^{-1}$ denote the Peclet number in the bulk and on the surface, with D and D_Γ the diffusion coefficients in the bulk and on the surface, respectively. The ad- and desorption term S is given according to the Langmuir sorption law,

$$S(c, c_\Gamma) = \text{Da}\, c\, (1 - c_\Gamma) - \text{Bi}\, c_\Gamma, \tag{22.10}$$

where $\text{Da} = k_a c_\Gamma^\infty V^{-1}$ and $\text{Bi} = k_d d V^{-1}$ are the Damköhler and the Biot number, respectively, with adsorption coefficient k_a and the desorption coefficient k_d.

22.2.2 *Diffuse-Interface Model*

In addition to the above description of the interface as a sharp hyper-surface, so-called diffuse-interface models can be used. Diffuse-interface models account for a partial mixing of the fluids at a small length scale. Therefore the interface is represented as a thin layer of finite thickness ϵ. An auxiliary phase-field function ϕ is used to indicate the phases. The phase-field function varies smoothly between distinct values in both phases, here $\phi = 1$ in the gas phase Ω_1 and $\phi = -1$ in the liquid phase Ω_2. The interface can be associated with an intermediate level set of the phase-field function (here $\phi = 0$). For a comprehensive overview on diffuse-interface models, we refer to another chapter of this book [5].

The simplest diffuse-interface model consists of the following coupled Navier–Stokes Cahn–Hilliard equations defined in $\Omega \times (0, T]$:

$$v(\phi) \left(\frac{\partial \mathbf{u}}{\partial t} + \mathbf{u} \cdot \nabla \mathbf{u} \right) + \nabla p - \nabla \cdot \left(\mathrm{Re}(\phi)^{-1} \left(\nabla \mathbf{u} + \nabla \mathbf{u}^T \right) \right)$$

$$= \mathrm{Fr}^{-1} v \mathbf{g} + \sigma \mu \nabla \phi + \frac{1}{2} |\nabla \phi| \mathbb{P}_\Gamma \nabla \sigma, \qquad (22.11a)$$

$$\nabla \cdot \mathbf{u} = 0, \qquad (22.11b)$$

$$\frac{\partial \phi}{\partial t} + \mathbf{u} \cdot \nabla \phi = \nabla \cdot (M(\phi) \nabla \mu), \qquad (22.11c)$$

$$\mu = \frac{3}{2\sqrt{2}} \left(\epsilon^{-1} W'(\phi) - \epsilon \Delta \phi \right). \qquad (22.11d)$$

Here μ is the chemical potential, M a mobility function, the function $W(\phi) = \frac{1}{4}(\phi^2 - 1)^2$ is a double-well potential and the surface projection $\mathbb{P}_\Gamma = \mathbb{I} - \mathbf{n}_\phi \otimes \mathbf{n}_\phi$ here involves the diffuse-interface normal $\mathbf{n}_\phi = \nabla \phi / |\nabla \phi|$. The Reynolds number and nondimensional density are defined as linear interpolations,

$$\mathrm{Re}(\phi) = \frac{1 + \phi}{2} \mathrm{Re}_2 \frac{\eta_2}{\eta_1} + \frac{1 - \phi}{2} \mathrm{Re}_2, \quad v(\phi) = \frac{1 + \phi}{2} \frac{\rho_1}{\rho_2} + \frac{1 - \phi}{2}.$$

The surfactant transport equations in the bulk and on the surface are extended to the whole domain Ω by the diffuse-domain approach, see also [5]. The resulting equations in $\Omega \times (0, T]$ are

$$\frac{\partial}{\partial t} (\chi(\phi)c) + \mathbf{u} \cdot \nabla (\chi(\phi)c) = \mathrm{Pe}^{-1} \nabla \cdot (\chi(\phi) \nabla c) - \delta(\phi) S(c, c_\Gamma), \qquad (22.12)$$

$$\frac{\partial}{\partial t} (\delta(\phi) c_\Gamma) + \mathbf{u} \cdot \nabla (\delta(\phi) c_\Gamma) = \mathrm{Pe}_\Gamma^{-1} \nabla \cdot (\delta(\phi) \nabla c_\Gamma) + \delta(\phi) S(c, c_\Gamma), \qquad (22.13)$$

where $\delta(\phi) = |\nabla\phi|/2$ and $\chi(\phi) = (1-\phi)/2$ are approximations to the surface Dirac delta function and the characteristic function of the liquid domain, respectively.

As an alternative, we also consider the following diffuse-interface model, where we replace (22.11) by an extension of the thermodynamically consistent model from [1] that additionally incorporates surfactants, see also [2, 28]. The model is given by

$$\frac{\partial}{\partial t}(\nu\mathbf{u}) + \left(\left(\nu\mathbf{u} - \frac{\rho_2 - \rho_1}{2\rho_2}M(\phi)\nabla\mu\right)\cdot\nabla\right)\mathbf{u} + \nabla p - \nabla\cdot\left(\mathrm{Re}^{-1}(\nabla\mathbf{u} + \nabla\mathbf{u}^T)\right)$$
$$= \mu\nabla\phi + \delta(\phi)\nabla\sigma + \mathrm{Fr}^{-1}\nu\mathbf{g} - \chi'(\phi)\left(G(c) - G'(c)c\right)\nabla\phi,$$
(22.14a)

$$\nabla\cdot\mathbf{u} = 0,$$
(22.14b)

$$\frac{\partial\phi}{\partial t} + \mathbf{u}\cdot\nabla\phi = \nabla\cdot(M(\phi)\nabla\mu),$$
(22.14c)

$$\mu + \nabla\cdot\left(\tfrac{2}{\pi}\epsilon\sigma\nabla\phi\right) = \tfrac{2}{\pi}\epsilon^{-1}\sigma W'(\phi) + \chi'(\phi)\left(G(c) - G'(c)c\right),$$
(22.14d)

where we note that $\chi'(\phi) = -\frac{1}{2}$. Here all variables and parameters are defined as before, except for the diffuse delta function, which is now

$$\delta(\phi) = \frac{2}{\pi}\left(\frac{\epsilon}{2}|\nabla\phi|^2 + \frac{1}{\epsilon}W(\phi)\right),$$

and for the double-well potential W, which in our model is given by the double-obstacle potential

$$W(\phi) = \begin{cases} \frac{1}{2}(1-\phi^2) & \text{if } |\phi| \leq 1, \\ +\infty & \text{else.} \end{cases}$$

In particular, (22.14d) abbreviates a variational inequality in this setting. The energy density $G(c)$ depends on the chosen sorption model, see [28], and in the case of the Langmuir sorption law (22.6) it holds that $G(c) - G'(c)c = -\mathrm{We}^{-1}\mathrm{E}\,c$.

22.3 Benchmark Without Surfactants

First, we consider the hydrodynamics of a clean surface without surfactants. In this case, the equations for c_Γ and c are dropped and the surface tension assumes the constant value $\sigma = \mathrm{We}^{-1}$. A first benchmark for two dimensional and axisymmetric Taylor flow hydrodynamics was presented in [8]. In this section, we recall the benchmark setting and results of [8], before we use the same setting and parameters to include surfactants in Sect. 22.4.

22.3.1 Setting and Parameters

We consider a single bubble in a two-dimensional channel. The initial bubble shape is prescribed as two half-circles connected by straight edges, the computational domain is $\Omega = [0, 1] \times [0, 10]$, see Fig. 22.1. Flow is created by imposing a pressure gradient between the channel inlet and outlet. A co-moving grid is employed such that the bubble remains in the center of the computational domain and evolves to a quasi-stationary state.

The hydrodynamic parameters are chosen from realistic Taylor flow experiments, see Table 22.1. The problem is essentially described by three different nondimensional numbers, the Reynolds number, the Weber number and the capillary number, only two of which are independent, see Table 22.2.

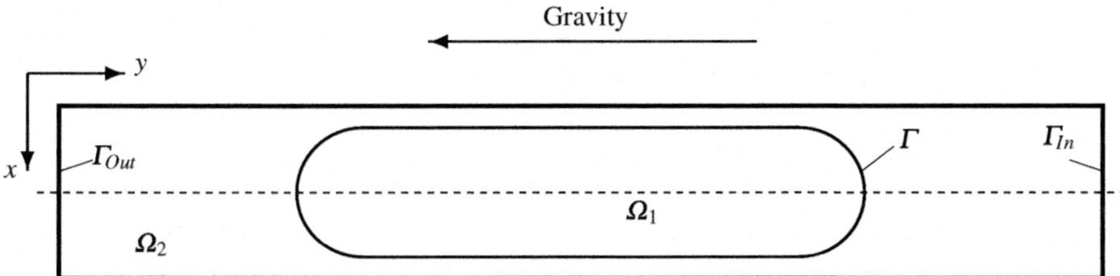

Fig. 22.1 Initial bubble setting and boundaries

Table 22.1 Physical parameters for the numerical benchmark

ρ_L	$1 \cdot 10^3$	$\frac{\text{kg}}{\text{m}^3}$	Density of liquid phase
ρ_G	$1 \cdot 10^{-3}$	ρ_L	Density of gas phase
η_L	$1 \cdot 10^{-2}$	Pa s	Dynamic viscosity of liquid phase
η_G	$1 \cdot 10^{-3}$	η_L	Dynamic viscosity of gas phase
σ_0	$5 \cdot 10^{-4}$	$\frac{\text{N}}{\text{m}}$	Surface tension coefficient
V	$1 \cdot 10^{-3}$	$\frac{m}{s}$	Characteristic velocity
d	$1 \cdot 10^{-3}$	m	Characteristic length, width of computational domain
L	$1 \cdot 10^{-2}$	m	Length of computational domain

Table 22.2 Dimensionless quantities for the numerical benchmark

Re_2	$\frac{\rho_L V d}{\eta_L}$	0.1	Reynolds no. in liquid phase
Ca	$\frac{\eta_L V}{\sigma_0}$	$2 \cdot 10^{\{-4,-3,-2,-1\}}$	Capillary no.
We	$Re_2\, Ca$	$0.1 \cdot Ca$	Weber no.

22.3.2 *Benchmark Results*

For the benchmark problems without surfactant three different scenarios where chosen [8], namely a 2D Taylor flow, the Bretherton problem, with varying capillary numbers, a 3D Taylor flow under the assumption of rotational symmetry, with varying capillary numbers, and a single 3D rotational symmetric Taylor bubble which was compared to experimental data. Note that rotational symmetry of the bubble can then be safely assumed due to the high surface tension.

Participating numerical codes were based both on a classical sharp-interface approach and also on a diffuse-interface model. For the sharp-interface model, two numerical codes using Arbitrary Lagrangian Eulerian (ALE) coordinates and a moving mesh were employed and for the diffuse-interface approach two different but similar mathematical models were compared. Data from all codes were compared to each other for varying capillary numbers in terms of velocity profiles, bubble shapes and pressure distribution.

By using a high-energy synchrotron radiation source [21], very precise measurements of the bubble shape were obtained. The different mathematical models and numerical codes obtained very good agreement to each other. The difference of the simulated bubble shapes to the experimental data was somewhat higher. While the exact reason for this slight disagreement is unclear as of yet, measurement errors in the experiment could explain part of this problem.

The width of the liquid film between the bubble and the capillary wall is one of the most sensible markers for Taylor flow. In a two-dimensional theoretical investigation, a relation between the capillary number and the film width has been derived [22]. For small capillary numbers ($\mathrm{Ca} \leq 3 \cdot 10^{-3}$) it was found that

$$h_{\mathrm{film}} = 0.669\,d\,\mathrm{Ca}^{2/3},$$

where d denotes the diameter of the 2D channel. For larger capillary numbers ($0.05 \leq \mathrm{Ca} \leq 100$) numerical simulations with the boundary element method led to the approximation [32]

$$h_{\mathrm{film}} = 0.209\,d\left(1 - \exp(-1.69 \cdot \mathrm{Ca}^{0.5025})\right).$$

Both formulas have been validated in the benchmark [8] for a range of capillary numbers $10^{-4} \leq \mathrm{Ca} \leq 1$. For high capillary numbers, all models showed perfect agreement with the film thickness approximations. For lower capillary numbers, the film width becomes very small and hard to resolve with the diffuse-interface models, because it requires a computationally expensive small interface thickness. For this reason only few results are shown in [8] for small Ca using diffuse-interface models.

In summary, in [8] it is concluded that despite their conceptual differences, the tested sharp- and diffuse-interface codes are well-applicable for the physically relevant parameter regime. Only diffuse-interface methods are restricted to ranges of moderately large capillary numbers to resolve the corresponding small film thickness and provide highly accurate results.

22.4 Benchmark with Surfactant

In the following we present an extension of the benchmark presented in the previous section to also include soluble surfactants. The basic setup and all hydrodynamic parameters are left unchanged. The additional parameters and conditions for the surfactants are stated below.

22.4.1 Setup

The benchmark setting is designed to simulate a single Taylor bubble in a vertical channel in a counter flow. As discussed earlier, stationary states are advantageous in particular for later comparison with experiments. Such a stationary equilibrium can be reached when buoyancy (upwards) and counter-flow (downwards) balance each other and the net velocity of the Taylor bubble becomes zero. To this end, the inflow velocity is adjusted during the time evolution towards the equilibrium state, see Sect. 22.4.1.2. Simultaneously, this ensures that the bubble remains in the center of the computational domain.

22.4.1.1 Initial Conditions

The initial bubble diameter is set to 0.7 units, the total bubble length is set to 4.7 units. The initial fluid is at rest, i.e. $\mathbf{u}(0) = 0$ in Ω. The initial surfactant concentration in the bulk is homogeneous $c(0) = c_0$. The initially constant surface surfactant concentration is chosen according to the equilibrium condition

$$\frac{\mathrm{Da}}{\mathrm{Bi}} = \frac{c_\Gamma}{c(1 - c_\Gamma)}. \tag{22.15}$$

22.4.1.2 Boundary Conditions

On the inflow boundary Γ_{In} the fluid velocity is prescribed by the following Dirichlet boundary condition

$$\mathbf{u}|_{\Gamma_{In}}(x, y, t) = -4u_{max}(t)x(1 - x)\mathbf{e}_y \quad x \in [0, 1], \tag{22.16}$$

where u_{max} is the maximum inflow velocity, and $\mathbf{e}_y$ denotes the unit vector in y-direction. The value of u_{max} is updated in every time step to reach a stationary bubble position according to the following law:

$$u_{max}^{n+1} = u_{max}^n + u_{bubble}^n, \tag{22.17}$$

where superscripts indicate time step indices and u_{bubble} is the bubble speed in the longitudinal direction, i.e. in the y-direction. We also set $u_{max}^0 = 0$.

The inflowing liquid contains a constant surfactant concentration, which leads to the Dirichlet boundary condition

$$c|_{\Gamma_{In}} = c_0. \tag{22.18}$$

On the outflow boundary Γ_{Out} a do-nothing (stress free) condition is given for the fluid velocity and the surfactant concentration, i.e.

$$\mathbf{n} \cdot \mathbb{S}(\mathbf{u}, p)|_{\Gamma_{Out}} = 0, \qquad \mathbf{n} \cdot \nabla c|_{\Gamma_{Out}} = 0, \tag{22.19}$$

where $\mathbf{n}$ denotes the outer unit normal on $\partial\Omega$. At the capillary wall a no-slip condition for the fluid is assumed

$$\mathbf{u}|_{\Gamma_W} = 0. \tag{22.20}$$

The flux of surfactant is obstructed by the capillary wall, hence, together with the no-slip condition for the fluid, a homogeneous Neumann condition is needed for the surfactant

$$\mathbf{n} \cdot \nabla c|_{\Gamma_W} = 0. \tag{22.21}$$

22.4.2 Surfactant-Related Parameters

Due to the wide variety of different surfactants with different properties, we freely choose the surfactant-related parameters. We use the fixed Damköhler and Biot numbers $Da = Bi = 0.01$. The other surfactant-related parameters are varied as follows. The surface elasticity $E \in \{0.5, 1.0\}$, the initial surfactant concentration $c_0 \in \{0.02, 0.08\}$ and the surface and bulk Peclet numbers are chosen to be equal $Pe = Pe_\Gamma \in \{1, 10\}$. The end time is specified as $T = 10$. The remaining parameters are: $Re = 0.1$, $We = 2 \cdot 10^{-3}$ and $Fr = 1 \cdot 10^{-4}$.

22.4.3 Numerical Methods

Four different solvers using different numerical approaches to the capillary problem at hand were used. In the following we give a short description of each solver.

22.4.3.1 Solvers Implementing the Sharp-Interface Model

The two sharp-interface solvers BGN and MooNMD use the Finite Element Method (FEM) to implement the sharp-interface model.

BGN

The solver BGN, realized within the finite element toolbox ALBERTA [51], is based on an implementation of the parametric finite element approximation introduced in [17]. In particular, it is based on an unfitted finite element method for two-phase flow [14, 19], where the bulk mesh, on which velocity, pressure and bulk surfactant are approximated, is totally independent of the polygonal interface mesh, on which the curvature and the surface surfactant are approximated.

In the bulk phase, the solver uses a standard P_2-P_1 Taylor–Hood element with a simple XFEM enrichment of the pressure space. In order to better capture the details around the phase interface, the bulk mesh is adapted to yield a fine resolution close to the interface, and a coarser resolution further away from it. As the bulk mesh is not fitted to the discrete interface, standard refinement and coarsening algorithms can be used to achieve the desired adaptation. In addition, an implicit tangential motion of the interface nodes leads to a nearly equidistributed discrete interface throughout [14]. As a consequence, no remeshings of interface or bulk meshes are needed in any of the computations.

An implicit Euler time stepping scheme, leading to linear subproblems to be solved at each time step, is employed. In the absence of surfactants, the time stepping scheme can be shown to be unconditionally stable [19]. The stability crucially relies on a variational treatment of curvature, and an appropriate coupling between bulk and interface approximations. The coupled linear systems arising at each time level are solved with the help of a Schur complement approach, which reduces the system to a standard saddle point problem arising from discretizations of Navier–Stokes problems. These saddle point problems are solved with a preconditioned GMRES iterative solver. More details on the solution procedure can be found in [19]. Finally, for the computations in this paper the symmetry of the planar problem was exploited. In particular, only half of the domain was used as the computational domain.

Variants of the BGN package have been successfully applied to dendritic growth [12], snow crystal growth [13], two-phase incompressible Navier–Stokes flow [14, 19], featuring insoluble [16] and soluble [17] surfactants, as well as to two-phase flow with a Boussinesq–Scriven surface fluid [18] and to the dynamics of fluidic membranes and vesicles [15, 20].

MooNMD

The MooNMD solver used here is branch based on the general purpose finite element solver MooNMD [39]. This package has been successfully applied, e.g. in the solution of the steady state and time dependent incompressible Navier–Stokes equations [36–38], large eddy simulations [35], free boundary value problems with capillary surfaces and ferrofluids [30, 43], multiphase problems with and without surface active agents [23, 24, 34], population balance systems [25, 40, 41] and convection-diffusion equations by stabilized finite elements [3, 46].

The MooNMD solver used here employs a Arbitrary Lagrangian Eulerian (ALE) finite element method [26, 27], such that the grid is always aligned to the interface and no cut-cells exist. The time discretization is done with a semi-implicit Euler scheme, that decouples the system (22.1) and (22.7) and allows for an iterative solution for each component.

A modified P_2-P_1 Taylor Hood finite element space is used for the Navier–Stokes equations, that allows pressure jump across the interface, but is continuous in the bulk phases.

For the surfactant a P_2 finite element space is used, in the bulk and on the surface as well.

The interface approximation is of second order, consisting of piecewise quadratic elements. Thus, the Navier–Stokes equation, the bulk transport and the surface transport equation are solved on a second order isoparametric grid at the interface. Further, the axisymmetry of the problem is employed and the computation is done only on one half of the domain using appropriate boundary conditions at the symmetry axis.

22.4.3.2 Solvers Implementing the Phase-Field Model

The two diffuse-interface solvers AMDiS and 2PStab were used for this study.

AMDiS

AMDiS is an adaptive finite element toolbox which is designed for fast implementation of arbitrary PDEs. The software has been already been used for benchmarks of two-phase flow [4, 6, 8]. Extensions for interfacial molecules[44], nanoparticles[7] and lipid bilayer membranes [10, 44] have been implemented.

The simple diffuse-interface model from (22.11)–(22.13) is used. A semi-implicit Euler time stepping algorithm is employed together with space discretization by a Taylor–Hood element for the Navier–Stokes equation, with P_1-elements for p and P_2-elements for $\mathbf{u}$, ϕ, μ, c, c_Γ. The occurring linear systems are solved with the direct solver UMFPACK.

We use $\epsilon = 0.0075$ and an adaptive grid with prescribed grid size $h = 1/128$ at the interface and $h = 1/32$ away from the interface. The mobility is kept constant at $M = 5 \cdot 10^{-5}$. Equidistant time steps of size 10^{-3} are used. Simulations until the end time $T = 10$ correspond to a computational time of $50h$.

2PStab

This code uses ideas from [29], where two-phase flow without surfactants is simulated, to obtain an energy stable time discretization for (22.14). Details will be provided elsewhere.

The code is written in C++ using the FEniCS package. The linear algebra backend is PETSc, and linear systems are solved using the direct solver MUMPS. The spatial discretization is represented using the library ALBERTA [51]. We use Taylor–Hood elements for the spatial discretization of the Navier–Stokes equation, i.e. P_2 elements for the velocity field $\mathbf{u}$ and P_1 elements for the pressure p. The further variables, i.e. ϕ, μ, c, c_Γ, are discretized using P_1 elements.

We set $\epsilon = 0.02$ and $M(\phi) = 10^{-5} \left(1 - 0.99\phi^2\right)$, approximating a degenerate mobility. To resolve the interfacial region, i.e. $|\phi| < 1$, we use triangles with diameter $h = 0.004$, while in the bulk regions we use $h = 0.025$. The time step size is $\tau = 5 \cdot 10^{-5}$.

22.4.4 Results

Different benchmark values are considered. Apart from the numerical values c, c_Γ and u_{max}, calculated by the codes, some of them are determined by a post process. Due to the different nature of the methods used here, the post process can differ likewise.

Here, the film width h_{film} is defined as the minimal distance between bubble and wall. For the sharp-interface schemes, h_{film} is computed directly from the discrete interface representation, while the diffuse-interface schemes use the zero level contour line of ϕ.

The bulk surfactant mass m_c and surface surfactant mass m_{c_Γ} are defined as the total amount of surfactant present in the bulk and on the surface, respectively. In case of the sharp-interface schemes the bulk surfactant exist in Ω_2 and the surface surfactant exists on Γ. The corresponding masses are calculated as

$$m_c = \int_{\Omega_2} c, \quad m_{c_\Gamma} = \int_\Gamma c_\Gamma.$$

In the case of the diffuse-interface schemes, the quantities c and c_Γ are defined in the whole domain Ω and, thus the masses are calculated as

$$m_c = \int_\Omega \frac{1}{2}(1 - \phi)c, \quad m_{c_\Gamma} = \int_\Omega \delta(\phi)c_\Gamma.$$

Having defined the benchmark values, we take a closer look at the results in the following.

Figure 22.2 shows the time evolution of the maximum inflow velocity u_{max} for the simulation parameters $(c_0, Pe, Pe_\Gamma, E) = (0.02, 1, 1, 1)$, exemplarily. All four codes show a similar behavior, after some oscillations u_{max} reaches a stationary state already before $t = 1$. According to the adaption law for u_{max}, recall (22.17), this also implies the stationarity of the bubble position in the channel. The sharp-interface results are very close to each other and so are the diffuse-interface results, that both predict a slightly smaller bubble velocity.

In Table 22.3 the final counter current flow velocity is shown for all chosen parameter combinations. Note that the variation of the flow velocity across the different numerical models dominates the variation across different parameters. The

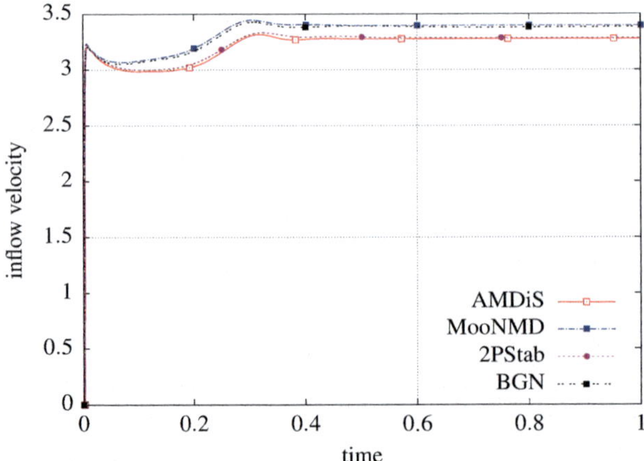

Fig. 22.2 Maximum inflow velocity over time (detail), for $(c_0, Pe, Pe_\Gamma, E) = (0.02, 1, 1, 1)$

Table 22.3 Maximum counter flow velocity u_{max} at the stationary state for different simulation parameters

	$Pe = Pe_\Gamma = 1$		$Pe = Pe_\Gamma = 10$	
	$E = 0.5$	$E = 1$	$E = 0.5$	$E = 1$
$c_0 = 0.02$				
AMDiS	3.27506	3.27336	3.27539	3.27391
MooNMD	3.39283	3.39284	3.39283	3.39284
2PStab ($t = 1$)	3.28185	3.28102	3.28248	3.28087
BGN	3.37914	3.37773	3.36149	3.35499
$c_0 = 0.08$				
AMDiS	3.26413	3.25311	–	–
MooNMD	3.39286	3.38990	3.39198	3.39203
2PStab ($t = 1$)	3.29367	–	–	–
BGN	3.35935	3.33733	3.33063	3.31528

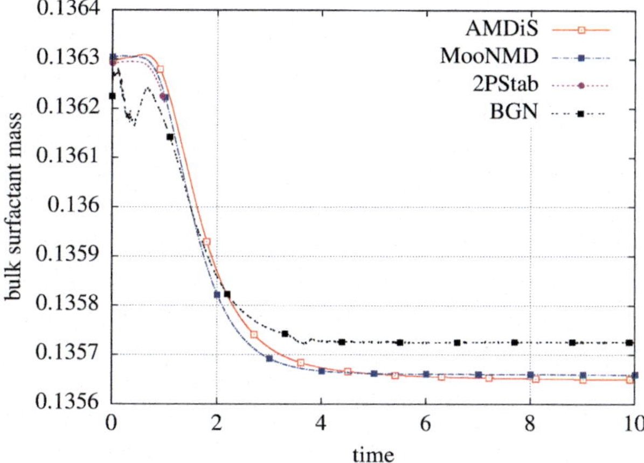

Fig. 22.3 Bulk surfactant mass over time, for $(c_0, Pe, Pe_\Gamma, E) = (0.02, 1, 1, 1)$

missing values in Table 22.3, in particular for $c_0 = 0.08, Pe = Pe_\Gamma = 10$, are caused by numerical problems due to high surfactant concentration at the bubble rear and the corresponding low surface tensions.

Next we look at the evolution of surfactant concentrations. The bulk surfactant mass and the surface surfactant mass over time are shown in Figs. 22.3 and 22.4, respectively. The bulk surfactant mass decreases over time and reaches a stationary state. It can be seen that the surfactant mass equilibrates slower than the inflow velocity and the stationary state is not reached until approximately $t = 5$. Agreement among the codes is particularly high between AMDiS and MooNMD, despite the conceptual differences between sharp- and diffuse-interface models. Due to the small time steps used, the simulation of 2PStab is limited to the end time $t = 1$.

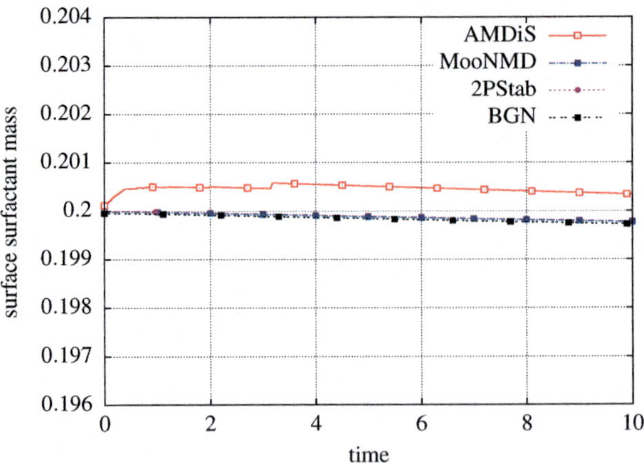

Fig. 22.4 Surface surfactant mass over time, for $(c_0, Pe, Pe_\Gamma, E) = (0.02, 1, 1, 1)$

In Fig. 22.4 we depict the temporal evolution of the surface surfactant mass. We observe a slow decrease of the surface surfactant mass over time, indicating that a discrete stationary solution is not yet found at $T = 10$. Although the continuous problem would eventually lead to a stationary state, for which conservation of mass holds trivially, the loss of exact divergence free flow in the discrete setting may lead to a slight, but steady, loss of surfactant mass over time. The deviation in the simulation using AMDiS at the beginning of the simulation cannot be explained, but in general the code shows the same overall behaviour as the other codes.

The stationary local distribution of surfactant in the bulk phase and on the surface is shown in Fig. 22.5, exemplarily for the stationary case with $(c_0, Pe, Pe_\Gamma, E) = (0.02, 1, 1, 1)$ (right) and $(c_0, Pe, Pe_\Gamma, E) = (0.02, 10, 10, 1)$ (left). The surface, the inner line, is scaled and translated a bit, in order to better distinguish it from the bulk. Note the different distribution due to different diffusion intensity. For both parameter regimes, surfactant is adsorbed at the bubble surface leading to a decrease in bulk surfactant next to the bubble. Along the bubble surface surfactant is transported with the flow to accumulate at the rear, where it is continuously desorbed to the bulk. In case of higher diffusion (left) local variations in surfactant concentration are more pronounced.

Figure 22.6 displays the stationary bulk surfactant distribution along the channel wall for $(c_0, Pe, Pe_\Gamma, E) = (0.02, 1, 1, 1)$. We find excellent agreement among the numerical models and codes.

Finally, we investigate the effect of surfactant on the film thickness. Figure 22.7 shows the film width over time for $(c_0, Pe, Pe_\Gamma, E) = (0.02, 1, 1, 0.5)$. The behavior of all four codes is similar and very close to each other. After some initial oscillations and interface waves, the bubble shape reaches a stationary state after a short time and the film width stays constant. The stationary film width is given for all codes and parameters in Table 22.4. All values are determined at $t = 10$, except

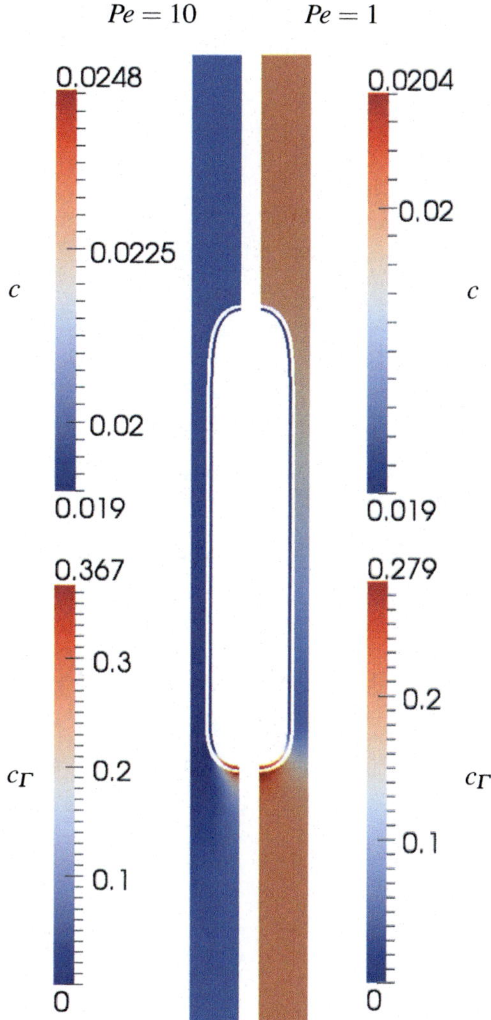

Fig. 22.5 Stationary surfactant distribution from MooNMD in the bulk phases and on the surface. For $(c_0, Pe, Pe_\Gamma, E) = (0.02, 1, 1, 1)$ (right), and for $(c_0, Pe, Pe_\Gamma, E) = (0.02, 10, 10, 1)$ (left). Note the different scalings

for the code 2PStab, for which they are taken at time $t = 1$, when the solution is already fairly stationary.

In general the diffuse-interface models predict a smaller film thickness which might be due to a minimal Cahn–Hilliard dynamics that leads to a slight retraction of the bubble. This explanation is also supported by the slightly smaller bubble lengths of diffuse-interface models, seen in Table 22.5.

Figure 22.8 shows a comparison of the film width with varying surface elasticity E (left) and with varying diffusion coefficients Pe, Pe_Γ (right). All codes show that increasing surface elasticity and surfactant diffusibility leads to larger film thickness. This is consistent with analytical predictions conducted in [49]. Due to the

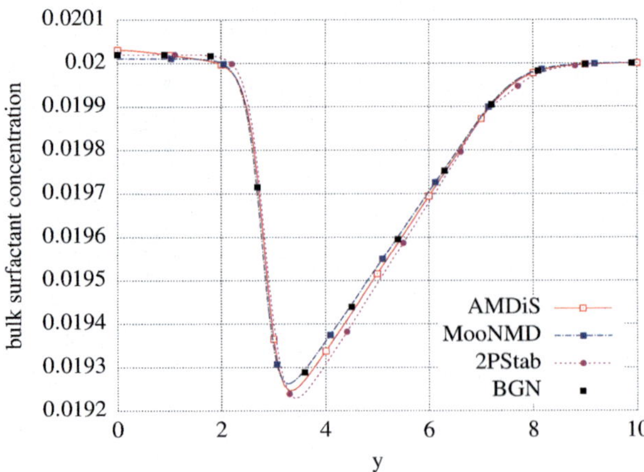

Fig. 22.6 Stationary bulk surfactant concentration along the channel wall at time $t = 10$ ($t = 1$ for 2PStab), for $(c_0, Pe, Pe_\Gamma, E) = (0.02, 1, 1, 1)$

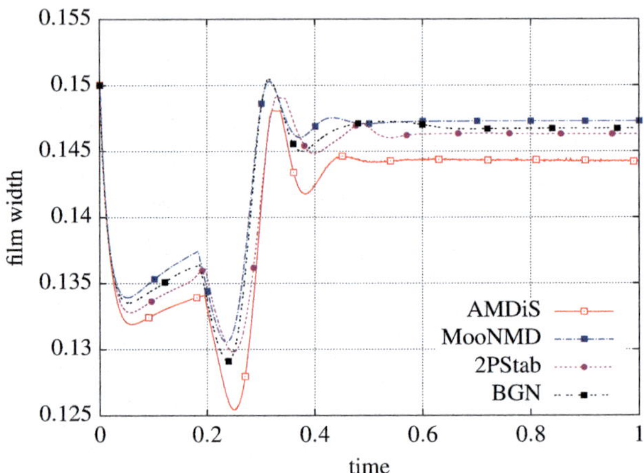

Fig. 22.7 Film width over time, for $(c_0, Pe, Pe_\Gamma, E) = (0.02, 1, 1, 0.5)$

chosen parameters, the change in film width across different parameters is minimal and the tiny differences between the codes may seem large. However, the differences in film thickness between the codes are smaller than 2% and the equal slope of the lines in Fig. 22.8 indicates good agreement between all four numerical codes with respect to changes in surfactant elasticity and diffusibility.

From Table 22.3 it can be seen that the bubble velocity is almost independent of the surfactant-related parameters, although the film thickness varies. This is in contrast to the results of [22] that predict an increasing bubble velocity with an increasing film width. This might be a unique behavior due to the presence of the surfactant that was not included in [22].

Table 22.4 Film width for different simulation parameters

| | $Pe = Pe_\Gamma = 1$ | | $Pe = Pe_\Gamma = 10$ | |
	$E = 0.5$	$E = 1$	$E = 0.5$	$E = 1$
$c_0 = 0.02$				
AMDiS	0.14408	0.14802	0.14304	0.14604
MooNMD	0.14724	0.15086	0.14601	0.14853
2PStab ($t = 1$)	0.14627	0.14957	0.14504	0.14739
BGN	0.14674	0.15053	0.14492	0.14789
$c_0 = 0.08$				
AMDiS	0.14960	0.14914	–	–
MooNMD	0.15121	0.15113	0.15037	0.15032
2PStab ($t = 1$)	0.14962	–	–	–
BGN	0.15101	0.14932	0.14845	0.15144

Table 22.5 Bubble length for different simulation parameters

| | $Pe = Pe_\Gamma = 1$ | | $Pe = Pe_\Gamma = 10$ | |
	$E = 0.5$	$E = 1$	$E = 0.5$	$E = 1$
$c_0 = 0.02$				
AMDiS	4.71900	4.72500	4.72000	4.73100
MooNMD	4.76351	4.77366	4.76632	4.77970
2PStab ($t = 1$)	4.73000	4.75000	4.74000	4.76000
BGN	4.75971	4.76951	4.76212	4.77465
$c_0 = 0.08$				
AMDiS	4.76200	4.80000	–	–
MooNMD	4.81922	5.05115	4.87267	4.97512
2PStab ($t = 1$)	4.80000	–	–	–
BGN	4.81032	4.82561	4.81045	4.84463

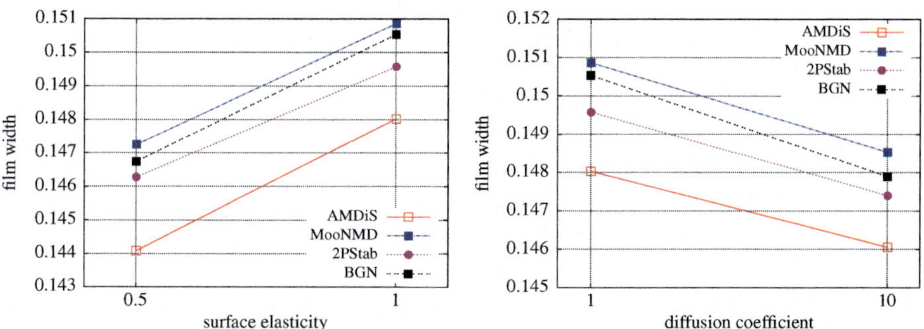

Fig. 22.8 Film width with surface elasticity for $(c_0, Pe, Pe_\Gamma) = (0.02, 1, 1)$ (*left*) and film width with diffusion coefficients for $(c_0, E) = (0.02, 1)$ (*right*)

22.5 Conclusion

We have presented a quantitative comparison of simulations based on diffuse- and sharp-interface models for two-phase flows with soluble surfactants. Therefore, we have extended the purely hydrodynamic benchmark study of Aland et al. [8] to a scenario with soluble surfactants. We considered a single Taylor bubble that assumes a stationary position in a channel flow as liquid inflow and buoyancy effects cancel each other out, which makes the scenario amenable to high resolution experimental imaging.

Several numerical simulations with different surfactant-related parameters were conducted with four numerical codes, two based on the diffuse-interface methods and two based on the sharp-interface methods. The codes were compared in terms of different benchmark quantities: the surfactant distribution, the film width between bubble and channel wall, the bubble length and the counter flow velocity required to keep the bubble stationary.

All four codes showed a quantitative agreement across all measured quantities. While bubble shapes and velocities were almost identical between the sharp-interface models, the diffuse-interface codes predicted a slightly smaller film thickness and bubble length which can be attributed to the Cahn–Hilliard dynamics leading to a minimal retraction of the bubble. The slight contraction of the bubble in the diffuse-interface models went along with an up to 3% smaller bubble velocity.

Furthermore, we analyzed the local distribution of surfactant in the system. Surfactant molecules are adsorbed at the bubble surface leading to a decrease in bulk surfactant next to the bubble. Along the bubble surface surfactant is transported with the flow to accumulate at the rear where it is then desorbed into the bulk. We found that all codes predicted almost identical values for the local surfactant distribution along the channel wall, in accordance with the characteristic decrease of the surfactant concentration in the film predicted in [49].

For increasing surfactant concentration and surface elasticity we found an increasing film thickness as predicted in [49]. Additionally, we observed an increase of the film width for higher surfactant diffusibility.

A further observation was that changes in the film width were not reflected in the counter flow velocity, which stayed nearly constant across all used parameters, contrary to the prediction of [22].

References

1. Abels, H., Garcke, H., Grün, G.: Thermodynamically consistent, frame indifferent diffuse interface models for incompressible two-phase flows with different densities. Math. Models Methods Appl. Sci. **22**(03), 1150013 (2012)
2. Abels, H., Garcke, H., Lam, K.F., Weber, J.: Two-phase flow with surfactants: diffuse interface models and their analysis. In: Advances in Mathematical Fluid Mechanics. Springer, Basel (2016); (a contribution in this book)

3. Ahmed, N., Matthies, G., Tobiska, L., Xie, H.: Discontinuous Galerkin time stepping with local projection stabilization for transient convection-diffusion-reaction problems. Comput. Methods Appl. Mech. Eng. **200**(21–22), 1747–1756 (2011). doi: 10.1016/j.cma.2011.02.003

4. Aland, S.: Time integration for diffuse interface models for two-phase flow. J. Comput. Phys. **262**, 58–71 (2014)

5. Aland, S.: Phase field models for two-phase flow with surfactants and biomembranes. In: Advances in Mathematical Fluid Mechanics. Springer, Basel (2016); (a contribution in this book)

6. Aland, S., Voigt, A.: Benchmark computations of diffuse interface models for two-dimensional bubble dynamics. Int. J. Numer. Methods Fluids **69**(3), 747–761 (2011)

7. Aland, S., Lowengrub, J., Voigt, A.: A continuum model of colloid-stabilized interfaces. Phys. Fluids **23**(6), 062103 (2011)

8. Aland, S., Boden, S., Hahn, A., Klingbeil, F., Weismann, M., Weller, S.: Quantitative comparison of Taylor flow simulations based on sharp-interface and diffuse-interface models. Int. J. Numer. Methods Fluids **73**, 344–361 (2013)

9. Aland, S., Lehrenfeld, C., Marschall, H., Meyer, C., Weller, S.: Accuracy of two-phase flow simulations: the Taylor flow benchmark. Proc. Appl. Math. Mech. **13**(1), 595–598 (2013)

10. Aland, S., Egerer, S., Lowengrub, J., Voigt, A.: Diffuse interface models of locally inextensible vesicles in a viscous fluid. J. Comput. Phys. **277**, 32–47 (2014). http://dx.doi.org/10.1016/j.jcp.2014.08.016

11. Angeli, P., Gavriilidis, A.: Hydrodynamics of Taylor flow in small channels: a review. Proc. Inst. Mech. Eng. C J. Mech. Eng. Sci. **222**(5), 737–751 (2008)

12. Barrett, J.W., Garcke, H., Nürnberg, R.: On stable parametric finite element methods for the Stefan problem and the Mullins–Sekerka problem with applications to dendritic growth. J. Comput. Phys. **229**(18), 6270–6299 (2010). doi:10.1016/j.jcp.2010.04.039. http://dx.doi.org/10.1016/j.jcp.2010.04.039

13. Barrett, J.W., Garcke, H., Nürnberg, R.: Numerical computations of faceted pattern formation in snow crystal growth. Phys. Rev. E **86**(1), 011604 (2012). doi:10.1103/PhysRevE.86.011604. http://dx.doi.org/10.1103/PhysRevE.86.011604

14. Barrett, J.W., Garcke, H., Nürnberg, R.: Eliminating spurious velocities with a stable approximation of viscous incompressible two-phase Stokes flow. Comput. Methods Appl. Mech. Eng. **267**, 511–530 (2013). doi:10.1016/j.cma.2013.09.023. http://dx.doi.org/10.1016/j.cma.2013.09.023

15. Barrett, J.W., Garcke, H., Nürnberg, R.: Numerical computations of the dynamics of fluidic membranes and vesicles. Phys. Rev. E **92**(5), 052704 (2015). doi:10.1103/PhysRevE.92.052704. http://dx.doi.org/10.1103/PhysRevE.92.052704

16. Barrett, J.W., Garcke, H., Nürnberg, R.: On the stable numerical approximation of two-phase flow with insoluble surfactant. M2AN Math. Model. Numer. Anal. **49**(2), 421–458 (2015). doi:10.1051/m2an/2014039. http://dx.doi.org/10.1051/m2an/2014039

17. Barrett, J.W., Garcke, H., Nürnberg, R.: Stable finite element approximations of two-phase flow with soluble surfactant. J. Comput. Phys. **297**, 530–564 (2015). doi:10.1016/j.jcp.2015.05.029. http://dx.doi.org/10.1016/j.jcp.2015.05.029

18. Barrett, J.W., Garcke, H., Nürnberg, R.: Stable numerical approximation of two-phase flow with a Boussinesq–Scriven surface fluid. Commun. Math. Sci. **13**(7), 1829–1874 (2015). doi:10.4310/CMS.2015.v13.n7.a9. http://dx.doi.org/10.4310/CMS.2015.v13.n7.a9

19. Barrett, J.W., Garcke, H., Nürnberg, R.: A stable parametric finite element discretization of two-phase Navier–Stokes flow. J. Sci. Comp. **63**(1), 78–117 (2015). doi:10.1007/s10915-014-9885-2. http://dx.doi.org/10.1007/s10915-014-9885-2

20. Barrett, J.W., Garcke, H., Nürnberg, R.: A stable numerical method for the dynamics of fluidic biomembranes. Numer. Math. **134**(4), 783–822 (2016). doi:10.1007/s00211-015-0787-5. http://dx.doi.org/10.1007/s00211-015-0787-5

21. Boden, S., dos Santos Rolo, T., Baumbach, T., Hampel, U.: Synchrotron radiation microtomography of Taylor bubbles in capillary two-phase flow. Exp. Fluids **55**(7), 1768 (2014). doi:10.1007/s00348-014-1768-7. http://dx.doi.org/10.1007/s00348-014-1768-7

22. Bretherton, F.: The motion of long bubbles in tubes. J. Fluid. Mech. **10**(02), 166–188 (1961)
23. Ganesan, S., Tobiska, L.: A coupled arbitrary Lagrangian–Eulerian and Lagrangian method for computation of free surface flows with insoluble surfactants. J. Comput. Phys. **228**(8), 2859–2873 (2009). doi:10.1016/j.jcp.2008.12.035
24. Ganesan, S., Tobiska, L.: Arbitrary Lagrangian–Eulerian finite-element method for computation of two-phase flows with soluble surfactants. J. Comput. Phys. **231**(9), 3685–3702 (2012). doi:10.1016/j.jcp.2012.01.018
25. Ganesan, S., Tobiska, L.: An operator-splitting finite element method for the efficient parallel solution of multi-dimensional population balance systems. Chem. Eng. Sci. **69**, 59–68 (2012)
26. Ganesan, S., Hahn, A., Held, K., Tobiska, L.: An accurate numerical method for computation of two-phase flows with surfactants. In: Eberhardsteiner, J., et al. (eds.), European Congress on Computational Methods in Applied Sciences and Engineering (ECCOMAS 2012), Vienna, Austria, pp. 10–14, September 2012
27. Ganesan, S., Hahn, A., Simon, K., Tobiska, L.: Finite element computations for dynamic liquid–fluid interfaces. In: Computational Methods for Complex Liquid-Fluid Interfaces. CRC Press, Boca Raton (2015)
28. Garcke, H., Lam, K., Stinner, B.: Diffuse interface modelling of soluble surfactants in two-phase flow. Commun. Math. Sci. **12**(8), 1475–1522 (2014)
29. Garcke, H., Hinze, M., Kahle, C.: A stable and linear time discretization for a thermodynamically consistent model for two-phase incompressible flow. Appl. Numer. Math. **99**, 151–171 (2016)
30. Gollwitzer, C., Matthies, G., Richter, R., Rehberg, I., Tobiska, L.: The surface topography of a magnetic fluid – a quantitative comparison between experiment and numerical simulation. J. Fluid Mech. **571**, 455–474 (2007)
31. Günther, A., Jhunjhunwala, M., Thalmann, M., Schmidt, M.A., Jensen, K.F.: Micromixing of miscible liquids in segmented gas-liquid flow. Langmuir **21**(4), 1547–1555 (2005). doi:10.1021/la0482406
32. Halpern, D., Gaver, D.: Boundary element analysis of the time-dependent motion of a semi-infinite bubble in a channel. J. Comput. Phys. **115**(2), 366–375 (1994)
33. Hayashi, K., Tomiyama, A.: Effects of surfactant on terminal velocity of a Taylor bubble in a vertical pipe. Int. J. Multiphase Flow **39**, 78–87 (2012)
34. Hysing, S., Turek, S., Kuzmin, D., Parolini, N., Burman, E., Ganesan, S., Tobiska, L.: Quantitative benchmark computations of two-dimensional bubble dynamics. Int. J. Numer. Methods Fluids **60**(11), 1259–1288 (2009). doi:10.1002/fld.1934
35. Iliescu, T., John, V., Layton, W.J., Matthies, G., Tobiska, L.: A numerical study of a class of LES models. Int. J. Comput. Fluid Dyn. **17**(1), 75–85 (2003). doi:10.1080/1061856021000009209
36. John, V.: Higher order finite element methods and multigrid solvers in a benchmark problem for the 3D Navier–Stokes equations. Int. J. Numer. Methods Fluids **40**(6), 775–798 (2002). doi:10.1002/fld.377
37. John, V.: Reference values for drag and lift of a two-dimensional time-dependent flow around a cylinder. Int. J. Numer. Methods Fluids **44**(7), 777–788 (2004). doi:10.1002/fld.679
38. John, V., Matthies, G.: Higher-order finite element discretizations in a benchmark problem for incompressible flows. Int. J. Numer. Methods Fluids **37**(8), 885–903 (2001). doi:10.1002/fld.195
39. John, V., Matthies, G.: MooNMD – a program package based on mapped finite element methods. Comput. Vis. Sci. **6**, 163–170 (2004). doi:10.1007/s00791-003-0120-1
40. John, V., Mitkova, T., Roland, M., Sundmacher, K., Tobiska, L., Voigt, A.: Simulations of population balance systems with one internal coordinate using finite element methods. Chem. Eng. Sci. **64**(4), 733–741 (2009). doi:10.1016/j.ces.2008.05.004; 3rd International Conference on Population Balance Modelling

41. Krasnyk, M., Mangold, M., Ganesan, S., Tobiska, L.: Numerical reduction of a crystallizer model with internal and external coordinates by proper orthogonal decomposition. Chem. Eng. Sci. **70**(0), 77–86 (2012). doi:10.1016/j.ces.2011.05.053; 4th International Conference on Population Balance Modeling
42. Kreutzer, M.T., Kapteijn, F., Moulijn, J.A., Heiszwolf, J.J.: Multiphase monolith reactors: chemical reaction engineering of segmented flow in microchannels. Chem. Eng. Sci. **60**(22), 5895–5916 (2005). doi:10.1016/j.ces.2005.03.022
43. Lavrova, O., Matthies, G., Mitkova, T., Polevikov, V., Tobiska, L.: Finite element methods for coupled problems in ferrohydrodynamics. In: Challenges in Scientific Computing – CISC 2002. Lecture Notes in Computational Science and Engineering, vol. 35, pp. 160–183. Springer, Berlin (2003)
44. Lowengrub, J., Allard, J., Aland, S.: Numerical simulation of endocytosis: Viscous flow driven by membranes with non-uniformly distributed curvature-inducing molecules. J. Comput. Phys. **309**, 112–128 (2016). doi:10.1016/j.jcp.2015.12.055. https://dx.doi.org/10.1016/j.jcp.2015.12.055
45. Marschall, H., Boden, S., Lehrenfeld, C., Falconi Delgado, C., Hampel, U., Reusken, A., Wörner, M., Bothe, D.: Validation of interface capturing and tracking techniques with different surface tension treatments against a Taylor bubble benchmark problem. Comput. Fluids **102**, 336–352 (2014)
46. Matthies, G., Tobiska, L.: A two-level local projection stabilisation on uniformly refined triangular meshes. Numer. Algorithms **61**, 465–478 (2012). doi:10.1007/s11075-012-9543-4
47. Muradoglu, M., Günther, A., Stone, H.A.: A computational study of axial dispersion in segmented gas-liquid flow. Phys. Fluids **19**(7) (2007). doi:10.1063/1.2750295
48. Olgac, U., Muradoglu, M.: Effects of surfactant on liquid film thickness in the Bretherton problem. Int. J. Multiphase Flow **48**, 58–70 (2013)
49. Ratulowski, J., Chang, H.C.: Marangoni effects of trace impurities on the motion of long gas bubbles in capillaries. J. Fluid Mech. **210**, 303–328 (1990)
50. Roy, S., Bauer, T., Al-Dahhan, M., Lehner, P., Turek, T.: Monoliths as multiphase reactors: a review. AIChE J. **50**(11), 2918–2938 (2004). doi:10.1002/aic.10268
51. Schmidt, A., Siebert, K.G.: Design of Adaptive Finite Element Software: The Finite Element Toolbox ALBERTA. Lecture Notes in Computational Science and Engineering, vol. 42. Springer, Berlin (2005)
52. Williams, J.L.: Monolith structures, materials, properties and uses. Catal. Today **69**(1–4) (2001). doi:10.1016/S0920-5861(01)00348-0

Chapter 23
Direct Numerical Simulations of Taylor Bubbles in a Square Mini-Channel: Detailed Shape and Flow Analysis with Experimental Validation

Holger Marschall, Carlos Falconi, Christoph Lehrenfeld, Rufat Abiev, Martin Wörner, Arnold Reusken, and Dieter Bothe

Abstract The Priority Program SPP 1506 "Transport Processes at Fluidic Interfaces" by the German Research Foundation DFG has established a benchmark problem for validation of two-phase flow solvers by means of specifically designed experiments for Taylor Bubble Flow. This chapter is devoted to results from Direct Numerical Simulations (DNS) of a single rising Taylor bubble and of Taylor flow in a square channel, where both bubble shape and flow pattern around the bubble have been thoroughly analyzed. Comparisons have been accomplished to highly resolved experiments providing detailed and local benchmark data for validation. An interesting three-dimensional backflow effect of technological relevance has

H. Marschall (✉) • D. Bothe
Fachbereich Mathematik, Mathematical Modeling and Analysis, Technische Universität
Darmstadt, Alarich-Weiss-Straße 10, 64287 Darmstadt, Germany
e-mail: marschall@mma.tu-darmstadt.de; bothe@mma.tu-darmstadt.de

C. Falconi
Institute for Chemical Technology and Polymer Chemistry, Karlsruher Institute für Technologie,
Engesserstraße 20, 76131 Karlsruhe, Germany
e-mail: carlos.falconi@kit.edu

C. Lehrenfeld
Institut für Numerische und Angewandte Mathematik, University of Göttingen, Engesserstr. 20,
76131 Karlsruhe, Germany
e-mail: lehrenfeld@math.uni-goettingen.de

R. Abiev
State Institute of Technology, St. Petersburg Technical University, Engesserstr. 20, 76131
Karlsruhe, Germany

M. Wörner
Institute of Catalysis Research and Technology, Karlsruher Institute für Technologie,
Engesserstraße 20, 76131 Karlsruhe, Germany
e-mail: martin.woerner@kit.edu

A. Reusken
Institut für Geometrie und Praktische Mathematik, RWTH Aachen University, Templergraben 55,
52056 Aachen, Germany
e-mail: reusken@igpm.rwth-aachen.de

© Springer International Publishing AG 2017 663
D. Bothe, A. Reusken (eds.), *Transport Processes at Fluidic Interfaces*,
Advances in Mathematical Fluid Mechanics, DOI 10.1007/978-3-319-56602-3_23

been revealed. A criterion to estimate its occurrence is deducted from thorough analysis of local simulation data.

23.1 Introduction

Taylor bubbles are elongated bubbles which almost completely fill the cross-sectional area of (commonly) straight channels—without wetting its confining walls but being surrounded by a thin liquid film. The flow of multiple subsequent Taylor bubbles in a channel is typically referred to as Taylor flow—sometimes also as bubble-train flow, segmented flow, or capillary slug flow. Taylor flow consists of a regular sequence of Taylor bubbles which are separated by liquid slugs.

Main advantages of Taylor flow in small (milli/micro-) channels have been already discussed in Chap. 19. Given mainly these advantages, gas-liquid Taylor flow in narrow channels is of great interest in, for instance, micro-process engineering [1] catalytically coated monolith reactors [2] material synthesis [3] and stimulus of biological cells [4]—to mention a few. Recent reviews of Taylor flow are given in [5–8] and the interested reader is referred to these articles and the references therein for further experimental and numerical studies on this subject.

It is well known that the hydrodynamics of Taylor bubbles in small channels is predominately determined by viscous (friction) and surface tension forces, with the inertial forces becoming important only at higher flow velocities. The relevant dimensionless groups are the Capillary number $Ca = \eta_L U_B / \sigma$ (ratio of viscous to surface tension force) and the Reynolds number $Re = \rho_L d_h U_B / \eta_L$, where U_B denotes the magnitude of the bubble velocity, d_h the hydraulic diameter of the channel, σ the surface tension coefficient, and ρ_L and η_L the liquid density and dynamic viscosity, respectively. Unlike in circular channels, Taylor bubble flow in square channels becomes more complex both regarding the bubble shape and the flow patterns being non-axisymmetric, as we set out in the remainder of this chapter. Nowaday Direct Numerical Simulations (DNS) enables us to gain detailed insights into local transport processes on the basis of local information as their direct outcome. For DNS of Taylor bubbles or Taylor flow this has two distinct implications: (i) the underlying code must feature fully 3D computations, and (ii) the numerical treatment of surface tension is an important issue, especially if the flow problems exhibit low capillary numbers Ca. Moreover, while many experimental and numerical studies on Taylor bubbles or Taylor flow consider circular channels, square channels are potentially of larger technological relevance for certain applications, but are less often studied.

This chapter is devoted to a coordinated numerical and experimental study on the shape of a Taylor bubble and local flow phenomena in Taylor flow under co-current upward laminar conditions in a square channel. We detail on the main results, reported originally by Marschall et al. [9] and Falconi et al. [10]. This work shall closely resemble and summarize the main aspects set out in [9, 10] and relevant findings. The studies have been accomplished within the framework of the Priority Program SPP 1506 "Transport Processes at Fluidic Interfaces" by the German Research Foundation. They are to be understood complementary: In the work of Marschall et al. [9] a combined experimental and numerical study on a single Taylor bubble in a square vertical mini-channel of 2.076 mm × 2.076 mm cross-sectional area is presented. The comparison focused on the non-axisymmetric bubble shape, where precise interface position profiles were measured in longitudinal cuts in lateral and diagonal direction by synchrotron radiation in conjunction with ultrafast radiographic imaging [11]. The influence of different numerical methods for interface resolving simulations and different surface tension force approximations is thoroughly investigated. The work of Falconi et al. [10] follows a similar combined experimental and numerical approach but with focus on local features of the velocity field in Taylor flow, which is measured within the liquid phase by micro Particle-Image-Velocimetry (µPIV) [12]. Such measurements are rare, especially when it comes to fully resolved velocity profiles within the liquid film [12] and highly accurate 3D interface position data [11]. As for the differences, in contrast to [9], which dealt with a single Taylor bubble, Taylor flow is considered in [10]. In [10] the liquid slug length is rather short so that the Poiseuille velocity profile is not fully developed and there is a notable interaction between neighboring Taylor bubbles, while in [9] the Taylor bubble was much longer. Moreover, the Ca was similar both in the study of Marschall et al. and of Falconi et al., however, the Reynolds number was higher, that is Re $= 17$ in [9] and Re $= 7$ in [10], respectively.

23.2 Mathematical Model and Numerical Methods

23.2.1 Continuum Model

The sharp interface model, i.e. the two-phase Navier-Stokes equations, provides the basis for all codes we worked with for this contribution. In sharp interface modeling the fluid interface $\Sigma(t)$ itself is a surface of discontinuity of zero thickness, which separates both fluid phase regions, for which we consider two incompressible, Newtonian, immiscible and isothermal fluids. For incompressible fluids, the transport equations for mass and linear momentum read

$$\nabla \cdot \mathbf{v} = 0 \qquad\qquad \text{in } \Omega \setminus \Sigma(t), \qquad (23.1)$$

$$\partial_t(\rho \mathbf{v}) + \nabla \cdot (\rho \mathbf{v} \mathbf{v}) = -\nabla p + \nabla \cdot \boldsymbol{\tau} + \rho \mathbf{g} \quad \text{in } \Omega \setminus \Sigma(t). \qquad (23.2)$$

Herein, the viscous stress tensor for a Newtonian fluid reads $\boldsymbol{\tau} = \eta \left(\nabla \mathbf{v} + (\nabla \mathbf{v})^{\mathsf{T}} \right)$. η denotes the dynamic fluid viscosity, ρ its density, and $\mathbf{g}$ is the gravitational acceleration. Moreover, the jump conditions at the interface read

$$\llbracket \mathbf{v} \rrbracket = \mathbf{0} \qquad \text{on } \Sigma(t), \tag{23.3}$$

$$\llbracket p\mathbf{I} - \boldsymbol{\tau} \rrbracket \cdot \mathbf{n}_\Sigma = \sigma\kappa\mathbf{n}_\Sigma \quad \text{on } \Sigma(t), \tag{23.4}$$

where a constant surface tension coefficient σ (i.e., no Marangoni effects), no phase change (due to evaporation or condensation) and no-slip at the interface has been assumed. Herein, $\mathbf{I}$ denotes the unity tensor, κ is twice the mean interface curvature, $\kappa = \nabla_\Sigma(-\mathbf{n}_\Sigma)$, where $\mathbf{n}_\Sigma$ denotes the unit normal at the interface.

Notably, there are some method-specific aspects to be considered: in interface capturing methods, (23.1) and (23.2) are solved in the so-called *one-field formulation*—inherently taking into account corresponding jump conditions according to (23.3) and (23.4). In the one-field approach specific marker functions are utilized in order to evaluate both density and viscosity fields locally. In Volume-of-Fluid methods this is accomplished by means of a phase indicator function resulting with a volume fraction field, whereas in Level-Set methods a signed distance function is used resulting with a level-set field. The one-field formulation of linear momentum transport reads as

$$\partial_t(\rho\mathbf{v}) + \nabla\cdot(\rho\mathbf{v}\mathbf{v}) = -\nabla p + \nabla\cdot\boldsymbol{\tau} + \rho\mathbf{g} + \mathbf{f}_\Sigma, \qquad \text{in } \Omega \tag{23.5}$$

where the singular surface tension force has been modified towards a volumetric interfacial force density $\mathbf{f}_\Sigma$. Effectively, the interfacial momentum jump condition is incorporated employing the Dirac distribution δ_Σ, i.e. $\mathbf{f}_\Sigma = \sigma\kappa\mathbf{n}_\Sigma\delta_\Sigma$, which is evaluated method-specifically using the corresponding marker function.

In interface tracking methods, the fluid interface itself is represented by a computational mesh boundary. Consequently, interface tracking inherently exhibits an explicit rather than an implicit interface representation as for interface capturing methods. The flow of each fluid phase is governed by a separate set of conservation equations in integral *Arbitrary Lagrangian Eulerian (ALE) formulation* [13, 14] being associated with separate fluid domains, viz.

$$\frac{d}{dt}\int_{V(t)} \rho\,dV + \int_{S(t)} \rho\,(\mathbf{v} - \mathbf{v}_S)\cdot\mathbf{n}\,dS = 0, \tag{23.6}$$

and

$$\frac{d}{dt}\int_{V(t)} \rho\mathbf{v}\,dV + \int_{S(t)} \rho\,(\mathbf{v} - \mathbf{v}_S)\,\mathbf{v}\cdot\mathbf{n}\,dS = \int_{S(t)} \boldsymbol{\sigma}\cdot\mathbf{n}\,dS + \int_{V(t)} \rho\mathbf{g}\,dV. \tag{23.7}$$

Herein, $\mathbf{v}_S$ represents the velocity of $S(t)$. Coupling of this set of conservation equations is achieved by directly enforcing the interfacial boundary conditions (23.3) and (23.4).

The two-phase hydrodynamics is typically governed by complementing above equations by one method-specific transport equation. Moreover, since the present Taylor bubble flow problem exhibits a low capillary number the numerical treatment of surface tension is an important issue. These aspects shall be briefly set out next on a per-method basis; for a detailed description the interested reader is referred to [9].

23.2.2 Numerical Methods

We perform DNS of Taylor bubbles/flow in a coordinated research effort using both the interface capturing (ICM) and the interface tracking (ITM) methods. While the first family of methods *captures* the position of a fluid interface from a marker field stored on a typically spatially fixed computational mesh, the latter family directly *tracks* the interface position by utilizing a moving mesh approach, in which the interface itself is represented either by a surface mesh aligned to a set of cell faces, or by a surface mesh which is moving relative to a fixed volume mesh. The latter approach, so-called Front tracking [15], is not considered here.

As for the ICM, we consider a Level-Set method [16–19] and two Volume-of-Fluid methods [20–22]; as for an ITM, we consider the Arbitrary Lagrangian Eulerian method [13, 14]. The methods are implemented in the Finite Element code DROPS and the Finite Volume codes FS3D, TURBIT-VOF and OpenFOAM (`interTrackFoam`),[1] respectively. The codes used in this coordinated research are very different in their underlying methodology. In what follows we briefly describe the numerical methods, particularly focusing on the different interface representation (see Fig. 23.1) and especially surface tension treatment, since the present Taylor bubble flow exhibits low capillary numbers and thus numerical treatment is a central aspect.

DROPS

The in-house code DROPS [23] is based on a level-set formulation. The scalar level-set function $\phi = \phi(x, t)$ is defined such that $\phi(x, t) < 0$ for $x \in \Omega_1(t)$, $\phi(x, t) > 0$ for $x \in \Omega_2(t)$ and $\phi(x, t) = 0$ for $x \in \Sigma(t)$. Ideally, the level set function is a signed distance function. In this setting, the interface $\Sigma(t)$ is given only implicitly as the zero-level of the level-set function. The interface motion is described by the linear hyperbolic level-set equation

$$\partial_t \phi + \mathbf{v} \cdot \nabla \phi = 0 \tag{23.8}$$

for $t \geq 0$ and $x \in \Omega$. With this, the density ρ and viscosity η can be expressed as jumping coefficients in terms of the level-set function—$\rho(\phi)$, $\eta(\phi)$. The effect

---

[1]OpenFOAM comprises of over 80 solvers to simulate specific problems in CCM and over 170 utilities. Hence, in order to avoid ambiguity, a closer specification of at least the used solver family is necessary, which we provide in the brackets.

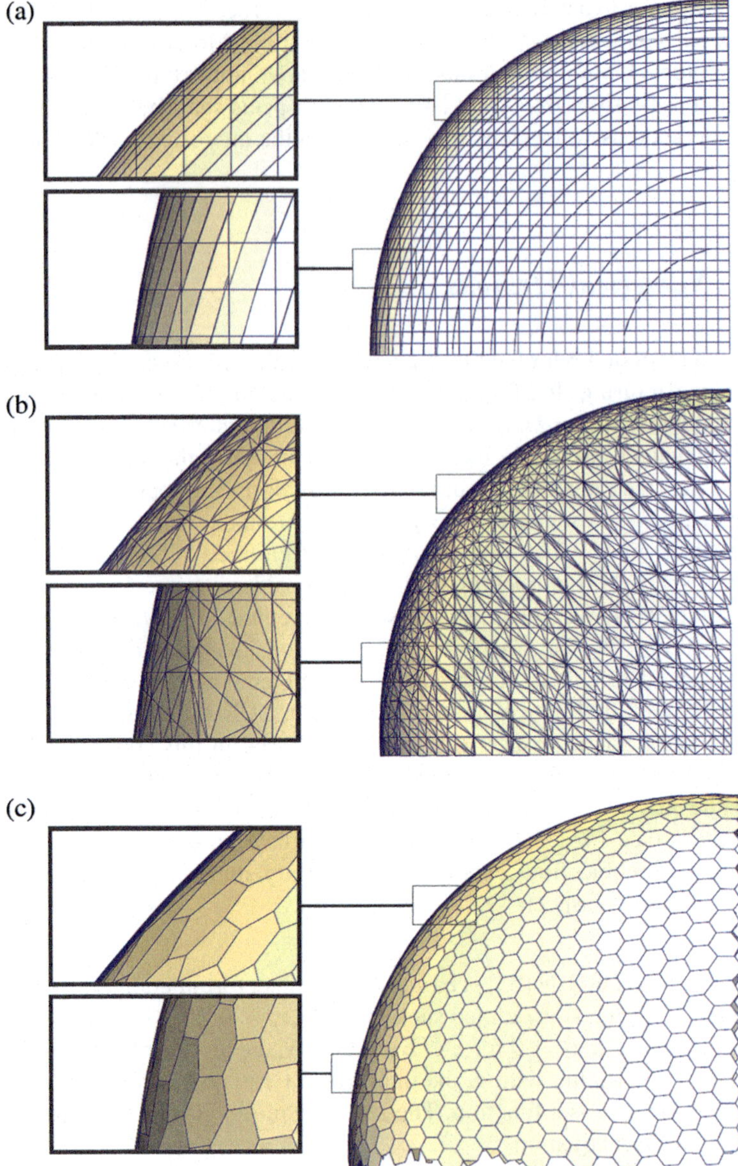

Fig. 23.1 Close-up view of different approach-specific interface representations at the Taylor bubble's rear (location of maximum interfacial curvature). Republished with permission of Elsevier, from [9]; permission conveyed through Copyright Clearance Center, Inc. (**a**) Piecewise planar interface approximation by quadrilaterals from PLIC (VOF-ICM, FS3D). (**b**) Piecewise planar interface approximation by triangles or quadrilaterals (LS-ICM, DROPS). (**c**) Polygonal interface representation by surface mesh (ALE-ITM, OpenFOAM/`interTrackFoam`)

of the surface tension is expressed in terms of a force localized at the interface. The localized surface tension force is given as $\mathbf{f}_\Sigma = \sigma \kappa \delta_\Sigma \mathbf{n}_\Sigma$ with δ_Σ the Dirac δ-function with support only on Σ.

The code is written in C++ and parallelized. An adaptive multilevel mesh hierarchy allows for an efficient resolution of the important flow features on unstructured tetrahedral meshes. For the spatial discretization of velocity and pressure, the LBB-stable Hood-Taylor finite element pair is used. To account for discontinuities in the pressure field, the pressure space is enriched using an extended finite element (XFEM) space and a modified Laplace-Beltrami technique is used to describe surface tension accurately [24]. For the discretization of the level-set function, continuous piecewise quadratics combined with streamline diffusion stabilization are used. For the time discretization, a backward Euler scheme is used where the nonlinearities are resolved using (modified) fixed point iteration schemes. For a detailed description the interested reader is referred to [24, 25] and the DROPS internet homepage [26].

FS3D & TURBIT-VOF

The in-house codes FS3D[27] and TURBIT-VOF[28, 29] are based on volume-of-fluid methods. The name FS3D means Free Surface 3D. The code is written in Fortran and parallelized using MPI and OpenMP, being actively developed at ITLR (Univ. Stuttgart) and MMA (Center of Smart Interfaces, TU Darmstadt). The code TURBIT-VOF is written in Fortran and has been developed at Karlsruhe Institute of Technology (KIT).

Both FS3D and TURBIT-VOF solve the two-phase Navier-Stokes equation in one-field formulation, cf. (23.5), for two incompressible Newtonian fluids on a structured staggered Cartesian mesh by a finite volume method. The interface position is captured implicitly introducing the phase indicator f_1 (for brevity f in the remainder) for one of the phases (e.g. phase 1) along with its corresponding transport equation,

$$\partial_t f + \mathbf{v} \cdot \nabla f = 0, \tag{23.9}$$

where

$$f(t, \mathbf{x}) := \begin{cases} 1 & \text{if } \mathbf{x} \in \Omega_1(t), \\ 0 & \text{otherwise.} \end{cases} \tag{23.10}$$

Note that TURBIT-VOF indeed solves for the locally volume-averaged equivalent of (23.1), (23.5) and (23.9), which we do not detail on here for the sake of brevity. In both approaches the interface is kept sharp during simulations by geometrically reconstructing and advecting the interface, adopting a Piecewise Linear Interface Calculation (PLIC) method according to [20, 22, 30] for FS3D and an in-house algorithm called EPIRA [28] for TURBIT-VOF, respectively. In FS3D the volumetric surface tension force density is incorporated as $\mathbf{f}_\Sigma = \sigma \kappa \delta_\Sigma \mathbf{n}_\Sigma$ employing the Dirac distribution $\delta_\Sigma \approx \|\nabla f\|$, where the term $\|\nabla f\|$ is to be understood in the sense of functions of bounded variation [31]. In this study, we use FS3D with the balanced Continuous Surface Force according to [32], which is based on generalized height functions for curvature estimation, following Popinet et al. [33]. In the volume-averaged single-field momentum formulation of TURBIT-VOF the surface tension term is $\mathbf{f}_\Sigma = \sigma \kappa a_\Sigma \mathbf{n}_\Sigma$, where within the volume-averaged framework the Dirac delta function δ_Σ has been approximated by the interfacial area density a_Σ. For advection a directional split algorithm is employed for FS3D, while TURBIT-VOF utilizes an unsplit advection algorithm. The underlying structured Cartesian grid supports staggered variable arrangement with the velocity field being stored on cell faces and the pressure field on cell centers. A divergence free velocity field is ensured at the end of each time step by a projection method. Further details can be found in [27] and [34], respectively.

OpenFOAM/interTrackFoam

OpenFOAM—Open Field Operation And Manipulation—is a free and Open Source C++ Class Library for Computational Continuum Mechanics (CCM) and Multiphysics [35–37]. OpenFOAM features efficient linear equation solvers with polyhedral cell support and is massively parallelized in domain decomposition mode. In this work we employ OpenFOAM's interface tracking method (`interTrackFoam` solver family). The method is based on the Space Conservation Law (SCL) [38], viz.

$$\frac{d}{dt} \int_{V(t)} dV - \int_{S(t)} \mathbf{v}_S \cdot \mathbf{n}\, dS = 0. \tag{23.11}$$

As the transport equations for the phase fraction or the level-set is central to the aforementioned interface capturing methods, so the SCL is central to the interface tracking method. It provides the relationship between the rate of change of the volume $V(t)$ and the corresponding velocity $\mathbf{v}_S$ of its bounding surface $S(t)$ causing this change.

The interface itself is represented by a computational mesh boundary, the movement of which is obtained as a part of the numerical two-phase flow solution, taking into account interfacial conditions by means of enforcing discretized forms of (23.3) and (23.4). For a detailed description the interested reader is referred to [39–42]. Using OpenFOAM, the original interface tracking methodology of Muzaferija and Perić [14] has been significantly extended by Tuković and Jasak [41] taking into account viscous and surface tension effects at the interface. Moreover, the Rhie-Chow interpolation practice has been improved (cf. [41] for details). The discretization methods underlying the interface tracking approach in OpenFOAM are a collocated (pseudo-staggered) Finite Area Method (FAM) and a collocated Finite Volume Method (FVM) on polyhedral meshes along with automatic mesh motion [35, 41–45]. Since only small deformations are present, Laplacian mesh motion as described in [35] is adopted in this study. The overall solution procedure is based on the iterative Pressure Implicit with Splitting of Operators (PISO) algorithm of Issa [46] for pressure-velocity coupling. The solution procedure is of second-order accuracy in space and time—in particular, it is worth noting that the surface tension calculation is of second order accuracy being based on a so-called force-conservative approach. Herein, the surface tension force is incorporated as area density using $\mathbf{F}^{\sigma}_{S_f} = \oint_{\partial S_f} \mathbf{m}\sigma\, dL \approx \sum_e (\sigma\mathbf{m})_e L_e$, viz. $(\kappa\sigma)_f = \sigma\kappa_f = \sigma \frac{1}{S_f} \mathbf{n}_f \cdot \sum_e \mathbf{m}_e L_e$, where L_e denotes the edge lengths of the polygonal control areas S_f with normals $\mathbf{n}_f$ and $(\sigma\mathbf{m})_e$ is the surface tension force per unit length.

23.3 Direct Numerical Simulation Results

In this section a thorough discussion on the fully-resolved data from Direct Numerical Simulations is provided. Focus is on the shape of a Taylor bubble in a square milli-channel and local flow phenomena in a Taylor flow in a square milli-channel, respectively. The computational setups of both cases are described in detail in [9, 10]. Thus, the interested reader is referred to these original journal papers. Both numerical studies are based on the experimental work [11, 12] and use the same channel geometry, i.e. a square vertical mini-channel of 2.076 mm $\times$ 2.076 mm cross-sectional area.

23.3.1 Detailed Shape of a Taylor Bubble in a Square Milli-Channel

Main focus in [9] has been on validation for the given case of a single rising Taylor bubble in a square milli-channel by means of the Taylor bubble shape and geometrical target quantities such as distances, curvatures and film thickness at locations where differences due to the different numerical surface tension treatment of the employed methods becomes evident.

A quantitative validation is achieved by examining the bubble's shape profiles in longitudinal and diagonal cuts through the flow domain, and by comparing with geometrical target quantities related to the bubble shape. Here, we detail on the first, as an example for the local benchmark data that have been gained within this study, while for the latter (geometrical target quantities) we refer to [9] where also a detailed discussion is provided. The bubble's shape profiles are depicted in Figs. 23.2 and 23.3. They have been obtained on diagonal and longitudinal cutting planes by intersection with the respective interfacial surface representation (iso-surface, PLIC-surface, or surface mesh representation, cf. Fig. 23.1). A corresponding close-up view is shown as an example for the longitudinal cutting plane in Fig. 23.3.

While for comparability of the bubble tip and rear profiles, the shapes have been vertically aligned to each other in Fig. 23.3 (both in Fig. 23.3a and b), this is intentionally not done in Fig. 23.2, where the shape profiles have been aligned to a common position of the bubble tips. From comparing solely the bubble shapes provided in Fig. 23.3 one might conclude, that all solvers show very similar results.

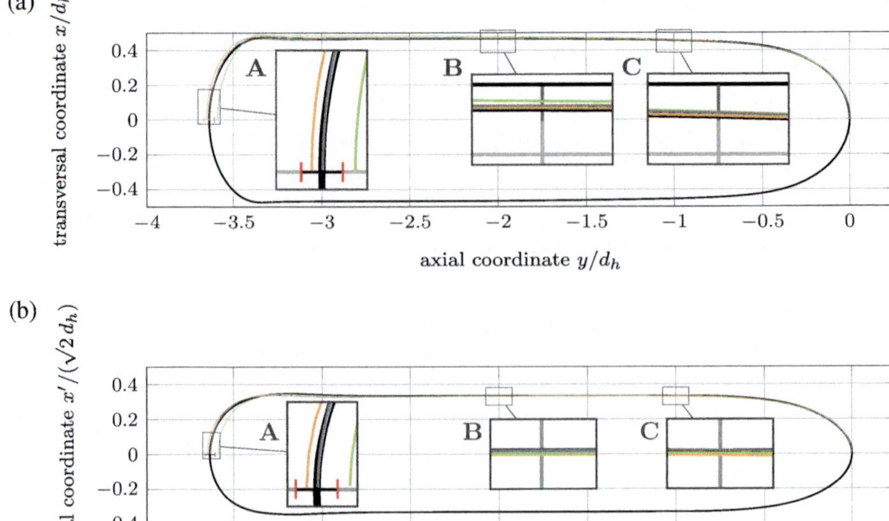

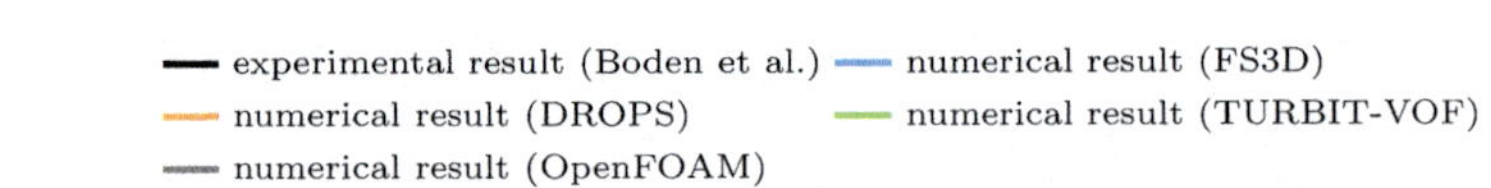

Fig. 23.2 Shape profile of Taylor bubble for distinct cutting planes through flow domain. Republished with permission of Elsevier, from [9]; permission conveyed through Copyright Clearance Center, Inc. (**a**) Longitudinal cut. (**b**) Diagonal cut

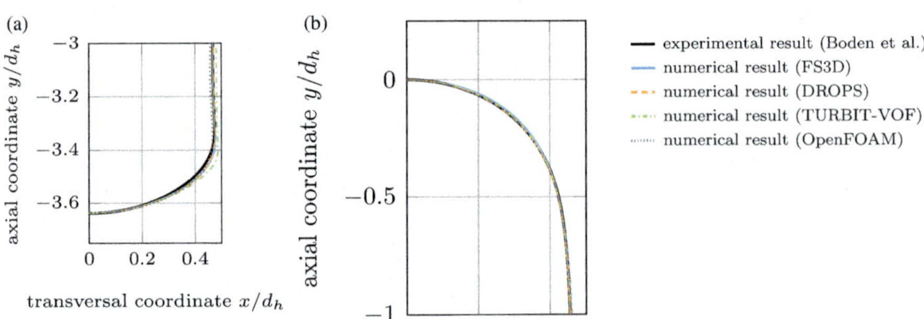

Fig. 23.3 Close-up view of Taylor bubble tip and rear for longitudinal cutting plane. Republished with permission of Elsevier, from [9]; permission conveyed through Copyright Clearance Center, Inc. (**a**) Bubble rear. (**b**) Bubble tip

In particular, the interfacial curvatures are well captured for the bubbles' tip part. Still a decent agreement with the experimental reference could be declared for its rear part. However, examining the bubble shapes as shown in Fig. 23.2—cp. the error bar in Inset A of Fig. 23.2a and b—a closer study of uncertainties and possible error sources is advisable. In [9] a thorough discussion on contributions to uncertainty and errors in the numerical and experimental methods is done in order to assess the reliability and accuracy of the employed two-phase flow solvers quantitatively. On this basis, for instance, the tolerable error between experimental and simulation results has been found to be 0.8% for the bubble length. One main contribution to this uncertainty in the numerical results (beside numerical errors) stems from the experimentally determined bubble volume, which is used to initialize the numerical simulations. An inaccurate initial bubble volume would directly result in erroneous numerical results, in particular, a wrong length of the Taylor bubble. Concluding, we find regarding the accuracy of the measurements and initialization procedures, and from the error bar depicted in Fig. 23.2a and b that all solvers do show similar results (though they employ very different approaches to surface tension calculation). A good agreement between numerical and experimental results can be declared, especially if either a force-balanced or a conservative approach to surface tension calculation has been employed.

23.3.2 *Local Flow Phenomena of a Taylor Flow in a Square Milli-Channel*

As a central objective in [10] and complementary to [9], quantitative local velocity field information are provided, e.g., local axial centerline velocity profiles in the liquid slugs for the particular rather complex Taylor flow in a square channel. On this basis an interesting flow phenomenon has been revealed, namely local backflow in the rear part of the liquid film. The results in [10] show that for a square channel the axial locations of this local backflow region differ in the lateral film and in the corner film. To the best of our knowledge, this behavior has not been reported in literature so far. It has been demonstrated, moreover, that (only) with the fully resolved local data gained by Direct Numerical Simulations it is possible to understand the underlying hydrodynamics and to derive a criterion for the occurrence of this phenomenon in practical flow systems. This is of significant technological relevance, since the local backflow during the passage of a Taylor bubble causes a temporal reversal of the sign of the wall shear stresses at a fixed position. This is of importance for heat and mass transfer in various applications with Taylor flow [4, 47–50].

Herein, $\tilde{z}$ denotes dimensionless distances—in particular in remainder of this chapter we use $\tilde{z}_B$ for the vertical distance from the rear meniscus of the bubble $\tilde{z}_B := z_B/L_B$, and $\tilde{z}_S$ for the liquid slug the vertical distance from the bubble front meniscus, viz. $\tilde{z}_S := z_S/L_S$. For a detailed quantitative analysis of the velocity field in the liquid slug and in the bubble region, local profiles of the axial and radial velocity are shown in Fig. 23.4 at distinct axial positions, where partly also experimental data are available for comparison [12]. As for the velocity field in the liquid slug, cf. Fig. 23.4a–c, the computed profiles of the axial velocity agree very well in the lateral near wall and corner regions with bypass flow, where they almost overlap for all three axial heights. Differences in the profiles of the axial velocity occur, however, in the channel center. These can partly be attributed to the different volumetric flow rates of the three codes. Also the rather large deviation between experimental data and numerical profiles close to the channel wall, both in the middle of the liquid slug (Fig. 23.4b) and close to the bubble nose (Fig. 23.4c), are attributed to uncertainties in the experimental data arising from the temperature dependence of the refractive index. For more details on the uncertainties the interested reader is referred to [10]. In summary, however, the deviation of the computational and experimental results for the local centerline axial velocity is about 3.5% for the three considered axial positions, while the deviation between the codes is less than 3%. Figure 23.4a and c also show profiles of the radial liquid velocity in a lateral and diagonal cut. In the middle of the liquid slug, the radial velocity is virtually zero; the respective profiles are therefore not included in Fig. 23.4b. The radial velocity profiles of all three codes only differ slightly in magnitude. Due to the recirculation pattern in the liquid slug (cf. Fig. 23.5a), the radial liquid velocity is non-zero close to the front and rear meniscus. Close to the bubble nose the liquid flows inward (i.e. toward the channel axis), while close to the bubble rear the liquid flow is outward. As for the velocity profile in the bubble region, cf. Fig. 23.4d–f, the deviation in the magnitude of the computed centerline axial velocity inside the bubble at the three considered axial positions is about 1.5–5% between the different codes. In the diagonal cut, the magnitude of the radial velocity is very small close to the walls and the profile for radial velocity exhibit a horizontal tangent for all three axial positions. In the lateral cut, the magnitude of the radial velocity close to the wall is larger than in the diagonal cut and the horizontal tangent exists only for $\tilde{z}_B = 0.75$ (Fig. 23.4d). The profiles for $\tilde{z}_B = 0.25$ and $\tilde{z}_B = 0.5$ show that liquid flows towards the wall. This is related to the axial profile of the bubble shape. Namely, the bubble diameter is increasing as $\tilde{z}_B$ decreases, and the bubble pushes liquid toward the wall, where it is redistributed laterally toward the channel corners. This results in a draining of the lateral liquid film. Examining the axial velocity at $\tilde{z}_B = 0.75$ and $\tilde{z}_B = 0.5$, the velocity profile in the liquid film is observed to be linear, which is of "Couette type" type as discussed by Meyer et al. [12]. At $\tilde{z}_B = 0.25$ there is a local backflow of liquid in the film, both in the experiment as well as in the three computations. The magnitude of this backflow is much larger in the diagonal cut than in the lateral cut; see the inset in Fig. 23.4f. This is consistent with the contour-plot of the vertical velocity shown in Fig. 23.5.

Figure 23.5 shows streamlines of the flow in a frame of reference moving with the bubble in a lateral and diagonal cut in combination with a contour-plot of the vertical velocity component in a fixed frame of reference (results of FS3D, Fig. 23.5a) and the bubble shape, the isosurface of and velocity vectors at the horizontal cross-sections $\tilde{z}_B = 0.17, 0.29$ and 0.52 (results of TURBIT-VOF, Fig. 23.5b). The velocity vectors indicate that the flow in the liquid film is upward at $\tilde{z}_B = 0.52$

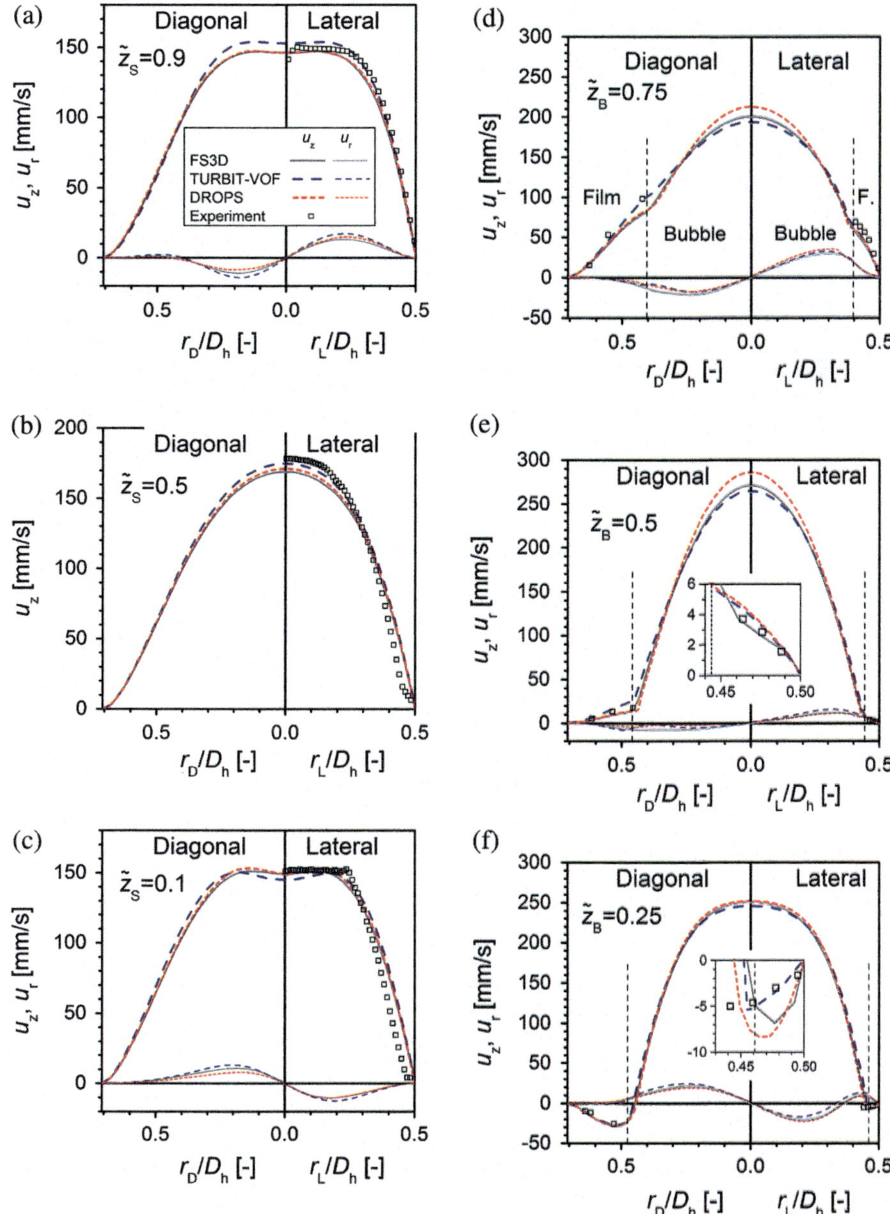

Fig. 23.4 Lateral and diagonal profiles of the axial and radial velocity in the liquid slug at different axial positions (*left*) and lateral and diagonal profiles of the axial and radial velocity in the bubble and the liquid film (*right*). The inset graphics in Fig. 23.4e and f show the lateral profiles of the axial velocity close to the wall. The *vertical dashed lines* denote the average bubble diameter of the three codes at the respective axial position. Reproduced from , with the permission of AIP Publishing. (**a**) $\tilde{z}_S = 0.9$. (**b**) $\tilde{z}_S = 0.5$. (**c**) $\tilde{z}_S = 0.1$. (**d**) $\tilde{z}_S = 0.75$. (**e**) $\tilde{z}_S = 0.5$. (**f**) $\tilde{z}_S = 0.25$

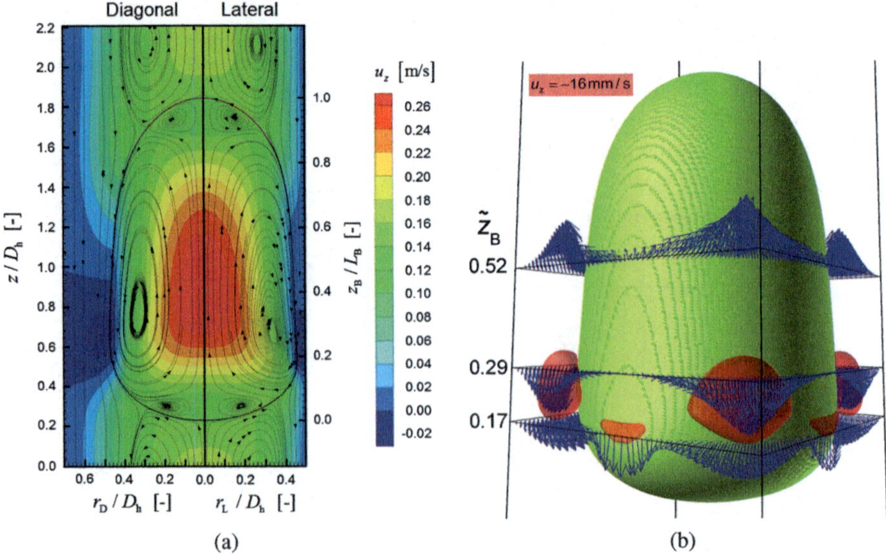

Fig. 23.5 Streamlines in the frame of reference moving with the bubble and vertical velocity in fixed frame of reference (color code) in a lateral and diagonal longitudinal cut (results of FS3D)—*left*. Visualization of the bubble shape, the isosurface of the vertical velocity, and the velocity field in three different horizontal cross-sections illustrating local backflow in the rear part of the liquid film (results of TURBIT-VOF)—*right*. Reproduced from , with the permission of AIP Publishing

and is downward at $\tilde{z}_B = 0.17$, and 0.29. For the two upper horizontal planes, the velocity in the lateral liquid film is quite small. In contrast large downward velocities exist in the lowest plane close to the bubble rear, where the lateral film is thinnest. From the volume enclosed by the isosurface for $v_z = -0.16$ m/s it is evident that the backflow region in the liquid film is much larger in the channel corners than at the channel sides. A close analysis of the local simulation data reveals that it is the different magnitude of the velocity in the lateral film and in the corner region which leads to azimuthal pressure differences in the lateral and diagonal liquid film which causes a slight deviation of the bubble from the rotational symmetry. This deviation is opposite in the front and rear part of the bubble and has the mentioned significant effects on the local flow field in the rear part of the liquid film. In Fig. 23.5a, the streamlines within the bubble indicate the presence of three toroidal vortices. The large central vortex in the body of the Taylor bubble drives two smaller toroidal vortices at the bubble front and bubble rear. The position of the center of the main vortex differs considerably in the lateral and diagonal cut. This shows that the flow in the bubble is not axisymmetric but three-dimensional, especially in the lower part of the main vortex. In the liquid slug there is one main toroidal vortex which rotates in the same direction as the main vortex in the bubble. In the region with bypass flow close to the channel walls the liquid velocity is rather low and one can identify regions in the rear part of the liquid film where the liquid is flowing downwards.

23.4   Summary and Conclusions

In a coordinated research effort we have thoroughly analysed local data from Direct Numerical Simulations of a rising Taylor bubble and Taylor flow in a square channel. Particular focus has been devoted to the bubble shape and flow pattern around the bubble. Achievements and findings of the work Marschall et al. [9] and Falconi et al. [10] can be summarized as follows.

- establishment of a detailed Taylor bubble validation benchmark for a comprehensive assessment and objective measure of accuracy and reliability of two-phase flow solvers. The basis for comparison are specifically designed high-resolution experiments providing accurate bubble shape profiles in longitudinal and diagonal cuts through the flow domain and sensitive geometrical target quantities. By this means the virtue and advantage of force-balanced and conservative numerical models for surface tension computation have become evident.
- in-depth description of local flow patterns and physical explanation of local liquid backflow in Taylor flow, which leads to a temporary reversal of wall shear stress during the passage of a bubble. Moreover, by means of a detailed analysis of the fully-resolved simulation data and employing a macroscopic mass balance of liquid flow, a useful estimate for the occurrence of backflow in the liquid film has been deducted, viz.

$$\frac{A_{B,\max}}{A} \cdot \frac{U_B}{J} > 1,$$

where $A_{B,\max}$ is the maximum cross sectional area of the bubble, A denotes the cross sectional area of the channel, J is the total superficial velocity and U_B the bubble velocity. Given the technological relevance, this correlation is seen useful for practical applications. Note, however, both ratios on the right hand side of the above criterion are dependent on the capillary number. In [10] we provide indicate on appropriate correlations for their approximative determination.

References

1. Jänisch, K., Baerns, M., Hessel, V., Ehrfeld, W., Haverkamp, V., Löwe, H., Wille, C., Guber, A.: Direct fluorination of toluene using elelement fluorine in gas/liquid microreactors. J. Fluor. Chem. **105**, 117 (2000)
2. Kreutzer, M.T., Kapteijn, F., Moulijn, J.A., Heiszwolf, J.J.: Phenylacetylene hydrogenation over [Rh(MBD)(PPh$_3$)^2]BF$_4$ catalyst in a numbered-up microchannels reactor. Ind. Eng. Chem. Res. **60**, 5895 (2005)
3. Günther, A., Jensen, K.F.: Multiphase microfluidics: from flow characteristics to chemical and material synthesis Lab Chip **6**, 1487 (2006)
4. El-Ali, J., Gaudet, S., Günther, A., Sorger, P.K., Jensen, K.F.: Cell stimulus and lysis in a microfluidic device with segmented gas-liquid flow. Anal. Chem. **77**, 3629 (2005)

5. Angeli, P., Gavriilidis, A.: Hydrodynamics of Taylor flow in small channels: a review. Proc. Inst. Mech. Eng. C J. Mech. Eng. Sci. **222**(5), 737–751 (2008)
6. Gupta, R., Fletcher, D., Haynes, B.: Taylor flow in microchannels: a review of experimental and computational work. J. Comput. Multiphase Flows **2**(1), 1–32 (2010)
7. Sobieszuk, P., Aubin, J., Pohorecki, R.: Hydrodynamics and mass transfer in gas-liquid flows in microreactors. Chem. Eng. Technol. **35**, 1346 (2012)
8. Talimi, V., Muzychka, Y.S., Kocabiyik, S.: A review on numerical studies of slug flow hydrodynamics and heat transfer in microtubes and microchannels. Int. J. Multiphase Flow **39**, 88 (2012)
9. Marschall, H., Boden, S., Lehrenfeld, C., Falconi Delgado, C.J., Hampel, U., Reusken, A., Wörner, M., Bothe, D.: Validation of interface capturing and tracking techniques with different surface tension treatments against a Taylor bubble benchmark problem. Comput. Fluids **102**, 336–352 (2014)
10. Falconi, C.J., Lehrenfeld, C., Marschall, H., Meyer, C., Abiev, R., Bothe, D., Reusken, A., Schlüter, M., Wörner, M.: Numerical and experimental analysis of local flow phenomena in laminar Taylor flow in a square mini-channel. Phys. Fluids **28**(1), 012109-1–012109-23 (2016)
11. Boden, S., dos Santos Rolo, T., Baumbach, T., Hampel, U.: Synchrotron radiation microtomography of Taylor bubbles in capillary two-phase flow. Exp. Fluids **55**(7), 1768 (2014)
12. Meyer, C., Hoffmann, M., Schlüter, M.: Micro-PIV analysis of gas-liquid Taylor flow in a vertical oriented square shaped fluidic channel. Int. J. Multiphase Flow **67**, 140 (2014)
13. Demirdžić, I., Perić, M.: Finite volume method for prediction of fluid flow in arbitrarily shaped domains with moving boundaries. Int. J. Numer. Methods Fluids **10**, 771–790 (1990)
14. Muzaferija, S., Perić, M.: Computation of free-surface flows using the finite volume method and moving grids. Numer. Heat Transf., Part B **32**, 369–384 (1997)
15. Unverdi, S.O., Tryggvason, G.: A front-tracking method for viscous, incompressible, multifluid flows. J. Comput. Phys. **100**, 25–37 (1992)
16. Osher, S., Sethian, J.A.: Fronts propagating with curvature-dependent speed: algorithms based on Hamilton-Jacobi formulations. J. Comput. Phys. **79**, 12–49 (1988)
17. Sethian, J.A.: Level Set Methods: Evolving Interfaces in Geometry, Fluid Mechanics, Computer Vision, and Material Science. Cambridge University Press, Cambridge (1996)
18. Sethian, J.A.: Level Set Methods and Fast Marching Methods: Evolving Interfaces in Computational Geometry, Fluid Mechanics, Computer Vision, and Materials Science. Cambridge University Press, Cambridge (1999)
19. Osher, S., Fedkiw, J., Ronald, P.: Level Set Methods and Dynamic Implicit Surfaces. Springer, New York (2002)
20. DeBar, R.B.: Fundamentals of the KRAKEN code. [Eulerian hydrodynamics code for compressible nonviscous flow of several fluids in two-dimensional (axially symmetric) region]. Technical report, California University, Livermore (USA). Lawrence Livermore Laboratory (1974)
21. Noh, W.F., Woodward, P.: SLIC (simple line interface calculation). Lecture Notes in Physics, vol. 59, pp. 330–340. Springer, Berlin (1976)
22. Hirt, C.W., Nichols, B.D.: Volume of fluid (VOF) method for the dynamics of free boundaries. J. Comput. Phys. **39**(1), 201–225 (1981)
23. Groß, S., Peters, J., Reichelt, V., Reusken, A.: The DROPS package for numerical simulations of incompressible flows using parallel adaptive multigrid techniques. Preprint 227, IGPM, RWTH Aachen (2002)
24. Groß, S., Reusken, A.: Finite element discretization error analysis of a surface tension force in two-phase incompressible flows. SIAM J. Numer. Anal. **45**, 1679–1700 (2007)
25. Groß, S., Reusken, A.: Numerical Methods for Two-phase Incompressible Flows. Springer, Berlin (2011)
26. The DROPS package, http://www.igpm.rwth-aachen.de/DROPS/ (2013)
27. Rieber, M.: Numerische Modellierung der Dynamik freier Grenzflächen in Zweiphasenströmungen. Fortschrittberichte VDI / 7. VDI-Verl. (2003)

28. Sabisch, W.: *Dreidimensionale numerische Simulation der Dynamik von aufsteigenden Einzel-blasen und Blasenschwärmen mit einer Volume-of-Fluid Methode.* PhD thesis, University Karlsruhe (2000)
29. Ghidersa, B.E., Wörner, M., Cacuci, D.G.: Exploring the flow of immiscible fluids in a square mini-channel by direct numerical simulation. Chem. Eng. J. **101**, 285 (2004)
30. Rider, W.J., Kothe, D.B.: Reconstructing volume tracking. J. Comput. Phys. **141**(2), 112–152 (1998)
31. Giusti, E.: Minimal Surfaces and Functions of Bounded Variation. Monographs in Mathematics, vol. 80. Birkhäuser, Boston (1984)
32. Francois, M.M., Cummins, S.J., Dendy, E.D., Kothe, D.B., Sicilian, J.M., Williams, M.W.: A balanced-force algorithm for continuous and sharp interfacial surface tension models within a volume tracking framework. J. Comput. Phys. **213**(1), 141–173 (2006)
33. Popinet, S.: An accurate adaptive solver for surface-tension-driven interfacial flows. J. Comput. Phys. **228**, 5838–5866 (2009)
34. Öztaskin, M.C., Wörner, M., Soyhan, H.S.: Numerical investigation of the stability of bubble train flow in a square minichannel. Phys. Fluids **21**, 042108-1–042108-17 (2009)
35. Jasak, H., Jemcov, A., Ž Tuković. OpenFOAM: a C++ library for complex physics simulations. In: International Workshop on Coupled Methods in Numerical Dynamics IUC, Dubrovnik, Croatia, 19–21 September 2007
36. Weller, H.G., Tabor, G., Jasak, H., Fureby, C.: A tensorial approach to computational continuum mechanics using object orientated techniques. Comput. Phys. **12**(6), 620–631 (1998)
37. The OpenFOAM CFD toolbox. http://www.openfoam.org, Jan 2013
38. Demirdžić, I., Perić, M.: Space conservation law in finite volume calculations of fluid flow. Int. J. Numer. Methods Fluids **8**, 1037–1050 (1988)
39. Muzaferija, S., Perić, M.: Computation of free-surface flows using interface-tracking and interface-capturing methods. In: Mahrenholtz, O., Markiewicz, M. (eds.) *Nonlinear Water Wave Interaction.* Computational Mechanics, Southampton (1998)
40. Demirdžić, I., Muzaferija, S., Perić, M., Schreck, E., Seidl, V.: Computation of flows with free surfaces. In: Scientific Computing in Chemical Engineering II, pp. 360–367. Springer, Berlin (1999)
41. Tuković, Ž., Jasak, H.: A moving mesh finite volume interface tracking method for surface tension dominated interfacial fluid flow. Comput. Fluids, **55**, 70–84 (2012)
42. Tuković, Ž., Jasak, H.: Simulation of free-rising bubble with soluble surfactant using moving mesh finite volume/area method. In: 6th International Conference on CFD in Oil & Gas, Metallurgical and Process Industries, SINTEF/NTNU, Trondheim, Norway, 10–12 June 2008
43. Jasak, H., Tuković, Ž.: Automatic mesh motion for the unstructured finite volume method. Trans. FAMENA **30**, 1–20 (2006)
44. Jasak, H.: Dynamic mesh handling in OpenFOAM. In: 48th AIAA Aerospace Sciences Meeting, Orlando, Florida (2009)
45. Menon, S.: A numerical study of droplet formation and behavior using interface tracking methods. Ph.D. thesis, University of Massachusetts Amherst (2011)
46. Issa, R.I.: Solution of the implicitly discretised fluid flow equations by operator-splitting. J. Comput. Phys. **62**, 40–65 (1986)
47. Yen, B.K.H., Günther, A., Schmidt, M.A., Jensen, K.F., Bawendi, M.G.: A microfabricated gas-liquid segmented flow reactor for high-temperature synthesis: the case of CdSe quantum dots. Angew. Chem. Int. Ed. **44**, 5447 (2005)
48. Hua, J., Erickson, L.E., Yiin, T.-Y., Glasgow, L.A.: A review of the effects of shear and interfacial phenomena on cell viability. Crit. Rev. Biotechnol. **13**, 305 (1993)
49. Mercier, M., Fonade, C., Lafforgue-Delorme, C.: How slug flow can enhance the ultrafiltration flux in mineral tubular membranes. J. Membr. Sci. **128**, 103 (1997)
50. Ratkovich, N., Chan, C.C.V., Berube, P.R., Nopens, I.: Experimental study and CFD modelling of a two-phase slug flow for an airlift tubular membrane. Chem. Eng. Sci. **64**, 3576 (2009)

Printed by Printforce, the Netherlands